THE BOOK ON MASKS

Your Comprehensive Guide to the Manipulative Psychology,
Malformed Philosophy, and Misrepresented Science
that Supercharged a Global Hysteria

For permission requests, write to philipbuckler@thebookonmasks.com, "Attention: Permissions."

This publication is designed to provide accurate and authoritative information regarding the subject matter covered. It is sold with the understanding that the author is not engaged in rendering medical, legal, or other professional services. The author and publisher shall not be liable for any loss of profit or any other commercial damages, including but not limited to special, incidental, consequential, personal, or other damages. Original illustrations by Philip Buckler. Cited illustration material by the listed authors and consistent with fair use.

ISBN: 979-8-9891699-0-0 (ebook)

ISBN: 979-8-9891699-1-7 (audiobook)

ISBN: 979-8-9891699-2-4 (Color Hardcover)

ISBN: 979-8-9891699-3-1 (B&W paperback)

ISBN: 979-8-9891699-4-8 (Color paperback)

Dedication

To everyone, everywhere, who has worn a mask against their will.

Table of Contents

Part 1: Mask Science – Efficacy and Effectiveness

Part 2: Mask Science - Side Effects

Part 3: Mask Psychology and Motivation

Part 4: Pro-Mask Philosophy is Bad Philosophy

Part 5: Under Color of Law — Legal Duels Over Compulsory Masking

Part 6: Fighting Back

Appendix: A Christian Case Against Compulsory Masking

Introduction

Mask Hysteria

In 2020, the world went mad for masks. Enough said.

The more distant we get from the year 2020 and the COVIDcrisis in general, the more tempting it becomes to view the subject matter of this book as dated, perhaps even irrelevant. I believe that would be a mistake. To be sure, there is an element of truth to such a belief, in the sense that new scientific studies on masks will be published, and new related case law will be handed down in the coming years. These may or may not find their way into future editions of *The Book on Masks* (which, themselves, may or may not be written). In the broader sense, however, good evidence, good science, and good arguments do not become bad evidence, bad science, and bad arguments with the passage of time. Bad legal decisions may be overturned, but good rulings which accurately conform to a Higher Law tend to stick around long after the judges that drafted them. Likewise, the bad science, bad reasoning, and destructive beliefs that fueled the COVIDcrisis will not look any better 100 years from now. If anything, the opposite is likely to be the case.

The science on masks is important, but the behavioral psychology harnessed to drive masking, and the opposing philosophies and belief systems at war beneath the surface, are as old as civilization. They have been relevant for every generation prior to 2020, and they will remain relevant for every generation after 2020. Masks were such an explosive issue because of how many deeper issues they tapped into. To get to the deeper issues, however, we need to first deal with the most public points of debate. You will find in this book the most comprehensive and thorough (and hopefully digestible) review of the science on masks yet written.

By the time you are halfway through this book, you will know far more about masks than every public health official and authority who eagerly forced one onto you, your friends, and your loved ones. If you read this book cover to cover, you will know far more about all aspects of facemasks than any of the "experts" who aggressively "recommended" that you wear one, or any of the authorities who were busily turning those "recommendations" into mandates with very real penalties that violated the Bill of Rights. You will also be equipped to stand in the forefront of pushing back on masks if you choose to do so.

My goal with this book is to not merely to provide readers with the single most comprehensive and effective arsenal of information available when it comes to all things mask-related (though this would be ambitious enough). My goal is to equip readers to recognize and counter the next manifestation of the masking mentality when history starts to rhyme again (as it inevitably will).

If you picked up this book because you saw compulsory masking for what it was from the very start, you can justly derive satisfaction from that. If you picked up this book because you came to question masks more gradually, cut yourself some slack. Escaping from Jonestown has always been harder than staying out of it in the first place. In either case, keep those critical faculties honed to a fine point and apply them consistently in other areas of your life.

This book is intended to be modular, and useable either for straight-through reading or as a handy reference volume. There are four main topics (five if you count the appendix).

- Mask Science
- Mask Psychology
- Mask Philosophy
- Mask Legalities

Other relevant topics, such as history, are distributed throughout. The appendix articulates just one of many possible sets of religious beliefs against wearing masks in the name of COVID. The sections are mutually supporting, refer to one another, and even include some overlap, but each can stand on its own. Feel free to read only those sections that interest you. If you are interested in the psychology used to push masking but not the physical science involved, go ahead and jump to that section. If you want to know more about the side effects of masks, you can start there instead. If you want to know more about the law involved, how some lawsuits against masks succeeded and why others failed, start with the legal section. If the idea of religious objections to wearing masks sounds either totally insane or intriguing, you are welcome to jump to the appendix straightaway.

INTRODUCTION

"Where's your public health degree?"

I do not have one and I do not need one. Neither do you. Based on the average performance of people with public health degrees during the COVIDcrisis, a public health degree might be as much of a liability as an asset. In his presentation to the fifteenth annual meeting of the National Science Teachers Association held in New York in 1966, Nobel Prize-winning physicist Richard Feynman said: "Science is the belief in the ignorance of experts."[1] In the same talk, he also said: "there is a considerable amount of intellectual tyranny in the name of science." Dr. Feynman explained:

> "When someone says, 'Science teaches such-and-such,' he is using the word incorrectly. Science doesn't teach anything; experience teaches it. If they say to you, 'Science has shown such and such,' you might ask, 'How does science show it? How did the scientists find out? How? What? Where?'

> "It should not be 'science has shown' but 'this experiment, this effect, has shown.' And you have as much right as anyone else, upon hearing about the experiments — but be patient and listen to all the evidence — to judge whether a sensible conclusion has been arrived at."

Good evidence does not become bad evidence with the passage of time, and bad evidence does not become good evidence no matter who cites it or how *many* people cite it. Having credentials does not make an expert. *Doing the work* makes an expert. The first working airplane was not built by "experts." Instead, the first airplane was built by a pair of bicycle mechanics who never went to college but who did a *lot* of independent reading. The medical field is no exception. I am an expert on masks because I did the work. You, too, can be a genuine expert on masks, and at the cost of considerably less time and effort.

For what it is worth, I have earned my living for over a decade by aerosolizing the contents of multiple strangers' mouths into my face on a daily basis — otherwise known as practicing dentistry. Dentists wear masks more than most MDs. But that is not why I am worth listening to when I talk about masks. I am worth listening to on this topic because I have spent more time and effort researching all aspects of masks than is typically spent researching all aspects of dentistry over the course of a two- or three-year postgraduate dental residency. I have very mixed feelings about that. Public health

---

1   Richard Feynman, "What is Science?" *The Physics Teacher*, Vol. 7, Issue 6, 1969, pp. 313-320, originally presented at the fifteenth annual meeting of the National Science Teachers Association in New York City, 1966; available online: http://www.feynman.com/science/what-is-science/

graduate programs do not even have so much as a single 1-credit class devoted to masks any more than do dental schools or medical schools. At the most, when you go through one of these programs, you get a few PowerPoint slides as part of the information firehose that you are drinking from. One of your clinical instructors then walks you through the basics, and then you practice the infection control rituals you have been taught without thinking much about them until they become a habit from which any deviation feels very uncomfortable. You assume that someone, somewhere, put out the effort to make certain it wasn't all a bunch of woo woo voodoo (or worse).

"You have a bias!"

I have strong and sincerely held beliefs about masks, obviously. No one writes a book like this without some passion for the subject. *Anyone* with an opinion one way or the other about masks could be accused of having a "bias." The important question is, *why* do I believe what I believe about masks? What is the *basis* for my beliefs? Motivated reasoning and confirmation bias are, unfortunately, part of human nature. If everyone has conclusions in search of answers, then the only real option is to weigh the arguments and evidence as best you can while respecting others' right to live according to their differing beliefs. As U.S. Supreme Court Justice Frank Murphy said in 1943: "It is in that freedom and the example of persuasion, not in force and compulsion, that the real unity of America lies."[2]

"The burden of proof is on you!"

Any claim to knowledge entails a burden of proof. The person who makes the claim bears the burden of showing why their claim is so. If someone says you should wear a mask, the burden of proof is on *them*, not you. If you counter that someone should *not* wear a mask, then you also have a burden of proof. All I really need to show is that the arguments and evidence in favor of masks are not sufficient to justify forcing everyone to wear one. In order to generate a moral or legal duty to wear a mask, showing by a preponderance of the best-quality clear and convincing evidence that masks are effective at mitigating the spread of respiratory viral infections is merely the *first* of several steps the pro-mask side of this debate needs to successfully make.

Each part of this book contains all of the references cited in that part. The reference citation style varies between parts. Personally, when I am reading, sometimes I like to check an author's references immediately, and sometimes I prefer to go back and do so later because immediate checking breaks the flow of my reading. I assume this is the same for other people. When I check an author's references, I like the process to be as convenient as possible. I do not like hunting back to an index, looking up references by number or alphabetical order, and then having to find my place again, or looking up

---

2   *West Virginia Board of Education v. Barnette et. al.*, 319 U.S. 624, 646 (1943) (Murphy, J., concurring), available online: https://www.loc.gov/item/usrep319624/

and downloading a separate document. I personally prefer footnotes, though I had to forgo making much use of them in the science sections because some of the citations are so heavily referenced that there would be several points at which whole pages would be nothing but redundant footnoted references. What I do *not* like in footnotes is looking down to see a sequence of "*Ibid.*" entries going back multiple layers to one that says "See [abridged title] *supra*" when the *supra* is forty notes or forty pages earlier. These are pet peeves of mine, and so this book does not have those things. The tradeoff is that avoiding this entails some redundancy in the footnotes and using more ink and paper.

I prefer to let people speak for themselves in their own words, and so I quote many sources at-length. It is very important to me that readers not have to take my word for anything, and that they be able to check my sources as quickly and easily as possible. I have therefore hyperlinked, cited, and referenced each section in a way that I as a reader tend to prefer, trying to find a good balance between readability and ease of reference without too much redundancy (often eschewing reference-formatting conventions to do so). I intentionally made use of as many publicly available sources as possible, though some of the linked sites (such as ResearchGate and archive.org) require free registration to access the relevant material, and others simply are not publicly available without paying (except possibly through the resources of one's local public or university library). For those sources which are not publicly available which any readers may wish to verify, I invite those readers to either contact me through this book's website, www.thebookonmasks.com (after browsing the extras section) or use the opportunity to develop their own skills as researchers. All hyperlinks were functional and accurate at the time of publication.

Additional material mentioned in this book, such as my (free and printable) poster-chart showing at a glance how the CDC's references stack up and balance against a more comprehensive body of scientific sources on masks, can be accessed at www.thebookonmasks.com/extras.

In a work this size, expect to find plenty of material that is new to you and some material you already knew, viewed from a different angle. Expect to find material you dispute, and even more material you agree with. Your time is valuable to me. I look forward to repaying your investment and earning a place on your bookshelf.

Sincerely,

Philip Traugott Buckler, DDS

Part 1:

Mask Science –
Efficacy and Effectiveness

Framing the Question

We can discover whether or not masks "work" as a public health measure by answering four basic questions:

I. What should we *expect* to observe if masks mitigate the spread of respiratory viruses and other pathogens?

II. What do we *actually* observe?

III. What is the *quality and quantity* of the evidence base supporting claims of mask efficacy?

IV. What *other important scientific aspects* of the problem need to be considered?

Absence of evidence is not evidence of absence, but finding evidence of absence is often pretty straightforward. Evidence of absence can come from direct observation, such as when you survey a refrigerator where you expect to find milk and, finding it to be empty, then conclude that you are out of milk. Evidence of absence can also come from incompatible positive findings. For example, if you establish that a man was physically present in California at a particular time, you can rule out the idea that he was physically present in New York at that same time. Contrary to certain tropes, it is entirely possible to prove a negative. Science often advances most effectively by falsifying hypotheses. A famous quote attributed to Thomas Edison encapsulates this approach: "If I find 10,000 ways something won't work, I haven't failed. I am not discouraged, because every wrong attempt discarded is another step forward."[3]

---

3   While it is certain that Edison said something to this effect (likely on multiple occasions), the exact wording is disputed, and variants appear in multiple biographies. For a more in-depth investigation of this quote, I recommend the work published in https://quoteinvestigator.com/2012/07/31/edison-lot-results/

Not all scientific evidence is created equal.

- The weakest types of scientific evidence are conjecture (also known as "*modeling*") and *expert opinion*.

- *Laboratory studies* are stronger, but hypotheses developed in controlled conditions need to be tested and verified by real-world observational studies and randomized controlled trials.

- Real-world *observational studies* are better than laboratory studies, but are often subject to a number of confounding biases. The following is only a partial list:

  o Selection bias occurs when there is a subtle loss of essential randomization, such as when an unusually high proportion of anxious people volunteer to participate in a study about the prevalence of anxiety in the general population.

  o Recall bias occurs when results of the study are highly dependent on the participant's ability to recall information, as when Solomon Asch's conformity study participants universally underestimated the extent to which they succumbed to group pressure (more on Asch's famous series of experiments in the section on mask psychology).

  o Surveillance bias occurs when looking more closely at a particular group finds more of a particular outcome, such as when testing a subset of the population twice as much as everyone else can make disease incidence look higher than it actually is.

  o Measurement bias occurs when data is collected in a systematically distorted matter, such as "cherry-picking" by discarding all the data that would falsify one's hypotheses.

  o Social Desirability bias occurs when responses are skewed in a direction that subjects find less personally embarrassing or believe will make them look better. "Why yes, now that you mention it, I *do* brush and floss seven days a week. Why do you ask, doctor?"

  o Publication bias occurs when studies are selectively published based on their results rather than on the inherent quality of the research they contain. This was especially rampant during COVID.

- *Randomized controlled trials* are usually held up as the gold standard of scientific studies, because they seek to eliminate as many confounding variables as possible. They are (on average) the strongest types of studies. They often require the most effort, as well.

- *Meta-analyses or systematic reviews* aggregate and assess the results of multiple individual lesser studies. Meta-analyses of randomized controlled trials are generally considered the highest-quality of evidence, but the outcomes of even these can be skewed or manipulated depending on which studies are selected for the review and analysis (garbage in, garbage out).

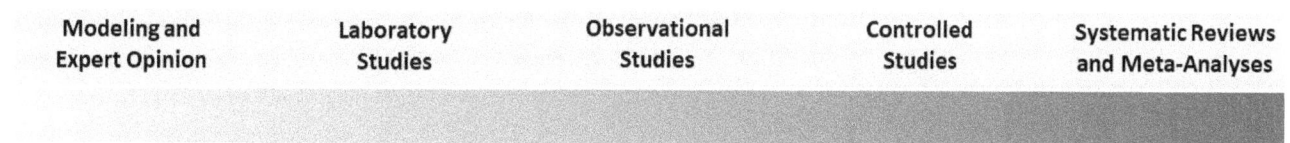

Modeling and Expert Opinion	Laboratory Studies	Observational Studies	Controlled Studies	Systematic Reviews and Meta-Analyses

Weaker Evidence   Stronger Evidence

Even within a category, the evidential quality of a study can vary dramatically depending on factors such as size, duration, observational detail, methodology, and the specific question being examined. Different sub-types of studies within each of these categories *also* vary considerably in methodology and overall strength. For example, there are multiple different sub-types of observational studies which can range from weak to quite strong. These range from case studies of single individuals, to prospective and retrospective cohort studies with thousands of participants.

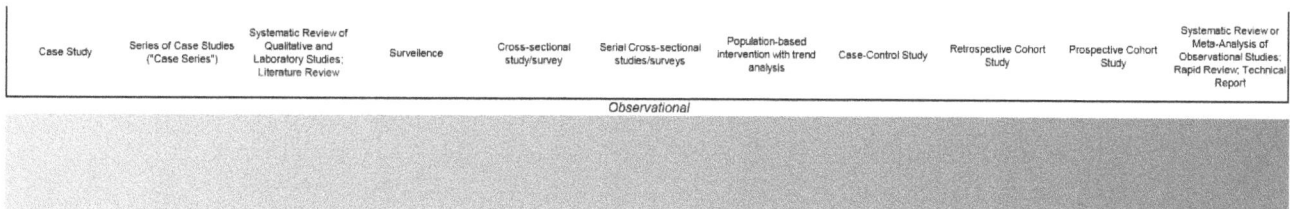

Case Study	Series of Case Studies ("Case Series")	Systematic Review of Qualitative and Laboratory Studies; Literature Review	Surveillance	Cross-sectional study/survey	Serial Cross-sectional studies/surveys	Population-based intervention with trend analysis	Case-Control Study	Retrospective Cohort Study	Prospective Cohort Study	Systematic Review or Meta-Analysis of Observational Studies; Rapid Review; Technical Report

Observational

Weaker Evidence   Stronger Evidence

Quite apart from extensive anecdotal evidence and independent research, there were more than 100 published studies by the middle of 2021, which, in combination, provided decisive evidence that masks do not mitigate the spread of respiratory viruses (or, for that matter, other pathogens) to any extent which would even come close to justifying mandates. This number more than doubles if you include studies which provide background evidence that would disfavor mask use even if masks could be shown to "work." To download a free printable pdf poster that provides a visual overview charting the category and content of more than 270 mask-related studies at a glance, visit www.thebookonmasks.com/extras.

What Should We Expect to Observe if Masks "Work"?

What would the scientific evidence, especially randomized controlled trials, need to show for us to conclude that masks "work" to mitigate the spread of respiratory viruses? If the hypothesis that masks are effective at mitigating the spread of respiratory viruses and other pathogens is true, then a preponderance of the best available scientific studies, especially randomized controlled trials, should show us the following:

- We should see masks provide ***clear and convincing benefits*** against post-operative infections when all members of surgical teams are masked compared to when some surgical team members are unmasked.
- We should see masks provide ***clear and convincing benefits*** against viral transmission within households when masks are worn compared to when they are not.
- We should see masks provide ***clear and convincing benefits*** against viral transmission when medical masks are worn compared to no masks at all among more densely packed sub-groups in the community, such as Hajj pilgrims or college students living in dorms.
- We should see masks provide ***clear and convincing benefits*** against respiratory viral infections among members of the general population who wear masks compared to those who do not.
- We should see ***clear and convincing benefits*** against viral respiratory infections among healthcare workers who wear top-quality medical masks like N95 respirators when compared to those healthcare workers who only wear standard medical-surgical masks.
- We should see independent systematic literature reviews and meta-analyses of multiple randomized controlled trials involving mask use consistently finding ***clear and convincing benefits*** against respiratory viral transmission from mask use.
- We should see studies purporting to show masks work generally being larger, longer, stronger, more rigorous, and more numerous than studies which show *no benefits* from wearing masks.

What Do We Actually Observe, Instead?

The greatest quantity and highest quality of scientific literature indicates that masks do nothing to mitigate the spread of respiratory viruses.

Randomized controlled trials show masks do not work

There are at least 25 Randomized Controlled trials examining the efficacy of facemasks. If it were true that our use of masks provides protection for others, we should expect to see a statistically significant difference in post-operative microbial infection rates when various surgical procedures are conducted with masked or non-masked medical personnel, yet randomized controlled trials from thousands of surgeries in hospital operating rooms show ___no clear and convincing benefits___ against post-surgical infections. The non-significant differences that *are* observed tend to show a slight *increase* in the rate of post-surgical infections with the fully masked surgical teams. [1, 2] If masks fail to make a difference in these post-surgical *bacterial* infection rates, there is no reason to suppose they will have an effect on much smaller respiratory viruses which are both more easily transmitted and harder to filter.

➤ A Swedish randomized controlled trial conducted over two years and including the same operating teams in the same operating rooms doing the same types of procedures on the same patient pool found that: "After 1,537 operations performed with face masks, 73 (4.7%) wound infections were recorded and, **after 1,551 operations performed without face masks, 55 (3.5%) infections occurred. This difference was not statistically significant (p > 0.05) and the bacterial species cultured from the wound infections did not differ in any way…**" [1] (emphasis added)

15

➢ Likewise, an Australian randomized controlled trial found that "**Surgical site infection rates did not increase when non-scrubbed operating room personnel did not wear a face mask.**" In the "Mask group," all non-scrubbed surgical staff wore a mask. In the "No Mask group," none of the non-scrubbed surgical staff wore masks. Out of 811 participants who completed the study: "a total of 83 (10.2%) surgical site infections were recorded; 46 (11.5%) were in the Masked group and 37 (9.0%) in the No Mask group. **The difference was not statistically significant.**" (emphasis added) [2]

Randomized controlled trials of respiratory viral transmission within households show **no clear and convincing benefits** against viral transmission when masks are worn vs. no masks. [3-10]

➢ A German Cluster randomized controlled trial of masking in 84 households totaling 302 participants found no decrease in the transmission of influenza when masks were worn: "In primary intention-to-treat analysis of all data, **the interventions did not lead to statistically significant reductions of SAR [secondary attack rate]**" (emphasis added) [3].

➢ In a 2011 Thailand cluster-randomized controlled trial involving 442 index (infected) children and 1,147 household members, secondary influenza attack rates between the handwashing and the handwashing + facemask group were identical. The Authors concluded: "**Influenza transmission was not reduced by interventions to promote hand washing and face mask use**" (emphasis added) [4].

➢ A cluster-randomized controlled trial was conducted in New York with 509 primarily Hispanic households (2,788 persons total) over a period of 19 months. It examined the effects of three interventions in three study cohorts: education, education + hand sanitizer, and education + hand sanitizer + facemasks. The authors found that: "**There were no significant differences in rates of URI [Upper Respiratory Infection], ILI [Influenza-like Illness], or influenza by intervention group**" (emphasis added) [5].

➢ A 2008 cluster randomized controlled trial by Cowling and colleagues compared three cohorts (groups): no intervention vs. facemasks vs. hand hygiene in 122 households. They found that: "**The laboratory-based or clinical secondary attack ratios did not significantly differ across the intervention arms.**" (**emphasis added**) [7] An even more well-known study by MacIntyre and colleagues one year later in 2009 [6] described Cowling's 2008 study as follows: "Another recent study that examined the use of surgical masks and handwashing for the prevention of influenza transmission also found no significant difference between the intervention arms."

➢ MacIntyre and colleagues' 2009 randomized controlled cohort study of 143 households included 286 adult participants and found that: "ILI [influenza-like illness] was reported in 21/94 (22.3%) in the surgical [mask] group, 14/92 (15.2%) in the P2 [respirator mask] group, and 16/100 (16.0%) in the control group, respectively**." "We concluded that household use of face masks is associated with low adherence and is ineffective for controlling seasonal respiratory disease"** (emphasis added) [6]. Did you catch that? In this study, too, the no-mask control group contracted influenza-like-illness no more frequently than either of the two masked groups.

➢ In 2009, Cowling and colleagues conducted another cluster randomized controlled trial involving 322 infected index cases and their household contacts. They found that: "Hand hygiene with or without facemasks seemed to reduce influenza transmission, but the differences compared with the control group were not significant." More importantly for the purposes of our analysis, they also reported that, **"No significant difference was found between the face-mask plus hand hygiene group and the hand hygiene group in RT-PCR–confirmed influenza virus infections in household contacts"** (emphasis added) [8]. Here, again, adding a facemask on top of hand hygiene made no difference.

➢ In a 2010 French cluster randomized controlled trial including 306 contacts in 105 households examining the efficacy of surgical facemasks for limiting influenza transmission, the authors reported: **"We observed a good adherence to the intervention. In various sensitivity analyses, we did not identify any trend in the results suggesting effectiveness of facemasks"** (emphasis added) [9]. In 2016, MacIntyre and colleagues (see below) described the French study as follows: "ILI [Influenza-Like Illness] was reported in 16.2% and 15.8% of contacts in the intervention and control arms, respectively, and the difference was not statistically significant" [10].

➢ The 2016 MacIntyre study quoted above was yet *another* randomized controlled trial [10] on masks. This particular randomized controlled trial was cited as evidence that masks do not work by 19 States in a joint lawsuit against the CDC's mask mandate.[4] This study was done in the same vein as the 2010 French study above, and is also noteworthy for how it characterizes previous randomized controlled trials on masks in a healthcare setting from 2011, 2013, and 2015:

> "Cloth and medical masks were originally developed as source control to prevent contamination of sterile sites by the wearer in operating theatres (OTs); however,

---

4   *State of Florida et al. v. Walensky et al.*, No. 22-cv-00718, Document 1 Initial Complaint, (Middle District of Florida, March 29, 2022), available online: https://legacy.myfloridalegal.com/webfiles.nsf/WF/GPEY-CCYHZH/$file/Mask%20Complaint%20as%20Filed.pdf

their effectiveness in preventing surgical site infections is yet to be proven" (emphasis added).

"Mask use as source control in healthcare settings has now been included in standard infection control precautions during periods of increased respiratory infection activity in the community, **yet there is no clinical efficacy evidence to support this recommendation**" (emphasis added).

The 2016 MacIntyre study examined the effects of medical masks as source control, using transmission of influenza-like illness from 245 index cases (half masked, half unmasked) to 597 household contacts. Though the authors repeatedly emphasized the lower number of infections in the masked group (two fewer infections, total, between the study cohorts), they nevertheless ultimately conceded: **"We did not find a significant benefit of medical masks as source control"** [10].

Randomized controlled trials among more densely packed sub-groups in the community, such as Hajj pilgrims, or college students living in dorms, likewise show *__no clear and convincing benefits__* against viral transmission when medical masks are worn vs. no masks [11-15].

➤ An interconnected trilogy of cluster Randomized Controlled Trials was conducted involving 7,851 total participants using tents of Hajj pilgrims as the cluster randomization units [11-13]. Though the 2014 pilot study of 164 participants [12] suggested a possible protective effect from medical masks, the larger follow-up studies with 7,687 participants found that masks were not associated with decreased risk for infections in Hajj pilgrims, *even when an infected index case was present within the same tent.* **"In intention-to-treat analysis, facemask use was neither effective against laboratory-confirmed vRTIs [Viral Respiratory Tract Infections] nor against CRI [Clinical Respiratory Infections], not even in per-protocol analysis"** (emphasis added) [11, 13].

➤ Two cluster-randomized controlled trials by Aiello et al. were published in 2010 and 2012 which looked at 2,475 student participants living in residence halls. These compared a control group to an intervention cohort with facemask only and a facemask-plus-hand-hygiene intervention cohort. In 2010, the authors reported that, "Both intervention groups compared to the control showed cumulative reductions in rates of influenza over the study period," however, they admitted in the same sentence: **"results did not reach statistical significance."** The 2012 study reiterates this point: **"Statistically significant findings were not observed for the face mask only group when compared to the control group"** (emphasis added) [14, 15]. This is consistent with the studies listed previously, which found that hand hygiene may have a beneficial effect, but that the addition of a facemask did not lead to a decrease in respiratory viral infections.

The *best* randomized controlled trial in the general population involving medical masks vs. no masks was Bundgaard et al.'s "Danish Facemask Study," which shows ***no clear and convincing benefits*** in the medical mask group against *SARS-CoV-2* and other respiratory viral infections [16].

➢ The "Danish Facemask Study" is important for being the first randomized controlled trial to examine SARS-CoV-2 infections in masked vs. non-masked civilian populations. It is also (by far) the highest-quality one to-date. 4,862 participants completed the study (2,392 masked; 2,470 non-masked). **The authors found no statistically significant difference in SARS-CoV-2 infection rates between the masked (1.8% infection rate) and non-masked (2.1% infection rate) study cohorts. The study cohort that reported wearing the mask "exactly as instructed" had a 2.0% SARS-CoV-2 infection rate. <u>There were also no statistically significant differences between the study cohorts in rates of other respiratory viral infections</u>.** In addition, the supplemental data found no statistically significant interaction in viral infection rates between wearers and non-wearers of eyeglasses, suggesting that eye protection measures like face shields are not beneficial. Also, the participant exit survey documented that the mere act of wearing a mask for two months produced a dramatic shift in *beliefs* about masks. (We will explore this in more detail in Part 3 when we look at mask psychology.)

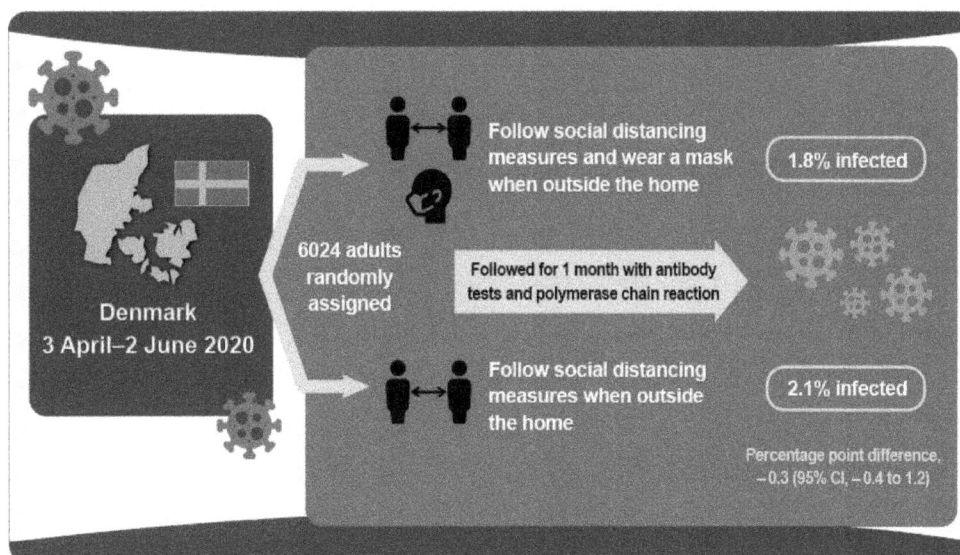

Does a recommendation to wear a surgical mask when outside the home reduce the wearer's risk for SARS-CoV-2 infection in a setting where masks were uncommon and not among recommended public health measures?

Denmark 3 April–2 June 2020

6024 adults randomly assigned

Follow social distancing measures and wear a mask when outside the home — 1.8% infected

Followed for 1 month with antibody tests and polymerase chain reaction

Follow social distancing measures when outside the home — 2.1% infected

Percentage point difference, −0.3 (95% CI, −0.4 to 1.2)

Only *one* randomized controlled trial [17] purports to show a statistically significant benefit from facemasks.

➤ The randomized controlled trial most favorable for facemask efficacy was conducted in Bangladesh in 2021, when **Abaluck and colleagues reported that cloth masks produced no statistically significant benefits, and that medical masks made an absolute difference of 1% in COVID infection rates** (8.6% COVID infection rate in the control cohorts vs. 7.6% in the mask cohorts) [17]. **This is a relative risk difference of 12%. Even taking Abaluck's study at face value, it is worth pointing out that a drug with 12% efficacy would not even be approved for use, much less mandated.**

Nevertheless, even the results above likely overstate the case for masks, because **Abaluck's study was seriously flawed, as described by two subsequent studies [18, 19] which re-analyzed the data** from that trial and found no benefit to facemasks. The first re-analysis noted that "the lack of blinding of the observers in the study contributed to highly significant differences between the population of the treatment and the control groups. These differences are much more significant than the measured differences in symptomatic seropositivity" [18].

In other words, the masked group and the control group were too dissimilar for any beneficial differences in COVID infections to be attributed to masks. The second re-analysis was equally, if not more, devastating for the Bangladesh Mask Study. The authors concluded "… all we can conclude is that there is a 95% chance that mask intervention would result in anything between 19,240 fewer positives and 18,500 MORE positives among every 100,000 people. The results therefore provide essentially no support even for the weak surrogate hypothesis that the mask intervention procedures reduce the seropositivity rate… **this would be much like flipping 201 coins, observing 101 'heads' and 100 'tails' and concluding that all coins are more likely to land on heads than tails**" [19]. Jefferson et al.'s 2023 Cochrane review of mask studies also rated the Bangladesh mask study as having a high risk of selection bias, detection bias, reporting bias, and performance bias [20]. In a later interview, the lead author of the Cochrane review stated the same criticisms of the Bangladesh Facemask study that others had voiced: "That was not a very good study because it was not a study about whether masks worked, it was a study about increasing compliance for wearing a mask" [21].

➢ Additionally, **according to the Bangladesh Mask Study author's own statements, the observed benefits of Facemasks were entirely concentrated in users over age 50**. This age-differentiated result from a mechanical intervention strongly suggests that the reported benefits were due to observation, sampling, and reporting biases, but even if these were not present, the results of Abaluck's study would provide **no evidence for benefits of masking anyone under age 50**.

Jason Abaluck
@Jabaluck

The reduction was larger in villages where we (randomly) used surgical masks than those where we used cloth masks; in surgical mask villages, we saw a 12% reduction in COVID overall and a 35% reduction among those aged 60+.

Figure 3: Effect on Symptomatic Seroprevalence by Age Groups, Surgical Masks Only

(a) Above 60 Years Old

60+ years old — 0.69% / 1.03% — Decrease of 34.7% p=0.001

(b) 50-60 Years Old

50-60 years old — 0.83% / 1.08% — Decrease of 23.0% p=0.011

(c) 40-50 Years Old

40-50 years old — 0.94% / 0.95% — No statistically significant decrease p=0.984

(d) Younger than 40 Years Old

<40 years old — 0.52% / 0.55% — No statistically significant decrease p=0.618

8:00 AM · Sep 1, 2021 · chirr.app

425 Retweets   **150** Quote Tweets   **1,664** Likes

If the theories behind compulsory mask-use and mask-guidance accurately reflect the reality of viral transmission, and if facemasks are truly effective at mitigating the spread of respiratory viruses, we should expect to see a preponderance of randomized controlled trials finding that healthcare workers who use higher-quality masks such as fitted N95 respirators experience *clear and convincing benefits* against viral respiratory infections when compared to those healthcare workers who only wear standard medical-surgical masks or non-fitted N95 respirators. What we *actually* observe from randomized controlled trials among healthcare workers is **no clear and convincing benefits** against respiratory viral infections from higher-quality N95 mask use over medical-surgical masks and non-fitted N95 respirators. [22-26]

➢ A randomized controlled trial conducted from 2020 to 2022 on 1009 healthcare workers which compared the efficacy of surgical masks to fit-tested N95 respirators found **no statistically significant difference in SARS-CoV-2 viral infection rates between healthcare workers who used medical masks and those who used fitted N95 respirators** [27]. These results were consistent with those of an earlier 2009 randomized controlled trial headed by the same author examining 446 nurses in Ontario hospitals: "Influenza infection occurred in 50 nurses (23.6%) in the surgical mask group and in 48 (22.9%) in the N95 respirator group." Notably, the earlier study *also* looked at coronavirus infection rates and had similar findings (4.3% rate of coronavirus infections in the surgical mask group vs. 5.7% in the N95 respirator group) [22]. Moreover, **the 2022 study was particularly notable because it *excluded* vaccinated healthcare workers and healthcare workers with prior SARS-CoV-2 infection, thus avoiding a strong potential confounder.**

Viral Infection Rates	Medical Masks	N95 (fitted)
Loeb et al (2009)	4.3% (coronaviruses) 25.9% (Influenza-like illness)	5.7% (coronaviruses) 22.6% (Influenza-like illness)
Loeb et al (2022)	10.5% (RT-PCR-confirmed SARS-CoV-2) 10.8% (Seroconversion-confirmed SARS-CoV-2) 5.4% (Acute respiratory illness) 0.6% (Lower respiratory infection or pneumonia)	9.3% (RT-PCR-confirmed SARS-CoV-2) 11.9% (Seroconversion-confirmed SARS-CoV-2) 6.1% (Acute respiratory illness) 0.6% (Lower respiratory infection or pneumonia)

➢ By far the most rigorous and extensive *single* randomized controlled trial comparing medical masks and N95 respirators to-date followed **over 4,000 healthcare workers** at 137 outpatient study sites in 7 medical centers across the United States over four flu seasons from 2011 to 2015. It was also partially sponsored by the CDC. "In this pragmatic, cluster randomized trial that involved multiple outpatient sites at 7 health care delivery systems across a wide geographic area over 4 seasons of peak viral respiratory illness, **there was no significant difference between the effectiveness of N95 respirators and medical masks in preventing laboratory-confirmed influenza among participants routinely exposed to respiratory illnesses in the workplace.** In addition, there were no significant differences between N95 respirators and medical masks in the rates of acute respiratory illness, laboratory-detected respiratory infections, laboratory-confirmed respiratory illness, and influenza-like illness among participants" (emphasis added) [26]. This study, by itself, is enough to show that the CDC and every other state public health agency either knew better or should have known better than to recommend respirator masks over surgical masks for the prevention of SARS-CoV-2 or any other respiratory viral infection.

➢ The most famous *series* of randomized controlled trials regarding masking in healthcare settings was conducted from 2011 to 2015 by MacIntyre and colleagues [23-25]. These compared the efficacy of medical masks vs. N95 masks, and double-layered cotton cloth masks vs. medical masks. They involved a total of 4,717 healthcare workers in China and Vietnam. In addition, the 2011 study included a convenience non-randomized non-masked control group of 481 healthcare workers. Contrary to current masking theory, MacIntyre's 2011 study, which among other things compared fit-tested and non-fit-tested N95s, found that **the non-fit-tested N95s outperformed the fit-tested N95s**: "After adjustment for clustering, non-fit-tested N95 masks were significantly protective compared to medical masks against CRI [Clinical Respiratory Infection], but **other outcomes were not significant between N95 and medical masks**" (emphasis added) [24]. Keep in mind that clinical respiratory infections include not just viruses, but bacteria and fungi as well, which are many times larger and thus easier to filter out).

- MacIntyre's 2013 study states: "The original purpose of medical masks was to prevent microbial contamination of wounds while worn by surgeons during surgery (hence their common name 'surgical masks'), yet **randomized controlled trials show no efficacy against wound contamination... Masks in community settings have no clearly proved efficacy**" (emphasis added) [25].

- MacIntyre's 2015 study describes the 2011 and 2013 MacIntyre studies as: **"two previous RCTs [Randomized Controlled Trials], in which no efficacy of medical masks could be demonstrated when compared with control or N95 respirators."** It goes on to state that: **"Observations during SARS suggested double-masking and other practices increased the risk of infection because of moisture, liquid diffusion and pathogen retention"** (emphasis added) [23].

- The most prominent finding of MacIntyre's 2015 study was that the best-case scenario for cloth mask use (trained healthcare workers who washed them daily) still produced a statistically significant **higher** rate of respiratory viral infection than any of the other study cohorts. "Cloth masks resulted in significantly higher rates of infection than medical masks, and also performed worse than the control arm…. The rates of all infection outcomes were highest in the cloth mask arm…. **the results caution against the use of cloth masks… the magnitude of difference raises the possibility that cloth masks cause an increase in infection risk in HCWs [healthcare workers]**" (emphasis added) [23]. If healthcare workers are warned against using cloth masks, forcing those same cloth masks on the general public cannot be justified. A 2020 *post-hoc* re-analysis found that the higher rate of viral infection in the case of cloth masks relative to medical masks disappeared for cloth masks which were disinfected in the hospital laundry, but this finding merely re-enforces the negative implications for universal community-level cloth masking [28].

    ○ Finally, MacIntyre's 2015 study also reported that: "**the rate of virus isolation in the no-mask control group in the first Chinese RCT was 3.1%, which was not significantly different to the rates of virus isolation in the medical mask arms in any of the three trials including this one**" (emphasis added) [23]. The values reported are reproduced in the table below for side-by-side comparison. It is noteworthy also that these results are within 1.0-1.3% of the SARS-CoV-2 infection rates found in Bundgaard et al.'s Danish Facemask Study from 2020.

Viral Attack Rates	Non-randomized no-mask control group (MacIntyre 2011)	Surgical/Medical Masks (MacIntyre 2011)	Surgical/Medical Masks (MacIntyre 2013)	Surgical/Medical Masks (MacIntyre 2015)
All Viruses	3.1%	2.6%	3.3%	3.3%

Systematic reviews and meta-analyses show masks do not work

Repeated, independent systematic literature reviews [29-36] and meta-analyses [37-42] including at least four done in the context of COVID and evaluating many of the mask-use randomized controlled trials cited earlier have also consistently found that masks provide **no clear and convincing benefits** against respiratory viral infections, and that N95s confer **no clear and convincing benefits** over surgical masks.

➤ A 2016 systematic review and meta-analysis compared the efficacy of surgical masks and N95 respirators. "**In the meta-analysis of the clinical studies, we found no significant difference between N95 respirators and surgical masks in associated risk of (a) laboratory-confirmed respiratory infection... (b) influenza-like illness... or (c) reported work-place absenteeism...**" (emphasis added) [41]. Importantly, this study also notes that N95s appear to have a benefit in laboratory settings, but that this does not seem to carry over to clinical settings.

➤ A 2017 meta-analysis that found a slight protective effect of N95 respirators against bacterial infections nevertheless concluded: "Compared to [medical] masks, N95 respirators conferred superior protection against CRI [clinical respiratory infections] and laboratory-confirmed bacterial, **but not viral infections or ILI [influenza-like illness]**" (emphasis added). The authors also stated: "Single-use medical masks are preferable to **cloth masks, for which there is no evidence**

of protection and which might facilitate transmission of pathogens when used repeatedly without adequate sterilization" (emphasis added) [39].

➤ A World Health Organization technical report published on the eve of COVID (September 19, 2019) [43] included the results of its own systematic literature review and meta-analysis [44] evaluating 10 randomized controlled trials on masks — all of which we have already reviewed [3-8, 10, 12, 14, 15]. In its own words, **the report's "Overall Result of Evidence on Face Masks" was that "there was no evidence that face masks are effective in reducing transmission of laboratory-confirmed influenza"** [43]. The report also specifically stated that "Reusable cloth face masks are not recommended." Yet in less than 9 months, despite no change in the balance of the evidence, those same reusable cloth masks would go from being "not recommended" to mandated across the globe.

Masks show no efficacy in 10 studies analyzed by the WHO

OVERALL RESULT OF EVIDENCE ON FACE MASKS

1. Ten RCTs were included in the meta-analysis, and there was no evidence that face masks are effective in reducing transmission of laboratory-confirmed influenza.

WHO, *Non-pharmaceutical public health measures for mitigating the risk and impact of epidemic and pandemic influenza.* 2019, World Health Organization Global Influenza Program WEP: Online.
https://www.who.int/publications/i/item/non-pharmaceutical-public-health-measuresfor-mitigating-the-risk-and-impact-of-epidemic-and-pandemic-influenza

Table 7. Description of studies included in the review of face masks

STUDY	STUDY DESIGN	STUDY PERIOD	POPULATION & SETTING	INTERVENTION	OUTCOME & FINDING	QUALITY OF EVIDENCE
Aiello AE, 2010 (20)	Cluster-randomized intervention trial	Nov 2006 – Mar 2007	1437 university hall residents (USA)	Mask; Mask + Hand hygiene; control	Significant reduction in ILI during weeks 4–6 in mask and hand hygiene group compared to control; No significant reduction in ILI in mask and hand group or mask-only group or control	Moderate
Aiello AE, 2012(23)	Cluster-randomized interventional trial	Nov 2007 – Mar 2008	1178 university hall residents (USA)	Mask; Mask + Hand hygiene; control	No significant reduction in rates of laboratory-confirmed influenza in mask and hand group or mask-only group or control group	Moderate
Barasheed O, 2014 (50)	Non-blinded cluster-randomized trial	Nov 2011 – Nov 2011	164 Australian pilgrims (Saudi Arabia)	Mask; control	No significant difference in laboratory-confirmed influenza in two arms; protective effect against syndromic ILI compared to controls (31% versus 53%, p = 0.04)	Moderate
Cowling BJ, 2008 (26)	Cluster-randomized intervention trial	Feb 2007 – Sep 2007	198 laboratory-confirmed influenza case and their household contacts	Mask; Hand hygiene; control	No significant reduction in the secondary influenza attack rate in control, mask or hand group	Moderate
Cowling BJ, 2009 (19)	Cluster-randomized intervention trial	Jan 2008 – Sep 2008	407 laboratory-confirmed influenza case and 794 household members	Mask; Mask + Hand hygiene; control	No significant difference in rates of laboratory-confirmed influenza in hand-only or mask and hand group	Moderate
Larson EL, 2010 (21)	Cluster-randomized intervention trial	Nov 2006 – Jul 2008	617 households	Mask + Hand hygiene; Hand hygiene; control	No significant reduction in rates of laboratory-confirmed influenza in control, hand, mask or hand group	Moderate
MacIntyre CR, 2009 (48)	Cluster-randomized intervention trial	Aug 2006 – Oct 2006 & Jun 2007 – Oct 2007	145 laboratory-confirmed influenza case and their adult household contacts	Surgical mask; P2 mask; control	No significant difference in rate of laboratory-confirmed influenza in control, face mask or P2 mask group	Moderate
MacIntyre CR, 2016 (49)	Cluster-randomized intervention trial	Nov 2013 – Jan 2014	245 ILI index case and 597 household contacts	Mask; control	Clinical respiratory illness, ILI and laboratory-confirmed viral infections were lower in the mask arm compared to control, but results were not statistically significant	Moderate
Simmerman JM, 2011 (22)	Cluster-randomized intervention trial	Apr 2008 – Aug 2009	465 laboratory-confirmed influenza case and their household contacts	Mask + Hand hygiene; hand hygiene; control	No significant reduction in rate of secondary influenza infection in control, hand, mask or hand group	Moderate
Suess (2012) (24)	Cluster-randomized intervention trial	Nov 2009 – Jan 2010 & Jan 2011 – Apr 2011	84 laboratory-confirmed influenza case and 218 household contacts	Mask; Mask + Hand; control	No significant difference in rate of laboratory-confirmed influenza in control, mask, mask or hand group	Moderate

ILI: influenza-like illness; USA: United States of America.

WORLD HEALTH ORGANIZATION

WHO, *Non-pharmaceutical public health measures for mitigating the risk and impact of epidemic and pandemic influenza.* **Annex: Report of systematic literature reviews.** 2019, World Health Organization: Online.
https://iris.who.int/handle/10665/329439 https://iris.who.int/bitstream/handle/10665/329439/WHO-WHE-IHM-GIP-2019.1-eng.pdf?sequence=1&isAllowed=y

➤ A 2020 meta-analysis made specifically in the context of COVID-19 concluded: "Compared with N95 respirators; the use of medical masks did not increase laboratory-confirmed viral (including coronaviruses) respiratory infection… or clinical respiratory illness." The authors went on to state that: "**There is no convincing evidence that medical masks are inferior to N95 respirators for protecting healthcare workers against laboratory-confirmed viral respiratory infections** during routine care and non–aerosol-generating procedures" (emphasis added) [37].

➤ A*nother* meta-analysis made specifically in the context of COVID-19 (and also one of the most thorough and straightforward) concluded: "Although mechanistic studies support the potential effect of hand hygiene or face masks, evidence from 14 randomized controlled trials of these measures did not support a substantial effect on transmission of laboratory-confirmed influenza." The study authors also stated: "**We did not find evidence that surgical-type face masks are effective in reducing laboratory-confirmed influenza transmission, either when worn by infected persons (source control) or by persons in the general community to reduce their susceptibility**" (emphasis added) [42].

➤ Another 2020 meta-analysis, concluded: "**There were no statistically significant differences in preventing laboratory-confirmed influenza, laboratory-confirmed respiratory viral infections, laboratory-confirmed respiratory infection and influenza-like illness using N95 respirators and surgical masks**… The use of N95 respirators compared with surgical masks is not associated with a lower risk of laboratory-confirmed influenza" (emphasis added) [38].

➤ Cochrane systematic reviews of surgical infection control trials published since 2005 have consistently concluded that: "**There was no statistically significant difference in infection rates between the masked and unmasked group in any of the trials**" (emphasis added) [33-35]. The conclusion of the 2023 Cochrane review, "Physical interventions to interrupt or reduce the spread of respiratory viruses," was no different [20]. The authors reviewed twelve of the Randomized Controlled trials already summarized [3, 6, 7, 9-12, 15-17, 23, 45], with over 275,000 total participants, and concluded with "moderate certainty": "Wearing masks in the community probably makes little or no difference to the outcome of influenza-like illness (ILI)/ COVID-19 like illness compared to not wearing masks." In a February 2023 interview, the review's lead author, Dr. Tom Jefferson, stated: "the evidence really didn't change from 2020 to 2023. There's still no evidence that masks are effective during a pandemic" [21]. What *did* change in 2020 was the politics around masks. In that same interview, Dr. Jefferson revealed that the 2020 edition of the Cochrane review on masks [46] — which included observational studies and also showed no benefit from masks — was ready for publication early in 2020. However, publication was mysteriously and unexpectedly delayed by 7 months for extra "peer review." That 7-month delay-period was when the mask hysteria really took off and ossified. Coincidentally, Bill Gates became one of Cochrane's major donors in 2016 [47].

➢ Earlier systematic reviews and studies were consistent with these, such as Bahli's 2009 conclusions after surveying trials from 1966 to 2007: "it is still not clear that whether wearing surgical face masks harms or benefit[s] the patients undergoing elective surgery" [29].

In 2012, Bin-Reza et al. conducted a systematic Literature Review which included 17 randomized controlled trials. The authors wrote: "None of the studies established a conclusive relationship between mask/respirator use and protection against influenza infection... Further, a simulation study found that strict adherence to guidance about personal protective equipment (which included masks and respirators) compromised normal ward functioning in a UK hospital setting" [30].

Two years later, a 2014 Danish systematic review of the mask-related scientific literature concluded: "[... **the studies unequivocally show that it does not have any effect on the incidence of postoperative infections whether the staff wears a mask or not**]" (emphasis added, translated from Danish) [31].

➢ Even a systematic literature review favorable towards masks, and which only included one randomized controlled trial with masks as an intervention [3], was still forced to concede (in a turn of phrase worthy of the most brilliant marketers) that "facemask use provided a **non-significant protective effect**" against influenza viral infections (emphasis added) [40]. Would you pay money to wear a piece of armor with a non-significant protective effect? Would you pay for a drug with a non-significant therapeutic effect?

➢ A 2019 literature review of "the evidence behind the ability of gloves, masks, gowns, drapes, head covers, footwear, and ventilation systems to prevent SSIs [surgical site infections]," drew attention to the following anecdote: "A study by Meleney and Stevens suggested that mask use may reduce the incidence of postoperative hemolytic streptococcus wound infections to 5%. The practice of wearing masks during surgery subsequently became more widespread. **Interestingly, the same authors refuted their initial findings 9 years later and reported infection rates with consistent mask use to be much higher than had been anticipated in their initial study. Recent studies comparing outcomes with or without surgical face masks have found little to no difference in SSIs** [surgical site infections]" [emphasis added] [36].

➢ As part of their own 2019 randomized controlled trial, Wang and colleagues conducted an additional internal meta-analysis of previous randomized controlled trials. At this point, I'm sure you are shocked to find out that they, too, found a total lack of efficacy from masks to prevent influenza or influenza-like illness [13].

➢ The verdict based on the totality of evidence from Randomized Controlled Trials did not change when Jefferson et al. revisited it in a 2020 Cold Spring Harbor Laboratory meta-analysis: "**Compared to no masks there was no reduction of influenza-like illness (ILI) cases or influenza for masks in the general population, nor in healthcare workers**" (emphasis added) [32].

Observational studies show masks do not work

There are also more than 40 observational studies covering situations from school districts to surgical suites which on their own provide strong evidence that masks do not provide a statistically significant infection control benefit even for bacteria, much less for respiratory viruses such as COVID-19. Many of these were in schools.

➤ A 2022 observational study in North Dakota was practically a naturally occurring randomized controlled trial, comparing two adjacent school districts with a combined total of over 23,000 students and congruent population demographics which differed only on mask policies, and covered a 5-month period overlapping both the delta and omicron waves of COVID. The authors concluded: "We observed no significant difference between student case rates while the districts had differing masking policies" [48].

➤ Another analysis of COVID-19 case data from all 50 states over a 9-month period from June 2020 to March 2021 concluded: "mask mandates and use are not associated with lower SARS-CoV-2 spread among US states" [49]. Post-mandate growth curves for non-mandate states, early mandate states, and late-mandate states were ultimately identical.

➤ A 2022 study by Chandra and Høeg [50] replicated a highly cited 2021 CDC publication by Budzn et al. [51]. The CDC's study used data from school districts in 520 counties and purported to show a negative association between school mask mandates and pediatric SARS-CoV-2 cases from July to October in 2021. Chandra and Høeg's replicating study found that not only did the CDC's study end at precisely the best chronological stopping point to make the CDC's case that masks work, but when the period of time and school counties were extended to provide a broader, longer, more representative sampling of 1,832 counties, the CDC's results disappeared or reversed, even when still using the CDC's own methodology. Drs. Chandra and Høeg were polite but unambiguous:

> "We replicated the CDC study and extended it to more districts and a longer period, employing seven times as much data.

> "… using the same methods and sample construction criteria as Budzyn et al., but a larger sample size and expanded time frame for analysis, we fail to detect a significant association between school mask mandates and pediatric COVID-19 cases.

"…while the presence of correlation does not imply causality, the absence of correlation can suggest causality is unlikely, especially if the direction of bias can be reasonably anticipated.

"…studies with significant findings are more likely to be published than those with non-significant or negative findings. This is particularly important in the context of the current pandemic where publishing studies that fit a certain narrative can become a self-fulfilling prophecy rather than an unbiased pursuit of truth" [50]

"While the presence of correlation does not imply causality, the absence of correlation can suggest causality is unlikely, especially if the direction of bias can be reasonably anticipated."

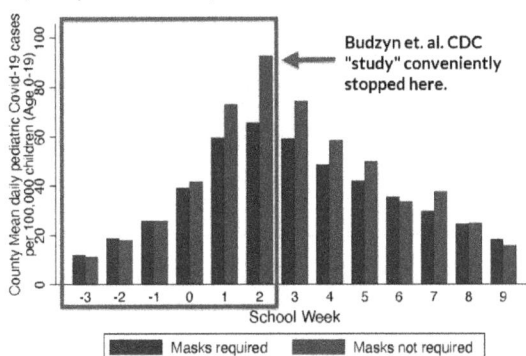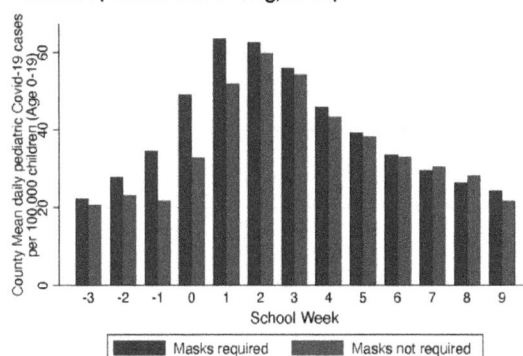

Chandra, A. and T.B. Høeg, *Lack of correlation between school mask mandates and paediatric COVID-19 cases in a large cohort.* Journal of Infection, 2022. **85**(6): p. 671-675. https://dx.doi.org/10.1016/j.jinf.2022.09.019

➤ Likewise, a retrospective study in Catalonia, Spain, comparing 3- to 11-year-olds, looked at over 599,000 children during the first trimester of the school year *September-December 2021*, especially non-masked 3- to 5-year olds compared with masked 6- to 11-year olds. They found that the incidence of COVID correlated with *age*, but not with *masks*. The authors reported: "**FCM [Face-Covering Mask] mandates in schools were not associated with lower SARS-CoV-2 incidence or transmission**, suggesting that this intervention was not effective" [52].

These results matched those obtained by a separate group of researchers the year earlier in the same Spanish province when evaluating schoolchildren from *September to December 2020* (see figure below). [53]

Alonso et al   *The Pediatric Infectious Disease Journal* • Volume 40, Number 11, November 2021

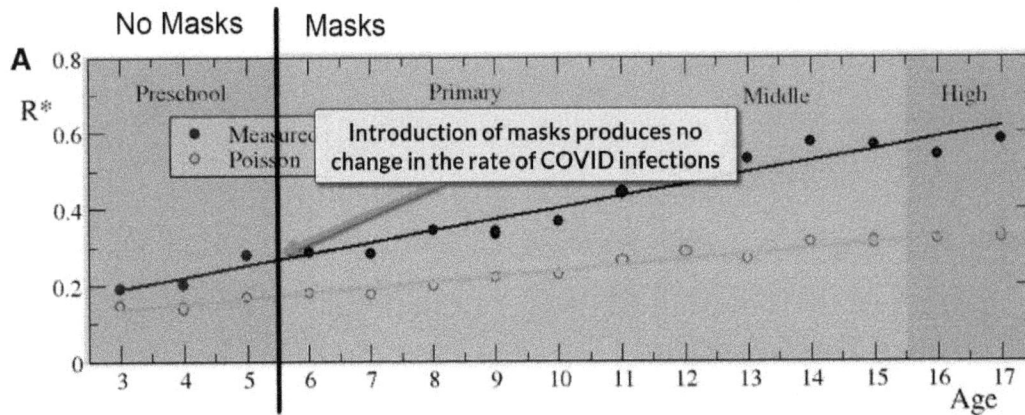

Alonso, S., et al., *Age-dependency of the Propagation Rate of Coronavirus Disease 2019 Inside School Bubble Groups in Catalonia, Spain.* The Pediatric infectious disease journal, 2021. **40**(11): p. 955-961. https://pubmed.ncbi.nlm.nih.gov/34321438

➤ These results are also consistent with those of a smaller (314 patients, 753 contacts) retrospective study of Catalonian Spanish households published in *The Lancet Infectious Diseases* in 2021, which reported that: "**We observed no association of risk of transmission with reported mask usage by contacts**, with the age or sex of the index case, or with the presence of respiratory symptoms in the index case at the initial study visit" [54].

➤ A 2022 study in Finland by authors affiliated with the Finnish Institute For Health and Welfare, Department of Health Security compared the COVID incidence among 10- to 12-year-olds in two Finnish cities: Helsinki, which did not have a school mask mandate for ages 10-12, and Turku, which did have such a mandate [55]. Here, again, COVID numbers varied wildly at any given time, but when all was said and done, there was no net benefit to masking school children.

> "According to our analysis, no additional effect seemed to be gained from this, based on comparisons between the cities and between the age groups of the unvaccinated children (10-12 years versus 7-9 years).

> "… one would expect to see some differences in the age-specific incidences if masking was an effective way to control transmission in schools."

➤ A U.S. study which compared case rates in masked vs. unmasked schools in Florida from October 2020 to April 2021 concluded: "**we do not see a correlation between mask mandates and COVID-19 rates among students in either adjusted or unadjusted analyses**" (emphasis added) [56].

➢ Yet another retrospective study compared COVID outcomes of 35 European countries over the fall and winter of 2020 to 2021 (from October 1, 2020 to March 31, 2021). [57] "Surprisingly, **weak positive correlations were observed when mask compliance was plotted against morbidity (cases/million) or mortality** (deaths/million) in each country... The positive correlation between mask usage and cases was not statistically significant, while the correlation between mask usage and deaths was positive and significant... **These findings indicate that countries with high levels of mask compliance did not perform better than those with low mask usage**" [57]. We will revisit this particular study at the conclusion of Part 2 when we talk about mask side effects.

➢ A 2023 observational study looking at U.S. data from all 50 states from January 2020 to July 2022 found that mask mandates were not associated with any difference in cumulative death rates. Though the authors reported a non-significant decrease in infection rates associated with mask mandates, even this was obtained only by excluding omicron infections and by using IHME infection estimates arrived at via modeling [58].

Bollyky, T.J., et al., *Assessing COVID-19 pandemic policies and behaviours and their economic and educational trade-offs across US states from Jan 1, 2020, to July 31, 2022: an observational analysis.* The Lancet, 2023.
https://dx.doi.org/10.1016/s0140-6736(23)00461-0

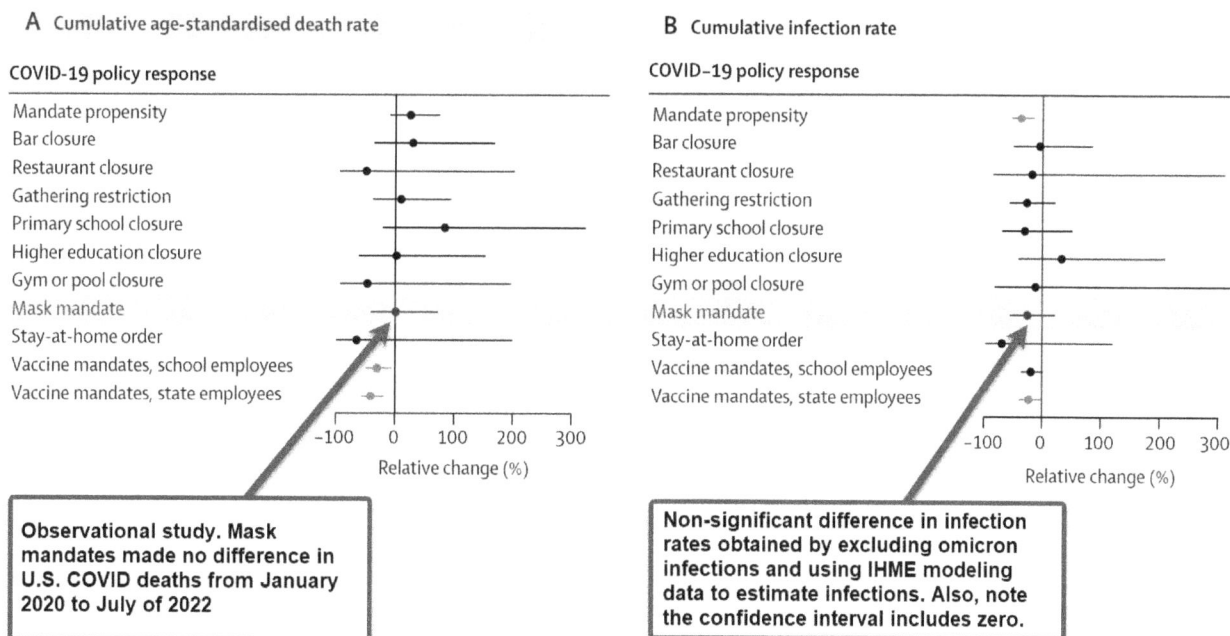

A Cumulative age-standardised death rate

B Cumulative infection rate

Observational study. Mask mandates made no difference in U.S. COVID deaths from January 2020 to July of 2022

Non-significant difference in infection rates obtained by excluding omicron infections and using IHME modeling data to estimate infections. Also, note the confidence interval includes zero.

➢ Moving on to observational studies on masks in healthcare settings, endophthalmitis is a bacterial infection of the tissue or fluids of the eyeball. It is a medical emergency which can easily result in permanent blindness. A 2021 study reviewing 483,622 intravitreal injections (shots given into the fluid of the eyeball through the front of the cornea) found that "physician face mask use did not affect the overall rate of postinjection endophthalmitis" [59]. There is no reason to think that masks will affect the transmission rate of a virus if they do not provide any statistically significant benefit against bacteria which are 10x the size. These study findings also undermined the rationale for wearing goggles and face shields to stop COVID infections.

➤ These findings that masks do nothing go back decades. In 1980, surgeon Neil Orr discontinued the use of masks in a surgical ward in England and found no increase in wound infections compared with the previous 4 years. Dr. Orr reported his findings in the *Annals of the Royal College of Surgeons*. "Nose and throat swabs were taken from all theatre personnel monthly or when they had a cold… **No restrictions in theatre were imposed on talking, movement, beards, or colds. In fact the theatre routine remained unchanged except that no one wore a mask.**" Not only did infections in the hospital ward *not* increase, Dr. Orr actually reported an overall *decrease* in wound infections when mask use was discontinued. Moreover, "The 8 infections which did occur bore no relation to the throat or nose cultures from the theatre team" [60]. In response to inquiries, Dr. Orr subsequently confirmed that this decrease in infections was not due to any changes in antibiotic regimens, because antibiotic regimens did not change during the trial [61].

➤ In 1996, a six month prospective observational trial was conducted at the Sheffield Northern General Hospital in England, comparing the rates of post-surgical infection when the surgical team wore open faceshields ("visors") vs. medical masks. Procedures performed this way included invasive thoracic surgeries [62]. The researchers found a 3.1% post-operative infection rate in the faceshield group, and a 4.4% post-operative infection rate in the mask group. This difference was not statistically significant, and is the *opposite* of what every COVIDcrisis theory supporting mask-use predicted.

➤ Similarly, Ruthman and colleagues reported in the *Illinois Medical Journal* that *no significant difference in emergency room infection rates* occurred when laceration repairs were done with or without caps and masks over the course of hundreds of procedures [63].

➤ A 1994 seroepidemiological case control study of respiratory virus infections among dental surgeons in Wales found that dentists generally had a higher prevalence of antibodies to influenza than the general population. What the study did *not* find, however, was any benefit from masks: "Wearing of masks or eye protection did not markedly reduce infection with these viruses among the dentists" [64].

A June 2020 survey of over 2,000 dentists published in the *Journal of the American Dental Association* [65] found a mere 1% rate of COVID infection at the time, and did not note any differences in outcomes based on PPE between "Medium Risk" dentists who wore respirator-type masks, and "High Risk" dentists who did not wear both face masks and eye protection.

In a 2021 study where dental treatment was performed on asymptomatic COVID patients, the authors reported that, "We also identified low copy numbers of SARS-CoV-2 virus in the saliva of several asymptomatic patients **but none in aerosols generated from these patients.**" The authors concluded that: "**dental treatment is not a factor in increasing the risk for transmission of SARS-CoV-2 in asymptomatic patients** and that standard infection control practices are sufficiently capable of protecting personnel and patients from exposure to potential pathogens" [66].

➢ A small study done in a Denver, Colorado pediatric hospital found: "gowning and masking did not appear to influence either illness or specific virus infection" [67].

➢ A small 2009 study of hospital workers by Jacobs et al. in Japan found no difference in respiratory viral infection rates between the masked and non-masked study cohorts [45].

➢ A 2001 German study concluded that "[**Surgical face masks worn by patients during regional anaesthesia did not reduce the concentration of airborne bacteria over the operation field in our study. Thus, they are dispensable**]" (emphasis added, translated from German) [68].

➢ A 2007 Saudi Arabian case-control study conducted on 250 medical personnel treating Hajj pilgrims reported: "**in our study regular use of facemasks offered no significant protection against ARI [Acute Respiratory Infection]. Our finding is in agreement, however, with the conclusion of the Centers for Disease Control and Prevention (CDC) in the USA which stated that surgical masks are not designed for use as particulate respirators and do not provide much protection against air-borne diseases** because they do not effectively filter small particles from the air or prevent leakage around the edge of the mask when the user inhales. **Furthermore, we found that intermittent use of surgical-type masks was actually associated with more than a 2.5-fold greater risk of infection**" (emphasis added). The study authors concluded: "The common practice among pilgrims and medical personnel of using surgical facemasks to protect themselves against ARI should be discontinued" [69].

➤ Another case-control study of 254 healthcare workers published in 2013 examining the 2009 spread of H1N1 influenza in Beijing concluded: "We were unable to demonstrate any impact of masks or hand washing in HCWs [Health Care Workers] against pandemic influenza" [70].

➤ A 2002 Danish study on the use of masks during percutaneous cardiac catheterization [71] got the same results as a 1989 study [72], which reported that caps and masks had no effect on post-operative infection rates and were unnecessary.

Catheterization and Cardiovascular Diagnosis 17:158–160 (1989)

Wearing of Caps and Masks Not Necessary During Cardiac Catheterization

Lawrence J. Laslett, MD, and Alisa Sabin

Although cardiac catheterization-related infections are rare, caps and masks are often worn to minimize this complication. However, documentation of the value of caps and masks for this purpose is lacking. We, therefore, prospectively evaluated the experience of 504 patients undergoing percutaneous left heart catheterization, seeking evidence of a relationship between whether caps and/or masks were worn by the operators and the incidence of infection. No infections were found in any patient, regardless of whether a cap or mask was used. Thus, we found no evidence that caps or masks need to be worn during percutaneous cardiac catheterization.

Key words: infection prevention and control; heart catheterization, adverse effects; percutaneous catheterization

➤ In a 2003 case-control study of 31 healthcare workers in a Toronto hospital who were exposed to the original SARS-CoV, half (3) of the resulting (6) infections occurred among the 13 healthcare workers who reported always wearing some kind of mask (including the 6 who wore N95s), and 3 infections occurred among the 18 who did not always wear a mask. Only 1 of the infections occurred among the sub-group of 8 who took "no precautions." [73] Congruently, a 2003 CDC MMWR report described 9 to 11 other healthcare workers in Toronto who were all infected by SARS-CoV when treating the same patient, despite the fact that: "… they all had worn the recommended personal protective equipment each time they entered the patient's room, including gown, gloves, PCM2000™ duckbill masks… and goggles with or without an overlying face shield" [74].

A 2004 retrospective study on the original SARS-CoV noted that of the 45 United States healthcare workers who had exposure to laboratory-confirmed SARS-CoV patients "without any mask use" during the course of a local outbreak, none were infected, despite the fact that: "An exposure was defined as any healthcare worker–patient interaction that occurred within droplet range (i.e., 3 feet)" [75]. Conversely, a separate 2004 study out of Hong Kong found that almost 100% of 72 study respondents who contracted SARS-CoV had done so despite wearing N95 masks [76].

Astute (or hostile) readers will note that the four studies just cited [73-76] are *consistent* with masks not working, but that masks not working is far from the only possible explanation for these study results. To this I say: "yes!" I would not have even bothered to include such weak studies in this section if the CDC had not used *even more* studies of exactly this type to argue in favor of masking. Hold on to your criticisms of these studies and apply them even-handedly when we review the evidence put forward by the CDC to argue that masks work.

➢ Masks were not mandated on flights to Canada until April 15, 2020. On January 22, 2020, a 15-hour flight carried 350 mostly unmasked passengers and one man later determined to have been symptomatic with COVID-19 from Wuhan to Guangzhou, then Guangzhou to Toronto. Despite this extremely close contact and conspicuous *lack* of masks for more than half a day, the only other passenger subsequently found by nasopharyngeal and throat swabs to be positive for COVID-19 was the man's wife [77]. On the other end of the spectrum, a November 2021 brief report in *Clinical Infectious Diseases* described multiple cases of SARS-CoV-2 transmission despite the carrier being masked, the exposed person being masked, or both persons being masked [78].

➢ On average, the highest risk setting for transmission of COVID-19 is within households, and 3 meta-analyses from 2020 and 2021 which included 54, 92, and 97 studies, respectively, found that the secondary attack rate averaged from 18% to 21% (95% Confidence Intervals ranged from 16% to 24%) [79-81]. This is worth noting, because a 2021 study of a COVID-19 delta variant outbreak in an Israeli hospital [82] found that despite a 96% vaccination rate and strict hospital-grade masking, "The calculated attack rate among all exposed patients and staff was 10.6% (16/151) for staff and **23.7% (23/97) for patients." In other words, the observed secondary attack rate of SARS-CoV-2 among fully masked and vaccinated patients was *as high or higher* than that observed in meta-analyses of intra-household transmission.**

➢ Based on those cited above and other studies, multiple reviews of laboratory studies and the observational literature have come to conclusions like those of the randomized controlled trial reviews and meta-analyses cited earlier.

A 1996 literature review published in the *British Journal of Theatre Nursing* said: "Current practices of operating room management and sterile technique are direct descendants of the elaborate principles of antisepsis and asepsis set down by [Dr. Joseph] Lister… the available clinical data suggests that the present generation of masks does not protect staff either from airborne bacteria or Hepatitis B virus (Ransjo 1986, Reingold 1988)… **it is now generally accepted that the wearing of masks for ward procedures is unnecessary** (Taylor 1980)… **Some studies suggest that surgical face masks might actually increase the incidence of surgical wound infection by increasing the shedding of facial skin** (Letts 1983). **Another hypothesis may be that by discarding masks individual nasal and oral droplets might be more likely to atomise and remain airborne** (Quesnel 1975)" (emphasis added) [83].

Another literature review from 2001 in the journal *Anaesthesia and Intensive Care* said: "A questionnaire-based survey, undertaken by Leyland' in 1993 to assess attitudes to the use of masks, showed that **20% of surgeons discarded surgical masks for endoscopic work**…. Equal numbers of surgeons wore the mask in the belief they were protecting themselves and the patient, **with 20% of these admitting that tradition was the only reason for wearing them**… **There is little evidence to suggest that the wearing of surgical face masks by staff in the operating theatre decreases postoperative wound infections. Published evidence indicates that postoperative wound infection rates are not significantly different in unmasked versus masked theatre staff**" (emphasis added) [84].

This continued into reviews in the 2010s. A 2014 literature review published in *The Journal of Bone and Joint Surgery* said: "**Surgical masks have not been shown to reduce rates of surgical site infection in the operating room**" (emphasis added) [85]. A year later, another author published in the *Journal of the Royal Society of Medicine* wrote uncontroversially: "… **overall there is a lack of substantial evidence to support claims that face-masks protect either patient or surgeon from infectious contamination**" (emphasis added) [86].

A living, rapid review published in the *Annals of Internal Medicine* in 2020 [87] and repeatedly updated until August 2022 stated: "**Randomized trials in community settings found possibly no difference between N95 versus surgical masks and probably no difference between surgical versus no mask in risk for influenza or influenza-like illness**" (emphasis added) [87]. Even in the case of those observational studies which may suggest some benefit from masks, "risk estimates were not statistically significant" [88].

The advent of SARS-CoV-2 did not somehow magically invalidate all the evidence on masks laboriously collected over the prior decades. Nevertheless, the political pressure for compulsory masking was so great that at least one article published in the journal *Medical Hypotheses* which questioned the efficacy of masks (citing, among other things, the WHO's and CDC's own recommendations from early 2020) was retracted by the editors following backlash [89]. Meanwhile, other far more deficient analyses promoting masking remained untouched.

➢ Multiple studies touted by the CDC to support mask-wearing turn out to show the opposite on closer examination. For example, a 2020 "population-based intervention with trend analysis" by Van Dyke et al. published in the CDC's August 2020 MMWR has been widely cited as evidence favoring the efficacy of facemasks [90]. The CDC leapt to put out a summary graphic which said: "Kansas implemented a mask mandate on July 3; some counties opted out. [In] counties with a mask mandate, new cases per 100,000 people decreased 6%. [In] counties without a mask mandate, new cases per 100,000 people increased 100%. CDC recommends everyone age 2 years and older wear masks in public."

However, that is not what this study, dubbed the "Kansas Mask Mandate Study" actually shows. In reality, over the six-week period evaluated from July 3, 2020 to August 21, 2020, COVID cases in the mask mandate counties actually *increased* far more quickly than they did in the non-mask mandate counties. The numbers used were publicly available, and I personally looked them up at the source. The *non-masked* Kansas counties began the study period with almost double the COVID cases of the interventionist mask mandate counties, but *finished* the study period with almost the same incidence of COVID cases per 100,000 people. Over the course of the study period, COVID cases in the mask mandate counties increased almost twice as quickly as they did in the *non*-mandate counties!

Kansas COVID Cases per 100,000 Population		
County Cohort	3 July 2020	21 August 2020
Mask-Mandate Counties	411	1,262
Non-Mask-Mandate Counties	825	1,271
Source: Kansas Dept of Health and Environment - COVID Summary		
https://www.coronavirus.kdheks.gov/160/COVID-19-in-Kansas		

The Kansas Mask Mandate Study authors reached their presented conclusion that masks slow viral spread by graphing a period of time where the rate of COVID acceleration (the *change* in the number of "new cases," known as "incidence") decreased slightly in the mask mandate counties, but was still increasing in the non-masked counties. Their conclusion says nothing about the actual *total* number of new cases of COVID in these counties (the velocity of COVID). Furthermore, as the study's own tables reveal, the Kansas counties that never mandated masks *still* ended the study with a *lower* overall COVID acceleration than the masked counties.

Instead of being straightforward like Bundgaard et al.'s randomized controlled trial Danish Mask Study, [16] the CDC's Kansas observational mask mandate study looked at the rate-of-change of the rate-of-change of COVID cases, and then cherry-picked a time period to get the results that best supported their desired conclusion. If the Kansas Mask Mandate Study authors had extended their study time period by just a week or two in either direction, they would have had to flip their conclusion, and as it is, they *still* could not completely hide the fact that the *non-masked* counties in their study did *better* overall. **Far from providing evidence in favor of mask mandates, the data underlying the Kansas Mask Mandate Study, if it does anything, actually provides evidence that mask mandates make things *worse*.**

The CDC's Deceptive Kansas Mask Mandate Study

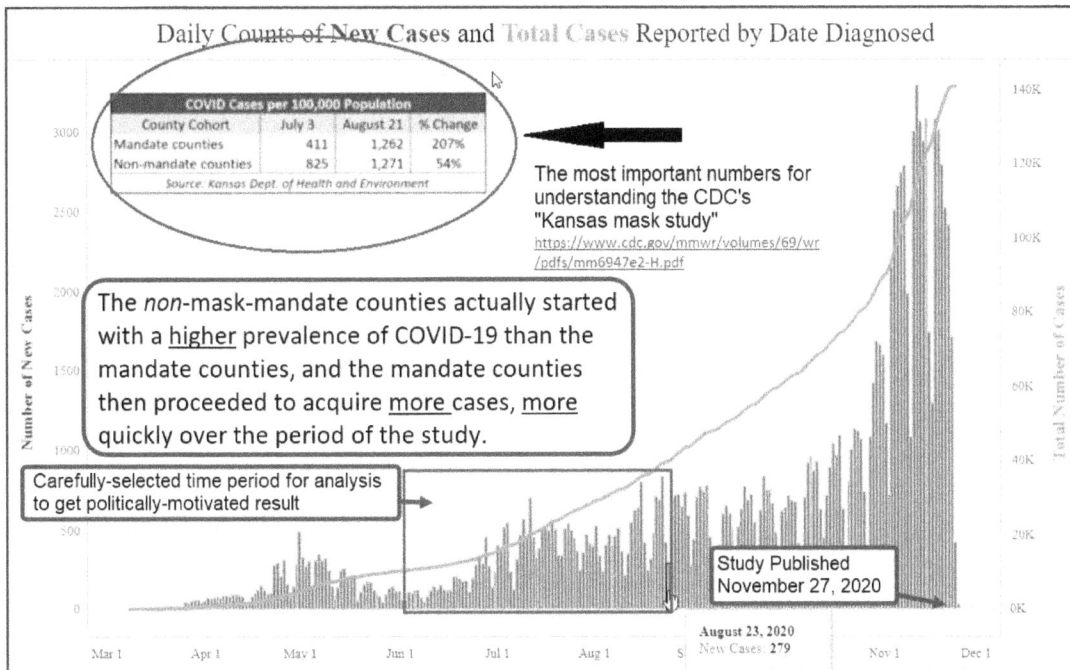

Daily Counts of New Cases and Total Cases Reported by Date Diagnosed

COVID Cases per 100,000 Population			
County Cohort	July 3	August 21	% Change
Mandate counties	411	1,262	207%
Non-mandate counties	825	1,271	54%
Source: Kansas Dept. of Health and Environment			

The most important numbers for understanding the CDC's "Kansas mask study"
https://www.cdc.gov/mmwr/volumes/69/wr/pdfs/mm6947e2-H.pdf

The *non*-mask-mandate counties actually started with a <u>higher</u> prevalence of COVID-19 than the mandate counties, and the mandate counties then proceeded to acquire <u>more</u> cases, <u>more</u> quickly over the period of the study.

Carefully-selected time period for analysis to get politically-motivated result

Study Published November 27, 2020

August 23, 2020
New Cases: 279

COVID cases in Kansas March-December 2020

Source: Kansas Department of Health and Environment - COVID Summary:
https://www.coronavirus.kdheks.gov/160/COVID-19-in-Kansas

TABLE. Confirmed COVID-19 infection 7-day rolling average case counts, rates, and percentage changes, by mask mandate status[*,†] and period — Kansas, June 1–August 23, 2020

Characteristic	Before executive order June 1–June 7	Executive order effective[§] July 3–9	After executive order August 17–23	% Change in incidence[¶] June 1–7 versus July 3–9	July 3–9 versus August 17–23
Mandated counties (N = 24)[*,]**		Mask mandate fails to significantly change incidence in the intervention counties.			
No. of daily cases[††]	60	333	310	N/A	N/A
Incidence[§§]	3	17	16	467	–6
Nonmandated counties (N = 81)[†,]**	Comparable incidence starting points		Non-mandate counties STILL doing better using these metrics		
No. of daily cases[††]	40	59	118	N/A	N/A
Incidence[§§]	4	6	12	50	100

Abbreviations: COVID-19 = coronavirus disease 2019; mandated = counties with a mask mandate; N/A = not applicable; nonmandated = counties without a mask mandate.
* Counties that as of August 11 did not opt out of the state mandate or adopted their own mask mandate shortly before or after the state mandate include Allen, Atchison, Bourbon, Crawford, Dickinson, Douglas, Franklin, Geary, Gove, Harvey, Jewell, Johnson, Mitchell, Montgomery, Morris, Pratt, Reno, Republic, Saline, Scott, Sedgwick, Shawnee, Stanton and Wyandotte. Total population in mask-mandated counties = 1,960,703 based on 2019 U.S. Census Bureau data.
† Counties that took no official action to opt out of the state mask mandate or adopted their own mask mandate shortly before or after the state mandate were considered to have a mask mandate in place. Counties were considered to not have a mask mandate in place if they took official action to opt out of the state mask mandate and did not adopt their own mask mandate or if their official action used only the language of guidance (e.g., "should" or "recommend"). Total population in non–mask-mandated counties = 952,611 based on 2019 U.S. Census Bureau data.
§ Week of governor's executive order (effective July 3, 2020).
¶ Change in incidence = [(incidence in period – incidence in previous period)/incidence in previous period] X 100.
** Data on county orders were collected through point-in-time surveys of local health department and other county officials and were supplemented with online searches for published orders and announcements on social media and local news sites. Text in the county orders was analyzed to determine whether mask mandates were in place as of August 11, 2020.
†† Seven-day rolling average number of new daily cases.
§§ Seven-day rolling average number of new daily cases per 100,000 population.

Van Dyke, M.E., et al., "Trends in County-Level COVID-19 Incidence in Counties With and Without a Mask Mandate - Kansas, June 1-August 23, 2020." *MMWR Morb Mortal Wkly Rep*, 2020. **69** (47): p. 1777-1781. https://www.cdc.gov/mmwr/volumes/69/wr/mm6947e2.htm

The authors of a July 2020 CDC MMWR report evaluated SARS-CoV-2 transmission among nearly 600 overnight campers in Georgia, where cloth masks were mandated for the staff but not for the campers [91] The cloth-masked staff subsequently experienced the highest attack rate of SARS-CoV-2 infection (56% compared to the campers' 33%-44%), and the authors reported this striking observation while simultaneously asserting that cloth masks have been shown to reduce the risk of infection from COVID-19 (citing only the CDC's MMWR Missouri Hair Salon no-control-group "study" to support that claim [92] — more on *that* particular embarrassment later).

➢ Another CDC MMWR cross-sectional report surveyed 169 Georgia K-5 schools during the strikingly short and specific 4-week time period of November 16 to December 11, 2020 [93]. Much of this paper was devoted to hypothesizing about why a statistically significant result for masking teachers and staff was observed, but not for students ("The 21% **lower incidence in schools that required mask use among students was not statistically significant compared with schools where mask use was optional**."), before conceding that "**the data from this cross-sectional study cannot be used to infer causal relationships**."

➢ Another "population-based intervention with trend analysis" which at the time of this writing is still cited by the CDC to support mask-use was conducted in the Massachusetts General Brigham healthcare network, and published in the *Journal of the American Medical Association* [94]. The study authors tracked SARS-CoV-2 PCR test positivity rates among the hospital healthcare workers from March 1st to April 30th, 2020. They observed that the percent of healthcare workers testing positive for SARS-CoV-2 declined after a universal mask mandate was imposed on all staff and patients in the hospital system on April 6th, and immediately assumed that masks were the cause.

The Massachusetts General Brigham Mask Study (Wang, X, et al., 2020)

Purported to show that masks mitigate COVID despite:

- Questionable association of timing between mask mandates and the infection curve.
- Total lack of a comparison group.
- The general surrounding population displaying an infection curve with the exact same timing and progress despite having no mask mandate.
- Masks being just one of many interventions implemented during the same time period.

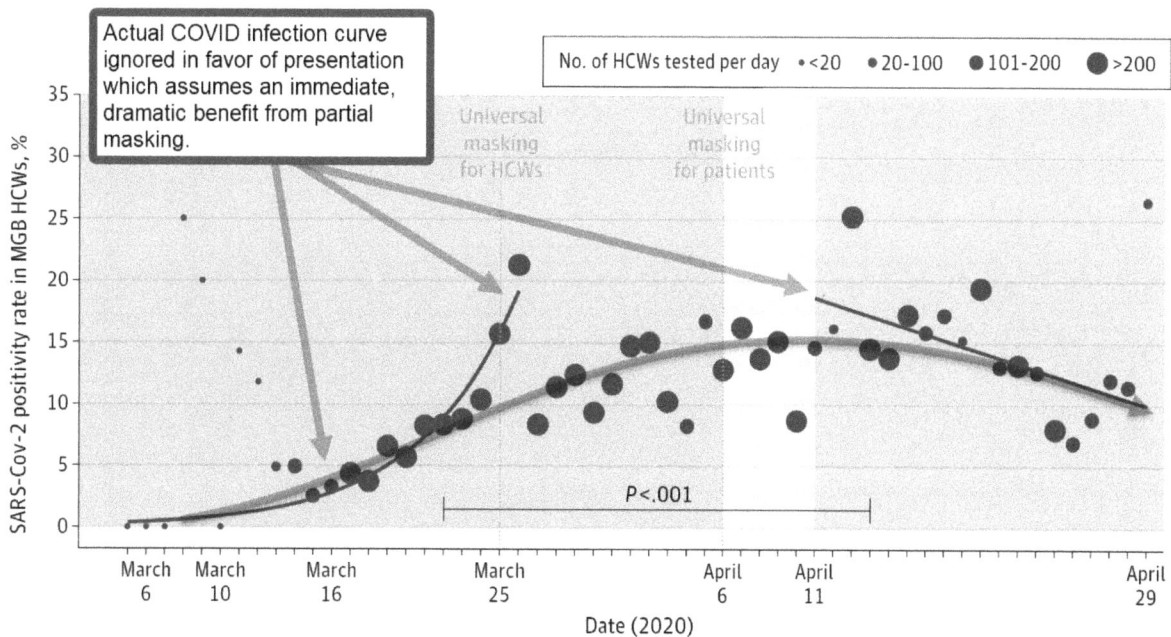

Wang, X., et al., "Association Between Universal Masking in a Health Care System and SARS-CoV-2 Positivity Among Health Care Workers." *Journal of the American Medical Association*, 2020. **324** (7): p. 703-4. https://jamanetwork.com/journals/jama/fullarticle/2768533

Unlike the Kansas Mask Mandate Study, the Massachusetts General Brigham Study lacks even the pretense of a control group, and relies entirely on chronological association. These fatal limitations did not prevent the authors from asserting: "we believe that our study provides definitive data on the value of universal masking in a healthcare setting during a pandemic, and that the results can be generalized to other settings even where social distancing is not possible." However, when the authors' findings are juxtaposed with publicly available data on COVID cases in Massachusetts as a whole during the same time frame (https://www.mass.gov/info-details/archive-of-covid-19-cases-in-massachusetts), what we see is two epidemic curves that follow an *identical* course. They even peak on the exact same day - April 13th. Moreover, it becomes obvious that Massachusetts Governor Baker's later masking order effective on May 6th — a full month after the hospital mandate — had no effect on the epidemic curve, either.

One might try to argue that the observed differences in SARS-CoV-2 test positivity rates between the Massachusetts General Brigham healthcare workers and Massachusetts general population — with the general population testing positive for COVID at a higher rate — should be ascribed to differences in PPE use. However, quite apart from the fact that masks were just one of many interventions aimed at stopping COVID imposed by the Massachusetts General Brigham hospital system during this same time period, there are at least two strong factors which, on their own, are adequate to explain any observed difference in test positivity rates.

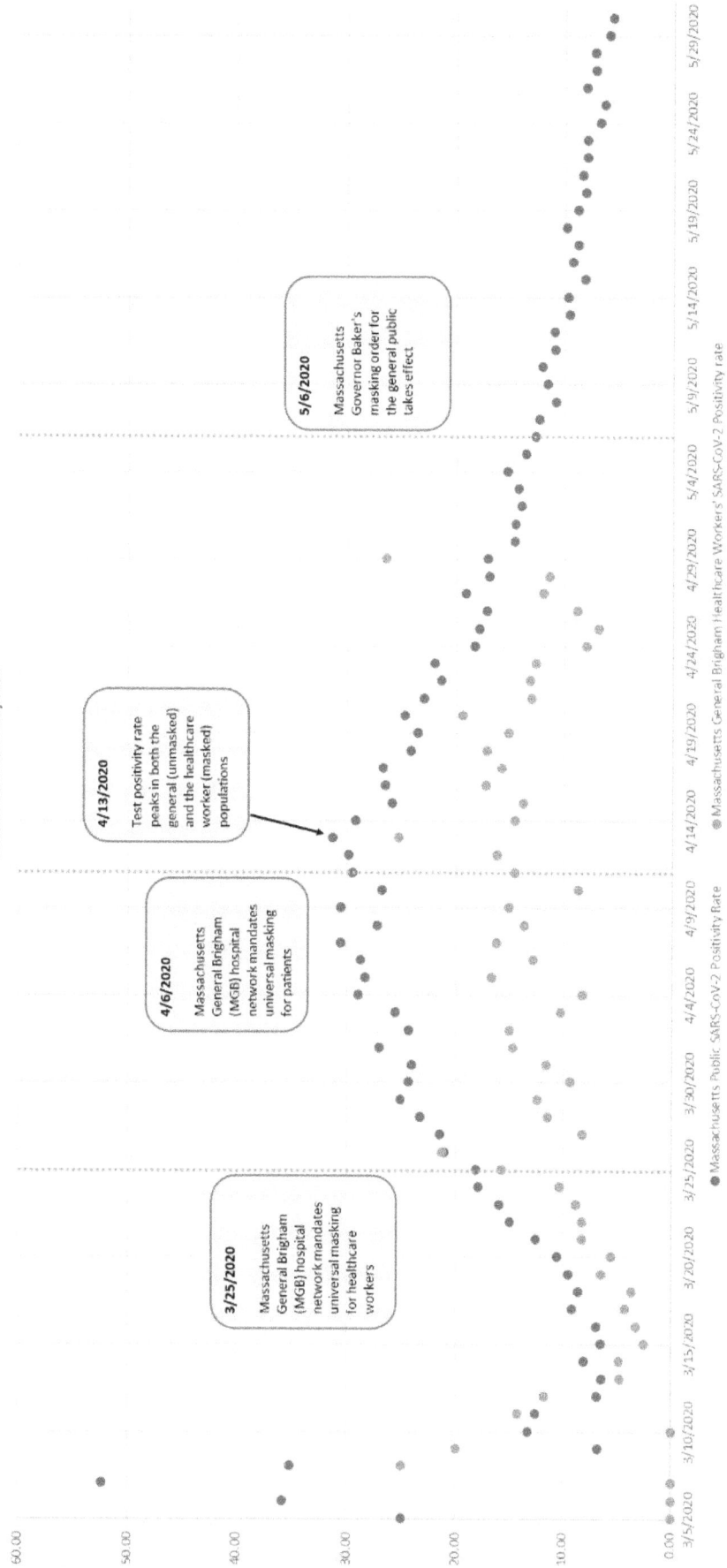

Massachusetts General Brigham Healthcare Workers' SARS-CoV-2 positivity rate vs. Massachusetts General Public SARS-CoV-2 positivity rate
5 March 2020 to 30 May 2020

3/25/2020

Massachusetts General Brigham (MGB) hospital network mandates universal masking for healthcare workers

4/6/2020

Massachusetts General Brigham (MGB) hospital network mandates universal masking for patients

4/13/2020

Test positivity rate peaks in both the general (unmasked) and the healthcare worker (masked) populations

5/6/2020

Massachusetts Governor Baker's masking order for the general public takes effect

● Massachusetts Public SARS-CoV-2 Positivity Rate
● Massachusetts General Brigham Healthcare Workers' SARS-CoV-2 Positivity rate

First, the general population had double-counting of cases. The MGB study authors eliminated duplicate positive results on the same person, whereas the Massachusetts health department did not do this. Second, we should expect the general population to have a higher positivity rate because the general population had a higher selection bias in favor of positive test results. During the time period studied, available tests were scarce. The hospital system had the resources to test all suspected cases among their employees, whereas the state as a whole could only test a small percentage. Thus, fewer of the less symptomatic cases among the general population were tested, which would produce a much higher overall positivity rate. **Thus, far from being evidence in favor of mask efficacy, when placed into the proper context, the Massachusetts General Brigham hospital study actually suggests that masks are *not* helpful.**

➤ In 2023, another hospital masking study was presented at the 33rd European Congress of Clinical Microbiology and Infectious Diseases in Copenhagen, Denmark. This study did everything that the Massachusetts General Brigham mask study authors neglected. Instead of looking at less than two months of data, the UK researchers followed COVID-19 infections at St. George's Hospital in southwest London for 40 weeks (nearly 10 months) in 2021-2022. The researchers compared the incidence of hospital-acquired Omicron COVID-19 infections when compulsory masking for staff and visitors was discontinued in the majority of wards at week 26 of the study period. The minority of wards which continued to require masks was used as a control group, and the researchers also compared the rate of COVID-19 in the hospital with that of the surrounding community. This later, longer, stronger study with multiple comparison groups found that the cessation of masking in hospitals was *not* associated with a statistically significant change in the rate of COVID-19 infections: "we found no evidence that a mask policy significantly impacts the rate of nosocomial [hospital-acquired] SARS-CoV-2 infection with the Omicron variant" [95].

Laboratory studies show masks do not work

The bulk of citations to support the efficacy of masking are mechanistic laboratory studies. Because they can be much more tightly controlled than real-world scenarios, and examine one or two features of masks in isolation, laboratory experiments by their nature are the most likely to find mask efficacy. However, even in this category of studies, there is an equal-or-greater weight of laboratory studies which call the efficacy of masks into question.

➤ By April 2 of 2020, very early in the events surrounding COVID, a study published in *The Lancet* observed that **SARS-CoV-2 remains viable for days longer on the inner and outer layers of**

medical masks than on wood, cloth, glass, stainless steel, or plastic [96]. The study found that while SARS-CoV-2 remained viable for less than three hours on tissue paper, and one or two days on cloth or glass, viable virus was recovered from the inner layer of medical masks after four days, and after up to seven days on the outer layer! This supplemental information was publicly available and easy to locate

SARS-CoV-2 remains viable longer on medical masks than on multiple other common surfaces.
Data published online April 2, 2020

Time	Virus titre (Log $TCID_{50}$/ml)									
	Paper		Tissue paper		Wood		Cloth		Glass	
	Mean	±SD	Mean	±SD	Mean	±SD	Mean	±SD	Mean	±SD
0 min	4.76	0.10	5.48	0.10	5.66	0.39	4.84	0.17	5.83	0.04
30 mins	2.18	0.05	2.19	0.17	3.84	0.39	2.84	0.24	5.81	0.27
3 hrs	U	-	U	-	3.41	0.26	2.21#	-	5.14	0.05
6 hrs	U	-	U	-	2.47	0.23	2.25	0.08	5.06	0.31
1 day	U	-	U	-	2.07#	-	2.07#	-	3.48	0.37
2 days	U	-	U	-	U	-	U	-	2.44	0.19
4 days	U	-	U	-	U	-	U	-	U	-
7 days	U	-	U	-	U	-	U			

SD = Standard Deviation

Time	Banknote		Stainless steel		Plastic		Mask, inner layer		Mask, outer layer	
	Mean	±SD	Mean	±SD	Mean	±SD	Mean	±SD	Mean	±SD
0 min	6.05	0.34	5.80	0.02	5.81	0.03	5.88	0.69	5.78	0.10
30 mins	5.83	0.29	5.23	0.05	5.83	0.04	5.84	0.18	5.75	0.08
3 hrs	4.77	0.07	5.09	0.04	5.33	0.22	5.24	0.08	5.11	0.29
6 hrs	4.04	0.29	5.24	0.08	4.68	0.10	5.01	0.50	4.97	0.51
1 day	3.29	0.60	4.85	0.20	3.89	0.33	4.21	0.08	4.73	0.05
2 days	2.47	0.23	4.44	0.20	2.76	0.10	3.16	0.07	4.20	0.07
4 days	U	-	3.26	0.10	2.27	0.09	2.47	0.28	3.71	0.50
7 days	U	-	U	-	U	-	U	-	2.79	0.46

Culture performed at room temperature: approximately 22 °C

$TCID_{50}$ = the median tissue culture infectious dose. A method of quantifying virus infectivity.

Supplement to: Chin, A.W.H. et al., "Stability of SARS-CoV-2 in different environmental conditions," *Lancet Microbe*, 2020. **1**(1): p. e10. https://pubmed.ncbi.nlm.nih.gov/32835322/

➢ In 2008, surgical masks were tested using NIOSH standards, and the results were published in the *American Journal of Infection Control*. The study authors reported that: "None of these surgical masks exhibited adequate filter performance and facial fit characteristics to be considered respiratory protection devices" [97].

In 2013, a team of researchers studying the protective effects of surgical masks admitted: "**Live, infectious virus was extracted from the air from behind all surgical masks tested.** This suggests that influenza virus can survive in aerosol particles that are able to bypass/penetrate a surgical mask" (emphasis added) [98].

➤ Consistent with the real-world observations and randomized controlled trials cited earlier which found that N95 respirators provide no statistically significant protective effect from respiratory viruses when compared with surgical masks, a 2022 laboratory study with a human subject using a bacteriophage virus as a proxy for SARS-CoV-2 conceded: "**even with the best fitting N95 mask there was still virus aerosol contamination of the nose** after 45min exposure to high virus aerosol load at close range in the absence of HEPA filtration" (emphasis added) [99].

This should have come as no surprise, because an earlier 2008 laboratory study had already found that N95 respirators perform at their worst when challenged by particles the size of a coronavirus [100]. Coronaviruses range in size from 0.06 to 0.14 µm (60 to 140 nm) [101]. The filter layer of N95 masks has an average pore size of about 30 µm (30,000 nm) or more, with the smallest pores still being at least 5 to 8 µm (5,000 to 8,000 nm) [102-104].

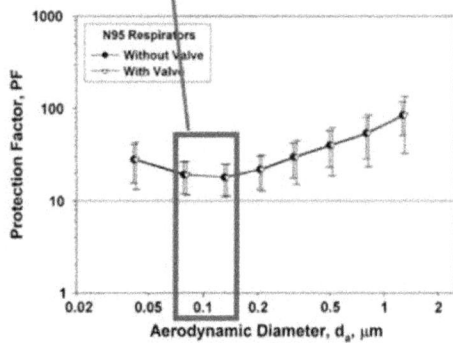

Fig. 3. The comparison of PFs against particles in bacterial and viral size ranges between respirators without (N95 Respirator C) and with (N95 Respirator D) exhalation valves. The tests were performed when the N95 respirators were donned on human subjects. Each data point represents an average and standard deviation of 36 observations for N95 Respirator C and 9 observations for N95 Respirator D.

"... **Most of the tested N95 respirators and surgical masks in this study were observed to perform at their worst against particles approximately between 0.04 and 0.2 µm, which includes the sizes of coronavirus and influenza virus.**"

Lee, S.A., S.A. Grinshpun, and T. Reponen, "Respiratory performance offered by N95 respirators and surgical masks: human subject evaluation with NaCl aerosol representing bacterial and viral particle size range." *The Annals of Occupational Hygiene*, 2008. **52**(3): p. 177-85. https://pubmed.ncbi.nlm.nih.gov/18326870/

Medical-surgical, cloth, and other types of masks perform even worse. The middle filter layers of medical-surgical masks have larger similar structures but average pore sizes ranging from 10 to 33 µm (10,000 to 33,000 nm) in diameter, but, importantly this filter layer is about one-half to one-quarter the thickness of the filter layer in an N95 [102]. (This is part of why surgical masks are easier to breathe through.) In polyurethane or cloth masks the pore sizes range from 100 to 500 µm (100,000 to 500,000 nm) [104, 105]. Suggesting that a respirator-type mask could protect against COVID was bad enough, but expecting a cloth or medical-surgical mask to slow down a respiratory virus is like expecting a wall perforated with football fields to stop a barbarian horde. Such suggestions should never have been taken seriously. Even assuming the viruses are carried on larger microdroplets in aerosols, they still have more clearance passing through masks than cars driving down 6-lane freeways or mosquitoes flying through chain-link fences.

Pore size isn't everything, but it might as well be.

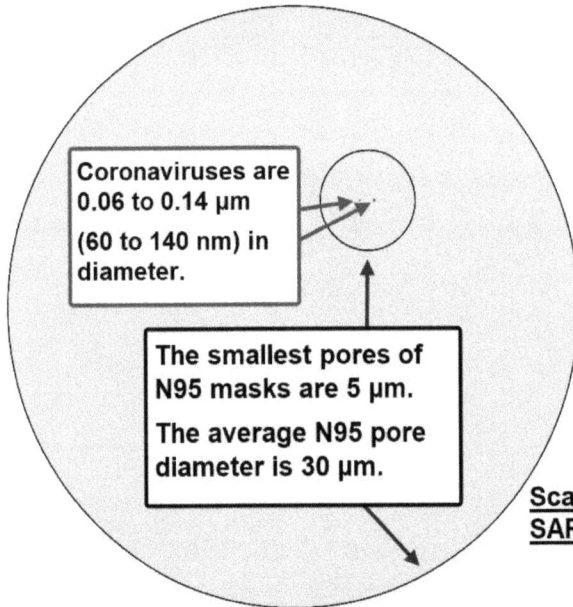

Coronaviruses are 0.06 to 0.14 μm (60 to 140 nm) in diameter.

The smallest pores of N95 masks are 5 μm.

The average N95 pore diameter is 30 μm.

Scale diagram comparing the size of SARS-CoV-2 to the pores of an N95.

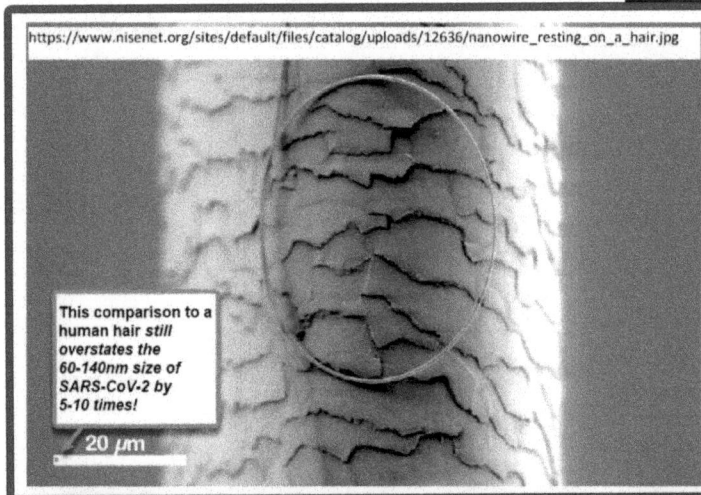

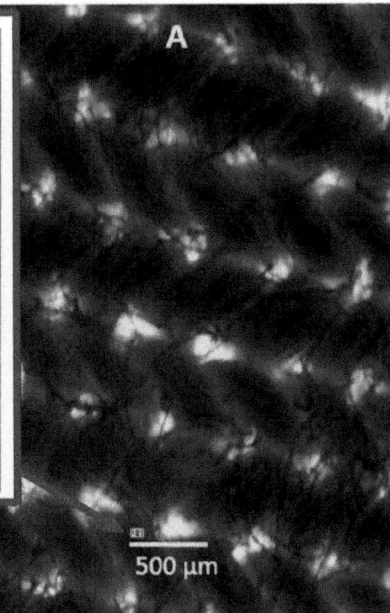

• Filter layer thickness, fiber diameter, and electrostatic charge make a difference, too, but pore size is the single most important determinant of a mask's filtration ability.

• Pointing out that virions can also be carried in larger droplets caught by masks does not defeat this basic point.

https://www.nisenet.org/sites/default/files/catalog/uploads/12636/nanowire_resting_on_a_hair.jpg

This comparison to a human hair *still overstates the* 60-140nm size of SARS-CoV-2 by 5-10 times!

20 μm

A

500 μm

Cloth mask pore size under bright-field microscopy.

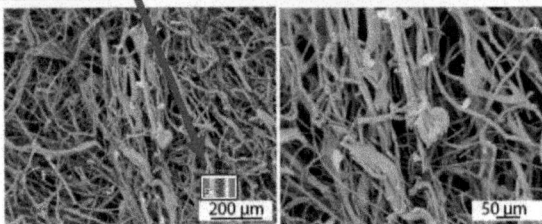

Neupane, B.B., et al., *Optical microscopic study of surface morphology and filtering efficiency of face masks.* PeerJ, 2019. 7: p. e7142, Figure 4.
https://dx.doi.org/10.7717/peerj.7142

200 μm

50 μm

3M 1860 N95 filter under scanning electron microscopy

Supplement to Yim, W., et al., "KN95 and N95 Respirators Retain Filtration Efficiency despite a Loss of Dipole Charge during Decontamination." *ACS Applied Materials & Interfaces*, 2020. **12**(49): p. 54473-54480, Figure S5B.
https://dx.doi.org/10.1021/acsami.0c17333

➤ Yet another 2013 experiment studying facemasks as source control found that after just two hours of use, the bacterial counts expelled from cloth and medical facemasks equaled or exceeded any initial reduction, and were on a trajectory to increase still further (the study did not examine counts beyond 2.5 hours) [106]. If masks produce this result with regard to bacteria, there is no reason to expect masks to perform better on viruses a fraction of the size.

Bacterial colony counts on culture plates 10-12 cm (roughly 4-5 inches) from the mouth

Table I. Colony counts after wearing fabric masks

Time	Average no. of bacteria (Mean ± SD)	p-value
Without Mask	5.36 + 4.38	
After 30 min	0.96 ± 1.06	< 0.001
After 1hr	2.33 ± 1.42	< 0.001
After 1.30 hr	3.23 ± 1.54	0.007
After 2hr	5.63 ± 4.02	0.67
After 2.30 hr	7.03 ± 4.45	0.019

Table II. Colony counts after wearing "Two – ply" disposable face masks

Time	Average no. of bacteria (Mean ± SD)	p-value
Without Mask	5.7 ± 2.99	
After 30 min	0.7 ± 0.87	<0.001
After 1hr	2.36 ± 1.03	<0.001
After 1.30 hr	4.16 ± 1.78	0.011
After 2hr	4.9 ± 1.98	0.161
After 2.30 hr	5.6 ± 2.21	0.951

Bacteria rapidly penetrate and reproduce within masks.
Bacteria expelled from masks equal or exceed non-masked condition within 2 to 2.5 hours.

Study did not extend past the 2.5 hour mark.

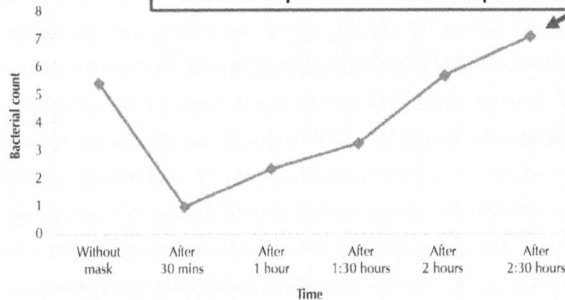

Figure 1. Average bacterial count plotted against time of wearing the fabric face mask

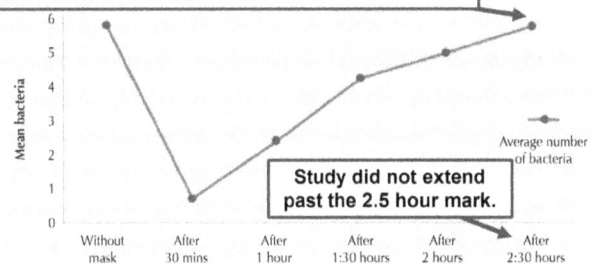

Figure 2. Average bacterial count plotted against time of wearing the disposable "Two – ply" mask

Kelkar, U.S., et al., *How effective are face masks in operation theatre? A time frame analysis and recommendations.* International Journal of Infection Control, 2013. 9(1).

➤ All these laboratory results are nothing new. Dr. W.H. Kellogg, M.D., Secretary and Executive Officer of the California State Board of Health observed that, "The masks, contrary to expectation, were worn cheerfully and universally, and also, contrary to expectation of what should follow under such circumstances, **no effect on the epidemic curve was to be seen**. Something was plainly wrong with our hypotheses" (emphasis added) [107]. Even using a bacterium more than ten times larger (and thus far easier to filter out) than SARS-CoV-2, Dr. Kellogg: "found that **with the element of aspiration introduced, as in the natural use of masks, even five layers did not give a sufficient reduction in count to make such a mask of value**" (emphasis added) [107]. Dr. Kellogg was writing in 1920 about the Spanish Flu, but his assessment is just as applicable to SARS-CoV-2 from 2020 to the present.

➢ One of the early lab studies on masks from 1975 was conducted in operating rooms using bacterial settle plates. The researchers were surprised to find that bacterial counts in the operating rooms: "were not influenced by the wearing of a face mask." [108]

As early as 1980, a team of researchers applied human albumin microspheres to the inner aspect of a surgeon's face mask before 20 major orthopedic procedures, the researchers were dismayed to find that: "Microspheres were retrieved from the wounds in all 20 experiments, thereby demonstrating wound contamination by the particles" [109].

A 1992 Swedish study examining air counts of colony-forming units (CFU) of bacteria over wounds *during live surgeries* found the same thing: "**the use of masks during operations does not influence the number of potentially pathogenic bacteria in the air close to the operation wound**" (emphasis added) [110].

Mask use and non-use did not impact levels of airborne bacteria around surgical sites.

Table I. *Air counts of bacteria (CFU) in the wound area during 14 operations on the thyroid and parathyroid glands*

Each operation included at least one 30 min period when personnel wore masks, and one during which they did not

Colony-forming units per cubic meter of air →	With masks (17 periods)		Without masks (28 periods)		Airborne bacterial levels were regularly sampled while masks were worn or not worn by _all_ members of the surgical team at different intervals during the surgeries.
	Mean CFU/m^3	Range of CFU/m^3	Mean CFU/m^3	Range of CFU/m^3	
Staph. epidermidis	6	0–32	5	0–18	
Corynebacterium spp.	2	0–6	1	0–6	
Propionibact. spp.	5	0–60	5	0–50	
α-Haemolytic strept.	0	0	1	0–3	
Other bacteria	2	0–16	1	0–10	

No statistically significant difference between masked and non-masked conditions during surgery.

Tunevall, T. Göran and Jörbeck, Hans, "Influence of wearing masks on the density of airborne bacteria in the vicinity of the surgical wound." European Journal of Surgery, 1992. **158**(5): p. 263-6.
https://pubmed.ncbi.nlm.nih.gov/1354489/

Another 1991 laboratory study in a surgical suite tried to determine if unmasked volunteers could contaminate bacterial settle plates. The results? "Oral microbial flora dispersed by unmasked male and female volunteers standing one metre from the table failed to contaminate exposed settle plates placed on the table… Further, volunteers were asked firstly to whisper and secondly to recite out loud with their mouths placed 15 cm from exposed culture plates for 5 min… **A small number of cfus [colony-forming units of bacteria] were found on settle plates situated on the operating table and the instrument tray when a volunteer recited out loud without** a mask. There was no contamination when up to four volunteers recited together standing one metre from the table… The numbers of airborne bacteria expelled from the nose and mouth are insignificant compared with the substantial numbers shed from the skin… These observations suggest that oral bacteria normally conveyed in droplets into the air during ordinary talking by non-scrubbed staff, who are not within the immediate vicinity of the operation site, do not pose an infection hazard and the wearing of masks is unnecessary" (emphasis added) [111]. Despite longstanding and numerous findings like these, activities like singing were frequently banned in the name of infection control during the COVIDcrisis.

A team of researchers in 2021 researching masks in the context of intravitreal ophthalmic injections found results that were congruent with the earlier studies: "In both the no talking and speech scenarios, **subjects wearing a tight-fitting face mask without tape had the highest bacterial growth—similar to, or even higher than, not wearing any mask**. It is possible that the tight-fitting face [mask] without tape may result in a greater amount of bacterial dispersion upward toward the subject's eye. Indeed, a recent study assessing respiratory droplet velocities with face mask use during simulated coughs reported that even with tight-fitting face masks, small openings can lead to leakage of droplets around the mask" [112].

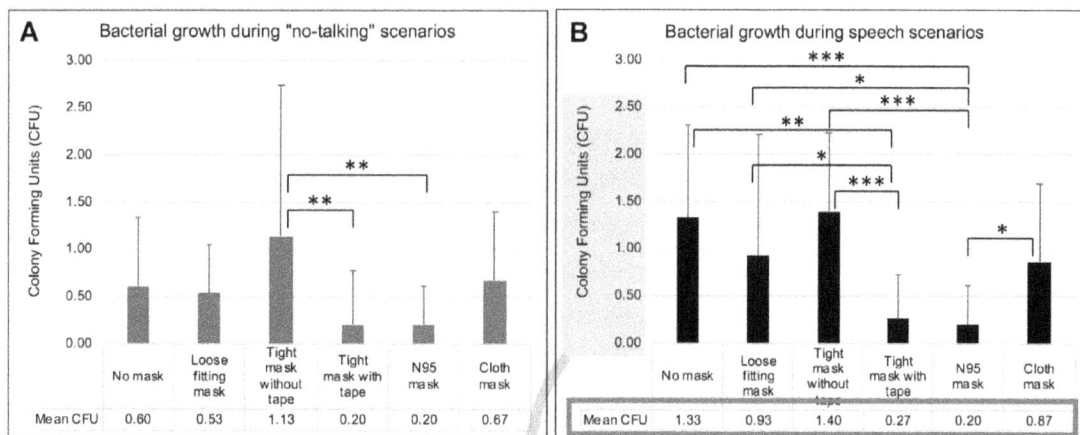

FIGURE 2. Mean bacterial growth based on colony forming units (CFUs) under various face mask conditions. A. No talking scenarios in which subjects were instructed to sit in silence for 2 minutes. B. Speech scenarios in which subjects were instructed to read a script for 2 minutes. Error bars represent standard deviations. *$P < .05$, **$P < .01$, and ***$P < .001$.

Patel, S.N., et al., *Bacterial Dispersion Associated With Various Patient Face Mask Designs During Simulated Intravitreal Injections.* American Journal of Ophthalmology, 2021. **223**: p. 178-183.

Note the scale of the chart. There is a difference of less than 1 mean colony-forming unit (CFU) between no mask and all other masks except for N95s and tight masks with tape, which produce a difference of only 1.13 CFUs *at most.*

These laboratory studies on masks combine to indicate that even under optimal conditions which do not reflect mask use in the real world, masks make, at best, a debatable and negligible difference in bacterial counts, producing no significant net effect on post-operative infection rates. Again, **if masks do not impede bacteria, there is no reason to think they will impede viruses which are less than one-tenth the size and mechanically far more difficult to exclude.**

➢ A light microscopy and filtration study published in 2019 highlights just two of the mechanisms by which any protective benefit from cloth masks may be offset on net balance: **stretching and washing associated with normal wearing causes a cumulative and rapid decline in the filtration capability of cloth masks** [105]. In this study, the researchers washed cloth masks four times and found that each washing resulted in about a 5% decline in filtration efficacy.

Importantly this study also re-highlighted one of the primary mechanical features which **makes a mockery of attempts to justify compulsory cloth mask use in the name of protection from a virus no larger than 0.14 μm: "The pore size of masks ranged from 80 to 500 μm**, which was much bigger than particular matter having [a] diameter of 2.5 μm or less" [105].

The effects of washing cloth masks described in this 2019 study take on even greater significance in light of a 2020 study from a team of researchers testing cloth masks made "**according to the CDC do-it-yourself instructions** for single- and double-layer t-shirt masks":

> "… wearing an unwashed single layer t-shirt (U-SL-T) mask while breathing **yielded a significant increase in measured particle emission rates compared to no mask,** increasing to a median of 0.61 particles/s. The rates for some participants (F1 and F4) exceeded 1 particle/s, representing **a 384% increase from the median no-mask value. Wearing a double-layer cotton t-shirt (U-DL-T) mask had no statistically significant effect on the particle emission rate, with comparable median and range to that observed with no mask"** (emphasis added) [113]."

In other words, the CDC instructions on how to make masks were useless *at best*. The observed particle emission rates of study participants during coughing and jaw movement were particularly noteworthy, with **the non-masked condition performing far better than the CDC-designed cloth masks**, comparable to the surgical masks, KN-95s, and even some individuals wearing N95s [113].

Additionally, the researchers elaborated on one of the mechanisms by which cloth masks may increase the incidence of infection above a non-masked baseline. They noted that their own previous work had established that: "'aerosolized fomites,' non-respiratory particles aerosolized from virus-contaminated surfaces such as animal fur or paper tissues, can also carry influenza virus and infect susceptible animals. **This observation raises the possibility that masks or other personal protective equipment (PPE), which have a higher likelihood of becoming contaminated with virus, might serve as sources of aerosolized fomites**" [113].

➤ Even the mechanistic laboratory studies cited as evidence to support the efficacy of masks by the CDC's *Science Brief: Community Use of Masks to Control the Spread of SARS-CoV-2* [114] are not unequivocally favorable to masks (especially not to cloth masks).

To support its masking recommendations, the CDC referenced a 2021 laboratory study of SARS-CoV-2, which reported that cloth and surgical masks were not significantly different in that both reduced viral RNA in fine aerosols by 48%. The CDC failed to mention, however, that the same study *also* reported that the facemask group had a *greater* number of infectious ("culture positive") fine aerosol samples than the non-mask group: "**None of the 75 fine-aerosol samples collected while not wearing face masks were culture-positive. Two (3%) of the 66 fine-aerosol samples collected from participants while wearing face masks were culture-positive**" (emphasis added) [115].

The CDC referenced a 2020 laboratory study as evidence that singing increases the risk of transmitting COVID by increasing the amount of exhaled respiratory particles, but neglected to mention that the same study said: "**SARS-CoV-2 could not be detected in the air samples collected while confirmed Covid-19 patients were singing and talking**" [116]. Oh reeeeallly? You don't say!

Another study referenced by the CDC in its Science Brief on masks, far from providing strong evidence to support cloth mask use, "showed that **cloth masks and other fabric materials tested in the study had 40–90% instantaneous penetration levels** against polydisperse NaCl aerosols" (emphasis added) [117] This instantaneous penetration "reached maximum (73–82%) at 100 nm" –coronavirus-size [117]

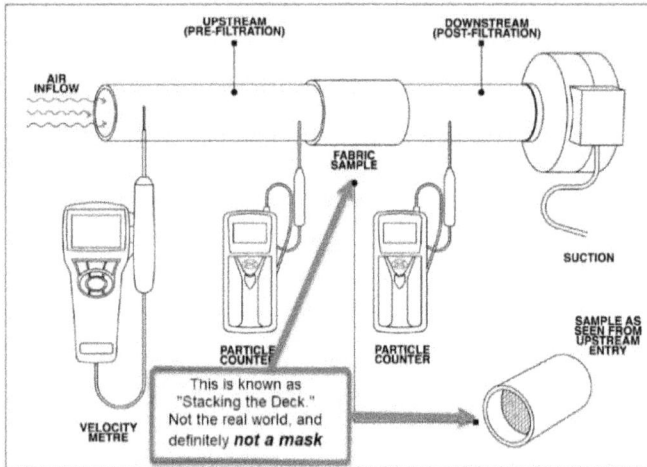

Figure 1 Diagram of experimental apparatus using two P-Trak Ultrafine Particle 8525 counters for simultaneous measurement and a TSI 9565 VelociCalc to measure face velocity.

O'Kelly, E., et al., *Ability of fabric face mask materials to filter ultrafine particles at coughing velocity.* BMJ Open, 2020. 10(9): p. e039424. https://dx.doi.org/10.1136/bmjopen-2020-039424

Multiple studies referenced to support cloth mask use by the CDC's Science Brief involve experimental setups which are impossible for mask-wearers to duplicate under functional real-world conditions. Even under such unrealistically favorable conditions, at least one study reported: "The average filtration efficiency of single layer fabrics and of layered combination was found to be 35% and 45%, respectively" [118]. That kind of efficacy is nowhere near enough to stop a virus.

Another 2020 study referenced in the CDC's Science Brief on masks highlighted this deficiency: **"publication of only the ideal filtration efficiency of materials in perfectly sealed settings can give mask wearers a false sense of security…"** [119].

Though referenced by the CDC Science Brief in a manner suggesting that it supports mask efficacy ("Multiple layers of cloth with higher thread counts have demonstrated superior performance compared to single layers of cloth with lower thread counts, in some cases filtering nearly 50% of fine particles less than 1 micron." [114]), this 2020 laboratory study actually found that "the filtration efficiency of loosely fitting masks/respirators against ultrafine particulates **can drop by more than 60% when worn compared to the ideal filtration efficiency of the base material**" (emphasis added) [119]. The researchers found that **when challenged by coronavirus-sized particles of 60 and 125nm under conditions more closely analogous to the real world, the filtration efficiency of medical masks, KN-95s, and even N95s was reduced to less than 20%** - *comparable to a basic #4 coffee filter.*

Even top-quality masks perform about as well as coffee filters against particles the size of SARS-CoV-2

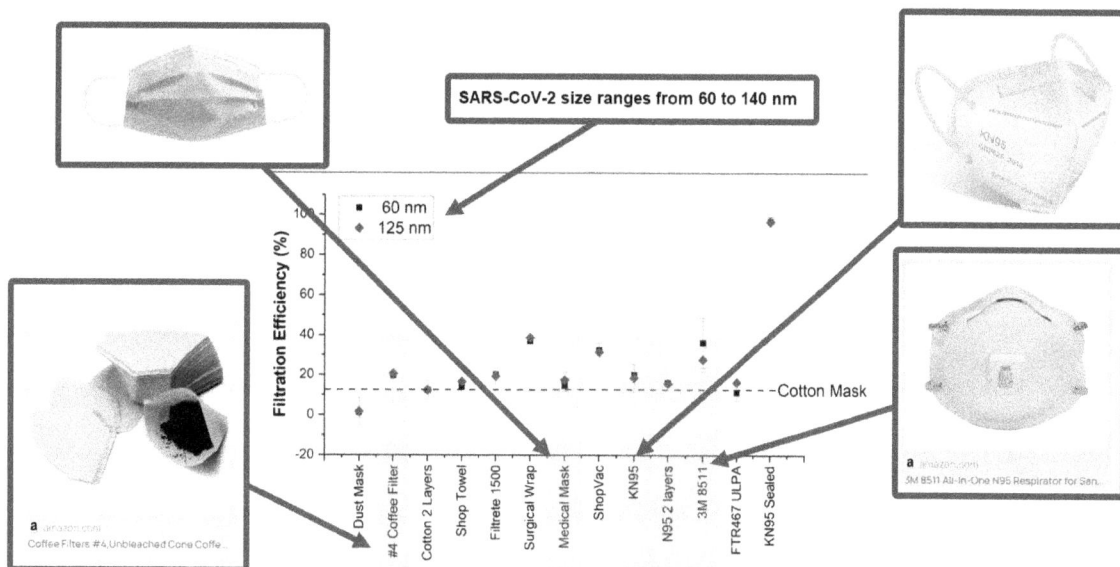

Figure 5. When inserted into a 2-layer cotton mask, all filter materials exhibited a significant drop in filtration efficiency compared to the measured base filtration efficiency. The 3M 8511 and KN95 respirators as well as the medical and dust masks were all tested as-received in mask form, whereas the remainder were tested through insertion between the two layers of the cotton mask. A KN95 respirator that was sealed to the headform before testing with thermoplastic adhesive (i.e. to demonstrate a leak-free fit) provided filtration efficiency very near to its base filtration efficiency, proving that the cause of efficiency reductions was related primarily to fit quality.

Hill, W.C., M.S. Hull, and R.I. Maccuspie, "Testing of Commercial Masks and Respirators and Cotton Mask Insert Materials using SARS-CoV-2 Virion-Sized Particulates: Comparison of Ideal Aerosol Filtration Efficiency versus Fitted Filtration Efficiency." *Nano Letters*, 2020. **20**(10): p. 7642-7647. https://dx.doi.org/10.1021/acs.nanolett.0c03182

> Similarly, another laboratory study found that when just 2% of mask materials' surface area was compromised, filtration efficiency declined by over 2/3 of the materials' maximum: "… these measurements showed that already very small leaks in the order of one percent of the total sample area can substantially reduce the overall filtration efficiency of a mask down to half or even less compared with the value of the material itself" [120].

Simple visual inspection reveals that non-respirator masks have more than 4-5x this amount of leakage. Additionally, the category of particle sizes used even for *this* study (≤2,500 nm) was many times larger than that of SARS-CoV-2 (140 nm), and would still tend to overestimate mask effectiveness against respiratory viruses [120].

Remember, because the Y-axis on this graph is *relative* filtration efficiency, "1.0" on this graph represents the mask's maximum filtration efficiency, which is *always* less than 100%.

A mere 2% surface area of leakage compromised mask filtration by more than 2/3 of the maximum! Most masks have at least 9-10% surface area leakage, if not more.

VelvetCotton
SurgicalMask4
solid: 0.03-2.5 µm particles
dashed: 10 µm particles

relative filtration efficiency

relative area of leakage / %

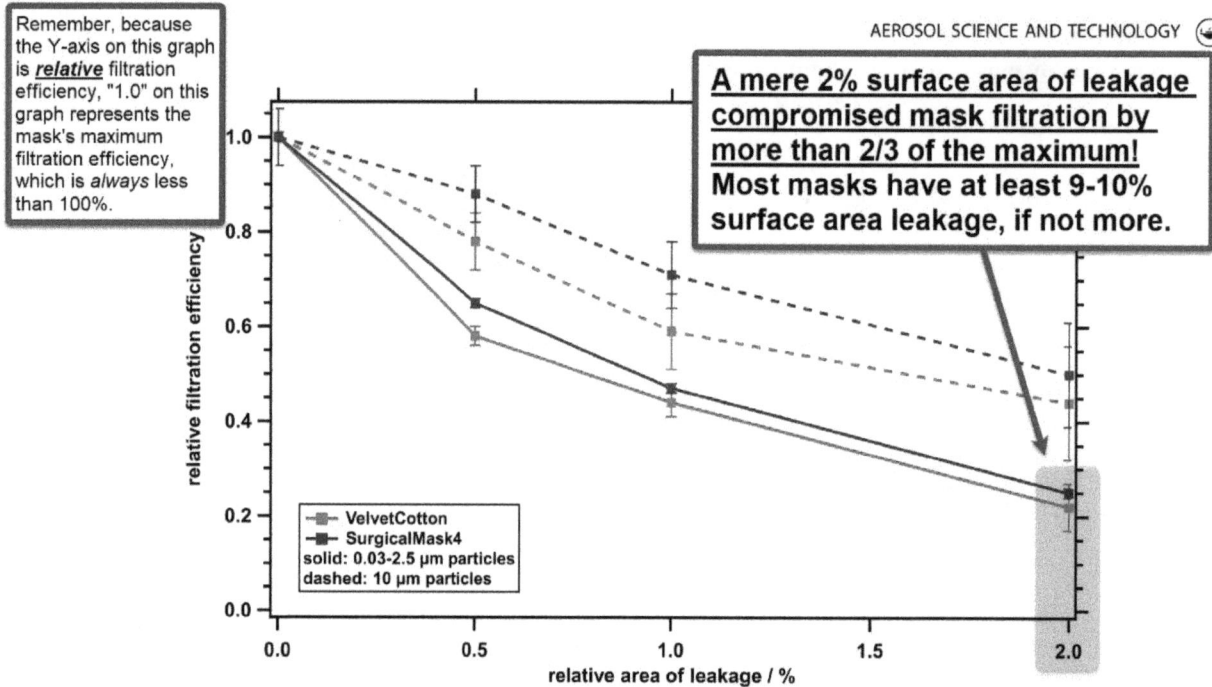

Figure 8. Filtration efficiency for velvet cotton (red) and surgical mask (blue) samples for $d_p \leq 2.5\,\mu m$ (solid line) and $d_p = 10\,\mu m$ (dashed line) versus relative leak area, normalized to the leak-free sample. Here, measurements of neutralized (*CPC setup*) and ambient aerosol (*SMPS/OPC setup*) were averaged, where available.

Drewnick, F., et al., *Aerosol filtration efficiency of household materials for homemade face masks: Influence of material properties, particle size, particle electrical charge, face velocity, and leaks.* Aerosol Science and Technology, 2021. **55**(1): p. 63-79.

Expert opinions; modeling studies; and qualitative & conceptual literature reviews predict that masks do not work

Political claims that "the Science is settled" in favor of masks, or that all experts agree masks "work," were false from inception. Pagan priestly divinations using the entrails of sacrificial animals have a better track record of accuracy than COVID-19 modeling studies, and when it comes to expert opinions, many experts and professional bodies have still been willing to oppose compulsory masking despite the political stigma that came with doing so. A handful of examples are presented here for the sake of completeness.

➤ Even or (especially) in healthcare settings, pronouncements on the dubious virtues or outright ineffectiveness of masks as infection control measures go back decades, especially in Scandinavian countries. As a 1993 expert opinion published in a Danish medical journal put it: "[There is therefore no reason from the point of view of preventing infection to maintain a general requirement for the use of surgical masks by others than those persons who may be situated within an arm's length of the operational field or instrument table]" [translated from Danish] [121]. In 2010, Swedish authors from the Karolinska institute noted with satisfaction that: "Anesthesia personnel are no longer required to wear disposable face masks in the operating room, a practice approved by our surgical colleagues." [122]

➤ Early in 2020, the *Journal of Paediatrics and Child Health* published a letter with eight signatories from departments of Infectious Diseases and Microbiology stating that: "There is no good evidence that facemasks protect the public against infection with respiratory viruses, including COVID-19" [123]. In August 2020, a Belgian MD/PhD and Honorary full professor at the School of Public Health publicly opined that: "It would appear that abstaining will often be more effective than wearing any 'mask' without discernment" [124]. German MD, Dr. Günter Kampf, wrote a letter to the editor published in the November 2020 *Proceedings of the National Academy of Sciences*, pointing out that: "data from Germany indicate that mandatory face masks in shops and public transport as a single measure did not accelerate the decline of new cases" [125].

➤ Even a published opinion which argued *for* the community use of facemasks by the general public admitted that the benefits of doing so were debatable and ran against the vast majority of official guidelines and recommendations going back decades prior to 2020 [126].

> **Prior to 2020, the majority of health authorities did not recommend general community use of masks.**

Table 1. Summary of the Earlier Recommendations on Medical Masks Use in the General Community Across Different Credible Health Authorities Prior to 6 April 2020

Source	Encourages Community Use of Face Masks?	Reasons/Further Notes Provided?	Suggestions on the Use of Masks for Healthy Individuals Under Alternative Circumstances?
WHO [2, 3]	No	– Improper use may hamper its use – No evidence to support the effectiveness against COVID-19 of mask-use in the community	Use masks when: – When the culture has been to use masks – When the local government encourages their use – Upon close contact with infected/suspected/high-risk individuals
United States [4]	No	– Spread of SARS-CoV-2 is mainly through close contact – Stockpiling of masks may place a burden on the supply to medical staff	Use masks when [5]: – In workplaces of and upon contact with infected/suspected/high-risk individuals
Canada [6]	No	– Improper use may increase infection risks – May induce a false sense of security that that played down other essential hygiene measures	Use masks when: – When the culture has been using masks – when the local government encourages their use – Upon close contact with infected/suspected/high-risk individuals
United Kingdom [7, 8]	Not explicit[a]	Nil	Use masks when: – Upon close contact with infected/suspected/high-risk individuals
Australia [9]	No	– little evidence supporting the widespread use of surgical masks in healthy people	Use masks when: – Upon close contact with infected/suspected/high-risk individuals
New Zealand [10]	No	– Cited as suggestions from WHO	Use masks when: – In workplaces of contact with infected/suspected/high-risk individuals
France [11]	No	– Facemasks cannot be worn at all times	Use masks when: – Upon prolonged close contact with an infected individual.
Italy [12]	No	– Citing as suggestions from WHO – Increased the risk of infection due to a false sense of security and greater contact between hands, mouth and eyes.	Use masks when: – Upon close contact with infected individuals
Spain [13]	No	– Worn by people who are sick. – An inadequate use of masks can contribute to a shortage of them in those situations for which they are indicated.	Use masks when: – Upon close contact with infected individuals
Germany [14]	No	– Citing as suggestions from WHO	Nil
Singapore [15]	No[b]	– Only for sick individuals	Nil
China [16, 17]	Yes	– The general community should make the judgment of mask-usage based the risk levels. – Masks are recommended in situations which include, going to medical institutions, in crowded open spaces, in a crowded or densely populated indoor environment, and close contact with people of quarantine at home	Masks are not required: – when you are at home (in isolation), engaging in outdoor activities or in well-ventilated indoor places
Hong Kong SAR, China [18, 19]	Yes	– Recommended when taking public transport or staying in crowded place, clinics or hospitals visits. – Face mask provides a physical barrier to fluids and large particle droplets. When used properly, surgical masks can prevent infections transmitted by respiratory droplets.	Nil
Macau SAR, China [20, 21]	Yes	– If it is necessary to go out, wear a mask at all times	Nil
South Korea [22]	Yes	– Wearing a mask can prevent infectious diseases	Nil.
Japan [23]	Yes and No	– If you wear a facemask in confined, badly ventilated spaces, it might help avoid catching droplets emitted from others	Masks are not required: – If you are in an open-air environment, the use of facemask is not very efficient.

> **Only in China, Hong Kong, South Korea, and Japan were masks even *recommended* prior to 2020.**

Tso, R.V. and B.J. Cowling, "Importance of Face Masks for COVID-19: A Call for Effective Public Education." Clinical Infectious Diseases, 2020. **71**(16): p. 2195-2198. https://academic.oup.com/cid/article/71/16/2195/5866410

➢ In a legal context, the Association of American Physicians and Surgeons (AAPS) states in the strongest terms: "The CDC mask mandate lacked medical justification, and cannot withstand a merely cursory logical or scientific analysis."[5] Likewise, America's Frontline Doctors (AFLDS) said: "it is the overwhelming weight of the available scientific and medical studies made on the subject of the wearing of masks, that masks simply don't work."[6]

Affirmative evidence for the negative

If masks were effective at decreasing the spread of respiratory viruses, we would expect to see a preponderance of the best scientific studies consistently finding ***clear and convincing evidence*** of benefits against respiratory viral transmission from mask use.

Instead, a thorough survey of the scientific literature — including independent systematic literature reviews and meta-analyses of multiple randomized controlled trials involving mask use — consistently reveals ***clear and convincing evidence of no benefits***. Additionally, those studies purporting to show that masks mitigate the spread of respiratory pathogens, especially viruses, are on average smaller, shorter, weaker, less rigorous, and less numerous than studies which show *no* benefits from wearing masks. A printable poster-sized pdf chart depicting this evidential disparity, including all the studies cited by the CDC in its *Science Brief: Community Use of Masks to Control the Spread of SARS-CoV-2* [114] and *The Science of Masking to Control COVID-19* [127] can be viewed and downloaded for free at www.thebookonmasks.com/extras.

One or two studies indicating non-efficacy for masks may be an anomaly, but more than 100 studies from dozens of researchers working independently in multiple countries over multiple decades is a decisive trend. As Drs. Ambarish Chandra and Tracy Beth Høeg put it: "while the presence of correlation does not imply causality, **the absence of correlation can suggest causality is highly unlikely, especially if the direction of bias can be reasonably anticipated.**" [128] This is not simply a *lack* of evidence that masks work. This is a *repeated* and longstanding pattern of *positive evidence* that masks do not work, and it is cumulatively *far stronger* than the *whole* body of evidence from 2020 to the present which has been put forward by officials and organizations with a conflict of interest in their increasingly desperate attempt to justify their mask mandates.

---

5   Brief for the Association of American Physicians and Surgeons (AAPS) as *Amicus Curiae*, *HFDF v. Biden* (USCA11 Case: 22-11287), Date Filed: 08/05/2022, page 16
https://aapsonline.org/judicial/aaps-amicus-mask-mandate-8-5-2022.pdf

6   Brief for America's Frontline Doctors (AFLDS) as *Amicus Curiae, HFDF v. Biden* (USCA11 Case: 22-11287), Date Filed: 08/08/2022, page 3
https://res.cloudinary.com/aflds/image/upload/v1660068993/aflds/Health_Freedom_Defense_Fund_Inc_et_al_v_Biden_etc_et_al_Americas_Frontline_Doctors_Amicus_Motion_Brief_Stamped_8_8_2022_21_cv_1693_MDFL_22_11287_CA_11_759748977c.pdf

If masks were actually *effective* at mitigating the spread of respiratory viruses and other pathogens, this *abundance* of negative findings and conspicuous *shortage* of positive findings would be *reversed*. The evidence supporting facemask use is not enough to surmount even the *empirical* hurdles — much less the *moral* and *legal* hurdles — necessary to justify mask mandates.

But What About the Evidence Supporting Masks?

Most of the time, instead of acknowledging, addressing, or attempting to rebut evidence like that which we just reviewed, apologists for mask mandates merely make broad or indirect references to CDC publications, recommendations, and guidelines to justify their positions, actions, and coercive mandates.

The CDC's flagship publication supporting mask use is its "Science Brief: Community Use of Masks to Control the Spread of SARS-CoV-2" [114]. At the time of this writing, it was last updated on December 6, 2021 to include a total of 90 references that give a first-glance impression of evidential strength [16, 17, 23, 28, 51, 77, 90, 92, 94, 115-119, 129-204]. Anyone who has actually *read* the studies the CDC cites in this *Science Brief*, however, knows otherwise. The evidence base put forward by the CDC to justify mask recommendations and mandates is shockingly, inexcusably weak. The CDC has acknowledged only a handful of studies from the peer-reviewed literature cited in the previous section. Instead, it proffers a body of evidence to support its mask-related recommendations and directives that is far smaller, far weaker, and fails to hold up under even mediocre cross-examination.

Dissecting the CDC's Science Brief on Masks

Many of the references provided by the CDC its *Science Brief* on masks do not even address the efficacy of masks to mitigate the spread of respiratory viruses. Thirty-three of them — more than a third — fall into this category [116, 129-131, 135, 138, 139, 145, 147-149, 153, 155, 158, 164, 167, 170, 175, 176, 179, 182, 184, 187-192, 194-198]. These 33 out of the CDC's 90 *Science Brief* citations provide additional background evidence and information which might *support* mask usage *if* masks work, but they do not themselves provide evidence that masks work.

In fact, eleven of these thirty-three references actually provide evidence that would *discourage* the use of masks *even if they worked* [116, 135, 138, 147, 149, 153, 164, 175, 179, 182, 198]. We will discuss many of these in more detail in the section on mask side effects, but for now, let's take a quick look at each of these eleven studies cited by the CDC in turn, and you can decide for yourself whether they favor or discourage the use of facemasks.

➢ One, which we looked at earlier, noted that "SARS-CoV-2 could not be detected in the air samples collected while confirmed Covid-19 patients were singing and talking" [116].

➢ Another of these studies was published in the *British Journal of Sports Medicine*, and reported: "**Our data indicate that wearing a cloth face mask significantly impaired participant performance during a CPET [Cardiopulmonary Exercise Test].** The observed significant differences in key performance variables (ie, reduced exercise time), physiological variables (e.g., VO_2max [maximum oxygen consumption], VE [minute ventilation], HR [heart rate], SpO_2 [oxygen saturation])[7] and perceptual variables (i.e., RPE [rating of perceived exertion], dyspnoea)[8] suggest that exercising while wearing a cloth face mask negatively impacted the exercise performance of our sample" (emphasis added) [147].

A 2020 study using healthy male volunteers reported that: "Assessment of the hemodynamic parameters showed that ffpm [FFP2/N95 masks] decreased avDO2 [arteriovenous O2 content difference] by 16.7 ± 11.2% compared to nm [no mask]. **The masks showed a marked effect on pulmonary parameters: VE [minute ventilation] for both sm [surgical masks] and ffpm [FFP2/ N95 masks] was significantly reduced by –12.0 ± 12.6% and –23.1 ± 13.6%, respectively, compared to nm [no mask]**" (emphasis added) [149].

Yet another of these studies reported: "wearing a treatment mask during the realization of a TM6 [six-minute walking test] does not modify the distance traveled but significantly and clinically increases dyspnea" (the sensation of difficulty breathing) [182].

A study on the population most optimally able to tolerate masks — healthy male volunteers in their mid-20s — reported: "In the healthy young men (age, 25.7 ± 3.5 years) in this study, the use of surgical face masks was associated with a significant increase in airway resistance, reduced oxygen uptake, and increased heart rate during continuous exercise" [164].

---

7   **VO_2max**: the maximum volume of oxygen your body uses while exercising at maximum intensity
VE: "minute ventilation" the volume of gas inhaled or exhaled from a person's lungs per minute.
HR: "heart rate"
SpO_2: "Saturation of peripheral oxygen" — blood oxygen saturation, usually measured through a pulse oximeter
8   **RPE**: "Rate of Perceived Exertion" — a subjective measure of the rate of effort put out during exercise. Similar to the pain scale "0 to 10" pain scale.

➢ Another study referenced by the CDC documented how facemasks impair non-verbal communication by hampering viewers' ability to read and recognize emotions. This effect was especially pronounced in younger children tested using pictures of masked faces: "for all groups, the percentage of correct responses is significantly reduced for the images with face masks compared to the images without face masks… the performance of toddlers is more affected by the use of a mask than the performance of both older children" [153].

➢ Four of these eleven studies concerned skin damage from masks. We shall look at all of these and more in greater detail when we cover the side effects masks have on skin. A research letter reporting on a survey of more than 1,200 members of the general population in Thailand found that many experienced new skin disorders or the exacerbation of pre-existing skin disorders when wearing masks, and only 1/3 reported wearing their masks comfortably [138].

A systematic review and meta-analysis of 35 dermatological studies by Spanish researchers reported that the rate of "cutaneous adverse events" related to mask use was over 50%, that 42% of children who used masks experienced irritant contact dermatitis, and that "In addition, five studies evaluated skin barrier function impairment due to mask wearing" [175].

A study conducted in Korea also reported cutaneous adverse events related to mask use in excess of 50%, and, "In patients with pre-existing dermatoses, the prevalence of skin lesions induced by wearing facial masks was 71.31%" [179]. A cross-sectional study on healthcare workers out of England reported: "more than half of the participants (454 cases; 54.5%) reported at least one adverse skin reaction related to a face mask." [198]

The other twenty-two background studies cited by the CDC in its *Science Brief* seek to address peripheral objections rather than make a positive case for mask-use based on efficacy [129-131, 139, 145, 148, 155, 158, 167, 170, 176, 184, 187-192, 194-197]. The information in these studies has value, but does not provide any actual evidence that masks "work." Rather, they constitute evidence which *could* favor mask use *if* masks "work." Additionally, when closely examined, most of these 22 references are hardly unequivocal endorsements of mask-use.

➢ Three studies showed that with intensive therapy and effort over multiple weeks, many autistic children can be trained not to rip off their masks or saturate them with saliva for short periods of up to 15 to 30 minutes [155, 167, 195].

➢ Two more studies discussed how facemasks affect people's ability to read and recognize emotions in a neutral or positive light. A 2020 study found no statistically significant differences between children' ability to accurately read emotions with or without a facemask [189], whereas another

2020 study documented that masks and sunglasses *both* significantly decreased children's' abilities to accurately read emotions, but opined: "while there may be some challenges for children incurred by others wearing masks, in combination with other contextual cues, masks are unlikely to dramatically impair children's social interactions in their everyday lives" [187].

➢ Three laboratory studies dealt with the effect of different activities on respiratory droplet generation. "Expiratory human activities such as breathing, coughing, sneezing or laughing result in droplet generation" [176], and "… speech or simple breathing could be a potent yet, until recently, unsuspected transport mechanism for pathogen transmission" [129]. Increases in speech volume increase the rate (but not the size distribution) of particles emitted during speech [131]. None of these three studies actually provide direct evidence about whether or not masks mitigate the transmission of respiratory pathogens. At best, they provide as much empirical justification for "no talking" mandates in the name of infection control as they do for mask mandates, and do nothing whatsoever to surmount the moral barriers associated with attempting such a thing.

➢ Another study simply surveyed perceptions about mask-use: "A small but non-negligible proportion of children reported discomfort and side effects that should be considered to ensure high adherence to mask wearing among school children" [130].

➢ The CDC *Science Brief*'s overview of the respiratory and cardiovascular side effects of masks is curated to make the strongest case that these side effects are mild and acceptable. It is, at best, an incomplete survey of the literature on this subject, and we will make the counter case in Part 2 of *The Book on Masks*, going into detail on many studies left out of the CDC's analysis. Here, we will simply review the studies on this topic cited by the CDC in its *Science Brief*.

Patients in a pediatric acute care unit were able to maintain full oxygen saturation without changes in end-tidal CO_2 ($EtCO_2$), peripheral oxygen saturation (SpO_2), Respiratory Rate (RR), and Heart Rate (HR) *when supplemental oxygen was administered through their masks* [145]. The CDC *Science Brief* referenced this study as though it somehow supports long-duration mask use by children in other settings.

Journal of PeriAnesthesia Nursing xxx (xxxx) xxx

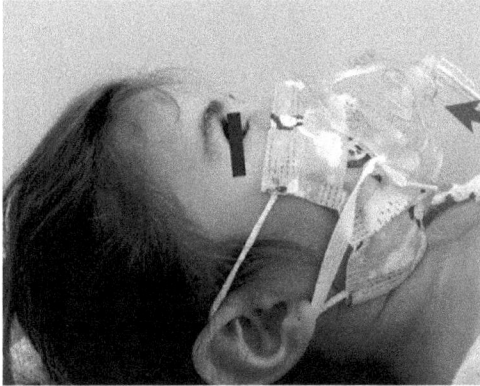

Providing 100% oxygen through a mask stacks the deck for any findings in favor of masks. It says nothing about the effects of masks on children in school or other real-life situations.

Figure 1. Oxygen treatment over the face mask. This figure is available in color online at www.jopan.org.

Dost, B., et al., "Investigating the Effects of Protective Face Masks on the Respiratory Parameters of Children in the Postanesthesia Care Unit During the COVID-19 Pandemic." *Journal of PeriAnesthesia Nursing*, 2021. https://europepmc.org/articles/PMC7877201

The CDC characterized a 2020 Italian study of masking children by saying: "A separate study observed no oxygen desaturation or respiratory distress after 60 minutes of monitoring among children less than 2 years of age when masked during normal play" [114]. While technically truthful, this summary is misleading. The children were, indeed, *monitored* for 60 minutes, but only actually wore the masks for 30 minutes of the 60-minute monitoring period [170]. Another study cited by the CDC likewise noted that children can wear masks for 30 minutes without sustaining significant changes in heart rate or oxygen saturation [197]. As we shall see in Part 2, heart rate and oxygen saturation are both fairly resistant to changes from increased levels of carbon dioxide.

According to the CDC, "Studies of healthy hospital workers, older adults, and adults with chronic obstructive pulmonary disease (COPD) reported no to minimal changes in oxygen or carbon dioxide levels while wearing a cloth or surgical mask either during rest or moderate physical activity" [114]. The five studies cited by the CDC to support this broad, conclusive assertion involved short durations of mask-wearing for up to 60 minutes by healthy subjects [135, 139, 158, 188, 192].

In one of these five studies, condominium residents in their 70s (excluding "those who had comorbid cardiac or respiratory conditions that could lead to dyspnea or hypoxia at rest") were able to maintain full peripheral oxygen saturation for an hour while wearing the lightest type of disposable, non-medical, 3-ply facemasks when at rest [139]. Another one of these five studies reported that "A small yet statistically significant difference in $EtCO_2$ [end-tidal carbon dioxide] increase was observed while wearing a mask" during a 5-minute walking test. $EtCO_2$ (end-tidal carbon dioxide) is the amount of carbon dioxide in exhaled air [135].

The remaining three studies in this cluster of the CDC's citations included findings that undercut conclusions about masks having acceptably mild side effects. After testing the most resilient group of subjects available — healthy, nonsmoking, young adult males who routinely exercised — one concluded, "short-term moderate-strenuous aerobic physical activity with a mask is feasible, safe, and associated with only minor changes in physiological parameters, particularly a mild increase in $EtCO_2$... The increase in $EtCO_2$ may be explained by the fact that re-breathing of the expired air which remains within the mask practically increases the dead space and may contribute to a mild hypercapnia" [148].

Another was a German study of "trained eleven-year-old boys," which reported that, "at the maximum performance level with surgical face mask, there was a significant reduction in running time as well as a significant increase in the subjective perception of exertion" [190]. This study was cited by the CDC to support its Science Brief's broad assertion that: "The safety of mask use during low to moderate levels of exercise has been confirmed in studies of healthy adults and adolescents" [114].

The fifth exercise study cited by the CDC was conducted in 2012, again on healthy subjects in their 20s. The authors wrote: "Surgical mask use for 1 h at a low-moderate work rate is not associated with clinically significant physiological impact or significant subjective perceptions of exertion or heat" [184]. The changes dismissed by the authors as not being *clinically* significant included statistically significant "increases in heart rate (9.5 beats/min; $p < 0.001$), respiratory rate (1.6 breaths/min; $p = 0.02$), and transcutaneous carbon dioxide (2.17 mm Hg; $p = 0.0006$)..."

The CDC Science Brief on masks characterizes these short-duration tests in young, healthy individuals as covering "most circumstances." In doing so, the CDC is being as selective with reporting these background studies as it is with studies which directly evaluate mask efficacy.

> *"J.A. Schumpeter said that the first thing a man will do for his ideals is lie. It is not necessary to lie, however, in order to deceive, when filtering will accomplish the same purpose."*
>
> - Thomas Sowell. *Intellectuals and Society,* 2011[9]

---

9   Thomas Sowell, *Intellectuals and Society*, New York: Basic Books, 2011, Kindle Edition, p. 128.

What about the CDC Science Brief's Remaining 57 references?

Eleven of the CDC Science Brief's remaining 57 references [16, 23, 28, 77, 90, 94, 115-119] were cited in our review of the evidence against masks, because upon close examination their results *directly contradict* the idea that masks are effective at mitigating respiratory virus spread. We also addressed Abaluck et al.'s Bangladesh Mask Study, which (even if taken at face value) still found no statistically significant benefit from cloth masks — meaning that the CDC's own best evidence for masks still invalidated virtually every mask mandate in the United States [17].

Of all scientific evidence, modeling studies (especially those based on speculative data) are the most prone to bias and manipulation. Such sources do not and *cannot* provide enough warrant to justify mandates. Nevertheless, 5 of the remaining references in the CDC's Science Brief are comprised of just such sources [140, 160, 162, 174, 186].

Fully twenty-two of the remaining studies are mechanistic laboratory studies [132-134, 137, 142, 144, 150, 156, 163, 166, 168, 169, 172, 177, 180, 185, 193, 199-202, 204]. These studies suggest that masks *should* decrease viral spread, but offer virtually no real-life evidence that they actually *do so*. If theories, models, and laboratory studies say something *should* work, but real-life observations and randomized controlled trials fail to confirm this, then the models and laboratory studies are clearly missing critical pieces of the puzzle.

➤ Successful use of a mask as a barrier for droplets large enough to be observed with the unaided eye does not necessarily imply that masks *also* stop infectious particles too small to be seen. Suggesting anything else is showmanship, not science.

Many of the laboratory studies cited by the CDC use photographs or video showing masks (as often as not on manikins) stopping or curtailing the spread of droplets large enough that they can be seen by the naked eye (sometimes assisted by lasers, thermal imaging, or other techniques). Such laboratory studies persuade through implication and unstated appeals to audience intuition. They ignore the crucial fact that infectious viral particles are so small that they cannot be seen with the naked eye, which makes direct photography at best an illustrative analogy rather than decisive evidence of effectiveness.

Citing to such studies is akin to showing that a common window screen can stop insects, and then claiming that window screens *also* meaningfully prevent air exchange, when, in fact, the entire point of using such a filter is to stop the large particles while letting the smaller ones pass through unimpeded [133, 134, 150, 201, 202]. Conversely, however, if masks *fail* to stop emissions large enough to be seen with the unaided eye, this is evidence that they do *not* stop particles which are measured in nanometers.

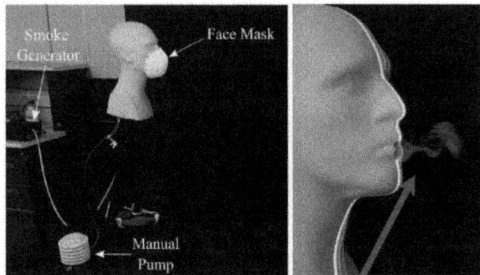

FIG. 1. Left—experimental setup for qualitative visualization of emulated coughs and sneezes. Right—a laser sheet illuminates a puff emerging from the mouth.

Visible smoke propelled by a cheap plastic foot pump is not a good proxy for a virus-laden cough.

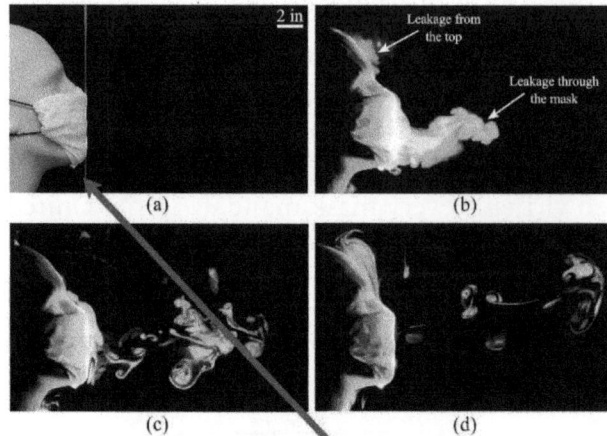

FIG. 3. (a) A face mask constructed using a folded handkerchief. Images taken at (b) 0.5 s, (c) 2.27 s, and (d) 5.55 s after the initiation of the emulated cough.

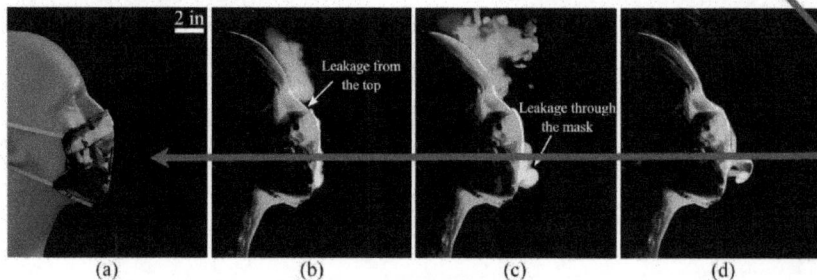

Masks on Mannequins' heads simply redirect whatever emissions do not get through.

FIG. 4. (a) A homemade face mask stitched using two-layers of cotton quilting fabric. Images taken at (b) 0.2 s, (c) 0.47 s, and (d) 1.68 s after the initiation of the emulated cough.

Verma, S., M. Dhanak, and J. Frankenfield, *Visualizing the effectiveness of face masks in obstructing respiratory jets.* Physics of Fluids, 2020. **32**(6): p. 061708. https://dx.doi.org/10.1063/5.0016018

➤ Laboratory studies purporting to show that masks filter microscopic particles often fail the applicability test as well. Simply looking at the experimental setup for many of these studies shows that while they may accurately describe the filtration properties of mask materials under optimal conditions, they are next-to-useless when describing real-world scenarios with masks [132, 156, 163, 169, 172, 185, 204]. Many of these are pictured here, and none are accurate proxies for real-world mask use. As we covered earlier, this fatal deficiency was highlighted by a 2020 study [119], which showed that under conditions far more akin to the real world, even surgical and respirator-type masks perform about as well as semi-transparent coffee filters. It takes a great deal of speculation, extrapolation, and chutzpah to parlay the laboratory studies cited by the CDC into recommendations, nevermind mandates.

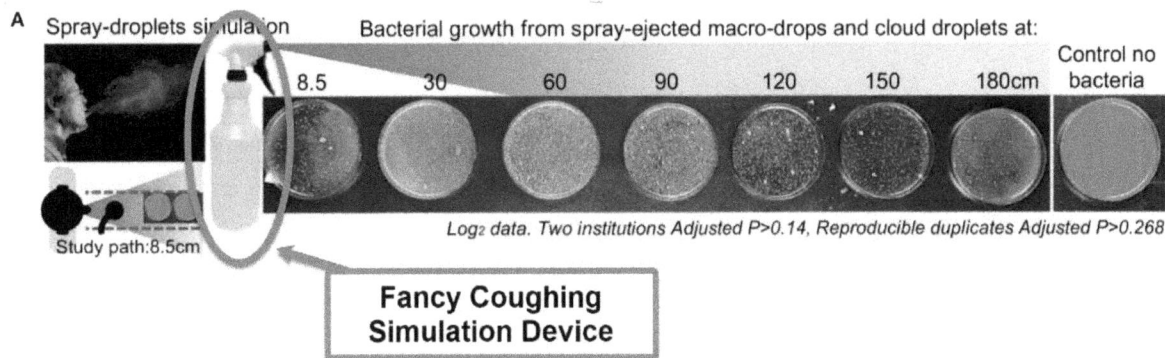

Rodriguez-Palacios, A., et al., *Textile Masks and Surface Covers—A Spray Simulation Method and a "Universal Droplet Reduction Model" Against Respiratory Pandemics.* Frontiers in Medicine, 2020. **7.**
https://dx.doi.org/10.3389/fmed.2020.00260

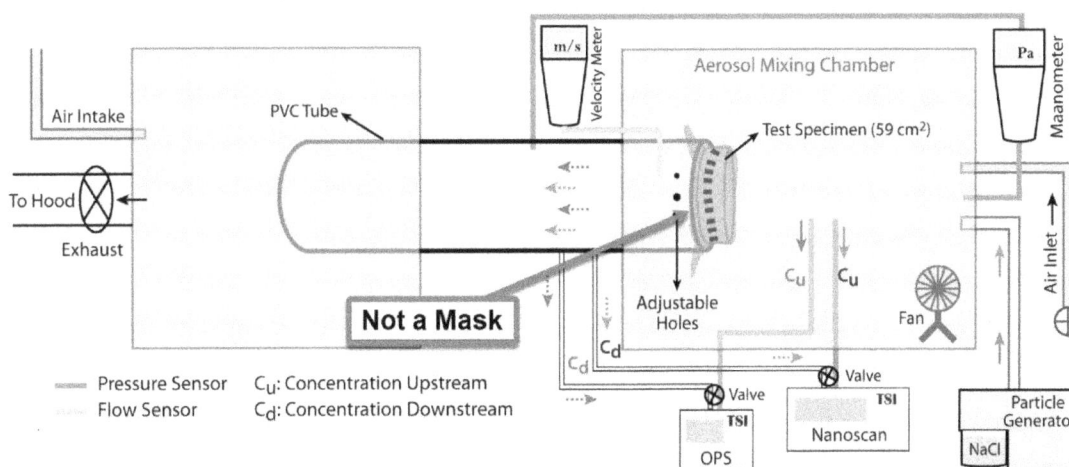

Figure 1. Schematic of the experimental setup. A polydisperse NaCl aerosol is introduced into the mixing chamber, where it is mixed and passed through the material being tested ("test specimen"). The test specimen is held in place using a clamp for a better seal. The aerosol is sampled before (upstream, C_u) and after (downstream, C_d) it passes through the specimen. The pressure difference is measured using a manometer, and the aerosol flow velocity is measured using a velocity meter. We use two circular holes with a diameter of 0.635 cm to simulate the effect of gaps on the filtration efficiency. The sampled aerosols are analyzed using particle analyzers (OPS and Nanoscan), and the resultant particle concentrations are used to determine filter efficiencies.

Konda, A., et al., *Aerosol Filtration Efficiency of Common Fabrics Used in Respiratory Cloth Masks.* ACS Nano, 2020. **14**(5): p. 6339-6347. https://www.ncbi.nlm.nih.gov/pmc/articles/PMC7185834/

---

10   I recall seeing something similar to this on a "Ty the Science Guy" video by industrial hygienist Tyson Gabriel and his colleagues, but when I returned to the original source (https://www.tyscienceguy.com/mask-documentary-series.html) two years later to see what (if any) similarities between the video's content and my own pictures existed, the website was no longer active. In any case, while this and the other "not a mask" pictures in this book were assembled by myself without any outside assistance or reference material, I wish to credit Tyson Gabriel and his colleagues with inspiring the general pattern of presentation.

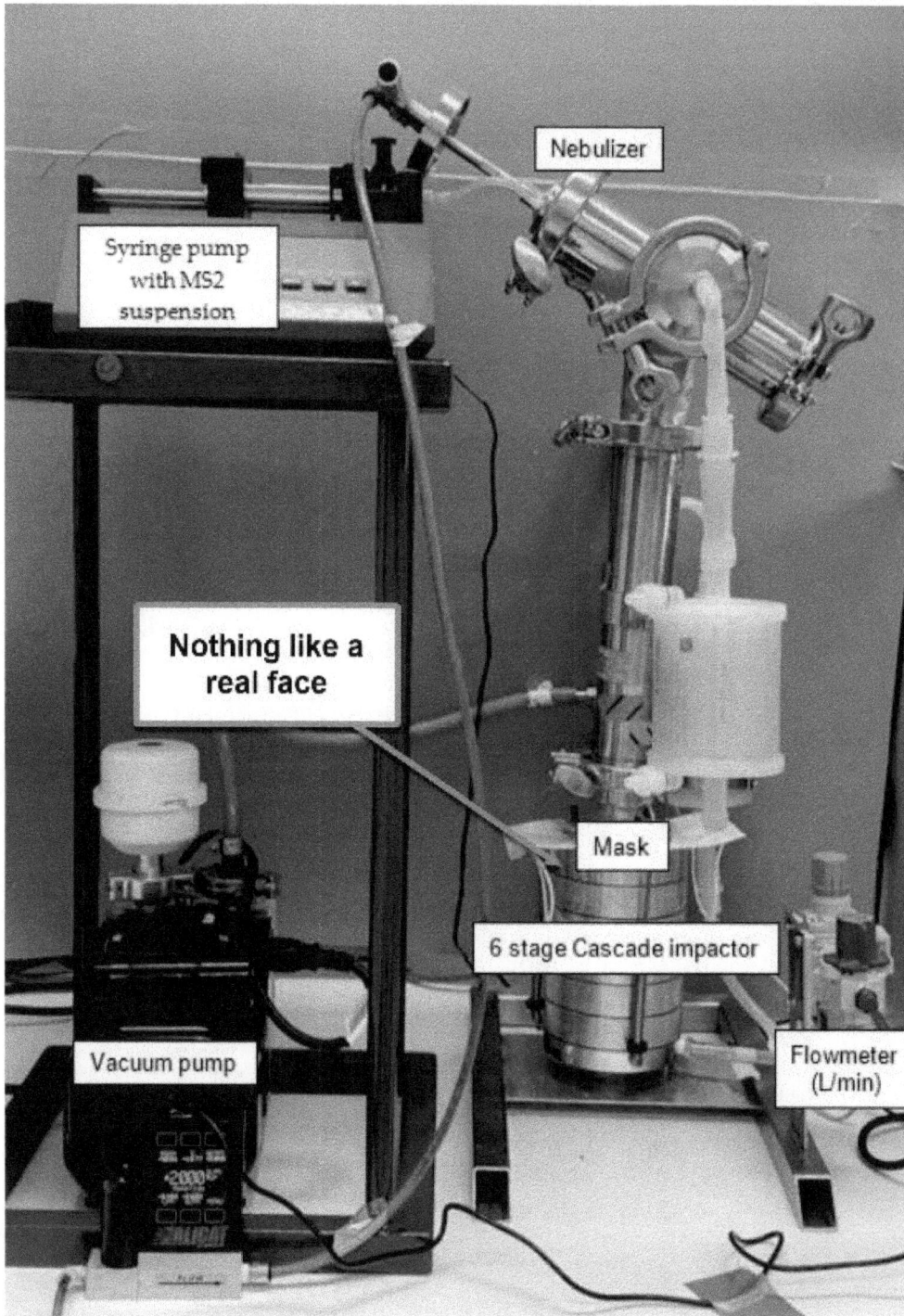

Figure 2. Mask testing rig, set up according to the ASTM F2101-14.

Whiley, H., et al., *Viral Filtration Efficiency of Fabric Masks Compared with Surgical and N95 Masks.* Pathogens, 2020. **9**(9): p. 762. https://dx.doi.org/10.3390/pathogens9090762

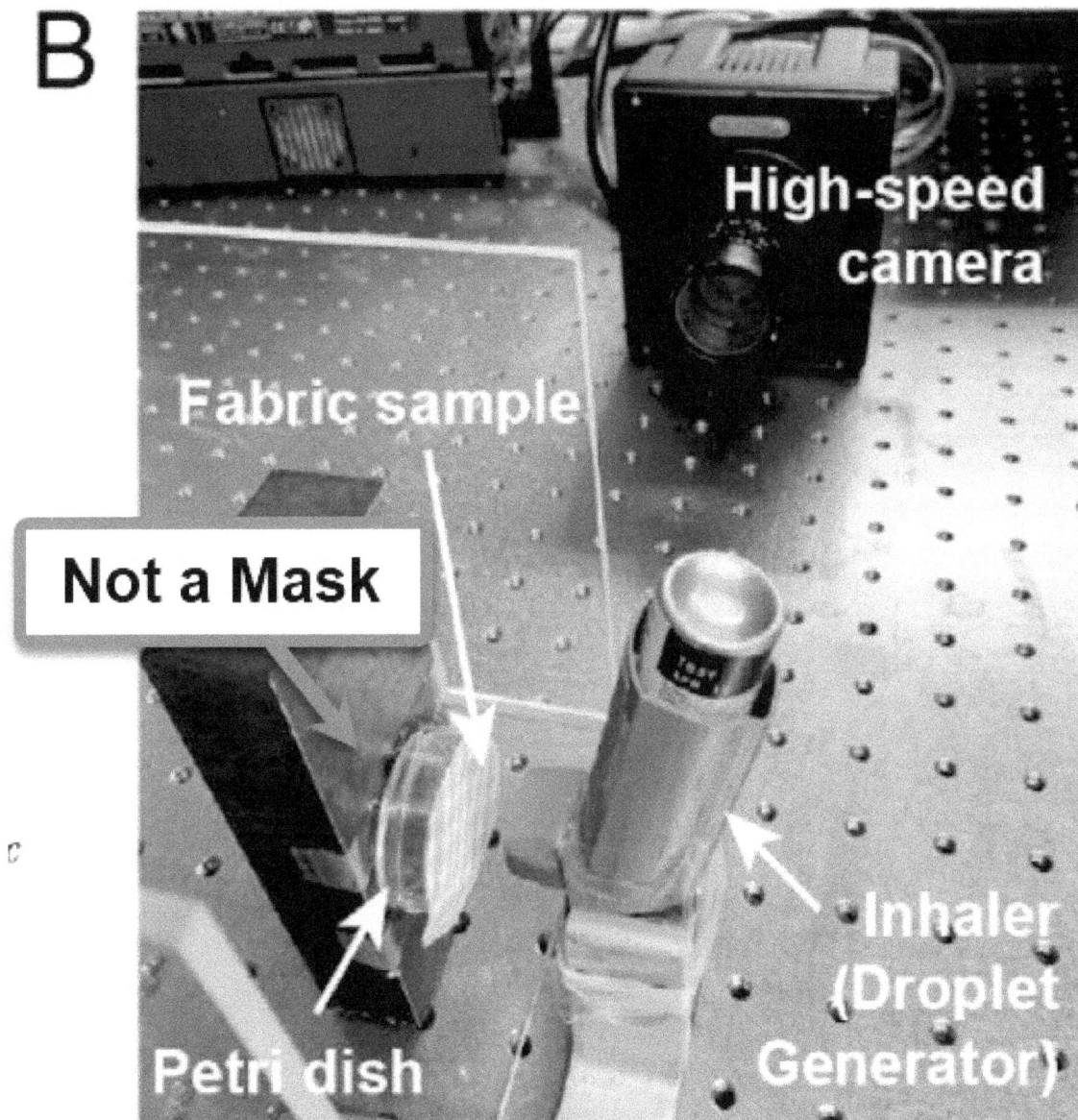

Aydin, O., et al., *Performance of fabrics for home-made masks against the spread of COVID-19 through droplets: A quantitative mechanistic study.* Extreme Mech Lett, 2020. 40: p. 100924. https://www.ncbi.nlm.nih.gov/pmc/articles/PMC7417273/

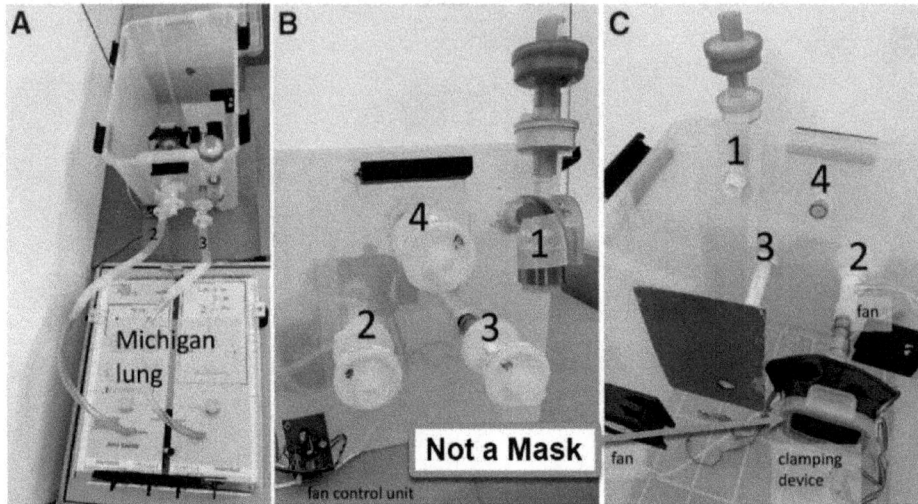

FIG. 1. **(A)** Shows how the Michigan lung is connected to port 2 (clamping device of the mask) and 3 (unfiltered reference port). **(B)** Shows the four connections on the front of the test box (1 = connection of the nebulizer head, 2 = connection to the clamping device, 3 = unfiltered reference port, 4 = port that enables pressure equalization during ventilation through the artificial lungs. The fans inside the boxes produce a homogeneous aerosol distribution after nebulization **(C)**.

Maurer, L., et al., *Community Masks During the SARS-CoV-2 Pandemic: Filtration Efficacy and Air Resistance.* Journal of Aerosol Medicine and Pulmonary Drug Delivery, 2021. **34**(1): p. 11-19.
https://dx.doi.org/10.1089/jamp.2020.1635

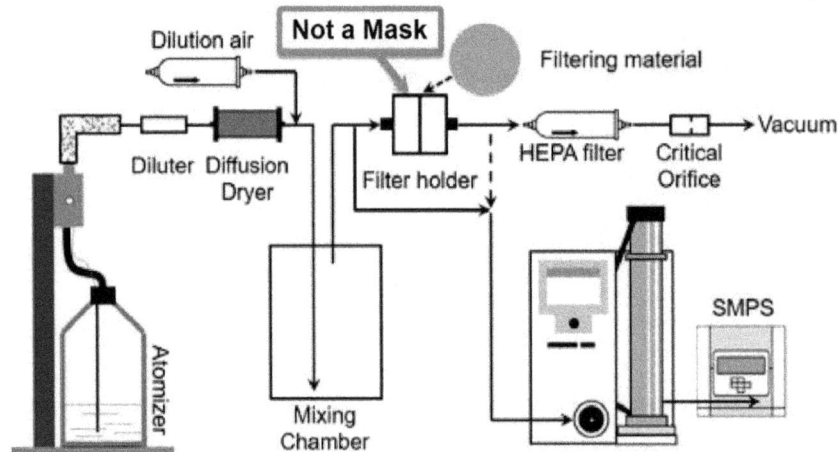

Fig. 1. A schematic diagram of the experimental setup of this study.

Hao, W., et al., *Filtration performances of non-medical materials as candidates for manufacturing facemasks and respirators.* Int J Hyg Environ Health, 2020. **229**: p. 113582.
https://www.ncbi.nlm.nih.gov/pmc/articles/PMC7373391/

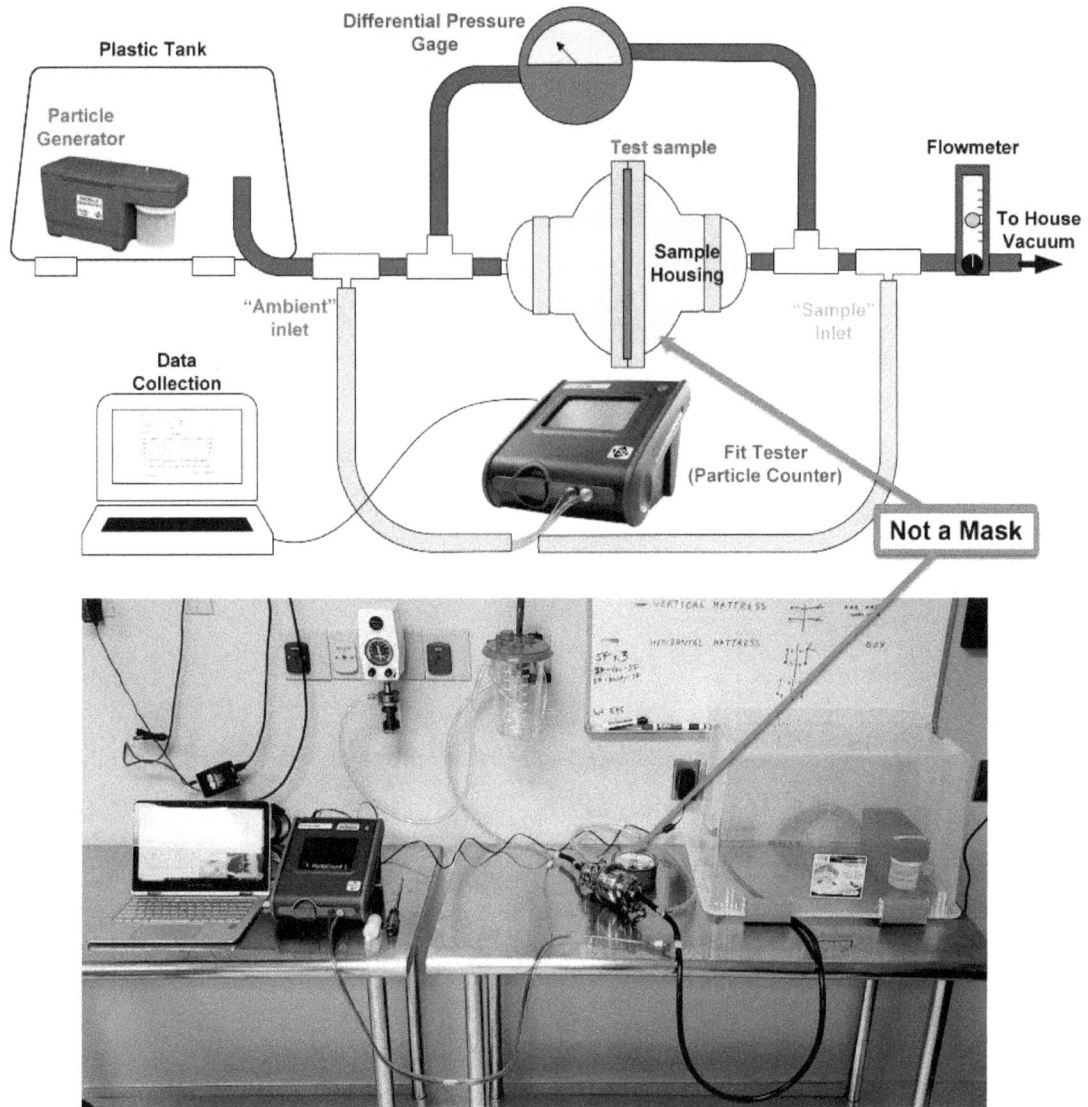

Fig 1. Filtration efficiency measurement setup. Diagram of filtration efficiency measurement setup, design utilizing hospital fit testing apparatus, particle generator, and house vacuum supply augmented with low-cost hardware.

https://doi.org/10.1371/journal.pone.0240499.g001

Long, K.D., et al., *Measurement of filtration efficiencies of healthcare and consumer materials using modified respirator fit tester setup*. PLOS ONE, 2020. **15**(10): p. e0240499.
https://dx.doi.org/10.1371/journal.pone.0240499

➤ Similarly, laboratory studies on improving mask fit or seal, which are cited by the CDC to support mask use, simply *assume* that the fit and filtration improvements described make a clinically or statistically significant real-world difference [142, 177]. The same applies to their laboratory study which reported that double-masking improves filtration efficiency by 10-15%, depending on which combination of masks is used [193].

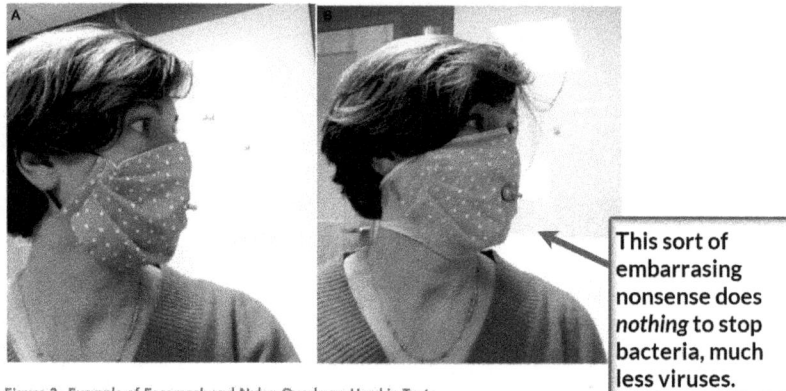

Figure 2. Example of Facemask and Nylon Overlayer Used in Tests
Facemask (Mask CS-1) worn as designed (A) and with a nylon overlayer (B) with tightly sealed grommet positioned at the philtrum of the upper lip. The grommet is used to sample air from inside the mask during testing. Note that: this mask could have been worn inside-out to ensure the folds faced down. However, for the purposes of this test, precautions against particle collection in folds were not considered necessary.

Mueller, A.V., et al., *Quantitative Method for Comparative Assessment of Particle Removal Efficiency of Fabric Masks as Alternatives to Standard Surgical Masks for PPE*. Matter, 2020. **3**(3): p. 950-962.
https://www.ncbi.nlm.nih.gov/pmc/articles/PMC7346791/

The dozens of observational and randomized-controlled trials which we reviewed earlier have found that this underlying assumption — i.e. that "improved or better-sealed masks = improved infection control" — is not borne out by real-world experience, even (or especially) in healthcare settings [22-26] [37-42] [62]. If going from a medical mask to a fitted N95 produces ***no clear and convincing benefits*** in controlling respiratory virus transmission, there is no good reason to think using an overlay of nylon hosiery to give a better fit to your surgical mask will do so in the way the CDC's study citations suggest.

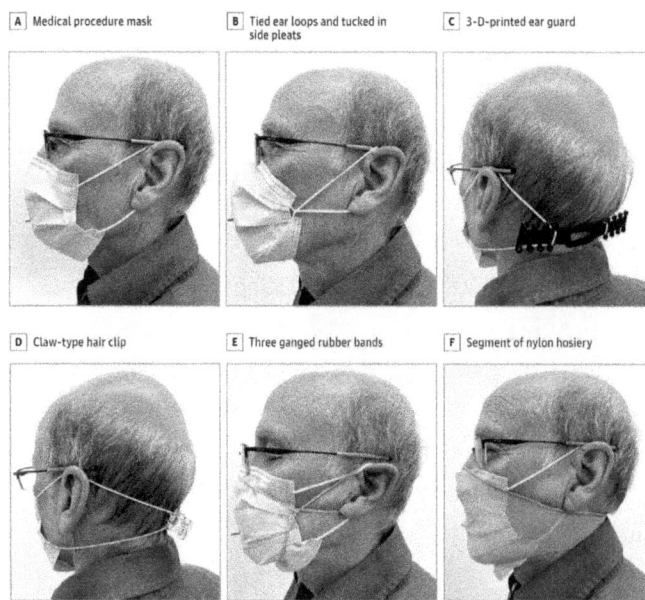

Clapp, P.W., et al., *Evaluation of Cloth Masks and Modified Procedure Masks as Personal Protective Equipment for the Public During the COVID-19 Pandemic*. JAMA Intern Med, 2021. **181**(4): p. 463-469. https://jamanetwork.com/journals/jamainternalmedicine/fullarticle/2774266

What about the CDC Science Brief's real-world studies?

After more than two *years* of motivated effort, the CDC's flagship Science Brief managed to scrounge up a mere 22 real-life studies on mask effectiveness that purportedly provide a scientific basis sufficiently robust to justify mask mandates. However, even light interrogation of these remaining studies reveals crippling deficiencies. Three of these we have already dealt with in detail — the Kansas Mask Mandate Study, the Massachusetts General Brigham Mask Study, and the Bangladesh Mask Study. [17, 90, 94] The Kansas Mask Mandate Study and the Massachusetts General Brigham Mask Study, in particular, so outrageously misrepresented the implications of their results and underlying data that insofar as they constitute any kind of evidence at all, they are most accurately categorized with studies providing evidence that masks do not work.

➢ Eight [51, 152, 154, 159, 161, 165, 171, 173] of the CDC's remaining 19 studies cited in its *Science Brief* are what the CDC euphemistically calls **"Population-based intervention with trend analysis."**

In other words: *Post hoc, ergo propter hoc.* "After this, therefore because [of] this." A rain dance could accurately be called: "a weather-based intervention with trend analysis." Rain dances look especially effective when the investigator forces people to do a rain dance every day of the year, and then confines their analysis to the rainy season. This is what the CDC is *repeatedly* guilty of doing with regard to mask evidence, engaging in statistical obfuscation and chicanery rather than objective scientific investigation seeking to find truth.

➢ All eight of these studies have one or more glaring deficiencies that made them unable to serve as the basis for a definitive recommendation, much less a mandate, by *any* informed public official of good conscience.

The July 1 to September 4, 2021 school study by Budzyn et al. [51] published in the CDC's MMWR was, as the authors stated: "an ecologic study, and causation cannot be inferred." As pointed out earlier, when the study was later replicated and extended by Drs. Ambarish Chandra and Tracy Beth Høeg, any and all benefits of mask mandates completely disappeared. Drs. Chandra and Høeg were too tactful to say this directly, but I have no such compunctions: the CDC authors carefully and deliberately chose which dataset they analyzed in order to get the result the CDC wanted for political reasons.

"While the presence of correlation does not imply causality, the absence of correlation can suggest causality is unlikely, especially if the direction of bias can be reasonably anticipated."

CDC cherry-picked time-period and dataset results (Budzyn et. al., 2021)

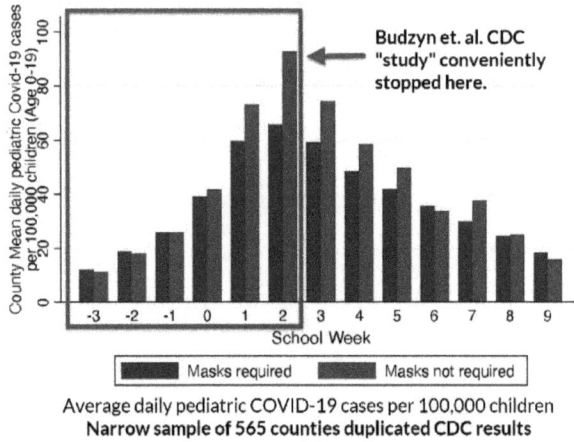

Budzyn et. al. CDC "study" conveniently stopped here.

Average daily pediatric COVID-19 cases per 100,000 children
Narrow sample of 565 counties duplicated CDC results

Expanded, *non*-cherry-picked time period and dataset (Chandra and Høeg, 2022)

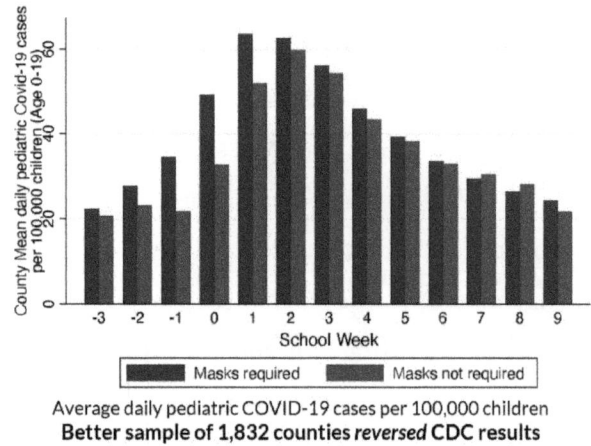

Average daily pediatric COVID-19 cases per 100,000 children
Better sample of 1,832 counties *reversed* CDC results

Chandra, A. and T.B. Høeg, *Lack of correlation between school mask mandates and paediatric COVID-19 cases in a large cohort.* Journal of Infection, 2022. **85**(6): p. 671-675. https://dx.doi.org/10.1016/j.jinf.2022.09.019

Another CDC MMWR study [154] looking at COVID case and death rates from March 1 to December 31, 2020, reported that mask mandates were associated with a <2% decrease in the case and mortality growth rates. Apart from the miniscule observed effect of masks, this CDC MMWR study includes the three-month period prior to June 2020, when overall test capacity had not reached the threshold for effective contact tracing, and stopped less than halfway through the winter season.

The same deficiencies described above are magnified by a smaller selection of states and an even shorter time period in another CDC MMWR report that looked at just 10 states (California, Colorado, Connecticut, Maryland, Michigan, Minnesota, New Mexico, New York, Ohio, and Oregon) from March 2020 to October 2020 and which associated decreased hospitalization rates with statewide mask mandates [161]. As you can see in the accompanying chart,[11] this study was published in February 2021 but completely excluded the much higher COVID peaks in November through January in all 10 of these states, which the draconian mask mandates did nothing to stop, and of which the authors could not have been unaware.

---

11   Data Sources: COVID cases and hospitalizations drawn from The Covid Tracking Project https://covidtrack-ing.com/data/download; State Populations used for per capita calculations were taken from The United States Census Bureau's numbers for 2020 https://www.census.gov/data/tables/time-series/demo/popest/2020s-state-total.html.

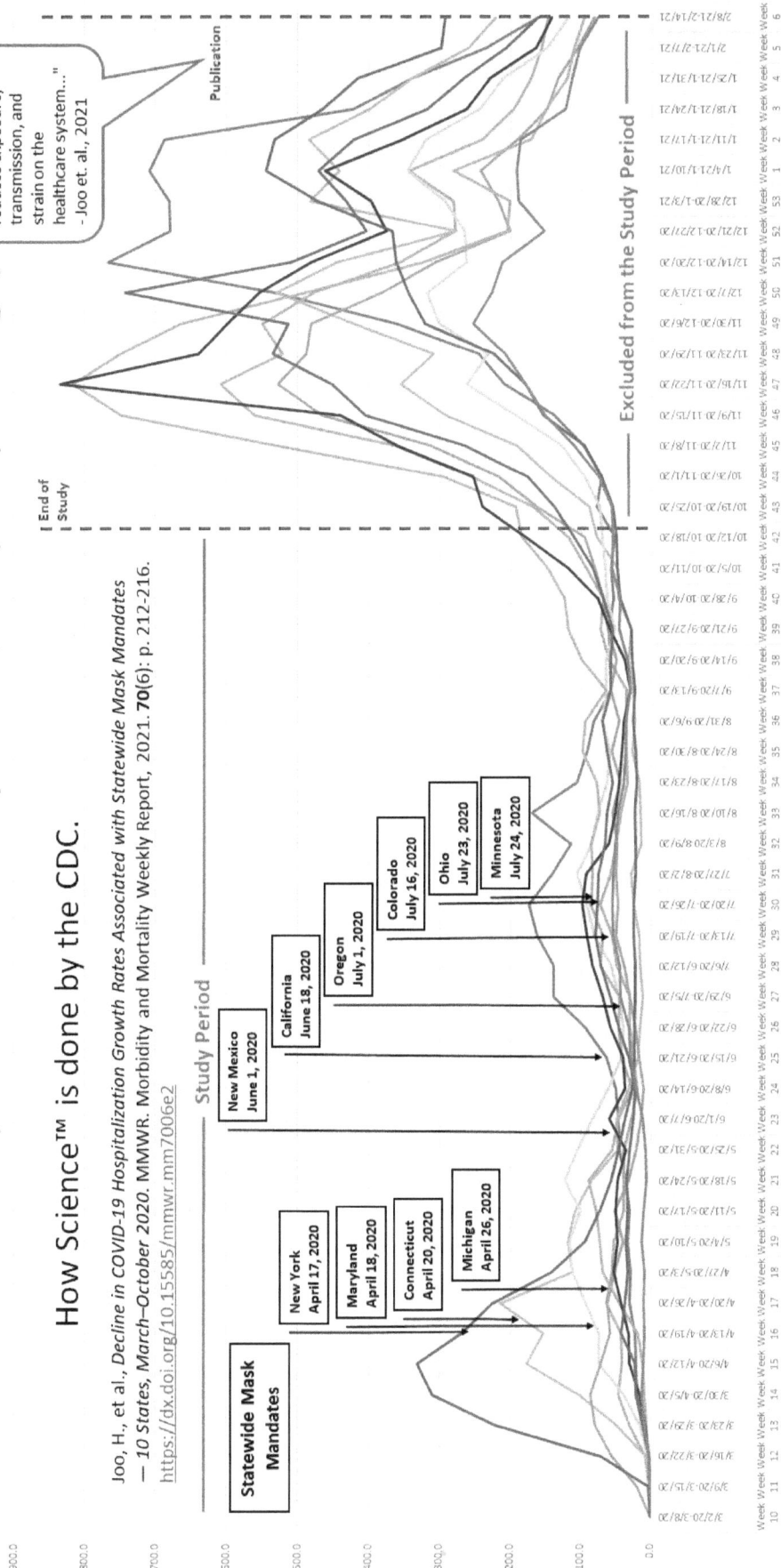

New Daily COVID-19 Cases per 100,000 (7-day average)

How Science™ is done by the CDC.

Joo, H., et al., *Decline in COVID-19 Hospitalization Growth Rates Associated with Statewide Mask Mandates — 10 States, March–October 2020.* MMWR. Morbidity and Mortality Weekly Report, 2021. **70**(6): p. 212-216. https://dx.doi.org/10.15585/mmwr.mm7006e2

"wearing a mask reduces exposure, transmission, and strain on the healthcare system..." - Joo et. al., 2021

Publication

End of Study

Study Period

Excluded from the Study Period

Statewide Mask Mandates

New York April 17, 2020

Maryland April 18, 2020

Connecticut April 20, 2020

Michigan April 26, 2020

New Mexico June 1, 2020

California June 18, 2020

Oregon July 1, 2020

Colorado July 16, 2020

Ohio July 23, 2020

Minnesota July 24, 2020

— California — Colorado — Connecticut — Maryland — Michigan — Minnesota — New Mexico — New York — Ohio — Oregon

Data Sources:
2020-2021 COVID case data from The Covid Tracking Project: https://covidtracking.com/data/download
State Populations taken from The United States Census Bureau state population estimates for 2020: https://www.census.gov/data/tables/time-series/demo/popest/2020s-state-total.html

77

The same applies to the CDC-cited study which used data from just 15 states and Washington DC over less than six weeks from April 8 to May 15, 2020 to conclude that masks work [171].

Compare and contrast these two studies cited by the CDC with just one of those we covered earlier: unlike the CDC's citations, Guerra et al. [49] looked at the data on *all* US States during the much longer period from June 2020 through March 2021, and encompassed the whole first winter wave of COVID, but found not even a small benefit from masks.

Moving on, the CDC's Science Brief cites Gallaway et al.'s study [152], published in a 2020 CDC MMWR, as strong evidence of beneficial changes in COVID-19 case rates in Arizona associated with mask use and other measures. The study authors themselves (to their credit) directly pointed out that: "the relationship between mitigation measures and changes in case counts are temporal correlations and should not be interpreted to infer causality." A simple visualization of the study time period in chronological context is enough to discredit any inferences of beneficial causality from masks in this sort of CDC study.[12]

---

12   Ian Miller, Twitter Post, *@ianmSC*, April 8, 2021, 2:54 p.m.
https://twitter.com/ianmSC/status/1380232494000132098

Ian Miller refutes the CDC with a single chart

> **Arizona mask mandates in place from June 17th did nothing to stop the fall and winter wave of COVID.**

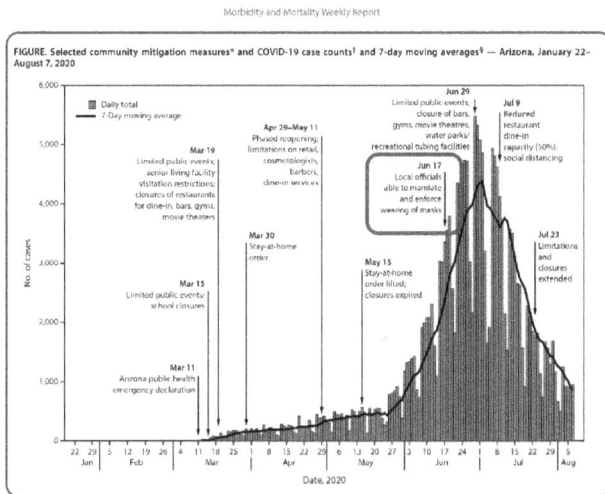

Gallaway, M.S., et al., "Trends in COVID-19 Incidence After Implementation of Mitigation Measures - Arizona, January 22-August 7, 2020," *MMWR. Morbidity and Mortality Weekly Report*, 2020. 69 (40): p. 1460-1463. https://dx.doi.org/10.15585/mmwr.mm6940e3

Ian Miller ✔
@ianmSC   **Subscribe**   ...

Two weeks ago, Arizona, to overwhelming local criticism, removed all county level mask mandates and rolled back nearly all restrictions on businesses. Nevada and California didn't, and yet their numbers are the exact same.

How many more times does this need to keep happening?

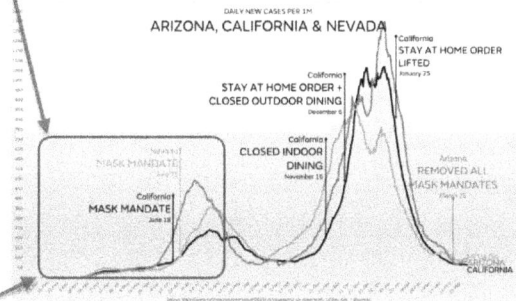

2:54 PM · Apr 8, 2021

💬 24   🔁 465   ♡ 1,038   🔖 67   ↑

The CDC tried to credit mask mandates and other mitigation measures in Arizona and other states with 2020's late-summer declines in COVID cases. In reality, mask mandates had no effect on seasonal COVID surges in any states, as Ian Miller highlighted by comparing Arizona, California, or Nevada.

Another CDC MMWR published in September 2021 looked at Arizona schools in Pima and Maricopa counties over a mere 45-day period from July 15 to August 31, 2021, and claimed to find that "after adjusting for potential described confounders, the odds of a school-associated COVID-19 outbreak in schools without a mask requirement were 3.5 times higher than those in schools with an early mask requirement" [159]. Yet this particular study does not even *look* at overall case numbers, but instead compares the *number of schools* with two or more cases, which means that schools with 2 cases or 200 cases get counted the same in determining the results. Neither does this study report the raw numbers of students and staff with COVID-19 infections which were traced to school spread, nor does it account for staff member vaccination rates. Furthermore, in this study, the CDC contact tracing guidelines *themselves* introduced a confounder, because CDC policy followed by the schools was to not even *consider* masked students as close contacts, effectively stacking the deck in favor of masks from the get-go. This leads to primary cases of COVID being over-reported in the un-masked districts and secondary cases being under-reported in masked districts.

Another "population-based intervention with trend analysis" still cited by the CDC looked at data from 183 countries from January to May 2020, and concluded that masks were negatively associated with COVID-19 mortality [165]. This is in contrast to a study we have already mentioned, which found that masks were associated with an increase in COVID deaths after evaluating a much longer, better-documented time period [57].

Similarly, another study which the CDC cited in its Science Brief, entitled "Face Masks Considerably Reduce COVID-19 Cases in Germany: A Synthetic Control Approach" [173] used nothing more than temporal association (and the total lack of a real control group) to claim that the fall "almost to zero" in new infections after facemasks were mandated in Germany between April 1 and April 10, 2020, was attributable to mask use. This ignored multiple critical variables, including the fact that cases had already started decreasing across all of Germany from April 2, 2020.

If it is fair game for the CDC to anchor its evidence favoring masks in chronological association with no control group, then it is more than fair to rebut it by pointing out that fanatical German mask mandates (including a medical mask mandate) had no effect on COVID spread in Germany at any time thereafter. This is extensively documented, especially by Ian Miller in his Substack (https://ianmsc.substack.com/) and excellent book, *Unmasked: The Global Failure of COVID Mask Mandates* [205] (these are essential mask-related reading with a great deal of original analysis and illustration).

Mandating medical-grade masks provided no benefits in Germany

Ian Miller ✔
@ianmSC

Subscribe ...

After 2.5 years of mask mandates proving to be completely useless, Germany is once again returning to mandates in a number of settings

No matter how many times they fail, politicians and "experts" won't just admit they were wrong — it's permanent commitment to what doesn't work

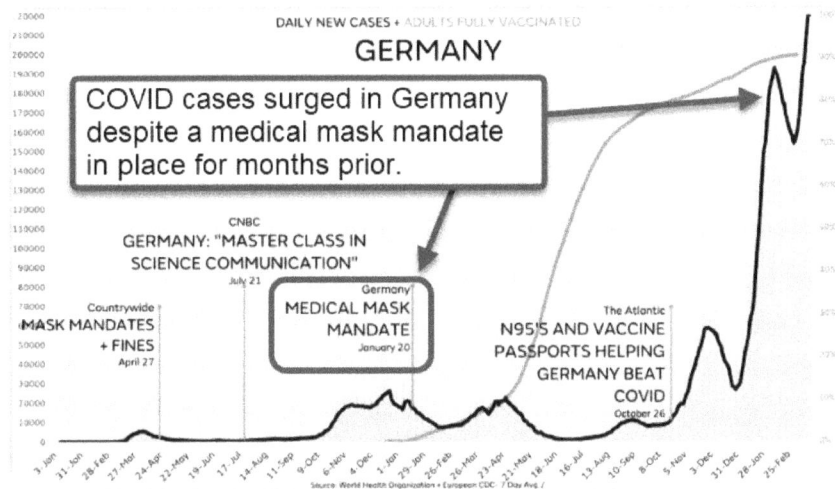

5:12 PM · Oct 3, 2022

Additionally, subsequent excess mortality numbers for 2020 and 2021 published by the World Health Organization showed that Sweden, unmasked and not locked down, ranked near the bottom of excess European mortality for 2020-2021 [206].

---

13   Ian Miller, Twitter Post, *@ianmSC*, October 3, 2022, 5:12 p.m.
https://twitter.com/ianmSC/status/1577044050661675008

2020-2021 Excess Mortality - WHO Report
(Normalized for Population Size)

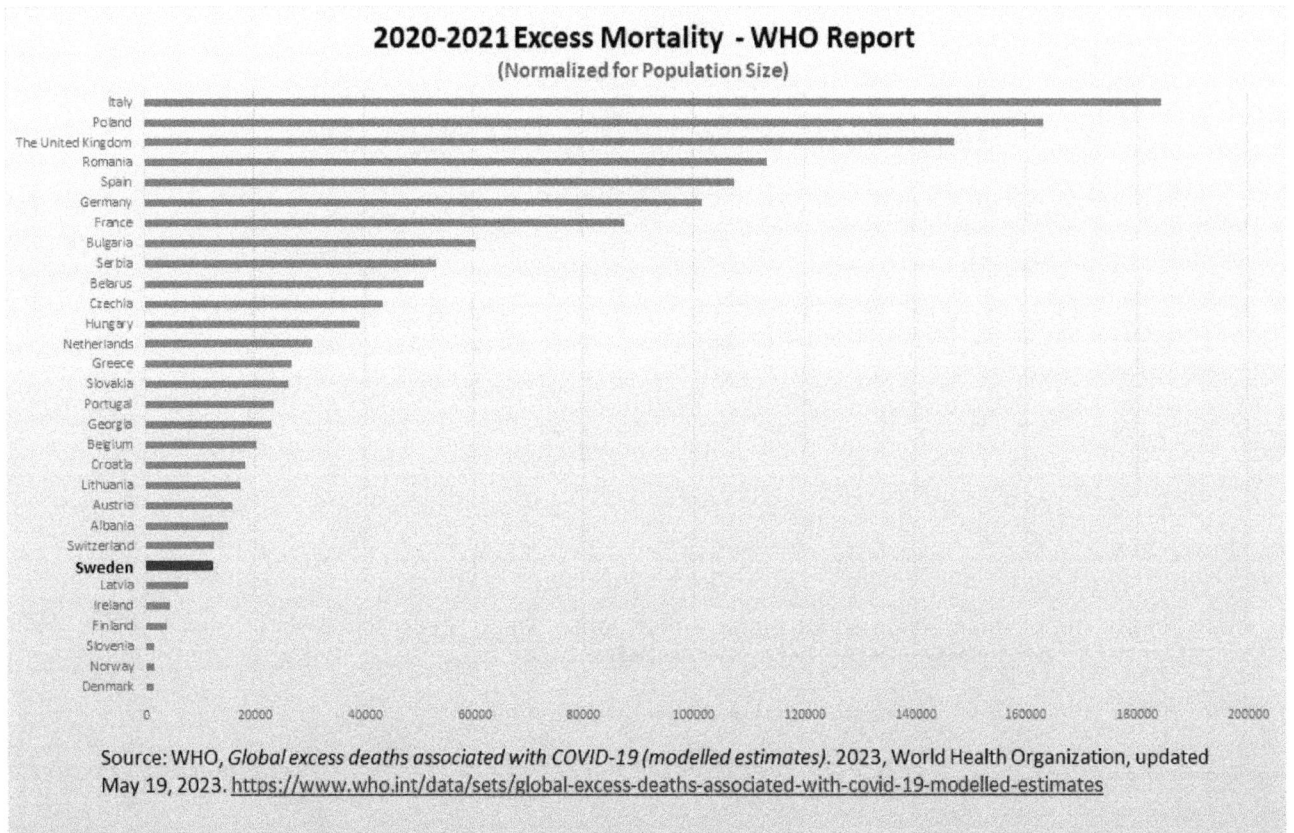

Source: WHO, *Global excess deaths associated with COVID-19 (modelled estimates)*. 2023, World Health Organization, updated May 19, 2023. https://www.who.int/data/sets/global-excess-deaths-associated-with-covid-19-modelled-estimates

Nevertheless, authorial disclaimers and fatal deficiencies have not stopped the CDC from citing the above 8 studies (six of which were published by the CDC itself) as strong evidence for a causal relationship between mask usage and lower transmission of COVID. It takes a special kind of brazenness to cite studies like this as evidence of mask efficacy powerful enough to justify mandates.

This leaves just 11 remaining references in the CDC's Mask Science Brief which we have yet to review. These consist of:

- 1 Meta-analysis of observational studies [141]

- 2 Literature Reviews [136, 143]

- 4 Retrospective Observational studies (2 Retrospective Cohort Studies [181, 203], 1 Case-Control Study [146], and 1 Cross-sectional study [183])

- 4 Reports of Surveillance Data [92, 151, 157, 178]

➢ Chu et al.'s meta-analysis of observational studies, published in June 2020 in *The Lancet*, is possibly the strongest study put forward by the CDC to support mask efficacy [141]. The authors combined the results of 29 observational studies (7 MERS, 18 SARS, and 4 COVID-19). The studies had a combined total of 3686 masked participants (with 163 infections) and 6,484 unmasked participants (with 546 infections). The authors calculated that masked individuals have a 3.1% risk of infection vs. 17.4% when unmasked. This generated a prevalent media sound bite that unmasked people were 5x more likely to catch COVID than those that wore masks. Even if many other studies, experience, and observation since 2020 had not shown that calculation to have completely failed the test of reality, it was *one* meta-analysis of *observational* studies, did not include a single one of the randomized controlled trials we looked at earlier, and even the authors rated their findings "low certainty." Also, it failed to include many observational studies which we reviewed earlier.

➤ The two remaining literature reviews cited in the CDC *Science Brief* are principally concerned with using a survey of existing literature to theorize about the potential effectiveness of cloth masks to stop viruses [136, 143]. Both rely primarily on laboratory filtration studies, but occasionally reference real-world studies such as MacIntyre's 2015 study comparing cloth masks to medical masks (the same 2015 MacIntyre study we looked at earlier which said: "This study is the first RCT [Randomized Controlled Trial] of cloth masks, and the results caution against the use of cloth masks" [23]).

The first of these two literature reviews is principally concerned with the decontamination of cloth mask materials and discusses four filtration mechanisms (electrostatic filtration, diffusion, interception, and impaction) [136]. The second focuses on summarizing 25 laboratory studies of cloth filtration efficiency, including the material, weave, thread count, testing methodology, aerosol used for testing, aerosol size, and flow rate [143]. Neither of these provides good evidence that cloth masks stop viruses or bacteria.

➤ As for the two retrospective cohort studies (as usual, published by the CDC MMWR), one was conducted using PCR testing among a convenience sample of 382 service members from the USS Theodore Roosevelt in April 2020, the study authors found that: "Service members who reported taking preventive measures had a lower infection rate than did those who did not report taking these measures (e.g., wearing a face covering, 55.8% versus 80.8%... respectively)" [181].

The other retrospective cohort study cited by the CDC was conducted by phone interview among 335 people in 124 families in Beijing from February 28 to March 27 of 2020 [203]. The authors reported that mask use was 79% effective in preventing secondary transmission of SARS-CoV-2 within a household, and that daily use of chlorine or ethanol-based disinfectants was 77% effective. The authors also reported that wearing a mask after illness onset of a family member was not significantly protective.

The case-control study cited by the CDC compared the characteristics of 211 cases of COVID-19 with those of 839 matched controls in Thailand from April 30 to May 27, 2020 [146]. 48.6% of the COVID-infected group reported not having worn a mask, whereas 60.7% of the matched uninfected case-controls also reported not wearing a mask. Yes, you read that right — a higher percentage of the *uninfected* group in this study reported not wearing a mask, but the CDC still cited this as evidence masks work because a higher percentage of the uninfected group also reported *always* wearing a mask. A much larger percentage of the infected group than the uninfected group reported wearing a mask "sometimes."

These three references, in particular, serve to highlight the CDC's outrageous double-standard when it comes to evidence regarding masks. The CDC attempts to dismiss the results of the Danish Facemask Randomized Controlled Trial [16] by saying, "The study was too small (i.e., enrolled about 0.1% of the population) to assess whether masks could decrease transmission from wearers to others." Yet in the same Science Brief, the CDC claims that masks are effective based on shorter, *un*controlled studies from Thailand, China, and an Aircraft Carrier [146, 181, 203], which together have a combined total of *less than half* the number of participants as the Danish Facemask Study, and are orders of magnitude smaller relative to the populations of their respective countries.

➢ The cross-sectional survey study [183] is *primarily a modeling study* using survey data collected over 8 weeks from June 3 to July 27, 2020. Again, compare and contrast this with the *non*-modeling studies cited earlier, covering case and death data over the course of more than 9 *months* across the United States, and encompassing an entire winter flu season [49]. Compare this also with any of the large-scale school studies showing no benefits from masking [48, 53, 55, 56, 128].

➢ The assortment of 4 reports of surveillance data [92, 151, 157, 178] are just as bad, if not worse.

The (in)famous Hair Salon Contact Tracing Study from May 2020 [92] totally lacked a control group. Half of the contacts (72) were not even tested (35 were not contacted, and 37 outright refused testing and were self-reporting during a time when people had strong incentives not to report themselves being sick because of quarantine and isolation expectations). Using this Hair Salon surveillance report as any kind of proof that masks prevent viral transmission is as bad as claiming that *not* wearing masks has a protective effect based on surveillance studies with the original SARS which we highlighted earlier, where 45 United States healthcare workers who had exposure to laboratory-confirmed SARS-CoV patients "without any mask use" were not infected

[75]. This Hair Salon "study" leaves many important questions unanswered, as Dr. Scott Atlas emphasized in 2021:

> "No other stylists working with the two infected hair stylists, Stylist A and Stylist B, all mingling without masks, developed COVID symptoms. Did they catch the infection? None were tested, so we don't know. Close at-home contacts of Stylist A became infected, but the two close at-home contacts of Stylist B did not develop COVID. Perhaps that implies that Stylist B was simply not a spreader of the infection, regardless of mask wearing." [207]

Similar to one of the CDC MMWR reports looked at earlier [154] (and using much the same data), a June 2020 Goldman Sachs Research analysis argued that mask use was associated with a decline in the growth of COVID cases and deaths of up to 3.7 percentage points [157]. The study predicted that widespread mask use would prevent a 5% drop in GDP and "substitute for renewed lockdowns." Wrong on all counts.

The third of these surveillance studies merely catalogs the published airline flights where secondary transmission of SARS-CoV-2 is believed to have occurred based on contact tracing [151]. "Three mass transmission flights without masking are contrasted to 5 with strict masking." The utility of the study is limited because any potential control group of unmasked flights was eliminated by airline mask mandates early in 2020, and as we have already seen, spending 15 hours on a plane near someone who has symptomatic COVID-19 was not necessarily enough to catch it even from the beginning, when natural immunity was at its least common [77]. Additionally, after the CDC's travel mask mandate was struck down in April 2022, there was no increase in COVID-19 infections on airplanes.

The last of this cluster of studies was simply a report on the results of surveillance contact tracing of in-school transmissions of SARS-CoV-2 in eight Massachusetts K-12 public school districts during the 2020-2021 school year [178]. The study authors reported that 435 index cases of COVID-19 produced just 29 possible or probable in-school transmissions despite having 1,771 school-based contacts. The CDC science brief highlights that the study authors "found an unadjusted secondary attack rate of 11.7% for unmasked versus 1.7% for masked interactions" [114]. Apart from the fact that this study leaves out roughly 280 other Massachusetts school districts, masks were mandated in Massachusetts schools throughout the study, so once again the CDC has built a crucial room of its evidential house on a study using no real control group. As the study authors themselves pointed out: "Notably, all reported classroom exposures were masked, so **these results do not directly inform the impact of masking within classrooms**" [178].

What About Other Studies Not Cited in the CDC Science Brief?

Other studies supposedly showing the efficacy of masks fare no better.

➤ A 2022 literature review by Alihsan et al. [208] was widely touted as evidence in favor of masks. German Federal Minister of Health Karl Lauterbach tweeted: "[For everyone who is still unsure whether masks protect against COVID: here is a new American mega study that evaluates over 1,700 studies. The benefit of the masks is very large, undisputed, and applies to many areas]" (translated from the original German).[14] Dr. Matthew K. Wynia, MD, MPH, Director of the Center for Bioethics and Humanities (who we will meet again in Part 4), also cited this study as evidence in favor of mask-wearing.[15] What Lauterbach neglected to mention was that the preprint study in question found over 1,700 studies in its initial search engine pass looking for evidence, but it only *included* 13 of them in its actual *analysis*.

---

14   Karl Lauterbach, Twitter Post, *@Karl_Lauterbach*, July 31, 2022, 3:59 p.m. https://twitter.com/Karl_Lauterbach/status/1553832826822344704

←   Post

Prof. Karl Lauterbach ✓
@Karl_Lauterbach

Für alle, die noch immer im Unklaren sind, ob Masken gegen COVID schützen: hier eine neue amerikanische Mega-Studie, die über 1.700 Studien auswertet. Der Nutzen der Masken ist sehr groß, unumstritten und gilt für viele Bereiche.

Translate post

medRχiv
THE PREPRINT SERVER FOR HEALTH SCIENCES

medrxiv.org
The Efficacy of Facemasks in the Prevention of COVID-19: ...
Facemasks have become a symbol of disease prevention in the context of COVID-19; yet, there still exists a paucity of ...

3:59 PM · Jul 31, 2022

1,171 Reposts   **414** Quotes   **6,140** Likes   **372** Bookmarks

15   Matthew Wynia, Twitter Post, *@MatthewWynia*, August 1, 2022, 9:23 a.m. https://twitter.com/MatthewWynia/status/1554095471383048192

Matthew Wynia
@MatthewWynia

Meta-analysis, in pre-print, and doesn't differentiate between high and low filtration masks, but still finds: "The probability of getting COVID-19 for mask wearers was 7% (97/1463, p=0.002), for non-mask wearers, probability was 52% (158/303, p=0.94)."

medRχiv
THE PREPRINT SERVER FOR HEALTH SCIENCES

medrxiv.org
The Efficacy of Facemasks in the Prevention of COVID-19: ...
Facemasks have become a symbol of disease prevention in the context of COVID-19; yet, there still exists a paucity of ...

9:23 AM · Aug 1, 2022

1 Retweet   **1** Quote Tweet

Anyone who cited this particular study as a serious reference supporting mask-use or mask mandates forfeited *any* claim to expertise by failing to do their due diligence, and may have been deliberately trying to deceive. If you want to know what "cherry picking" in professional literature looks like, this preprint is a prime example. The Alihsan et al. literature review purports to be a "systematic review" of the efficacy of facemasks in preventing COVID-19 transmission, but with the exception of one random case report on SARS from 2004 [209] (which did not even meet the authors' own inclusion criteria because it was a 1-person case study where one SARS patient transmitted SARS to no one else), this review limited its analysis to 12 articles published from April 2020 to August 2020. Do you think a little thing like deliberately picking *April to August of 2020* might skew the authors' literature findings just a tiny bit?

The 13 studies included in the Alihsan et al. analysis consist of: 1 retrospective cohort study [203]; 2 case-control studies [210, 211]; 6 surveillance [92], survey [210, 212, 213], or anecdote [214, 215] studies; 3 contact tracing studies [216-218]; and 1 detailed write-up of hospital policies [219]. Even the CDC only bothered to cite two of these thirteen in its own *Science Brief* on masks ([92, 203]), for good reason. Nine of the thirteen do not even have a control group, and at least two of them (insofar as they provide any evidence at all) run as much *against* mask efficacy as they do in favor of it.

In total, the 13 studies include 247 SARS-CoV-2 infections and a total of 1551 participants (1076 in healthcare and 475 in community settings). Remember, the CDC tried to dismiss the Bundgaard et al. Danish facemask study as being "too small" despite a total of more than 4,800 participants completing just that ONE randomized controlled trial – more than twice as many as all 13 of the studies cited by Alihsan and colleagues in this so-called "systematic review" *combined*.

A table listing all thirteen studies with comments is presented below.

The 13 studies cited by the Alihsan et al. (2022) "Literature Review"	
Canova et al., 2020	Swiss contract tracing study. No control group. 21 healthcare workers exposed to one COVID case. *None wore masks. None tested positive.* 3 reported mild respiratory symptoms. "These results indicate that routing short clinical examinations and short physical contacts did not place the healthcare workers at risk sufficient for them to have acquired SARS-CoV-2." This study contains no evidence favoring mask-use.
Caruhel et al., 2021	Anecdotal one-page article talking about how 20 tracheotomies were successfully performed by personnel wearing gas masks. Obviously, there was no non-masked control group.
Çelebi et al., 2020	Case-control study of healthcare workers (including custodial staff) who had a positive COVID PCR test. 47 cases and 134 controls were analyzed out of 703 total staff surveyed/tested. Results were based off the finding that 33 of the 47 COVID cases had eaten for >15 minutes in the presence of other healthcare workers without wearing masks. This is *extremely* weak association at best and cannot show causality. Also, this summary ignored the fact that, according to the study results, simply *touching* a suspected or confirmed COVID patient was associated with 3x to 5x *more* risk of infection than intubating or collecting a respiratory sample from a suspected or confirmed COVID patient! This *alone* should render any conclusions suspect.
Chaovavanich et al., 2004	This is a 1-person case study from 2004 where a single SARS patient failed to transmit SARS to anyone. According to Alihsan et al.'s own inclusion criteria, this study should have been left out. No control group (obviously).
Chang et al., 2020	This study simply recorded a 1% COVID secondary infection rate (and no tertiary infections) in the absence of mask usage in a hospital. This observed transmission rate is even lower than that found by Bundgaard et al., and provides no evidence that masks work to stop respiratory viruses. If anything, it suggests the opposite.
Guo et al., 2020	Case-control study based on a survey of 26 COVID-infected orthopedic surgeons in Wuhan from 8 hospitals. The same deficiencies as Çelebi et al., 2020, but worse. It also excluded orthopedic surgeons who worked in COVID wards, an omission which skews the results in favor of masks because the COVID-ward surgeons always wore masks but had higher infection rates. Also unermining the study's findings in favor of masks, the authors found that "insufficient" PPE supply was *not* associated with a statistically-significant difference in infection rates. The study authors stated it best: "The case-control study design was used to test the possible link between the exposures and the outcome, but not to confirm the causal relationship, and recall bias could have occurred because of its nature."
Hendrix et al., 2020	The CDC's "Hair Salon Study" with no control group. These results are also consistent with generally low transmission rates from minimally symptomatic individualls, as shown by Luo et al.'s August 2020 contact tracing.
Kang et al., 2020	This study's inclusion in the literature review was based on contact-tracing case reports from page 4. Of the two people infected with COVID, one wore a mask, and one did not, making this study useless to show that masks work.
Nir-Paz et al., 2020	After opening with the statement that: "there is no evience to support viral transmission during flights," this anecdotal article then goes on to credit the lack of COVID transmission on a 15-person flight (4 crew, 11 passengers) to everyone wearing masks, despite the fact that multiple passengers took of their masks for extended periods of time to eat. No control group.
Rolland et al., 2020	Long-term care facility study conducted via questionnaire, and covering the period March-May 2020. Found that mask use was equally prevalent among facilities which had and had not been contaminated with COVID (pg 3 of the study). The authors said nothing about masks in their discussion. If anything, this study suggests masks made no difference, and including it in this analysis is not justified.
Wang et al., 2020	Retrospective cohort study cited by the CDC in its *Science Brief: Community Use of Makss to Control the Spread of SARS-CoV-2.*" Looked at 335 people in 124 familites with at least one COVID case. Found that mask use was more prevalent in the households that reported no secondary transmission. Found that chlorine and ethanol use for disinfection was equally prevalent in households that reported no secondary infection. Found an overall secondary attack rate of 23%. Result contradicted by dozens of later studies, but this is the closest thing to a justifiably citeable study for a literature review in this entire list.
Wang et al., 2020	Cross-sectional survey of a convenience sample of 92 healthcare workers (31 infected, 61 uninfected) carried out from 1 January 2020 to 29 February 2020 at Zhongnan Hospital of Wuhan University. Inclusion criteria was having worked at the hospital for 14 days or more since the first case of COVID was reported there. Participants were asked not whether they did or did not wear a mask, but whether they wore it "correctly." Only 2 of the 92 healthcare owrkers admitted not wearing a mask correctly, and those two happened to be part of the 31 that became infected. If anything, this study could just as easily be cided as evidence against masks working, since everyone that got infected was wearing a mask, meaning that there was no real control group).
Xu et al., 2020	April 2020 published report of policies and procedures in a Chinese general hospital. Required 100% masking of staff and reported that 73-74% of those staff wore their masks correctly. Reported that as of 31 January 2020, there were no hospital-acquired COVID-19 infections among staff in the hospital, which it attributes to its policies and procedures. **Including this study in any systematic literature review as evidence for masks is a bad joke**. This is basicaly an outline of policies and procedures which are supposed to be evidence-based, but do not themselves constitute real evidence. Moreover, the policies and procedures it describes lacked (and still lack) an empirical evidential basis. It also cuts off at the end of January 2020. No one in 2022 should be citing this. It does not include any reports of COVID cases in patients or staff. It was written by bureaucrats for bureaucrats so that bureaucrat policy makers could have some published reference to point at as justification for their own masking policies.

➤ Another 2022 study published in the *New England Journal of Medicine* purported to show that lifting mask mandates in Massachusetts was associated with 44.9 more cases of COVID per 1000 students [220]. However, this study suffers from the same defects of cherry-picked time period and dataset as the earlier Budzyn et al. study published in the CDC's MMWR [51]. More importantly, this study's own charts clearly show that the school districts being compared experienced even greater differences in COVID case numbers *during the period when all of them were subject to the same universal masking policy*. This strongly suggests that any differences in COVID-19 incidence after the mask mandates were lifted were more than likely caused by whatever factors caused the dramatic differences in COVID-19 incidence when the mask mandates were still in place.

Cowger, T.L., et al., *Lifting Universal Masking in Schools — Covid-19 Incidence among Students and Staff.* New England Journal of Medicine, 2022. https://dx.doi.org/10.1056/nejmoa2211029

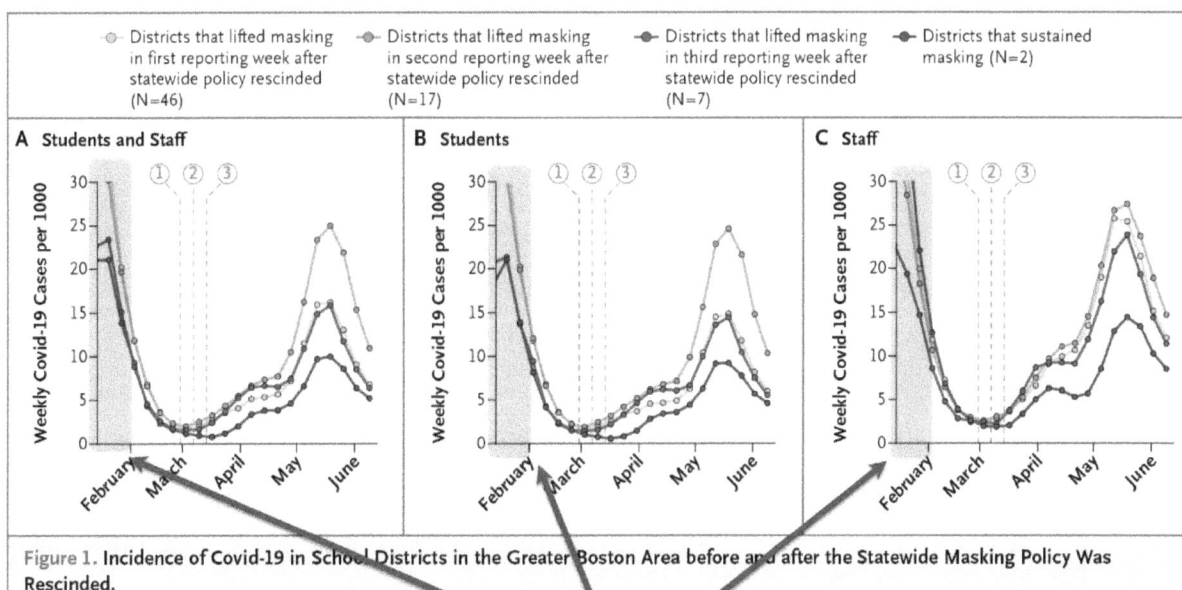

Figure 1. Incidence of Covid-19 in School Districts in the Greater Boston Area before and after the Statewide Masking Policy Was Rescinded.

The authors' own graphs can't hide the fact that the study groups had even **greater** differences in COVID cases while still under the **same** universal masking policy!

Wrapping up the CDC's case for masks

If anything, the foregoing discussions give the CDC *too much* credit, because the CDC was putting its best foot forward when picking and choosing which of its MMWR publications to cite in its *Science Brief* on masks. Three persevering researchers — Tracy Høeg, Vinay Prasad, and Alyson Haslam — doggedly accomplished the mind-numbing feat of tracking down, carefully reviewing, cataloging, and analyzing every CDC MMWR publication pertaining to masks from 1978 to 2023 [221]. They published their findings in July 2023, and unlike the CDC, they made it *easy* for reviewers to check their work.

The team's first notable finding was that the CDC *had* no MMWR publications prior to 2020 which pertained to masks — all of the CDC's MMWR writings on masks came from 2020 and later. In total (not including five guidance documents and a standalone search result), the team found 77 CDC MMWR studies on masks: 30 from 2020, 33 from 2021, and 14 from 2022. Of those 77 studies, 58 included conclusions that masks were effective, but only 23 actually assessed mask effectiveness. It gets worse. Of those 23 studies looking at mask effectiveness, only 12 were statistically significant, and only 12 had anything resembling a control group. (Not every MMWR publication with statistically significant results had a control group, or vice versa.) In four of the twelve MMWR studies where something like a control group was present, the masked group actually had *more* cases of COVID than the comparator group, but the studies *still* concluded that masks were effective and therefore essential! Høeg et al. found that the *only* one of the CDC's 77 MMWR studies pertaining to masks in which causal language was used appropriately was a *mannequin* study with highly questionable applicability to real life. Similarly, the only one of the 77 MMWR studies to even *mention* data that suggested masks might not work was a study involving masks and *influenza*. Not a single one of the CDC's MMWR publications pertaining to masks were randomized trials or cited randomized data.

The studies cited and referenced by the CDC never came anywhere *close* to satisfying the burden of proof necessary to sustain mask recommendations, let alone *mandates*. The CDC's aggregate conclusions flew in the face of mountainous contrary evidence. Characteristic deficiencies of studies set up to justify masking include (but are not limited to): lack of a control group; small size; short duration (with suspiciously convenient time periods selected for analysis); selective presentation; uncontrolled and un-addressed confounders; and suspiciously unusual selection criteria or endpoints. As Ian Miller, author of one of the few accurate books on masks, aptly commented: "No matter how much respect you've lost in the CDC over the past few years, it's truly not enough."[16]

---

16   Ian Miller, "The CDC Repeatedly And Purposefully Put Out Misleading, Low Quality Studies To Push Masks," *Unmasked* (Blog), July 15, 2023, online: https://ianmsc.substack.com/p/the-cdc-repeatedly-and-purposefully

What are we forgetting? Why masks do not work.

The evidence base cited to support the efficacy of masks fails to consider important aspects of the problem. Filtration is just one of *many* mechanisms by which masks act out their real-world effects. If susceptible individuals have enough viral exposure to reach the infection threshold with or without a mask, then mask wearing will not make a significant difference in the overall infection rate. Even if it was undisputed that mask filtration properties provide infection control benefits, **there are multiple additional effects associated with mask-wearing that could, on net balance, easily negate or even outweigh any benefits from filtration**, potentially increasing the overall risk of disease transmission depending on circumstances.

➢ Masks facilitate cross-contamination by providing an additional surface and microbial growth medium, especially if skin irritation or repeated adjustments result in more frequent touching, transport, and improper storage [96, 104, 106, 222-234]. We will discuss the infection control issues with masks in more detail when we cover mask side effects in Part 2.

➢ The increases in temperature and humidity on skin surfaces underneath masks enhance microbial growth, expanding the pool of microbes which can be dispersed [230, 235-238]. Normally, exhaled respiratory emissions dry quickly in free air, but when trapped in masks, microbe-laden droplets stay warm and damp. This may explain some of the study findings that we looked at earlier: how bacteria emissions through surgical masks increase in volume over just two hours of use until they equal or surpass unmasked emissions [106]. We will cover this topic in much greater detail in Part 2, as well.

➢ Masks can serve as sources and launchpads for aerosolized fomites — non-respiratory particles which have been contaminated [113]. As we saw earlier, multiple CDC-designed and mandate-compliant homemade cloth masks emit substantially more airborne particles than a non-masked baseline [113].

➤ Similar to how the aerator on your faucet or the nozzle on a spray bottle works, masks act like nebulizers, increasing the proportion of respiratory emissions composed of fine aerosols [113, 239]. A 2020 laboratory study found that when no mask was worn, 35% of particles emitted were in the smallest measured size range of 0.3 to 0.5 μm (300 to 500nm, still much larger than SARS-CoV-2). When a surgical mask was worn, however, this fraction increased from 35% to 46% of emitted particles, and increased still further to 60% of particles when a KN95 mask was worn [113]. Even assuming that a mask blocks 50% of emissions under perfect conditions, we already know that having just 1-2% of a mask's area compromised and leaking can cut that filtration benefit by half or more [120]. If a mask functioning at half or less of its theoretical filtration also creates a relative increase in the wearer's aerosol emissions of 25-30% or more, the net result becomes either no net benefit or even a net harm. So much for the slogan "my mask protects you, your mask protects me."

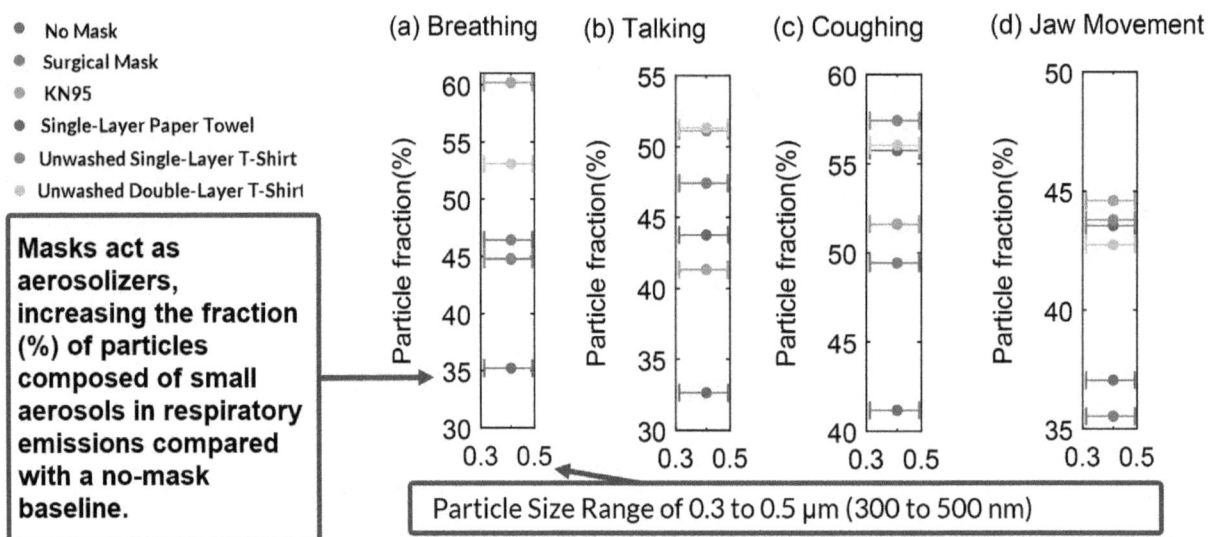

Adapted from Asadi, S., et al., "Efficacy of masks and face coverings in controlling outward aerosol particle emission from expiratory activities." *Scientific Reports*, 2020. **10**(1). https://dx.doi.org/10.1038/s41598-020-72798-7

➤ As noted earlier, SARS-CoV-2 remains viable longer on medical masks than on most other surfaces [96].

➤ Mask-associated rhinitis (nasal inflammation), laryngeal, or respiratory irritation can increase respiratory secretions [240, 241]. More on this in Part 2.

➢ The increased resistance to airflow from masks may necessitate more vigorous inhalation or expiration, thereby generating more scary bioaerosols [149, 236, 242-251].

➢ Another hypothesis based on *in vivo* laboratory findings and observational studies suggests that deep re-inhalation of hypercondensed droplets or pure virions caught in facemasks can worsen prognosis and might be linked to long-term effects of COVID-19 infection (this has been dubbed, the "Foegen effect" theory) [252].

➢ Mask abrasion on skin often releases more microbes into the air than untouched skin [108, 110, 253].

➢ Long-duration mask-wearing damages skin, compromising skin's ability to act as a barrier to all microbes, not just viruses [138, 179, 198, 230, 235-238, 244, 254-263]. More on skin damage from masks in Part 2.

➢ Wearing a mask can cause counterproductive behavioral changes such as risk-compensation, with wearers becoming less careful because of perceived protection [227, 231]. This psychological side effect of mask-wearing was also noted in the 2023 Cochrane Review on masks[17] [20]. More on the psychological side effects of masks in Part 2.

➢ The overwhelming weight of evidence is that surgical masks make no statistically significant difference against infections [1-11, 13-16, 20, 32, 35, 42, 46, 59, 60, 64, 69, 70, 72], and when directly compared to surgical masks, cloth masks perform worse [23, 28]. Additionally, stretching and washing associated with normal wearing causes a cumulative and rapid decline in the filtration capability of cloth masks [105]. Study findings of higher respiratory infection rates in cloth-mask-wearers than when either no masks are worn [91] or when surgical or respirator-type masks are worn [23, 28], raise the strong possibility that widespread cloth mask usage during COVID actually *increased* respiratory infections slightly above a non-masked baseline. Similarly, a case-control observational study on healthcare providers treating Hajj pilgrims found that *intermittent* medical mask use was associated with an increased chance of acute respiratory infection [69], as did a case-control study out of Thailand cited by the CDC [146].

➢ There is a final point you can easily test for yourself the next time you cough. Compare the experience of coughing without covering your mouth to the experience of coughing into a cupped hand, cloth or good paper towel. You will notice that your cough is more productive when you

---

17   Specifically, on page 34 (pdf page 36) of that review. Jefferson, T., et al., "Physical interventions to interrupt or reduce the spread of respiratory viruses," *Cochrane Database of Systematic Reviews*, 2023. **2023**(1). https://dx.doi.org/10.1002/14651858.cd006207.pub6

cover your mouth. Similar to how covering your mouth with your cupped hand or coughing into a cloth can help make your cough more productive, masks create this same plunger-like back pressure and turbulence, dislodging more of those scary respiratory secretions held deep in the lungs. There is no net public health benefit to a filter that catches half of what you cough up if it makes you cough up twice as much material from deeper in your lungs in the first place.

Part 1 Conclusion:
Over 100 years of science

The first part of this book has been concerned primarily with the empirical debate about the efficacy of masks in mitigating the spread of respiratory pathogens — especially viruses. In the next section on mask science, we will cover numerous additional studies which provide background information about well-documented non-trivial side effects and tradeoffs that would discourage mask use and compulsory masking even if masks were effective at mitigating the transmission of respiratory pathogens.

Every society already has a very effective method of avoiding infectious disease transmission. This method has been tried, tested, verified, and established over centuries of experience. It is called "staying away from obviously sick people." Compulsory masking of apparently healthy people, even if it worked, would only address a small subset of risk — lethal spread of COVID from asymptomatic individuals. Asymptomatic individuals are only a fraction of total COVID cases [101, 264, 265]. Additionally, *multiple* meta-analyses — incorporating more than 150 contact tracing studies [79-81] — have shown symptomatic carriers of COVID to be 6 to 20 times more contagious than asymptomatic carriers, and that the chances of catching COVID through workplace exposure, even in a *healthcare* workplace setting, are less than 1/5 the odds of doing so through exposure in the home.

Table 1. Summary Table of the Pooled SAR and R_{obs} for the Exposure Locations Considered in this Study

SAR = Secondary Attack Rate (The lower the better)

Setting	Pooled SAR (%)	95% CI	Pooled R_{obs}	95% CI
Households	21.1	17.4–24.8	0.96	0.67–1.32
Social gatherings with family and friends	5.9	3.8–8.1	0.38	0.18–0.64
Travel	5.0	0.3–9.8	...	...
Healthcare	3.6	1.0–6.9	1.18	0.65–2.04
Workplace	1.9	0.0–3.9	...	...
Casual close contacts	1.2	0.3–2.1	...	...

Where values are missing, there were not enough data available to estimate a pooled value.

Abbreviations: CI, confidence interval; R_{obs}, observed reproduction numbers; SAR, secondary attack rate.

Symptomatic SARS-COV-2/COVID-19 Secondary Attack (Transmission) Rates

Severe Acute Respiratory Syndrome Coronavirus 2 Setting-specific Transmission • CID 2021:73 (1 August) • e759

Thompson, H.A., et al., _Severe Acute Respiratory Syndrome Coronavirus 2 (SARS-CoV-2) Setting-specific Transmission Rates: A Systematic Review and Meta-analysis._ Clinical Infectious Diseases, 2021. **73**(3): p. e754-e764. https://dx.doi.org/10.1093/cid/ciab100

Study	Country	Exposure Location	Index cases	Secondary cases	Total contacts		Secondary Attack rate
Asymptomatic							
Jiang et al	China	combined	8	1	174		0.01 (95% CI: 0.00-0.03)
Han et al	China	combined	18	0	41		0.00 (95% CI: 0.00-0.09)
Chaw et al	Brunei	household	19	2	32		0.06 (95% CI: 0.01-0.21)
Ohen et al	China	combined	30	6	146		0.04 (95% CI: 0.02-0.09)
Zhang et al	China	combined	83	1	119		0.01 (95% CI: 0.00-0.05)
Park et al	South Korea	household	97	0	4		0.00 (95% CI: 0.00-0.60)
Liu et al	China	combined	147	24	914		0.03 (95% CI: 0.02-0.04)
Luo et al	China	combined	347	1	305		0.00 (95% CI: 0.00-0.02)
Beta-Binomial Summary $\gamma = 0.047$							**0.02 (95% CI: 0.01-0.03)**

Compare this to the _Symptomatic_ Secondary Attack (Transmission) Rate!

Thompson, H.A., et al., _Severe Acute Respiratory Syndrome Coronavirus 2 (SARS-CoV-2) Setting-specific Transmission Rates: A Systematic Review and Meta-analysis._ Clinical Infectious Diseases, 2021. **73**(3): p. e754-e764. https://dx.doi.org/10.1093/cid/ciab100

Moreover, the purported lethality of COVID-19 was vastly overstated in media right from the very beginning. A French study published in May 2020 found that the mortality rate of patients hospitalized with SARS-CoV-2 was comparable to the mortality rate of patients hospitalized with other coronaviruses (in fact, this study found that the OC43 and NL63 coronaviruses had double the mortality of SARS-CoV-2) [266]. The authors concluded, "there does not seem to be a significant difference between the mortality rate of SARS-CoV-2 in OECD countries and that of common coronaviruses." In January 2021, a meta-analysis by Stanford epidemiologist Dr. John Ioannidis incorporating data from nearly 50 countries suggested that COVID-19 had an Infection Fatality Rate (IFR) of ~0.15% [267]. In other words, a bad flu. A subsequent analysis in preprint by October 2022 which Dr. Ioannidis coauthored, found even this modest previous estimate was far too high: "At a global level, pre-vaccination IFR may have been as low as 0.03% and 0.07% for 0-59 and 0-69 year old people, respectively" [268].

According to this data using 40 seroprevalence studies from 38 countries, the age-stratified median infection fatality rate for COVID-19 was estimated to be:

- 0.0003% at 0-19 years
- 0.003% at 20-29 years
- 0.011% at 30-39 years
- 0.035% at 40-49 years
- 0.129% at 50-59 years
- 0.501% at 60-69 years

Compulsory measures cannot be justified to control the risk from a virus which the CDC's own seroprevalence publications admit was successfully contracted and cleared by more than half of the United States population as early as February 2022 [269].

Compulsory masking was initially motivated not by science and carefully considered data-driven public policy, but by primal fears of contamination, unexamined intuition, a desire to be (or at least appear to be) helpful, a desire to make others feel better, an "it can't hurt" attitude, and in some cases a pathological risk-aversion, a conscious desire to manipulate others through behavioral psychology, and/or desperate *post-hoc* self-justification.

It is worth repeating that from a public policy standpoint, it is not actually necessary to prove that masks do not "work". All that is really necessary is to show that the balance of scientific evidence does not favor masks strongly enough to justify forcing all citizens to wear one. This is a *much* lower burden of proof than most people realize, and it has been met since the time of the Spanish Flu [107]. In a multiple-choice public health policy test, mandatory masking was and remains the *wrong answer*. It was the wrong answer *before* 2020, and it is even *more* clearly the wrong answer from 2020 onwards. Claims or assertions that compulsory masking is a "least restrictive means" to further a compelling public health interest, or that compulsory masking is somehow required by "the common good" or even "military necessity" simply never matched reality.

In the wake of the Spanish Flu, the U.S. Navy Surgeon General conducted an extensive investigation which included evaluating whether masks had been beneficial. His conclusion still holds true more than 100 years later:

EPIDEMIOLOGICAL AND STATISTICAL
DATA, U. S. NAVY, 1918

———

Reprinted from Annual Report of Surgeon General, U. S. Navy
For the Fiscal Year 1919

UNIV OF CALIF
MEDICAL SCHOOL

WASHINGTON
GOVERNMENT PRINTING OFFICE
1920

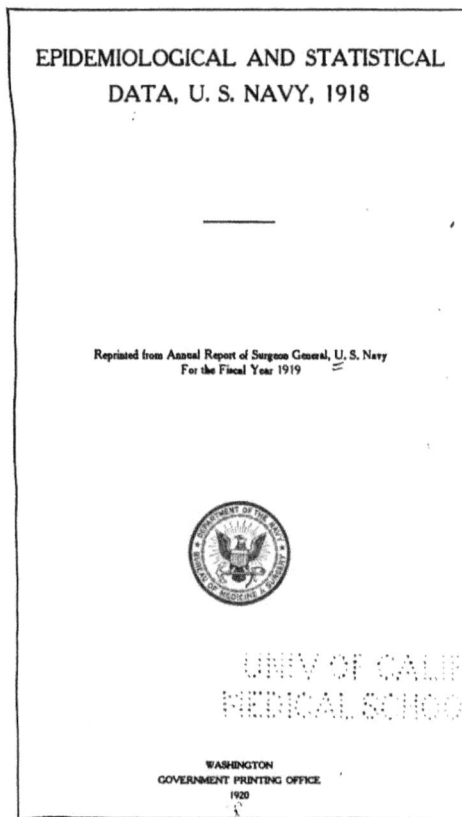

"**No evidence was presented which would justify compelling persons at large to wear masks during an epidemic.** The mask is designed only to afford protection against a direct spray from the mouth of the carrier of pathogenic microorganisms; and assuming that it affords such protection, the probability that the microorganisms will eventually be carried into the mouth or nose by the fingers is very great if the mask is worn for more than a brief period of time. Masks of improper design, made of wide-mesh gauze, which rest against the mouth and nose, become wet with saliva, soiled with the fingers, and are changed infrequently, may lead to infection rather than prevent it, especially when worn by persons who have not even a rudimentary knowledge of the modes of transmission of the causative agents of communicable diseases."[18]

---

18   "Epidemiological and Statistical Data, US Navy, 1918," Reprinted from the Annual Report of the Surgeon General, US Navy, (Washington, DC: Government Printing Office, 1919) 434. (emphasis added), available online: https://www.google.com/books/edition/Epidemiological_and_Statistical_Data/hBpFAQAAMAAJ?hl=en&gbpv=0

Part 2:

Mask Science - Side Effects

Mask side effects are not all in your head

When bans on mask mandates finally began to appear in 2021, courts that previously had no objections to mask *mandates* passed by emergency fiat suddenly had a problem with *bans* on mask mandates passed by legislatures. In October 2021, district judge Robert Pratt enjoined a ban on school mask mandates passed by the Iowa state legislature and signed by Governor Kim Reynolds. Not only are masks effective, said Judge Pratt, but: "Nor is there any evidence that mask-wearing is linked to psychological or emotional harm, except maybe in circumstances in which 'parents or community members perpetuate false claims that masks are harmful.'"[19]

"Masks save lives," went the narrative that none were permitted to question. Any non-fatal side effect was trivial by comparison — just a minor inconvenience. At best, raising the issue of mask side effects when objecting to mandates was seen as whining and making excuses for not being willing to do one's civic duty by masking up. But the fact of the matter is that not only have masks been decisively shown to lack any real efficacy in stemming the spread of COVID and other respiratory viruses, masks also have numerous non-trivial side-effects. (A free, poster-sized pdf chart listing some of these side effects can be found on this book's website, at www.thebookonmasks.com/extras.) Like any medical intervention, the side effects from masks vary in severity from person to person, but that does not make them any less real.

The human body is a marvel of engineering. It is designed to maintain function under a wide range of adverse conditions, compensating, adapting and even thriving, depending on the type of stress.

---

19   Order Granting Plaintiffs' Motion for a Preliminary Injunction, *Arc of Iowa v. Reynolds*, Case No. 4:21-cv-00264, Document 60, p. 18 (Southern District of Iowa October 8, 2021), available online: https://thearc.org/wp-content/uploads/2021/10/Order-Granting-Motion-for-Preliminary-Injunction-IA.pdf, main case documents: https://thearc.org/resource/the-arc-of-iowa-v-reynolds/

Thus, it is no surprise that most people can function with masks on, even for hours at a time, day-in and day-out. But there is a very real physical and psychological cost, and the higher that cost gets, the harder it becomes to justify compulsory masking. The extensively documented side effects from masks would weigh heavily against compulsory masking even if the balance of evidence had shown that masks were effective at stopping disease transmission. We already dealt with 33 of these studies when we looked at the studies cited by the CDC in its *Science Brief.* After reviewing the scientific literature on mask side effects, I have found it helpful to sort the adverse effects from mask-wearing into seven broad categories. There is some unavoidable overlap between categories, as when headaches caused by masks make it harder to focus and think straight, when heightened CO_2 levels or impaired ventilation through masks triggers anxiety, or when bacterial and fungal colonies growing in the warm, moist environment inside masks produce an odor. Depending on the individual, masks impose:

- Respiratory and Cardiovascular Burdens due to increased dead air space and resistance to airflow during ventilation cycles. This has direct detrimental effects on oxygenation, blood carbon dioxide, and can change other respiratory variables depending on the affected individual [135, 147-149, 164, 182, 184, 223, 230, 236-239, 242-249, 251, 261, 270-297].
- Infection Control Hazards through cross-contamination, risk compensation, pathogen retention, and enhanced bacterial and fungal growth because of increased temperature and humidity under the mask [96, 104, 106, 222-233].
- Social, Cognitive, and Psychological Burdens, including adverse user experiences, anxiety, sleep disturbance, concentration difficulties, relational impairments and communication impairments [113, 153, 236, 237, 259-261, 277, 296, 298-305].
- Headaches ranging from minor to migraine, along with symptoms like nausea, dizziness, and cognitive impairment [244, 259-261, 271, 277, 282, 284, 296, 306-309].
- Skin Damage, including irritation, itch, ulceration, allergic reactions, and exacerbation of issues caused by bacteria (acne, or "maskne," for example) [138, 175, 179, 198, 230, 235-238, 244, 254-256, 258-260, 262, 263, 310, 311].
- Heightened Fatigue, including increased exertion while wearing masks, more rapid fatigue, dizziness, drowsiness, and outright exhaustion [147, 149, 236, 237, 242-244, 248, 259, 261, 275, 277, 281, 282, 284, 296, 312].
- Other side effects that do not readily fit into the previous six categories, including increased difficulty recognizing faces [299, 302, 313], nasal inflammation [240], vocal disorders [238, 241, 259, 299], bad odor [149, 236, 237, 302, 305], intraocular pressure increase [314], eye irritation [235, 237, 259, 302, 311, 315], glasses fogging [238, 299], the cost-effectiveness of masks (or lack thereof) [316], a correlation with increased fatality rates from COVID

that warrants further investigation to rule out causation [57, 252], and any other potential drawbacks relevant to their widespread use [317-319].

Many studies document side effects from several of these categories, and so appear in multiple sections of this discussion. When mask side effects are observed, the effects are almost invariably more pronounced for medical-surgical masks than for cloth masks, and worse still for respirator-type masks like N95s [138, 149, 175, 179, 198, 236, 237, 239, 242, 250, 254, 256, 259, 260, 276, 277, 282, 284, 292, 293, 296, 312]. The "better" the mask, the worse the side effects. Because side effects are less central to the arguments surrounding masks than the efficacy of masks, we will not be going into quite the study-by-study depth that we did in the previous section, though we will still be covering this aspect of masking in more than enough detail.

One of the major problems with mask studies looking at detailed physiology is that most of the studies conducted were of short duration and did not include many participants when compared to studies of mask efficacy. However, many studies on mask side effects still found statistically significant differences between masked and non-masked conditions, and if small, short-duration studies found these differences, it is reasonable to expect stronger effects when masks are worn for longer periods of time. This is consistent with the results of those studies that *did* look at longer-durations of mask-wearing.

As with other studies on masks, there was an overt publication bias when it came to studies on mask side effects. In 2020 and 2021, even *hypothesizing* that the side effects from masks might make compulsory masking unjustifiable was enough to get one study retracted from a journal specifically devoted to medical *hypotheses* [89]. Authors who published any findings of side effects from masks that might seriously challenge public health pronouncements about the safety and efficacy of masks could (at the very least) expect to have to respond to multiple sniping letters to the editor [320, 321]. A 2021 study which looked at the effects of mask-wearing on children and found that mask-wearing dramatically increased the carbon dioxide content of inhaled air was retracted over the authors' objections after being subjected to a level of scrutiny that none of the CDC's MMWR publications on masks would have survived [288].

In 2023, a team of researchers with two previously published studies on mask side effects under their belt nevertheless had their systematic review and meta-analysis of mask side effects summarily retracted in spite of their objections [276]. Perhaps the nearly 200 other non-retracted studies the team cited (many of which I have personally read) were not enough of an evidence base. Perhaps their study was retracted for noting that nearly 40% of the major long-COVID symptoms overlap with mask-related complaints and then suggesting: "It is possible that some symptoms attributed to long-COVID-19 are predominantly mask-related." Perhaps their study was retracted for (quite rightly) saying that the efficacy of masks against viral transmission was still unproven. Perhaps their study was retracted for offending authors of other studies on masks by having the temerity to express the truthful (or at least highly defensible) opinion that:

> "In contrast to our study, most of the recent systematic reviews only analyzed a few outcome threshold values without considering comprehensive effects... Therefore, their recommendations e.g., masks are harmless and safe for everybody etc. appears to be superficial, non-medical, non-holistic, and misleading." [276]

I can only speculate about the real reasons for this study's retraction, because the journal's retraction notice provided no explanation other than to say that "concerns were raised regarding the scientific validity of the article." I'll bet there were. Apparently, lesser measures like precisely listing the concerns raised and letting the authors publish errata were off the table. The retraction notice provided no additional detail about the particular "concerns" raised. Studies endorsing masks were not subjected to anywhere near this level of scrutiny, and far worse studies on masks with equally bold and less-supported conclusions remain proudly published, not just in the CDC's MMWR, but in journals like the *Journal of the American Medical Association* (for example, the Massachusetts General Brigham masking study [94]). I can attest from personal experience that *JAMA*'s staff filtered out public comments critical of pro-masking studies, no matter how tactfully worded or well-referenced.

Respiratory and cardiovascular side effects of facemasks

Apart from their efficacy (or lack thereof) in stopping airborne pathogens, the most obvious and hotly debated side effect of masks is their impact on wearers' ability to breathe, and the extent to which mask-wearing does or does not affect blood oxygen saturation and carbon dioxide levels. When perception of exertion or dyspnea (the sensation of having difficulty breathing) from wearing masks has been measured, people almost universally notice that they do not breathe as easily while wearing a mask as they do without one [135, 147, 182, 223, 230, 237, 238, 242-245, 247, 248, 261, 275, 279, 282, 284, 285, 294-296]. In healthy individuals at rest, these sensations are usually mild, but during work, play, or exercise, they get noticeably worse. Anyone who has tried their usual exercise routine while wearing a mask knows this to be an indisputable fact. Since the amount of CO_2 exhaled by a typical healthy man increases by 7.5x or more during even moderate exercise (from about 220 ml/min of CO_2 to over 1,650 ml/min) [322], it is not surprising that the difficulty of exercising while wearing a mask scales up faster than the difficulty of wearing one during light activities or at rest, and the effects of mask-wearing are most obvious during exercise.

These respiratory and cardiovascular side effects are not limited to healthcare masks, and they were noticed long before COVID. A 2018 pre-COVID study using elevation training masks to simulate increased altitude reported that even in the case of "healthy resistance trained men" in their early 20s, three out of twenty-five subjects were unable to complete the prescribed series of exercises for the study because the effects from the masks were too severe [275]. The study authors also speculated that the training masks might actually *hinder* long-term respiratory and cardiovascular adaptation. Yet another pre-COVID study from 2015 noted that even highly ventilated fencing masks produced a measurable difference in nasal airflow during training [247]. The authors concluded that "when compared to no mask, exercising *with* a mask decreases the body's ability to increase airflow through the nose to keep up with increased oxygen demand." (Having some fencing experience myself in high school and afterwards, I can personally attest that the effect described was mild, but noticeable.) A 2009 study on respirator mask side effects in healthcare workers found that more than half of the study participants stopped wearing their masking ensembles within less than 8 hours, citing a number of issues including visual, auditory, and vocal problems, heat, pressure, itching, nausea, dizziness, difficulty concentrating, and having the masks physically interfere with their work [259]. On another personal note, I have found that when doing dentistry, N95 respirator masks get in the way far more than do standard medical masks. One's magnifying loupes never sit quite right on the dang things, and when they do, they break the seal or mess with your vision through the magnifying lenses. I, personally, find that I practice much higher quality dentistry when wearing a medical mask according to the pre-COVID standards, and I will not wear an N95 when providing dental treatment.

In the politically contaminated atmosphere pervading scientific publication during the COVIDcrisis, it was borderline impossible for a study which found that masks affected wearer oxygen or carbon dioxide to get published and stay that way unless the study also concluded that any side effects were comparatively mild and more than acceptable in light of the benefits mask-wearing had for infection control. Many study authors also clearly believed in masks and genuinely favored their use. Despite this, dozens of studies from both before and during the COVIDcrisis documented that masks do, indeed, have a very real, measurable, and detrimental effect on wearers' ability to oxygenate [147, 149, 164, 237, 242, 243, 245, 270, 273, 275, 280, 281, 283, 284, 287, 293, 295] and expel carbon dioxide [148, 236, 239, 242, 244-246, 271-274, 278-280, 282, 283, 286-288, 290, 293, 295]. The reason for this is simple: masks increase the effort required to breathe by increasing airflow resistance and the volume of dead air space for each breath. Expired air with a high concentration of carbon dioxide is trapped within the mask's additional dead air space to be re-inhaled on the next breath. This increases the effort required for ventilation, and at the same time decreases the efficiency and benefit from each breath. Many chronic respiratory diseases like emphysema and chronic obstructive pulmonary disease create similar effects (if not always through the same mechanism). In the case of masks, the effect on carbon dioxide levels is the largest and most noticeable, but the effect on oxygen does not lag far behind.

Not counting water vapor, normal fresh air is 78% Nitrogen, 21% Oxygen, just under 1% Argon, and about 0.04% (just over 400 ppm) Carbon dioxide. Carbon dioxide is a gas necessary for human life. It is a waste product for humans, but is essential for the plants that humans depend on. Like anything in medicine, it is the *dose* that makes the poison. *Within a certain range*, carbon dioxide is beneficial, but *too much* carbon dioxide can have intoxicating or toxic effects. The higher the level of carbon dioxide in the air you breathe, the steeper becomes the chemical gradient your body has to work against with every breath to offload its waste carbon dioxide.

Tests done on dogs in the late 1980s corroborated that the toxic effects from *too much* carbon dioxide were not simply due to a lack of oxygen. When the non-oxygen components in air were replaced by carbon dioxide, creating air that was 80% carbon dioxide, death occurred very rapidly even though oxygen availability remained at normal atmospheric levels of 20-21% [322]. Dogs exposed to this kind of atmosphere stopped all respiratory movements within just 1 minute, and complete circulatory collapse followed just a few minutes later. Upping the oxygen concentration to 50% and leaving just 50% carbon dioxide was not much better. Dogs exposed to this kind of atmosphere struggled harder to breathe for just 1-2 minutes before their breathing and consciousness gradually dropped off until they died an hour and a half later.

In humans, carbon dioxide has been used therapeutically for short durations to help counteract the effect of oxygen deprivation. This is because cerebral blood flow increases by three to six percent for every 1 mm Hg[20] increase in the blood partial pressure of CO_2 [271]. Inhaling 2.5% to 3.5% CO_2 for up to ten minutes can increase cerebral blood flow by up to 100%, helping the brain maintain oxygenation. However, extended duration exposure to carbon dioxide levels of 2-5% (20,000-50,000 ppm) causes dizziness, sweating, dyspnea, and headaches (likely from the blood vessel dilation just mentioned). Cases of CO_2 intoxication and poisoning are rare but well-documented (often they occur when large amounts of dry ice are improperly handled within enclosed spaces). When CO_2 levels climb to 6-10%, dizziness gets worse, the heart starts racing (tachycardia), and people start to hyperventilate if they were not already doing so. Concentrations of CO_2 over 9% act on the central nervous system like a narcotic. At CO_2 levels of 11-17%, irresistible drowsiness, muscle twitching, and loss of consciousness occur, with convulsions, coma, and death occurring if levels over 17% are prolonged [322]. One study done on psychiatric patients in 1954 found that 30% CO_2 causes loss of consciousness in just 24 to 28 seconds [322].

Just as the human body can adjust to high altitudes, acclimation to higher-than-normal levels of CO_2 has also been shown to occur [322]. In the early 20th century, multiple studies were conducted in which small groups of participants were housed for days at a time in artificially induced atmospheres of 1%

---

20   Note: mm Hg is a unit of pressure measurement. Blood pressure of 120/80 means that your systolic blood pressure is 120 mm Hg and your diastolic blood pressure is 80 mm Hg. The dictionary definition is that 1 mm Hg is equal to the pressure exerted by a column of mercury 1 millimeter high at 0°C (32°F) under earth's gravitational acceleration of 9.8m/s².

      Some sources give slightly varying values for what CO_2 levels are normal in humans. Obviously, blood carrying oxygen and nutrients *to* tissues has a lower concentration of CO_2 than blood carrying waste products and CO_2 *from* tissues. My source for the normative 35-45 mm Hg for blood CO_2 in humans is:

Messina Z, Patrick H. "Partial Pressure of Carbon Dioxide," [Updated September 26, 2022]. In: StatPearls [Internet]. Treasure Island (FL): StatPearls Publishing; 2023 Jan-. Available from: https://www.ncbi.nlm.nih.gov/books/NBK551648/

to 5% CO_2 — twenty to one hundred times the normal atmospheric concentration! A 1967 review coming out of the School of Aerospace Medicine, published by the National Technical Information Service, listed ten such studies [323]. Three of the listed studies involved durations of 30 days or more. One of the longest was a 1963 Russian study where two men lived for 30 days in an atmosphere of 2.0% CO_2 (20,000 ppm — 40x to 50x normal). Cognitive tests were either not performed or not commented on, but a "marked deterioration in ability to perform exercise" was noted throughout the study [323]. Most of these studies on long-duration CO_2 exposure only included a handful of from 2 to 8 subjects, but the two largest and longest among these studies had dozens, and were conducted by the U.S. Navy during the 1940s and 1950s.

The highest CO_2 concentrations tested on humans for a period of days ranged from just over 5% to 6.75%. This was done in a series of six experiments conducted by the U.S. Navy in 1944, simulating extreme conditions within submarines. The results were published three years later in 1947 [324]. The most extreme test in this series had 4 men kept for 72 hours at 85°F (29.4°C) and 75% humidity, as CO_2 in their environment gradually reached 5.4% and oxygen levels dropped to half of normal — 10.5%. In the sixth and largest of this series of tests, a group of 77 men experienced CO_2 levels naturally climbing to just over 5% while oxygen levels dropped to 13% over the course of 50 hours. The Navy researchers observed that to maintain basic oxygenation under these extreme conditions, subjects had to increase their per-minute ventilation (the total amount of air passing in and out of their lungs) by 2.5x-3x. There was also a surprisingly subtle rise in heart rate of approximately 10 beats per minute. These changes in heart rate and respiration occurred almost identically even in trials where the oxygen levels were kept at near-atmospheric levels, confirming that the changes in breathing were more a response to the change in CO_2 than changes in oxygen. The researchers also noted that as CO_2 levels peaked, many of the men found it necessary to breathe through their mouths even at rest (mouth-breathing greatly lowers airflow resistance compared to breathing through the nose). After two days in these conditions, hand-arm steadiness declined dramatically and swaying when standing increased noticeably, but, remarkably, performance on cognitive test scores for simple computations declined by only about 10%. At the close of the experiment, a quarter of the men were suffering from headaches which resolved after a day or two in normal air. Incredibly, only one of the 77 men had to be removed early, though that one man was removed because of severe headache, nausea, vomiting, and a progressively increasing high blood pressure.

Nine years later, in 1953, the U.S. Navy conducted the longest trial of heightened CO_2 on humans, using the decommissioned WWII submarine USS Haddock. The experiment was rather dramatically named "Operation Hideout" [323]. Many of the results were formally published in a series of articles in the *Journal of Applied Physiology* beginning ten years later [325]. In this study, 21 men were housed in an atmosphere of 1.5% CO_2 (15,000 ppm — more than 30x atmospheric levels) for 42 days.

The men apparently experienced no symptoms or performance degradation in cognitive tests, motor functions, or the physical exercises that could be performed aboard the submarine [323]. As expected, their oxygen requirements remained constant. The men's rate of breathing initially rose slightly (by 1-2 breaths per minute) and then gradually declined back to normal over several weeks. Similarly, their blood pH initially dropped (a process known as respiratory acidosis that occurs when carbon dioxide levels in the blood increase), but then gradually shifted back towards normal in just under a month.[21]

So how were the men's bodies compensating long term for the much higher CO_2 levels they had to work against to offload their metabolic CO_2? The primary noticeable change was that their per-minute ventilation immediately jumped by 34-39%, and it stayed that way until they returned to a normal atmosphere. Since the men's overall lung capacity did not change, and their respiratory rate was returning to normal, this increase in ventilation was accomplished by using more of their existing lung capacity. The men underwent an increase in their tidal volume — the amount of air that moves in and out of the lungs during a normal breath. Calculated tidal volume rose from an average of just over 800 ml to just over 1000 ml. Tests revealed that most of this additional volume came from inhaling more deeply, but the men also exhaled a bit more fully, as well.

However, increasing tidal volume comes with its own tradeoffs. In addition to the respiratory muscles having to work harder, when you increase tidal volume, your natural dead air space goes up, too. On average, the men gained 68 to 81 ml of dead air space, partially offsetting the increased tidal volume. But the marvel of engineering that is the human body has an additional mechanism to compensate for this drawback, too. Over time, the body can finesse the perfusion of different parts of the lungs to improve its gas exchange efficiency by decreasing blood flow to the under-ventilated dead space lung alveoli while at the same time increasing blood flow to the areas where ventilation has improved. The men's kidneys also helped compensate for the increased CO_2 in their environment by excreting more CO_2 in the urine while retaining more bicarbonate. This increased the amount of bicarbonate in the men's blood, which helped buffer the acidity of the additional CO_2 the men needed to carry in their blood to maintain the chemical gradient that allowed them to continue to offload CO_2 into the surrounding air.

Nevertheless, the Navy researchers found that they had something of an astonishing mystery on their hands. As we just saw, the men's oxygen consumption stayed constant. This meant the total amount of CO_2 the men were getting rid of should have stayed constant as well, but it didn't. Despite living

---

21   In the 2-man Russian experiment mentioned earlier, blood pH took 15-20 days to return to normal. H.A. Glatte, Jr. and B.E. Welch, *Carbon Dioxide Tolerance: A Review.* 1967, U.S. Department of Commerce, National Technical Information Service: School of Aerospace Medicine, Brooks Air Force Base, Texas. https://apps.dtic.mil/sti/citations/ADA017159 citing S.G. Zharov et al. "Effect on man of prolonged exposure to atmosphere with high carbon dioxide content." Aviation and space medicine (Moscow), 1963, pp. 182-185.

with CO_2 levels many times normal, the total amount of CO_2 the men were excreting actually *dropped* by 16% for the first 23 days of the experiment! In fact, when CO_2 levels were returned to normal on day 42, the men were *still* excreting 5% less CO_2 than they had been at the start of the experiment. So where were the men's bodies storing multiple *days'* worth of extra CO_2 during the more than month-long experiment, and how did they get rid of it later? Getting rid of the extra CO_2 they had stored didn't seem to be a problem once conditions returned to normal. As soon as conditions returned to normal, the men's respiratory excretion of CO_2 jumped to 10% *above* normal and stayed that way for several weeks — an amount of time proportionate to the amount of CO_2 stored earlier. Based on this result and parallel findings in guinea pig studies, the researchers surmised that the men's bodies converted the excess CO_2 to forms that could be stored in their bones, and then converted it back to gaseous CO_2 for excretion when conditions later permitted.

The Navy researchers' findings were not yet done. They made another observation with special relevance for universal compulsory masking. Before the main 1.5% CO_2 exposure began, the researchers exposed the subjects to 5% CO_2, and they did so again 40 days into the trial. As expected, the men's respiratory centers recalibrated over the course of the experiment, becoming far less sensitive to CO_2. After 40 days of living in an atmosphere of 1.5% CO_2, all of the men exhibited a much greater tolerance for 5% CO_2. What has special relevance for universal masking, though, was that a third of the men had a much lower ventilatory response to 5% CO_2 in the first place, and these were the men with naturally larger tidal volumes. All of the men were able to adapt when they had to *live* in an environment of heightened CO_2, but a third of the men's respiratory systems started out more naturally acclimated. In humans, individual responses and tolerance for CO_2 vary greatly, just as with other gases or medications. This implies that to the extent that it affects CO_2, universal masking will be much harder on some people than on others, and this is true even of fit, healthy people. Normal partial pressure for CO_2 in humans ranges from 35 to 45 mm Hg (roughly 4.5%-6.0%), and a 2017 review of the scientific literature on carbon dioxide poisoning noted that the blood gas concentration of CO_2 among subjects who experienced symptoms ranged from 41.8 to 64.6 mm Hg [326]. This overlap between the range of normal CO_2 levels and the range of symptomatic CO_2 levels means that one person's CO_2 normal can be enough to produce signs and symptoms in someone else. Reviews as early as 1983 were noting:

> "Certain 'types' of individuals are especially responsive to exercise, heat, elevated CO_2, lactate production, and oxygen deficient atmospheres. These individuals have been referred to as 'hyperventilators,' and their ventilatory responses seem to exceed the normal physiological demand." [301]

Just breathing 5% CO_2 for 15 to 20 minutes or less can be enough to artificially trigger a panic attack. In studies done on adults in 1988 [327] and 2001 [328], exposure to 5% CO_2 triggered a panic attack in one-third to one-half of patients with previously diagnosed panic disorder, but *also* triggered panic attacks in one out of ten people from the control group who didn't even *have* any anxiety disorders to begin with.[22]

The evidence we have seen so far would suggest that those people without panic disorders who still suffered panic attacks from heightened CO_2 levels were from that part of the population more naturally sensitive to CO_2. The 2001 study also evaluated the *severity* of the panic attacks, and found that the subjects *without* panic disorder actually had the most extreme reactions once they *did* panic. Similarly, in a smaller study involving children and adolescents, 3/14 (20%) of the control group terminated the CO_2 flow early due to anxiety, compared to 7 of 18 (38%) of the children with anxiety disorders [329].

Like the sailors in Operation Hideout, workers in occupations who are regularly exposed to concentrations of carbon dioxide up to 1.5% (15,000 ppm) such as brewers, miners, submariners, and astronauts, show no obvious ill effects. Applying a safety factor of three, the U.S. National Institute for Occupational Safety and Health (NIOSH) has set the time-weighted average (TWA) CO_2 exposure concentration for an 8-hour workday in a 40-hour work week at 0.5% (5,000 ppm), and the 15-minute short-term exposure limit (STEL) at 3.0% (30,000 ppm) [330]. In 2008, NASA reaffirmed its 0.7% (7,000 ppm) Spacecraft Maximum Allowable Concentration of CO_2 for 7 to 180-day durations that was established in 1996 [331]. The U.S. Navy set CO_2 exposure limits for submarines carrying female crew at 0.8% (8,000 ppm). The Navy's setting of this particular value was in part based on a 2012 finding that female rats started to show a statistically significant decrease in litter size when living with CO_2 levels elevated to 3% [332].[23]

It is worth noting that this study only looked at the effects of long-term exposure to heightened CO_2 on pregnancy, appetite, bodyweight, and gross visceral malformations. Like the early studies, it was not intended — and did not attempt — to evaluate more subtle effects like anxiety, learning, immune function, or memory.

As part of their 2023 scoping literature review on the potential effects of heightened CO_2 from mask-wearing [239], Kisielinski and colleagues surveyed studies documenting the effects of chronically elevated CO_2 in animals, including the 2012 Navy study on pregnant female rats just mentioned.

---

22   Control group participants were pre-screened to rule out pre-existing undiagnosed anxiety disorders insofar as possible.

23   They found that the highest measured concentration of CO_2 in the experiments where defects did not appear was 2.5%, so they applied an interspecies uncertainty factor of three. 2.5% divided by three, rounded down, is 0.8%.

These studies conducted prior to the COVIDcrisis found that, at least in mice, rats, and guinea pigs, chronically elevated CO_2 levels much lower than those in the early experiments produced noticeable negative neurological and reproductive effects. Especially notable were a pair of studies on rats and mice published in 2014 [333, 334]. In both of these 2014 studies, oxygen levels were kept within atmospheric norms, but CO_2 levels were increased to 3,000 ppm (0.3% — 7x atmospheric conditions). The rat study [333] found that when rat pups were exposed to CO_2 at 0.3% concentration 24 hours a day for 7 weeks, starting *in utero*, the resultant rat pups demonstrated increased anxiety along with reduced memory and spatial learning when tested at 6 weeks. Examination of the rat pups' brain tissue revealed that when compared to their normal-CO_2 counterparts, the 0.3%-CO_2 adolescent rats had neuron counts only about 2/3 as high in the prefrontal cortex and hippocampus — an area of the brain that plays an essential role in learning, memory, and mediating emotions, and which is among the first areas to be affected in Alzheimer's disease.

In the study on adolescent mice [334], a team with many of the same researchers as the rat study found that, like the rats, when placed into a water maze, adolescent mice which had been exposed 24 hours a day for 7 weeks to 0.3% CO_2 were able to *swim* as well as mice which had been kept in normal air, but whereas the normal-CO_2 mice were able to find the maze's lifesaving platform more quickly during exercises on subsequent days — finishing in half their original time after four trials — the high-CO_2 mice were not able to do this, and took just as much time to find the platform in the final challenge as they had needed on the first attempt [334]. Here, too, the researchers documented heightened anxiety in the high-CO_2 mice,[24] as well as a clear loss of neuron density in both the hippocampus and prefrontal cortex.[25]

It is worth noting that the rat study included a *third* group of rats exposed to CO_2 at just 0.1% concentration (1,000 ppm — 2.5x atmospheric conditions). With just one-third the chronic CO_2 level of the 0.3%-CO_2 rats, this additional group of rats displayed elevated anxiety compared to their normal-CO_2 counterparts, but quite a bit *less* anxiety than the 0.3%-CO_2 rats. Importantly, the 0.1%-CO_2 rats showed no apparent loss of memory function, and were still able to learn at a

---

24   To measure rat and mouse anxiety, researchers employ what is known as an open-field test. Essentially, rats or mice are placed into a standardized, gridded, open box about 1 square meter in area, and their behavior is recorded for a set amount of time. Mice that spend more time hugging the walls, less time exploring, less time in the more exposed grid squares, and who cover less ground when exploring are more anxious. The 3,000ppm-CO_2, mice were far less (barely over half) as likely to spend time in open areas, and they covered less ground in total when exploring compared to their normal-CO_2 counterparts.

25   The paper on mice did not include neuron counts, but included pictures of tissue slides where the loss of neuron density through apoptosis (programmed cell death) was clear, and other quantified values (such as changes in the counts of cells that stained positive with a dye corresponding to apoptosis) between the normal-CO_2 and 0.3%-CO_2 mice were very similar to differences between the normal-CO_2 and 0.3%-CO_2 rats.

rate competitive with their normal-CO_2 peers, though microscopy still revealed a slight reduction in neuron density in their hippocampi and prefrontal cortices.

In humans, a number of subtle and not-so-subtle effects of CO_2 on higher cognitive functions were documented in the two decades leading up to the COVID crisis. For example, a 2016 double-blind study [335] tested 24 professional-grade employees, including architects, engineers, programmers, and managers, performing their normal work activities in office conditions of fresh-air CO_2 (0.05%, 500ppm), moderate indoor CO_2 of 0.09% (900 ppm) and high indoor CO_2 of 0.14% (1,400ppm). The workers spent several days in their temporary office without leaving for lunch. At the end of each day, the office workers were tested in 9 cognitive domains: basic activity level, applied activity level, focused activity level, task orientation, crisis response, information seeking, information usage, breadth of approach, and strategy. Some domains like information seeking and task orientation showed no effect from heightened CO_2, but other domains, especially information usage, breadth of approach, and strategy, showed pronounced declines when the high-indoor CO_2 level of 1,400 ppm was compared to the fresh air of 500-600 ppm CO_2. The researchers concluded that, "On average, a 400-ppm increase in CO_2 was associated with a 21% decrease in a typical participant's cognitive scores across all domains." A 2018 literature review examining the effects of small increases in CO_2 on human cognitive performance described more studies in which performance on complex tasks decreased as CO_2 levels increased to between 0.1% and 0.25% (1,000ppm to 2,500ppm) [336].

These studies are relevant to universal masking because masks indisputably affect the amount of CO_2 wearers breathe in. Predictably, the "better" the masks and the less healthy the individuals wearing them, the stronger this effect becomes. Exercise just worsens the effect. Even though the dead air space within masks is only about 100ml [278, 290], because it is the relatively small tidal volume that determines how much fresh air is taken in with each breath, the dead air space within masks can still have a substantial impact on how much CO_2 people are actually breathing in. (Also, the dead air space in some respirator-type masks can be over 250 ml [297].) As a rule of thumb, adult human males have a respiratory tidal volume of 500ml, females 400ml, and children much less depending on age. Even if the air in a room is fresh from the outside, with 500 ppm CO_2 or less, if 100 ml of tidal volume is taken up by air being rebreathed from underneath a mask which has five-*thousand* ppm CO_2, the average concentration of CO_2 in each breath jumps to 1,400 ppm for a man and just over 1,600 ppm for a woman. This amount goes up proportionately if the ambient air is not fresh outside air, or if the masks are trapping *more* than 5,000 ppm CO_2 for rebreathing (which they often are).

At the end of June 2021, a two-page research letter published in *JAMA Pediatrics* created a sensation [288]. In this letter, a team of researchers led by Harald Walach reported the results from their testing of carbon dioxide levels underneath surgical and FFP2 (KN95) masks worn by children.

They found the average CO_2 concentrations under both types of masks to be just over 13,000 ppm (1.3% CO_2), and the youngest children were being subjected to the highest CO_2 values, with the mask environment of one 7-year-old measured at 25,000 ppm (2.5%). This level of CO_2 underneath masks puts wearers' net exposure squarely at or above the range for which cognitive effects were observed in the pre-COVID studies that we just looked at. Since this finding very much went against the dogma that masks were well-tolerated and had minor, clinically insignificant effects on inhaled CO_2, the 2021 Walach paper was subjected to a barrage of criticisms and summarily removed just 16 days after publication — despite the authors' objections and responses.[26] No CDC MMWR publication on masks would have survived this kind of treatment. If this first study was not enough to earn Dr. Walach and his colleagues a place in the hall of unsung COVIDcrisis heroes, they responded to this affront by putting forward the ultimate rebuttal: they published an even *stronger* study in another journal less than 12 months later in 2022 which validated their earlier findings by replicating them [290]!

Walach and colleagues were not the only researchers who found that masks dramatically increased the amount of rebreathed CO_2. Another 2022 study [278] found resting CO_2 concentrations under surgical masks reached 4,800 ppm for adults and 6,400 ppm for children, with concentrations under FFP2 respirator-type masks rising to 9,000 ppm and just under 13,000 ppm, respectively. These concentrations of CO_2 were generally lower than those found by Walach et al., but still 10x or more above atmospheric conditions. Importantly, this study [278] also found that the concentration of CO_2 under masks increased as the number of breaths per minute increased. Another study done outdoors in Spain in 2020 [280] found *higher* concentrations of CO_2 under surgical masks than Walach and colleagues did - more than 14,000 ppm at rest and over 17,000 ppm during short periods of exercise! Going back even further, a successful 2005 PhD thesis study described CO_2 levels of 21-24 mm Hg (roughly corresponding to 3.0%; 30,000 ppm) underneath surgical masks, producing shifts of up to 5.5 mm Hg in arterial CO_2 [272]. The bottom line is that those studies which examine CO_2 concentrations under masks universally show dramatic increases. Masks can and do create microenvironments where the concentration of CO_2 being breathed by wearers exceeds the normal concentration of CO_2 in the atmosphere by ten times (often more). The net effect is to raise wearers' overall CO_2 exposure to levels where physical, psychological, and cognitive effects have been documented in both humans and lab animals. The effect is proportionately worse for women and worst of all for children because of their smaller respiratory tidal volumes.

---

26   Dr. Walach made his response publicly available (along with links to other documents in the process) on his website within 3 days of the retraction.
Harald Walach, "Unsere Impfstudie und unsere Kindermaskenstudie — Erläuterungen," July 19, 2021, online: https://harald-walach.de/2021/07/19/impfstudie-kindermaskenstudie-erlauterungen/

As we saw earlier, the body's goal is to maintain internal equilibrium (homeostasis). Under a variety of stresses and conditions, the body works to keep oxygen saturation as close to 100% as possible, and the concentration of CO_2 within a narrow range of about 35-45 mm Hg. CO_2 levels that are either too high or too low *both* cause problems and can be diagnostic indicators of disease.[27] Even during exercise, the concentration of exhaled CO_2 in healthy people does not vary more than a few points, and usually still stays within this range. But here, too, there is substantial individual variation. As one study on masks noted: "A critical threshold for clinically significant hypercapnia/ hypoxemia[28] is not defined in the current guidelines as it differs greatly between individuals and depends on the respective baseline value" [242]. A 2008 study on athletically trained volunteers [337] found that in a pool of just 45 well-trained volunteers, the measured CO_2 in exhaled breaths during exercise varied from 35.1 mm Hg to 62.6 mm Hg. Interestingly, this study also noted that the highest normal CO_2 levels clustered in the third of participants who also had the highest tidal volumes and lowest respiratory rates — very similar to the Navy's 1953 submarine study (of which the 2008 authors appear to have been unaware).

The adjustments people's bodies make to the immediate heightened CO_2 levels, increased breathing resistance, and increased dead air space from masks can and do vary widely from individual to individual, and this variation is reflected in the findings of the many studies. Some people compensate with a higher number of more shallow breaths, while others naturally take fewer, deeper breaths. Heart rates may stay the same or show a small increase (though at least one study [147] found a small *decrease* in heart rates when cloth masks were worn). Oxygen saturation often drops [135, 147, 237, 242, 245, 270, 275, 280, 281, 292, 293], but not always [148, 164, 182, 184, 242, 246, 274, 287, 291, 294, 295]. The changes seen most often are an increase in the concentration of CO_2 in the blood or in exhaled breath [135, 148, 184, 242, 245, 271, 274, 279-281, 287, 291, 292, 295] and a feeling of dyspnea (difficulty breathing) [135, 147, 182, 223, 230, 237, 238, 242-245, 247, 248, 261, 275, 279, 281, 282, 284, 285, 294-296].

Keep in mind when we discuss broad trends in how masks affect particular respiratory and cardiovascular variables that the only complete constant is the dramatically increased concentration of CO_2 in the air inside masks.[29]

For any other given variable, there are multiple studies which show everything from no change to a dramatic shift. The point is that simply because a given study finds the exhaled CO_2, respiratory rate,

---

27   In certain forms of pulmonary hypertension, the end tidal partial pressure of CO_2 actually goes *down* substantially — Into the low 30s and high 20s mm Hg.

28   Hypercapnia is excessively high CO_2 in the bloodstream, while hypoxemia refers to excessively low oxygen.

29   Outside of masks that supply air to the wearers in some way, I know of no study which has found the CO_2 content of air inside masks to be anywhere near normal.

blood pressure, heart rate, etc. keep within the range of normal or only change slightly when wearing masks does not necessarily mean that the body is not having to put in extra work to make sure those values *stay* that way. A shift in CO_2 pressure of just a few mm Hg in arterial blood or expired air is often still an indicator that something is going on. An increase in exhaled CO_2 generally means either that the body is producing more CO_2, that the body is having a harder time clearing the CO_2 it is already producing, or that more CO_2 is being *in*haled. Even when the body is dealing with an increased environmental load of CO_2, the overall *amount* of CO_2 exhaled with each breath may very well *not* increase because the amount of oxygen being metabolized and the amount of metabolic CO_2 being created has not changed. Remember that in the 42-day Navy study, the men's CO_2 excretion actually *decreased* under high CO_2 conditions. Pointing to this or that respiratory or cardiovascular measurement in any given study which did not undergo significant changes when masks were worn misses the broader, more important point that people's bodies have to significantly change *something* to compensate for the additional burden whenever they wear masks.

Cardiovascular and Respiratory Side Effects of Masks

Note: This chart does not include the many additional qualitative survey studies in which people reported increased dyspnea, dizziness, and other breathing difficulties while wearing masks.

Legend:
- Statistically significant change (↑ or ↓)
- No change or change not statistically significant
- Not measured or not reported

Measurement columns (for both tables): O₂ Saturation ↓ (%) · O₂ Partial Pressure ↓ (mm Hg) · CO_2 Partial Pressure ↑ (mm Hg) · Respiration ↑ or ↓ (breaths/min) · Respiration ↓ (tidal volume – mL) · Heart Rate ↑ or ↓ (beats/min) · Blood Pressure ↑ (mm Hg) · ↑ Dyspnea or Perceived Exertion

Study	Year	Mask Type	Participants
Beder et al.	2008	Medical/Surgical	Healthy Volunteers
Roberge et al.	2012	Medical/Surgical	Healthy Volunteers
Porcari et al.	2016	Elevation Trng Mask	Athletic Volunteers
Jagim et al.	2018	Elevation Trng Mask	Athletic Volunteers
Person et al.	2018	Medical/Surgical	Healthy Volunteers
Fikenzer et al.	2020	Medical/Surgical	Healthcare Workers
Georgi et al.	2020	Cloth	Healthcare Workers
Georgi et al.	2020	Medical/Surgical	Healthcare Workers
Lässing et al.	2020	Medical/Surgical	Healthy Volunteers
Liu et al.	2020	Cloth	Healthy Volunteers
Liu et al.	2020	Medical/Surgical	Healthy Volunteers
Mo et al.	2020	Medical/Surgical	Elderly COPD patients
Pifarré et al.	2020	Medical/Surgical	Healthy Volunteers
Bar-On et al.	2021	Medical/Surgical	Healthy Volunteers
Dirol et al.	2021	Medical/Surgical	Healthy Volunteers
Driver et al.	2021	Cloth	Healthy Volunteers
Epstein et al.	2021	Medical/Surgical	Healthy Volunteers
Hirai et al.	2021	Medical/Surgical	Elderly COPD patients
Toprak and Bulut	2021	Medical/Surgical	Healthy Pregnant Women
Zhang et al.	2021	Medical/Surgical	Healthy Volunteers
Sukul et al.	2022	Medical/Surgical	Healthy Volunteers

Study	Year	Mask Type	Participants
Kao et al.	2004	FFP3/N95	End-stage renal disease
Kim et al.	2013	FFP3/N95	Healthy Volunteers
Roberge et al.	2014	FFP3/N95	Pregnant & Non-pregnant
Tong et al.	2015	FFP3/N95	Pregnant Volunteers
Goh et al.	2019	FFP2/KN95	Healthy Children
Bharatendu et al.	2020	FFP3/N95	Healthcare Workers
Fikenzer et al.	2020	FFP2/KN95	Healthcare Workers
Georgi et al.	2020	FFP2/KN95	Healthcare Workers
Kyung et al.	2020	FFP3/N95	Volunteers with COPD
Liu et al.	2020	FFP2/KN95	Healthy Volunteers
Epstein et al.	2021	FFP3/N95	Healthy Volunteers
Toprak and Bulut	2021	FFP2/KN95	Healthy Pregnant Women
Sukul et al.	2022	FFP2/KN95	Healthy Volunteers

Those studies which concluded that masks were well-tolerated and had minimal respiratory and cardiovascular side effects tended to be short, limited to more restful conditions, or looked at only one or two variables that were naturally unlikely to shift much even under high CO_2 loads. Additionally, as we saw when looking at the studies cited by the CDC to support mask use, methods could easily include procedures that stacked the deck in favor of making masks look less burdensome, such as using the thinnest possible non-medical masks for elderly wearers [139], or pumping 100% oxygen through masks being worn by children [145].

The small size of many of these studies — often just 10 to 25 individuals — hindered more subtle results from reaching statistical significance. A 2010 study [283] observed a clear decrease in respiratory rate and a 25% increase in tidal volume when N95 masks were worn, but the results did not achieve statistical significance because the study only had 10 healthcare worker for subjects. A 2021 study on 16 male volunteers found the 4 mm Hg increase in exhaled CO2 from medical masks during exercise to be non-statistically significant, but found that the 7 mm Hg increase produced by N95s *was* statistically significant [148]. Short studies which looked at mask-wearing periods of 9 to 15 minutes [242, 274] concluded that masks were well-tolerated and any changes in blood gases were "measurable but clinically irrelevant" [242]. Based on studies minutes long with a few dozen healthy people, decision makers deemed mask respiratory and cardiovascular side effects to be trivial, and forced masks on the rest of the population for hours on end. In the selective application of a bad precautionary principle, the most severe infections with SARS-CoV-2 were used to justify universal forced masking, but the most severe effects of masks were ignored or denied altogether.

Like every medical intervention, the *average* results for masks have prominent individual exceptions that usually cannot be predicted in advance. Even those studies which found no average effect had a substantial number of outliers in the results. For example, a 2010 study on 10 healthcare workers wearing N95s which found no *average* significant changes in oxygen saturation, expired CO_2, or tidal volume, had at least two of the subjects who *did* show substantial increases of CO_2 in their expired air (reaching >50 mm Hg) [283]. Recall the broad variations observed in athletic subjects during exercise, and the otherwise-healthy man who had to be removed from the Navy's CO_2 experiment for severe headache, nausea, vomiting, and a progressively increasing high blood pressure.

In healthy individuals, the measured impact of medical-surgical masks ranges from no significant changes during rest or exercise [242] to a 3.6% drop in oxygen saturation at rest [292] and a 5.5% drop in oxygen saturation during intense exercise [280]. In a 2008 study, 53 surgeons found that their pulse oximetry readings declined by an average of 1.5% over the course of about four hours when wearing surgical masks, from between 97-98% to 96% [270]. Similarly, the average impact of surgical masks on expired or blood gas CO_2 in healthy people ranges from no significant change at rest [149, 242] to a 4.5 mm Hg increase in CO_2 at a slow (4 km/hr) walk, and an increase of 8.4 mm Hg at a brisk walk of 7 km/hr [135]. Remember, these averages do not include the greater changes measured in outlying individual cases. The average measured impact of FFP2-KN95 or "better" masks on oxygen saturation and CO_2 covers a similar overall range, but the effects show up more consistently in studies looking at averages, and when these "better" masks are directly compared with medical-surgical masks in the same studies, their measured effects are roughly one-and-a-half to two times worse [148, 236, 237, 278, 288, 290, 292].

The effect of surgical masks on oxygen saturation at rest has been reported at just under 1% in elderly patients with COPD [294] and just over 2% [293] in healthy pregnant patients. Authors of a 2004 study reported that when they tested the effects of N95 respirators on adults with end-stage renal disease, arterial carbon dioxide levels did not significantly change, but the patients' oxygen partial pressures declined by 9 points from 101.7 mm Hg to 92.7 mm Hg, and this was *after* excluding those patients in especially poor health who had oxygen saturation levels below 92% even *without* masks. The largest study looking at masks and pregnancy to-date was done in 2021 and involved 297 pregnant women [293]. Carbon dioxide levels were not evaluated, but the researchers observed a significant average decline in oxygen saturation of 2.2% for surgical masks and 4.1% with FFP2 masks after a *non*stress test.[30] In fact, 25 out of the 148 women in the FFP2 mask group (17%) had their overall oxygen saturation levels drop below 92% following the nonstress test.

A common objection raised by mask apologists is that the observed respiratory and cardiovascular shifts from masks might be *statistically* significant, but are too small to be *clinically* significant. It is true that not every distinct, measurable change has a real-world practical effect, but physiologic changes (signs) combined with concurrent symptoms are the very *definition* of clinically significant, and masks produce both of those in spades. Additionally, the argument that statistical significance is not the same as clinical significance goes both ways. To the extent one insists on discounting the clinical relevance of statistically significant mask side effects, one must also discount the clinical relevance of any studies purporting to show statistically significant mask efficacy and effectiveness against respiratory infections.

In one 2020 study [245], 7 out of 97 patients with chronic obstructive pulmonary disease (COPD) could not even complete a 10-minute rest and 6-minute walk test while wearing N95 masks. This failure rate under such mild testing would be worth noting on its own, but the most important finding of this study may be the average major shifts in blood gas that were *not* observed. The changes seen in those COPD patients who failed to complete the study procedures were well within the range of changes seen in healthy people wearing masks. Their average oxygen saturation declined by just 1.5% and their average end-tidal CO_2 increased by just 5.3 mm Hg, yet such low-magnitude changes called "clinically insignificant" in other studies still reflected adverse experiences severe enough that the subjects were unable to complete the study procedures.

Several studies suggest that the body makes compensating changes for mask-wearing on the cellular level. In a 2015 study on healthy pregnant women [287], per-minute ventilation actually *declined* by 25%. In fact, oxygen consumption, tidal volume, and CO_2 production *all* significantly decreased while wearing N95 masks during low-intensity activities. Despite this, maternal and fetal heart rates and oxygen saturation did not significantly change. A 2022 study that used

---

30   Nonstress tests are prenatal tests that measure the baby's heart rate and response to movement. For the mother, they involve wearing a belt monitor while sitting in a reclining chair and keeping track of the baby's movements. They are not a form of exercise.

real-time mass spectrometry to analyze exhaled volatile organic compounds while subjects were wearing masks, reported significant changes in exhaled levels of alcohols, ketones, aromatics, aldehydes, and other metabolites after just 15 to 30 minutes [292]. What these findings imply is that the body compensates for the additional burden imposed by mask-wearing at the level of cellular metabolism and by sending strong signals to the brain to curtail exertions while wearing masks. This would certainly help explain why the universal reports of increased difficulty breathing while wearing masks often seem disproportionate to the observed shifts in basic respiratory and cardiovascular measurements.

Another common intuitive objection is that SARS-CoV-2 is much larger than molecules of oxygen and carbon dioxide, so to the extent masks are effective at preventing *air* exchange, they must also be effective at preventing *SARS-CoV-2* exchange. We will talk more about the counterproductive infection control side effects of masks shortly, but for now, the problem with this objection is that when carried to its logical conclusion, it just makes masks look worse. Masks do not ultimately prevent viral particles from escaping any more than they ultimately prevent CO_2 from escaping. At the best, if masks affect the flow of SARS-CoV-2 like they affect the flow of air, not only do masks ultimately *fail* to prevent viral escape (enough air must actually be exchanged with the surrounding environment to avoid passing out), but they have the "bonus" effect of creating a pocket of concentrated viral particles with the concentrated CO_2 between the mask and the face for repeated re-inhalation, and this cluster is simply released every time the mask is removed or adjusted.

A related objection is that by the nature of how statistical significance is defined, we should expect to see one in every twenty respiratory or cardiovascular measurements show a statistically significant impact from mask-wearing even if no causal relationship is actually present. The problem with this objection is that if observed effects from masks were really the result of statistical noise, we should expect to see the findings randomly distributed. That is not the case with studies on masks, where the statistically significant measurements consistently turn up in the same areas, accompanied by symptoms. This indicates we are observing something beyond mere statistical background noise. You do not need to take my word for this, either. Other published literature reviews have likewise concluded that masks produce hypercapnia (increased CO_2) in wearers, especially during exercise [244, 248, 273].

The effect of masks on blood oxygen levels suggests a further mechanism by which mask-use could nullify any filtration benefits. In order to enter lung epithelial cells, SARS-CoV-2 must bind to angiotensin-converting enzyme II (ACE2) proteins on the lung cell surfaces. One laboratory study found that mouse and human lung cells in hypoxic (low oxygen) conditions dramatically increase the number of these ACE2 proteins that lung cells express on their surface, providing more binding sites for SARS-CoV-2 to latch onto. "After 24 h in 12% oxygen, lung Ace2 mRNA expression was increased almost 5-fold" [338]. Normal oxygen levels in the air are around 21%, and masks certainly do not produce this kind of artificial 43% drop in oxygen levels in the human body, but you don't need a 5-fold increase in SARS-CoV-2 binding sites to make a difference. Even as little as a 5-10% increase in viral binding sites in lung tissue could conceivably be enough to counter *any* marginal, *theoretical*, filtration benefits ascribed to masks. Other laboratory studies suggest that certain types of immune system cells (CD4+ T-helper 1 cells), do not perform or proliferate as effectively under hypoxic conditions [339, 340]. This line of research does not yet have enough evidence to say for certain *if* or *how much* of a real-world difference small changes in oxygenation can make in viral infection rates. But then again, as we have already seen, much of the evidence cited to force masks onto people was just this kind of laboratory evidence, so it is only fair to count this kind of evidence *against* compulsory masking when that is the direction it points in.

Apart from the influence of psychology on blood pressure, masks generally have a statistically insignificant effect on blood pressure, though if there is an effect, it is to create a slight increase [147, 244, 276, 279, 297]. The same is true for heart rate, though again, individual study results vary considerably [147, 149, 164, 236, 237, 244-246, 250, 270, 282, 284, 291], and plenty of studies *have* found a significant effect on heart rate [147, 149, 164, 236, 237, 244, 246, 270, 282, 284, 297]. The increases in heart rate from mask-wearing ranged from 3 to 11 beats per minute, but are typically on the order of 5 to 9 bpm, getting more towards the upper end of that range the longer the masks were worn. A 2008 study which looked at medical masks worn over a four-hour period found about a 6 bpm difference at the four-hour mark between masks and no masks [270].

Recall that in the early Navy studies, while heightened CO_2 produced a modest (but sometimes dramatic) temporary increase in respiratory rate (especially at higher levels of 2.0% to 5.0% CO_2), long-term respiratory rates gradually worked their way back towards normal. The primary way subjects' bodies compensated for high environmental CO_2 was by increasing tidal volume in each breath.

In the more extreme shorter-duration trials, mouth breathing helped by decreasing airflow resistance in and out of the lungs. When looking at the effect of masks on respiratory rate, some studies have found decreases in respiratory rate, [149, 246, 250, 283], while other studies [184, 236, 242-246, 277, 295, 297] have found no change or an increase in the number of breaths per minute. When Kisielinski and colleagues pooled studies in their meta-analysis of this effect, they were surprised to find that, on average, masks produced no net difference in the rate of breathing [276]. The reason for these varied findings remains an open question. Possibly, this overall finding resulted from the different ages and populations being examined in the pooled studies. People's breathing characteristics change over the course of their lives. Children breathe about twice as quickly as adults, and even adult breathing patterns change with age. Different age groups may exhibit different changes in breathing as a response to wearing masks. In one study out of Germany, younger adults compensated for wearing masks by breathing more slowly and deeply, while adults over 60 increased their breathing rate and expended more conscious effort [292]. The authors of this study noted that changes in respiratory rate and heart rate from mask-wearing were only seen in older adults and were the most pronounced with respirator-type masks [292].

When we dig deeper, we find that the effects of masks on ventilation directly contrast with the adaptations to high CO_2 seen in early non-masked studies. Though at least one 2010 study found a distinct increase in tidal volume of about 25% when wearing N95 masks during light walking [283] — similar to the tidal volume increase seen in the Navy's 1953 submarine experiment — this finding turned out to be the exception rather than the rule, and the small number of just 10 participants prevented it from reaching statistical significance. In contrast, recall the 2015 study with pregnant women using N95 masks which found a 23% *decrease* in tidal volume during light activities [287]. 2020 and 2021 studies found the same effect, with cloth masks [147], surgical masks [291], and N95 masks [149] *all* producing statistically significant *decreases* in respiratory tidal volume despite the *increased* amount of CO_2 which wearers were inhaling with each breath. The authors of the 2021 study using surgical masks suggested that the decrease in respiratory tidal volumes seen with mask use is most likely due to the increase in airflow resistance produced by the masks [291]. This increase in airflow resistance with every breath was not something that participants in early CO_2 studies were subjected to. Similarly, studies that looked at *total ventilation* during periods of mask wearing *also* found a significant net decrease [147, 149, 164, 250, 251, 271, 287, 291] compared to a no-mask condition.

In their meta-analysis, Kisielinski and colleagues calculated this decrease in total ventilation from mask use to be 19% for medical masks and 24% for N95s [276]. These effects of *decreased* tidal volume and *decreased* overall ventilation when wearing masks are in direct contrast to the findings in the *non*-masked high-CO_2 Navy studies, where overall ventilation *increased* during acclimation to compensate for the increase in CO_2. One of the most detailed 2020 studies frankly stated that both surgical masks and respirator-type masks "significantly reduce pulmonary parameters at rest" [149]. The authors concluded: "Medical face masks have a marked negative impact on cardiopulmonary capacity that significantly impairs strenuous physical and occupational activities. In addition, medical masks significantly impair the quality of life of their wearer." With multiple studies documenting significant decreases in total per-minute ventilation of 10%, 20%, or more while wearing masks, it is no wonder people felt like masks were reducing the amount of air they were getting. That is *exactly* what masks do!

Double masking

As the COVIDcrisis progressed, it became clear to more and more people that masks were not producing the promised benefits. Public health authorities responded by doubling down and recommending *two* masks. "CDC Says Double-Masking Offers More Protection Against the Coronavirus," proclaimed America's National Public Radio in February 2021.[31] Dr. Fauci and multiple YouTubers were happy to demonstrate on video how to "properly" double-mask.[32] At the time of this writing, double mask recommendations have yet to be repudiated. The July 2023 memorandum released by the U.S. Under Secretary of Defense for Personnel and Readiness which rescinded the "Consolidated Department of Defense Coronavirus Disease 2019 Force Health Protection Guidance," *still* recommended masks for

---

31   Laurel Wamsley, "CDC Says Double-Masking Offers More Protection Against the Coronavirus," *npr. org*, Feburary 10, 2021, online: https://www.npr.org/sections/coronavirus-live-updates/2021/02/10/966313710/cdc-now-recommends-double-masking-for-more-protection-against-the-coronavirus

32   Maura Hohman, "Dr. Fauci demonstrates how to wear 2 masks correctly only TODAY," *TODAY*, February 11, 2021, available online: https://www.today.com/health/dr-fauci-shows-how-wear-2-masks-correctly-today-t208765

respiratory viruses, and said: "A good practice is to wear a disposable mask underneath a cloth mask for added protection if this does not interfere with breathing."[33]

No studies have shown *any* real-world infection control benefits from wearing two masks. Only a handful of studies have looked at the side effects of double-masking [241, 259, 277, 282, 284, 309, 312], and fewer still looked in detail at things like heart rate and oxygenation. Those which *did* look at whether double-masking interferes with breathing, however, showed dramatic results. One of these studies was done on an all-male group of 20 healthy oral surgeons in Italy, and compared the effects of various length surgeries to an unmasked pre-surgery baseline [284]. It found that double-masking produced dramatic effects on both heart rate and blood pressure. Surgeons' heart rates that had started out ranging from 51-69 bpm at rest jumped to 71-95 bpm after performing a 20-minute surgery. By the end of four-*hour* double-masked surgeries, the double-masked surgeons' heart rates were running from 86 to 110 bpm. This increase is more than three times the increase in heart rates observed under multi-day conditions of 5% CO_2 and 13% oxygen during the U.S. Navy's WWII submarine experiments. For the double-masked Italian oral surgeons, oxygen saturation which had been hovering comfortably at 98% pre-surgery was down to 94% after 20 minutes and 91% at the four-hour mark. People can certainly *live* at 91% oxygen saturation, but values like this are only "normal" in severe chronic diseases like COPD and lung cancer, not health. Unsurprisingly, the surgeons in this study also reported dramatic increases in shortness of breath, light-headedness, and headaches the longer they worked while double-masked.

Another study conducted in Turkey found that the double masked participants had more of an aggravation in headaches when compared to the groups of single-masked participants [277]. In a pre-COVID 2013 study [282], nurses that wore surgical masks over their N95s did not show significant changes in oxygen levels and heart rate compared to just wearing an N95, but had higher CO_2 levels, nausea, and visual difficulties. One possible explanation for the different study findings is that the Italian oral surgeons in the 2021 study were wearing FFP2 masks under their medical masks instead of N95s. FFP2 masks are substantially easier to breathe through than N95s, and so adding a medical mask would make a bigger relative difference. Regardless, this earlier 2013 study also found that wearing N95 masks for an entire 12-hour shift had statistically significant negative effects on both physiologic measures and subjective symptoms:

---

33   Department of Defense Office of the Undersecretary of Defense for Personnel and Readiness, "Force Health Protection Guidance — Coronavirus Disease 2019 and Other Infectious Respiratory Diseases," July 26, 2023, https://media.defense.gov/2023/Jul/28/2003270467/-1/-1/0/FORCE-HEALTH-PROTECTION-GUIDANCE-CORONAVIRUS-DISEASE-2019-AND-OTHER-INFECTIOUS-RESPIRATORY-DISEASES-SIGNED-JULY-26-2023.PDF

"Over time, nurses' CO_2 levels became significantly elevated, from a statistical standpoint, compared with beginning-of-shift baseline measures; perceived exertion, perceived shortness of air; and complaints of headache, lightheadedness, and difficulty communicating also increased over time. CO_2 levels increased from a baseline average of 32.4 at the beginning of the shift to 41.0 at the end of each shift."

I say this as a healthcare provider with expertise on the subject: *nobody* should be double-masking!

Bringing it all together

So how do we reconcile the early studies on CO_2 exposure, which report effective long-term adaptation to heightened CO_2 levels, with the later studies on CO_2 exposure that found detrimental effects from much lower increases? What is the significance for masking? I believe most of the contradictions are more apparent than real, or can be explained by differences in the methodology being used and the populations being tested. The most obvious difference is that the early CO_2 studies were not conducted using masks. The additional burdens on breathing that mask-wearing imposes (airflow resistance, increased dead air space, and lower oxygenation) likely lower the concentration of CO_2 needed to produce overt effects. Additionally, the U.S. Navy's tests from the 1940s and 1950s were conducted on healthy adult males in military physical condition, and so they represent a best-case scenario for CO_2 tolerance. By their nature, these studies do not generalize well to children whose brains are still developing, the elderly, pregnant women, and people with chronic respiratory deficiencies.

Most of the early studies only included CO_2 exposures up to 30 minutes, and were primarily concerned with basic mechanical effects. They were not designed or intended to look at more subtle issues. Unlike the later studies on mice, none of the early studies evaluated the effects of heightened CO_2 on human reproduction, immune system function, anxiety, and long-term memory formation. When it came to cognitive and performance tests in the largest high-CO_2 studies, the Navy researchers said they deliberately set up the experiment "to minimize the effect of rapid learning and irregular adaptation to test conditions" [324]. If people were tested on tasks they already learned, that would at least partially conceal any effect of high CO_2 on memory. More complex problem-solving eschewed by these early experiments was what later studies observed to be most affected by higher CO_2 levels in the workplace [335, 336]. Also, even the early studies *did* note some performance decreases. The incentives faced by the researchers also played a role in the findings. In the early studies, researchers were trying to find the upper limits under which people could perform simple essential tasks, whereas in more recent studies, the researchers were seeking to determine the lowest levels at which effects could be observed when performing more complex tasks.

Studies which looked only at basic parameters like heart rate, blood pressure, breaths per minute, the amount of CO_2 being exhaled, and oxygen saturation, tended to underestimate the impact masks were having on users because these things are not as consistently and strongly affected by changes in oxygenation and carbon dioxide as are things like blood carbon dioxide concentration, respiratory tidal volume, complex mental tasks, memory, and anxiety.

When discussing mask side effects, always bear in mind the wide range of individual differences. If your body is signaling your brain that it is having a more difficult time achieving full oxygenation, then unless you have a strong basis to believe otherwise, you can reasonably assume your body is providing your brain with far more detailed, comprehensive, accurate up-to-the-moment information about your individual situation and needs than any white lab coat or politician who is making recommendations and issuing mandates based on the averages from a handful of studies. One thing I *can* say with complete confidence is that masks *do* elevate the wearers' carbon dioxide and make oxygenation more difficult, and the "better" the mask is, the stronger this effect becomes. Even assuming that there is a genuine conflict in some of the studies, this alone should have been enough to rule out universal compulsory masking.

I want to be clear that I am not simultaneously arguing for the contradictory positions that the human body is a marvel of engineering able to function under much higher CO_2 and lower-oxygen conditions, but that any burden from increased CO_2 and decreased oxygen is too much. The relevant question is not whether people *can* do this, but what additional burdens compulsory masking imposes, and what tradeoffs the body has to make to bear those burdens. What I *am* arguing is that the physical tradeoffs involved are real, significant, and much harder on some people than others. Even short-duration studies hinted at this, and surveys looking at the numerous side effects people reported from long periods of mask use bore this out. These inescapable tradeoffs were denigrated and ignored by authorities, decision-makers, and public health "experts" at every level during the COVIDcrisis.

The human body really is an engineering masterwork. It is adaptive and resilient enough that even a doubling of atmospheric CO_2 would not be harmful, but compulsory masking went far beyond this. Mask-wearing as practiced during the COVIDcrisis managed to achieve the worst of both worlds, yielding all the short-term drawbacks of CO_2 extremes, while at the same time being too intermittent, irregular, and imposing too many additional burdens to allow the kind of altitude-like acclimation seen in early studies on adaptation to high CO_2. If you only spend a minority of your hours at high-altitude, you are not going to acclimate, and a similar issue occurred with masks. The increased airflow resistance and dead air space from wearing masks not only made oxygenation more difficult, it magnified the impact from increases in CO_2 by preventing wearers' bodies from using their usual methods of compensation via increasing overall ventilation. Compulsory masking imposed conditions of chronic respiratory disease on 90% of the population in the name of inhibiting a transient respiratory infection which was no real threat to more than *99%* of the population. Nice work, "experts"!

Germ reservoirs — infection control problems with masks

Given that infection control was the rationale on which masks were mandated, it may seem counterintuitive to raise the issue of infection control as a problem with masks. Yet we have already seen in Part 1 of this book, that by April 2nd, 2020, scientific data had been published showing that SARS-CoV-2 remains stable and viable for *days* longer on the inner and outer surfaces of medical masks than it does on paper, tissue paper, wood, cloth, glass, banknotes, stainless steel, or plastic [96]. This alone should have been enough to preclude any mask mandates. Forcing people to go about covering both of their airways with a medium that *preserves* the respiratory viruses you are trying to control is inexcusably bad public health policy.

The fact of the matter is that masks create microenvironments with elevated temperature and humidity that are ideal for incubating and growing all kinds of microbes [149, 226, 230, 235-238, 244, 248, 259, 276, 285]. In Part 1, we briefly looked at a 2013 study which found that the amount of bacteria expelled from cloth and medical facemasks after just 120 to 150 minutes of use equaled or exceeded any initial reduction when they were first put on [106]. When this 2013 study's final data points were collected at the 150-minute mark, bacteria counts expelled from masks were on a trajectory to increase still further past unmasked levels [106]. If masks perform that poorly with bacteria, we should expect them to perform far worse with respiratory viruses one tenth the size of a bacillus or less.

In 2014, researchers in a Bangkok, Thailand, hospital cultured the bacteria and fungi found on surgical masks worn by hospital staff in multiple departments over the course of a single shift, and compared this content to the ambient air [228]. They found an average of 47 colony-forming units of bacteria and 15 colony-forming units of fungi per ml of material on the inside of the masks. The exterior mask material contained a higher average of 166 colony-forming units of bacteria and 33 colony-forming units of fungi per ml. By comparison, the average concentration of colony-forming units of bacteria and fungi in samples collected from the ambient air in the hospital ranged from 38 and 23 per cubic meter of air in the operating rooms to 575 colony-forming units of bacteria and

262 colony-forming units of fungi per cubic meter in the out-patient department air. This makes masks look pretty good until we remember the units of measurement involved. 1 cubic meter is 1 *million* milliliters, meaning that the lowest concentration of bacteria and fungi harvested from the material in the masks worn all day was still *thousands* of times higher by volume than in the most bacteria- and fungi-laden air.

A 2018 study found that the bacterial contamination on surgical masks worn in the operating room more than doubled between the second and fourth hour the surgical masks were worn [232]. The researchers also concluded that bacteria cultured from the outward-facing surface of the masks were more likely from the surgeons than from their environment, and that the surgical masks *themselves* could be the source of bacterial shedding. This 2018 study acknowledged the then-uncontroversial position that "a direct correlation between mask and SSIs [surgical site infections] has not been proven in the literature." Another 2018 laboratory study published in the *Journal of Virological Methods* looked at whether contaminated personal protective equipment could, itself, serve as a source for viral infections, and characterized surgical and N95 masks as "personal bioaerosol samplers" [222].

Masks quickly pick up bacteria and fungi, and provide an ideal greenhouse environment for their growth, which can then be passed on through direct contact when masks are put on, removed, touched, and adjusted. The less comfortable masks are, the more often they are touched, removed, and adjusted [230, 285], undermining or reversing any supposed infection control benefits from filtration. A 2020 German study which looked at contamination of masks during dental treatments concluded: "The surgical mask seems to provide excellent conditions for the survival of oral or dermal bacteria" [226]. "Touching the outer surface of the mask should be avoided at any time. After touching or removing the mask, the hands must be disinfected" [226]. We all know this is *not* how masks were used during the COVIDcrisis. In a 2020 study done using volunteers from an Italian university's department of oral surgery, participants were observed to touch or adjust N95 respirators 25 times in just one hour, compared to 8 touches and adjustments for standard surgical masks [230]. N95s were touched, adjusted, and removed due to discomfort so frequently that the researchers concluded: "it is better to wear a surgical mask correctly than an N95 which, due to the discomfort, causes displacements with the hands and temporary withdrawals of the mask from the face." So much for "better" masks.

Taiwan was praised for its early lack of COVID cases, much of which was wrongly attributed to their universal mask use [341]. As in other countries, however, closer investigation suggested that the way masks were being used in practice was at best a net neutral and likely a net negative when it came to infection control. To start, the sale of masks was severely rationed. Adults in Taiwan were limited to two surgical masks per week for most of February 2020, upped to nine masks every two weeks in April [341]. A survey of Taiwanese university students found that masks were donned and doffed several times throughout the day because of work, and stored in bags, pockets, and on desks when not in use [223]. When ten of these students performed basic office tasks in a preset environment while wearing masks coated with dye, the area "contaminated" through touching averaged 530 cm² (slightly less than the footprint of a sheet of printer paper) after less than ten minutes.

A 2022 study found that more than half of surveyed mask-users wore the same masks for two days or more [104], and sought to determine the extent to which microbes could colonize masks after several days of use (defining a day of use for a mask as being worn for at least 3-6 hours). The researchers cultured bacteria and fungi from masks that had been used for 1, 2, and 3 days. Unlike the 2014 study out of Thailand, in this 2022 study, it was the *face* side of the masks that had the higher bacterial colony counts, with the face sides of masks producing 168 colonies on agar plates compared to just 36 from the outer side. The researchers also found no significant differences in colony counts between different types of masks or masks worn in different commuting environments, such as private vehicles, public transportation, walking, and cycling. Interestingly, while bacterial accumulation seemed to reach an upper limit by day three, fungal counts (though much lower at about 4-6 colonies per side) continued to increase. Most of the bacterial and fungal species retrieved from the masks examined were non-pathogenic, but some pathogenic species of both bacteria and fungi were present [104]. Commensal *Staphylococcus aureus* was common, and *Staphylococcus saprophyticus* (which causes urinary tract infections) and *Bacillus cereus* (a bacillus that causes food poisoning) were also present. Fungal species which colonized the masks included *Microsporum* and *Trichophyton*, which cause fungal infections like ringworm and jock itch. Another study done in 2021 looking solely at fungal growth on masks found that 70% of the 50 randomly sampled masks worn by out-patients had one or more species of fungus growing on them (42% of the 50 had this growth on the inner layer), including 7 out of 8 N95 masks [233].

Defenders of compulsory masking may be inclined to point to some of these findings as evidence that masks are doing a good job of filtering contaminants from the air. However, to the extent that masks are successful at catching larger microbes, every microbe captured is now being held against your face in a nice warm, humid nursery, and the smallest of the captured bacteria are still ten times the size of a virus. Studies which specifically look for the presence of respiratory viruses on masks are remarkably rare. The authors of one such study in 2019 said: "To our knowledge this is the first study examining the presence of respiratory viruses on the outer surface of used medical masks" [224]. This 2019 study found that only 14 out of 99 masks managed to pick up *any* detectable viruses, despite being worn by healthcare workers in *hospitals* for more than 6 hours straight. Just *one* out of 49 masks worn for less than six hours had any detectable virus. Thus, out of 148 total masks tested, only 15 managed to pick up any detectable respiratory viruses, and of those 15, only *2* were influenza — the closest analogue of SARS-CoV-2.[34]

The overall infection control picture of masks that emerges when these puzzle pieces are put together is one of medical devices that are of questionable net benefit even under ideal circumstances when used by trained healthcare providers and replaced every two hours. Take a look at the filter in your home's vacuum cleaner and HVAC system. One side is cleaner than the other, but do you genuinely believe that strapping either side against your face would provide a net infection control benefit? Of course, this analogy is imperfect. HVAC and vacuum cleaner air filters only have to deal with airflow that goes in one direction. Masks, on the other hand, are two-way highways, not one-way lanes. What is expelled from a mask can also be inhaled from one. Picture the N95 that your healthcare provider had to wear all day for a week before they could get a new one during the early months of the COVIDcrisis. Now picture the cloth or medical mask left in your hot car during the summer for a couple of weeks. Is this speculative? Somewhat. Part of the purpose of this section is to shed light on some of the possible reasons for the demonstrated worldwide failure of masks to do anything to stop respiratory viruses. But any speculation in this section is still bounded by far better data than anything put forward to justify compulsory masking. COVIDcrisis masking was worse than useless for the purposes of "source control" or any other microbial control. Rather than stemming from any rational multifaceted scientific analysis, COVIDcrisis compulsory masking policies were instead based on delusion, fear, knee-jerk authoritarianism, wishful thinking, and a desire to be seen "doing something."

---

34   The full breakdown was that out of 15 positives: 7 were adenovirus, 2 bocavirus, 1 human metapneumovirus, 2 influenza, 2 respiratory syncytial virus, and 1 parainfluenza virus.

Not all harms are physical — social, cognitive, and psychological side effects of masks

"Sweat and panic about the next time."

"Forgetfulness, tiredness, hunger for breath."

"Dizziness, stress, depression, anger."

"Inattentiveness in traffic."

"Panic attacks due to PTSD."

"Weeping fits."

"Child trauma comes up."

"Lack of patience with my children."

"An unbelievable anger arises."

"Feeling of loss of control, fear, sadness, aggression."

"I'm the stupid lemming who lets everything be done to himself, just so that I can participate in public life."

"Increased aggressive behavior in road traffic."

"Despair because coercion doesn't stop."

These are just a few of the free-form comments about wearers' experiences with masks that were submitted to an exploratory cross-sectional survey conducted in Germany over the first two weeks of June 2020 [302]. A wide selection of these free-form comments was published with the results. These comments could have just as easily been written by people from any country subjected to mask mandates. Reports of respiratory symptoms like coughs, dizziness, and shortness of breath from mask-wearing were common, as were headaches and skin rashes (more on those shortly), but cognitive and psychological effects predominated or figured prominently in nearly half the submitted comments. In the mask psychology section of this book, we will look at the behavioral psychology that was exploited to maximize compliance by manipulating, pressuring, and coercing people into wearing masks, as well as exploring some of the implications of the deindividuating and anonymizing effect masks have on wearers. The present subsection is dedicated to elaborating on some of the less subtle cognitive and psychological side effects that came with mask-wearing for many people. The chart below lists just a few of the studies documenting these side effects.

Cognitive, Psychological, and Interpersonal Side Effects From Masks

		Morgan 1983	Li et al. 2005	Radonovich et al. 2009	Wong et al. 2013	Rebmann et al. 2013	Carbon 2020	Liu et al. 2020	Mheidly et al. 2020	Prousa 2020	Ramirez et al. 2020	Cheok et al. 2020	Gori et al. 2021	Rosner 2021	Koseoglu et al. 2021	Silke et al. 2021
User Experience	Increased Overall Sensation of Discomfort	✓	✓		✓		✓		✓	✓	✓	✓			✓	✓
	Pain Spots	✓	✓						✓	✓	✓					
Emotion	Increased Anxiety	✓														✓
	Claustrophobia	✓							✓							
	Irritability, Anger or Aggression								✓	✓	✓					✓
	Feeling of helplessness or powerlessness								✓							
	Sadness								✓							✓
Sleep	Sleep Disturbance								✓						✓	✓
Cognition	Impaired concentration and/or cognition	✓			✓		✓	✓	✓	✓				✓		✓
Communication	Communication Impairment	✓		✓		✓		✓	✓		✓	✓				✓
	Increased difficulty reading facial expressions	✓					✓						✓			
	Decreased empathy		✓													

Whether it was the sight of masked faces triggering perpetual fear of contamination and contagion, dismay at how quickly and easily we and others fell into compliant lockstep, or anxiety about what was coming after compulsory masking had established a psychological foot in the door for future compliance with more extreme demands, no one thing contributed more to the atmosphere of uncontrolled anxiety during the COVIDcrisis than masks. Apart from the constant visual communication of danger and abnormality that universal masking creates, we have already seen that immediate increases in CO_2 can artificially induce anxiety in both humans and lab animals, and that masks create just such increases in CO_2. This strongly suggests that at least some of the widespread jump in anxiety during the COVIDcrisis had an underlying physical cause, but anxiety is just one of multiple cognitive and psychological side effects that mask wearing produces in many users.

In one cross-sectional survey, 78% of users from the general public reported increased discomfort from mask-wearing that averaged between 4 and 5 on a scale of 1 to 10 [299]. Many of these discomforts were physical, but just as many were mental and emotional. Adverse emotional experiences that directly resulted from mask-wearing and compulsory masking ran the gamut from increased anxiety, claustrophobia, increased irritability, anger, aggression, and decreased empathy towards others, to sadness, depression, a sense of helplessness or powerlessness, and increased restlessness [261, 301, 302, 304]. One commentary published in the *International Journal of Environmental Research and Public Health* pointed out that compulsory masking implicated and either directly attacked or subverted three universal, fundamental needs for optimal wellbeing: autonomy, psychological relatedness, and competence [303].

Sleep disturbances were another commonly reported side effect of mask-use [261, 277, 296, 302]. Side-effects like headaches, itching, lack of sleep, dizziness, emotional upset, and impaired ventilation all tend to make it harder to think straight, so it comes as no surprise that when asked, one-quarter [260] to one-half [261] of mask-wearers reported some degree of impaired cognition and increased difficulty focusing or concentrating while wearing masks [237, 260, 261, 282, 296, 302]. One study [260] specified that out of the quarter of participants who experienced impaired cognition, more than half experienced this after four hours of mask use, though 12% experienced it within the first hour.

Communication impairments and nuisances from mask-wearing were a known issue with masks going back decades before the COVIDcrisis [236, 259, 282, 301]. Forcing *everyone* to wear a mask did nothing to make these issues go away [261, 299, 300, 302]. Facemasks hinder interpersonal communication by making it harder to read social and emotional cues in others' faces [300]. A 2020 German study found that people's ability to read neutral and fearful faces was the same with and without masks, but that the ability to accurately read emotions like sadness, anger, happiness, and disgust declined by anywhere from 10% to nearly 50% [298]. The CDC cited studies which found (or opined) that masks made no statistically significant difference in people's ability to read facial expressions [187, 189], but other studies found that masks made reading and interpreting emotions and facial expressions more difficult [298, 300], especially for children who were just learning how to do so [153, 187]. A 2013 randomized controlled trial found that when doctors wore facemasks during consultations, patients perceived them as being less empathetic [304].

While masks had the overall effect of greatly increasing anxiety, they could also have a temporary emboldening effect for some risks while being worn. Risk compensation refers to how people adjust their behavior in response to perceived risks, behaving less carefully when they feel safe, and more cautiously when they feel at risk. Just like wearing a seat belt or having car insurance can lead to more aggressive driving, so mask-wearing can create a false sense of security that has the potential to increase the risk of transmitting infections. The evidence on risk compensation while wearing masks is mixed, but in a 2020 behavioral survey, "participants indicated they would stand, sit, or walk closer" to strangers when either one of them was wearing a mask, and this effect was stronger for those who believed masks were protective [227]. One study using smart device location surveillance data from the first six months of the COVIDcrisis concluded that Americans in areas with mask mandates increased their visits to public locations and spent less time at home when compared to areas without mask orders [231]. The detailed 2023 Cochrane review on masks [20] also mentioned this, citing further studies suggesting that mask-use can create "possible risk compensation behavior leading to an exaggerated sense of security."

In October 2020, the German University of Witten/Herdecke opened a registry to which parents, teachers, and doctors could submit their observations of the effects masks were having on their children [261]. Within a week, more than 20,000 respondents provided data on 25,930 children. A minority of children experienced no apparent impairments from the mask wearing, but over two-thirds of both children and parents reported impairments directly caused by masks. The most common side effects reported were increased irritability (60%), headache (53%), difficulty concentrating (50%), general fatigue or malaise (37-42%), and impaired learning (38%). Less-common observations were decreased sleep (31%), sleeping more than usual (25%), playing less (15%), or having increased restlessness (8%). Even allowing for self-reporting bias whereby those with negative experiences are more likely to respond to surveys asking about negative experiences, there is no way to honestly characterize the many adverse psychological effects of masks and compulsory masking as "trivial" or "rare."

Masks cause headaches

For many people, the psychological turmoil involved with mask-wearing and mask mandates was made worse by pain from the physical headaches that their masks caused or exacerbated [224, 237, 244, 245, 259-261, 271, 277, 282, 284, 296, 306-309]. Headaches, of course, themselves often have a number of secondary effects like sleep disruption, fatigue, rapid breathing, dizziness, nausea, and heart palpitations [277], so even when masks did not *directly* generate these symptoms, the first order side-effects of masks such as headaches often did so indirectly. We have already seen that an increase in CO_2 can generate headaches through cerebral vasodilation,[35] but in plenty of people, things like wearing bands around the head can compress superficial nerves and cause headaches as well.

---

35   Bharatendu et al.'s 2020 study used Transcranial Doppler Ultrasound to monitor blood flow in the middle cerebral artery and found that wearing N95 masks induced a 12% increase in the average blood flow velocity within just 5 minutes.

Rates of Headache by Associated Mask Type	Cloth	Medical/Surgical	FFP2/KN95	FFP3/N95	Additional Notes
Lim et al., 2006				37%	Of the 37% who developed headaches, 60% required medication, 8% lost workdays.
Chughtai et al., 2019		6%			
Bharatendu et al., 2020				80%	Migraine-intensity headaches in 25%.
Georgi et al., 2020		17%			9-minute exercise trial. Reported headache results for all mask-types in aggregate.
Hajjij et al., 2020				62%	33% of the participants experienced *de novo* headaches, while 29% experienced exacerbations of pre-existing headaches.
Ong et al., 2020				81%	Headaches were graded as mild by 71% of respondents. 23% of those with headaches experienced other symptoms, including nausea and vomiting; 7% of respondents took sick leave for their headaches.
Ramirez et al., 2020		57%	74%		Of those who developed headaches, the most common complaint was a lack of concentration (67%).
Zaheer et al., 2020				28%	Most headaches were moderate intensity. 2/3 of headaches were bilateral.
Köseoğlu et al., 2021		45%	58%		73% of those who **double-masked** developed new or worsening headaches. Percentages include exacerbations of pre-existing headaches (surgical masks 26% *de novo*, 19% exacerbations; filtering masks 50% *de novo*, 8% exacerbations).
Rosner, 2021			71%		59% of respondents wore surgical masks and 41% wore N95s. Headache data was reported as an aggregate number.
Scarano et al., 2021			21%-31%		Study on **double-masked oral surgeons, FFP2 masks covered by surgical masks**; liklihood of headache varied with length of operation; from 21% for surgeries up to 20 minutes, to 31% for surgeries 2-4 hours in duration.
Silke et al., 2021	53%				Headache frequency reported as an aggregate of all mask types. Most (65%) of the children wore cloth masks, 21% wore surgical masks, 2% wore FFP masks, mask type was not specified for 11%.
Kisielinski et al., 2023			62%		**Meta-analysis**, reported 62% incidence of headaches for general multi-hour mask use, up to 70% with N95 masks.

Not everyone developed headaches from wearing masks, but many people did. Reported rates of headache associated with surgical masks range from 6% [224] to 57% [296]. Reported rates of headache associated with N95 masks range from 28% [309] to 81% [308]. The only study I know of which looked at rates of headache for cloth masks reported aggregated rates of headache for cloth and surgical masks to be 53% in children [261]. Kisielinski et al.'s meta-analysis pegged the overall rate of headaches at 62% for generalized multi-hour mask use [276].

When sufferers were asked about what *they* thought was causing their headaches, masks or PPE were cited by anywhere from 40% [309] to nearly 100% [261] of study participants, with others listing masks as a contributing factor but attributing the primary cause of their headaches to things like lack of sleep [309]. While masks tend to exacerbate pre-existing headaches of all types, headaches arising *de novo* from mask-use are most often characterized in the literature as compressive headaches [277, 296, 306, 309], were typically bilateral, and located in areas compressed by masks or by the elastic bands that held them in place [277, 308]. A 2020 study on healthcare workers in a Casablanca hospital found that the majority of the headaches associated with N95 respirator-masks were frontally located, 40% involved both sides of the head, and a quarter involved the whole cranium [306].

As you would expect, headaches from masks typically developed over the course of several hours, and the longer the masks were worn, the more headaches occurred [260, 296, 308]. Frequency and duration of the mask-related headaches also varied greatly, from just a few days each month on upward. They tended to be longest and most frequent in people who were prone to suffering from headaches even before masks entered the picture [277, 307, 308]. The intensity of mask-related headaches was usually mild and manageable with over-the-counter painkillers [308], but they could often be moderate or worse. One study reported that 25% of N95 headaches in healthcare workers reached migraine intensity [271]. In another study, two-thirds of those healthcare workers with pre-existing migraines reported that their migraines worsened with mask use [277]. Two studies (one from 2006 and another from 2020) reported that of those healthcare workers who did develop mask-related headaches, 7-8% lost workdays as a result [307, 308].

Skin damage from masks

We have already seen how the warm, damp environment underneath masks forms an ideal greenhouse for microbes, but apart from simple infection control issues, this has implications for skin conditions like acne and certain types of dermatitis. Masks can also directly damage the skin, whether through physical trauma like pressure ulcers [175, 198, 254, 256, 258], chemical irritation [175, 179, 254, 255, 263, 311], or repeated abrasion [138, 175, 254, 255, 258, 263]. Facial acne [138, 175, 179, 198, 244, 256, 260, 276, 311] and itching [138, 175, 179, 198, 235, 236, 238, 244, 256, 258, 259, 262, 263, 276, 311] were the two most common effects of long-term mask use on skin, but a substantial minority of users also experienced rashes [138, 175, 179, 198, 256, 258, 311], various forms of dermatitis [175, 179, 254, 255, 263, 276, 311], and abrasions that progressed to fissures, scaling, oozing, and/or crusting [175, 179, 254, 255, 258, 263]. Some wearers' skin responded to the increased temperature and humidity inside masks by becoming more greasy [138], while for others, it felt more dry and tight [138, 175, 179, 237, 254, 258, 311]. A small minority of users even experienced changes in pigmentation of the areas where they wore their masks [175, 198, 256].

A 2020 cross-sectional survey of 833 healthcare workers and non-healthcare workers in Thailand found that 55% of wearers reported adverse skin reactions from their masks, of which acne was the most common (399 cases - 40% of the participants). 18% of respondents in this study experienced rashes, and 16% reported itching [198]. Other effects included changes in pigmentation in 3.6% of wearers, and pressure-related skin injuries in 2% [198]. Another 2021 study conducted in South Korea found that 59% of mask-wearers developed some sort of adverse skin reaction [179]. A 2021 systematic literature review and meta-analysis put the overall rate of adverse mask-related side effects on skin at 58% [175]. Those with pre-existing skin conditions were especially vulnerable to adverse skin reactions from masks [138, 175, 179, 244, 254, 255, 263, 311], and suffered disproportionately.

Skin Reactions to Masks
(Literature Reviews and Meta-analyses are bolded)

Skin Reactions to Masks	**Kisielinski et al., 2023**	Rosner, 2021	Park et al., 2021	**Montero-Vilchez et al., 2021**	**Kisielinski et al., 2021**	Chaiyabutr et al., 2020	Zuo et al., 2020	Veraldi et al., 2020	Techasatian et al., 2020	Szepietowski et al., 2020	Matusiak et al., 2020	Liu et al., 2020	Lan et al., 2020	Hua et al., 2020	Bothra et al., 2020	Bhalla et al., 2009	Radonovich et al., 2009	Foo et al., 2006	Li et al., 2005
Increased Temperature		✓			✓	✓			✓			✓	✓		✓			✓	
Increased Humidity / Sweating		✓			✓	✓			✓			✓	✓		✓		✓		
Itch		✓		✓	✓	✓	✓	✓		✓	✓	✓		✓	✓			✓	
Rash				✓	✓	✓	✓	✓		✓				✓					✓
Tenderness or pain					✓		✓		✓					✓	✓				
Acne		✓	✓	✓	✓	✓	✓	✓		✓									✓
Change in Pigmentation					✓					✓									
Pressure Damage / Ulceration					✓					✓			✓	✓			✓		✓
Skin Dryness and/or tightness				✓	✓	✓	✓	✓						✓			✓		
Abrasions / Desquamation					✓	✓	✓	✓	✓					✓		✓	✓		
Oozing / Crusting / Fissures / Scaling				✓	✓				✓					✓		✓	✓		
Dermatitis (Allergic, Irritant, Friction) / Urticaria (hives)				✓	✓	✓	✓	✓	✓							✓	✓		
Aggravation of pre-existing skin conditions (Psoriasis, Eczema, Dermatitis, Urticaria, etc.)					✓	✓			✓							✓	✓		

Under normal pre-COVID working conditions, occupational contact dermatitis in healthcare workers hovered in the region of 30-33%, but under the "new normal" of COVID Personal Protective Equipment, this rate rose to 75% or more [310]. Though it included all PPE and did not isolate the effects of masks, one study found skin damage in 97% of nearly 550 surveyed healthcare providers subjected to the "new normal" in anti-COVID PPE standards [258]. The nasal bridge, often covered by both masks and goggles, was the most commonly affected site, at 83% [258]. (As a side note, one of the colleagues with whom I shared an office in 2020 developed a long-lasting and painful-looking sore on the bridge of his nose from the combined weight of his N95 and dental loupes.) Affected skin also included the border areas and areas rubbed by the mask straps (especially behind the ears) [254].

As usual, the type of masks worn and the amount of time they were worn made a large difference. Itching and other issues were more frequent the longer masks were worn [138, 175, 179, 198, 262]. A 2020 study reported that 69% of participants who wore N95 masks for less than 6 hours per day had skin damage compared to 82% of participants who wore N95 masks for more than 6 hours each day [258]. Here, too, the effects of the "best" respirator-type masks tended to be more adverse than those of cloth and surgical masks [235]. Respirator-type masks can contain formaldehyde, which kills microbes but is also a known strong irritant and allergen; cloth and surgical masks also contain other preservatives that can induce irritant or topical allergic reactions [254, 311].

With a couple of exceptions [238, 262], the general trend seen in so many other areas held true, with N95 masks producing more side effects than medical-surgical masks, and medical-surgical masks producing more side effects than cloth masks.[36] In a 2020 cross-sectional survey study of 404 healthcare workers in China, 58% of those who wore N95 masks reported dermatological symptoms compared to 40% of those who wore medical masks [311]. A 2021 study of 303 healthcare workers in South Korea found that 64% of those who wore respirator-type masks developed adverse skin reactions compared to 54% of those who wore surgical masks [179]. In a mixed 2020 study of healthcare workers and non-healthcare workers in Thailand, 58% of those who wore surgical masks had adverse skin reactions compared to 48% of those who wore cloth masks [198]. Similarly, in a larger 2021 survey study of more than 1,200 members of the general population in Thailand, surgical masks showed consistently higher incidences than cloth masks of acne (46% vs. 39%) and itching (33% vs 25%), though the incidence of rash was not significantly different (17.4% vs. 15.7%) [138]. A rough rule of thumb is that when going from cloth masks to surgical masks and from surgical masks

---

36   The exceptions came from two Polish cross-sectional survey studies of students conducted by overlapping groups of authors. One survey of 2,300 Polish students found no statistically significant difference in the rates of itching between cloth, surgical, and respirator masks, and another found that when compared to surgical masks, cloth masks had higher rates of difficulty breathing, warming/sweating, and itch.

to respirator-type masks, expect the percentage of wearers experiencing adverse skin reactions to jump by 10 to 20% for each "upgrade" in mask type.

One set of authors described iatrogenic (meaning intervention- or treatment-induced) dermatitis from masks as "a pandemic within a pandemic" [254]. In a research letter to the editor, Italian dermatologists reported mask-induced worsening in 20 out of 43 of their patients with seborrheic dermatitis over just the three-month period between December 2019 and February 2020 [263]. The dermatologists suggested that the increased temperature and humidity underneath masks caused a change in the skin's microbiome, with proliferation of the *Malassezia* yeasts that are a major contributing factor to seborrheic dermatitis [263]. Though *Malassezia* fungus was not reported in the two studies which cultured fungi off masks that we reviewed earlier [228, 233], those studies did not include cultures from the skin or masks of patients suffering from seborrheic dermatitis, and as we have seen, masks certainly do facilitate fungal growth, as well as pick up fungi from the air.

Depending on mask type and how long they were worn for, masks have been reported as inducing flair ups of acne in up to 69% [179] of wearers [138, 198, 256, 260]. Reported incidence of mask-induced rash ranged from 12% to 36% of users [138, 179, 198, 256, 311]. For a large minority of wearers, the itching and discomfort from masks was a chronic distraction. Reported incidence of mask-induced itching in the literature is generally between 15% [198, 311] to 33% [138] but ranged from less than 10% [259] all the way up to 58% [179]. Kisielinski et al.'s 2023 meta-analysis put the overall incidence of acne at 38%, skin irritation at 36%, and itching at 26% [276]. In a survey of 2,300 Polish students, 20% reported itching from facemasks [262], and 10% of this subgroup (2% of all mask-wearers surveyed) reported *constant* itching while wearing masks. This was also the only study which evaluated itch intensity, and the majority of participants reported their mask-induced itching as being "moderate" (as opposed to "mild," "severe," or "very severe"), though 12% of those with mask-induced itch reported it as being "severe" or "very severe" [262]. In yet another example of real-life mask use undermining the theory behind mask mandates, at least half of those in this survey who reported itching also reported removing their masks to scratch [262].

Tired yet? — mask fatigue

Fatigue has a built-in element of subjectivity. It reflects a physical reality, but also depends on individual perceptions and internal experience. Many things contribute to it. Most of the side effects from masks that we have reviewed tend to increase fatigue. If one or more of the physical side effects from masks were not enough — increased carbon dioxide; decreased oxygen ventilation; increased resistance when breathing; headaches; skin damage; physical discomfort from increased temperature and humidity — then the disrupted sleep and psychological demoralization from being forced to wear masks for hours at a time against one's will would often do the trick. Widespread heightened fatigue as a by-product of mask-wearing was inevitable.

Write-in survey responses describe "fatigue," "drowsiness," and "exhaustion" from mask-wearing [302]. As mentioned earlier, 37-42% of respondents to the German survey study conducted at the University of Witten/Herdeke attributed an observed increase in drowsiness, tiredness, fatigue or malaise in children and adolescents to their long-duration mask-wearing [261]. To be sure, a few studies found no statistically significant difference in fatigue or perceived exertion when wearing masks [184, 243, 246, 275]. However, the majority of studies reported an increase in participant fatigue or perceived exertion when masks were worn, and others found that "better" respirator-type masks resulted in more fatigue than cloth or surgical masks [147, 149, 236, 242, 244, 261, 276, 277, 291, 295, 296, 302, 312]. As you would expect, double-masking increased wearers' perceived exertion and fatigue still further [282, 284]. Often, fatigue accompanied mask-related headaches [277].

The fact of the matter is that doing things simply takes more effort while wearing a mask than without one. A 2021 study done with healthy young adults in Texas [147] found that cloth masks imposed a 14% reduction in exercise time and a 29% reduction in participants' maximum oxygen uptake when compared to an unmasked state. Participants also stated that exercising was more difficult in the cloth masks ("harder than last time, I fatigued quicker"; "running was harder than normal"). Another 2021 cardiopulmonary function study done in China, also on healthy young subjects [291], reported that participants rated their perceived exertion 19% higher when wearing surgical masks than without (4.78 on a scale of 1 to 10 compared to 5.69 out of 10). In still another 2021 study conducted by faculty members at a Turkish medical school [295], the average fatigue reported by healthy volunteers performing a six-minute walk test without masks was 3.15 out of 10. With surgical masks on, the average fatigue rating for the same test rose to 4.64 out of 10 — a 47% increase. These mask-related increases in fatigue, large as they are, were still smaller than the 2020 results from a study conducted in the Department of Cardiology at the University of Leipzig, Germany. In the German study, subjects performing an incremental exertion test whose unmasked fatigue averaged just 2.7 on a scale of 1 to 10 found that their fatigue when masked averaged 5.8 out of 10 for surgical masks and 6.5 out of 10 with N95 masks [149]. Keep in mind that most of these tests were short (less than 30 minutes) and performed on healthy adults. The two studies which observed the strongest effects from masks — the German [149] and Turkish [277] studies, also had the oldest participants (averaging 38 years and 40 years old, respectively), whereas the subjects in the Texas and China studies which found smaller effects averaged just 23 and 27 years old [147, 291]. If anything, then, these studies likely understate the cumulative effect of masks on fatigue, especially in older, less fit, or unwell individuals.

After reviewing 109 studies in 2021 (44 of which were sufficiently quantitative to permit mathematical grouping and analysis), Kisielinski and colleagues postulated the existence of a Mask-Induced Exhaustion Syndrome (MIES) which can affect healthy people in addition to the unwell [244]. Subsequent study findings have supported this [290, 292]. The authors of the German study looking at exhaled metabolites when wearing masks said: "Our study has not only provided support for MIES [Mask-Induced Exhaustion Syndrome], but has also advanced the existing hypothesis by extending the side-effects onto exhaled metabolites" [292]. The additional fatigue you felt after wearing a mask all day at work or in school was not just an illusion; rather, it was another red-ink entry on your body's ledger tallying the cost of wearing a mask. Whether at work or at play, people consistently noticed that they got more tired more quickly when wearing masks.

Where does this all stop? other side effects from masks

Masks have a grab bag of further side effects that range from incidental to annoyances to real impairments. A 2021 study found that a group of glaucoma patients experienced a statistically significant but slight increase of 1 to 2 mm Hg in their intraocular pressure while walking and wearing respirator-type masks [314]. According to the American Academy of Ophthalmology, intraocular pressure is normally 10 to 20 mm Hg.[37] Your guess is as good as mine what (if any) clinical significance this particular change may have (hopefully, none).

Moving up the scale from incidental to annoyance to impediment, anything that covers the mouth muffles or attenuates the voice. Masks made it harder to recognize people and to communicate, as well as changing the content of communication, incentivizing shorter and simpler exchanges with less humor or playfulness [313]. As one study respondent in Germany reported, "Everything is so unreal. Even friends are hard to recognize" [302]. A 2020 UK study published in the *International Journal of Audiology* [313] reported that:

> "With few exceptions, participants reported that face coverings negatively impacted hearing, understanding, engagement, and feelings of connection with the speaker. Impacts were greatest when communicating in medical situations. People with hearing loss were significantly more impacted than those without hearing loss. Face coverings impacted communication content, interpersonal connectedness, and willingness to engage in conversation; they increased anxiety and stress, and made communication fatiguing, frustrating and embarrassing —

---

37   Dan T. Gudgel, "Eye Pressure," May 24, 2022, *American Academy of Ophthalmology: Eye Health A-Z*, online: https://www.aao.org/eye-health/anatomy/eye-pressure

both as a speaker wearing a face covering, and when listening to someone else who is wearing one."

Meanwhile, in Singapore, 18% of respondents to another cross-sectional survey reported they had a harder time recognizing people wearing masks, and 46% reported that masks made it harder to communicate [299]. In a 2009 study, one out of 27 healthcare providers reported that their medical mask negatively affected their voice [259]. 12% of surveyed Polish students reported that their speech was slurred when wearing masks [238], and in one Chilean study, questionnaire results suggested that up to 26% of healthcare workers experienced a mask-related voice abnormality as compared to previous data of 7% for the general population [241].

Bad odors from masks are documented in multiple studies [149, 236, 237, 302, 305]. While the chemicals used to manufacture and preserve masks can irritate the nose in the same way they irritate the skin, over time, the bacterial and fungal growth on masks creates its own uniquely distasteful fragrance of used mask.

Glasses often fog up when masks are worn because air (and anything carried in that air) is deflected up and into the eyes [238, 299]. Having your glasses fogged while walking around is an irritation, but it can be a real handicap when it interferes with things like reading or your ability to see precisely where on your patient you are trying to place an IV or which tooth you are drilling on. While adjusting the fit and seal of the mask can help with this, in non-respirator masks, improvement comes because the air is simply being redirected back out to the sides or downward, which is hardly a solution as far as the theory behind compulsory masking is concerned.

Moving beyond baseline annoyance, air deflected upward out of a mask also dries the eyes, and eye irritation is a consistent side-effect for a minority of mask-wearers [235, 237, 259, 302, 311, 315]. In the literature, reports of eye irritation from mask-wearing are generally in the region of 5-6% [235, 311], up to 10% or more of users for some types of N95 respirator-masks [235, 259]. Cross-sectional survey write-in responses describe "Itchy eyes, for hours," and "Burning eyes (contact lenses)" as a result of compulsory mask-wearing [302]. In one publication, ophthalmologists reported seeing an increase in ocular symptoms at their practice from patients and staff who wore masks regularly [315]. Taping the borders of masks reduced air leakage, but dry eyes remained a problem because the addition of tape interfered with the normal lubricating movements of the lower eyelid [315].

Even if we set aside all the psychological and moral arguments against compulsory masking, assuming for the sake of argument that masks provide some net medical benefit to their users or those around them would *still* not be enough to automatically make masking good public policy because of the sheer volume of resources allocated to masking and the resultant garbage. When the failure of masks (including double-masking) became apparent to most wearers, the CDC switched to emphasizing "high-quality" or respirator-type masks like N95s, but a 2017 study estimated that the net incremental cost of upgrading from medical masks to N95s was $490-$1,230 for each clinical respiratory infection prevented, even *after* factoring in the cost of healthcare, medications, and lost working hours from illness [316]. Based on a survey of mask use and a literature review of over 130 publications (including those of the World Health Organization), a 2021 study estimated that the UK, US, Australia, India, Sri Lanka, and Singapore cumulatively generated over 6 billion used masks weighing over 16,000 tons every *week* [319]. This is an annual total of more than 300 billion masks weighing over 800,000 tons. This does not include the mask output of other developed countries in Europe, or the more than 100 billion masks put out by China in 2020 *alone* [318]. Soil and waterway pollution from masks falls under pollution involving plastics, which is practically its own separate sub-genre of environmental science that we will not cover in this book, but literature reviews of multiple studies from 2021 and 2022 have confirmed that masks release thousands of tons of microplastics into the environment with use and disposal [318, 319].

Melt-blown polypropylene forms the middle filter layer in most medical masks, while spun-bond polypropylene and higher density polyethylene are often integral to the facial and outer layers [102, 103, 318]. (respirator-type masks often have a second middle layer composed of cellulose and polyester or cotton [103, 319].) Polypropylene and polyethylene are used in thousands of products without causing harm. However, certain forms can be irritating for many people when they lodge in mucous membranes. Ironically, masks supposed to protect from respiratory secretions were also responsible for *increasing* those very secretions in many people. Anecdotal reports of nasal congestion and other forms of mucosal irritation as a direct result of mask-wearing are common [302]. A group of nasal and allergy specialists in Germany reported on a series of cases where polypropylene fibers from filtering facepiece (respirator-type) masks were found in multiple patients who presented with severe irritant rhinitis (inflammation inside the nose), sneezing, itching, and rhinorrhea (runny rose) even in the absence of a true allergic reaction to the masks' component microplastics [240]. After three mask-free days, the polypropylene fibers found in the patients' noses had naturally cleared by nearly 90%, and their signs and symptoms correspondingly subsided. Microplastics from masks can and do get into the nose, causing inflammation and increased respiratory secretions in a segment of the populace. This is yet another one of the many mechanisms by which any filtration benefits ascribed to masks based on laboratory studies can be nullified or reversed on net balance during real life use.

In late 2021, researchers at the UK's Castle Hill Hospital collected lung tissue left over after thoracic surgery on thirteen patients. They found that eleven of these thirteen patients had varying amounts of microplastics in their lungs [317]. The researchers actually identified a total of 39 different types of microplastics present, but by far the most abundant at over 40% of the total were polypropylene and polyethylene — the same plastics in masks. While these findings are *consistent* with microplastic fibers from masks being permanently inhaled, they do not quite *prove* this is happening as a result of perpetual mask use. We should expect to find some microscopic representatives of many substances from our environment in our lungs, and microplastics have been identified in tissue samples from human lungs as early as the 1990s. Ideally, we would need to compare lung tissue from people who have never worn masks, people who wore masks throughout the COVIDcrisis, healthcare providers who wear masks as part of their jobs, and industry workers who manufacture masks. In any case, the amount of microplastic found by the UK researchers was small, averaging 1-3 pieces per gram of lung tissue, and I recommend not losing sleep visualizing mask microplastics piling up in your lungs. Still, these findings raise questions which need answers. What proportion (if any) of microplastics found in lung tissue come from mask use? How quickly do they build up? What effects (if any) do they have once ensconced in the lungs?

On a related note, in 2023, researchers at Jeonbuk National University in South Korea evaluated the off-gassing of Toxic Volatile Organic Compounds from different types of masks [342]. While they found 15 types of hazardous chemicals present, the concentrations were "very low, nearly 100-1000 times less than the recommended exposure limit set by NIOSH." However, the "best" masks tested — the respirator-type masks — off-gassed concentrations of these compounds that were many times higher when fresh out of the packaging. The concentration of off-gassed compounds increased dramatically at body temperature or warmer, but declined by about 80% during the first thirty minutes after unwrapping. The researchers recommended letting respirator-type masks sit for at least 30 minutes after taking them out of the packaging before putting them on.

Finally, we have already dealt with studies cited by the CDC that were presented as purporting to show benefits from large-scale masking at a population level. We have seen how these studies only looked at short periods of 3 months or less, usually during the late spring and summer of 2020 when mask mandates coincided with the decline phase of COVID's natural cycle in the areas surveyed. Some of them, like the Kansas Mask Mandate study [90], turn out to provide evidence *against* masks working when closely scrutinized. A subsequent analysis of the same Kansas counties over the same time period as that published by the CDC expanded the dataset to include case fatality rates [252]. Depending on how the COVID-related death rate was defined, Kansas counties with mask mandates turned out to have had case fatality rates from COVID that were 46% to 83% higher on average than the Kansas counties which did not mandate masks (representing an absolute difference in reported case fatality rates of 0.6% to 0.9%).

We have likewise seen how many other studies which looked at longer time periods found no benefit from mask use. A 2022 study went beyond this "no effect" conclusion after looking at mask use in 35 European countries over the six-month period from October 1ˢᵗ, 2020, to March 31ˢᵗ, 2021, covering the first European late fall, winter, and early spring with COVID [57]. This period was selected for analysis because the COVID vaccine rollout was still in its infancy and unlikely to be a confounding factor, with only three of the countries studied having vaccination rates over 20% by March 31ˢᵗ (Serbia, Hungary, and the UK). However, when the data from these 35 European countries was sifted, not only did mask use *not* corelate with lower COVID cases, it correlated with a small but statistically significant *increase* in COVID deaths (the increase in COVID *cases* associated with mask use was not statistically significant). Given the time period analyzed in this study, the correlation between higher mask use and higher COVID deaths is unlikely to be explained by countries being hit harder by COVID responding with more masking. Many of the European countries like France, Italy, Portugal and Spain went into the fall and winter of 2020 with mask mandates which had *already* been entrenched for months. Germany had gone even further and mandated FFP2 respirator-type masks. In every country, mask use was widespread, and COVID reporting was more uniform and systematized than in early 2020. Small countries vs. large countries, Western Europe vs. Eastern Europe, no matter how the author looked at the data, he was unable to find a comparison where more masks correlated with fewer cases or deaths from COVID. While this observational correlation certainly does not prove that masks *cause* increased COVID deaths, at the very least, it should prompt follow-up studies, and provides further strong evidence that *masks do not work*.

Part 2 Conclusion:
There is nothing "minor" or "inconsequential" about it

The side effects from masks are numerous and non-trivial. They involve detrimental respiratory and cardiovascular effects, headaches, fatigue, skin damage, odors, and more. Masks also present substantial social, cognitive, and psychological side effects in many people. A number of these side effects directly or indirectly undermine the stated rationale behind compulsory masking. In particular, contact transmission of any pathogen is a risk that masks increase. Despite rarely picking up any viruses even when worn for hours in hospitals, masks quickly pick up and help incubate bacteria and fungi, especially when worn over multiple days, while any SARS-CoV-2 virions a mask does manage to pick up then remain viable for days longer than they would on many other surfaces.

Keep this discussion in mind later in this book when we review the legal duels surrounding masking, especially when reading dogmatic assertions by some judges that wearing a mask is a "minimal inconvenience"[38] and "may be an inconvenience or annoyance, but it is a trivial imposition"[39] (with the implication that no one should be complaining). Making a sandwich for someone is trivially easy, too, a "minor inconvenience," you might say, but you wouldn't accept that as an excuse from someone for *forcing* you to make them a sandwich, even though people need food to survive. In any case, the numerous, well-documented, side effects of masks are in direct conflict with expansive descriptions of health put forward by bodies like the World Health Organization, which define health as a state of mental, physical, and social well-being rather than a simple absence of disease. In aggregate, the side effects from masks would make compulsory masking unjustifiable even if masks were 100% effective.

---

38   *Machovec v. Palm Beach County*, No. 2020CA006920AXX — Filing #110806335, 27 July 2020 (15th Judicial Circuit, Florida), p. 11 available online: https://clearinghouse.net/doc/109346/ & https://www.floridacivilrights.org/wp-content/uploads/2020/07/Order-Denying-Temporary-Injunction.pdf

39   *Stewart et al. v. Justice et al.* No. 3:2020cv00611 - Document 45 (S.D.W. Va. 2021), p. 10, available online: https://law.justia.com/cases/federal/district-courts/west-virginia/wvsdce/3:2020cv00611/230251/45/.

But then, if masks were 100% effective, universal compulsory masking *still* would not be needed or justified. Given the repeated failure of masks to demonstrate *any* real-world efficacy against viral infections like SARS-CoV-2, mask side effects rendered compulsory masking completely inexcusable.

Parts 1 & 2 References

1. Tunevall, T.G., *Postoperative wound infections and surgical face masks: a controlled study.* World Journal of Surgery, 1991. **15**(3): p. 383-7; discussion 387-8. https://pubmed.ncbi.nlm.nih.gov/1853618/

2. Webster, J., et al., *Use of face masks by non-scrubbed operating room staff: a randomized controlled trial.* ANZ J Surg, 2010. **80**(3): p. 169-73. https://pubmed.ncbi.nlm.nih.gov/20575920/

3. Suess, T., et al., *The role of facemasks and hand hygiene in the prevention of influenza transmission in households: results from a cluster randomised trial; Berlin, Germany, 2009-2011.* BMC Infectious Diseases, 2012. **12**(1): p. 26. https://dx.doi.org/10.1186/1471-2334-12-26

4. Simmerman, J.M., et al., *Findings from a household randomized controlled trial of hand washing and face masks to reduce influenza transmission in Bangkok, Thailand.* Influenza and Other Respiratory Viruses, 2011. **5**(4): p. 256-267. https://dx.doi.org/10.1111/j.1750-2659.2011.00205.x

5. Larson, E.L., et al., *Impact of Non-Pharmaceutical Interventions on URIs and Influenza in Crowded, Urban Households.* Public Health Reports, 2010. **125**(2): p. 178-191. https://dx.doi.org/10.1177/003335491012500206

6. MacIntyre, C.R., et al., *Face mask use and control of respiratory virus transmission in households.* Emerg Infect Dis, 2009. **15**(2): p. 233-41. https://pubmed.ncbi.nlm.nih.gov/19193267/

7. Cowling, B.J., et al., *Preliminary Findings of a Randomized Trial of Non-Pharmaceutical Interventions to Prevent Influenza Transmission in Households.* PLoS ONE, 2008. **3**(5): p. e2101. https://dx.doi.org/10.1371/journal.pone.0002101

8.   Cowling, B.J., et al., *Facemasks and hand hygiene to prevent influenza transmission in households: a cluster randomized trial.* Ann Intern Med, 2009. **151**(7): p. 437-46. https://pubmed.ncbi.nlm.nih.gov/19652172/

9.   Canini, L., et al., *Surgical Mask to Prevent Influenza Transmission in Households: A Cluster Randomized Trial.* PLoS ONE, 2010. **5**(11): p. e13998. https://dx.doi.org/10.1371/journal.pone.0013998

10.   Macintyre, C.R., et al., *Cluster randomised controlled trial to examine medical mask use as source control for people with respiratory illness.* BMJ Open, 2016. **6**(12): p. e012330. https://dx.doi.org/10.1136/bmjopen-2016-012330

11.   Alfelali, M., et al., *Facemask versus No Facemask in Preventing Viral Respiratory Infections During Hajj: A Cluster Randomised Open Label Trial.* SSRN Electronic Journal, 2019. https://papers.ssrn.com/sol3/papers.cfm?abstract_id=3349234

12.   Barasheed, O., et al., *Pilot Randomised Controlled Trial to Test Effectiveness of Facemasks in Preventing Influenza-like Illness Transmission among Australian Hajj Pilgrims in 2011.* Infect Disord Drug Targets, 2014. **14**(2): p. 110-6. https://www.researchgate.net/publication/282747527_Pilot_Randomised_Controlled_Trial_to_Test_Effectiveness_of_Facemasks_in_Preventing_Influenza-like_Illness_Transmission_among_Australian_Hajj_Pilgrims_in_2011

13.   Wang, M., et al., *A cluster-randomised controlled trial to test the efficacy of facemasks in preventing respiratory viral infection among Hajj pilgrims.* J Epidemiol Glob Health, 2015. **5**(2): p. 181-9. https://europepmc.org/article/MED/25922328

14.   Aiello, A.E., et al., *Mask use, hand hygiene, and seasonal influenza-like illness among young adults: a randomized intervention trial.* J Infect Dis, 2010. **201**(4): p. 491-8. https://academic.oup.com/jid/article/201/4/491/861190

15.   Aiello, A.E., et al., *Facemasks, Hand Hygiene, and Influenza among Young Adults: A Randomized Intervention Trial.* PLoS ONE, 2012. **7**(1): p. e29744. https://dx.doi.org/10.1371/journal.pone.0029744

16.   Bundgaard, H., et al., *Effectiveness of Adding a Mask Recommendation to Other Public Health Measures to Prevent SARS-CoV-2 Infection in Danish Mask Wearers : A Randomized Controlled Trial.* Ann Intern Med, 2021. **174**(3): p. 335-343.

https://www.acpjournals.org/doi/full/10.7326/M20-6817?rfr_dat=cr_pub++0pubmed&url_ver=Z39.88-2003&rfr_id=ori%3Arid%3Acrossref.org

17. Abaluck, J., et al., *Impact of community masking on COVID-19: A cluster-randomized trial in Bangladesh*. Science, 2021. **0**(0): p. eabi9069. https://www.science.org/doi/abs/10.1126/science.abi9069

18. Chikina, M., W. Pegden, and B. Recht, *A note on sampling biases in the Bangladesh mask trial*. arXiv pre-print server, 2021. https://arxiv.org/abs/2112.01296

19. Fenton, N., *The Bangladesh Mask study: a Bayesian perspective*. 2022. https://www.researchgate.net/publication/360320982_The_Bangladesh_Mask_study_a_Bayesian_perspective

20. Jefferson, T., et al., *Physical interventions to interrupt or reduce the spread of respiratory viruses*. Cochrane Database of Systematic Reviews, 2023. **2023**(1). https://dx.doi.org/10.1002/14651858.cd006207.pub6

21. Demasi, M., *EXCLUSIVE: Lead author of new Cochrane review speaks out; A no-holds-barred interview with Tom Jefferson*. 2023: Substack: Maryanne Demani, reports, . Online: https://maryannedemasi.substack.com/p/exclusive-lead-author-of-new-cochrane?publication_id=1044435&post_id=100756872&isFreemail=true

22. Loeb, M., et al., *Surgical mask vs N95 respirator for preventing influenza among health care workers: a randomized trial*. Jama, 2009. **302**(17): p. 1865-71. https://pubmed.ncbi.nlm.nih.gov/19797474/

23. MacIntyre, C.R., et al., *A cluster randomised trial of cloth masks compared with medical masks in healthcare workers*. BMJ Open, 2015. **5**(4): p. e006577. https://www.ncbi.nlm.nih.gov/pmc/articles/PMC4420971/

24. MacIntyre, C.R., et al., *A cluster randomized clinical trial comparing fit-tested and non-fit-tested N95 respirators to medical masks to prevent respiratory virus infection in health care workers*. Influenza Other Respir Viruses, 2011. **5**(3): p. 170-9. https://pubmed.ncbi.nlm.nih.gov/21477136/

25. MacIntyre, C.R., et al., *A randomized clinical trial of three options for N95 respirators and medical masks in health workers*. Am J Respir Crit Care Med, 2013. **187**(9): p. 960-6. https://pubmed.ncbi.nlm.nih.gov/23413265/

26. Radonovich, L.J., Jr., et al., *N95 Respirators vs Medical Masks for Preventing Influenza Among Health Care Personnel: A Randomized Clinical Trial.* Jama, 2019. **322**(9): p. 824-833. https://pubmed.ncbi.nlm.nih.gov/31479137/

27. Loeb, M., et al., *Medical Masks Versus N95 Respirators for Preventing COVID-19 Among Health Care Workers : A Randomized Trial.* Ann Intern Med, 2022. https://pubmed.ncbi.nlm.nih.gov/36442064/

28. MacIntyre, C.R., et al., *Contamination and washing of cloth masks and risk of infection among hospital health workers in Vietnam: a post hoc analysis of a randomised controlled trial.* BMJ Open, 2020. **10**(9): p. e042045. https://www.ncbi.nlm.nih.gov/pmc/articles/PMC7523194/

29. Bahli, Z.M., *Does evidence based medicine support the effectiveness of surgical facemasks in preventing postoperative wound infections in elective surgery?* J Ayub Med Coll Abbottabad, 2009. **21**(2): p. 166-70. https://www.ayubmed.edu.pk/JAMC/PAST/21-2/Zahid.pdf

30. Bin-Reza, F., et al., *The use of masks and respirators to prevent transmission of influenza: a systematic review of the scientific evidence.* Influenza Other Respir Viruses, 2012. **6**(4): p. 257-67. https://www.ncbi.nlm.nih.gov/pmc/articles/PMC5779801/

31. Carøe, T., *[Dubious effect of surgical masks during surgery].* Ugeskrift for laeger, 2014. **176**(27): p. V09130564. http://europepmc.org/abstract/MED/25294675

32. Jefferson, T., et al., *Physical interventions to interrupt or reduce the spread of respiratory viruses. Part 1 - Face masks, eye protection and person distancing: systematic review and meta-analysis.* 2020, Cold Spring Harbor Laboratory.https://dx.doi.org/10.1101/2020.03.30.20047217

33. Lipp, A. and P. Edwards, *Disposable surgical face masks: a systematic review.* Can Oper Room Nurs J, 2005. **23**(3): p. 20-1, 24-5, 33-8. https://pubmed.ncbi.nlm.nih.gov/16295987/

34. Lipp, A. and P. Edwards, *Disposable surgical face masks for preventing surgical wound infection in clean surgery.* Cochrane Database of Systematic Reviews, 2014. https://dx.doi.org/10.1002/14651858.cd002929.pub2

35. Vincent, M. and P. Edwards, *Disposable surgical face masks for preventing surgical wound infection in clean surgery.* Cochrane Database of Systematic Reviews, 2016. https://dx.doi.org/10.1002/14651858.cd002929.pub3

36. Yu, J., et al., *Evidence-based Sterility: The Evolving Role of Field Sterility in Skin and Minor Hand Surgery.* Plast Reconstr Surg Glob Open, 2019. **7**(11): p. e2481. https://www.researchgate.net/publication/337551632_Evidence-based_Sterility_The_Evolving_Role_of_Field_Sterility_in_Skin_and_Minor_Hand_Surgery

37. Bartoszko, J.J., et al., *Medical masks vs N95 respirators for preventing COVID-19 in healthcare workers: A systematic review and meta-analysis of randomized trials.* Influenza Other Respir Viruses, 2020. **14**(4): p. 365-373. https://pubmed.ncbi.nlm.nih.gov/32246890/

38. Long, Y., et al., *Effectiveness of N95 respirators versus surgical masks against influenza: A systematic review and meta-analysis.* J Evid Based Med, 2020. **13**(2): p. 93-101. https://pubmed.ncbi.nlm.nih.gov/32167245/

39. Offeddu, V., et al., *Effectiveness of Masks and Respirators Against Respiratory Infections in Healthcare Workers: A Systematic Review and Meta-Analysis.* Clin Infect Dis, 2017. **65**(11): p. 1934-1942. https://pubmed.ncbi.nlm.nih.gov/29140516/

40. Saunders-Hastings, P., et al., *Effectiveness of personal protective measures in reducing pandemic influenza transmission: A systematic review and meta-analysis.* Epidemics, 2017. **20**: p. 1-20. https://pubmed.ncbi.nlm.nih.gov/28487207/

41. Smith, J.D., et al., *Effectiveness of N95 respirators versus surgical masks in protecting health care workers from acute respiratory infection: a systematic review and meta-analysis.* Cmaj, 2016. **188**(8): p. 567-574. https://pubmed.ncbi.nlm.nih.gov/26952529/

42. Xiao, J., et al., *Nonpharmaceutical Measures for Pandemic Influenza in Nonhealthcare Settings—Personal Protective and Environmental Measures.* Emerging Infectious Diseases, 2020. **26**(5): p. 967-975. https://dx.doi.org/10.3201/eid2605.190994

43. WHO, *Non-pharmaceutical public health measures for mitigating the risk and impact of epidemic and pandemic influenza.* 2019, World Health Organization Global Influenza Program WEP: Online. https://www.who.int/publications/i/item/non-pharmaceutical-public-health-measuresfor-mitigating-the-risk-and-impact-of-epidemic-and-pandemic-influenza; Alternate: https://web.archive.org/web/20231005052047/https://www.who.int/publications/i/item/non-pharmaceutical-public-health-measuresfor-mitigating-the-risk-and-impact-of-epidemic-and-pandemic-influenza; Additional alternate: https://resourcecentre.savethechildren.net/pdf/9789241516839-eng.pdf/

44.   WHO, *Non-pharmaceutical public health measures for mitigating the risk and impact of epidemic and pandemic influenza. Annex: Report of systematic literature reviews.* 2019, World Health Organization: Online. https://iris.who.int/handle/10665/329439 https://iris.who.int/bitstream/handle/10665/329439/WHO-WHE-IHM-GIP-2019.1-eng.pdf?sequence=1&isAllowed=y; Alternate: https://resourcecentre.savethechildren.net/pdf/who-whe-ihm-gip-2019.1-eng.pdf/; Alternate: https://www.scribd.com/document/683622235/World-Health-Organization-Global-Influenza-Programme-Non-pharmaceutical-public-health-measures-for-mitigating-the-risk-and-impact-of-epidemic-and-pand

45.   Jacobs, J.L., et al., *Use of surgical face masks to reduce the incidence of the common cold among health care workers in Japan: a randomized controlled trial.* Am J Infect Control, 2009. **37**(5): p. 417-419. https://www.ajicjournal.org/article/S0196-6553(08)00909-7/fulltext

46.   Jefferson, T., et al., *Physical interventions to interrupt or reduce the spread of respiratory viruses.* Cochrane Database of Systematic Reviews, 2020. **2020**(11). https://dx.doi.org/10.1002/14651858.cd006207.pub5

47.   Cochrane, *Cochrane announces support of new donor.* 2016. Online: https://www.cochrane.org/news/cochrane-announces-support-new-donor

48.   Sood, N., et al., *Association between School Mask Mandates and SARS-CoV-2 Student Infections: Evidence from a Natural Experiment of Neighboring K-12 Districts in North Dakota.* 2022, Research Square Platform LLC. https://dx.doi.org/10.21203/rs.3.rs-1773983/v1

49.   Guerra, D. and D.J. Guerra, *Mask mandate and use efficacy in state-level COVID-19 containment.* 2021, Cold Spring Harbor Laboratory. https://dx.doi.org/10.1101/2021.05.18.21257385

50.   Chandra, A. and T.B. Høeg, *Lack of correlation between school mask mandates and paediatric COVID-19 cases in a large cohort.* Journal of Infection, 2022. **85**(6): p. 671-675. https://dx.doi.org/10.1016/j.jinf.2022.09.019

51.   Budzyn, S.E., *Pediatric COVID-19 Cases in Counties With and Without School Mask Requirements—United States, July 1–September 4, 2021.* MMWR. Morbidity and Mortality Weekly Report, 2021. **70**. https://www.cdc.gov/mmwr/volumes/70/wr/mm7039e3.htm

52. Coma, E., et al., *Unravelling the Role of the Mandatory Use of Face Covering Masks for the Control of SARS-CoV-2 in Schools: A Quasi-Experimental Study Nested in a Population-Based Cohort in Catalonia (Spain)*. 2022. https://papers.ssrn.com/sol3/papers.cfm?abstract_id=4046809

53. Alonso, S., et al., *Age-dependency of the Propagation Rate of Coronavirus Disease 2019 Inside School Bubble Groups in Catalonia, Spain*. The Pediatric infectious disease journal, 2021. **40**(11): p. 955-961. https://pubmed.ncbi.nlm.nih.gov/34321438

54. Marks, M., et al., *Transmission of COVID-19 in 282 clusters in Catalonia, Spain: a cohort study*. The Lancet Infectious Diseases, 2021. https://dx.doi.org/10.1016/s1473-3099(20)30985-3

55. Juutinen, A., et al., *Use of face masks did not impact COVID-19 incidence among 10–12-year-olds in Finland*. 2022, Cold Spring Harbor Laboratory. https://dx.doi.org/10.1101/2022.04.04.22272833

56. Oster, E., et al., *COVID-19 mitigation practices and COVID-19 rates in schools: Report on data from Florida, New York and Massachusetts*. medRxiv, 2021. https://www.medrxiv.org/content/10.1101/2021.05.19.21257467v1

57. Spira, B., *Correlation Between Mask Compliance and COVID-19 Outcomes in Europe*. Cureus, 2022. **14**(4). https://www.cureus.com/articles/93826-correlation-between-mask-compliance-and-covid-19-outcomes-in-europe

58. Bollyky, T.J., et al., *Assessing COVID-19 pandemic policies and behaviours and their economic and educational trade-offs across US states from Jan 1, 2020, to July 31, 2022: an observational analysis*. The Lancet, 2023. https://dx.doi.org/10.1016/s0140-6736(23)00461-0

59. Patel, S.N., et al., *The Impact of Physician Face Mask Use on Endophthalmitis After Intravitreal Anti–Vascular Endothelial Growth Factor Injections*. American Journal of Ophthalmology, 2021. **222**: p. 194-201. https://dx.doi.org/10.1016/j.ajo.2020.08.013

60. Orr, N.W., *Is a mask necessary in the operating theatre?* Ann R Coll Surg Engl, 1981. **63**(6): p. 390-2 https://www.ncbi.nlm.nih.gov/pmc/articles/PMC2493952/pdf/annrcse01509-0009.pdf

61. Orr, N., *Is a mask necessary in the operating theatre?* Ann R Coll Surg Engl, 1982. **64**(3): p. 205. https://www.ncbi.nlm.nih.gov/pmc/articles/PMC2493986/

62. Norman, A., *A comparison of face masks and visors for the scrub team. A study in theatres.* The British journal of theatre nursing: NATNews: the official journal of the National Association of Theatre Nurses, 1995. **5**(2): p. 10-13. https://www.scribd.com/document/552633554/1995-Norman-A-Comparison-of-Face-Masks-and-visors-faceshields

63. Ruthman, J.C., et al., *Effect of cap and mask on infection rates in wounds sutured in the emergency department.* IMJ Ill Med J, 1984. **165**(6): p. 397-9. https://pubmed.ncbi.nlm.nih.gov/6146589/
https://www.scribd.com/document/552632881/1984-Ruthman-Effect-of-Cap-and-Mask-on-Infection-rates

64. Davies, K.J., et al., *Seroepidemiological study of respiratory virus infections among dental surgeons.* Br Dent J, 1994. **176**(7): p. 262-5. https://www.nature.com/articles/4808430

65. Estrich, C.G., et al., *Estimating COVID-19 prevalence and infection control practices among US dentists.* J Am Dent Assoc, 2020. **151**(11): p. 815-824. https://pubmed.ncbi.nlm.nih.gov/33071007/

66. Meethil, A.P., et al., *Sources of SARS-CoV-2 and Other Microorganisms in Dental Aerosols.* J Dent Res, 2021: p. 220345211015948. https://journals.sagepub.com/doi/10.1177/00220345211015948?url_ver=Z39.88-2003&rfr_id=ori:rid:crossref.org&rfr_dat=cr_pub%20%20 0pubmed

67. Murphy, D., et al., *The use of gowns and masks to control respiratory illness in pediatric hospital personnel.* The Journal of Pediatrics, 1981. **99**(5): p. 746-750. https://dx.doi.org/10.1016/s0022-3476(81)80401-5

68. Lahme, T., et al., *[Patient surgical masks during regional anesthesia. Hygenic necessity or dispensable ritual?].* Der Anaesthesist, 2001. **50**(11): p. 846-851. https://doi.org/10.1007/s00101-001-0229-x

69. Al-Asmary, S., et al., *Acute respiratory tract infections among Hajj medical mission personnel, Saudi Arabia.* Int J Infect Dis, 2007. **11**(3): p. 268-72. https://www.researchgate.net/publication/6882796_Acute_respiratory_tract_infections_among_Hajj_medical_mission_personnel_Saudi_Arabia

70. Zhang, Y., et al., *Factors associated with the transmission of pandemic (H1N1) 2009 among hospital healthcare workers in Beijing, China.* Influenza Other Respir Viruses, 2013. **7**(3): p. 466-71. https://pubmed.ncbi.nlm.nih.gov/23078163/

71. Sjøl, A. and H. Kelbaek, *[Is use of surgical caps and masks obsolete during percutaneous heart catheterization?].* Ugeskrift for laeger, 2002. **164**(12): p. 1673-1675. http://europepmc.org/abstract/MED/11924291

72. Laslett, L.J. and A. Sabin, *Wearing of caps and masks not necessary during cardiac catheterization.* Catheterization and Cardiovascular Diagnosis, 1989. **17**(3): p. 158-160. https://doi.org/10.1002/ccd.1810170306

73. Scales, D.C., et al., *Illness in intensive care staff after brief exposure to severe acute respiratory syndrome.* Emerg Infect Dis, 2003. **9**(10): p. 1205-10. https://pubmed.ncbi.nlm.nih.gov/14609453/

74. Ofner, M., et al., *Cluster of severe acute respiratory syndrome cases among protected health-care workers-Toronto, Canada, April 2003.* MMWR: Morbidity & Mortality Weekly Report, 2003. **52**(19): p. 433-433. https://www.cdc.gov/mmwr/preview/mmwrhtml/mm5219a1.htm

75. Park, B.J., et al., *Lack of SARS transmission among healthcare workers, United States.* Emerg Infect Dis, 2004. **10**(2): p. 244-8. https://wwwnc.cdc.gov/eid/article/10/2/03-0793_article

76. Lau, J.T.F., et al., *SARS Transmission among Hospital Workers in Hong Kong.* Emerging Infectious Diseases, 2004. **10**(2): p. 280-286. https://dx.doi.org/10.3201/eid1002.030534

77. Schwartz, K.L., et al., *Lack of COVID-19 transmission on an international flight.* Canadian Medical Association Journal, 2020. **192**(15): p. E410-E410. https://dx.doi.org/10.1503/cmaj.75015

78. Klompas, M., et al., *Transmission of Severe Acute Respiratory Syndrome Coronavirus 2 (SARS-CoV-2) From Asymptomatic and Presymptomatic Individuals in Healthcare Settings Despite Medical Masks and Eye Protection.* Clinical Infectious Diseases, 2021. **73**(9): p. 1693-1695. https://dx.doi.org/10.1093/cid/ciab218

79. Madewell, Z.J., et al., *Household Transmission of SARS-CoV-2.* JAMA Network Open, 2020. **3**(12): p. e2031756. https://dx.doi.org/10.1001/jamanetworkopen.2020.31756

80. Madewell, Z.J., et al., *Factors Associated With Household Transmission of SARS-CoV-2.* JAMA Network Open, 2021. **4**(8): p. e2122240. https://dx.doi.org/10.1001/jamanetworkopen.2021.22240

81. Thompson, H.A., et al., *Severe Acute Respiratory Syndrome Coronavirus 2 (SARS-CoV-2) Setting-specific Transmission Rates: A Systematic Review and Meta-analysis.* Clinical Infectious Diseases, 2021. **73**(3): p. e754-e764. https://dx.doi.org/10.1093/cid/ciab100

82. Shitrit, P., et al., *Nosocomial outbreak caused by the SARS-CoV-2 Delta variant in a highly vaccinated population, Israel, July 2021.* Eurosurveillance, 2021. **26**(39). https://dx.doi.org/10.2807/1560-7917.es.2021.26.39.2100822

83. McCluskey, F., *Does wearing a face mask reduce bacterial wound infection? A literature review.* Br J Theatre Nurs, 1996. **6**(5): p. 18-20, 29. https://pubmed.ncbi.nlm.nih.gov/8850864/ https://www.scribd.com/document/552637864/1996-McCluskey-Does-Wearing-a-Face-Mask-Reduce

84. Skinner, M.W. and B.A. Sutton, *Do anaesthetists need to wear surgical masks in the operating theatre? A literature review with evidence-based recommendations.* Anaesth Intensive Care, 2001. **29**(4): p. 331-8. https://pubmed.ncbi.nlm.nih.gov/11512642/

85. Salassa, T.E. and M.F. Swiontkowski, *Surgical attire and the operating room: role in infection prevention.* J Bone Joint Surg Am, 2014. **96**(17): p. 1485-92. https://pubmed.ncbi.nlm.nih.gov/25187588/; https://www.scribd.com/document/552638974/2014-Salassa-Surgical-Attire-and-the-Operating

86. Da Zhou, C., P. Sivathondan, and A. Handa, *Unmasking the surgeons: the evidence base behind the use of facemasks in surgery.* J R Soc Med, 2015. **108**(6): p. 223-8. https://pubmed.ncbi.nlm.nih.gov/26085560/

87. Chou, R., et al., *Masks for Prevention of Respiratory Virus Infections, Including SARS-CoV-2, in Health Care and Community Settings.* Annals of Internal Medicine, 2020. **173**(7): p. 542-555. Original: https://dx.doi.org/10.7326/m20-3213 Final update (Update 11): https://doi.org/10.7326/L22-0235

88. Chou, R., et al., *Update Alert 11: Epidemiology of and Risk Factors for Coronavirus Infection in Health Care Workers.* Annals of Internal Medicine, 2022 https://doi.org/10.7326/L22-0235

89. Vainshelboim, B., *Facemasks in the COVID-19 era: A health hypothesis.* Medical hypotheses, 2021. **146**: p. 110411-110411.
https://www.ncbi.nlm.nih.gov/pmc/articles/PMC7680614/

90. Van Dyke, M.E., et al., *Trends in County-Level COVID-19 Incidence in Counties With and Without a Mask Mandate - Kansas, June 1-August 23, 2020.* MMWR Morb Mortal Wkly Rep, 2020. **69**(47): p. 1777-1781.
https://www.cdc.gov/mmwr/volumes/69/wr/mm6947e2.htm

91. Szablewski, C.M., et al., *SARS-CoV-2 Transmission and Infection Among Attendees of an Overnight Camp — Georgia, June 2020.* MMWR. Morbidity and Mortality Weekly Report, 2020. **69**(31): p. 1023-1025. https://dx.doi.org/10.15585/mmwr.mm6931e1

92. Hendrix, M.J., et al., *Absence of Apparent Transmission of SARS-CoV-2 from Two Stylists After Exposure at a Hair Salon with a Universal Face Covering Policy - Springfield, Missouri, May 2020.* MMWR Morb Mortal Wkly Rep, 2020. **69**(28): p. 930-932.
https://www.cdc.gov/mmwr/volumes/69/wr/mm6928e2.htm?s_cid=mm6928e2_w

93. Gettings, J., et al., *Mask use and ventilation improvements to reduce COVID-19 incidence in elementary schools—Georgia, November 16–December 11, 2020.* Morbidity and Mortality Weekly Report, 2021. **70**(21): p. 779.
https://www.ncbi.nlm.nih.gov/pmc/articles/PMC8158891/

94. Wang, X., et al., *Association Between Universal Masking in a Health Care System and SARS-CoV-2 Positivity Among Health Care Workers.* Journal of the American Medical Association, 2020. **324**(7): p. 703-4. https://jamanetwork.com/journals/jama/fullarticle/2768533

95. Patterson, B., R. Mehra, and A. Breathnach, *Unmasking the mask: a time-series analysis of nosocomial COVID-19 rates before and after removal,* in *33rd European Congress of Clinical Microbiology and Infectious Diseases: ECCMID 2023; "abstract 5979,"* Published April 6, 2023, Presented April 17, 2023, ECCMID Presentation Number 2738, Session Code PS089: Copenhagen, Denmark. https://escmid.reg.key4events.com/AbstractList.aspx?e=17&header=0&ai=17204&preview=1&aig=-1 or
https://achern-weiss-bescheid.de/wp-content/uploads/2023/04/5979masks.pdf

96. Chin, A.W.H., et al., *Stability of SARS-CoV-2 in different environmental conditions.* Lancet Microbe, 2020. **1**(1): p. e10. https://pubmed.ncbi.nlm.nih.gov/32835322/

97.   Oberg, T. and L.M. Brosseau, *Surgical mask filter and fit performance*. Am J Infect Control, 2008. **36**(4): p. 276-82. https://pubmed.ncbi.nlm.nih.gov/18455048/

98.   Makison Booth, C., et al., *Effectiveness of surgical masks against influenza bioaerosols*. Journal of Hospital Infection, 2013. **84**(1): p. 22-26. https://dx.doi.org/10.1016/j.jhin.2013.02.007

99.   Landry, S.A., et al., *Fit-tested N95 masks combined with portable HEPA filtration can protect against high aerosolized viral loads over prolonged periods at close range*. The Journal of Infectious Diseases, 2022. https://dx.doi.org/10.1093/infdis/jiac195

100.   Lee, S.A., S.A. Grinshpun, and T. Reponen, *Respiratory performance offered by N95 respirators and surgical masks: human subject evaluation with NaCl aerosol representing bacterial and viral particle size range*. Ann Occup Hyg, 2008. **52**(3): p. 177-85. https://pubmed.ncbi.nlm.nih.gov/18326870/

101.   Wiersinga, W.J., et al., *Pathophysiology, Transmission, Diagnosis, and Treatment of Coronavirus Disease 2019 (COVID-19)*. JAMA, 2020. **324**(8): p. 782. https://dx.doi.org/10.1001/jama.2020.12839

102.   Du, W., et al., *Microstructure analysis and image-based modelling of face masks for COVID-19 virus protection*. Communications Materials, 2021. **2**(1). https://dx.doi.org/10.1038/s43246-021-00160-z

103.   Yim, W., et al., *KN95 and N95 Respirators Retain Filtration Efficiency despite a Loss of Dipole Charge during Decontamination*. ACS Applied Materials & Interfaces, 2020. **12**(49): p. 54473-54480. https://dx.doi.org/10.1021/acsami.0c17333

104.   Park, A.-M., et al., *Bacterial and fungal isolation from face masks under the COVID-19 pandemic*. Scientific Reports, 2022. **12**(1). https://dx.doi.org/10.1038/s41598-022-15409-x

105.   Neupane, B.B., et al., *Optical microscopic study of surface morphology and filtering efficiency of face masks*. PeerJ, 2019. **7**: p. e7142. https://dx.doi.org/10.7717/peerj.7142

106.   Kelkar, U.S., et al., *How effective are face masks in operation theatre? A time frame analysis and recommendations*. International Journal of Infection Control, 2013. **9**(1). https://www.ijic.info/article/view/10788

107. Kellogg, W.H. and G. Macmillan, *An Experimental Study of the Efficacy of Gauze Face Masks.* American Journal of Public Health, 1920. **10**(1): p. 34-42. https://dx.doi.org/10.2105/ajph.10.1.34

108. Ritter, M.A., et al., *The operating room environment as affected by people and the surgical face mask.* Clinical Orthopaedics and Related Research®, 1975. **111**: p. 147-150. https://journals.lww.com/clinorthop/Citation/1975/09000/The_Operating_Room_Environment_as_Affected_by.20.aspx

109. Ha'eri, G.B. and A.M. Wiley, *The efficacy of standard surgical face masks: an investigation using "tracer particles".* Clin Orthop Relat Res, 1980(148): p. 160-2. https://corona-blog.net/wp-content/uploads/2020/10/3.Haeri-GB-Wiley-AM-1980.pdf

110. Tunevall, T.G. and H. Jörbeck, *Influence of wearing masks on the density of airborne bacteria in the vicinity of the surgical wound.* European Journal of Surgery, 1992. **158**(5): p. 263-6. https://pubmed.ncbi.nlm.nih.gov/1354489/ https://www.scribd.com/document/552635830/1992-Tunevall-Influence-of-Wearing-Masks-on-the-density-of-airborne-bacteria

111. Mitchell, N.J. and S. Hunt, *Surgical face masks in modern operating rooms--a costly and unnecessary ritual?* J Hosp Infect, 1991. **18**(3): p. 239-42. https://www.journalofhospitalinfection.com/article/0195-6701(91)90148-2/pdf https://www.scribd.com/document/552634740/1991-Mitchell-Are-Surgical-Masks-in-the-or-Unnecessary

112. Patel, S.N., et al., *Bacterial Dispersion Associated With Various Patient Face Mask Designs During Simulated Intravitreal Injections.* American Journal of Ophthalmology, 2021. **223**: p. 178-183. https://dx.doi.org/10.1016/j.ajo.2020.10.017

113. Asadi, S., et al., *Efficacy of masks and face coverings in controlling outward aerosol particle emission from expiratory activities.* Scientific Reports, 2020. **10**(1). https://dx.doi.org/10.1038/s41598-020-72798-7

114. CDC.gov, *Science Brief: Community Use of Masks to Control the Spread of SARS-CoV-2 | CDC.* 2021. https://www.cdc.gov/coronavirus/2019-ncov/downloads/science-of-masking-full.pdf also at https://web.archive.org/web/20230930222431/https://www.cdc.gov/coronavirus/2019-ncov/downloads/science-of-masking-full.pdf

115. Adenaiye, O.O., et al., *Infectious Severe Acute Respiratory Syndrome Coronavirus 2 (SARS-CoV-2) in Exhaled Aerosols and Efficacy of Masks During Early Mild Infection.* Clinical Infectious Diseases, 2021. https://dx.doi.org/10.1093/cid/ciab797

116. Alsved, M., et al., *Exhaled respiratory particles during singing and talking.* Aerosol Science and Technology, 2020. **54**(11): p. 1245-1248. https://dx.doi.org/10.1080/02786826.2020.1812502

117. Rengasamy, S., B. Eimer, and R.E. Shaffer, *Simple Respiratory Protection—Evaluation of the Filtration Performance of Cloth Masks and Common Fabric Materials Against 20–1000 nm Size Particles.* The Annals of Occupational Hygiene, 2010. **54**(7): p. 789-798. https://dx.doi.org/10.1093/annhyg/meq044

118. O'Kelly, E., et al., *Ability of fabric face mask materials to filter ultrafine particles at coughing velocity.* BMJ Open, 2020. **10**(9): p. e039424. https://dx.doi.org/10.1136/bmjopen-2020-039424

119. Hill, W.C., M.S. Hull, and R.I. Maccuspie, *Testing of Commercial Masks and Respirators and Cotton Mask Insert Materials using SARS-CoV-2 Virion-Sized Particulates: Comparison of Ideal Aerosol Filtration Efficiency versus Fitted Filtration Efficiency.* Nano Letters, 2020. **20**(10): p. 7642-7647. https://dx.doi.org/10.1021/acs.nanolett.0c03182

120. Drewnick, F., et al., *Aerosol filtration efficiency of household materials for homemade face masks: Influence of material properties, particle size, particle electrical charge, face velocity, and leaks.* Aerosol Science and Technology, 2021. **55**(1): p. 63-79. https://www.tandfonline.com/doi/pdf/10.1080/02786826.2020.1817846?src=getftr

121. Jepsen, O.B., et al., *[Importance of surgical masks for peroperative asepsis].* Ugeskr Laeger, 1993. **155**(25): p. 1940-2. https://pubmed.ncbi.nlm.nih.gov/8317057/

122. Sellden, E. and H.C. Hemmings, *Is Routine Use of a Face Mask Necessary in the Operating Room?* Anesthesiology, 2010. **113**(6): p. 1447-1447. https://doi.org/10.1097/ALN.0b013e3181fcf122

123. Isaacs, D., et al., *Do facemasks protect against COVID -19?* Journal of Paediatrics and Child Health, 2020. **56**(6): p. 976-977. https://dx.doi.org/10.1111/jpc.14936

124. De Brouwer, C., *Wearing a Mask, a Universal Solution Against COVID-19 or an Additional Health Risk?* 2020. https://papers.ssrn.com/abstract=3676885

125. Kampf, G., *Protective effect of mandatory face masks in the public-relevant variables with likely impact on outcome were not considered.* Proc Natl Acad Sci U S A, 2020. **117**(44): p. 27076-27077. https://www.pnas.org/content/117/44/27076

126. Tso, R.V. and B.J. Cowling, *Importance of Face Masks for COVID-19: A Call for Effective Public Education.* Clin Infect Dis, 2020. **71**(16): p. 2195-2198. https://academic.oup.com/cid/article/71/16/2195/5866410

127. CDC.gov. *The Science of Masking to Control COVID-19.* 2021 11/16/2020 [cited 2020; Available from: https://www.cdc.gov/coronavirus/2019-ncov/downloads/science-of-masking-full.pdf. CDC direct pdf link closed in late 2023, but can also be found at https://web.archive.org/web/20230930222431/https://www.cdc.gov/coronavirus/2019-ncov/downloads/science-of-masking-full.pdf

128. Chandra, A. and T.B. Høeg, *Revisiting Pediatric COVID-19 Cases in Counties With and Without School Mask Requirements—United States, July 1—October 20 2021.* Available at SSRN 4118566, 2022. https://papers.ssrn.com/sol3/papers.cfm?abstract_id=4118566

129. Abkarian, M., et al., *Speech can produce jet-like transport relevant to asymptomatic spreading of virus.* Proceedings of the National Academy of Sciences, 2020. **117**(41): p. 25237-25245. https://dx.doi.org/10.1073/pnas.2012156117

130. Ammann, P., et al., *Perceptions towards mask use in school children during the SARS-CoV-2 pandemic: the Ciao Corona Study.* 2021, Cold Spring Harbor Laboratory. https://dx.doi.org/10.1101/2021.09.04.21262907

131. Asadi, S., et al., *Aerosol emission and superemission during human speech increase with voice loudness.* Scientific Reports, 2019. **9**(1). https://dx.doi.org/10.1038/s41598-019-38808-z

132. Aydin, O., et al., *Performance of fabrics for home-made masks against the spread of COVID-19 through droplets: A quantitative mechanistic study.* Extreme Mech Lett, 2020. **40**: p. 100924. https://www.ncbi.nlm.nih.gov/pmc/articles/PMC7417273/

133. Bahl, P., et al., *Face coverings and mask to minimise droplet dispersion and aerosolisation: a video case study.* Thorax, 2020. **75**(11): p. 1024. http://thorax.bmj.com/content/75/11/1024.abstract

134. Bandiera, L., et al., *Face Coverings and Respiratory Tract Droplet Dispersion*. 2020, Cold Spring Harbor Laboratory. https://dx.doi.org/10.1101/2020.08.11.20145086

135. Bar-On, O., et al., *Effects of Wearing Facemasks During Brisk Walks: A COVID-19 Dilemma*. The Journal of the American Board of Family Medicine, 2021. **34**(4): p. 798-801. https://dx.doi.org/10.3122/jabfm.2021.04.200559

136. Bhattacharjee, S., et al., *Last-resort strategies during mask shortages: optimal design features of cloth masks and decontamination of disposable masks during the COVID-19 pandemic*. BMJ Open Respiratory Research, 2020. **7**(1): p. e000698. https://dx.doi.org/10.1136/bmjresp-2020-000698

137. Brooks, J.T., et al., *Maximizing Fit for Cloth and Medical Procedure Masks to Improve Performance and Reduce SARS-CoV-2 Transmission and Exposure, 2021*. MMWR Morb Mortal Wkly Rep, 2021. **70**(7): p. 254-257. https://www.cdc.gov/mmwr/volumes/70/wr/mm7007e1.htm

138. Chaiyabutr, C., et al., *Adverse skin reactions following different types of mask usage during the COVID-19 pandemic*. Journal of the European Academy of Dermatology and Venereology, 2021. **35**(3). https://dx.doi.org/10.1111/jdv.17039

139. Chan, N.C., K. Li, and J. Hirsh, *Peripheral Oxygen Saturation in Older Persons Wearing Nonmedical Face Masks in Community Settings*. Jama, 2020. **324**(22): p. 2323-2324. https://www.ncbi.nlm.nih.gov/pmc/articles/PMC7600049/

140. Chernozhukov, V., H. Kasahara, and P. Schrimpf, *Causal impact of masks, policies, behavior on early covid-19 pandemic in the U.S.* J Econom, 2021. **220**(1): p. 23-62. https://www.ncbi.nlm.nih.gov/pmc/articles/PMC7568194/

141. Chu, D.K., et al., *Physical distancing, face masks, and eye protection to prevent person-to-person transmission of SARS-CoV-2 and COVID-19: a systematic review and meta-analysis*. Lancet, 2020. **395**(10242): p. 1973-1987. https://www.ncbi.nlm.nih.gov/pmc/articles/PMC7263814/

142. Clapp, P.W., et al., *Evaluation of Cloth Masks and Modified Procedure Masks as Personal Protective Equipment for the Public During the COVID-19 Pandemic*. JAMA Intern Med, 2021. **181**(4): p. 463-469. https://jamanetwork.com/journals/jamainternalmedicine/fullarticle/2774266

143. Clase, C.M., et al., *Forgotten Technology in the COVID-19 Pandemic: Filtration Properties of Cloth and Cloth Masks—A Narrative Review.* Mayo Clinic Proceedings, 2020. **95**(10): p. 2204-2224. https://dx.doi.org/10.1016/j.mayocp.2020.07.020

144. Davies, A., et al., *Testing the efficacy of homemade masks: would they protect in an influenza pandemic?* Disaster Med Public Health Prep, 2013. **7**(4): p. 413-8. https://www.ncbi.nlm.nih.gov/pmc/articles/PMC7108646/

145. Dost, B., et al., *Investigating the Effects of Protective Face Masks on the Respiratory Parameters of Children in the Postanesthesia Care Unit During the COVID-19 Pandemic.* Journal of PeriAnesthesia Nursing, 2021. https://europepmc.org/articles/PMC7877201

146. Doung-Ngern, P., et al., *Case-Control Study of Use of Personal Protective Measures and Risk for SARS-CoV 2 Infection, Thailand.* Emerging Infectious Diseases, 2020. **26**(11): p. 2607-2616. https://dx.doi.org/10.3201/eid2611.203003

147. Driver, S., et al., *Effects of wearing a cloth face mask on performance, physiological and perceptual responses during a graded treadmill running exercise test.* British Journal of Sports Medicine, 2021: p. bjsports-2020-1. https://dx.doi.org/10.1136/bjsports-2020-103758

148. Epstein, D., et al., *Return to training in the COVID-19 era: The physiological effects of face masks during exercise.* Scandinavian Journal of Medicine & Science in Sports, 2021. **31**(1): p. 70-75. https://dx.doi.org/10.1111/sms.13832

149. Fikenzer, S., et al., *Effects of surgical and FFP2/N95 face masks on cardiopulmonary exercise capacity.* Clin Res Cardiol, 2020. **109**(12): p. 1522-1530. https://www.ncbi.nlm.nih.gov/pmc/articles/PMC7338098/

150. Fischer, E.P., et al., *Low-cost measurement of face mask efficacy for filtering expelled droplets during speech.* Sci Adv, 2020. **6**(36). https://dx.doi.org/10.1126/sciadv.abd3083

151. Freedman, D.O. and A. Wilder-Smith, *In-flight transmission of SARS-CoV-2: a review of the attack rates and available data on the efficacy of face masks.* Journal of Travel Medicine, 2020. **27**(8). https://dx.doi.org/10.1093/jtm/taaa178

152. Gallaway, M.S., et al., *Trends in COVID-19 Incidence After Implementation of Mitigation Measures — Arizona, January 22–August 7, 2020.* MMWR. Morbidity and Mortality Weekly Report, 2020. **69**(40): p. 1460-1463. https://dx.doi.org/10.15585/mmwr.mm6940e3

153. Gori, M., L. Schiatti, and M.B. Amadeo, *Masking Emotions: Face Masks Impair How We Read Emotions.* Frontiers in Psychology, 2021. **12**(1541). https://www.frontiersin.org/article/10.3389/fpsyg.2021.669432

154. Guy, G.P., et al., *Association of State-Issued Mask Mandates and Allowing On-Premises Restaurant Dining with County-Level COVID-19 Case and Death Growth Rates — United States, March 1–December 31, 2020.* MMWR. Morbidity and Mortality Weekly Report, 2021. **70**(10): p. 350-354. https://dx.doi.org/10.15585/mmwr.mm7010e3

155. Halbur, M., et al., *Tolerance of face coverings for children with autism spectrum disorder.* Journal of Applied Behavior Analysis, 2021. **54**(2): p. 600-617. https://dx.doi.org/10.1002/jaba.833

156. Hao, W., et al., *Filtration performances of non-medical materials as candidates for manufacturing facemasks and respirators.* Int J Hyg Environ Health, 2020. **229**: p. 113582. https://www.ncbi.nlm.nih.gov/pmc/articles/PMC7373391/

157. Hatzius J, S.D.R.I., *Goldman Sachs | Insights - Face Masks and GDP.* 2020. https://www.goldmansachs.com/insights/pages/face-masks-and-gdp.html

158. Hopkins, S.R., et al., *Face Masks and the Cardiorespiratory Response to Physical Activity in Health and Disease.* Annals of the American Thoracic Society, 2021. **18**(3): p. 399-407. https://dx.doi.org/10.1513/annalsats.202008-990cme

159. Jehn, M., *Association Between K–12 School Mask Policies and School-Associated COVID-19 Outbreaks—Maricopa and Pima Counties, Arizona, July–August 2021.* MMWR. Morbidity and Mortality Weekly Report, 2021. **70**. https://www.cdc.gov/mmwr/volumes/70/wr/mm7039e1.htm

160. Johansson, M.A., et al., *SARS-CoV-2 Transmission From People Without COVID-19 Symptoms.* JAMA Network Open, 2021. **4**(1): p. e2035057. https://dx.doi.org/10.1001/jamanetworkopen.2020.35057

161. Joo, H., et al., *Decline in COVID-19 Hospitalization Growth Rates Associated with Statewide Mask Mandates — 10 States, March–October 2020.* MMWR. Morbidity and Mortality Weekly Report, 2021. **70**(6): p. 212-216. https://dx.doi.org/10.15585/mmwr.mm7006e2

162. Karaivanov, A., et al., *Face masks, public policies and slowing the spread of COVID-19: evidence from Canada.* Journal of Health Economics, 2021: p. 102475. https://www.nber.org/system/files/working_papers/w27891/w27891.pdf

163. Konda, A., et al., *Aerosol Filtration Efficiency of Common Fabrics Used in Respiratory Cloth Masks.* ACS Nano, 2020. **14**(5): p. 6339-6347. https://pubs.acs.org/doi/10.1021/acsnano.0c03252

164. Lässing, J., et al., *Effects of surgical face masks on cardiopulmonary parameters during steady state exercise.* Sci Rep, 2020. **10**(1): p. 22363. https://www.nature.com/articles/s41598-020-78643-1

165. Leffler, C.T., et al., *Association of Country-wide Coronavirus Mortality with Demographics, Testing, Lockdowns, and Public Wearing of Masks.* Am J Trop Med Hyg, 2020. **103**(6): p. 2400-2411. https://www.ncbi.nlm.nih.gov/pmc/articles/PMC7695060/

166. Leung, N.H.L., et al., *Respiratory virus shedding in exhaled breath and efficacy of face masks.* Nature Medicine, 2020. **26**(5): p. 676-680. https://dx.doi.org/10.1038/s41591-020-0843-2

167. Lillie, M.A., et al., *Increasing passive compliance to wearing a facemask in children with autism spectrum disorder.* Journal of Applied Behavior Analysis, 2021. **54**(2): p. 582-599. https://dx.doi.org/10.1002/jaba.829

168. Lindsley, W.G., et al., *Efficacy of face masks, neck gaiters and face shields for reducing the expulsion of simulated cough-generated aerosols.* 2020, Cold Spring Harbor Laboratory. https://dx.doi.org/10.1101/2020.10.05.20207241

169. Long, K.D., et al., *Measurement of filtration efficiencies of healthcare and consumer materials using modified respirator fit tester setup.* PLOS ONE, 2020. **15**(10): p. e0240499. https://dx.doi.org/10.1371/journal.pone.0240499

170. Lubrano, R., et al., *Assessment of Respiratory Function in Infants and Young Children Wearing Face Masks During the COVID-19 Pandemic.* JAMA Network Open, 2021. **4**(3): p. e210414. https://dx.doi.org/10.1001/jamanetworkopen.2021.0414

171. Lyu, W. and G.L. Wehby, *Community Use Of Face Masks And COVID-19: Evidence From A Natural Experiment Of State Mandates In The US.* Health Affairs, 2020. **39**(8): p. 1419-1425. https://dx.doi.org/10.1377/hlthaff.2020.00818

172. Maurer, L., et al., *Community Masks During the SARS-CoV-2 Pandemic: Filtration Efficacy and Air Resistance.* Journal of Aerosol Medicine and Pulmonary Drug Delivery, 2021. **34**(1): p. 11-19. https://dx.doi.org/10.1089/jamp.2020.1635

173. Mitze, T., et al., *Face masks considerably reduce COVID-19 cases in Germany.* Proceedings of the National Academy of Sciences, 2020. **117**(51): p. 32293-32301. https://dx.doi.org/10.1073/pnas.2015954117

174. Moghadas, S.M., et al., *The implications of silent transmission for the control of COVID-19 outbreaks.* Proceedings of the National Academy of Sciences, 2020. **117**(30): p. 17513-17515. https://dx.doi.org/10.1073/pnas.2008373117

175. Montero-Vilchez, T., et al., *Skin adverse events related to personal protective equipment: a systematic review and meta-analysis.* Journal of the European Academy of Dermatology and Venereology, 2021. **35**(10): p. 1994-2006. https://dx.doi.org/10.1111/jdv.17436

176. Morawska, L., et al., *Size distribution and sites of origin of droplets expelled from the human respiratory tract during expiratory activities.* Journal of Aerosol Science, 2009. **40**(3): p. 256-269. https://dx.doi.org/10.1016/j.jaerosci.2008.11.002

177. Mueller, A.V., et al., *Quantitative Method for Comparative Assessment of Particle Removal Efficiency of Fabric Masks as Alternatives to Standard Surgical Masks for PPE.* Matter, 2020. **3**(3): p. 950-962. https://www.ncbi.nlm.nih.gov/pmc/articles/PMC7346791/

178. Nelson, S.B., et al., *Prevalence and risk factors for in-school transmission of SARS-CoV-2 in Massachusetts K-12 public schools, 2020-2021.* 2021, Cold Spring Harbor Laboratory. https://dx.doi.org/10.1101/2021.09.22.21263900

179. Park, S.J., et al., *Adverse skin reactions due to use of face masks: a prospective survey during the COVID-19 pandemic in Korea.* Journal of the European Academy of Dermatology and Venereology : JEADV, 2021. **35**(10): p. e628-e630. https://europepmc.org/articles/PMC8447351

180. Parlin, A.F., et al., *A laboratory-based study examining the properties of silk fabric to evaluate its potential as a protective barrier for personal protective equipment and as a functional material for face coverings during the COVID-19 pandemic.* PLOS ONE, 2020. **15**(9): p. e0239531. https://dx.doi.org/10.1371/journal.pone.0239531

181. Payne, D.C., et al., *SARS-CoV-2 Infections and Serologic Responses from a Sample of U.S. Navy Service Members — USS Theodore Roosevelt, April 2020*. MMWR. Morbidity and Mortality Weekly Report, 2020. **69**(23): p. 714-721. https://dx.doi.org/10.15585/mmwr.mm6923e4

182. Person, E., et al., *[Effect of a surgical mask on six minute walking distance]*. Rev Mal Respir, 2018. **35**(3): p. 264-268. https://pubmed.ncbi.nlm.nih.gov/29395560/

183. Rader, B., et al., *Mask-wearing and control of SARS-CoV-2 transmission in the USA: a cross-sectional study*. Lancet Digit Health, 2021. **3**(3): p. e148-e157. https://www.ncbi.nlm.nih.gov/pmc/articles/PMC7817421/

184. Roberge, R.J., J.H. Kim, and S.M. Benson, *Absence of consequential changes in physiological, thermal and subjective responses from wearing a surgical mask*. Respir Physiol Neurobiol, 2012. **181**(1): p. 29-35. https://www.sciencedirect.com/science/article/abs/pii/S1569904812000341?via%3Dihub

185. Rodriguez-Palacios, A., et al., *Textile Masks and Surface Covers—A Spray Simulation Method and a "Universal Droplet Reduction Model" Against Respiratory Pandemics*. Frontiers in Medicine, 2020. **7**. https://dx.doi.org/10.3389/fmed.2020.00260

186. Rothamer, D.A., et al., *Strategies to minimize SARS-CoV-2 transmission in classroom settings: Combined impacts of ventilation and mask effective filtration efficiency*. 2021, Cold Spring Harbor Laboratory. https://dx.doi.org/10.1101/2020.12.31.20249101

187. Ruba, A.L. and S.D. Pollak, *Children's emotion inferences from masked faces: Implications for social interactions during COVID-19*. PLOS ONE, 2020. **15**(12): p. e0243708. https://dx.doi.org/10.1371/journal.pone.0243708

188. Samannan, R., et al., *Effect of Face Masks on Gas Exchange in Healthy Persons and Patients with Chronic Obstructive Pulmonary Disease*. Annals of the American Thoracic Society, 2021. **18**(3): p. 541-544. https://dx.doi.org/10.1513/annalsats.202007-812rl

189. Schneider, J., et al., *The Role of Face Masks in the Recognition of Emotions by Preschool Children*. JAMA Pediatrics, 2021. https://dx.doi.org/10.1001/jamapediatrics.2021.4556

190. Schulte-Körne, B., et al., *Einfluss einer Mund-Nase-Maske auf die objektive körperliche Leistungsfähigkeit sowie das subjektive Belastungsempfinden bei gut-trainierten, gesunden*

Jungen. Wiener Medizinische Wochenschrift, 2021.
https://dx.doi.org/10.1007/s10354-021-00851-9

191. Shaw, K.A., et al., *The impact of face masks on performance and physiological outcomes during exercise: a systematic review and meta-analysis.* Applied Physiology, Nutrition, and Metabolism, 2021. **46**(7): p. 1-11. https://dx.doi.org/10.1139/apnm-2021-0143

192. Shein, S.L., et al., *The effects of wearing facemasks on oxygenation and ventilation at rest and during physical activity.* PLOS ONE, 2021. **16**(2): p. e0247414.
https://dx.doi.org/10.1371/journal.pone.0247414

193. Sickbert-Bennett, E.E., et al., *Fitted Filtration Efficiency of Double Masking During the COVID-19 Pandemic.* JAMA Internal Medicine, 2021.
https://dx.doi.org/10.1001/jamainternmed.2021.2033

194. Singh, L., A. Tan, and P.C. Quinn, *Infants recognize words spoken through opaque masks but not through clear masks.* Developmental Science, 2021. **24**(6).
https://dx.doi.org/10.1111/desc.13117

195. Sivaraman, M., J. Virues-Ortega, and H. Roeyers, *Telehealth mask wearing training for children with autism during the COVID-19 pandemic.* Journal of applied behavior analysis, 2021. **54**(1): p. 70-86. https://europepmc.org/articles/PMC7753388

196. Slimani, M., et al., *Effect of a Warm-Up Protocol with and without Facemask-Use against COVID-19 on Cognitive Function: A Pilot, Randomized Counterbalanced, Cross-Sectional Study.* International Journal of Environmental Research and Public Health, 2021. **18**(11): p. 5885. https://dx.doi.org/10.3390/ijerph18115885

197. Smith, J., A. Culler, and K. Scanlon, *Impacts of Blood Gas Concentration, Heart Rate, Emotional State, And Memory in School-Age Children with And without The Use of Facial Coverings in School during The COVID-19 Pandemic.* The FASEB Journal, 2021. **35**.
https://www.ncbi.nlm.nih.gov/pmc/articles/PMC8239941/

198. Techasatian, L., et al., *The Effects of the Face Mask on the Skin Underneath: A Prospective Survey During the COVID-19 Pandemic.* Journal of Primary Care & Community Health, 2020. **11**: p. 215013272096616. https://dx.doi.org/10.1177/2150132720966167

199. Ueki, H., et al., *Effectiveness of Face Masks in Preventing Airborne Transmission of SARS-CoV-2.* mSphere, 2020. **5**(5). https://www.ncbi.nlm.nih.gov/pmc/articles/PMC7580955/

200. Van Der Sande, M., P. Teunis, and R. Sabel, *Professional and Home-Made Face Masks Reduce Exposure to Respiratory Infections among the General Population.* PLoS ONE, 2008. **3**(7): p. e2618. https://dx.doi.org/10.1371/journal.pone.0002618

201. Verma, S., M. Dhanak, and J. Frankenfield, *Visualizing the effectiveness of face masks in obstructing respiratory jets.* Physics of Fluids, 2020. **32**(6): p. 061708. https://dx.doi.org/10.1063/5.0016018

202. Viola, I.M., et al., *Face Coverings, Aerosol Dispersion and Mitigation of Virus Transmission Risk.* IEEE Open Journal of Engineering in Medicine and Biology, 2021. **2**: p. 26-35. https://dx.doi.org/10.1109/ojemb.2021.3053215

203. Wang, Y., et al., *Reduction of secondary transmission of SARS-CoV-2 in households by face mask use, disinfection and social distancing: a cohort study in Beijing, China.* BMJ Global Health, 2020. **5**(5): p. e002794. https://dx.doi.org/10.1136/bmjgh-2020-002794

204. Whiley, H., et al., *Viral Filtration Efficiency of Fabric Masks Compared with Surgical and N95 Masks.* Pathogens, 2020. **9**(9): p. 762. https://dx.doi.org/10.3390/pathogens9090762

205. Miller, I., *Unmasked: The Global Failure of COVID Mask Mandates.* 2022: Post Hill Press

206. WHO, *Global excess deaths associated with COVID-19 (modelled estimates).* 2023, World Health Organization, updated May 19, 2023. https://www.who.int/data/sets/global-excess-deaths-associated-with-covid-19-modelled-estimates

207. Scott W. Atlas, M.D., *A Plague on Our House: My Fight at the Trump White House to Stop COVID from Destroying America.* 2021, Kindle edition: Bombardier Books

208. Alihsan, B., et al., *The Efficacy of Facemasks in the Prevention of COVID-19: A Systematic Review.* 2022, Cold Spring Harbor Laboratory. https://dx.doi.org/10.1101/2022.07.28.22278153

209. Chaovavanich, A., et al., *Early containment of severe acute respiratory syndrome (SARS); experience from Bamrasnaradura Institute, Thailand.* Journal of the Medical Association of Thailand = Chotmaihet thangphaet, 2004. **87**(10): p. 1182-1187. http://europepmc.org/abstract/MED/15560695

210. Guo, X., et al., *Survey of COVID-19 Disease Among Orthopaedic Surgeons in Wuhan, People's Republic of China.* J Bone Joint Surg Am, 2020. **102**(10): p. 847-854. https://pubmed.ncbi.nlm.nih.gov/32271208/

211. Çelebi, G., et al., *Specific risk factors for SARS-CoV-2 transmission among health care workers in a university hospital.* American Journal of Infection Control, 2020. **48**(10): p. 1225-1230. https://dx.doi.org/10.1016/j.ajic.2020.07.039

212. Wang, Y., et al., *Super-factors associated with transmission of occupational COVID-19 infection among healthcare staff in Wuhan, China.* Journal of Hospital Infection, 2020. **106**(1): p. 25-34. https://dx.doi.org/10.1016/j.jhin.2020.06.023

213. Rolland, Y., et al., *Guidance for the Prevention of the COVID-19 Epidemic in Long-Term Care Facilities: A Short-Term Prospective Study.* The journal of nutrition, health & aging, 2020. **24**(8): p. 812-816. https://dx.doi.org/10.1007/s12603-020-1440-2

214. Nir-Paz, R., et al., *Absence of in-flight transmission of SARS-CoV-2 likely due to use of face masks on board.* Journal of Travel Medicine, 2020. **27**(8). https://dx.doi.org/10.1093/jtm/taaa117

215. Caruhel, J.-B., et al., *Military gas mask to protect surgeons when performing tracheotomies on patients with COVID-19.* BMJ Military Health, 2021. **167**(3): p. 214-214. https://dx.doi.org/10.1136/bmjmilitary-2020-001547

216. Kang, Y.-J., *Lessons Learned From Cases of COVID-19 Infection in South Korea.* Disaster Medicine and Public Health Preparedness, 2020. **14**(6): p. 818-825. https://dx.doi.org/10.1017/dmp.2020.141

217. Canova, V., et al., *Transmission risk of SARS-CoV-2 to healthcare workers -observational results of a primary care hospital contact tracing.* Swiss Med Wkly, 2020. **150**: p. w20257. https://smw.ch/article/doi/smw.2020.20257

218. Chang, M.C., J. Hur, and D. Park, *Strategies for the Prevention of the Intra-Hospital Transmission of COVID-19: A Retrospective Cohort Study.* Healthcare, 2020. **8**(3): p. 195. https://dx.doi.org/10.3390/healthcare8030195

219. Xu, C., et al., *Application of refined management in prevention and control of the coronavirus disease 2019 epidemic in non-isolated areas of a general hospital.* Int J Nurs Sci, 2020. **7**(2): p. 143-147. https://www.ncbi.nlm.nih.gov/pmc/articles/PMC7141553/pdf/main.pdf

220. Cowger, T.L., et al., *Lifting Universal Masking in Schools — Covid-19 Incidence among Students and Staff.* New England Journal of Medicine, 2022. https://dx.doi.org/10.1056/nejmoa2211029

221. Høeg, T.B., A. Haslam, and V. Prasad, *An analysis of studies pertaining to masks in Morbidity and Mortality Weekly Report: Characteristics and quality of all studies from 1978 to 2023.* 2023, Cold Spring Harbor Laboratory. https://dx.doi.org/10.1101/2023.07.07.23292338

222. Blachere, F.M., et al., *Assessment of influenza virus exposure and recovery from contaminated surgical masks and N95 respirators.* Journal of Virological Methods, 2018. **260**: p. 98-106. https://dx.doi.org/10.1016/j.jviromet.2018.05.009

223. Chao, F.-L., *Adolescents' face mask usage and contact transmission in novel Coronavirus.* Journal of Public Health Research, 2020. **9**(1). https://pubmed.ncbi.nlm.nih.gov/32582579/

224. Chughtai, A.A., et al., *Contamination by respiratory viruses on outer surface of medical masks used by hospital healthcare workers.* BMC Infectious Diseases, 2019. **19**(1). https://dx.doi.org/10.1186/s12879-019-4109-x

225. Fisher, E.M., et al., *Validation and Application of Models to Predict Facemask Influenza Contamination in Healthcare Settings.* Risk Analysis, 2014. **34**(8): p. 1423-1434. https://dx.doi.org/10.1111/risa.12185

226. Gund, M., et al., *Contamination of surgical mask during aerosol-producing dental treatments.* Clin Oral Investig, 2020: p. 1-8. https://europepmc.org/article/PMC/PMC7590255

227. Luckman, A., et al., *Risk Compensation during COVID-19: The Impact of Face Mask Usage on Social Distancing.* OSF Preprints, 2020. https://doi.org/10.31219/osf.io/rb8he

228. Luksamijarulkul, P., N. Aiempradit, and P. Vatanasomboon, *Microbial Contamination on Used Surgical Masks among Hospital Personnel and Microbial Air Quality in their Working Wards: A Hospital in Bangkok.* Oman Medical Journal, 2014. **29**(5): p. 5. https://www.omjournal.org/articleDetails.aspx?coType=1&aId=564

229. Monalisa, A.C., et al., *Microbial contamination of the mouth masks used by post-graduate students in a private dental institution: An In-Vitro Study.* IOSR J Dent Med Sci, 2017. **16**: p. 61-7. https://www.ncbi.nlm.nih.gov/pmc/articles/PMC7269059/

230. Scarano, A., F. Inchingolo, and F. Lorusso, *Facial Skin Temperature and Discomfort When Wearing Protective Face Masks: Thermal Infrared Imaging Evaluation and Hands Moving the Mask.* International Journal of Environmental Research and Public Health, 2020. **17**(13): p. 4624. https://dx.doi.org/10.3390/ijerph17134624

231. Yan, Y., et al., *Risk compensation and face mask mandates during the COVID-19 pandemic.* Scientific Reports, 2021. **11**(1). https://dx.doi.org/10.1038/s41598-021-82574-w

232. Zhiqing, L., et al., *Surgical masks as source of bacterial contamination during operative procedures.* J Orthop Translat, 2018. **14**: p. 57-62. https://www.ncbi.nlm.nih.gov/pmc/articles/PMC6037910/

233. Keri, V.C., et al., *Pilot study on burden of fungal contamination in face masks: need for better mask hygiene in the COVID-19 era.* Le Infezioni in Medicina, 2021. **36**: p. 25-49. https://www.researchgate.net/profile/Vishakh-Keri/publication/357221512_Pilot_study_on_burden_of_fungal_contamination_in_face_masks_need_for_better_mask_hygiene_in_the_COVID-19_era/links/61c1ebd0abfb4634cb3363c9/Pilot-study-on-burden-of-fungal-contamination-in-face-masks-need-for-better-mask-hygiene-in-the-COVID-19-era.pdf

234. Schnirman, R., et al., *A case of legionella pneumonia caused by home use of continuous positive airway pressure.* SAGE Open Medical Case Reports, 2017. **5**: p. 2050313X1774498. https://dx.doi.org/10.1177/2050313x17744981

235. Hua, W., et al., *Short-term skin reactions following use of N95 respirators and medical masks.* Contact Dermatitis, 2020. **83**(2): p. 115-121. https://dx.doi.org/10.1111/cod.13601

236. Li, Y., et al., *Effects of wearing N95 and surgical facemasks on heart rate, thermal stress and subjective sensations.* International Archives of Occupational and Environmental Health, 2005. **78**(6): p. 501-509. https://dx.doi.org/10.1007/s00420-004-0584-4

237. Liu, C., et al., *Effects of wearing masks on human health and comfort during the COVID-19 pandemic.* IOP Conference Series: Earth and Environmental Science, 2020. **531**: p. 012034. http://dx.doi.org/10.1088/1755-1315/531/1/012034

238. Matusiak, Ł., et al., *Inconveniences due to the use of face masks during the COVID-19 pandemic: A survey study of 876 young people.* Dermatologic Therapy, 2020. **33**(4). https://dx.doi.org/10.1111/dth.13567

239. Kisielinski, K., et al., *Possible toxicity of chronic carbon dioxide exposure associated with face mask use, particularly in pregnant women, children and adolescents – A scoping review.* Heliyon, 2023. **9**(4): p. e14117. https://dx.doi.org/10.1016/j.heliyon.2023.e14117

240. Klimek, L., et al., *A new form of irritant rhinitis to filtering facepiece particle (FFP) masks (FFP2/N95/KN95 respirators) during COVID-19 pandemic.* World Allergy Organization Journal, 2020. **13**(10): p. 100474. https://dx.doi.org/10.1016/j.waojou.2020.100474

241. Heider, C.A., et al., *Prevalence of Voice Disorders in Healthcare Workers in the Universal Masking COVID -19 Era.* The Laryngoscope, 2021. **131**(4). https://dx.doi.org/10.1002/lary.29172

242. Georgi, C., et al., *The Impact of Commonly-Worn Face Masks on Physiological Parameters and on Discomfort During Standard Work-Related Physical Effort.* Deutsches Arzteblatt international, 2020. **117**(40): p. 674-675. https://www.ncbi.nlm.nih.gov/pmc/articles/PMC7838380/

243. Kao, T.W., et al., *The physiological impact of wearing an N95 mask during hemodialysis as a precaution against SARS in patients with end-stage renal disease.* J Formos Med Assoc, 2004. **103**(8): p. 624-8. https://www.researchgate.net/publication/8371248_The_physiological_impact_of_wearing_an_N95_mask_during_hemodialysis_as_a_precaution_against_SARS_in_patients_with_end-stage_renal_disease

244. Kisielinski, K., et al., *Is a Mask That Covers the Mouth and Nose Free from Undesirable Side Effects in Everyday Use and Free of Potential Hazards?* International Journal of Environmental Research and Public Health, 2021. **18**(8): p. 4344. https://dx.doi.org/10.3390/ijerph18084344

245. Kyung, S.Y., et al., *Risks of N95 Face Mask Use in Subjects With COPD.* Respiratory Care, 2020. **65**(5): p. 658-664. https://dx.doi.org/10.4187/respcare.06713

246. Roberge, R.J., J.-H. Kim, and J.B. Powell, *N95 respirator use during advanced pregnancy.* American Journal of Infection Control, 2014. **42**(10): p. 1097-1100. https://dx.doi.org/10.1016/j.ajic.2014.06.025

247. Passali, D., et al., *Effects of a Mask on Breathing Impairment During a Fencing Assault: A Case Series Study.* Asian Journal of Sports Medicine, 2015. **6**(3). https://dx.doi.org/10.5812/asjsm.23643

248. Johnson, A.T., *Respirator masks protect health but impact performance: a review.* Journal of Biological Engineering, 2016. **10**(1). https://dx.doi.org/10.1186/s13036-016-0025-4

249. Zimmerman, N.J., et al., *Effects of respirators on performance of physical, psychomotor and cognitive tasks.* Ergonomics, 1991. **34**(3): p. 321-334. https://doi.org/10.1080/00140139108967316

250. Mapelli, M., et al., *"You can leave your mask on": effects on cardiopulmonary parameters of different airway protection masks at rest and during maximal exercise.* Eur Respir J, 2021 https://erj.ersjournals.com/content/58/3/2004473.long

251. Lee, H.P. and D.Y. Wang, *Objective Assessment of Increase in Breathing Resistance of N95 Respirators on Human Subjects.* The Annals of Occupational Hygiene, 2011. **55**(8): p. 917-921. https://doi.org/10.1093/annhyg/mer065

252. Fögen, Z., *The Foegen effect: A mechanism by which facemasks contribute to the COVID-19 case fatality rate.* Medicine, 2022. **101**(7): p. e28924. https://journals.lww.com/md-journal/Fulltext/2022/02180/The_Foegen_effect__A_mechanism_by_which_facemasks.60.aspx

253. Chamberlain, G.V. and E. Houang, *Trial of the use of masks in the gynaecological operating theatre.* Annals of the Royal College of Surgeons of England, 1984. **66**(6): p. 432-3. https://www.ncbi.nlm.nih.gov/pmc/articles/PMC2494468/

254. Bhatia, R., et al., *Iatrogenic dermatitis in times of COVID-19: a pandemic within a pandemic.* J Eur Acad Dermatol Venereol, 2020. **34**(10): p. e563-e566. https://pubmed.ncbi.nlm.nih.gov/32495393/

255. Bothra, A., et al., *Retroauricular dermatitis with vehement use of ear loop face masks during COVID-19 pandemic.* Journal of the European Academy of Dermatology and Venereology, 2020. **34**(10): p. e549-e552. https://onlinelibrary.wiley.com/doi/full/10.1111/jdv.16692

256. Foo, C.C.I., et al., *Adverse skin reactions to personal protective equipment against severe acute respiratory syndrome - a descriptive study in Singapore.* Contact Dermatitis, 2006. **55**(5): p. 291-294. https://dx.doi.org/10.1111/j.1600-0536.2006.00953.x

257. Jacek C. Szepietowski, Ł.M.M.S.P.K.K. and B.-B. Rafał, *Face Mask-induced Itch: A Self-questionnaire Study of 2,315 Responders During the COVID-19 Pandemic | Abstract | Acta Dermato-Venereologica.* Acta DermatoVenereologica, 2020. **100**. https://www.medicaljournals.se/acta/content/abstract/10.2340/00015555-3536

258. Lan, J., et al., *Skin damage among health care workers managing coronavirus disease-2019.* Journal of the American Academy of Dermatology, 2020. **82**(5): p. 1215-1216. https://dx.doi.org/10.1016/j.jaad.2020.03.014

259. Radonovich, L.J., et al., *Respirator Tolerance in Health Care Workers.* JAMA, 2009. **301**(1): p. 36. https://dx.doi.org/10.1001/jama.2008.894

260. Rosner, E., *Adverse Effects of Prolonged Mask Use among Healthcare Professionals during COVID-19.* Journal of Infectious Disease and Epidemiology, 2021. **6**(3): p. 130. https://clinmedjournals.org/articles/jide/journal-of-infectious-diseases-and-epidemiology-jide-6-130.php?jid=jide

261. Schwarz, S., et al., *Corona children studies "Co-Ki": First results of a Germany-wide registry on mouth and nose covering (mask) in children.* Research Square, 2021. https://doi.org/10.21203/rs.3.rs-124394/v2

262. Szepietowski, J.C., et al., *Face Mask-induced Itch: A Self-questionnaire Study of 2,315 Responders During the COVID-19 Pandemic.* Acta Derm Venereol, 2020. **100**(10): p. adv00152. https://www.medicaljournals.se/acta/content/abstract/10.2340/00015555-3536

263. Veraldi, S., L. Angileri, and M. Barbareschi, *Seborrheic dermatitis and anti-COVID-19 masks.* Journal of Cosmetic Dermatology, 2020. **19**(10): p. 2464-2465. https://onlinelibrary.wiley.com/doi/10.1111/jocd.13669

264. Byambasuren, O., et al., *Estimating the extent of asymptomatic COVID-19 and its potential for community transmission: systematic review and meta-analysis.* 2020, Cold Spring Harbor Laboratory. https://dx.doi.org/10.1101/2020.05.10.20097543

265. Sah, P., et al., *Asymptomatic SARS-CoV-2 infection: A systematic review and meta-analysis.* Proceedings of the National Academy of Sciences, 2021. **118**(34): p. e2109229118. https://dx.doi.org/10.1073/pnas.2109229118

266. Roussel, Y., et al., *SARS-CoV-2: fear versus data.* Int J Antimicrob Agents, 2020. **55**(5): p. 105947. https://www.ncbi.nlm.nih.gov/pmc/articles/PMC7102597/pdf/main.pdf

267. Ioannidis, J.P.A., *Reconciling estimates of global spread and infection fatality rates of COVID-19: An overview of systematic evaluations.* European Journal of Clinical Investigation, 2021. https://dx.doi.org/10.1111/eci.13554

268. Pezzullo, A.M., et al., *Age-stratified infection fatality rate of COVID-19 in the non-elderly informed from pre-vaccination national seroprevalence studies.* 2022, Cold Spring Harbor Laboratory. https://dx.doi.org/10.1101/2022.10.11.22280963

269. Clarke, K.E.N., et al., *Seroprevalence of Infection-Induced SARS-CoV-2 Antibodies — United States, September 2021–February 2022.* MMWR. Morbidity and Mortality Weekly Report, 2022. **71**(17): p. 606-608. https://dx.doi.org/10.15585/mmwr.mm7117e3

270. Beder, A., et al., *Preliminary report on surgical mask induced deoxygenation during major surgery.* Neurocirugía, 2008. **19**(2): p. 121-126. https://dx.doi.org/10.1016/s1130-1473(08)70235-5

271. Bharatendu, C., et al., *Powered Air Purifying Respirator (PAPR) restores the N95 face mask induced cerebral hemodynamic alterations among Healthcare Workers during COVID-19 Outbreak.* Journal of the Neurological Sciences, 2020. **417**: p. 117078. https://dx.doi.org/10.1016/j.jns.2020.117078

272. Butz, U., *Rückatmung von Kohlendioxid bei Verwendung von Operationsmasken als hygienischer Mundschutz an medizinischem Fachpersonal [Rebreathing of carbon dioxide of surgical staff using hygienic masks],* in *Fakultät für Medizin der Technischen Universität München, Munich, Germany, 2005.* 2005, Institute for Anaesthesiology at the Technical University of Munich: Universitätsbibliothek der Technischen Universität München p. 55. https://mediatum.ub.tum.de/doc/602557/602557.pdf

273. Chandrasekaran, B. and S. Fernandes, *"Exercise with facemask; Are we handling a devil's sword?" - A physiological hypothesis.* Med Hypotheses, 2020. **144**: p. 110002. https://www.ncbi.nlm.nih.gov/pmc/articles/PMC7306735/

274. Goh, D.Y.T., et al., *A randomised clinical trial to evaluate the safety, fit, comfort of a novel N95 mask in children.* Scientific Reports, 2019. **9**(1). https://dx.doi.org/10.1038/s41598-019-55451-w

275. Jagim, A.R., et al., *Acute Effects of the Elevation Training Mask on Strength Performance in Recreational Weight lifters.* The Journal of Strength & Conditioning Research, 2018. **32**(2). https://journals.lww.com/nsca-jscr/Fulltext/2018/02000/Acute_Effects_of_the_Elevation_Training_Mask_on.22.aspx

276.   Kisielinski, K., et al., *Physio-metabolic and clinical consequences of wearing face masks-Systematic review with meta-analysis and comprehensive evaluation.* Front Public Health, 2023. **11**: p. 1125150. https://www.ncbi.nlm.nih.gov/pmc/articles/PMC10116418/

277.   Köseoğlu Toksoy, C., et al., *Headache related to mask use of healthcare workers in COVID-19 pandemic.* The Korean Journal of Pain, 2021. **34**(2): p. 241-245. https://dx.doi.org/10.3344/kjp.2021.34.2.241

278.   Martellucci, C.A., et al., *Inhaled CO2 concentration while wearing face masks: a pilot study using capnography.* medRxiv, 2022: p. 2022.05.10.22274813. https://www.medrxiv.org/content/medrxiv/early/2022/05/11/2022.05.10.22274813.full.pdf

279.   Mo, Y., et al., *Risk and impact of using mask on COPD patients with acute exacerbation during the COVID-19 outbreak: a retrospective study.* 2020, Research Square. https://dx.doi.org/10.21203/rs.3.rs-39747/v1

280.   Pifarré, F., et al., *COVID-19 and mask in sports.* Apunts Sports Medicine, 2020. **55**(208): p. 143-145. https://www.sciencedirect.com/science/article/pii/S2666506920300250

281.   Porcari, J.P., et al., *Effect of Wearing the Elevation Training Mask on Aerobic Capacity, Lung Function, and Hematological Variables.* Journal of sports science & medicine, 2016. **15**(2): p. 379-386. https://pubmed.ncbi.nlm.nih.gov/27274679 https://www.ncbi.nlm.nih.gov/pmc/articles/PMC4879455/

282.   Rebmann, T., R. Carrico, and J. Wang, *Physiologic and other effects and compliance with long-term respirator use among medical intensive care unit nurses.* American Journal of Infection Control, 2013. **41**(12): p. 1218-1223. https://dx.doi.org/10.1016/j.ajic.2013.02.017

283.   Roberge, R.J., et al., *Physiological impact of the N95 filtering facepiece respirator on healthcare workers.* Respir Care, 2010. **55**(5): p. 569-77. http://rc.rcjournal.com/content/55/5/569.short

284.   Scarano, A., et al., *Protective Face Masks: Effect on the Oxygenation and Heart Rate Status of Oral Surgeons during Surgery.* Int J Environ Res Public Health, 2021. **18**(5). https://pubmed.ncbi.nlm.nih.gov/33670983/

285. Smart, N.R., et al., *Assessment of the Wearability of Facemasks against Air Pollution in Primary School-Aged Children in London.* International Journal of Environmental Research and Public Health, 2020. **17**(11): p. 3935. https://dx.doi.org/10.3390/ijerph17113935

286. Smith, C.L., J.L. Whitelaw, and B. Davies, *Carbon dioxide rebreathing in respiratory protective devices: influence of speech and work rate in full-face masks.* Ergonomics, 2013. **56**(5): p. 781-790. https://dx.doi.org/10.1080/00140139.2013.777128

287. Tong, P.S.Y., et al., *Respiratory consequences of N95-type Mask usage in pregnant healthcare workers—a controlled clinical study.* Antimicrobial Resistance & Infection Control, 2015. **4**(1). https://dx.doi.org/10.1186/s13756-015-0086-z

288. Walach, H., et al., *Experimental Assessment of Carbon Dioxide Content in Inhaled Air With or Without Face Masks in Healthy Children.* JAMA Pediatrics, 2021. https://dx.doi.org/10.1001/jamapediatrics.2021.2659

289. Zhu, J., et al., *Evaluation of rebreathed air in human nasal cavity with N95 respirator: a CFD study.* Trauma Emerg. Care, 2016. **1**(2): p. 15-18. https://www.oatext.com/pdf/TEC-1-106.pdf

290. Walach, H., et al., *Carbon dioxide rises beyond acceptable safety levels in children under nose and mouth covering: Results of an experimental measurement study in healthy children.* Environmental Research, 2022. **212**: p. 113564. https://www.sciencedirect.com/science/article/pii/S001393512200891X

291. Zhang, G., et al., *Effect of Surgical Masks on Cardiopulmonary Function in Healthy Young Subjects: A Crossover Study.* Frontiers in Physiology, 2021. **12**. https://www.frontiersin.org/articles/10.3389/fphys.2021.710573

292. Sukul, P., et al., *Effects of COVID-19 protective face masks and wearing durations on respiratory haemodynamic physiology and exhaled breath constituents.* European Respiratory Journal, 2022. **60**(3): p. 2200009. https://dx.doi.org/10.1183/13993003.00009-2022

293. Toprak, E. and A.N. Bulut, *The effect of mask use on maternal oxygen saturation in term pregnancies during the COVID-19 process.* Journal of Perinatal Medicine, 2021. **49**(2): p. 148-152. https://www.researchgate.net/publication/347948092_The_effect_of_mask_use_on_maternal_oxygen_saturation_in_term_pregnancies_during_the_COVID-19_process https://doi.org/10.1515/jpm-2020-0422

294. Hirai, K., et al., *Effect of surgical mask on exercise capacity in COPD: a randomised crossover trial.* European Respiratory Journal, 2021. **58**(4): p. 2102041. https://dx.doi.org/10.1183/13993003.02041-2021

295. Dirol, H., et al., *The physiological and disturbing effects of surgical face masks in the COVID-19 era.* Bratisl Lek Listy, 2021. **122**(11): p. 821-825. https://doi.org/10.4149/BLL_2021_131

296. Ramírez-Moreno, J.M., et al., *Mask-associated de novo headache in healthcare workers during the Covid-19 pandemic.* 2020, Cold Spring Harbor Laboratory. https://dx.doi.org/10.1101/2020.08.07.20167957

297. Kim, J.-H., S.M. Benson, and R.J. Roberge, *Pulmonary and heart rate responses to wearing N95 filtering facepiece respirators.* American journal of infection control, 2013. **41**(1): p. 24-27. https://www.sciencedirect.com/science/article/abs/pii/S0196655312007171

298. Carbon, C.-C., *Wearing face masks strongly confuses counterparts in reading emotions.* Frontiers in Psychology, 2020. **11**: p. 2526. https://www.ncbi.nlm.nih.gov/pmc/articles/PMC7545827/

299. Cheok, G.J.W., et al., *Appropriate attitude promotes mask wearing in spite of a significant experience of varying discomfort.* Infect Dis Health, 2021. https://www.idhjournal.com.au/article/S2468-0451(21)00002-X/fulltext

300. Mheidly, N., et al., *Effect of Face Masks on Interpersonal Communication During the COVID-19 Pandemic.* Frontiers in Public Health, 2020. **8**. https://dx.doi.org/10.3389/fpubh.2020.582191

301. Morgan, W.P., *Psychological Problems Associated with the Wearing of Industrial Respirators: A Review.* American Industrial Hygiene Association Journal, 1983. **44**(9): p. 671-676. https://doi.org/10.1080/15298668391405544

302. Prousa, D., *Studie zu psychischen und psychovegetativen Beschwerden mit den aktuellen Mund-Nasenschutz-Verordnungen [Study of psychological and psycho-vegetative complaints by the current mouth and nose protection regulations in Germany].* 2020. https://www.psycharchives.org/handle/20.500.12034/2751

303. Scheid, J.L., et al., *Commentary: Physiological and Psychological Impact of Face Mask Usage during the COVID-19 Pandemic.* Int J Environ Res Public Health, 2020. **17**(18). https://pubmed.ncbi.nlm.nih.gov/32932652/

304. Wong, C.K.M., et al., *Effect of facemasks on empathy and relational continuity: a randomised controlled trial in primary care.* BMC Family Practice, 2013. **14**(1): p. 200. https://dx.doi.org/10.1186/1471-2296-14-200

305. Little, A.C., B.C. Jones, and L.M. Debruine, *Facial attractiveness: evolutionary based research.* Philosophical Transactions of the Royal Society B: Biological Sciences, 2011. **366**(1571): p. 1638-1659. https://dx.doi.org/10.1098/rstb.2010.0404

306. Hajjij, A., et al., *Personal Protective Equipment and Headaches: Cross-Sectional Study Among Moroccan Healthcare Workers During COVID-19 Pandemic.* Cureus, 2020. **12**(12): p. e12047. https://pdfs.semanticscholar.org/f37c/09eb6873f6dc13c3fdf4f9308da09107c3a4.pdf

307. Lim, E.C.H., et al., *Headaches and the N95 face-mask amongst healthcare providers.* Acta Neurologica Scandinavica, 2006. **113**(3): p. 199-202. https://dx.doi.org/10.1111/j.1600-0404.2005.00560.x

308. Ong, J.J.Y., et al., *Headaches Associated With Personal Protective Equipment - A Cross-Sectional Study Among Frontline Healthcare Workers During COVID-19.* Headache, 2020. **60**(5): p. 864-877. https://pubmed.ncbi.nlm.nih.gov/32232837/

309. Zaheer, R., et al., *Association of Personal Protective Equipment with De Novo Headaches in Frontline Healthcare Workers during COVID-19 Pandemic: A Cross-Sectional Study.* Eur J Dent, 2020. **14**(S 01): p. S79-s85. https://www.researchgate.net/publication/348001702_Association_of_Personal_Protective_Equipment_with_De_Novo_Headaches_in_Frontline_Healthcare_Workers_during_COVID-19_Pandemic_A_Cross-Sectional_Study

310. Lin, P., et al., *Adverse skin reactions among healthcare workers during the coronavirus disease 2019 outbreak: a survey in Wuhan and its surrounding regions.* British Journal of Dermatology, 2020. **183**(1): p. 190-192. https://doi.org/10.1111/bjd.19089

311. Zuo, Y., et al., *Skin reactions of N95 masks and medical masks among health-care personnel: A self-report questionnaire survey in China.* Contact Dermatitis, 2020. **83**(2): p. 145-147. https://dx.doi.org/10.1111/cod.13555

312. Shenal, B.V., et al., *Discomfort and Exertion Associated with Prolonged Wear of Respiratory Protection in a Health Care Setting.* Journal of Occupational and Environmental Hygiene, 2012. **9**(1): p. 59-64. https://dx.doi.org/10.1080/15459624.2012.635133

313. Saunders, G.H., I.R. Jackson, and A.S. Visram, *Impacts of face coverings on communication: an indirect impact of COVID-19.* International Journal of Audiology, 2021. **60**(7): p. 495-506. https://www.researchgate.net/publication/346715295_Impacts_of_face_coverings_on_communication_an_indirect_impact_of_COVID-19_View_supplementary_material_Impacts_of_face_coverings_on_communication_an_indirect_impact_of_COVID-19

314. Janicijevic, D., et al., *Intraocular pressure responses to walking with surgical and FFP2/N95 face masks in primary open-angle glaucoma patients.* Graefes Arch Clin Exp Ophthalmol, 2021: p. 1-6. https://www.ncbi.nlm.nih.gov/pmc/articles/PMC8023773/

315. Moshirfar, M., W.B. West, and D.P. Marx, *Face Mask-Associated Ocular Irritation and Dryness.* Ophthalmology and Therapy, 2020. **9**(3): p. 397-400. https://dx.doi.org/10.1007/s40123-020-00282-6

316. Mukerji, S., et al., *Cost-effectiveness analysis of N95 respirators and medical masks to protect healthcare workers in China from respiratory infections.* BMC Infectious Diseases, 2017. **17**(1). https://dx.doi.org/10.1186/s12879-017-2564-9

317. Jenner, L.C., et al., *Detection of microplastics in human lung tissue using µFTIR spectroscopy.* Science of The Total Environment, 2022. **831**: p. 154907. https://dx.doi.org/10.1016/j.scitotenv.2022.154907

318. Rathinamoorthy, R. and S. Raja Balasaraswathi, *Mitigation of microfibers release from disposable masks - An analysis of structural properties.* Environ Res, 2022. **214**(Pt 4): p. 114106. https://doi.org/10.1016/j.envres.2022.114106

319. Selvaranjan, K., et al., *Environmental challenges induced by extensive use of face masks during COVID-19: A review and potential solutions.* Environmental Challenges, 2021. **3**: p. 100039. https://www.sciencedirect.com/science/article/pii/S2667010021000184

320. Fikenzer, S. and U. Laufs, *Response to Letter to the editors referring to Fikenzer, S., Uhe, T., Lavall, D., Rudolph, U., Falz, R., Busse, M., Hepp, P., & Laufs, U. (2020). Effects of surgical and FFP2/N95 face masks on cardiopulmonary exercise capacity. Clinical research in cardiology: official journal of the German Cardiac Society, 1-9. Advance online publication.* Clin Res

Cardiol, 2020. **109**(12): p. 1600
https://doi.org/10.1007/s00392-020-01704-y.

321. Fikenzer, S. and U. Laufs, *Response to Letter to the editors of Hopkins et al.: Effects of surgical and FFP2/N95 face masks on cardiopulmonary exercise capacity: the numbers do not add up.* Clin Res Cardiol, 2020. **109**(12): p. 1607. https://doi.org/10.1007/s00392-020-01749-z

322. Langford, N.J., *Carbon Dioxide Poisoning.* Toxicological Reviews, 2005. **24**(4): p. 229-235. https://doi.org/10.2165/00139709-200524040-00003

323. Glatte, H.A., Jr. and B.E. Welch, *Carbon Dioxide Tolerance: A Review.* 1967, U.S. Department of Commerce, National Technical Information Service: School of Aerospace Medicine, Brooks Air Force Base, Texas. https://apps.dtic.mil/sti/citations/ADA017159

324. Consolazio, W.V., et al., *Effects on Man of High Concentrations of Carbon Dioxide in Relation to Various Oxygen Pressures During Exposures as Long as 72 Hours.* American Journal of Physiology-Legacy Content, 1947. **151**(2): p. 479-503. https://journals.physiology.org/doi/abs/10.1152/ajplegacy.1947.151.2.479

325. Schaefer, K.E., et al., *Respiratory acclimatization to carbon dioxide.* Journal of Applied Physiology, 1963. **18**(6): p. 1071-1078. https://journals.physiology.org/doi/abs/10.1152/jappl.1963.18.6.1071

326. Permentier, K., et al., *Carbon dioxide poisoning: a literature review of an often forgotten cause of intoxication in the emergency department.* International Journal of Emergency Medicine, 2017. **10**(1). https://dx.doi.org/10.1186/s12245-017-0142-y

327. Gorman, J.M., et al., *Ventilatory physiology of patients with panic disorder.* Arch Gen Psychiatry, 1988. **45**(1): p. 31-9. https://jamanetwork.com/journals/jamapsychiatry/article-abstract/494205

328. Gorman, J.M., et al., *Physiological Changes During Carbon Dioxide Inhalation in Patients With Panic Disorder, Major Depression, and Premenstrual Dysphoric Disorder.* Archives of General Psychiatry, 2001. **58**(2): p. 125. https://dx.doi.org/10.1001/archpsyc.58.2.125

329. Pine, D.S., et al., *Ventilatory Physiology of Children and Adolescents With Anxiety Disorders.* Archives of General Psychiatry, 1998. **55**(2): p. 123. https://dx.doi.org/10.1001/archpsyc.55.2.123

330. CDC, *NIOSH pocket guide to chemical hazards - carbon dioxide*. Last accessed July 21, 2023: Online. https://www.cdc.gov/niosh/npg/npgd0103.html

331. James, J.T., *Spacecraft Maximum Allowable Concentrations for Selected Airborne Contaminants: Volume 5 "Chapter 7 - Carbon Dioxide"*. 2008, National Aeronautics and Space Administration (NASA). https://nap.nationalacademies.org/read/12529/chapter/10

332. Howard, W.R., Wong, B., Okolic, M., Bynum, K.S., James, R.A. , *The Prenatal Development Effects of Carbon Dioxide (CO2) Exposure in Rats (Rattus Norvegicus)*. Defense Technical Information Center, Fort Belvoir, VA, 2012. https://apps.dtic.mil/sti/pdfs/ADA583166.pdf

333. Kiray, M., et al., *Effects of carbon dioxide exposure on early brain development in rats*. Biotechnic & Histochemistry, 2014. **89**(5): p. 371-383. https://doi.org/10.3109/10520295.2013.872298 Published version not available without payment. Preprint version: https://www.researchgate.net/publication/259984078_Effects_of_carbon_dioxide_exposure_on_early_brain_development_in_rats

334. Uysal, N., et al., *Effects of exercise and poor indoor air quality on learning, memory and blood IGF-1 in adolescent mice*. Biotechnic & Histochemistry, 2014. **89**(2): p. 126-135. https://doi.org/10.3109/10520295.2013.825318 Published version not available without payment. Preprint version: https://www.researchgate.net/publication/256835687_Effects_of_exercise_and_poor_indoor_air_quality_on_learning_memory_and_blood_IGF-1_in_adolescent_mice

335. Allen, J.G., et al., *Associations of Cognitive Function Scores with Carbon Dioxide, Ventilation, and Volatile Organic Compound Exposures in Office Workers: A Controlled Exposure Study of Green and Conventional Office Environments*. Environ Health Perspect, 2016. **124**(6): p. 805-12. http://dx.doi.org/10.1289/ehp.1510037

336. Azuma, K., et al., *Effects of low-level inhalation exposure to carbon dioxide in indoor environments: A short review on human health and psychomotor performance*. Environment International, 2018. **121**: p. 51-56. https://dx.doi.org/10.1016/j.envint.2018.08.059

337. Bussotti, M., et al., *End-tidal pressure of CO2 and exercise performance in healthy subjects*. Eur J Appl Physiol, 2008. **103**(6): p. 727-32. https://pubmed.ncbi.nlm.nih.gov/18521623/

338.   Sturrock, A., et al., *Hypoxia induces expression of angiotensin-converting enzyme II in alveolar epithelial cells: Implications for the pathogenesis of acute lung injury in COVID-19.* Physiological Reports, 2021. **9**(9). https://dx.doi.org/10.14814/phy2.14854

339.   Shehade, H., et al., *Cutting Edge: Hypoxia-Inducible Factor 1 Negatively Regulates Th1 Function.* The Journal of Immunology, 2015. **195**(4): p. 1372-1376. https://doi.org/10.4049/jimmunol.1402552

340.   Westendorf, A., et al., *Hypoxia Enhances Immunosuppression by Inhibiting CD4+ Effector T Cell Function and Promoting Treg Activity.* Cellular Physiology and Biochemistry, 2017. **41**(4): p. 1271-1284. https://dx.doi.org/10.1159/000464429

341.   Yi-Fong Su, V., et al., *Masks and medical care: Two keys to Taiwan's success in preventing COVID-19 spread.* Travel Med Infect Dis, 2020. **38**: p. 101780. https://www.ncbi.nlm.nih.gov/pmc/articles/PMC7270822/

342.   Ryu, H. and Y.-H. Kim, *Measuring the quantity of harmful volatile organic compounds inhaled through masks.* Ecotoxicology and Environmental Safety, 2023. **256**: p. 114915. https://www.sciencedirect.com/science/article/pii/S0147651323004190

Part 3:

Mask Psychology and Motivation

Wear. A. Mask.

"Symbolism is a primitive but effective way of communicating ideas. The use of an emblem or flag to symbolize some system, idea, institution, or personality, is a short cut from mind to mind."

- Robert H. Jackson, Supreme Court Justice and Chief Prosecutor in the Nuremberg Trials, in: *WEST VIRGINIA STATE BOARD OF EDUCATION et al. v. BARNETTE et al.*, June 14, 1943

> *"The thing most likely to guide a person's behavioral decisions isn't the most potent or instructive aspect of the whole situation; instead, it's the one that is most prominent in consciousness at the time of decision."*
>
> — Robert Cialdini, *Influence, New and Expanded,* 2021[40]

---

40   Robert B. Cialdini, *Influence, New and Expanded* (p. 375). HarperCollins. Kindle Edition. 2021.

Mask speech

In March and April 2020, almost overnight, the world developed a mania for facemasks. People who had never put on a mask in their life were suddenly wearing them when walking alone outside.[41] I was living in El Paso, Texas, at the time, and lost count of the number of people I saw wearing facemasks, even when walking alone outside, driving in their cars, or walking with family members. As late as December 2022, there were still plenty of people wearing facemasks in outdoor areas like grocery store parking lots.

Intuitively, one would expect most mask-wearing to be motivated by altruism, politeness, reasonable fear, a sensibly prudent desire to avoid giving or receiving an infection, or some combination thereof. Certainly, these were (and still are) the most commonly ascribed motivations for mask wearing in publications from the *Los Angeles Times*[42] to editorials in the *British Medical Journal*.[43] The least altruistic motivation attributed to mask-wearing was completely understandable crippling anxiety in the face of a new Spanish Flu.

Paired with this was a widespread self-righteous, vitriolic refusal to credit similar beneficent motivations to those who resisted mask-wearing. "To wear a face mask is an expression of deep goodwill toward your fellow humans," opined one writer, continuing, "My glasses slip off the bridge of my nose because they can't hold to the cloth. And still, I wear a mask. Why? Because I'm not a selfish jerk." [44]Another author in *Psychology Today* semi-jokingly proposed a new diagnosis for people who refused to wear masks: "antisocial masking disorder" characterized by "consistent irresponsibility," "lack of remorse," "recklessness," and "violation of the physical or emotional rights of others" (exactly *which*

41   I resided in El Paso, Texas, at the time, and lost count of the number of people I saw wearing masks, even when alone outside or in their cars.

42   Mariel Garza, "Editorial: To wear a mask or not to wear a mask. It's no longer a question," *The Las Angeles Times*, April 1, 2020. https://www.latimes.com/opinion/story/2020-04-01/editorial-to-wear-a-mask-or-not-to-wear-a-mask-its-still-a-question

43   Amy Price, *Why I wear a mask indoors and out.* BMJ, 2021: p. n1055. https://dx.doi.org/10.1136/bmj.n1055

44   Elle Silver, "Americans Refuse to Wear Face Masks Because of Selfish Individualism," *Medium*, June 25, 2020. https://medium.com/the-silver-mine/americans-refuse-to-wear-face-masks-because-of-selfish-individualism-d0e4b93a140

physical or emotional rights were never specified)[45]. Letters, opinions, and editorials castigating the "brainless"[46] behavior by "childish,"[47] non-maskers with "weapons-grade selfishness"[48] were aired prominently and widely. Those who refused to wear masks were denied services and expelled from stores.[49] Healthcare providers who questioned masking were treated as committing malfeasance or worse. Dr. Steven LaTulippe, M.D., was one of the few physicians with enough courage to go on record opposing compulsory mask-wearing, and had his license summarily revoked by the Oregon Medical Board for propagating such dangerous heresy.[50]

In a nod to George Orwell's *1984*, *Dilbert* author and social commentator Scott Adams coined the term "loserthink" to refer to dysfunctional habits of thinking.[51] Loserthink mind-reading was rampant in editorial after editorial where writers speculated on the reasons behind many people's supposedly baffling refusal to fulfill their "civic duty"[52] and "moral obligation"[53] to wear masks. Some suggested reasons, like differing risk assessments or allusions to psychological factors, got at part of the truth. Others were wildly off the mark, such as one academic's assessment that, "Anti-maskers are not afraid of losing their freedom, they're afraid of facing the reality of their own mortality."[54]

45   Steven Reidbord, "Antisocial Masking Disorder: It feels good to diagnose, but it's better to understand." *Psychology Today*, May 25, 2020,
https://www.psychologytoday.com/us/blog/sacramento-street-psychiatry/202005/antisocial-masking-disorder

46   "Letters to the Editor: Refusing to wear a mask is the most brainless, selfish way to assert your liberty" *Los Angeles Times*, June 20, 2020. https://www.latimes.com/opinion/story/2020-06-20/refusing-to-wear-a-mask-brainless-selfish

47   Simon Ljunggren, "Anti-mask crowd needs to put aside childish whining," *Bozeman Daily Chronicle*, July 19, 2020. https://www.bozemandailychronicle.com/opinions/letters_to_editor/anti-mask-crowd-needs-to-put-aside-childish-whining/article_ec8b01ce-efb5-5fee-b6cb-f7be4c256786.html/

48   Fiona Jones, "'Refusing to wear a facemask is not a philosophy, it's just selfish,' says James O'Brien," *LBC Leading Britain's Conversation*, July 24, 2020. https://www.lbc.co.uk/radio/presenters/james-obrien/facemask-refusers-are-selfish/

49   I personally experienced this. Apart from the routine hassling and dirty looks, among other things I was denied services at my bank until I put on a mask, and was confronted by a police officer and Walmart manager at the self-checkout and ordered to leave without being able to purchase my groceries when I refused to put on a mask. (Whereupon, I drove to another Walmart close by, entered without a mask, and went through the entire shopping trip again – this time successfully getting through the self-checkout without being accosted or ever putting on a mask.)

50   Steven LaTulippe, "COVID-19 and Medical Board Tyranny," *Journal of American Physicians and Surgeons*, 2022, Volume 27, Number 4. https://www.jpands.org/vol27no4/latulippe.pdf
Ralph Ellis, "Doctor Who Claimed Masks Hurt Health Loses License," *WebMD*, September 21, 2021 https://www.webmd.com/covid/news/20210920/doctor-who-claimed-masks-hurt-health-loses-license

51   See generally: Adams, Scott, *Loserthink: How Untrained Brains Are Ruining America*, Penguin Publishing, 2019.

52   Scullen, Patrick, "Take it from a veteran: It is our civic and patriotic duty to wear face masks when around other people," *Cleveland.com*, July 2, 2020, https://www.cleveland.com/letters/2020/07/take-it-from-a-veteran-it-is-our-civic-and-patriotic-duty-to-wear-face-masks-when-around-other-people.html

53   Ahmed-Zaid, Said, "Wearing a mask during pandemic is not just a health decision, it's a moral imperative," *Idaho Statesman*, July 19, 2020, Online: https://www.idahostatesman.com/living/religion/article244282522.html

54   Carlos Alberto Sánchez, "Wearing a Mask Makes Us Face Our Own Mortality (OPINION)," *San José State University SJSU Scholar Works*, January 27, 2021, Online: https://scholarworks.sjsu.edu/public-voices-2020/47/ (last

Fear of mortality had nothing to do with it. Politicians and other authority figures helped fan natural tension into open mutual hostility by consistently putting the worst possible spin on refusal to wear masks. With a lack of self-awareness which will make them barn-sized targets for criticism in high-school papers for generations to come, many of the same people who mandated masks then went on to decry the politicization of mask-wearing which was a direct result of their own actions.

These superficial and often self-flattering popular assessments sidestepped the real issues and motives underlying most people's decision to put on a mask or resist doing so. While fear of infection for either oneself or others was certainly a primary motivation for some, overall, it played a relatively small part in most people's decision to wear a mask. As we will see, psychological studies looking at possible motivations for mask-wearing often ruled out many possibilities by their experimental design. For example, the only directly motivation-oriented questions used in a 2021 Swiss study of mask-wearing offered a binary choice between altruism and risk-aversion — a desire to protect oneself or a desire to protect others — excluding other possible motives from consideration.[55] While it will come as no surprise to anyone that greater risk-aversion heavily predicted mask-wearing, a desire to avoid infection or to protect others were far from the only reasons people might have to wear or not wear a mask.

A 2020 Japanese study[56] cast a wider net by looking at six potential psychological drivers of mask-wearing: altruism, self-interest, perceived severity of the disease, personal relief from anxiety, a "do whatever you can" attitude, and conformity to a perceived social norm, framed in question form as, "When you see other people wearing masks, do you think that you should wear a mask?" Contrary to expectation and popular rhetoric: "the expectation of risk reduction (personal or collective) explained only a small portion of mask usage." Instead, the researchers found that "conformity to the social norm was the most prominent driving force for wearing masks." Behavioral Psychologist Robert Cialdini summarizes the study, "although multiple reasons were measured… only one made a major difference in mask-wearing frequency: seeing other people wearing masks." [57]

Mask-wearing is a potent form of communication. The authors of the 2020 Japanese study not only recognized the communicative effect of masks, they explicitly endorsed using the

---

accessed January 8, 2023)

55   Ankush Asri, et al., *Wearing a mask—For yourself or for others? Behavioral correlates of mask wearing among COVID-19 frontline workers.* PLOS ONE, 2021. **16**(7): p. e0253621. https://dx.doi.org/10.1371/journal.pone.0253621 Survey questions listed in the supplementary appendix pages 1-4.

56   Kazuya Nakayachi, et al., *Why Do Japanese People Use Masks Against COVID-19, Even Though Masks Are Unlikely to Offer Protection From Infection?* Front Psychol, 2020. **11**: p. 1918. https://www.ncbi.nlm.nih.gov/pmc/articles/PMC7417658/

57   Robert B. Cialdini, *Influence, New and Expanded* (p. 132). HarperCollins. Kindle Edition. 2021.

psychological "nudge" effect it produced, stating, "… wearing masks can be a symbol of collective confrontation against a pandemic… we encourage policymakers to apply the effects of the social norm on the wearing of masks to promote collective efforts to combat COVID-19."[58] Similarly, researchers writing in the academic journal *Anxiety, Stress, & Coping* acknowledged that "mask-wearing within certain contexts represents a sign of conformity with accepted social norms," and described "following mask-wearing directives as a way to express communal responsibility for combating the pandemic."[59]

The social psychologists Richard Thaler and Cass Sunstein highlighted the social influences inherent in mask-wearing in their book, *Nudge*:

> "Social influences come in two basic categories. The first involves information. If many people do something or think something, their actions and their thoughts convey information about what is best for you to do or think. If people are picking up after their dogs, buckling their seatbelts, driving under the speed limit, saving for retirement, treating people equally, or wearing masks, you might think that is the right thing to do. The second involves peer pressure. If you care about what other people think about you (perhaps in the mistaken belief that they are paying some attention to what you are doing…), then you might go along with the crowd to avoid their wrath or curry their favor. In some places during the COVID-19 pandemic, you'd get a cold stare or worse, if you weren't wearing a mask in a public place— and in some places, you'd get a cold stare or worse if you were wearing a mask."[60]

This gets at one of the core issues at the heart of much resistance to masking. More often than not, people who refused to wear masks did *not* want to communicate that wearing masks was the right thing to do, and resisted because they consciously or intuitively recognized that this was *exactly* what wearing masks communicated. They recognized the message inherent in mask-wearing.

"'It's a garment that has a strong visual impact,' says Lauren Fajardo, one of the owners of Cuban fashion brand Dador. 'It is also a way to express yourself. I don't even have to talk for someone to

58   Kazuya Nakayachi, et al., *Why Do Japanese People Use Masks Against COVID-19, Even Though Masks Are Unlikely to Offer Protection From Infection?* Front Psychol, 2020. **11**: p. 1918.
https://www.ncbi.nlm.nih.gov/pmc/articles/PMC7417658/
59   Sidney A. Saint and David A. Moscovitch, "Effects of mask-wearing on social anxiety: an exploratory review." *Anxiety, Stress, & Coping*, 2021. **34**(5): p. 487-502. https://dx.doi.org/10.1080/10615806.2021.1929936
60   Richard Thaler and Cass R. Sunstein, *Nudge* (pp. 65-66). Penguin Publishing Group. 2008 (2021 Kindle Edition).

see what I'm trying to say with my face mask.'"[61] Since 2020, the symbolic speech and expressive conduct inherent in mask-wearing has been widely and explicitly acknowledged, even exploited, by people on both sides of the political spectrum. On May 12, 2020, governor of New York Andrew Cuomo declared:

> "This mask says I respect the nurses and the doctors… This mask says, I respect the essential workers who get up every day, and drive the bus, or drive the train, or deliver the food, or keep the lights on so that I can stay home and I can stay safe. It says I respect others. And I respect you. And that is a statement that we should all be willing to make any day. But especially in the middle of this."[62]

As early as July 2020, Minnesota Governor Tim Walz complained that masks had become "a political statement rather than a public health statement." [63]Both of these governors' statements affirmed that masks are speech no matter what angle they are viewed from.

In 2020, President Trump reportedly refused to wear a mask because doing so would "send the wrong message," and was widely criticized for this stance by the mainstream media.[64] When 94-year-old Queen Elizabeth II did not wear a mask during her public appearance in October 2020, commentators worried that by *not* wearing one, The Queen was potentially "sending the wrong message."[65] The authors of a 2020 perspective published in the *New England Journal of Medicine* wrote: "It is also clear that masks serve symbolic roles. Masks are not only tools, they are also talismans[.]"[66] "It says to those around us that, whatever our vaccine status, we value community safety," wrote Dana

61   Andrea Rodriguez, *The Associated Press* "In Latin America, face masks become a form of expression," *The Seattle Times: World*, May 6, 2020, online:

https://www.seattletimes.com/nation-world/world/in-latin-america-face-masks-become-a-form-of-expression/

62   Andrew Cuomo, "Video, Audio & Rush Transcript: Amid Ongoing COVID-19 Pandemic, Governor Cuomo Urges New Yorkers to Wear Their Masks Out of Respect For the Nurses And Doctors Who Are Fighting to Save Lives," *New York Governor's Press Office,* May 12, 2020, https://www.governor.ny.gov/news/video-audio-rush-transcript-amid-ongoing-covid-19-pandemic-governor-cuomo-urges-new-yorkers

63   Cathy Wurzer, "Walz says a mask mandate is the right call, but he's waiting for more GOP support," *MPRnews,* July 17, 2020, online: https://www.mprnews.org/story/2020/07/17/gov-walz-hoping-for-more-buyin-on-masks-before-mandate

64   Mollie Mansfield, "NOT A GOOD LOOK Trump 'thinks wearing mask would send wrong message before election and believes US death toll is exaggerated,'" *The U.S. Sun: US News*, May 7, 2020, online: https://www.the-sun.com/news/794835/trump-mask-wrong-message-us-death-toll-exaggerated/

65   Mikhaila Friel, "The 94-year-old Queen sent the wrong message to fans by not wearing a mask, according to experts," *Insider*, Oct 20, 2020, online: https://www.insider.com/queen-sent-wrong-message-no-mask-experts-say-2020-10

66   Michael Klompas, et al., *Universal Masking in Hospitals in the Covid-19 Era.* New England Journal of Medicine, 2020. **382**(21): p. e63. https://dx.doi.org/10.1056/nejmp2006372

Stevens, in *The Atlantic*. [67] Publications from a children's book titled "Wearing a Mask Says I Love You"[68] to political cartoons such as Ben Garrison's "The Masks Speak…"[69] clearly recognized the fact that masks are a form of speech.

Even Dr. Fauci acknowledged the symbolic communication inherent in mask-wearing, and he did so very early on in a May 2020 interview, when he said: "I want to make it be a symbol for people to see that that's the kind of thing you should be doing."[70] Dr. Fauci reaffirmed this a year later in May 2021 during an appearance on *Good Morning America* when he described the CDC's ever-changing mask policies: "I am now much more comfortable in… in people seeing me indoors without a mask. Before the CDC made the recommendation change, I didn't want to look like I was giving mixed signals."[71]

67   Stevens, Dana, "Excuse Me If I'm Not Ready to Unmask," *The Atlantic*, May 19, 2021. Available Online: https://www.theatlantic.com/ideas/archive/2021/05/mask-wearing-cdc-guidelines/618916/ (last accessed 1/6/2023)

68   Jen Welter and Brooke Foley, *Wearing a Mask Says I Love You*, Critter Fitter Publishing: Florida (2020), https://www.amazon.com/Wearing-Mask-Says-Critter-Fitter-ebook/dp/B08CY75V9T/ref=sr_1_1?crid=C270S-JXRANWN&keywords=wearing+a+mask+says+I+love+you+critter+fitters+Jen+Welter&qid=1662921774&sprefix=wearing+a+mask+says+i+love+you+critter+fitters+jen+welter%2Caps%2C75&sr=8-1

69   Ben Garrison, "The Masks Speak…" Political Cartoon, Grrrgraphics (2020) https://grrrgraphics.com/the-mask-speaks/

70   Anthony Fauci interview with Jim Sciutto, 27 May 2020, quoted in Veronica Stracqualursi, "Fauci says he wears a mask to be a symbol of what 'you should be doing,'" *CNN*, Online: https://www.cnn.com/2020/05/27/politics/fauci-coronavirus-wear-masks-cnntv/index.html
See also: https://www.usatoday.com/story/news/politics/2020/05/27/coronavirus-fauci-says-he-wears-mask-symbol-what-do/5266189002/

71   Interview with Morley Safer on ABC's, *Good Morning America*, published online March 18, 2021, "Dr. Anthony Fauci discusses confusion over CDC's new mask guidance," https://www.youtube.com/watch?v=p3lgd2vM5_M (quote starts at 2:30 of the clip)

Even before COVID, the symbolic and communicative properties of mask-wearing were ingrained in the societal subconscious. The very word "mask" connotes pretense and concealment. Popular and scientific literature both reflect this. In his 2005 book, *False Alarm: The Truth About the Epidemic of Fear*, Dr. Marc Siegel observed: "Respiratory masks and other paraphernalia meant to shield us actually spread panic more effectively than any terrorist agent by sending the message that something is in the offing."[72]

One of the many studies we reviewed in Part 1 of this book was a 2009 masking study of hospital workers in Japan. In addition to finding no protective benefits against upper respiratory infections from wearing surgical masks, the authors of this 2009 study further noted that "Wearing a face mask may also communicate to the other party that the individual with the mask is either (1) infectious or (2) thinks the person they are talking with is infectious."[73] Similarly, the author of a 2018 research article discussed his clinical findings regarding the subtype of process addiction known as "mask dependency." He quoted one mask-wearer who was quite clear about what wearing a mask communicates: "When I went out, I always wore a mask as an indication of my intention to refuse to interact with people."[74] A 2021 literature review of the effects of mask-wearing on social anxiety noted that among individuals with social anxiety disorder, there are "certain safety behaviours that resemble the function of mask-wearing, including tendencies to 'hide your [their] face' and 'wear clothes or makeup to hide blushing.'"[75] Safety behaviors are a normal part of reducing or coping with anxiety and things that cause it, but when taken to extremes, they can become maladaptive and ultimately worsen the anxiety they are meant to alleviate.

When questioned by Members of the British Parliament, David Halpern, the psychologist heading up the UK's Behavioural Insights Team (infamously known as the "Nudge" Unit), not only drew attention to both the communicative and symbolic effect of masks, but also pointed out the self-contradictory nature of the surrounding mandates, given that the (marginally) more effective viral control measure of handwashing was not mandated. During a Public Administration and Constitutional Affairs Committee hearing on January 19, 2021, Halpern stated:

> "It took a long time to get people in masks. Our view early on was that masks
> are effective, not least because of the signal they create — and of course the

72   Marc Siegel, M.D., *False alarm : the truth about the epidemic of fear.* 2006, Hoboken, N.J.: John Wiley and Sons

73   Joshua L. Jacobs, et al., *Use of surgical face masks to reduce the incidence of the common cold among health care workers in Japan: a randomized controlled trial.* Am J Infect Control, 2009. **37**(5): p. 417-419. https://www.ajicjournal.org/article/S0196-6553(08)00909-7/fulltext

74   Noboru Watanabe, *Mask Dependent.* Stress Science Research, 2018: p. 15-20. https://www.jstage.jst.go.jp/article/stresskagakukenkyu/33/0/33_2018006/_article/-char/ja/

75   Sidney A. Saint and David A. Moscovitch, "Effects of mask-wearing on social anxiety: an exploratory review." *Anxiety, Stress, & Coping*, 2021. **34**(5): p. 487-502. https://dx.doi.org/10.1080/10615806.2021.1929936

underlying evidence. And of course there were… I mean, I gave you an example on handwashing where we did a number of trials and we found this was more effective. Um, there was… the evidence would suggest it was more effective and it wasn't taken up."[76]

Based on the results of her interviews with government officials, in her book *A State of Fear*, Journalist Laura Dodsworth elaborated on the underlying psychological mechanism involved, where she also highlighted Halpern's admission:

"… behavioural psychologists love masks. They absolutely love them. They believe they promote collectivism, the feeling that we are all 'in it together'. This attitude, clearly gleaned from my interviews with the SPI-B advisors, was confirmed when David Halpern answered MPs' questions… How does the signal work? Normative pressure is enhanced and sustained when we wear masks in public. They are a visible indicator that there is danger present all around, in the air we breathe and in the people we meet. Masked faces prime you to think of danger. We become walking billboards for disease and danger. They keep fear in our faces. Literally."[77]

Ms. Dodsworth's analysis has strong evidential support from popular and scientific sources. It is far from some fringe take. Most people will intuitively recognize the truth in her words. Even the Director-General of the World Health Organization, Tedros Ghebreyesus, explicitly made this connection when he proclaimed on August 3rd, 2020, "the mask has come to represent solidarity." He went on to urge healthcare workers, in particular, to "show us your solidarity in following national guidelines and safely wearing a mask — whether caring for patients or loved ones, riding on public transport to work, or picking up essential supplies… By wearing a mask, you're sending a powerful message to those around you that we are all in this together."[78]

---

76   David Halpern, Public Administration and Constitutional Affairs Committee hearing on 19 January 2021, Start Time Tuesday 19 January 2021 9:30 a.m., "09:30:15 Witness(es): Dr David Halpern, Behavioral Insights Team; Professor Steve Reicher, University of St Andrews," published online by parliamentlive.tv, https://parliamentlive.tv/Search?Keywords=&Member=&MemberId=&House=&Business=&Start=19%2F01%2F2021&End=19%2F01%2F2021 (quoted passage starts at 9:42:45 of the proceedings)

77   Laura Dodsworth, *A State of Fear: How the UK government weaponised fear during the Covid-19 pandemic*, London: Pinter and Martin Ltd (2021

78   Tedros Adhanom Ghebreyesus, "WHO Director-General's opening remarks at the media briefing on COVID-19 - 3 August 2020," *World Health Organization*, August 3, 2020, online: https://www.who.int/director-general/speeches/detail/who-director-general-s-opening-remarks-at-the-media-briefing-on-covid-19---3-august-2020 also at https://www.youtube.com/watch?v=TywepFySz_Q

Masks enhanced and perpetuated the anxieties they were supposed to alleviate. Psychiatrist Mark McDonald succinctly assessed the communicative and psychological effect of masking for COVID-19, saying, "Perhaps the greatest tool government has wielded in instigating and maintaining fear is the mask…. Their function has been purely symbolic, an emblem of fear, anxiety, and compliance."[79] Psychiatrist and bioethicist Aaron Kheriaty wrote, "masks became a potent symbol of physical purity and civic spirit, as well as a symbol of mutual mistrust, during the pandemic."[80]

Robert Freudenthal, a psychiatrist in the UK National Health Service, writes about masking:

> "It suggests that, at its core, interpersonal relationships are the true drivers of illness, and therefore our interconnectedness and relational lives, rather than being the very essence of our humanity, become a risk that should be managed and ideally avoided. Masking gives off the message 'I am an infection risk. You are an infection risk. We are to be avoided. Don't get close. I am better off away from you. Stay away.'"[81]

These properties of mask-wearing were common knowledge prior to COVID. In a 2018 Keynote Address to the National Academy of Medicine, health journalist (and World Economic Forum Global Health Security Advisory Board Member[82]) Laurie Garrett answered a question about the scientific evidence underlying masking by describing a study out of Japan, "It seemed like the major efficacy of a mask is that it causes alarm in the other person[.]"[83]

Times of declared emergency are often more dangerous to fundamental liberties than they are to life and limb. In 1943, in the middle of World War II, the United States Supreme Court courageously upheld the First Amendment when it ruled in *West Virginia State Board of Education et al. v. Barnette et al.* that saluting the American flag constituted a form of symbolic speech protected

---

(the entire segment starting at 6:45 assumes the symbolic effect of seeing someone wearing a mask, the precise quote starts at 7:02 in the full video)

79   Mark McDonald, *United States of Fear: How America Fell Victim to a Mass Delusional Psychosis*. New York • Nashville: Bombardier Books (2021)

80   Kheriaty, Aaron. *The New Abnormal: The Rise of the Biomedical Security State* (pp. 13-14). Regnery Publishing. Kindle Edition.

81   Freudenthal, Robert, "The True Meaning of Masking," *Brownstone Institute Articles*, October 30, 2021. https://brownstone.org/articles/the-true-meaning-of-masking/ last accessed December 31, 2022.

82   As listed in her website bio: https://www.lauriegarrett.com/about/ (last accessed January 1, 2023).

83   Laurie Garrett, "From the 1918 Influenza Pandemic to 2009 H1N1 Pandemic to Now: Is the World Ready to Respond to the Next Outbreak?" Published on the National Academy of Medicine YouTube channel, December 7, 2018, https://www.youtube.com/watch?v=sSExbHTS3nE&t=2504s (quoted remark timestamp 41:58)

by the First Amendment, and could not be compelled in violation of conscience — even during a national wartime emergency. In delivering the court's ruling, Justice Robert H. Jackson stated, "Symbolism is a primitive but effective way of communicating ideas. The use of an emblem or flag to symbolize some system, idea, institution, or personality, is a short cut from mind to mind."[84] Masks are indisputably such an emblem and form of speech, and in the context of COVID, mask-wearing became a perpetual salute. (This has critical implications under the First Amendment, which we will explore in detail in Part 5.)

Masks are broad-spectrum symbolic declarations of caution, fear, and danger — of infectiousness or infectibility. Everyone walking around with a mask on is effectively broadcasting that other people should wear a mask. Everyone walking around without a mask broadcasts the opposite. In particular, when perceived experts and medical authority figures like healthcare providers wear masks, it constitutes visual approval and social proof for these medical interventions and modified quarantine measures, as well as tacit endorsement of their moral and legal legitimacy. This social modeling and social proof speak and persuade more powerfully than any words.

Behavioral psychologist Robert Cialdini perceptively observed that "People have a natural tendency to think a statement reflects the true attitude of the person who made it. What is surprising is that they continue to think so even when they know the person did not freely choose to make the statement."[85] The universal principle of "ignore what they say and watch what they do,"[86] means that even acts of protest like wearing masks with anti-mask messages printed on them simply served to make protesting wearers self-contradictory hypocrites in the eyes of many viewers — after all, their convictions about masks being useless were not strong enough to prevent them from wearing one.

---

84   *West Virginia Board of Education v. Barnette et al.*, 319 U.S. 624, 632 (1943), available online: https://www.loc.gov/item/usrep319624/

85   Robert B. Cialdini, *Influence, New and Expanded* (p. 316). HarperCollins. Kindle Edition. 2021.

86   Variant on an old proverb. Reported as "Don't listen to what people say; watch what they do." Steven D. Levitt and Stephen J. Dubner, *Think Like a Freak*, HarperCollins: Toronto (2014) p. 112

Key Takeaways

🔑 Mask-wearing is an unusually potent and persuasive form of symbolic communication, as attested to by behavioral psychologists, psychiatrists, leaders on both sides of the political spectrum, and regular people from all walks of life.

🔑 The *stated* primary motive for mask wearing, infection control, was not the *actual* primary motive for most wearers.

Masks R Us:
Social influences on mask-wearing

Behavioral psychology was weaponized to drive masking. In fact, compulsory masking has been one of the most visible and effective weaponizations of behavioral psychology in history. Even if one were to assume for the sake of argument that masks provide all of the infection control benefits erroneously ascribed to them, those benefits would still be dwarfed by their virulent communicative impact and toxic psychological effects. There are three psychologists, in particular, whose influential work helps explain not only *why* compulsory masking worked the way it did, but also how to combat it and similar manipulations in the future. These three are Solomon Asch and his conformity studies, Stanley Milgram and his studies on obedience to authority, and Philip Zimbardo with his work on the situational factors that lead otherwise good people to abuse power or do evil things. Compulsory masking occupies the space where these three bodies of work converge.

In 1951, Solomon Asch conducted the first of his seminal conformity experiments.[87] Subjects were recruited for what was presented to them as a simple group visual perception experiment. Subjects were part of a group shown straight lines anywhere from 2 to 10 inches long and then asked to select the line of matching length from three adjacent choices. The catch was that each member of the group had to state their selection publicly, and apart from the actual research subject, every other member of the group giving answers was one of Asch's accomplices, who had been instructed to confidently state the same wrong answers for certain questions, thereby creating passive pressure to conform to an artificial "consensus."

---

87   Solomon E. Asch, "Effects of group pressure upon the modification and distortion of judgments." *Groups, Leadership and Men: Research in Human Relations*, Harold Guetzkow, ed. United States: Carnegie Press (1951), p. 177-190, available online: https://www.google.com/books/edition/Groups_Leadership_and_Men/aPRGAAAAMAAJ?hl=en

Asch's experimental design trapped a subject between "two contradictory and irreconcilable forces — the evidence of his own experience of an utterly clear perceptual fact and the unanimous evidence of a group of equals." What happens when a person is confronted by a group consensus of their peers which their own two eyes tell them is obviously wrong? Asch's control group showed a virtual absence of errors when matching the lines, and he initially predicted that relatively few people would conform to group opinion by making demonstrably false statements. What Asch actually *found*, however, was that fully three quarters of subjects acquiesced to the incorrect group answer at least once, and 30% of the subjects conformed to the wrong group answers the *majority* of the time.

In the face of entirely passive but unanimous group pressure, *only 26%* of test subjects in Asch's initial experiment were able to maintain their complete independence, and continued to give the correct answer. Even for those individuals who were able to maintain their independence, the act of contradicting even a non-antagonistic group majority was a clear source of substantial stress. Those who maintained their autonomy and gave the correct answers were careful to avoid badmouthing the group consensus even though their answers inescapably implied that the group was wrong. One of the strongest subjects who had maintained his autonomy would only go so far as to affirm the *possibility* that the group was wrong when pressed, saying, "You're probably right, but you may be wrong!" When the nature of the experiment was finally disclosed to him, he described himself as "exultant and relieved," adding that, "at times I had the feeling: 'to heck with it, I'll go along with the rest.'"

In conducting exit interviews with his subjects and analyzing his experimental results, Asch found that those who succumbed to group influence and gave the wrong answer generally fell into three categories: those whose *actions* were distorted, those whose *judgements* were distorted, and those whose *perceptions* were distorted. Those whose actions were distorted did not think their judgement was wrong, but yielded to the group out of a need not to appear different, inferior, or cause friction with the group. Those whose *judgements* were distorted thought that because everyone else in the group saw the lines a certain way, their own ability to judge the length of the lines was most likely in error, and so joined the group, denying the evidence of their own eyes. Finally, and most remarkably, the third subset of those who yielded to group pressure reported being unaware of any inconsistency between the group verdict and their own perceptions. In a way, the discomfort of seeing something different from the group consensus caused these subjects to actually *see* what the other group members merely *said* they saw. Their perceptions *changed* to fit the group descriptions.

Solomon Asch's experiments demonstrated that even when we have personal, undeniable factual knowledge, the mere condition of being surrounded by peers who demonstrate a consensus — even when no threats or direct interaction are involved — creates pressure to conform which only a minority of people resist completely, and which *no one* resists without effort. Moreover, most people have no conscious awareness when this is occurring. When Solomon Asch inverted his experiment, using a majority of test subjects, but adding a single accomplice who consistently dissented by giving wrong answers, the majority did not take the dissenter seriously. They even found him amusing: "contagious laughter spread through the group at the droll minority." However, when Asch increased the number of accomplices giving the wrong answer to at least three, he found that "the attitude of derision in the majority turns to seriousness and increased respect."[88] While subsequent modifications to Asch's experimental template highlighted variations of outcome based on circumstance, gender, and culture, Asch's initial core findings have been repeatedly replicated and validated. Keep in mind, too, that the pressure to conform increases with the stakes. Asch's experiments involved a trivial disagreement: matching the length of lines on paper. Dissent from the group could not reasonably be characterized as harming others, the way refusing to wear a mask often was.

Solomon Asch's findings are directly applicable to mask-wearing. As the authors of a 2022 study published in *BMC Public Health* stated, "behavior is contagious. When governments impose face mask regulations, this enforces a social norm that subsequently stimulates the advocated behavior because people want to conform to the group standard."[89] This study surveyed nearly 7,000 students from 10 countries, and found that by far the strongest predictor of mask-wearing was the stringency of mask mandates. Mandates got people to wear masks. Conversely, even though fear for personal safety correlated with mask-wearing, concern for others had no substantial impact on mask-wearing behavior. The study also found that international students were significantly more likely to report wearing masks, possibly because their social status as foreign nationals was less secure, forcing them to attune more closely to the visible group consensus.[90] All of these findings are consistent with results from Japan and other countries, and they point to social science rather than medical science as the real driving imperative for masking. Masking was a form of conspicuous caring, and mask *mandates* were *coerced* conspicuous caring.[91] In public health lingo, masks were a vector for transmitting infectious,

88   Solomon E. Asch, "Effects of group pressure upon the modification and distortion of judgments." *Groups, Leadership and Men: Research in Human Relations*, Harold Guetzkow, ed. United States: Carnegie Press (1951), p. 177-190, p. 189, available online: https://www.google.com/books/edition/Groups_Leadership_and_Men/aPRGAAAAMAAJ?hl=en

89   Annelot Wismans et al., "Face mask use during the COVID-19 pandemic: how risk perception, experience with COVID-19, and attitude towards government interact with country-wide policy stringency." *BMC Public Health*, 2022. **22**(1). https://dx.doi.org/10.1186/s12889-022-13632-9

90   Annelot Wismans et al., "Face mask use during the COVID-19 pandemic: how risk perception, experience with COVID-19, and attitude towards government interact with country-wide policy stringency." *BMC Public Health*, 2022. **22**(1). https://dx.doi.org/10.1186/s12889-022-13632-9

91   I was first introduced to this particularly apt term in Kevin Simler and Robin Hanson's excellent book *The Elephant in the Brain: Hidden Motives in Everyday Life*, Oxford University Press: New York (2018).

obsessive, debilitating, fear-based groupthink, and decisions to wear a mask were rarely as altruistic or selfless as propaganda made them out to be.

Solomon Asch's study yielded further important findings. He also observed that once a subject had yielded to the group, they continued to yield more than previously independent subjects. Asch explained,

> "[H]aving once committed himself to yielding, the individual finds it difficult and painful to change his direction. To do so is tantamount to a public admission that he has not acted rightly. He therefore follows the precarious course he has already chosen in order to maintain an outward semblance of consistency and conviction."

This is another lever of influence activated by masking, and we will examine it in more depth later.

The conformity effect created by a group extends even to morality, because people apprehend moral realities and experience moral convictions similarly to how they apprehend and experience other aspects of the external world. In a 2012 study, 80% of tested college students shifted their views of torture from being morally unacceptable to a more favorable stance after being convinced that the majority of their peers in a laboratory group had a more favorable view of the use of torture in interrogations.[92]

As mentioned earlier, there are two components of group conformity: informational and social normative. Group behavior communicates information and constitutes its own type of highly influential evidence, known as *social proof*. Communicating this social proof is known as *social modeling*. Robert Cialdini describes how this works:

> "We determine what is correct by finding out what other people think is correct… We view an action as correct in a given situation to the degree that we see others performing it."[93]

---

92   Nicholas P. Aramovich, Brad L. Lytle, and Linda J. Skitka, "Opposing torture: Moral conviction and resistance to majority influence," *Social Influence*, 2012. 7(1): p. 21-34. https://doi.org/10.1080/15534510.2011.640199, available online: https://www.tandfonline.com/doi/full/10.1080/15534510.2011.640199
This study was also highlighted by Psychologist Robert Cialdini in his book: *Influence, New and Expanded* (p. 131). HarperCollins. Kindle Edition.
93   Robert B. Cialdini, *Influence, New and Expanded* (p. 130). HarperCollins. Kindle Edition. 2021.

"Following the advice or behaviors of the majority of those around us is often seen as a shortcut to good decision-making. We use the actions of others as a way to locate and validate a correct choice. If everybody's raving about a new restaurant, it's probably a good one that we'd like too. If the great majority of online reviewers is recommending a product, we'll likely feel more confident clicking the purchase button."[94]

Who we witness taking actions also makes a difference. If we see someone we view as an expert or authority on the subject doing something, we are more likely to follow their example than we would be if it is a peer or someone we dislike. This made the sight of healthcare providers wearing masks an especially potent form of social proof.

In his study, Asch also noted that "the majority effect grows stronger as the situation diminishes in clarity."[95] Thus, conformity is higher when we are personally in a state of uncertainty, and the effect of social proof becomes stronger the more people we witness doing something. As shown by Asch's and subsequent experiments, group behavior is so powerfully persuasive that it can go so far as to *create* doubt and uncertainty where none previously existed and none would normally exist. Simply hearing a decisive pronouncement by their peers led many of Asch's subjects to doubt the evidence of their own two eyes when viewing something less than six feet away — evidence whose accuracy at judging object size and distance they otherwise took for granted and depended on every day when navigating and interacting with the world around them.

Using group consensus as a shortcut to effective decision making is not entirely irrational. It is an intuitive form of playing the odds, and most of the time has a net beneficial survival value. In fact, as professional risk analyst Douglas Hubbard has pointed out, in certain contexts such as when people are spending their own money in prediction markets,

"[A]ggregating opinions is better at forecasting than almost any of the individual participants in the market are… participants have an incentive not only to consider the questions carefully but even, especially when a lot of money is involved, to expend their own resources to get new information to analyze about

---

94   Robert B. Cialdini, *Influence, New and Expanded* (p. 157). HarperCollins. Kindle Edition. 2021.

95   Solomon E. Asch, "Effects of group pressure upon the modification and distortion of judgments." *Groups, Leadership and Men: Research in Human Relations*, Harold Guetzkow, ed. United States: Carnegie Press (1951), p. 177-190, p. 189, available online: https://www.google.com/books/edition/Groups_Leadership_and_Men/aPRGAAAAMAAJ?hl=en

the investment."[96]

Often, this flexible willingness to re-evaluate (and possibly modify) beliefs previously held with certainty, based simply on social context or social modeling, is one of humanity's greatest strengths. On an individual level, emulating others can be a great spur to self-improvement and even rapid relief from fear. Memorable exceptions to this general rule notwithstanding, children at the dental office who witness a sibling or peer behaving in a non-anxious manner will more often than not be less anxious themselves once they climb into the chair. Psychologist Albert Bandura found that more than half (55%) of a study group of nursery-school-aged children who were initially terrified of dogs, were willing to climb into a playpen with a dog and remain there, petting and scratching the dog, even when alone, after just four sessions of positive modeling from a peer. Just 13% of the control group which did not get the benefit of positive peer modeling were willing to do this. All of us can recall childhood and adult experiences that confirm these experimental findings, as well as personal examples of fears that are resistant to calming peer influences.[97]

Key Takeaways

- Solomon Asch's behavioral psychology experiments on group conformity demonstrate the powerful influence that even passive pressure from peer group modeling can have on individual behavior and beliefs.
- Behaviors like mask-wearing are more contagious than any virus.

---

96   Douglas W. Hubbard, *How to Measure Anything*, Third Edition. (p. 347). Wiley. Kindle Edition. 2014.

97   For me, personally, one stubborn fear that resists all peer modeling influence is my fear of heights.

Deluding the masses

Herd wisdom is far from infallible, however. It works best when individuals have the freedom to choose whether or not to follow the herd, but even then, mass delusions are regular occurrences. Charles MacKay catalogued many of these in his 1841 three-volume book set on *The Madness of Crowds*. Christopher Booker and Richard North added to this list in their 2007 book *Scared to Death*. Though individual readers are certain to disagree with these authors about particular issues listed as delusions, they are also certain to find far more points of agreement.

On a cultural and scientific level, social proof and modeling are simultaneously necessary for — and obstacles to — innovation, moral development, and the general advancement of knowledge. Virtually every current scientific tenet was a minority belief at one time. A dedicated minority with moral or empirical insight can use social influence processes to alter society for the better. The Western eradication of slavery happened through just such persistent efforts by a dedicated minority. However, the strength inherent in this mental flexibility includes an unavoidable corresponding vulnerability. The same process can normalize acceptance of false "facts" or morally degenerate behavior. Change can occur with astonishing rapidity in either direction, depending on the sources of influence. Ancient wisdom literature is replete with instructions highlighting the double-edged nature of this elemental human characteristic. The Ancient Egyptian *Wisdom of Ptahhotep*, one of the oldest books in the world, instructs: "If you are weak, follow a man of excellence and all your conduct will be good before god."[98] The Biblical book of Proverbs says, "Walk with the wise and become wise, for a companion of fools suffers harm,"[99] and "Do not make friends with a hot-tempered person, do not associate with one easily angered, or you may learn their ways and get yourself ensnared."[100] The common thread in these aphorisms is an understanding that while we cannot completely control the influence others have on us, we *can* choose our company with conscious awareness of this fact.

---

98   *The Wisdom of Ptahhotep*. Papyrus Prisse column 7, lines 7-9. 2003 University College London. Available online: https://www.ucl.ac.uk/museums-static/digitalegypt/literature/ptahhotep.html

99   Proverbs 13:20, *NIV*

100   Proverbs 22:24-25, *NIV*. See also 1 Corinthians 15:33: "Do not be misled: 'Bad company corrupts good character.'" (*NIV*)

Badness and fear are contagious, but so are goodness and courage. Ancient readers were urged to carefully and intentionally select their social models, keeping in mind the effects these will have on their own actions and beliefs.

The other paramount driving impulse for group conformity comes from a primal social need. In his book *The Lucifer Effect: Understanding How Good People Turn Evil*, psychologist Philip Zimbardo explains, "other people are more likely to accept us when we agree with them than when we disagree, so we yield to their view of the world, driven by a powerful need to belong, to replace differences with similarities."[101] At bottom, the social need to conform is not simply a desire for approval, a desire to receive assistance, or a need to grease the wheels of collaboration. Whereas the information conveyed by social proof and social modeling may or may not help us survive and thrive as individuals, the group *itself* can, and often *does*, become a direct threat to our individual survival. Remaining part of the main group (or at least not being perceived by that group as a threat that needs to be neutralized or eliminated) is a matter of brutal survival that we all recognize on a visceral level. We are, very reasonably, extremely hesitant about anything which could jeopardize our status in this regard. History is littered with the bodies of those who were thrust into an outgroup, often against their will, based on physical, religious, social and political differences.

The results of Solomon Asch's experiments predict that even if everyone knew for a fact going into 2020 that masks were useless against respiratory viruses, once the practice of mask-wearing took hold, only a minority of people would resist wearing a mask in all situations, and even then, doing so would require conscious effort and involve a great deal of personal discomfort. Passive social pressure, alone, would be strong enough to get about three quarters of people to wear masks at least some of the time, and about a third of people would wear masks all or most of the time. But mask-wearing during COVID was much greater than this. In Solomon Asch's experiments, pressure from the group was mostly implicit and indirect, rather than overt, and certainly not aggressive in any physical sense. What happens when indirect, passive, pressure to *conform* becomes direct pressure to *obey*? Ten years after Asch published his initial findings, Stanley Milgram began conducting another groundbreaking series of experiments to answer this question. Unless painful self-knowledge can be considered an injury, Milgram's genius was to acquire this knowledge without injuring anyone.

Stanley Milgram's subjects believed that they were participating in an experiment to determine how learning and memory were affected by the use of punishment. In reality, they participated in an experiment to see how far ordinary men and women will go in obedience to authority. Subjects believed they were acting as a "teacher," administering tests of word pairs to a "learner," and flipping

---

101   Philip Zimbardo. *The Lucifer Effect*, New York: Random House Publishing Group, 2007, Kindle Edition, p. 264

switches to give increasingly severe shocks with every wrong answer. The shocks started at 15 volts and went all the way up to 450 volts in 15-volt increments, with one dedicated switch per shock. The panel of switches had additional labels describing the voltages, running from Slight Shock, Moderate Shock, and Strong Shock, to Very Strong Shock, Intense Shock, and Extreme Intensity Shock. The 375-volt switch was marked with "Danger, Severe Shock." The final two 435- and 450-volt switches were simply marked with XXX, possibly because a skull and crossbones would have been too over the top. Before starting, the "teacher" subjects were given a sense of what they were doing to the "learner," as well as further convincing proof of the experiment's veracity, when they received a 45-volt shock administered by flipping the third switch. For comparison purposes, home electric outlets in the United States are roughly 120 volts, with double that going to large appliances.[102] As Milgram's experiment progressed, the learner-victim would verbally complain at 120 volts, and demand to be released from the experiment at 150 volts. At 285 volts, he let out an agonized scream. At 300 volts, the learner-victim would refuse to answer any further questions, and after 330 volts, the shocks would not even produce grunts or groans — only ominous silence.[103]

Thirty-nine psychiatrists who reviewed the experimental procedure beforehand predicted that just over one in a thousand subjects would be willing to go all the way up to the final 450-volt switch marked XXX. What Milgram actually *found*, however, was that when persistently instructed to do so by an authority figure in a lab coat, 65% of men *and* women were willing to administer successively stronger electric shocks to a screaming, pleading victim, strapped into a chair, begging to be let out — not only going all the way up to 450 volts, but adding two additional 450-volt shocks for good measure before the lab-coated experimenter declared the procedure completed. The only things faked were the shocks and harm to the "victim."

So, what escalating sequence of prods from an authority figure could *possibly* be compelling enough in the moment that 65% of people were willing to administer shocks to a begging, pleading victim, going all the way up to 450 volts long after the subject had gone silent and presumably lost consciousness?

Prod 1: "Please continue," or "Please go on."

Prod 2: "The experiment requires that you continue."

Prod 3: "It is absolutely essential that you continue."

---

102   Basic North American home circuit breakers will usually trip at 15 amps of current, so outlets typically cap at 12 amps.

103   Stanley Milgram, *Obedience to Authority*, 1974, New York: Harper and Row, p. 4, available online: https://archive.org/details/obediencetoautho0000milg/page/n7/mode/2up

Prod 4: "You have no other choice, you must go on."

No imprecations. No threats of job loss, jail time, or reputational harm. Just persistent instructions to administer the tests and shocks coupled with reassurance that the subject would suffer no permanent tissue damage and assertions that the authority figure accepted full responsibility for the subject's actions. The experimenter's tone was firm, but not impolite, and if the subject still refused after the fourth prod, the experiment was terminated and the subject was immediately debriefed.

Were the subjects gleeful as they flipped each switch to shock the learner after he started to demand to quit at 150 volts? Not at all! While some completed the whole sequence mechanically or with minimal signs of distress, a greater number protested vehemently. They looked sad; they shook with tension; they laughed nervously; they expressed concern for the victim. Some balled their fists, and one made a half-hearted offer to trade places with the "learner." Yet they *still* complied and administered the shocks up to 450 volts, with two more 450-volt "shocks" for good measure. When subjects could dimly see their learner-victim writhing from the shocks, they often averted their eyes. Compliance decreased in variants where the subjects were in closer proximity to the "victim" of their shocks, but remained substantial. In one variant where the subject was within reach of the strapped down "victim" and had to force the victim's hand onto a metal plate in order to administer the shock, over one quarter still obeyed to the very end.

Later researchers who had misgivings about the accuracy and implications of Milgram's findings wanted to see what would happen if the victim was authentic.[104] Their experiments with a group of college students not only verified Milgram's findings, but, as Philip Zimbardo later wrote, they illustrated "how powerful the obedience effect can be when legitimate authorities exercise their power within their power domains." [105] In these new experiments, "a cute, fluffy puppy" was substituted for the hidden learner-victim in Milgram's experiments. The cute puppy was kept on an electrified floor and could be clearly seen responding to very real electric shocks. Subjects were student volunteers being guided by a course instructor. The students were offered course credit for participating, but before the experiment began, they "were informed that, by the mere act of showing up for the experiment they had fulfilled their course credit requirement for participation in experiments. Thus, course credit was not contingent on shocking the puppy." [106]

---

104   Charles L. Sheridan and Richard G. King, "Obedience to authority with an authentic victim," in *Proceedings of the annual convention of the American Psychological Association*. 1972. American Psychological Association. https://cynlibsoc.com/clsology/pdf/women-torture-puppies.pdf

105   Philip Zimbardo. *The Lucifer Effect*, New York: Random House Publishing Group, 2007, Kindle Edition, p. 275.

106   Charles L. Sheridan and Richard G. King, "Obedience to authority with an authentic victim," in *Proceedings of the annual convention of the American Psychological Association*. 1972. American Psychological Association.

Though the actual intensities of the shocks were much lower than labeled, the final 10 shocks were all 800 volts at 1 milliamp of current, and every switch flipped clearly caused muscle spasms and increasing pain to the helpless puppy. The puppy could only howl, squeal, run and jump in futile attempts to escape the electrified grid. As in Milgram's experiments, the same sequence of verbal prods was used, and the experiment was immediately ended and the subject debriefed after refusing the fourth prod.[107] Despite being clearly able to see the pain they were inflicting, 54% of male subjects and 100% of female subjects still went through the entire sequence of shocks, up to the maximum voltage. The authors had no choice but to conclude that their findings were "in consonance with the view that Milgram's findings may correctly be taken at face value… Milgram's findings have proven remarkably robust in the face of a variety of procedural variations."[108] As with Milgram's subjects, inflicting such pain on a helpless victim clearly distressed the students. Some of the girls even broke down crying, but they still ran the hurting puppy through the entire sequence of shocks. Additionally, this puppy-shocking experiment was designed to allow two different ways to disobey. Subjects could directly refuse the experimenter's orders, but they could also deceive the experimenter by saying that the puppy had solved the problem correctly and so did not need to be shocked. Half the boys who refused to obey chose this alternative method over direct confrontation.

As a side note, Milgram's own experiments found comparable male and female obedience rates, though the puppy-shock study authors also reported personal correspondence with another professor whose unpublished Milgram-type study using adolescent girls found "virtually identical" 100% obedience. Two possible explanations for these differences in experimental outcomes are the ages of the subjects involved, and the nature of the learner-victim as a puppy instead of a human. Milgram's male and female subjects were fully mature, whereas the female subjects in the follow-up studies were adolescents and early college students — i.e., teenagers.[109] Alternatively, as Dr. Zimbardo points out in *The Lucifer Effect*: "The typical finding with *human* 'victims,' including Milgram's own findings, is that there are no male-female gender differences in obedience" (*emphasis added*).[110]

https://cynlibsoc.com/clsology/pdf/women-torture-puppies.pdf

107   Charles L. Sheridan and Richard G. King, "Obedience to authority with an authentic victim," in *Proceedings of the annual convention of the American Psychological Association*. 1972. American Psychological Association. https://cynlibsoc.com/clsology/pdf/women-torture-puppies.pdf

108   Charles L. Sheridan and Richard G. King, "Obedience to authority with an authentic victim," in *Proceedings of the annual convention of the American Psychological Association*. 1972. American Psychological Association. https://cynlibsoc.com/clsology/pdf/women-torture-puppies.pdf

109   Although he does not directly address Milgram-type experiments, see generally: Dr. Leonard Sax, *Why Gender Matters*, Second Edition (Harmony/Rodale, 2017), for more examples of inherent differences between males and females with social manifestations.

110   Philip Zimbardo. *The Lucifer Effect*, New York: Random House Publishing Group, 2007, Kindle Edition, p. 276.

Dr. Philip Zimbardo was a close contemporary of Stanley Milgram (they actually went to high school together). Dr. Zimbardo's own most famous study, the Stanford Prison Experiment, produced results so extreme that it was terminated early, and future iterations were banned. Dr. Zimbardo's fascinating account of the experiment, as well as a deep-dive into the implications and applications of the behavioral principles at play, are contained in his book *The Lucifer Effect: Understanding How Good People Turn Evil.* His book is essential reading for anyone who wants to psychologically inoculate themselves against subtle or overt situational factors influencing them to do things they later regret.

The goal of the Stanford Prison Experiment was to use simulated prison conditions to gain insight into the psychology of imprisonment, and the extent to which situational and systemic forces and social roles influence individuals' beliefs and actions. As Dr. Zimbardo put it: "our research will attempt to differentiate between what people bring into a prison situation from what the situation brings out in the people who are there."[111] Normal, healthy, intelligent, well-adjusted college students were randomly assigned to the roles of "guard" or "prisoner." Every participant went into the experiment with the conscious knowledge that they were playacting, that the experiment was only two weeks long, and that they could quit at any time. All that subjects had to do to get out of the experiment was to verbally assert their right to stop participating. In his book, Dr. Zimbardo narrates how he briefed guard participants on what they could and could not do when performing their roles in this experiment:

> "[T]here were limits to what could be accomplished in an experiment using only a 'mock prison.' The prisoners knew they were being imprisoned for only the relatively short time of two weeks, unlike the long years most real inmates serve. They also knew that there were limits to what we could do to them in an experimental setting, unlike real prisons, where prisoners can be beaten, electrically shocked, gang-raped, and sometimes even killed. **I made it clear that we couldn't physically abuse the 'prisoners' in any way.**

> "I also made it evident that, despite these constraints, we wanted to create a psychological atmosphere that would capture some of the essential features characteristic of many prisons I had learned about recently.

> "'We cannot physically abuse or torture them,' I said. 'We can create boredom. We can create a sense of frustration. We can create fear in them, to some degree.

---

111   Philip Zimbardo. *The Lucifer Effect*, New York: Random House Publishing Group, 2007, Kindle Edition, p. 33.

We can create a notion of the arbitrariness that governs their lives, which are totally controlled by us, by the system, by you, me, Jaffe. They'll have no privacy at all, there will be constant surveillance—nothing they do will go unobserved. They will have no freedom of action. They will be able to do nothing and say nothing that we don't permit. We're going to take away their individuality in various ways. They're going to be wearing uniforms, and at no time will anybody call them by name; they will have numbers and be called only by their numbers. In general, what all this should create in them is a sense of powerlessness. We have total power in the situation. They have none. The research question is, what will they do to try to gain power, to regain some degree of individuality, to gain some freedom, to gain some privacy? Will the prisoners essentially work against us to regain some of what they now have as they freely move outside the prison?'

"I indicated to these neophyte guards that the prisoners were likely to think of this all as 'fun and games' but it was up to all of us as prison staff to produce the required psychological state in the prisoners for as long as the study lasted. We would have to make them feel as though they were in prison; we should never mention this as a study or an experiment…

"[M]y initial interest was more in the prisoners and their adjustment to this prisonlike situation than it was in the guards. The guards were merely ensemble players who would help create a mind-set in the prisoners of the feeling of being imprisoned." (**emphasis added**) [112]

Dr. Zimbardo and everyone else involved got more than they bargained for as reality and illusion rapidly became indistinguishable. Within less than a week, everyone had so internalized their assigned roles that the "guards" (especially those on the night shift) were engaging in every prisoner abuse and humiliation they could think of which didn't violate the letter of the rules from higher-ups. This included (but was not limited to) forced pushups, sleep disruption and deprivation, control of toilet use, collective punishments imposed for the purpose of turning prisoners against their fellows who would not obey, and even stripping "prisoners" naked for tearing the ID numbers off the front of their prisoner uniforms. The "prisoners," for their part, were responding so authentically that they often seemed to have completely forgotten they had the ability to quit. Although part of their

---

112   Philip Zimbardo. *The Lucifer Effect*, New York: Random House Publishing Group, 2007, Kindle Edition, pp. 54-55.

refusal to quit may have been because doing so would have made them see themselves as weak, or "quitters," a turning point was reached when one prisoner falsely asserted that no one was allowed to quit. In the atmosphere created by the experiment, this claim was so plausible that it affected all the others dramatically. One "prisoner," Glen, reported afterwards, "The actual belief that 'we are really prisoners' was real — one couldn't escape without truly drastic action followed by a series of unknown consequences."[113] Another student-prisoner even became something of a snitch in order to get improved treatment.

The Stanford Prison Experiment demonstrated, under relatively controlled conditions, what many students of history, including America's Founding Fathers, already knew from painful experience: even normal and "good" people will rapidly begin to abuse power under the right conditions. It also demonstrated the difficulty of predicting ahead of time how any given person will act in such situations. One of the most humble and introspective Stanford Prison Experiment "prisoner" participants articulated this in a later televised interview when challenged by the most abusive "guard" who asked, "Well, you in that position, what would you have done?" The former "prisoner" responded, slowly and carefully, "I don't know. I can't tell you that I know what I'd do."[114] Even such a founding luminary of the American Revolution as John Adams was not immune to these forces. In 1798, he signed the Alien and Sedition Acts passed by Congress in direct violation of the First Amendment. In his correspondence with Thomas Jefferson nearly 20 years later, Adams, perhaps thinking back to this or other episodes, soberly reflected: "Power always thinks it has a great Soul, and vast Views, beyond the Comprehension of the Weak; and that it is doing God Service, when it is violating all his Laws."[115]

The revelatory thing about Asch's, Milgram's, and Zimbardo's experiments is that they involve ordinary people like you and I. They dispel the notion that some uniquely pathological personality is a required ingredient for evil. Milgram himself exhibited no doubt about the implications of his work. In a 1979 interview for CBS News' *Sixty Minutes*, he stated: "If a system of death camps were set up in the United States of the sort we had seen in Nazi Germany, one would be able to find sufficient personnel for those camps in any medium-sized American town."[116]

---

113   Philip Zimbardo. *The Lucifer Effect*, New York: Random House Publishing Group, 2007, Kindle Edition, p. 71.

114   This exchange was televised on NBC's *Chronolog* program, episode titled "819 Did a Bad Thing." Part of the exchange is reproduced in Dr. Zimbardo's book on pages 193-194.

115   From John Adams to Thomas Jefferson, 2 February 1816," *Founders Online*, National Archives, https://founders.archives.gov/documents/Adams/99-02-02-6575 .

116   Quoted from an interview with Morley Safer on CBS News, *Sixty Minutes*, March 31, 1979 in *Obedience to authority: Current perspectives on the Milgram paradigm*. 2000, Mahwah, New Jersey: Lawrence Erlbaum Associates, Inc., Taylor & Francis e-Library, 2009. ed.

From 2020 onward, with the exception of a few imperfect havens like Sweden, South Dakota, and Florida, our worldwide experiences with facemasks (and the COVID-19 response in general) exist in the zone where Solomon Asch's, Stanley Milgram's, and Philip Zimbardo's findings overlapped and amplified one another. In the context of COVID, masks effectively turned wearers into actor-accomplices in a Solomon Asch compliance experiment.[117] Mandates and the involvement of doctors exacerbated the impact by incorporating the influence of authority figures on obedience as described in Stanley Milgram's work.[118] When petty authorities at every level found most people complying, many doing so enthusiastically, and their superiors either standing back or endorsing their actions, the processes for the abuse of power described by Philip Zimbardo went into overdrive. The COVIDcrisis response in general, and compulsory masking and vaccination in particular, were in effect a series of nationwide, uninformed, non-consensual experiments, with vicious, often at least temporarily successful attempts to virtually eliminate any kind of control group. I was just one of many who independently recognized this. In January 2021, A Voltaire-caliber online wit with the twitter handle *el gato malo* ("the bad cat") provided what is probably the most succinct summary to-date, accompanied by a helpful Venn Diagram: "Asch showed us that people ignore their own perceptions to fit into a group. Milgram showed us that people will do things they know to be wrong if pressured by authority figures. Stanford showed us how badly people handle power, especially if egged on. COVID validated them all."[119] Two days after posting this remark, *el gato malo* was deplatformed ("de-catformed") and banned from Twitter.

117   Solomon E. Asch, "Effects of group pressure upon the modification and distortion of judgments." *Groups, Leadership and Men: Research in Human Relations*, Harold Guetzkow, ed. United States: Carnegie Press (1951), p. 177-190, available online: https://www.google.com/books/edition/Groups_Leadership_and_Men/aPRGAAAAMAAJ?hl=en Solomon E. Asch, *Studies of independence and conformity: I. A minority of one against a unanimous majority.* Psychological monographs: General and applied, 1956. **70**(9): p. 1, available online: https://cynlibsoc.com/clsology/pdf/independence-and-conformity.pdf

118   Milgram, S., *Obedience to Authority*. 1974, New York: Harper and Row. https://archive.org/details/obediencetoautho0000milg/page/n7/mode/2up

119   Original punctuation cleaned up. El gato malo (pseud.), Twitter Post, *@boriquagato*, January 5, 2021, 3:30 p.m., https://twitter.com/boriquagato/status/1346539611061854208, last accessed December 30, 2022. Also re-posted on the commentator's Substack, El gato malo (pseud.), "bad cats and bluebirds," *bad cattitude* (blog), December 25, 2022: https://boriquagato.substack.com/p/bad-cats-and-bluebirds (El gato malo's Twitter account was reinstated on Christmas, December 25, 2022.)

el gato malo @boriquagato · Jan 5, 2021 · · ·

asch showed us that people ignore their own perceptions to fit into a group.

milgram showed us that people will do things they know to be wrong if pressured by authority figures.

stanford showed us how badly people handle power, esp if egged on.

covid validated them all.

ASCH CONFORMITY EXPERIMENT

MILGRAM AUTHORITY EXPERIMENT

COVID HEALTH POLICY

STANFORD PRISON EXPERIMENT

ılıl   ○ 15   ⮌ 209   ♡ 513   ⬆

el gato malo   · · ·
@boriquagato

note that 2 out of 3 of these experiments had previously been banned from being repeated over severe ethical concerns.

but hey, we abandoned all the other science and ethics of the last 100 years in 2020, why not that too?

3:30 PM · Jan 5, 2021

ılıl View Tweet analytics

19 Retweets   **1** Quote Tweet   **186** Likes

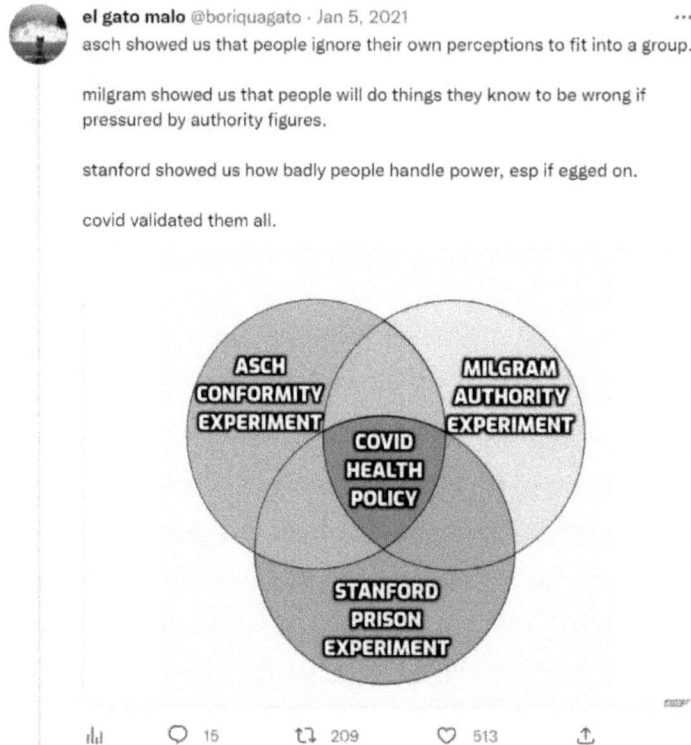

People are not wired simply to recognize and conform to *existing* social trends, but also to recognize and adopt *emerging* social trends. As Robert Cialdini put it,

> "Researchers have identified a consequential quirk in human perception. When we notice a change, we expect the change will likely continue in the same direction when it appears as a trend. This simple presumption has fueled every financial-investment bull market and real-estate bubble on record."[120]

---

120   Robert B. Cialdini, *Influence, New and Expanded* (p. 188). HarperCollins. Kindle Edition. 2021.

Before mass panic set in and authorities in most areas made masks mandatory, only a small minority of people who met two predisposing conditions were putting on masks. These people, 1) believed masks worked, and 2) feared SARS-CoV-2 more than they feared standing out from the group. They were a minority that has always existed, and whom most of us regarded as aberrations to be pitied rather than taken seriously, especially as early proclamations by public health and other authority figures initially ran counter to wearing masks.[121] However, as fear and uncertainty regarding the virus increased, so did the number of those wearing masks. The snowball started rolling downhill.

Key Takeaways

🔑 Stanley Milgram and his successors' experiments showed that most people will go farther in obedience to authority than most people realize, including inflicting direct harm on innocent third parties simply because an authority figure insists that they do so.

🔑 Philip Zimbardo's Stanford Prison Experiment showed how quickly people can slip into roles defined for them, turning even illusory roles into reality, and how prone even normal people are to abuse power and authority under the right circumstances.

🔑 Masks were an essential and highly-effective prop for getting people to internalize their new subordinate "good citizen in a pandemic" roles during the COVIDcrisis. This, in turn, enhanced compliance with all COVID-related directives issued by authorities, who more-often-than-not followed the same behavioral patterns as the abusive "guards" in the Stanford Prison Experiment.

---

121   Two of the most famous such figures were NIH Director Anthony Fauci and Surgeon General Jerome Adams. "Right now, in the United States, people should not be walking around with masks… When you're in the middle of an outbreak, wearing a mask might make people *feel* a little bit better, and it might even block a droplet, but it is not providing the perfect protection that people think that it is. And often, there are unintended consequences…" Fauci, Anthony *60 Minutes Overtime*, March 8, 2020. *60 Minutes* YouTube channel:
https://www.youtube.com/watch?v=PRa6t_e7dgI Last accessed December 30, 2022.
On February 29, 2020 Surgeon General Jerome Adams tweeted: ""Seriously people - STOP BUYING MASKS! They are NOT effective in preventing general public from catching #Coronavirus" (John Bacon, "Seriously people — STOP BUYING MASKS!': Surgeon general says they won't protect from coronavirus," *USA Today*, March 2, 2020. https://www.usatoday.com/story/news/healthcare/2020/03/02/seriously-people---stop-buying-masks-surgeon-general-says-they-wont-protect-from-coronavirus/112244966/
(Adams later deleted the tweet.)

Engineering an avalanche

> *"I am afraid, and therefore unquestioningly obedient."*
> — Václav Havel, *The Power of the Powerless*, 1978

Expanding on Asch's work, Cialdini describes three conditions that maximize social proof's persuasive force: "when we are unsure of what is best to do (uncertainty); when the evidence of what is best to do comes from numerous others (the many); and when that evidence comes from people like us (similarity)."[122] All of these converged in masking for COVID. As with Asch's experimental majority, the more the minority consensus in favor of masking grew, the more seriously we took it. Fear and uncertainty about the risk from SARS-CoV-2 and its transmission became nearly universal. Many others like ourselves were modeling mask-wearing, and this number increased daily, creating its own strong social proof that was only heightened when many authority figures like doctors and other healthcare providers began masking up as well, especially when they went to greater extremes by adding further accoutrements like goggles and a face shield. More people began to wear masks, generating yet more fear, more social proof, and more validation for those already wearing masks. The psychological avalanche acquired a self-perpetuating, self-enhancing critical mass almost overnight.

As if all this was not enough, many public health officials, politicians, and news sources engaged in deliberate efforts and messaging designed to heighten fear still further. Social psychologists have spent the last century meticulously studying the influence of individual and group behavior on belief. Internal governmental reports about COVID-related public health measures show that working knowledge gained from the last century of behavioral studies was consciously applied to the public. For example,

---

122   Robert B. Cialdini, *Influence, New and Expanded* (p. 143). HarperCollins. Kindle Edition. 2021.

in an October 2020 report, the United Kingdom's Scientific Pandemic Insights Group on Behaviors advised that in order to increase young people's compliance with COVID-19 preventative behaviors:

> "Communications should draw upon social norms of effective adherence by emphasising what other peers are doing (descriptive e.g. your peers are switching to socialising online) and approved perceptions of behaviours (injunctive e.g. your peers think you should start socialising online)… Communications targeting young people should where possible be delivered by trusted, non-governmental sources e.g. charities, celebrities, sports clubs, commercial brands."[123]

A 17-page German Department of the Interior ("Bundesinnenministerium") paper leaked in March 2020 recommended that the government should intentionally *increase* the population's fear of SARS-CoV-2.[124] The paper endorsed three specific strategies for doing this: 1) Stress the breathing problems of COVID-19 patients to deliberately trigger primal fears of death by suffocation; 2) Terrify and guilt children by persuading them that they are responsible for whether or not one of their parents and grandparents catches COVID and "dies in agony"; and 3) Emphasize the unknown long-term health effects caused by SARS-CoV-2 infection, including the theoretical possibility of sudden death or irreversible damage, then let people's imaginations run wild filling in the blanks.

One straightforward and overt use of behavioral rules of influence to push masks involved the rule of liking. The rule of liking states that people we like influence our behavior. Many people like celebrities. Put these two facts together in 2020, and you get celebrities and influencers across the globe modeling

---

123   *Scientific Pandemic Insights Group on Behaviors (SPI-B)*, "Increasing adherence to COVID-19 preventative behaviours among young people," U.K. Government Scientific Advisory Group for Emergencies, October 22, 2020: Online at https://www.gov.uk/government/publications/spi-b-increasing-adherence-to-covid-19-preventative-behaviours-among-young-people-22-october-2020 (last accessed January 1, 2023).

124   Bundesinnenministerium. Wie Wird Covid 19 unter Kontrolle Bekommen. Originally Available online: https://www.bmi.bund.de/SharedDocs/downloads/DE/veroeffentlichungen/2020/corona/szenarienpapier-covid19.html (last accessed on 20 December 2020) this document was no longer available on the German governmental web address in December 2022, but can still be found at https://fragdenstaat.de/dokumente/4123-wie-wir-covid-19-unter-kontrolle-bekommen/ as of the time of this writing. It was also reported on in April 1, 2020, by Arne Semsrott: "Corona strategy of the Ministry of the Interior: If you want to avert danger, you have to know it." *FragDenStaat*, April 1, 2020. https://fragdenstaat.de/blog/2020/04/01/strategiepapier-des-innenministeriums-corona-szenarien/ (last accessed on December 31, 2022). The section quoted is found on page 13.

face masks.[125] In one representative example from July 2020, a 30-second advertisement from the Alliance for Aging Research included masked pictures of everyday individuals, people in scrubs, and actresses like Jennifer Aniston, Natalie Portman, and Gal Gadot, alongside pictures of Republican and Democrat politicians, including both Donald Trump and Joe Biden wearing masks.[126] Celebrity monologues castigating people who refused to wear masks were given plenty of airtime.

Once we started putting on masks, a more subtle interior effect began to play out. Compulsory masking hijacked an ongoing process at work in every one of us throughout our lives (often at an unconscious level). We all know intuitively that our behavior and beliefs affect each other. We try to behave according to our beliefs, and when we consistently behave in a certain way, we find that our beliefs support those behaviors. The aphorism "fake it 'til you make it" refers to our conscious efforts to steer this process, usually against our in-the-moment preferences. Dr. Zimbardo comments:

> "Get people to perform good actions, and they will generate the necessary underlying principles to justify them. Talmudic scholars are supposed to have preached not to require that people believe before they pray, only to do what is needed to get them to begin to pray; then they will come to believe in what and to whom they are praying."[127]

This underlying mechanic is part of being human. It is neither inherently good nor bad, and can work in either direction, guided by our conscious choices. This dynamic also predicts that the simple act of wearing a mask makes people more likely to believe they *should* wear a mask. As it turns out, we have good evidence from nothing less than a randomized controlled trial demonstrating that this is indeed how mask wearing influenced our beliefs in practice. The Danish Facemask Study by Bundgaard et al. did not just look at the impact of mask-wearing on viral transmission rates. It also examined the impact of mask-wearing on people's beliefs about whether they *should* wear masks.[128]

---

125   Examples were this was trumpeted in the media are too numerous to count, an article published by *Insider* is representative: Zoë Ettinger, "20 photos of dressed-down celebrities wearing face masks," *Insider*, May 27, 2020, online: https://www.insider.com/photos-celebrities-face-masks-just-like-us-2020-5

126   Alliance for Aging Research, "Wear a Mask," July 28, 2020, online: https://www.youtube.com/watch?v=7iUxagedpWA

127   Philip Zimbardo. *The Lucifer Effect*, New York: Random House Publishing Group, 2007, Kindle Edition, p. 449

128   These results are contained in Supplement Table 5 of Henning Bundgaard, et al., *Effectiveness of Adding a Mask Recommendation to Other Public Health Measures to Prevent SARS-CoV-2 Infection in Danish Mask Wearers : A Randomized Controlled Trial.* Ann Intern Med, 2021. **174**(3): p. 335-343. https://www.acpjournals.org/doi/full/10.7326/M20-6817?rfr_dat=cr_pub++0pubmed&url_ver=Z39.88-2003&rfr_id=ori%3Arid%3Acrossref.org

As we saw when we looked at the Danish Facemask Study in Part 1, the two study cohorts, masked and non-masked, were carefully randomized at the start, so there is no reason to assume a lopsided distribution of beliefs regarding mask-wearing going into it. At the end of the study, however, participant exit polling revealed a remarkable thing. Despite the ongoing marketing blitz for masking, just 21% of the control group that had *not* worn masks favored the use of facemasks for future epidemics and fully half of the control group said that masks should *not* be used for future epidemics. Among the group that *had* routinely worn facemasks for two months, however, the proportion of people who believed that masks should *not* be used for future epidemics dropped from one-half to one-third, and the portion who said that facemasks *should* be used for future epidemics nearly *doubled*, going from 21% to 37%! Both groups were exposed to the same social environment over the same time period, so this difference between the two groups cannot simply be attributed to changes in popular attitudes towards masks as 2020 progressed. This dramatic 16-18% shift in the study participants' beliefs about masks is greater than the popular vote margin of most presidential elections throughout America's history. No strong social pressure or pseudo-legal coercion was necessary. All it took was the mere *act* of wearing masks for less than two months. Now imagine the impact on belief from more than two *years* of routine facemask-use, especially in countries where mask-wearing was compulsory, social pressure to wear masks was high, and the drive for masking was magnified by pseudo-legal coercion and the use of other behavioral psychology tools.

Anyone inclined to think that masks are a good idea might see these randomized controlled study results as more evidence which favors mask mandates — reasoning that if you force people to habitually do the "right thing," eventually they'll do it voluntarily. Everyone inclined to value independent thought and freedom of belief, on the other hand, should find this very concerning, and remain on their guard against other attempts to manipulate their belief through behavior. The mutually causal relationship between behavior and belief means that mandatory masking constituted not merely compelled symbolic speech in violation of the First Amendment, but also a brute-force attempt to compel *belief*. If you are thinking to yourself, dear reader, that wearing masks for a couple of years did not affect your beliefs about whether you *should* wear masks, I sincerely hope you are right, and you may very well be part of that minority of the population of whom this is true. Unfortunately, that is demonstrably untrue when applied to the population as a whole. The more people wear masks, the more they believe they *should* be wearing masks.

Key Takeaways

🔑 The power of social influence is maximized in situations when we see many people like us doing the same thing, especially when we are individually uncertain about what we ought to do.

🔑 Fear and behavioral psychology were deliberately exploited to get people to wear masks and conform to other COVIDcrisis measures.

🔑 Because behavior and belief influence each other, the mere act of wearing masks for an extended period of time had a dramatic impact on people's beliefs about whether they *should* wear masks.

The slippery slope is not always a fallacy

Behaviorally, mask-wearing was a highly effective foot in the door that opened wearers to complying with greater impositions down the road. The classic experiment looking at this mechanic of human nature was done in 1966 by Jonathan Freedman and Scott Fraser. Freedman and Fraser found that people who consented to putting up a 3x3-inch card promoting safe driving in the corner of one of their windows subsequently became more than twice as likely to consent to having a large, obnoxious safe driving billboard erected in their front yard. Fewer than 20% of those in the control group who were *not* psychologically primed by putting up a 3x3 inch card in their window agreed to put the large sign on their lawn compared to over 55% of the experimental subjects.[129] Freedman and Fraser's study is justly famous. In his own work on behavioral psychology, Robert Cialdini dwells on some of the implications:

> "Freedman and Fraser's findings tell us to be very careful about agreeing to trivial requests because that agreement can influence our self-concepts. Such an agreement can not only increase our compliance with very similar, much larger requests but also make us more willing to perform a variety of larger favors that are only remotely connected to the little favor we did earlier. It's this second kind of influence concealed within small commitments that scares me. It scares me enough that I am rarely willing to sign a petition anymore, even for a position I support. The action has the potential to influence not only my future behavior but also my self-image in ways I may not want. Further, once a person's self-

---

129   Jonathan L. Freedman and Scott C. Fraser, *Compliance without pressure: the foot-in-the-door technique.* J Pers Soc Psychol, 1966. 4(2): p. 195-202. https://web.mit.edu/curhan/www/docs/Articles/15341_Readings/Influence_Compliance/Freedman_Fraser_Foot-in-the-door.pdf

image is altered, all sorts of subtle advantages become available to someone who wants to exploit the new image."[130]

 "Once people have been induced to take actions that shift their self-images to that of, let's say, public-spirited citizens, they are likely to be public spirited in a variety of other circumstances where their compliance may also be desired. And they are likely to continue their public-spirited behavior for as long as their new self-images hold."[131]

Don't be selfish. Be a good person. Be a good citizen. Care about others. Be part of the solution. Put on a mask. Sound familiar?

Cialdini also describes how basic tools of psychology like the foot-in-the-door and gradual successive steps were used on Americans captured during the Korean War to get almost all of them to collaborate with their Chinese captors in one way or another. The most skilled of the interrogators started with small, innocuous, eminently reasonable and truthful concessions, such as simply getting the POWs to agree that "the United States is not perfect." Once the minor concession had been made, more substantive ones were pushed, until a prisoner might find himself describing a problem with America in writing. Once he put something in writing, why not sign it? After all, if it weren't true, you wouldn't have written it. If you're not willing to criticize your country, how about criticizing yourself? You wouldn't be a prisoner if you hadn't done *something* wrong, even if only an honest mistake. Confession and self-criticism are marks of humility, after all, and humility is a virtue. Of course, any written statements could then be used as tangible evidence to help persuade other prisoners to greater cooperation, or to help get the author to internalize the label of "collaborator," leading to even more extensive acts of collaboration. Essays by these new "collaborators" could produce an even greater return on investment by being recorded and broadcast.

The point is that the "Slippery Slope Fallacy" is not always a fallacy, and even seemingly reasonable concessions are sometimes better rejected, even if refusal seems irrational at first glance. In Stanley Milgram's experiments, each new shock was just 15 volts higher than the previous one. Progressing in gradual successive steps ("just a little bit more") is #7 on Philip Zimbardo's list of 10 lessons from the Milgram studies on how to get good people to do evil things. The more innocuous each increment, the harder it becomes to spot the process as a whole from within. Dr. Zimbardo describes how "gradual escalation of abuses from minimal to extreme" occurred at Abu Ghraib, culminating in the infamous and justly condemned night shift abuses. Dr. Zimbardo also describes how conformity

---

130   Robert B. Cialdini, *Influence, New and Expanded* (p. 313). HarperCollins. Kindle Edition. 2021.
131   Robert B. Cialdini, *Influence, New and Expanded* (p. 340). HarperCollins. Kindle Edition. 2021.

and socialized obedience to authority — two integral characteristics of compulsory masking and the COVID response in general — were foundational elements in building that abusive environment. In light of all this, consider a short 2002 article published in the *British Medical Journal*. It describes how prisoners held without trial in Guantanamo Bay, Cuba, were forced to wear surgical masks, earmuffs, and blindfolds for extended periods of time as part of a degrading standardized procedure designed to make them more psychologically submissive.[132]

Like the interplay between behavior and belief, the ability of even small acts to shift a person's self-image or self-concept is not inherently bad. We each have the individual ability to determine whether a foot-in-the-door followed by gradual successive steps leads us downwards or upwards, depending on which trivial requests we choose to comply with. A conscious, intentional awareness of how this process works can alert us so that we do not blindly follow any given set of railroad tracks laid out for us by others who may not have our best interests in mind. Our free will permits us to intentionally choose to continue taking steps along a gradually ascending path, doing the right thing when fear, discouragement, or social pressure would otherwise stop us. It also permits us to arrest and reverse a downward slide that, to all appearances, seems inexorable.

In his discussion of this process, Dr. Zimbardo envisions a "slow ascent into goodness step by step," as a kind of "reverse-Milgram" progression. By implication, enlightened moral authority figures could help direct this process. There is, however, one major vulnerability that was on full display during the response to COVID. Any reverse-Milgram authority figures assisting progress into goodness will be just as vulnerable to behavioral psychology and irrationality (sometimes moreso) than the people they seek to lead. Along with many other episodes of history, the COVIDcrisis provided another working example of how and why outsourcing our moral judgements to ostensibly beneficent authorities — especially distant politicians with no personal knowledge of our individual situations — creates more risk and harm than any natural hazard, viral or otherwise.

---

132   Dyer, O., *Prisoners' treatment is "bordering on torture," charity says. British Medical Journal*, 2002. **324**(7331): p. 187. https://www.ncbi.nlm.nih.gov/pmc/articles/PMC1122119/

Key Takeaways

🔑 The use of gradual successive steps to reach a particular end state is a powerful behavioral psychology tool that can be used on an individual or societal level for good or ill. This is especially true of small acts that make subtle changes to how we see ourselves — our internal self-concepts.

🔑 Compulsory masking was a step that seemed minor and justifiable to most people, but which actually had major import, and paved the way for greater acts of compliance down the road.

🔑 We need to be vigilant about where small concessions we make can lead, consciously define our personal boundaries ahead of time, and resist any changes that move us towards a destination we do not want to reach. In particular, we must avoid outsourcing our moral judgements.

Sheeple and science-deniers

Another basic compliance technique used to maximum effect (and with zero subtlety) in support of masking throughout the COVIDcrisis was the use of labeling and altercasting. Altercasting is a persuasion or compliance technique within the broader labeling theory of behavior, whereby people take on the characteristics they are described as having — whether good or bad.

When you altercast, you describe someone as possessing characteristics you want them to display. Want someone to be more cooperative? Compliment them on how well they work with others. Want a child to pick up after themselves? Let them know that their teacher complimented how neat and organized they are compared to the other children. Want to help someone with their diet? Praise them as having behaved characteristically when they exhibit self-discipline in exercise or food choices. In other words, give people a label you want them to live up to.

Dr. Zimbardo's Stanford Prison Experiment illustrates how powerful labeling and altercasting can be, even in an experimental setting where everyone knows at some level that they are play-actors getting paid for the roles they have been assigned. Dr. Zimbardo explains:

> "Our sense of identity is in large measure conferred on us by others in the ways
> they treat or mistreat us, recognize or ignore us, praise us or punish us. Some
> people make us timid and shy; others elicit our sex appeal and dominance.
> In some groups we are made leaders, while in others we are reduced to being
> followers. We come to live up to or down to the expectations others have of
> us. The expectations of others often become self-fulfilling prophecies. Without
> realizing it, we often behave in ways that confirm the beliefs others have about
> us. Those subjective beliefs can create new realities for us. We often become who
> other people think we are, in their eyes and in our behavior."[133]
> "Give someone an identity label of the kind that you would like them to have as

---

133   Philip Zimbardo. *The Lucifer Effect*, New York: Random House Publishing Group, 2007, Kindle Edition, p. 321.

someone who will then do the action you want to elicit from them. When you tell a person that he or she is helpful, altruistic, and kind, that person is more likely to do helpful, altruistic, and kind behaviors for others."[134]

Dr. Cialdini lists still more examples:

"[W]hat those around us think is true of us importantly determines what we ourselves think. For example, one study found that a week after hearing they were considered charitable people by their neighbors, people gave much more money to a canvasser from the Multiple Sclerosis Association. Apparently, the mere knowledge that others viewed them as charitable caused the individuals to make their actions congruent with that view."[135]

Admittedly, there are some caveats. The label has to have some plausible relationship to reality, and it works best if the person sees that trait or characteristic as being desirable. Even better if it's already part of their self-concept or if you can get people to label *themselves*. At its best, sincere altercasting reinforces and enhances good attributes people already possess or are striving to develop. In another application of the technique inhabiting more of a moral grey zone, Dr. Cialdini describes how life-insurance agents make profitable use of altercasting by highlighting parents' role as protectors to make them more willing to purchase life-insurance for their families. At the opposite extreme, negative labels can produce the opposite effect if the targets internalize them. "You're such a troublemaker!" or "You don't care about anyone other than yourself!" can turn into self-fulfilling prophecies through labeling and altercasting. A great deal of self-help literature is devoted to helping people remove the self-fulfilling negative labels they carry internally.

Mask-wearing was a classic case of altercasting good citizenship. For COVID, the roles of protector, "good person," and "good citizen" were effectively altercast onto virtually everyone in society, with detailed instructions about the performance expected of those roles — including wearing a mask. This provided an artificial island of certainty in a sea of uncertainty and fear, with predictable results on compliance. Masks became symbolic "self-labeling of helpfulness." Once bestowed, these positive labels could later be threatened and publicly withdrawn from any who refused to wear masks or get vaccinated, creating yet another disincentive to push back on the pervading narrative. Conversely,

---

134   Philip Zimbardo. *The Lucifer Effect*, New York: Random House Publishing Group, 2007, Kindle Edition, p. 451.

135   Robert B. Cialdini, *Influence, New and Expanded* (p. 317). HarperCollins. Kindle Edition. 2021.

those who maintained their autonomy and refused to wear masks were regularly vilified, labeled as inexcusably selfish at best, and deliberately committing negligent homicide at worst. Mask non-compliance (no matter how little of it there was) became an easy scapegoat when scary COVID numbers went up, whereas declines in COVID cases coinciding with mask-mandates were endlessly hyped as proof that masks work. *Post hoc, ergo propter hoc.* After this, therefore because of this.

Scorn and frustration embodied in negative labels like "brainless," "selfish," "childish," and "science-denier," were flung at resisters, who responded in kind to the "cowards," "sheeple," and "Branch COVIDians." In his book, Dr. Zimbardo describes what happened in variants of Milgram-style shock experiments where different labels were applied to the "victims:"

> "Those labeled in a more dehumanizing way, as 'animals,' were shocked most intensively, and their shock level increased linearly over ten trials. It also climbed higher and higher over trials, up to an average of 7 out of the maximum of 10 for each group of participants. Those labeled 'nice' were given the smallest amount of shock, while the unlabeled, neutral group fell in the middle of these two extremes."[136]

The more effectively a role is played, the more powerful and immersive its effect becomes, potentially to the point where idea and reality merge (bringing these two things closer together is, after all, part of the goal of altercasting in the first place). During the Stanford Prison Experiment, a visit from an actual prison chaplain dramatically increased the sense of authenticity for the participants. Dr. Zimbardo described how the priest's effective roleplaying deepened the illusion of the experiment, which was already nearly off the rails by that point:

> "The priest's visit highlights the growing confusion here between reality and illusion, between role-playing and self-determined identity. He is a real priest in the real world with personal experience in real prisons, and although he is fully aware that ours is a mock prison, he so fully and deeply enacts his assumed role that he helps to transform our show into reality."[137]

---

136   Philip Zimbardo. *The Lucifer Effect*, New York: Random House Publishing Group, 2007, Kindle Edition, p. 309.

137   Philip Zimbardo. *The Lucifer Effect*, New York: Random House Publishing Group, 2007, Kindle Edition, p. 105.

With compulsory masking, whole populations were press-ganged into exactly this kind of highly immersive mutual roleplaying, blurring the lines between pandemic illusion and viral reality until the two were virtually indistinguishable.

Like the priest in Dr. Zimbardo's Stanford Prison Experiment, the performance of healthcare workers in wearing masks and adhering to other COVID-related displays had an especially potent immersive effect. Healthcare providers' portrayal of anything related to health and disease is especially credible because of their perceived expertise, but people with healthcare degrees are, on the whole, no less vulnerable to behavioral psychology than anyone else. (Personally, I believe many are even more vulnerable than average because of habits they had to form in order to successfully progress through the many-layered medical training process and a hierarchy which rewards things like unquestioning deference to "experts.") A lot of people with healthcare or public health degrees who had less special knowledge about masks than most painters or industrial workers suddenly found themselves altercast on a grand scale by the media (or on an intimate scale by friends and family) as experts on all things related to infection control, including masks. This segues into how the even more powerful behavioral psychology rule of commitment and consistency was harnessed to fuel and perpetuate mask-wearing, because healthcare workers of all kinds locked themselves into positions on masks by engaging in very public displays of mask use or outright endorsements which gradually escalated. Once they did this, the psychological rule of commitment and consistency immediately took hold like a mental straightjacket.

Key Takeaways

🔑 How we see ourselves, and how we act based on that self-concept, are influenced by the roles and labels that other people assign to us.

🔑 Widespread labeling and altercasting were used to enhance compliance with mask wearing.

🔑 As illustrated in the Stanford Prison Experiment, playing even artificial roles can effectively merge illusion and reality. When a real-life role is incorporated into this process, as was the case with the prison chaplain recruited by Dr. Zimbardo to assist the Stanford Prison Experiment, and with healthcare workers wearing masks, the effect is greatly magnified.

The Matrix creates its own glitch

Generally, good personal consistency is a healthy, desirable, and even vital trait. Consistency is an essential precondition for logic and rationality, and we all think highly of people who consistently adhere to their principles, especially in situations where doing so involves a high personal cost. Hypocrisy, on the other hand, is by definition inconsistency between one's professed beliefs and one's behavior. Internal inconsistencies and contradictions within a story help us detect false statements and better get at the truth. It is natural (and usually beneficial) that, as Cialdini puts it, "Once we make a choice or take a stand, we encounter personal and interpersonal pressures to think and behave consistently with that commitment."[138] Much of this pressure stems from the self-image we create or reinforce by our actions, but the binding internal strength of some commitments is greater because they tap into our self-image more effectively than others. Commitments become strongest when they are "active, public, effortful, and freely chosen."[139] Mask-wearing meets the first three out of four of these criteria for *everyone*, but fulfilled all four in the case of everyone who put on a mask voluntarily. Wearing a mask is an affirmative act, highly visible, takes effort, and was embraced wholeheartedly by a substantial portion of the population.

When it comes to mask-wearing, there is no middle ground or way to conceal your symbolic speech or your public stance that accompanies it. You either wear a mask or you don't. The public nature of mask-wearing made the consistency principle even more powerful than it would be otherwise, especially for those who started wearing masks prior to the mandates out of genuine belief or instinctive concession to simple social pressure. Because of how visible it is, mask wearing (and non-wearing) cannot help being not just a symbolic public *statement*, but also a public *commitment*. Public commitments of this type are the most difficult and painful to reverse. As Cialdini says, "It's hardly surprising that people try to avoid the look of inconsistency. For appearances' sake, the more public a stand, the more reluctant we are to change it."[140] When choices and commitments are made publicly, the power of the internal drive for consistency

138   Robert B. Cialdini, *Influence, New and Expanded* (p. 294). HarperCollins. Kindle Edition. 2021.
139   Robert B. Cialdini, *Influence, New and Expanded* (p. 314). HarperCollins. Kindle Edition. 2021.
140   Robert B. Cialdini, *Influence, New and Expanded* (p. 322). HarperCollins. Kindle Edition. 2021.

is multiplied. Consequently, public commitments are the most likely to be lasting commitments. Many individual public commitments to masking will never be reversed, at least not explicitly.

There is at least one way, however, in which the zeal for mask-wearing that culminated in overt propaganda and widespread mandates created its own Achilles heel, and helped to loosen the mental vice-grip of the masking mentality even as outward compliance increased. In order to maintain a behavior in the absence of oversight and external pressure, a person needs to have internalized that behavior and incorporated the reason for it into their self-image. As Robert Cialdini points out, "the less detectable outside pressure such a reason contains, the better."[141] In other words, for behaviors like mask-wearing to really stick, people need to engage in them as voluntarily as possible (or at least *believe* they are doing so). The more overt and detectable the external pressures and incentives applied to induce a behavior become, the less likely people are to internalize the behavior. When the visible group consensus endorsing masks was supported by mandates with a credible, eagerly exercised threat of force, the pressure to conform went off the scale. The quarter of the population who Asch's experiments predict would have been able to resist the simple passive influence of seeing everyone else wearing masks was reduced to less than 1 in 100 in many places.[142] The stronger the pressure got, however, the more visible and noticeable it became, and the less internal commitment practitioners needed.

On their own, excesses and extremes have a tendency to make individuals pause and reevaluate what is going on. Mask mandates and other highly coercive measures like denying essential services (including medical care)[143] to the unmasked certainly were excessive and extreme, but another major psychological exit ramp this created was more subtle. The draconian mandates and penalties provided everyone wearing a mask with a very plausible and sensible reason for their mask-wearing — a reason which did not warp their self-concept into one that would keep them wearing masks when the mandates were lifted. The self-concept "I wear a mask to stay out of trouble," is perfectly compatible with other self-concept beliefs

---

141   Robert B. Cialdini, *Influence, New and Expanded* (p. 340). HarperCollins. Kindle Edition. 2021.

142   El Paso, Texas, was one of these places.

143   This includes my own personal experience. Several additional examples are listed below. A major effect of these mandates was deterrence from even seeking medical care in the first place.

Emily Czachor, "Woman With 'Broken Finger' Outraged Hospital Kicked Her Out for Refusing to Wear Mask," *Newsweek*, July 6, 2020, available online:

https://www.newsweek.com/woman-broken-finger-outraged-hospital-kicked-her-out-refusing-wear-mask-1515738

Rob Russell, "Veteran Who Can't Wear a Mask 'cuz PTSD Kicked Out of a VA Hospital for Not Wearing a Mask," *Granite Grok*, August 3, 2021, online: https://granitegrok.com/mg_manchester/2021/08/veteran-who-cant-wear-a-mask-cuz-ptsd-kicked-out-of-a-va-hospital-for-not-wearing-a-mask

Joseph v. Becerra, 22-cv-40-wmc, Document 1, p. 2,6 (W.D. Wis. Nov. 29, 2022), available online: https://storage.courtlistener.com/recap/gov.uscourts.wiwd.48728/gov.uscourts.wiwd.48728.1.0.pdf more case documents available at: https://www.courtlistener.com/docket/62627567/joseph-mark-v-becerra-xavier/

such as "I do not believe masks work," and "Wearing a mask does not make me a better person." It can also slowly and quietly supplant other beliefs that would lead a person to continue wearing a mask, such as mistaken beliefs about the proven effectiveness or moral necessity of mask-wearing. For a large fraction of the population, the forceful imposition of mask mandates provided an obvious explanation for their mask-wearing that helped them avoid internally committing to the practice. Because it was clear (or at least plausible) to themselves and others that they had not made a public commitment to mask-wearing so much done so to stay out of trouble, this critical segment of the population remained open to taking off their masks. The coercion and control used to push masks were so overt and forceful that even people who were genuinely committed to mask-wearing at some point were able to ascribe their behavior to external pressure rather than causes which might threaten their self-image. When the direction of social pressure to conform finally shifted, and masks were no longer being pushed, the coercive methods used to push masks provided people with a built-in face-saving excuse for changing their habits.

Despite this hidden silver lining side effect of the coercive methods used to push masks, the behavioral forces of commitment and consistency impelling people to wear masks were often overwhelming. Social media campaigns with convenient profile picture templates and frames like the #IWearAMaskBecause campaign[144] were another stroke of psychological Machiavellianism. By inducing people to not only make a public commitment to wearing a mask, but to commit to their *reasons* for wearing a mask, campaigns like this got participants to embrace *personal* responsibility for wearing their masks, thus internalizing the behavior and the underlying mentality. Some of the most public mask advocates who cannot scrub their personal histories or are unwilling to admit a mistake, will no doubt go to their graves insisting that the mask mandates were necessary and saved thousands of lives, or that mandating masks was still the right thing to do even if they had no medical benefits. As Cialdini explains, our innate compulsion for consistency is what drives this behavior, but it can easily go from being an asset to a vulnerability.

> "Sometimes it is not the effort of hard, cognitive work that makes us shirk thoughtful activity but the harsh consequences of that activity. Sometimes it is the cursedly clear and unwelcome set of answers provided by straight thinking that makes us mental slackers. There are certain disturbing things we simply would rather not realize. Because it is a preprogrammed and mindless method of responding, automatic consistency can supply a safe hiding place from troubling

---

144   Li Cohen, "As US nears 300,000 COVID-19 deaths, #IWearAMaskBecause hashtag urges fellow Americans to remain vigilant," *CBS News*, December 5, 2020, online:
https://www.cbsnews.com/news/covid-cdc-wear-a-mask-because-twitter-hashtag/

realizations. Sealed within the fortress walls of rigid consistency, we can be impervious to the sieges of reason."[145]

"[A]lthough consistency is generally good—even vital—there is a foolish, rigid variety to be shunned. We need to be wary of the tendency to be automatically and unthinkingly consistent, for it lays us open to the maneuvers of those who want to exploit the mechanical commitment and consistency sequence for profit."[146]

The bottom line is this: consistency must always remain a secondary value at most. The moment we give consistency precedence over a primary value like truth, it mutates from a virtue and asset into a fault and liability. Self-awareness and a personal commitment to truth as the superseding value serve as the most effective defense against having our default impulse to remain consistent used against us, as it was turned against so many of us in the case of masking.

Key Takeaways

🔑 People's natural impulse to maintain internal and external consistency was taken advantage of to drive masking by getting people to publicly commit to their support for the practice and to their reasons for wearing masks.

🔑 The heavy-handed coercion and pressure used to push masking created a psychological escape route for many people by giving them a reason to wear masks that allowed them to stop wearing masks without feeling inconsistent or threatening their self-image.

🔑 Consistency must remain a secondary value, subordinate to higher primary values like truth.

---

145   Robert B. Cialdini, *Influence, New and Expanded* (p. 296). HarperCollins. Kindle Edition. 2021.
146   Robert B. Cialdini, *Influence, New and Expanded* (p. 350). HarperCollins. Kindle Edition. 2021.

The rule of reciprocation

The sheer number and power of behavioral psychology rules and principles deployed to get people to wear masks has been so great that it is practically impossible to pick out any single one for the title of "most effective." That being said, if I personally had to put down money on the issue, I would give serious consideration to the rule of reciprocation for the top spot. This hardwired psychological rule simply states "we should provide to others the kind of actions they have provided us."[147] While most of us feel gratitude at a gift, we also experience great discomfort if we are unable to reciprocate or repay in-kind. To get rid of the resulting feeling of discomfort, we may even wind up repaying several times over to eliminate any hint of indebtedness. This usually unconscious psychological rule can enforce even *uninvited* debts because our sense of obligation to repay does not require that we *asked* for the gifts or favors. The Japanese have a specific term for favors like this, which connotes the discomfort and trouble they cause for the recipient: *arigata-meiwaku*, meaning an unwanted favor, or unwelcome kindness.

Robert Glover, a psychologist specializing in relationships, describes the dysfunctional dynamic that arises when people misuse the rule of reciprocation (especially in romantic or intimate relationships). People can create "covert contracts" using unwanted or unsolicited favors as an indirect attempt to get needs met. In other words, giving primarily in order to get. This strategy sometimes works in the short run, but in the long run, resentment tends to build on both sides of the relationship. The "giving" partner does not get their needs met reliably because they are rarely communicated clearly. At the same time, the recipient partner develops a feeling of emotional indebtedness and comes to resent being involuntarily put in that position, especially because the terms of the contract are rarely clear even to the "giving" partner.[148] As psychologists Henry Cloud and Robert Townsend put it in their bestselling book, *Boundaries: When to Say Yes, How to Say No*:

> Caring for someone so that they'll care back for us is simply an indirect means
> of controlling someone else. If you've ever been on the "receiving" end of that

---

147   Robert B. Cialdini, *Influence, New and Expanded* (p. 44). HarperCollins. Kindle Edition. 2021.
148   See generally. Glover, Robert A. *No More Mr. Nice Guy*, 2000. Barnes and Noble Digital.

kind of maneuver, you'll understand. One minute you've taken the compliment, or favor—the next minute you've hurt someone's feelings by not figuring out the price tag attached.[149]

While reciprocal exchange in loving relationships is a profound source of joy, and even unwanted favors are usually well-intentioned, unscrupulous manipulators or compliance specialists can and do take advantage of this hardwired psychological rule of influence. During the COVIDcrisis, the rule of reciprocation was unscrupulously used to turn everyone's mask-wearing into one giant exchange of favors. The fiendishly clever propaganda phrase: "my mask protects you, your mask protects me" was a psychological coup. Not only did it invoke the powerful social influence rule of reciprocation to drive mask-wearing, it recast and promoted the society-wide application of a dysfunctional and manipulative relationship tactic as an act of virtue. The response to COVID in general, and masking in particular, recycled a dysfunctional relationship dynamic to gain compliance by turning our best motivations against us. Many people consciously understood or at least intuited that this was occurring, and very reasonably resented and resisted being forced into an abusive relationship. Meanwhile, others so fully embraced this facet of the masking mentality that when confronted with an unmasked face, they behaved like scorned lovers.

Key Takeaways

Humans have a hardwired need to repay favors that they receive, even unasked, unwanted favors. This "rule of reciprocation" was exploited to increase compliance with mask-wearing through slogans like "my mask protects you, your mask protects me."

---

149   Henry Cloud and John Townsend. *Boundaries: When To Say Yes, How to Say No* (p. 56). Zondervan. Kindle Edition. 2017.

The rule of unity —
"we're all in this together."

Not to be left out, the social influence rule of unity was also invoked directly and perpetually throughout the COVIDcrisis using society-wide jingoistic appeals, like "we're all in this together" and, "do your part, wear a mask." Simply put, the rule of unity is that people generally behave most favorably towards those they see as sharing an identity in common with them. [150]This rule isn't limited to superficial physical characteristics (although those can certainly play a role), but is especially relevant to commonalities of ideology, cultural outlook, and shared experiences. Shared suffering, in particular, bonds people into "we" groups like nothing else does. Apart from the obvious visible categorization and symbolic communications inherent in mask-wearing, the artificial but very real shared suffering that mask-wearing entails created a unique sense of in-group cohesion, particularly for sincere wearers. This shared suffering naturally increased positive feelings of mask-wearers towards others who were wearing masks, creating a new group identity. Unfortunately, the negative rule of unity is that people behave worst towards those they see as having an identity which is adversarial to their own. As the in-group "we" effect from mask-wearing became stronger, resentment of those who refused to participate in the unifying form of group suffering grew proportionately, especially for those who were so affected by the propaganda campaign that they experienced visceral fear at the sight of an unmasked face. Those who refused to wear masks experienced their own corresponding version of this bonding effect with each other, centered around the social suffering, opprobrium, and ostracism inflicted on them by especially Pharisaic mask wearers and enforcing authorities.

In practice, social indicators provide people with as much information about the status of an epidemic as deaths and hospitalizations do. At the close of the Philadelphia Yellow Fever epidemic of 1793 (which was proportionately far worse than COVID, killing 10% of the city's population over the course of about twelve weeks), John Adams' son, Thomas Adams, wrote to his mother from the city

---

150   Robert Cialdini devotes an entire chapter of his book, *Influence, New and Expanded* (HarperCollins: Kindle edition, 2021), to this rule, starting on page 363.

to reassure her, "The idea of danger is dissipated in a moment when we perceive thousands walking in perfect security about their customary business, & no ill consequences ensuing from it."[151] In the case of COVID, compulsory masking forcefully suppressed this essential social proof, enabling the sense of emergency to drag out far longer than the time it took even the Black Plague to rip through communities. Compulsory masking was like refusing to allow the defendant in a criminal case to present any exculpatory evidence. With no visible alternatives, group pressure and groupthink produced self-reinforcing mass hysteria. A 2021 article in the *International Journal of Environmental Research and Public Health* on mass hysteria in the context of COVID was very direct in explaining how mass anxiety can grip a population, as well as how beneficial and essential dissenting minorities become during these times:

> "Once anxiety has spread and the majority of a group behaves in a certain way, there is the phenomenon of conformity, i.e., social pressure makes individuals behave in the same way as other members of the group. In the end, there may be a phenomenon that has been called emergent norms: when a group establishes a norm, everyone ends up following that norm. For example, if a group decides to wear masks, everyone agrees to that norm.

---

151   Adams, T.B., *Thomas Boylston Adams to Abigail Adams, 24 November 1793*, A. Adams, Editor. 1793, University of Virginia Press: Founders Online, National Archives https://founders.archives.gov/documents/Adams/04-09-02-0263. https://founders.archives.gov/documents/Adams/04-09-02-0263

See also, *Hurley v. Irish-American Gay, Lesbian and Bisexual Group of Boston, Inc.,* 515 U. S. 557, 561 (1995), available online https://supreme.justia.com/cases/federal/us/515/557/case.pdf.

quoted in *NetChoice v. Paxton* 49 F.4th 439, 458 (5th Cir. 2022), available online: https://casetext.com/case/netchoice-llc-v-paxton-2 ("The Court concluded that the parade was a 'form of expression' that receives First Amendment protection.")

"[W]hile anyone in a hysteria related to public health may voluntarily close their own business, wear a mask, or stay at home, in a minimal state, no one can use coercion to force others who are healthy and do not succumb to the hysteria to close their businesses, wear masks, or quarantine. A minority can just ignore the collective panic and continue to live their normal lives, because they are free to do so. Such a minority can be an example and a wake-up call to those that do succumb to the collective hysteria or are close to doing so. This minority may be especially attractive to borderline cases. Suppose that a small group of people during a collective health hysteria continues to go shopping, to work, to socialize, and breathe freely and does not fall ill (massively and fatally). Having this example, the anxiety of observers may fall. Observers may follow the example, and the group of hysterics shrinks." [152]

Mask-wearing naturally incorporates so many hardwired rules of influence and compliance that it serves as an extremely potent behavioral psychology device even before fear of any virus gets thrown into the mix. Many of these psychological rules were consciously or intuitively invoked and amplified to get people to internalize their mask-wearing compliance. Others were organically swept up along the way. Masking by its very nature maximized the mutually causal relationship between behavior and belief, "for, as we know from dissonance theory," writes Dr. Zimbardo, "beliefs follow behavior." [153] Mask-wearing was a "foot in the door" — a seemingly small initial act (after all, "it's just a mask") which primed wearers for more extreme acts of compliance down the road. The seemingly trivial nature of mask-wearing was also used as a rhetorical bludgeon to browbeat those who objected into compliance. Once people started wearing masks, the behavioral psychology rules of reciprocation, unity, commitment, and consistency gave the behavior and implicit underlying ideology its own self-perpetuating, positive-feedback momentum. Simultaneously, the influential power of positive and negative labeling, as well as the rule of liking, were intentionally used to further increase mask-wearing compliance. Despite all this, a small, almost paradoxical measure of saving grace was that, though mask mandates maximized compliance and genuinely convinced a lot of people through simple habituation, they also helped *undermine* the mask-related mass psychosis by providing a psychological exit ramp for many of those people who had not completely internalized the masking mentality.

---

152   Philipp Bagus, José Antonio Peña-Ramos, and Antonio Sánchez-Bayón, *COVID-19 and the Political Economy of Mass Hysteria.* International Journal of Environmental Research and Public Health, 2021. **18**(4): p. 1376. https://dx.doi.org/10.3390/ijerph18041376

153   Philip Zimbardo. *The Lucifer Effect*, New York: Random House Publishing Group, 2007, Kindle Edition, p. 449.

Key Takeaways

🔑 Humans have a natural tendency to group themselves, and to favor members of their own personal subgroups over "outsiders." In addition, shared suffering bonds group members more closely.

🔑 Compulsory masking artificially created two new sub-groups, the masked and the unmasked. The shared suffering from wearing masks, combined with official censure of non-masking, created a self-reinforcing cohesion within that part of the population who willingly wore masks.

🔑 Compulsory masking also extended the COVIDcrisis long past its natural termination point by artificially suppressing the social proof necessary to end the crisis (unmasked individuals going about their lives without harm).

Individual mask psychology

> *"Behavior change happens mostly by speaking to people's feelings."*
>
> — John Kotter, author of *Leading Change*[154]

Facemasking has been nothing less than weaponized behavioral psychology, and even those who recognized or suspected the implications right away still found themselves conforming all-to-often. Masking was and remains one of the biggest psychological lynchpins of what future generations will rightly see as an epoch-defining global hysteria. So far, we have investigated some specific reasons that people wear masks by looking at the empirical case for masks as an infection control tool, the communicative effect of masks, and the powerful social influences that led the majority of people to put on masks starting in 2020. Fear, altruism, and a sense of duty were deliberately exploited to gain increased compliance, but for many people these were only part of the complex motivations involved. At an individual level, there are multiple reasons why one person dons a mask while another adamantly refuses to do so.

Wearing a mask produces what Philip Zimbardo describes as a deindividuating or anonymizing effect. For individuals with social anxiety, this effect can provide a sense of comfort and protection that has nothing to do with infection control. [155]Prior to 2020, mask addiction was recognized in Japan in at least some professional circles as a sub-type of behavioral addiction or process dependency to which

154   John Kotter, quoted in Deutschman, Alan, "Change or Die," FastCompany.com, May 1, 2005, available online: https://www.fastcompany.com/52717/change-or-die (last accessed January 7, 2023).

155   One lawyer put it more dramatically but no less accurately when he said that "masking tends to erode or destroy a citizen's manifested personality, his identity, his sense of, and the public projection of, his innate *self*." *Green v. Alachua County*, 323 So. 3d 246, Appellant's Initial Brief, Filing #109556371 E-Filed 06/29/2020 (Fla. Dist. Ct. of App., 2021), p. 40, https://1dca.flcourts.gov/content/download/876478/file/2020-1661_Brief_1043158_RC03202D20Initial20Brief20on20Merits.pdf

patients with Social Anxiety Disorder were especially vulnerable. [156]While the American *Diagnostic and Statistical Manual of Mental Disorders, Fifth Edition* (DSM-5) does not yet include behavioral addictions as a separate category,[157] it does include some behavioral addictions like gambling disorder and internet gaming disorder under separate entries. To one degree or another, behavioral addictions are recognized in both the US and Japan, and in Japan are they are generally categorized as either process addictions (e.g., gambling, internet use, shopping, etc.), or addiction to human relations (e.g., sex addiction or love addiction).[158]

Japanese Psychologist Yuzo Kikumoto is credited with popularizing (possibly even coining) the phrase "mask dependency" in his prescient 2011 book, *Date Mask Addiction: People Standing at the Entrance to an Unrelated Society.* "[159]Wearing a mask hides your identity and makes you feel safe," Kikumoto wrote nearly 10 years before COVID. One of his patients explained to him that "Hiding her face makes her feel better because she doesn't have to be seen by other people." Another said simply, "it somehow calms me down."[160] These sentiments were mirrored during COVID in a May 2021 op-ed for the *New York Daily News*, with one author writing:

> "[F]rom the time I was a child, I worked to hide myself in public. I hunched
> over. I kept my eyes down, arranged my features into neutral nonentity…
> Making myself as invisible as possible became second nature… Then came

---

156 Noboru Watanabe, *Mask Dependent.* Stress Science Research, 2018: p. 15-20.
https://www.jstage.jst.go.jp/article/stresskagakukenkyu/33/0/33_2018006/_article/-char/ja/

157

> "… groups of repetitive behaviors, which some term behavioral addictions, with such subcategories as 'sex addiction,' 'exercise addiction,' or 'shopping addiction,' are not included because at this time there is insufficient peer-reviewed evidence to establish the diagnostic criteria and course descriptions needed to identify these behaviors as mental disorders."

American Psychiatric Association, *Diagnostic and Statistical Manual of Mental Disorders* (5th ed., 2013), p. 481, https://archive.org/details/diagnostic-and-statistical-manual-of-mental-disorders-dsm-5-pdfdrive.com/page/481/mode/2up

158 Kuroki, T., et al., *Current viewpoints on DSM-5 in Japan.* Psychiatry and Clinical Neurosciences, 2016. **70**(9): p. 371-393. https://dx.doi.org/10.1111/pcn.12421

159 Kikumoto, Yuzo, *Date Mask Addiction: People Standing at the Entrance to an Unrelated Society*, Fusosha, Japan. June 1, 2011.
https://www.amazon.co.jp/dp/B0087KZDVQ/ref=olp-opf-redir?aod=1
https://kikiwell.com/free/hon

160 Kikumoto, Yuzo, *Date Mask Addiction: People Standing at the Entrance to an Unrelated Society*, Fusosha, Japan. June 1, 2011.
https://www.amazon.co.jp/dp/B0087KZDVQ/ref=olp-opf-redir?aod=1
https://kikiwell.com/free/hon

Quotes machine-translated using Adobe OCR and Google Translate.

COVID and masks… Now, poof, courtesy of a bit of fabric, no need to worry about my expression, that I look too open, too available for interruption."[161]

There is a price to pay for this transient relief from social anxiety, however. As we saw in Part 2, a 2013 randomized controlled trial conducted in Hong Kong showed that when we wear masks, we come across as less empathetic to other people. The disconnecting effect produced by masks works on viewers as well as wearers. When doctors wear masks while taking to their patients, patients perceive them as being less empathetic.[162] In his own work, Dr. Zimbardo emphasizes another effect of masks:

> "In addition to the power of rules and roles, situational forces mount in power with the introduction of uniforms, costumes, and masks, all disguises of one's usual appearance that promote anonymity and reduce personal accountability. When people feel anonymous in a situation, as if no one is aware of their true identity (and thus that no one probably cares), they can more easily be induced to behave in antisocial ways."[163]

This makes one especially loathsome manipulation tactic that received wall-to-wall media support particularly ironic. In yet another transparent attempt to increase compliance (using variants of labeling and altercasting), Antisocial Personality Disorder (ASPD) was publicly and repeatedly portrayed as being linked with refusal to wear a mask or comply with other anti-COVID measures. "Sociopathic traits linked to not wearing a mask or social distancing during pandemic: study,"[164] read one typically overstated headline. Another trumpeted: "Refusing to wear a face mask can be a sign

161   Neuman, Kate, "Me and my mask: Why I'm not willing to give it up yet," *New York Daily News*, May 19, 2021, available online: https://www.nydailynews.com/opinion/ny-oped-me-and-my-mask-20210519-hs3x7zbswvgo-5c3js7m4kpdvsa-story.html (last accessed January 6, 2023)

162   Carmen K. M. Wong, et al., *Effect of facemasks on empathy and relational continuity: a randomised controlled trial in primary care.* BMC Family Practice, 2013. **14**(1): p. 200. https://dx.doi.org/10.1186/1471-2296-14-200

163   Philip Zimbardo. *The Lucifer Effect*, New York: Random House Publishing Group, 2007, Kindle Edition, p. 219.

164   Stieg, Cory, "Sociopathic traits linked to not wearing a mask or social distancing during pandemic: study," *CNBC*, September 15, 2020, available online:
https://www.cnbc.com/2020/09/02/study-refusal-to-wear-face-mask-associated-with-psychopathy-traits.html

of serious personality disorder, docs warn." [165]As was typical during COVID, anyone who bothered to *read* the study being cited would find these characterizations of the results misleading at best.[166]

The study touted by media as providing evidence of sociopathic traits in anti-maskers simply involved administering part of a personality test to 1578 Brazilian adults, combined with three yes/no questions asking whether respondents thought it was necessary to engage in social distancing, handwashing, and/or wearing a facemask to control COVID. [167]The authors themselves were careful to *avoid* saying that the personality characteristics of *any* of their surveyed groups even deviated from the range of what is normal, much less suggest pathology. At the very least, the omission means that the numerous histrionic headlines spawned from it badly mischaracterized the study's results. Instead, they simply make much of the marginal differences in personality traits which they did find, while ignoring commonalities like the groups' identical degrees of irresponsibility. The real takeaway from the study which ought to have been emphasized was the truly marginal nature of observed personality differences between the participants who refused COVID containment measures and those who eagerly embraced them.

165   Williams, Terri-Ann, "MASKING THE ISSUE: Refusing wear a face mask can be a sign of serious personality disorder, docs warn," *The U.S. Sun*, October 27, 2020, available online: https://www.the-sun.com/news/1696946/refusing-face-mask-sign-serious-personality-disorder/

166   Miguel, F.K., et al., *Compliance with containment measures to the COVID-19 pandemic over time: Do antisocial traits matter?* Personality and individual differences, 2021. **168**: p. 110346. https://www.sciencedirect.com/science/article/pii/S0191886920305377

167   An important caveat that somehow got missed in all the breathless reporting is that the study, as published, did not even isolate personality traits that correlate with attitudes towards masking, specifically. Out of the 1578 participants surveyed, the only 29 participants who we *know* eschewed masking *also* eschewed handwashing and social distancing. 29 out of 1578 also makes for a very lopsided sample comparison. What is particularly telling is that the authors of this personality study *could* have included a mask-specific analysis, but chose not to do so. The actual number of respondents who rated masking as "not necessary" are contained in the raw data which the study authors did not provide in the study itself and have not, to my knowledge, made public.

    The other two study groups, who rated one or two of the three COVID-containment measures as "not necessary," had a combined total of 347 people. Only the study authors know how many of these 347 people rated masks as unnecessary. It is safe to assume, however, that the number who did so is greater than 29, meaning that anywhere from 30 to 347 people who eschewed masking were lumped in with two of the other comparison groups. It is possible the study authors simply were not interested in isolating personality traits associated with refusing masks. It is equally possible that an analysis of the personality traits associated with rating masks not important was omitted because there were no statistically significant personality differences between those individuals and the rest of the study participants.

2021 - Miguel et. al. Survey Results Correlating Personality Characteristics with how people rate the importance of social distancing, masks, and/or handwashing.

(The choice was binary: "either important" or "not important")

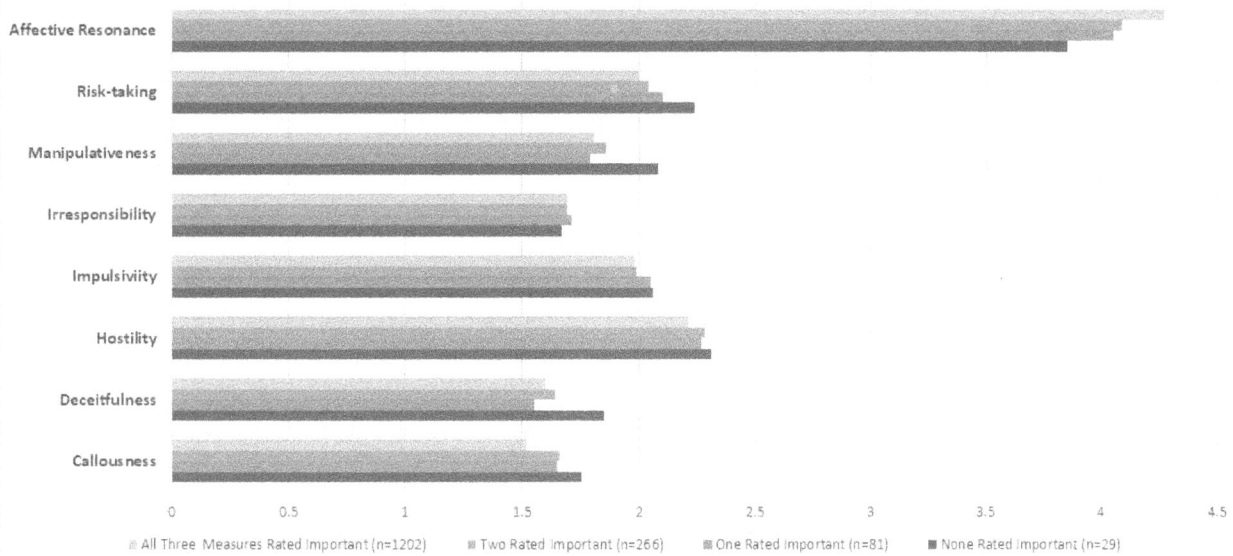

Affective Resonance · Risk-taking · Manipulativeness · Irresponsibility · Impulsivity · Hostility · Deceitfulness · Callousness

0   0.5   1   1.5   2   2.5   3   3.5   4   4.5

■ All Three Measures Rated Important (n=1202)   ■ Two Rated Important (n=266)   ■ One Rated Important (n=81)   ■ None Rated Important (n=29)

Estimated Marginal Means (EMM) from PID-5 and ACME Questionnaires, values reported on page 5 of

Miguel, F.K., et al., *Compliance with containment measures to the COVID-19 pandemic over time: Do antisocial traits matter?* Personality and individual differences, 2021. **168**: p. 110346.
https://www.sciencedirect.com/science/article/pii/S0191886920305377

Key Takeaways

🔑 Wearing masks creates an isolating, anonymizing effect that can provide temporary relief from social anxiety.

🔑 Prior to 2020, mask addiction was recognized in at least some professional circles as a subtype of behavioral addiction or process dependency.

🔑 Despite slanted media presentations attempting to link it to antisocial personality traits, refusal to wear a mask did not imply any psychological abnormality or pathology.

Rational caution vs. obsession

Fears of disease and contagion long pre-date germ theory, but fear of physical sickness, *per se*, is not always the main driver. In a 2020 article published in the *Journal of Obsessive-Compulsive and Related Disorders*, Dr. Frederick Aardema, a psychologist specializing in obsessive-compulsive and related disorders, held up a more historically minded, almost psychoanalytic, lens through which fears of COVID might be understood.[168] Dr. Aardema differentiates between exaggerated, almost phobic reactions to potential infection, and obsessive-compulsive disorders centered around a more primal dread of contamination. One of the chief signs that something is an obsessive disorder, rather than a phobia or rational-but-excessive fear, is that obsessive disorders tend to be highly selective and idiosyncratic. As an example, Dr. Aardema cites the case of one of his patients who routinely wore gloves and avoided shaking hands out of obsessional fear of Hepatitis C "but showed no heightened concerns about the coronavirus."[169]

The line between rational caution regarding illness and obsessive fear of contamination can get very blurry in practice, but the response to COVID, including compulsory masking, clearly hurtled across this line with the pedal to the metal. Many of the measures claiming to be for the control of COVID, especially masking regulations, were *highly* idiosyncratic, having more in common with obsessive disorders than any rational infection control measure. A prime example of this were requirements to wear facemasks when entering a restaurant, but which permitted removal of masks when seated at a table. The same idiosyncrasy applies to every exception that allowed mask removal for brief periods, including while eating and drinking on airplanes or anywhere else. The tendency to treat cloth masks, medical masks, and N95 respirators as being equally capable of meeting mandate requirements is another example. Masking outdoors was *never* rational, and touching a mask to pull

---

168   Frederick Aardema, *COVID-19, obsessive-compulsive disorder and invisible life forms that threaten the self.* Journal of obsessive-compulsive and related disorders, 2020. **26**: p. 100558-100558.
https://pubmed.ncbi.nlm.nih.gov/32834943 or https://www.ncbi.nlm.nih.gov/pmc/articles/PMC7324330/

169   Also, as Dr. Aardema explains, obsessive disorders are not limited to pathogens or other potential causes of harm. They can include compulsions related to symmetry- or re-checking things.

it back enough to sip a drink or take a bite of food, then immediately replacing it, was worse than useless from an infection control standpoint. Idiosyncrasies apparent when viewing masks solely as a public health measure vanish, however, when masking is viewed as an obsessive means of social signaling, psychological manipulation, social control, and coerced symbolic speech.

Dr. Aardema suggests that justified concerns to avoid contagion shift towards obsessional fears of contamination the more a pathogen is personified. Many counter-productive COVID-related infection control or "safety" measures implicitly personified SARS-CoV-2. When it is personified, a pathogen acquires by implication the power to impart mental and spiritual corruption. Efforts to avoid this deeper threat, which spawns a primal dread, are often out of all proportion to the physical danger. Dr. Aardema quotes one particularly self-aware 19th century commentator who describes this primitive sensation of dread: "I have no particular apprehension of contracting small-pox or any other disease I can specify. It is an overpowering feeling that I shall be defiled in some mysterious way, that presses me with a force I cannot resist…" The many COVID curfews were an obvious example of this, as though the virus will intentionally "get" naughty citizens who stay out too late or socialize too much. From a rational public health standpoint, curfews simply forced citizens to cram their activities into fewer hours every day, ensuring that more people were out and about and interacting at the same time. Closing bars which serve food but allowing restaurants which serve alcohol to operate was another example of this sort of irrational idiosyncrasy (we will go into more detail on some of these in Part 5).

A codified scientistic veneer, especially when it has intuitional appeal, multiplies the staying power and persuasiveness of sub-rational fears. Perceptive 19th and early 20th-century observers drew attention to this. In an 1866 lecture delivered at the Royal Institution in London, simply titled "Superstition," the Reverend Charles Kingsley pointed out that man has a unique capacity to "organize his folly; erect his superstitions into a science; and create a whole mythology out of his blind fear of the unknown."[170] In his well-timed 2019 book, *How Fear Works*, psychologist Frank Furedi spells out this difference between science and fear-based Science™. On the one hand, "the term 'Research shows' has the character of a

---

170 Kingsley, Charles, "Superstition: A Lecture Delivered at the Royal Institution, April 24, 1866", *Fraser's Magazine For Town and Country*, volume 73 (January to June, 1866), p. 705. Available online at: https://ia801804.us.archive.org/14/items/sim_frasers-magazine_1866-06_73_438/sim_frasers-magazine_1866-06_73_438.pdf (last accessed January 5, 2023)

See also his companion lecture: Charles Kingsley, "Science: A Lecture Delivered at the Royal Institution.", *Fraser's Magazine For Town and Country*, volume 74, July to December, 1866, p. 15. Available online at: https://ia801700.us.archive.org/31/items/sim_frasers-magazine_1866-07_74_439/sim_frasers-magazine_1866-07_74_439.pdf (last accessed January 5, 2023)

Frank Furedi, *How Fear Works* (p. 275). Bloomsbury Publishing. Kindle Edition.

ritualistic incantation. It is the modern equivalent of the mandate offered by the Holy Scripture."[171] Simultaneously,

> "[T]he absence of evidence does not inhibit advocates from asserting their beliefs. And in many cases the absence of evidence is reinterpreted as proof that the threat is actually far greater than previously believed… arguments about threats are often founded on a prior conviction, rather than discoveries made by science. That is why it is so difficult to distinguish the scientific from the moral in the language of fear."[172]

There is also a correlation between affluence and anxiety: "In Western societies, it is usually the wealthiest, the most economically and socially secure individuals, who tend to be most concerned for their safety."[173] Masking provided visual validation for a lot of this free-floating anxiety — and not in a good way.

Though authorities constantly appealed to science and experts to justify universal compulsory masking and other totalitarian forms of COVIDcrisis overreach, what they meant by "science" was very different from the methodology for discovering truths about the physical world. In reality, what they were referring to was The Science™, which was nothing more than a euphemism for the whole framework of collective underlying beliefs that supported their policies and impositions — including compulsory masking. One of America's greatest 20th and 21st century minds, Thomas Sowell, used his characteristic dry wit to describe the real processes involved when The Science™ is invoked to guide public policy:

> "[E]xperts are often called in, not to provide factual information or dispassionate analysis for the purpose of decision-making by responsible officials, but to give political cover for decisions already made and based on other considerations entirely. The shifting of socially consequential decisions from systemic processes, involving millions of people making mutual accommodations—at their own costs and risks—to experts imposing a master plan on all would be problematic even if the experts were free to render their own best judgment. In situations where experts are simply part of the window dressing concealing arbitrary and even corrupt decisions by others, reliance on what 'all the experts' say about a

---

171   Frank Furedi, *How Fear Works* (p. 127). Bloomsbury Publishing. Kindle Edition.
172   Frank Furedi, *How Fear Works* (p. 120). Bloomsbury Publishing. Kindle Edition.
173   Frank Furedi, *How Fear Works* (p. 22). Bloomsbury Publishing. Kindle Edition.

given issue is extremely risky. Even where the experts are untrammeled, what 'all the experts' are most likely to agree on is the need for using expertise to deal with problems."[174]

Dr. Furedi's and Dr. Sowell's analyses were repeatedly borne out during the COVIDcrisis. Very real socio-legal threats and punishments were used to actively *prevent* people from making voluntary mutual accommodations regarding masking and other COVID-related preventative measures. At the same time, vague hand-waving appeals to The Science™ were used as moralizing truncheons to end discussion and quell dissent.

---

174   Thomas Sowell, *Intellectuals and Society*, New York: Basic Books, 2011, Kindle Edition, pp. 26-27.

Sacred Values

Public health was a *de facto* quasi-religion for at least a few people even before COVID, but *during* COVID, public health fully metastasized into a secular religion, with mask-wearing and vaccination as the chief performative rituals. "In Fauci We Trust," popped up on numerous products in what was, at best, a semi-satirical allusion to America's national motto, "In God We Trust." Widespread acceptance of medical reasons for not wearing a mask as being legitimate and acceptable, coupled with a simultaneous denial of the legitimacy and acceptability of religious reasons for refusal, testified to this transformation. To categorically elevate something's importance *over* religion is to enshrine it as a form of religion. For example, to permit dental treatment (which by definition requires the removal of masks) while denying the removal of masks on religious grounds is to take for granted that matters of dentistry are more important than matters of religion.

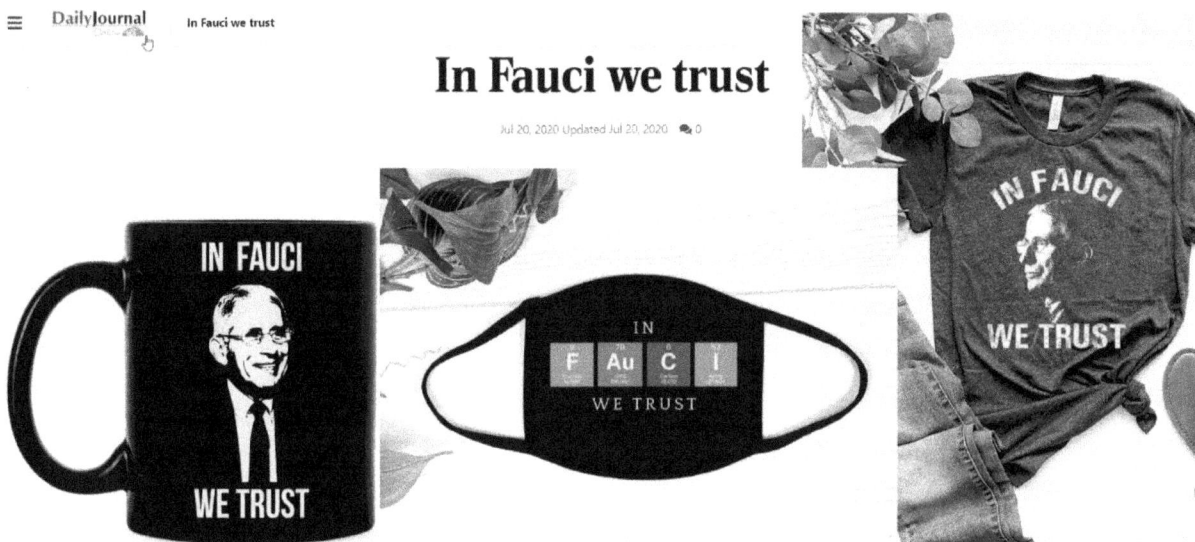

Author's collage of "In Fauci We Trust" products and headlines from various sources.

The extent to which public health had been exalted was made luminously apparent in May 2020 when the statue of Christ the Redeemer in Rio de Janeiro, Brazil, was lit up by a projector to appear as if Jesus was wearing a mask, with #MascaraSalva ("Mask Saves") emblazoned on his chest. This was followed in May 2021 by "Vaccine Saves," after the vaccination campaign in Brazil commenced in January with two healthcare workers symbolically receiving their shots at the foot of the statue of Christ.[175]

Charitable viewers might be inclined to see this as an act of symbolic analogy, or an invocation seeking divine blessing rather than any real worship of public health. But even if we assume such an interpretation to be true, this type of symbolism possessed a very real cutting edge, and testifies to the religious-level fervor that often drove masking.

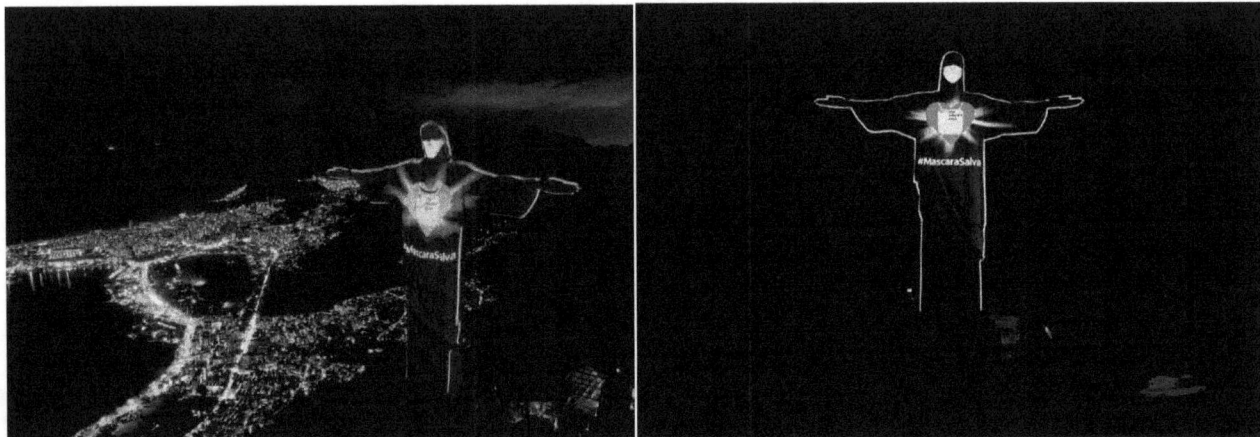

Kim, Allen, "Face mask projected onto Brazil's Famous Christ the Redeemer Statue," *CNN Travel*, May 4, 2020
https://edition.cnn.com/travel/article/rio-redeemer-statue-trnd/index.html

Shumaker, Lisa, "Brazil's Christ the Redeemer statue lights up for vaccine equality,"*Reuters*, May 17, 2021,
https://www.reuters.com/world/americas/brazils-christ-redeemer-statue-lights-up-vaccine-equality-2021-05-17/

A late 2020 study in the *Journal of Social Psychology* found that even *theoretical* intellectual challenges to COVID elimination approaches were met with responses more typically seen when "sacred values" (e.g., religious beliefs) are questioned. In the study, researchers crafted two identical proposals, with

---

175   Shumaker, Lisa, "Brazil's Christ the Redeemer statue lights up for vaccine equality," *Reuters*, May 17, 2021,
https://www.reuters.com/world/americas/brazils-christ-redeemer-statue-lights-up-vaccine-equality-2021-05-17/

the only difference being that one was framed as questioning anti-COVID measures, and one was framed as supporting them. One proposal hypothesized that the human suffering caused by COVID-19 elimination efforts outweighed the suffering that would occur if those measures were abandoned (i.e., that the prevention or cure might be worse than the disease), and vice versa. Both proposals contained the same amount of empirically validated information. The only difference was in how the proposal was framed — in what hypothesis was to be investigated. In theory, the two proposals should have received identical or nearly identical ratings from reviewers. What *actually* happened was that not only did participants rate the study proposal framed as questioning anti-COVID measures as "less accurate, less methodologically sound, and less valuable to society," but they also rated the *researchers*, personally, as being less competent and less trustworthy![176]

Tensions over masking bear a strong resemblance to tensions over religious beliefs because masking in general, and *compulsory* masking in particular, implicate many beliefs that people hold sacred. Under American law, "religious beliefs" are not defined as being limited to traditional organized religions. Rather, "Religious beliefs include theistic beliefs as well as non-theistic 'moral or ethical beliefs as to what is right and wrong which are sincerely held with the strength of traditional religious views.'"[177] In other words, the law classifies "'ultimate ideas' about 'life, purpose, and death'" — including ideas about death from disease and illness — as religious if they are "held with the strength of traditional religious views." In its guidance documents derived from courtroom precedents (precedents covered in detail in Part 5 of this book), the United States Equal Employment Opportunity Commission (EEOC) explains that:

> "Religion includes not only traditional, organized religions such as Christianity, Judaism, Islam, Hinduism, Sikhism, and Buddhism, but also religious beliefs that are new, uncommon, not part of a formal church or sect, only subscribed to by a small number of people, or that seem illogical or unreasonable to others… A belief is "religious" for Title VII purposes if it is "religious" in the person's "own scheme of things," i.e., it is a 'sincere and meaningful' belief that 'occupies a place in the life of its possessor parallel to that filled by . . . God.'"[178]

---

176   Maja Graso, Fan Xuan Chen, and Tania Reynolds, *Moralization of Covid-19 health response: Asymmetry in tolerance for human costs.* Journal of experimental social psychology, 2021. **93**: p. 104084-104084. https://pubmed.ncbi.nlm.nih.gov/33311735 or https://www.ncbi.nlm.nih.gov/pmc/articles/PMC7717882/

177   U.S. Equal Employment Opportunity Commission Office of Legal Counsel, "Section 12: Religious Discrimination," in *Compliance Manual on Religious Discrimination*, January 15, 2021, available online: https://www.eeoc.gov/laws/guidance/section-12-religious-discrimination#_ftnref26

178   U.S. Equal Employment Opportunity Commission Office of Legal Counsel, "Section 12: Religious Discrimination," in *Compliance Manual on Religious Discrimination*, January 15, 2021, available online:

During the COVIDcrisis, all of the organic and artificial social pressure, nudges, manipulation, and coercive measures combined to overpower even strong-willed individuals. For those who resisted all these pressures and still refused to put on a mask, some stronger, predominantly internal force was needed to counterbalance these external pressures. This internal force consisted of bedrock beliefs, often instinctually grasped in the heat of the moment. In this highly charged context, refusal to wear a mask was an expression of one's deepest beliefs about authority and the rights of the individual over the collective. It was a declaration of self-ownership, of non-slavery, of the same type of sincerely held religious belief that led Quakers such as William Penn to refuse to uncover their heads as a show of deference to authority.[179] Any symbolic political elements were important, and certainly implicated the Free Speech clause of the First Amendment, but were nevertheless subsidiary to even more bedrock principles that easily meet the threshold for religious beliefs recognized and protected by the First Amendment's Free Exercise Clause. By the same token, the decision to *wear* a mask could also be based on similarly held sacred values, and doubtless was for many people.

Many authors, including Dr. Zimbardo, have written about how individuals are constantly engaged in two-way exchanges with society. They are shaped by it, but also shape it. They adapt to its norms, rules, and prescriptions, but they also define and change them. [180]More than 40 years ago, Václav Havel, a Czech statesman known for his courageous opposition to Communism, described the simple dynamic created when people succumb to a totalitarian system of mandated, manipulative compliance: "They conform to a particular requirement and in so doing they themselves perpetuate that requirement."[181]

> "[They] create through their involvement a general norm and, thus, bring pressure to bear on their fellow citizens. And further: so they may learn to be comfortable with their involvement, to identify with it as though it were something natural and inevitable and, ultimately, so they may—with no external urging—come to treat any non-involvement as an abnormality, as arrogance, as

https://www.eeoc.gov/laws/guidance/section-12-religious-discrimination#_ftnref26

179   "The Quakers, William Penn included, suffered punishment rather than uncover their heads in deference to any civil authority. Braithwaite, The Beginnings of Quakerism (1912) 200, 229-230, 232-233, 447, 451; Fox, Quakers Courageous (1941) 113."
West Virginia Board of Education v. Barnette et al., 319 U.S. 624, 633 n 13 (1943), available online:
https://www.loc.gov/item/usrep319624/

180   Philip Zimbardo. *The Lucifer Effect*, New York: Random House Publishing Group, 2007, Kindle Edition, p. 266

181   Václav Havel, *The Power of the Powerless*. International Journal of Politics October, 1978.
https://www.nonviolent-conflict.org/resource/the-power-of-the-powerless/

an attack on themselves, as a form of dropping out of society."[182]

This process described by Václav Havel is *exactly* what happened with masks during COVID. People reacted so strongly to differences in masking or non-masking, not primarily because of any perceived infection control risk, but because the masked or unmasked person's mere *presence* implied a negation of their sacred values and implicitly reproached their own actions. Masking is so contentious because it taps directly into people's most basic beliefs and assumptions, putting irreconcilable conflicts between them very literally in-your-face.

Someone wearing a mask, who has adopted the beliefs that masks work and are necessary to protect against SARS-CoV-2, that all rules should be presumed lawful until proven otherwise, or that their mask-wearing is a manifestation of them being a good person, will naturally perceive an unmasked face as a form of assault on their worldview, their character, their body, or all three at once, and react accordingly. A similar effect and reaction can occur within someone refusing to put on a mask when they are ordered to do so. For many people during COVID, refusal to put on a mask in the face of intense social pressure, directives from authority figures, and pseudo-legal pressure to conform, was an important symbolic declaration of their deepest beliefs about authority and the rights of the individual over the collective — a collective selfishly asserting ownership over their body by demanding they put on a face covering. In light of this, is it any wonder that 2021 was the worst year on record for altercations on airplanes, and that 4,290 of the 5,981 incidents of "unruly" behavior during air travel were mask-related?[183]

Making masks mandatory also attached executive and public health authorities' sense of power and prestige to the act, making masking into an even more prominent ideological symbol than before. Many institutions, especially those accustomed to a high degree of control, such as correctional facilities and the Armed Forces,[184] not only drank their own Kool-Aid with regard to the evidence and ideology behind masking, but also came to view compulsory masking as representative of their overall authority. Any challenge to masking, in turn, was seen as an existential threat to that authority. Anything less than total compliance was deemed subversive and unacceptable.

---

182   Václav Havel, *The Power of the Powerless.* International Journal of Politics October, 1978. https://www.nonviolent-conflict.org/resource/the-power-of-the-powerless/

183   Brief of *Amici Curiae* 313 Airline Workers in Support of Appellant Urging Reversal, *Wall v. Centers for Disease Control & Prevention*, No. 22-11532, p. 22 (11th Circuit Court of Appeals, July 5, 2022) Available online: https://lucas.travel/wp-content/uploads/2022/07/Amicus-Brief-of-Airline-Workers-FILED.pdf

184   Often conflating *real* military readiness with military readiness defined by administrative fiat.

In the Stanford Prison experiment, the "guards" instituted a rule that prisoners must "eat at mealtimes and only at mealtimes."[185] When a new "prisoner," Prisoner 416, refused to eat, the "guards" saw this act of disobedience as an existential threat to their authority, because "such refusal could trigger further challenges to their authority from the others, who until now had traded rebellion for docility." To protect their "authority" the "guards" escalated their abusive behavior against the prisoners even further. The guards successfully turned several of the other prisoners against the new rebel by imposing collective punishment (taking away everyone's ability to have visitors) and blamed it on Prisoner 416's refusal to obey orders. Depressingly, the rest of the "prisoners" fell for this tactic to one extent or another, leaving their most effective resister isolated at a time when their collective support might have reversed the downward trajectory of the entire experiment (or at least their experiences within it). Instead of getting angry with the "guards" for unjustly and illogically making an arbitrary connection between their own visiting privileges and someone else's refusal to eat, most of the other "prisoners" got angry with the "troublemaker." The guards also intuitively made use of the psychological tools described earlier. For example, they ordered the other prisoners to tell Prisoner 416 *why* he was selfish, thus getting the other "prisoners" to publicly commit to their compliance in the same manner as the #IWearAMaskBecause campaign got people to publicly commit to their compliance with masks.[186]

Once the practice of mask-wearing took hold, deference to authority was just one of the psychological forces which helped perpetuate it. Confirmation bias (the propensity for people to look for what confirms their beliefs and ignore what contradicts their beliefs), motivated reasoning (selective skepticism, not applying the same standard of proof to things we already agree with), post-hoc justification, and the sunk cost fallacy all kicked in and further hampered honest investigation of the evidence. *Never* underestimate the power of confirmation bias and motivated reasoning — conclusions wandering in search of arguments. Many people who pushed masks and other anti-COVID measures would pay a high psychic price in self-image alone to admit that they were wrong. Politicians, judges, and public health officials are even less likely to admit that they were wrong for the same reason. The cost is high. Admitting a mistake to oneself is hard enough. Publicly doing so is even more rare, and takes a degree of moral courage matching or exceeding that required to never succumb to social pressures in the first place. In particular, people who coerced or pressured others into wearing masks will have a strong internal incentive to justify themselves and to use the alternative escape route retrospectively minimizing the extent to which they coerced and pressured others. This means that a

185   Philip Zimbardo. *The Lucifer Effect*, New York: Random House Publishing Group, 2007, Kindle Edition, p. 125.

186   These methods were not something the "guards" made up on the fly. By that point in the experiment, they'd had prior practice. Prisoner 416 was actually brought in to replace Prisoner 819, who Dr. Zimbardo sent home after finding him past the point of tears under similar treatment. The full details of these encounters, as well as the "guards'" and "prisoners'" accounts of what was going through their heads at the time, are covered in Chapter 6 of Zimbardo's book. For those who do not feel like reading, just search YouTube for "Prisoner 819 did a bad thing."

large portion of the administrative class who initiated, perpetuated, and enforced coerced masking is unlikely to ever admit, even to themselves, the extent to which they did so or that they were wrong to do so.

People who refuse to wear masks are not immune to these same self-justification incentives. I do not claim to be immune to these things either, and that is part of my point: *no one is!* These features of human nature are held in common by all of us. If no one is immune to these things, ask yourself these questions when trying to decide who is right: "who is trying to persuade you, who is trying to control you, and which side of the argument has been willing to sacrifice more of *their own material well-being* for what they believe?"

Key Takeaways

🔑 Public health during COVID took on the hallmarks typically associated with a religion.

🔑 Compulsory masking for COVID was, and remains, so controversial because it implicates core sacred values and beliefs.

🔑 Once mask-wearing was made compulsory, many authorities came to view refusal to comply as a personal affront, as well as an existential threat to their overall legitimacy, and responded accordingly.

Mask psychology conclusion — the experts know less than you think

> *"…the masses are much less likely than the elites to think that they should be overruling people whose stake and whose relevant knowledge for the issue at hand are far greater than their own. Moreover, the masses are less likely to have the rhetorical skills to conceal from others, or from themselves, that this is what they are doing."*
>
> Thomas Sowell, *Intellectuals and Society,* 2011[187]

Academics, professionals, intellectuals, and the intelligentsia tend to confuse lack of special knowledge or a lower knowledge per capita with having adequate knowledge to justify decision-making dictates. An enthusiastic amateur may often surpass an expert on a particular matter even within that expert's broader field, and when it comes to issues outside of their expertise, experts are no better off than the rest of us.

Herein lies a strong clue to answering the rhetorical question of why doctors wear facemasks. The ritual of mask-wearing for infection control came into use late in the 19th century, when germ theory was in its infancy, and doctors were implementing every practice they could think of which intuition suggested might help decrease germ transmission. Medical schools and public health programs do not offer classes in masks. Masks are typically just one small portion of classes on infection control — a few PowerPoint slides or at most a 1-hour lecture. There is simply too much other material to cover in these programs to permit going into depth on such a niche issue. As we've already seen, the psychology harnessed by mask-wearing is enough on its own to make the practice self-perpetuating

---

187   Thomas Sowell, *Intellectuals and Society*, New York: Basic Books, 2011, Kindle Edition, pp. 29-30.

in the absence of scientific evidence. These psychological forces are strong enough to perpetuate masking even when the practice is *contrary* to the clear and convincing weight of scientific evidence. Generations of doctors have been taught by respected mentors and other authority figures that they should wear masks to prevent disease transmission, and the longer a doctor has been in practice, the longer he or she has routinely worn a mask. Given that doctors as a group are no more immune to behavioral psychology than any other portion of the population, imagine the effect on belief of decades-long, rigorous, ritualistic, socially enforced, and socially validated mask-wearing. In fact, doctors were routinely penalized for *not* wearing masks in many situations. Sound familiar? The real reasons doctors have been wearing masks are largely the same reasons everyone else started wearing masks for COVID.

The methods used to push facemasks and other COVID-19 mitigation measures upon the general population were direct descendants of trailblazing psychological research on obedience, compliance, and social influence conducted by Solomon Asch, Stanley Milgram, Philip Zimbardo, their contemporaries, and their successors. Everything from gradual successive steps to diffusion of responsibility. "15 days to flatten the curve." Essential vs. nonessential. Now the masks. Now the vaccine, but you still have to wear the masks. The people making the scientific recommendations were not the ones issuing the actual orders, so officials' pet scientists could say that they were just issuing recommendations, and the officials could claim to simply be following The Science™. The people issuing punitive citations for noncompliance were not the same ones who imposed the regulations. The people conducting inspections without warrants were not the same ones enforcing the inspectors' findings by closing down businesses or arresting the owners for trying to earn a living under the "new normal." Collective punishments were decreed in the name of "safety" and "public health." Neighborhoods and businesses were locked down. The Stanford Prison Experiment's dynamics were writ-large, as the people being hurt by lockdowns were incessantly told to blame their objecting and non-compliant neighbors instead of the people who were actually hurting them.

The moralizing dimension of mask mandates was conveyed by euphemistic language, which turned compliance with abuse into a form of virtue. (People who didn't wear their masks right were not showing enough "mask discipline" for example.) Contrary to the strident and vitriolic moralizations of politicians, pundits, and televised public health officials, compliance with mask-wearing was never a route to ending the COVIDcrisis, only to deepening it. Compulsory mask-wearing was ideology masquerading as public health. It was coerced symbolic speech under the pretext of infection control. Voluntarily giving an inch with masks drastically increased compliance with other COVID-control measures, including lockdowns, and led directly to giving a mile with compulsory vaccinations. It is not "just a mask," and it never *was*.

Key Takeaways

🔑 One of the fundamental errors at the heart of compulsory masking was the false belief that individual rights are granted by the government, or "society."

🔑 The liberties that each individual human naturally possesses exist independently of whether they are enforceable or recognized by the government.

🔑 America's Founders repudiated the view that individual rights turn on utilitarian numbers games.

🔑 The individual rights of personal security and self-preservation do not extend to forcing everyone else to wear a mask or stay at home as a pre-emptive self-defense measure. No aggregate community right of personal security and self-preservation does this either.

🔑 There is no individual or collective "right" to safety or public health which can be used to override *actual* individual rights like those recognized in the Bill of Rights.

Part 4:

Pro-Mask Philosophy is Bad Philosophy

The Malformed Philosophical Beliefs and Assumptions
that Underlie Compulsory Masking

Why include this section at all?

"Your right to swing your fist ends with my face!" ["Therefore, you must wear a mask!"]

"There is no Constitutional Right to infect others." ["Therefore, you must wear a mask!"]

"People who refuse to wear masks are selfish." ["Therefore, you must wear a mask!"]

"Your liberties are not absolute." ["Therefore, you must wear a mask!"]

"We're all in this together." ["Therefore, you must wear a mask!"]

"Refusing to wear a mask violates others' rights." ["Therefore, you must wear a mask!"]

- Rhetoric, slogans, and sophistry masquerading as philosophy.[188]

---

188   Rhetoric of this type saturated public debate during the COVIDcrisis. The quotes given are generic rather than taken from any particular source, but they encapsulate some of the major sound bites used to shut down debate and justify the draconian measures imposed on populations across the world in the name of stopping COVID. Precise examples of this are provided below for the sake of completeness:

"[Y]our right to swing your fist ends at someone else's nose."
Dr. Tom Frieden, "Making – and muddling – the case for mask and vaccine mandates," *CNN*, February 25, 2022. Online at: https://www.cnn.com/2022/02/25/health/mask-vaccine-mandates-analysis-frieden/index.html

"We do not have a constitutional or protected right to infect others."
Judge John Kastrenakes, *Machovec v. Palm Beach County*, No. 2020CA006920AXX — Filing #110806335, 27 July 2020, (15th Judicial Circuit, Florida), p.12, available online: https://clearinghouse.net/doc/109346/ & https://www.floridacivilrights.org/wp-content/uploads/2020/07/Order-Denying-Temporary-Injunction.pdf

"People who refuse to wear masks are selfish and stupid."
Mike Barnhart, "Selfish and stupid," *The Register-Guard*, August 1, 2020. Online at:
https://www.registerguard.com/story/opinion/letters/2020/08/01/selfish-and-stupid/112782328/

"Our fundamental liberties are *not* absolute."
Phyllis Feng, "Mandated Masks: Which Comes First, The Individual or the Group?" *Affinity Magazine*, September

If I moved straight from explaining the science, and how the psychological magic trick worked, to the legal duels over masking, I would be ignoring the most important facet of compulsory masking. The Science™ provided a pretext and a cloak, the psychology involved explains much of *how* most of the world came to wear masks so religiously, and the legalities provided a means of implementation and enforcement. All these aspects are very important, and include some overlap, but they do not address the ideological *source* of the infection, which must be exposed and dealt with if any other remedies are to have real staying power.

In the science sections, I explained why the belief that masks work is false. In the psychology section, I dealt with some of the mechanisms by which many people came to hold the belief that they should wear masks. My goal was to help readers rebut the conspicuous errors and arm themselves against being deceived by similar tactics in the future. The manipulative power of the tactics and techniques used goes down dramatically when targets become aware of what is going on. Principled self-awareness and critical thinking form one of the most effective broad-spectrum defenses. It helps us know when to turn off our autopilot and carefully evaluate both what is going on around us and what we are being told. It helps us decide when to say "yes," and when to dig in our heels and not give an inch. It also helps us discern when we are most likely being used as a means to an end. (Hint: the more influence tactics and shoddy science employed, the more likely manipulation is involved.)

At the most, however, showing *how* someone came to hold a belief can only show that someone lacks good *warrant* for that belief. Showing that someone lacks a good foundation for their belief is not at all the same thing as showing the belief itself to be false. If you believe the earth is a sphere because you read it in a fortune cookie, that might not be a good *reason* for believing that the earth is a sphere, but your inadequate *basis* for holding that particular belief does not affect whether the belief *itself* is true or false. Trying to prove someone's belief or course of action erroneous by showing how they *arrived* at that belief or course of action is what is known as a textbook genetic fallacy. In this section, I address the false core beliefs that underlay compulsory masking. These false beliefs include the idea that if masks work, people must have a moral duty to wear them, and that it then becomes morally acceptable to compel

27, 2020, online:
https://affinitymagazine.us/2020/09/27/mandated-masks-which-comes-first-the-individual-or-the-group/

"We're all in this together. #WearAMask"
American Hospital Association, "Sample Social Media Messages," *Wear A Mask Campaign*, August 2020, online:
https://www.aha.org/wearamask#wamsoc

"My right to life and liberty is restricted when the letter writer refuses to wear a mask. Her liberty is not infringed one bit by that same requirement."
Sheryl Feinman, "Medical masks are not new," *The Tampa Bay Times*, June 27, 2020, online:
https://www.tampabay.com/opinion/2020/06/27/does-wearing-a-mask-violate-our-freedoms-heres-what-readers-say/

their use. I hesitated about writing this more philosophically oriented section for months. I consider myself a well-informed layman, not a teacher. As such, any teaching in these chapters is meant for the purpose of supporting passionate arguments — not for browbeating or talking down to readers. But well-informed laymen can (and *should*) be competent to accurately recognize and call out nonsense when it starts coming from someone with a degree. Of themselves, degrees confer no special knowledge. They are merely a proxy indicator of specialized knowledge in narrow fields. Thomas Sowell spent an entire book describing the fallacies, arrogance, and ignorance of intellectuals and professionals who operate on the mistaken, self-serving premise that a few people with superior *per capita* knowledge in specific fields can and should override everyone else's decision-making power for the greater good:[189]

"If no one has even one percent of the knowledge currently available, not counting the vast amounts of knowledge yet to be discovered, the imposition from the top down of the notions in favor among elites, convinced of their own superior knowledge and virtue, is a formula for disaster… The ignorance, prejudices, and groupthink of an educated elite are still ignorance, prejudice, and groupthink."[190]

More often than not, the credential held by a sermonizer is unrelated to the topic being lectured on, which doesn't stop the credentialed from making sweeping self-refuting pronouncements like Stephen Hawking's famous philosophical assertion that "philosophy is dead."[191] Even a degree in philosophy provides no guarantee that one's overall argument is valid, though it is usually a decent proxy for verbal skill. Well-informed laymen are perfectly capable of putting out solid arguments if they put in the time and effort. Besides, never forget that we locked down, mandated masks (and worse) not based on well-reasoned arguments, but based on sound bites, slogans, and moral browbeating made plausible mostly by carefully cultivated and totally unnecessary terror. The real source of the mental infection — an infection which mask-wearing only exacerbated — came through primarily as the hidden premises underlying sound bites, rhetorical slogans, and loaded questions. Crucial assumptions, axioms, and logical fallacies were occasionally mentioned and even directly debated on the periphery, but closer to the propulsive core which guided COVID policy, these remained, for the most part, unspoken or only implicit. Fear, force, and rhetorical sleight-of-hand

189   Thomas Sowell, *Intellectuals and Society*, New York: Basic Books, 2011, Kindle Edition.
190   Thomas Sowell, *Intellectuals and Society*, New York: Basic Books, 2011, Kindle Edition, pp. 21-22.
191   Stephen Hawking and Leonard Mlodinow, *The Grand Design*, London: Bantam Press (2010), p. 1.

let proponents of unexamined, false, and destructive philosophical assumptions use non sequiturs to jump our societal center-of-mass onto railroad tracks headed straight off a cliff.

If law ceases to reflect objective moral reality, it loses legitimacy. [192]Legal reforms based on bad or mistaken philosophical assumptions tend to be fragile and short-lived, unless they get it right by accident during trial-and-error. (One thinks of Supreme Court Justice Oliver Wendell Holmes Jr.'s famous observation that "The life of the law has not been logic: it has been experience.")[193] If the unstated underlying moral beliefs and bad reasoning supporting compulsory masking are not stated directly and explicitly dealt with, then no matter which way the other debates surrounding facemasks eventually shake out, the same basic issue will recur over and over with slightly different decorations every time enough people with power become convinced that a particular practice will save or lengthen some lives if universalized. Before we turn to details of the legal and constitutional cases against compulsory masking, it is essential that the bad philosophical assumptions undergirding compulsory masking be exposed and addressed.

---

192   If you hold to a positivist theory of law that denies valid laws must satisfy certain minimum moral standards, you may take issue with this premise. I hope and pray such legal theories never achieve dominance in America.

193   Oliver Wendell Holmes, Jr., *The Common Law*, Boston: Little, Brown and Company, (1881), p. 1, available online: https://archive.org/details/commonlaw0000oliv_m6f7/page/n23.

Thomas Sowell deemed this quote important enough to include it no less than four times in his book: *Intellectuals and Society*.

Rights and "rights": a fundamental confusion

My brother-in-law and sister are two of the best and most intelligent people I know. Not only are both of them MDs (psychiatrists), but they gave up hundreds of thousands of dollars in foregone income to do medical missions in Africa for years at a time. They are also engaged, loving parents. Why bring up my immediate family? Because doing so provides a perfect example of how good people could — and did — get essential matters of principle badly wrong when it came to COVID. Like families around the world, we engaged in heated arguments over the response to COVID. Direct experience vividly brought to life the real-world implications of what had previously been cordial (or entirely unnoticed) abstract disagreements about fundamental principles. The governmental response to COVID split families along lines which often turned out to run thicker than blood. In my case, most of the vehemence was between myself and my sister. My brother-in-law got dragged in because of his eloquence. His primary contribution to our inter-sibling debate was to articulate, in succinct writing, one of the most foundational errors underlying the measures taken during COVID:

> "First people must realize that 'civil liberties' only exist with a civilization, and not in a libertarian sense; American civil liberties only exist within the American schema... US citizens unfortunately commonly think that by virtue of being US citizens we cannot be told that we can't do something; the patent absurdity of this is manifest by the fact that our rights only exist by virtue of being US citizens, which exist only with the US schema, which only exists within the context of having a Federal government which exists to protect those rights.

A simpler way of saying this is "you are only free because you submit to a government that mandates that freedom."[194]

In other words, your rights, your liberties, your freedoms, are not something you possess because you are a human and all humans have certain unalienable rights. Instead, your rights depend for their existence on your "society," the government, and those of your neighbors who at the moment are scared witless and posting to social media that anyone who doesn't stay home and wear a mask is committing negligent homicide. This pernicious but widespread error confuses and conflates the *existence* of rights with one's ability to *exercise* them. It is also a view that America's Founders rightly, explicitly, and vigorously repudiated. The philosophy undergirding the American Revolution presupposed the existence of certain individual rights which have an objective existence *independent* of any government.

For something to exist objectively means that it exists on its own, independent of any person's (or any group's) perception or opinion. Individuals and societies may disagree about what is right or wrong, but right and wrong are not just social constructs. A knowable realm of objective morality exists externally to ourselves. Likewise, each of us has individual rights that exist objectively, independent of government and independent of what other people or even other societies may think. To say otherwise is to affirm a form of sociocultural relativism whereby morality and your rights are dependent on the society you live in. Sociocultural relativism renders it impossible to criticize *any* society's morals because if sociocultural relativism is true, then there is no objective external standard by which to judge. Morality becomes nothing more than the personal preference of the most powerful or most persuasive. Sociocultural relativism makes the Aztec institution of human sacrifice and the Spanish conquests equally acceptable. It would also validate the defense raised by German doctors on trial after WWII in Nuremberg — after all, everything they did under the Third Reich was perfectly legal (even mandated) according to their government at the time. But some things really are right and some things really are wrong. Even if every human on the planet were to fail to perceive a moral truth, whether through innocent ignorance or willful blindness, that moral truth still exists. It has been a source of bitter humor to see people deny the existence of objective moral values and duties one moment, and then turn around the next moment and assert that it is objectively wrong to refuse to wear a mask.

---

194   Personal correspondence, April 2, 2020. Note: Just as I would not want my own position on an issue mis-characterized, I do not want to mischaracterize my brother-in-law's or sister's positions on this one. I do not think that this quote was taken out of context, but I provided my sister and brother-in-law an advance copy of this book, along with an offer to host any comments, response, clarification, or rebuttal they would like to make. I will host any public written response they provide as a download available on this book's website at www.thebookonmasks.com/extras.

America was founded on the belief that, regardless of what your immediate government happens to say, it is self-evident — *axiomatic* — that "all men are created equal, that they are endowed by their Creator with certain unalienable Rights, that among these are Life, Liberty and the pursuit of Happiness." Americans are blessed enough to live in a country founded by individuals who correctly believed that our individual rights have objective existence, independent of the government, and that these rights are inalienable. Generations of Americans since that time have grappled with working out the practical implications of this belief. What we call "civil liberties" are simply the best way the framers of our Constitution could find to articulate those fundamental rights and principles and enshrine them into law. They did their best to ensure that the government they designed was perpetually forced to recognize these rights, and constrained from perpetrating the kind of free-reign tyrannies which governments across the world plunged headlong into during the COVIDcrisis. The fact that most other governmental entities around the world (and most of even America's authorities) were grossly derelict in their duty to reaffirm and vigorously protect those liberties does nothing to defeat the contention that individual civil liberties exist independent of government.

James Wilson, who holds the distinction of being the only man to sign the Declaration of Independence, sign the Constitution, *and* serve as a Justice on the United States Supreme Court, asked and answered rhetorically:

> "What was the primary and the principal object in the institution of government?... was it to acquire new rights by a human establishment? Or was it, by a human establishment, to acquire a new security for the possession or the recovery of those rights, to the enjoyment or acquisition of which we were previously entitled by the immediate gift, or by the unerring law, of our all-wise and all-beneficent Creator? The latter, I presume, was the case..."[195]

Your Natural God-given rights as an individual do not disappear simply because you cross a national border, or because you live in a society with laws that violate those rights. The purpose of government is to *protect* the rights of individuals, not to confer rights on individuals only insofar as doing so benefits the collective "society." Even if one chooses to accept the slogan "liberty is not absolute" as a face-value argument (which, on its own, it is not), the natural rejoinder is that the ability of politicians and public health officials to curtail liberty is even *less* absolute.

---

195   Kermit L. Hall and Mark David Hall, eds., *Collected Works of James Wilson*, 2 vols. (Indianapolis, Ind.: Liberty Fund Press, 2007), p. 1053-1054. Online at: http://files.libertyfund.org/files/2072/Wilson_4140.01_LFeBk.pdf and http://files.libertyfund.org/files/2074/Wilson_4140.02_LFeBk.pdf

In the same lecture just quoted, James Wilson also cited one of the foremost authorities on English Common Law, Sir William Blackstone. English Common Law was embodied in the *Magna Carta* of 1215, with roots and influences stretching back even further to Anglo-Saxon, Roman, and Mosaic law. William Blackstone was the preeminent English jurist of his time, and his magisterial *Commentaries on the Laws of England* was essential reading for every lawyer and legal theorist of the time. His work has been quoted by numerous American Founders including, John Jay, John Adams, Alexander Hamilton, and the first Supreme Court Chief Justice John Marshall. Later American Statesman like Abraham Lincoln also quoted Blackstone, as have numerous Supreme Court rulings throughout the 20th and 21st Centuries.[196] One survey of more than 15,000 founding-era documents from 1760 to 1805 found that Blackstone was the second-most cited secular author overall, and the most-cited by the 1790s, when the Bill of Rights was passed. [197]"The absolute rights of every Englishman,' wrote Sir William Blackstone, 'are, in a political and extensive sense, usually called their liberties.'"[198]

This makes Blackstone's *Commentaries* a natural starting point for examining the absolute, unalienable rights that got trampled by the COVID response in general and by compulsory masking in particular. In his *Commentaries,* Blackstone distinguishes between absolute rights that every human possesses by nature, and relative rights that arise from the specific circumstances of a society. Blackstone begins his *Commentaries* by defining these absolute rights in Book 1 ("Of the Rights of Persons"), Chapter 1 ("Of the Absolute Rights of Individuals"):

---

196   Miles, Albert S. (2000). "Blackstone and his American Legacy". Australia & New Zealand Journal of Law and Education. 5 (2). ISSN 1327-7634 — page 57, available online:
https://www.anzela.edu.au/assets/anzjle_5.2_-_4_albert_s_miles,_david_l_dagley__christina_h_yau.pdf
197   The French Philosopher Baron de Montesquieu (who among other things articulated the theory of separation of powers) was quoted slightly more frequently than Blackstone over the period 1760 to 1805.
Lutz, Donald S. "The Relative Influence of European Writers on Late Eighteenth-Century American Political Thought." *The American Political Science Review* 78, no. 1 (1984): 189–97. https://doi.org/10.2307/1961257 .
198   William Blackstone, *Commentaries on the Laws of England Volume I,* Clarendon Press, Oxford (1765), p. 127 (available online: https://archive.org/details/bim_eighteenth-century_commentaries-on-the-law_blackstone-sir-william_1766/page/126

> "The absolute rights of every Englishman (which, taken in a political and extensive sense, are usually called their liberties)…"

Quoted in Kermit L. Hall and Mark David Hall, eds., *Collected Works of James Wilson*, 2 vols. (Indianapolis, Ind.: Liberty Fund Press, 2007), p. 1054. Online at: http://files.libertyfund.org/files/2072/Wilson_4140.01_LFeBk.pdf and http://files.libertyfund.org/files/2074/Wilson_4140.02_LFeBk.pdf

"[T]hese may be reduced to three principal or primary articles; the right of personal security, the right of personal liberty; and the right of private property[.]"[199]

"The right of personal security consists in a person's legal and uninterrupted enjoyment of his life, his limbs, his body, his health, and his reputation."[200]

"[P]ersonal liberty consists in the power of locomotion, of changing situation, or removing one's person to whatsoever place one's own inclination may direct; without imprisonment or restraint, unless by due course of law… it cannot ever be abridged at the mere discretion of the magistrate, without the explicit permission of the laws. Here again the language of the great charter is, that no freeman shall be taken or imprisoned, but by the lawful judgment of his equals, or by the law of the land… The confinement of the person, in any wise, is an imprisonment. So that the keeping a man against his will in a private house, putting him in the stocks, arresting or forcibly detaining him in the street, is an imprisonment."[201]

"The third absolute right, inherent in every Englishman, is that of property: which consists in the free use, enjoyment, and disposal of all his acquisitions, without any control or diminution, save only by the laws of the land."[202]

You do not need to be a biologist to tell a man from a woman; you do not need to be a veterinarian to tell a dog from a cat, and you do not need to be a visionary or someone with special legal training to see that during the COVIDcrisis, all three of these elemental rights were repeatedly and flagrantly

---

199   William Blackstone, *Commentaries on the Laws of England Volume I,* Clarendon Press, Oxford (1765), p. 129, available online: https://archive.org/details/bim_eighteenth-century_commentaries-on-the-law_blackstone-sir-william_1766/page/128

200   William Blackstone, *Commentaries on the Laws of England Volume I,* Clarendon Press, Oxford (1765), p. 129, available online: https://archive.org/details/bim_eighteenth-century_commentaries-on-the-law_blackstone-sir-william_1766/page/128

201   William Blackstone, *Commentaries on the Laws of England Volume I,* Clarendon Press, Oxford (1765), pp. 134-136, available online: https://archive.org/details/bim_eighteenth-century_commentaries-on-the-law_blackstone-sir-william_1766/page/134

202   William Blackstone, *Commentaries on the Laws of England Volume I,* Clarendon Press, Oxford (1765), p. 138, available online: https://archive.org/details/bim_eighteenth-century_commentaries-on-the-law_blackstone-sir-william_1766/page/138

violated at the panic- or power-driven whims of magistrates, governors, mayors, judges, public health officials, and petty bureaucrats of every type. In many places, these basic human rights are still being routinely violated on the same or similar pretexts.

Blackstone's jurisprudence, while foundational, was nevertheless imperfect and drew criticism. On the one hand, proto-utilitarians like his contemporary Jeremy Bentham went after Blackstone's work from the starting point that: "it is the greatest happiness of the greatest number that is the measure of right and wrong."[203] (More on the problems with utilitarianism later.) On the other hand, while Blackstone affirmed absolute individual rights, his definitions and elaborations included caveats which weakened those rights to the point where, in practice, they were absolute in name only:

> "But every man, when he enters into society, gives up a part of his natural liberty, as the price of so valuable a purchase; and, in consideration of receiving the advantages of mutual commerce, obliges himself to conform to those laws, which the community has thought proper to establish.[204]

> "The law, which restrains a man from doing mischief to his fellow citizens, though it diminishes the natural, increases the civil liberty of mankind." [205]

America's Founders learned from painful lived experience that generalized caveats and circumscription of individual rights would inevitably be abused by rulers who claimed the authority to make law, whether in the form of Parliament, King George III, or their deputies. The Founding Fathers' cornerstone written responses were the *Declaration of Independence, The Constitution*, and the *Bill of Rights*, which explicitly codified a broader, stronger view and protection of Natural Rights than Blackstone articulated. James Wilson countered Blackstone's assertion that men gave up some of their natural rights when they entered society by using a rhetorical question to point out (with the

---

203   Jeremy Bentham, *A Fragment on Government; or A Comment on the Commentaries*, Dublin (1776), p. i, available online: https://archive.org/details/bim_eighteenth-century_a-fragment-on-government_bentham-jeremy_1776_0/page/n1

204   William Blackstone, *Commentaries on the Laws of England Volume I*, Clarendon Press, Oxford (1765), p. 125, available online: https://archive.org/details/bim_eighteenth-century_commentaries-on-the-law_blackstone-sir-william_1766/page/124

205   William Blackstone, *Commentaries on the Laws of England Volume I*, Clarendon Press, Oxford (1765), p. 125-126, available online: https://archive.org/details/bim_eighteenth-century_commentaries-on-the-law_blackstone-sir-william_1766/page/124

medieval theologian Thomas Aquinas) that even in a state of nature, men never possessed the "right" to do mischief to anyone in the first place.[206]

America's Founders asserted collective liberty to protect individual liberty. Ironically, a major objection to the Bill of Rights was the contention that it actually *undermined* the notion of limited government and individual rights because the very existence of a document reserving specific fundamental rights to individuals implied that everything *not* listed was somehow surrendered. Alexander Hamilton raised this objection in *Federalist* 84:

> "… the people surrender nothing; and as they retain everything they have no need of particular reservations… bills of rights, in the sense and to the extent in which they are contended for, are not only unnecessary in the proposed Constitution, but would even be dangerous. They would contain various exceptions to powers not granted; and, on this very account, would afford a colorable pretext to claim more than were granted. For why declare that things shall not be done which there is no power to do? Why, for instance, should it be said that the liberty of the press shall not be restrained, when no power is given by which restrictions may be imposed?"[207]

The belief that liberties (civil or otherwise) only *exist* because one lives in a particular society, or that one's right to do something only comes from the government, are just two of the false premises underlying compulsory masking. While *some* peripheral rights do indeed come into being as a product of citizenship, Americans and all people across the globe also possess certain inalienable fundamental rights which were trampled during COVID. These rights exist objectively, are God-given (or for the non-theist, Natural), inalienable, and inhere in all members of mankind, regardless of national borders.

---

206   I am indebted to Hadley Arkes for his essay, "A Natural Law Manifesto" which provided me with a jumping off point for hours of edifying research, reading, and writing as I followed up the original sources he quotes and references, including James Wilson's essay "Of the Natural Rights of Individuals," Thomas Aquinas' *Ethics*, and the particular salience of *Federalist* 84. At the time of this writing, Hadley Arkes' essay is available online at https://jameswilsoninstitute.org/about/a-natural-law-manifesto

207   "The Federalist No. 84, [28 May 1788]," *Founders Online,* National Archives, https://founders.archives.gov/documents/Hamilton/01-04-02-0247. [Original source: *The Papers of Alexander Hamilton*, vol. 4, *January 1787–May 1788,* ed. Harold C. Syrett. New York: Columbia University Press, 1962, pp. 702–714.]

A more moderate iteration of the mistaken belief above, that "the government exists to temper individuals' actions so as to assure that the public is protected,"[208] is partially true (if by "the public" one means "other individuals"), but this does not extend to things like compulsory masking. In his book, *Principles For A Free Society: Reconciling Individual Liberty With The Common* Good, legal scholar Richard Epstein (who has no connection whatsoever to the Epstein that owned an island prior to not hanging himself) says something very similar: "social obligations have to be structured to limit the risk of unbridled self-interest threatening the welfare of all individuals."[209] However, as he explains, there are some serious qualifiers to this concept of social obligation:

> "The proper standard of the common good does not only authorize state action; it also hems it in."[210]

> "[T]he forces that link individual liberty to the common good are far stronger than those that seemingly drive them apart… a sound legal order is one that responds to the fragility of knowledge by giving no one absolute control and power. It seeks the dispersion of power across individuals and social groups."[211]

A proper understanding of "the common good" serves to *limit* government action as much or more than it permits the government to act. Government exists to secure individuals' rights against one another and from outside aggression. Outside aggression or unbridled self-interest may originate from other individuals, groups of individuals, or the government itself. An essential role of government is to temper the actions of the mob so that individuals are not unjustly deprived of their rights by mob action (say, for example, hypothetically speaking, during a mass panic). This is true even, or *especially*, when the size of the mob grows to include a majority of the population. As The Father of the Constitution, James Madison, said in his 1792 *Essay on Property*:

---

208   This phrase is also reproduced nearly verbatim from elsewhere in my brother-in-law's correspondence quoted earlier: "the government exists to temper their actions so as to assure that the public is protected."

209   Richard A. Epstein. *Principles For A Free Society: Reconciling Individual Liberty With The Common Good*, Reading, Massachusetts: Perseus Books (1998), p. 19, available online: https://archive.org/details/principlesforfre0000epst/page/18/mode/2up

210   Richard A. Epstein. *Principles For A Free Society: Reconciling Individual Liberty With The Common Good*, Reading, Massachusetts: Perseus Books (1998), p. 4, available online: https://archive.org/details/principlesforfre0000epst/page/4/mode/2up

211   Richard A. Epstein. *Principles For A Free Society: Reconciling Individual Liberty With The Common Good*, Reading, Massachusetts: Perseus Books (1998), p. 9-10, available online: https://archive.org/details/principlesforfre0000epst/page/8/mode/2up

"Government is instituted to protect property of every sort; as well [as] that which lies in the various rights of individuals, as that which the term particularly expresses. This being the end of government, that alone is a just government which impartially secures to every man whatever is his own."[212]

Ironically, much of the destruction of *actual* individual rights during the COVIDcrisis was done under the cover of rhetoric which either asserted *non-existent* rights or over-extended real ones. Thomas Sowell succinctly summarized this process back in 2011:

"Much advocacy by intellectuals involves assertions of 'rights,' for which no basis is asked or given. Neither constitutional provisions, legislative enactments, contractual obligations, nor international treaties are cited as bases for these 'rights.' Thus there are said to be 'rights' to 'a living wage,' 'decent housing,' 'affordable health care,' and numerous other benefits, material and psychic. That such things may be desirable is not the issue. The real issue is why such things are regarded as obligations—the logical corollary of rights—upon other people who have agreed to no such obligation to provide these things. If someone has a right, someone else has an obligation."[213]

Foremost among the newly discovered "rights" invoked during the COVIDcrisis were "health," especially "public health," and open-ended "safety." "Health is a fundamental human right," proclaimed World Health Organization Director-General Tedros Ghebreyesus in 2017, three years prior to COVID.[214] Everyone subjected to any manner of discrimination or duress for the purpose of coercing their mask wearing or vaccination will no doubt be relieved to find out that, in the same article, Ghebreyesus specified that, "the right to health also means that everyone should be entitled to control their own health and body... free from violence and discrimination." Apparently, a lot of people, including Ghebreyesus himself, missed that part of their own memo. The right to control one's own body

---

212   "For the National Gazette, 27 March 1792," *Founders Online*, National Archives, https://founders.archives.gov/documents/Madison/01-14-02-0238 . [Original source: *The Papers of James Madison*, vol. 14, *6 April 1791–16 March 1793*, ed. Robert A. Rutland and Thomas A. Mason. Charlottesville: University Press of Virginia, 1983, pp. 266–268.
213   Thomas Sowell, *Intellectuals and Society*, New York: Basic Books, 2011, Kindle Edition, p. 108.
214   Tedros Adhanom Ghebreyesus, "Health is a fundamental human right," World Health Organization, December 10, 2017. Online: https://www.who.int/news-room/commentaries/detail/health-is-a-fundamental-human-right (last accessed January 14, 2023)

certainly exists, but it does not come from any general open-ended right to optimal health. Saying we have a right to health is like saying that we have a right not to get sick — it's just malformed.

The basis on which health and safety could be declared "rights" was, of course, never articulated (it was simply taken as an *a priori* given). A frequently used tactic was (and is) to simply insert these newly created "rights" into a list of actual rights, thereby giving them a veneer of legitimacy via simple association. People who objected, or who asked why, exactly, the right to health and safety would demand measures like compulsory universal mask-wearing, found themselves on the receiving end of that classic rebuttal favored by those who resent having their positions challenged: vitriolic *ad hominem* attacks.

We have seen there is an unquestionable individual right of personal security which all humans possess, and this includes the right of self-preservation. As America's 5th President and last Founding Father, James Monroe, said when addressing Congress in 1818, "The right of self defense never ceases. It is among the most sacred, and alike necessary to nations and to individuals[.]"[215] The American Second Amendment takes this as a given, for the right to self-preservation includes the right of self-defense, which, in turn, entails the right to possess the *means* of self-defense. However, to say that the rights of personal security and self-preservation imply the two much broader (and often conveniently vague) rights to health or safety is to indefensibly overextend them.

My right to property includes the right to work and dispose of my earnings without being robbed, but this does not imply that others have a legal or moral obligation to provide me with employment or forego competing with me for a job. My right of personal security does not include the right to preemptively disarm one of my neighbors (much less *every* neighbor) who has not done me any harm and has not displayed both a means and intention of imminently doing so. In a similar vein, my right to not be assaulted or robbed does not imply that my neighbor has a general moral or legal obligation to purchase a weapon, act as my bodyguard, or even install a security camera on his property. It also does not imply that my neighbor has a moral or legal obligation to protect himself, even if his doing so may benefit me indirectly. To try to use the right of personal security or self-preservation so broadly and preemptively is a gross overextension. Lockdowns and mask mandates went even further. Lockdowns and mask mandates implicitly assumed a broad, nonsensical right not to get sick, which was somehow enforceable on everyone else, superseding multiple preexisting *actual* rights to personal security, personal liberty, and personal property. Even presuming for the sake of argument that masks work, my right to personal security does not entail that other people have

---

215   James Monroe, Second Annual Message to Congress, Delivered November 16, 1818, transcript available online: https://millercenter.org/the-presidency/presidential-speeches/november-16-1818-second-annual-message

a moral or legal obligation to wear a mask on the rationale that doing so might indirectly benefit myself or someone else I care about.

My neighbor's right of personal security entails the right to not have something forced *into* his body, *onto* his body, or taken *from* his body against his will. In the same way, my right of personal security includes no component by which I can justify violating my neighbor's absolute right of personal property by forcing him to give up his job, stay at home, or wear a mask based merely on the presence of some disease in the community which he may or may not have. This remains true no matter how frightening or dangerous such a disease may be to me personally, how widespread it may be in the community, or how artfully and incessantly details about that disease happen to be presented. My neighbor is not responsible for my health, and vice versa. Similarly, we do not mandate blood and bone marrow donations for good reason. Your body is your most fundamental property, and your right to its use exceeds any claim which "society" may exert. The same principles that preclude involuntary servitude or mandating blood, bone marrow, tissue and organ donations also preclude mandating masks, vaccines, or any number of other medical interventions. My right to not be assaulted with a particular bioweapon (either naturally occurring or man-made) does not imply that my neighbor has an obligation to protect me by taking every possible precaution to protect himself. This is especially true if doing so involves him making use of a medical intervention he strongly objects to.

But someone may respond that they do not consent to being infected with a virus, and therefore *you* have a moral and legal duty to wear a mask. Apart from being a non sequitur, this objection is misconceived. Refusing to consent to being infected with any particular pathogen at large in a world with many trillions of them is like refusing consent to be exposed to water vapor in a fog or rainforest. In the case of a respiratory virus, one might as well demand to only breathe air that has never been in someone else's lungs. I have a right to take shelter or avoid places where I am more likely to encounter such environmental hazards, but in no way does this create an obligation for my neighbors to do so as well. My right to personal security may include a right to not be knocked down by someone else, but it certainly does not include a "right" to demand that others do everything possible to keep me from tripping and falling. My right to take whatever personal measures I think may lower my risk of disease do not extend to deciding what personal measures others take, and vice versa, even if I or someone else stands to benefit. It is the difference between asserting one's right not to be stabbed, and claiming a "right" to force others to do everything possible to keep you from being cut by a sharp object.

What we really have is a right to not be infected through deliberate action, which perhaps extends to criminal negligence in certain specific and *very limited* contexts, such as (maybe) when someone else is *demonstrably* ill. But one cannot establish negligence simply by asserting that it has occurred, nor even by proving that the accused did not take every possible preventative measure. One also cannot establish negligence by appealing to what people suffering from an artificially induced panic or borderline obsession might consider "reasonable" (regardless of whether such sufferers happen to constitute a majority of the population). Showing that someone has a 1 or 2% chance of being sick does not even come close to reaching this threshold of proof. Unlike vaccination, masks cannot contribute to primary immunity even in theory. Any theoretical risk benefits attributable to masking are thus interchangeable with other common COVID-19 containment methods such as social distancing, handwashing, and quarantining of the actually sick.

Key Takeaways

- One of the fundamental errors at the heart of compulsory masking was the false belief that individual rights are granted by the government, or "society."
- The liberties that each individual human naturally possesses exist independently of whether they are enforceable or recognized by the government.
- America's Founders repudiated the view that individual rights turn on utilitarian numbers games.
- The individual rights of personal security and self-preservation do not extend to forcing everyone else to wear a mask or stay at home as a pre-emptive self-defense measure. No aggregate community right of personal security and self-preservation does this either.
- There is no individual or collective "right" to safety or public health which can be used to override actual individual rights like those recognized in the Bill of Rights.

"Safety" is not a right

"Safety" is like any number of desirable conditions, such as clean water and clean air, to which the economic law of diminishing returns inescapably applies.

> "Despite the political appeal of categorical phrases like 'clean water' and 'clean air,' there are in fact no such things, never have been, and perhaps never will be. Moreover, there are diminishing returns in removing impurities from water or air. Reducing truly dangerous amounts of impurities from water or air may be done at costs that most people would agree were quite reasonable. But, as higher and higher standards of purity are prescribed by government, in order to eliminate ever more minute traces of ever more remote or more questionable dangers, the costs escalate out of proportion to the benefits…
>
> "Depending on what the particular impurity is, minute traces may or may not pose a serious danger. But political controversies over impurities in the water are unlikely to be settled at a scientific level when passions can be whipped up in the name of non-existent 'clean water.' No matter how pure the water becomes, someone can always demand the removal of more impurities. [216]

Open-ended commitment to "safety" is even more simpleminded than an open-ended commitment to "clean" water and air, because all safety measures, by their very nature, involve risk trade-offs. The fewer entry and exit points a building has, the easier it is to prevent unauthorized access or theft, but that same lack of portals instantly becomes a liability in the case of a fire or mass shooter. There is a wealth of evidence showing that Vitamin D enhances immune function,[217]

216   Thomas Sowell, *Basic Economics: Fifth Edition*, Basic Books: New York (2015), Kindle Edition, pp. 417-418

217   For just four such examples of this evidence *see*:

1) Lorenz Borsche, Bernd Glauner, and Julian Von Mendel, "COVID-19 Mortality Risk Correlates Inversely with Vitamin D3 Status, and a Mortality Rate Close to Zero Could Theoretically Be Achieved at 50 ng/mL 25(OH)D3: Results of a Systematic Review and Meta-Analysis." *Nutrients*, 2021. 13(10): p. 3596.

but the sunlight necessary to produce Vitamin D can also cause an increase in some types of cancer.[218] This sort of tradeoff applies to masks, as well. At the end of our review of mask science in Part 1, we examined some of the tradeoff mechanics associated with mask use that, on net balance, easily negate or outweigh any theoretical advantages that might be expected from looking solely at their filtration properties.

A Mercatus Center research publication: "The Unintended Consequences of Safety Regulation," lists safety tradeoffs in several domains, including automobile safety. In the years immediately after the September 11 terrorist attacks, fear of flying and a desire to avoid the hassle caused by TSA security measures led more people to drive. If nobody flew, deaths from air travel would go to zero, but since flying is safer per mile than driving, the result of avoiding perceived dangers from terrorists in the air was a net increase in deaths because of automobile accidents.[219] Compulsory automobile insurance laws increased reckless driving, leading to more traffic fatalities. Even air bags had tradeoffs: "While air bags reduced fatalities by 24 percent among adults, they increased fatalities by 34 percent among children under the age of 10."[220] This may have represented a net decrease in deaths, but it was only achieved by shifting traffic fatality risk from adults to children.

In his book, *The Failure of Risk Management,* author and data-driven risk management specialist Douglas Hubbard describes how many risk management programs are often based on subjective intuition that makes astrology look good by comparison. Such programs can be quite effective at reducing *legal* risk of fines or litigation (which is often the primary unspoken driving force behind their implementation). At the same time, however, they provide no benefit whatsoever, and may even have a net detrimental impact, on risks involving injury or illness. In their book *Playing by the Rules: How Our Obsession With Safety is Putting Us All at Risk,* authors Tracy Brown and Michael Hanlon list even more examples of counterproductive risk tradeoffs and dumb rules implemented

https://dx.doi.org/10.3390/nu13103596

2) J.J. Cannell et al., "Epidemic influenza and vitamin D." *Epidemiology and Infection*, 2006. 134(6): p. 1129-1140. https://dx.doi.org/10.1017/s0950268806007175

3) Yonghong Li et al., "Assessment of the Association of Vitamin D Level With SARS-CoV-2 Seropositivity Among Working-Age Adults." *JAMA Network Open*, 2021. 4(5): p. e2111634. https://dx.doi.org/10.1001/jamanetworkopen.2021.11634

4) Shaun Sabico et al., "Effects of a 2-Week 5000 IU versus 1000 IU Vitamin D3 Supplementation on Recovery of Symptoms in Patients with Mild to Moderate Covid-19: A Randomized Clinical Trial." *Nutrients*, 2021. 13(7): p. 2170. https://dx.doi.org/10.3390/nu13072170

218   c.f. Martin-Gorgojo, A., Y. Gilaberte, and E. Nagore, *Vitamin D and Skin Cancer: An Epidemiological, Patient-Centered Update and Review.* Nutrients, 2021. **13**(12): p. 4292. https://dx.doi.org/10.3390/nu13124292

219   Sherzod Abdukadirov, *The Unintended Consequences of Safety Regulation.* George Mason University Mercatus Center, June 14, 2013, Online at: https://papers.ssrn.com/sol3/papers.cfm?abstract_id=2343923

220   Sherzod Abdukadirov, *The Unintended Consequences of Safety Regulation.* George Mason University Mercatus Center, June 14, 2013, Online at: https://papers.ssrn.com/sol3/papers.cfm?abstract_id=2343923

in the name of "safety."[221] Moreover, what is true of every constituent part is not necessarily true of the whole. Simply because any *particular* accident could have been prevented does not mean *every* accident can be prevented, which makes the common bureaucratic safety goal of "zero preventable harm" a classic fallacy of composition.

All of the authors cited above share one thing in common: they are primarily concerned with physical safety, rather than concrete-but-intangible injuries and harms like a loss of essential individual liberty.[222] (In Part 5, we will look more at how such a loss can rise to the level of a legally recognized irreparable injury.) Our "safety" ended on the day each one of us was conceived, and the human mortality rate is everywhere, ultimately, 100%. As one commentator and data analyst bluntly but accurately observed: "There is no such thing as a respiratory infection that doesn't kill some frail elderly."[223] The COVIDcrisis provides an excellent illustration of how compulsively running to the government for safety from any given threat is, itself, a very risky *modus operandi*. In an essay presciently titled: "Is Progress Possible? Willing Slaves of the Welfare State," C. S. Lewis anticipated our gain-of-function future even while he admonished those in his day who succumbed to fear of atomic warfare:

> "I care far more how humanity lives than how long. Progress, for me, means increasing goodness and happiness of individual lives. For the species, as for each man, mere longevity seems to me a contemptible ideal…
>
> "I am more concerned by what the Bomb is doing already. One meets young people who make the threat of it a reason for poisoning every pleasure and evading every duty in the present. Didn't they know that, Bomb or no Bomb, all men die (many in horrible ways)? There's no good moping and sulking about it…

---

221   Tracy Brown and Michael Hanlon, *Playing by the Rules: How Our Obsession With Safety is Putting Us All at Risk*, Sourcebooks: Illinois (2014).

222   Philosophically, there is a distinction between concrete objects and abstract objects. Concrete objects have causal powers, and include such tangible entities as persons, physical objects, and events (or the intangible terms or names that denote such things), whereas abstract objects are causally impotent numbers, classes, states, qualities, and relations. The definition provided above is my best understanding after listening to Philosopher William Lane Craig's verbal summary of the issue during one of this *Defenders* series of lectures on Christian Theology, Philosophy, and Apologetics. The specific clip I refer to can be found on YouTube at the time of this writing: https://www.youtube.com/watch?v=FuKMRNvtiB4

223   Mathew Crawford, "The Omicron Hypothesis, Part 1," *Rounding the Earth Newsletter*, January 20, 2022, online at: https://roundingtheearth.substack.com/p/the-omicron-hypothesis-part-1

"We shall grow able to cure, and to produce, more diseases - bacterial war, not bombs, might ring down the curtain - to alleviate, and to inflict, more pains, to husband, or to waste, the resources of the planet more extensively. We can become either more beneficent or more mischievous. My guess is we shall do both; mending one thing and marring another, removing old miseries and producing new ones, safeguarding ourselves here and endangering ourselves there.[224]

During the COVIDcrisis, rhetoric, panic, and bad reasoning created an open-ended commitment to safety that was more costly than any virus. Compulsory masking was just one manifestation of this destructive philosophical starting point. However desirable "safety" may be, it must never be elevated to the position of a primary value. It must remain subordinate to the individual liberties recognized in the Bill of Rights. Safety is not, never was, and never will be, a human right in the sense it was alleged to justify violations of *actual* human rights during the response to COVID. There is a right to personal security, but that is not the same as a general open-ended right to safety. Thomas Sowell pointed out the distinction with his usual insightful brevity:

"Rights from government interference — "Congress shall make no law," as the Constitution says regarding religion, free speech, etc. — may be free, but rights to anything mean that someone else has been yoked to your service involuntarily... For society as a whole, nothing is a right—not even bare subsistence, which has to be produced by human toil."[225]

Likewise, health is neither a right nor an obligation. A "right" or "freedom" which can be used to impose an authoritarian obligation on everyone else to put something into or onto their bodies is not a real right or freedom. During COVID, this false conception of "rights" was used to tar people who tried to assert their *actual* rights with labels like "selfish," "entitled," "dangerous," "anti-social," or worse.

---

224   Clive Staples Lewis, *God in the dock: essays on theology and ethics*, Grand Rapids, Michigan: William B. Eerdmans Publishing Company (1970), pp. 311-312. Available online: https://archive.org/details/godindockessayso0000lewi/page/312/mode/2up

225   Thomas Sowell, *The Vision of the Anointed*, Basic Books: New York (1995), 2019 Kindle Edition reprint, p. 100.

Key Takeaways

- The open-ended pursuit of safety is neither rational nor realistic, because there will always be some further risk that can be reduced, the law of diminishing returns applies, and all safety measures involve tradeoffs which may not be beneficial on net balance.
- Physical and psychological harms are not the only harms that count. Intangible harms like the loss of fundamental rights and liberties are often just as bad or worse.
- Mere safety or longevity should not be our superseding values, either as individuals or societies.
- False, overbroad definitions of "rights" were used to justify mandating masks and other anti-COVID measures. Ironically, resistance to things like compulsory masking was actually based on a more limited conception of individual rights.

Mis-defining "freedom"

At their heart, debates over freedom in the context of COVID-19 (or for that matter, any other pathogen) have turned on radically different definitions of freedom. "Anti-maskers infringe upon the freedom of everyone else," complained an opinion piece in *The Salt Lake Tribune*.[226] "'Anti-Maskers' Actually Infringe Upon My Freedom" insisted a double-masked author published in online health forum *The Mighty*.[227] At the height of the panic, these opinions may have been held by a majority, and were given pseudo-legal force through emergency mandates and widespread social pressure.

Thomas Sowell insightfully described the conflict between visions of the world which define "freedom" differently. This *Conflict of Visions* lies at the heart of many bitter disputes over morality and policy, including public health, masks, and vaccines:

> "The constrained vision is a tragic vision of the human condition. The unconstrained vision is a moral vision of human intentions, which are viewed as ultimately decisive."[228]

> "Freedom, as well as justice, is defined differently by the two visions… In the constrained vision, freedom is a process characteristic—the absence of externally imposed impediments. [Thomas] Hobbes applied this concept of freedom both to man and to inanimate things: A man was not free if chained or restricted by prison walls, and water was not free if hemmed in by river banks or by the walls

226   Paul Gibbs, "Anti-maskers infringe on the freedom of everyone else," *The Salt Lake Tribune*, August 23, 2021, online at: https://www.sltrib.com/opinion/commentary/2021/08/23/paul-gibbs-anti-maskers/

227   Emily Filmore, "'Anti-Maskers' Actually Infringe Upon My Freedom," *The Mighty*, December 22, 2020. Online: https://themighty.com/topic/corona-virus-covid-19/anti-maskers-freedom-covid-19/
This author had also self-professed to only "risking" exposure once by December of 2020, when she took a walk outside, to go to the park.

228   Thomas Sowell, *A Conflict of Visions*, Basic Books: New York (2007), p. 27, Kindle edition

of a container. But where the lack of movement was due to internal causes—a man "fastened to his bed by sicknesse" or a stone that "lyeth still"—that was not considered by Hobbes to be a lack of freedom. The same concept of freedom continues to characterize the constrained vision today. Freedom to [Friedrich] Hayek means "freedom from coercion, freedom from the arbitrary power of other men," but not release from the restrictions or compulsions of "circumstances." In the unconstrained vision, however, freedom is defined to include both the absence of direct, externally imposed impediments and of the circumstantial limitations which reduce the range of choice[.]"[229]

An insistence that everyone has a moral and/or legal obligation to put on a mask to stop a disease contains the implicit presumption of an individual or collective "right" to the unconstrained version of freedom. Such a presumption is also latent in appeals to a "social contract" or duty to "society" by which everyone can be forced to wear a mask in the name of a broader right to "public health" or a "freedom" to not be infected. These competing visions of freedom can — and often do — overlap, but are just as often mutually exclusive.

To those holding the constrained vision, pathogens at large in the community (however they were introduced) simply form part of the general "circumstances" that make up the tragedy of the human condition. Within this vision, the spread of disease only acquires a moral or legal component when it incorporates a *mens rea* and *actus reus*– a guilty mind and guilty act. Any form of criminal negligence under the constrained definition is limited to *extremely* narrow, well-defined circumstances. Failing to take one or more interchangeable preventative measures such as mask-wearing or vaccination (especially in the absence of signs or symptoms of illness) does not come close to meeting that standard, regardless of whether or not such measures have been proven effective.

By contrast, to those holding the more expansive unconstrained vision, where human intention and actions are ultimately determinative, the progress of disease through a community has an inextricable moral component. Moral or legal failures which easily rise to the level of criminal negligence are to be *presumed* unless proven otherwise by the exercise of all possible preventative measures — including those that have only a *possible* benefit. The unconstrained vision, with its broader definitions of freedom, thus tends to presume a generalized, open-ended moral (and often *legal*) — duty to avoid becoming infected or allowing anyone else to pick up pathogens that might have used one's body as a roadhouse during some part of their lifecycle.

---

229   Thomas Sowell, *A Conflict of Visions*, Basic Books: New York (2007), p. 98-99, Kindle edition

When people use the terms "personal responsibility" or "individual responsibility" in arguments over masking, those with constrained and unconstrained visions mean two different things. Those with the constrained view mean something more along the lines that people have individual responsibility for their own health, constrained by circumstances that necessitate choosing between imperfect tradeoffs, On this view, extension of responsibility for things like health tends to be limited to the members of one's intimate circle and involves voluntary mutual accommodations. When those with a more unconstrained view use phrases like "individual responsibility," they are describing a responsibility borne by every individual to do what is within their power to *ensure* their personal health and that of every member of their community.

During COVID, people who held the unconstrained vision of freedom accused those who resisted putting on masks of having a false and overbroad concept of rights and freedom. "Not only do American citizens not have the right to refuse masks, they have a positive duty to wear them," chided an opinionated author published by the *Washington Post* in August 2020.[230] At the same time, people who resisted compulsory masking in the name of individual liberty were manifesting the much more fundamentally *narrow* definition of freedom found in the constrained vision. In another of the many bitter ironies to emerge from the COVIDcrisis, the broader concept of freedom was used as a justification for overriding the actual rights and liberties of those who limited their definition of individual rights to the much narrower domain of taking personal protective measures without the option of using personal or third-party coercion to force others to do the same.

---

230   Helena Rosenblatt, "No, there isn't a constitutional right to not wear masks," *The Washington Post*, August 20, 2020, online: https://www.washingtonpost.com/outlook/2020/08/20/no-there-isnt-constitutional-right-not-wear-masks/

Key Takeaways

🔑 Two different definitions of freedom and responsibility were competing in the debates over compulsory masking.

🔑 The narrower view of freedom and responsibility holds that individuals are responsible for their own welfare and have a right to freedom from coercion and the arbitrary power of others, but that freedom from circumstantial and personal limitations is not a right.

🔑 Compulsory masking was based on the broader, unconstrained definitions of freedom and responsibility. These hold that freedom means people have a right to certain desirable outcomes, or at least to have roughly the same opportunity to be free from specific obstacles like fear or risk of disease, and that everyone has a corresponding obligatory personal responsibility to do what they can to help achieve these broader collective freedoms.

"The General Welfare" and "The Public Interest"

At this point, proponents of compulsory masking are apt to object that there is no need to locate a right to public health in the Bill of Rights because both this "right" *and* the power to enforce it are built into the General Welfare Clause of the Constitution.[231] This clause states, "The Congress shall have Power To lay and collect Taxes, Duties, Imposts and Excises, to pay the Debts and provide for the common Defence and general Welfare of the United States."[232] People who put forward this or similar arguments seem to forget that the powers granted to the Federal Government by the Constitution are intentionally "few and defined," to "be exercised principally on external objects, as war, peace, negotiation, and foreign commerce," as James Madison explained.[233] James Madison made clear in *Federalist* 41 that the "general welfare" clause *cannot* be legitimately interpreted as any kind of "an unlimited commission to exercise every power which may be alleged to be necessary for the common defense or general welfare," because all the specific items "alluded to by these general terms" were immediately laid out in the same article, "not even separated by a longer pause than a semicolon[.]"[234] Notably, "public health" in either a general or a narrow sense is nowhere to be found in that list of powers delegated to the Federal government.

---

231   Article 1, Section 8.

232   Constitution Annotated: Analysis and Interpretation of the U.S. Constitution, https://constitution.congress.gov/browse/article-1/section-8/ (accessed January 21, 2023)

233   "*The Federalist* Number 45, [26 January] 1788," *Founders Online,* National Archives, https://founders.archives.gov/documents/Madison/01-10-02-0254. [Original source: *The Papers of James Madison*, vol. 10, *27 May 1787–3 March 1788*, ed. Robert A. Rutland, Charles F. Hobson, William M. E. Rachal, and Frederika J. Teute. Chicago: The University of Chicago Press, 1977, pp. 428–432.

234   "*The Federalist* Number 41, [19 January] 1788," *Founders Online*, National Archives, https://founders.archives.gov/documents/Madison/01-10-02-0237. [Original source: *The Papers of James Madison*, vol. 10, *27 May 1787–3 March 1788*, ed. Robert A. Rutland, Charles F. Hobson, William M. E. Rachal, and Frederika J. Teute. Chicago: The University of Chicago Press, 1977, pp. 390–398.]

Students of history may try to rebut this citation by pointing out that James Madison signed the Vaccine Act of 1813 into law.[235] However, all this act actually did was authorize the President to appoint an agent to preserve vaccine matter and send it to anyone who requested some through the United States' mail free of any postage charges. The claim that this piece of legislation "effectively established a national vaccine institution"[236] is a massive overstatement and mischaracterization. It did not establish any federal health programs, nor did it involve mandates or coercion.[237] James Madison's signature on the Vaccine Act of 1813 no more conferred legitimacy on overextensions of the General Welfare Clause than John Adams' signature on the Alien and Sedition Acts legitimized their blatant violations of the First Amendment. It is a testament to the Founders' integrity that they avoided or corrected such violations as often as they did. Pointing to the 1813 Vaccine Act in isolation also ignores the fact that during James Madison's second presidential term in 1816, Congress killed another proposed House bill that would have used general Federal Revenue to create and fund a federal bureaucracy for the purpose of freely distributing smallpox vaccine throughout the United States. Thus, Americans of the founding generation directly considered and rejected measures to deal with smallpox that were taken for granted during the COVID crisis.[238]

In any event, the 1813 Vaccine Act produced a fiasco nine years later in North Carolina, which came to be referred to as the "Tarboro Tragedy." The vaccine agent appointed, Baltimore physician

---

235   Twelfth Congress. Session II, Chapter 37. 27 February 1813. "An Act to encourage Vaccination," repealed by "An Act to Repeal the Act to Encourage Vaccination," ch. 50, 3 Stat. 677 (1822). Reprinted in Annals of the Congress of the United States, Twelfth Congress — Second Session, Gales and Seaton (1853), p. 1336-1337
Available online at: https://digital.library.unt.edu/ark:/67531/metadc30356/

236   As the often-cited writer below has asserted.
Rohit K. Singla, "Missed Opportunities: The Vaccine Act of 1813 (1998 Third Year Paper)," Harvard University Digital Access to Scholarship at Harvard, p. 46. Available online: https://dash.harvard.edu/handle/1/10015266

237   As far as the Constitutionality, even the oft-cited writer above outright conceded:
        "The 1813 Act marked one of the first times Congress stepped squarely outside the text of
        [Section] 8 of Article 1 [of the Constitution]." He also notes that the primary lobbyist for the
        1813 Vaccine Act, Dr. James Smith, "seems to have become almost obsessed with maintaining his
        vaccine institution and distributing vaccine freely to the entire nation. He propagated the same
        sample of cow-pox virus received in 1801 for over twenty years, continuously vaccinating new
        individuals to maintain the strain, and spent twenty-five years promoting universal vaccination."

238   Deliberations found in Annals of the Congress of the United States, Fourteenth Congress - First Session, Gales and Seaton (1853), p. 719, 1408, 1412, and 1457. Available online at:
https://digital.library.unt.edu/ark:/67531/metadc30356/?q=H.%20OF%20R
Further details of the Bill are found in Rohit K. Singla, "Missed Opportunities: The Vaccine Act of 1813 (1998 Third Year Paper)," Harvard University Digital Access to Scholarship at Harvard, p. 46. Available online:
https://dash.harvard.edu/handle/1/10015266
H.R. 73 as amended, 14th Cong., 1st Sess. (1816) (on file with National Archives: House Documents, Box 14A.B1, Original House Bills). The unamended bill is micro-formed on Early American Imprints, Second Series, Fiche 39298 (Readex).

Dr. James Smith, accidentally mailed smallpox scabs to his auxiliary receiving agent, Dr. John Ward. The label was marked with "*Variol.,*" an abbreviation for the Latin term for smallpox,[239] in Smith's handwriting. Dr. Ward was either unable to read Dr. Smith's handwriting, or did not recognize the abbreviation. Either way, Dr. Ward proceeded to inoculate (*variolate*) his patients with what he mistakenly thought was harmless cowpox matter. This mistake produced a small epidemic that led to at least 60 cases of smallpox and 10 deaths by late 1821.[240] To his credit, Dr. Smith immediately published and widely distributed a warning to physicians who had recently received his vaccines that the batch they received might be tainted. This understandably produced a public uproar. Less to either of their credit, Dr. Smith and Dr. Ward then proceeded to try to throw one another under the bus. Dr. Smith pointed out that smallpox had already been in the area, his sending had been in response to a request for assistance, and that any physician competent to vaccinate should have known the difference between smallpox and cowpox scabs without the assistance of any label. Dr. Ward riposted by pointing to the persistent presence of smallpox in Baltimore despite more than a decade of Dr. Smith's enthusiastic vaccinations. Congress responded to the popular outrage and disgust by repealing the 1813 Vaccine Act.[241]

All these historical details are relevant, but the main point here is that even if the "general welfare" clause were interpreted as broadly as possible, it is not in any way synonymous with — or reducible to — public health. The good of others, and the common good, cannot be reduced to public health because the general welfare encompasses a myriad of other physical components, as well as distinct spiritual and moral dimensions of just as much importance and immediacy. Individual liberty, freedom of religion, freedom of conscience, and the primacy of the Bill of Rights are much larger parts of the general welfare than the incidence of any disease. Additionally, the rights of personal property, due process, freedom of association, and the presumption of innocence (to name just a few), are all more important.

Forcing masks onto everyone in the name of public health effectively elevated physical harm from pathogens above all other types of harm. In the name of an imprecise, speculative benefit against one virus out of thousands, compulsory masking inflicted immediate, broad, and severe harm to others by striking at their psychology, critical thinking, and fundamental liberties. Sacrificing preeminent transcendent values and principles in the pursuit of transient risk reduction did far greater harm to the "general welfare" than the aggregate harm to "society" from the worst diseases ever could. Public health

---

239   Latin: *variolous*

240   Rohit K. Singla, "Missed Opportunities: The Vaccine Act of 1813 (1998 Third Year Paper)," Harvard University Digital Access to Scholarship at Harvard, p. 67. Available online: https://dash.harvard.edu/handle/1/10015266

241   Rohit K. Singla, "Missed Opportunities: The Vaccine Act of 1813 (1998 Third Year Paper)," Harvard University Digital Access to Scholarship at Harvard, p. 67. Available online: https://dash.harvard.edu/handle/1/10015266

can and should take a back seat when it infringes on other non-physical (yet no-less-concrete) interests.[242] The myopic fixation on public health and mask mandates was far more harmful to all of these essential components of the general welfare than any number of individuals going unmasked could possibly have been, *even if masks worked.*

As U.S. District Court Judge Terry Doughty correctly observed when striking down the Federal Head Start Mask and Vaccine Mandate, "The public interest is served by maintaining the constitutional structure and maintaining the liberty of individuals who do not want to take the COVID-19 vaccine. This interest outweighs Government Defendants' interests."[243] The 5th Circuit Court of Appeals upheld his ruling on the same reasoning, "The public interest is also served by maintaining our constitutional structure and maintaining the liberty of individuals to make intensely personal decisions according to their own convictions — even, or perhaps particularly, when those decisions frustrate government officials." [244]Yet even if these courts had ruled otherwise,[245] those rulings would still not alter the basic truth that public health is a subsidiary good which *cannot* and *should not* trump other even more basic goods. The Declaration of Independence appealed to a Natural Law higher than those of any governing official: "the Laws of Nature and of Nature's God." It did this to recognize and lay claim to individual rights of "Life, Liberty, and the Pursuit of Happiness," not "Life, Public Health, and the Pursuit of Safety."

---

242   Philosophically, concrete objects can be physically tangible or intangible. The definition of concrete objects includes having causal power, in contrast to abstract objects, which lack causal power.

243   *Louisiana v. Becerra*, 577 F. Supp. 3d 483 (W.D. La. 2022). Case 3:21-cv-04370-TAD-KDM, Document 128, Filed 09/21/22, Page 26
https://libertyjusticecenter.org/wp-content/uploads/2021/12/2022-9-21-Brick-v.-Biden.pdf

244   *BST HOLDINGS v. Occupational Safety & Health Admin.*, 17 F.4th 604 (5th Cir. 2021). Case: 21-60845, Document: 00516091902, Page: 20, Date Filed: 11/12/2021
https://attorneygeneral.utah.gov/wp-content/uploads/2021/11/BST-Holdings-Stay-opinion.pdf

245   Early in the COVIDcrisis, plenty of lower courts certainly did rule otherwise — more on this in Part 5.

Key Takeaways

- The writings of America's Founders and their historical precedents teach that the General Welfare Clause of the U.S. Constitution does not extend to anything and everything that may benefit Americans as a whole.

- Public health is just one of many equally or even more important things, including individual liberty, that make up the "general welfare."

- The individual right to control over highly personal decisions about which medical interventions to use or to forego is an indispensable component of the general welfare.

Slick rhetoric turns non-action into assault

> *"Law has this much to learn from medicine: first, do no harm."*
>
> — Richard A. Epstein, *Principles for a Free Society,* 1998[246]

During COVID, organized medicine failed to follow its own most important principles. The first and most important of these is to do no harm.[247] Imagine, for a moment, that you own a small business. One day, mobsters move into your area. You mind your own affairs, hoping to be left alone, but a few weeks later, some of these mobsters show up at your business. They tell you in no uncertain terms that leaving you alone counts as a form of protection, and therefore you now owe them routine payments to maintain this "protection."

The added expense of paying mobsters *not* to wreck your business slowly begins to bankrupt you. This, in turn, generates creeping despair but produces renewed interest in spiritual matters. So, you begin attending a local house of worship. It comes as a shock when, several months later, one of your new co-religionists pulls you aside pulls you aside to say that you are not representing the religion well, because you are not a good enough person. Your actions, like being rude to non-believers, lying to get out of volunteering, and failure to donate or participate in ministry is turning others away from the faith. Therefore, it will be *your* fault when they are lost for eternity.

---

246   Richard A. Epstein. *Principles For A Free Society: Reconciling Individual Liberty With The Common Good*, Reading, Massachusetts: Perseus Books (1998), p. 39, available online: https://archive.org/details/principlesforfre0000epst/page/38/mode/2up

247   *Primum non nocere*, a Latin medical aphorism of debated origin, but with general parallels going back to the Hippocratic Oath.

The mobsters and co-religionist described in the imaginary scenario above are using a false equivalency to make unjustified demands. Ideally, we would all refuse to accept bad reasoning like this from anyone. We should likewise refuse to accept it from politicians and public health officials. Non-action is not a form of protection, and neither is it any form of assault. It is another false equivalency to say that inaction is "imposing risk" on oneself or others. Inactions like refusing to wear a mask are not doing something "to" anybody. At the worst, it is refusing to do something "for" somebody, which is a crucial distinction.

A common fallacy or rhetorical trick is to use a single sentence that everyone agrees with to imply or jump to an entirely unwarranted conclusion. Florida Circuit Court Judge John Kastrenakes provided a prime example in the case of facemasks when he dismissively shot down an early legal challenge to compulsory masking by acerbically pointing out: "we do not have a constitutional or protected right to infect others."[248] Like many slogans, this slick piece of verbal virtuosity[249] does four things with just one sentence. First, it neatly conflates (and equates) non-action with a form of assault. It also sidesteps actual justification, in the same way that someone can say, "Well, that's just your opinion" as a dodge to avoid having to defend their *own* opinion. Third, it fabricates an implied right to personal or public health out of thin air. Lastly, it presumes masks actually *work*.

Often, arguments for compulsory public health measures simply start from an *a priori* false framing that assumes rights and choices are privileges or gifts from authorities, and that non-actions can "impose risk." Thus, you get broad, unqualified assertions like those of Dr. Matthew K. Wynia, Director of the Center for Bioethics and Humanities, which categorically elevate matters of public health above matters even of religion, like "avoiding mandates to uphold the ideals of personal and religious freedom is not worth the risk to others[.]"[250]

Saying that "we do not have a Constitutional right to infect others" or that unmasked and unvaccinated individuals are "imposing a risk" on everyone else smuggles in an outrageously false equivalency. Even if masks worked, an individual's failure or refusal to wear a mask would not actively "impose a risk" of death by disease on others, any more than failure to give charitably "imposes a risk" of destitution on potential recipients. "Silence is violence," is another example of such a false-equivalency slogan. Apart from being catchy, these slogans work because they contain a germ of truth. Sins of omission

---

248   *Machovec v. Palm Beach County*, No. 2020CA006920AXX — Filing #110806335, 27 July 2020, (15th Judicial Circuit, Florida), p.12, available online: https://clearinghouse.net/doc/109346/ & https://www.floridacivilrights.org/wp-content/uploads/2020/07/Order-Denying-Temporary-Injunction.pdf

249   To my knowledge, credit to Thomas Sowell for coining this term.

250   Matthew K. Wynia, Thomas D. Harter, Jason T. Eberl, "Why A Universal COVID-19 Vaccine Mandate Is Ethical Today", Health Affairs Blog, November 3, 2021. DOI: 10.1377/hblog20211029.682797. Online at: https://www.healthaffairs.org/do/10.1377/forefront.20211029.682797/ (last accessed January 31, 2023).

(failure to do what one has a moral duty to do) and criminal negligence do exist, and contrast with sins of commission (actively doing something which one has a moral duty *not* to do). However, these slogans completely fail to carry their burden of proof to show that a sin of omission or criminal negligence has actually occurred, and they include a further cheat by falsely equating the asserted sin of *o*mission with a sin of *co*mmission.

Attempting to justify mask mandates using slogans like, "We do not have the right to infect others," is an intellectual sleight-of-hand that lumps assault with a deadly bioweapon into the same bucket as the inescapable risk we all incur simply by existing in a world saturated with living microorganisms. Similarly, this type of rhetoric conflates risks that are conceivable or theoretically possible, with risks that are scientifically *probable*. Just as risk of death is an inescapable corollary of living, anyone's and everyone's mere existence imposes a plethora of risks *and* benefits on everyone else. Simply because another person would benefit from some action does not magically make *withholding* that particular action the same as inflicting harm. To isolate one theoretical risk and use that as an excuse to infringe others' absolute rights is using rhetorical alchemy to transmute a duty to do no harm into an open-ended duty to assist. A duty to do no harm does not imply a duty to assist, much less an *open-ended* duty to assist. The response to COVID in general, and mask mandates in particular, twisted a duty to do no harm into an open-ended duty to assist, enforced by immoral, coercive mandates and manipulative psychology.

Key Takeaways

- Slogans like "we do not have a Constitutional right to infect others" were used to push compulsory masking and other COVIDcrisis impositions. These slogans were rhetorical dodges which allowed users to get away with intellectual sleight of hand, like conflating sins of omission with sins of commission, or turning a duty to do no harm into an open-ended duty to assist.
- Refusing to do something *for* someone is not the same as actively doing something *to* them. Even if masks worked, refusing to wear a mask would not "impose" risk on anybody.
- To retain legitimacy, laws must reflect a higher moral reality, as moral luminaries from St. Augustine to St. Thomas Aquinas to Martin Luther King Jr. have consistently held. Instead of reflecting this higher moral reality, mask mandates violated it.

The flawed hidden moral reasoning behind compulsory masking

Many people throughout history, and especially in 2020, have found the false-equivalency and non sequitur reasoning described above to be highly persuasive. But why? Fear in the moment is not, on its own, an adequate explanation. Most popular-level diatribes against those labeled "anti-maskers" and "anti-vaxxers" simply take for granted the foundational moral strength of compulsory masking, and make no detailed attempt to describe or defend it. Most people advocating compulsory masking and compulsory vaccination simply take it for granted that if those measures are effective (or even merely *possibly* effective) in reducing disease transmission, then it is moral to force them on others. I.e., if wearing a mask reduces transmission of COVID, then it is moral to force everyone to do it. But as it stands, this is a non sequitur, meaning that the conclusion does not logically follow. Science may be able to tell you what will *happen* if you do something, but it cannot not tell you if you *should* do something.[251]

---

251   The most cogent summary of five things science cannot tell you of which I am aware was provided in a famous 1998 debate on the existence of God between Dr. William Lane Craig and Dr. Peter Atkins:

Dr. Atkins: Do you deny that science cannot account for everything?

Dr. Craig: Yes I do deny that science can account for everything.

Dr. Atkins: So, what can't it account for?

Dr. Craig: Well, had you brought that up in the debate I had a number of examples that I was going to give. I think there are a good number of things that cannot be scientifically proven, but that we're all rational to accept.

Dr. Atkins: Such as?

Dr. Craig: Let me List five:

Logical and mathematical truths cannot be proven by science. Science presupposes logic and math, so that to try to prove them by science would be arguing in a circle.

Metaphysical truths, like there are other minds other than my own or that the external world is real or that the past was not created five minutes ago with an appearance of age are rational beliefs that cannot be scientifically proven.

Ethical beliefs about statements of value are not accessible by the scientific method. You can't show by science whether the Nazi scientists in the camps did anything evil as opposed to the scientists in western democracies.

Aesthetic judgments, number four, cannot be accessed by the scientific method because the beautiful, Like the good, cannot be scientifically proven.

And finally, most remarkably, would be science itself. Science cannot be justified by the scientific method. Science is permeated with unprovable assumptions.

If we assume for the sake of argument that masks work, all that follows is that fewer people would catch COVID if everyone wore a mask. The bridge that gets people from the initial premise that "masks work" to the conclusion that "it is moral to force everybody to wear one," is a primitive, intuitive form of consequentialism or utilitarianism (usually unexamined). New York governor Andrew Cuomo espoused this in 2020 when he said: "This is about saving lives, and if everything we do saves just one life, I'll be happy."[252] This is yet another one of the primary underlying philosophical drivers behind the arguments for compulsory masking.

For example, in the special theory of relativity, the whole theory hinges on the assumption that the speed of light is constant in a one-way direction between any two points A and B. But that strictly cannot be proven. We simply have to assume that in order to hold to the theory.

Dr. Atkins: But you're missing the whole point.

Dr. Craig: So none of these beliefs can be scientifically proven and yet they are accepted by all of us, and we're rational in doing so.

William Lane Craig in debate with Peter Atkins, "What is the evidence for/against the existence of God?" Carter Presidential Center, Atlanta, Georgia, April 3, 1998, video 1:03:32, transcript p. 22.

Recording Online: https://www.youtube.com/watch?v=mEoznzPSguI

Transcript online:

https://www.reasonablefaith.org/media/debates/what-is-the-evidence-for-against-the-existence-of-god

252   Andrew Cuomo, quoted in "Video, Audio, Photos & Rush Transcript: Governor Cuomo Signs The 'New York State on Pause' Executive Order," governor.ny.gov, March 20, 2020, Online at: https://www.governor.ny.gov/news/video-audio-photos-rush-transcript-governor-cuomo-signs-new-york-state-pause-executive-order (last accessed January 31, 2023)

Three additional examples included the mayor of the New Mexico city of Taos as he imposed a curfew in the name of infection control: ""If it saves just one life, it's worth it to me. Anything we can do at this point to save lives, we have to do it[.]"

T.S. Last, "Taos Town Council imposes nighttime curfew," Albuquerque Journal, April 3, 2020, Online at: https://www.abqjournal.com/1440107/taos-town-council-imposes-curfew-in-effort-to-slow-coronavirus-spread.html

Rev. Marek Zabriskie of Hartford, Conneticut said, "I thought to myself, 'We need to protect our people. If it saves just one life, it is worth it." The wardens and vestry and I conferred. We were all in agreement."

Marek Zabriskie, "With churches closed, new ways to worship," *Hartford Courant*, March 29, 2020, Online at: https://www.courant.com/opinion/op-ed/hc-op-zabriskie-churches-closed-coronavirus-0329-20200329-f4qbavqgq-jgjblktwxltmvmcyu-story.html

Elizabeth, New Jersey police department used drones to warn residents of fines up to $1,000 for violating COVID "guidelines." "'If this plan saves one life, then it is worth it,' the department said."

Associated Press, "NJ, Fla. Police use 'talking' drones to enforce COVID-19 social distancing," Police1 by LEX-IPOL, April 9, 2020, Online at:

https://www.police1.com/coronavirus-covid-19/articles/nj-fla-police-use-talking-drones-to-enforce-covid-19-social-distancing-eHRd8M2slg5LkM4y/

"If it saves just one life, it's worth it," sounds good as a slogan when everyone is scared, but involves multiple hidden premises which are taken for granted, but rarely (if ever) explicitly stated or defended.[253]

The unexamined chain of reasoning usually runs something like this:

1. "X works" — doing X achieves some desirable end, such as preserving, sustaining, or improving life for oneself or others.
2. If X works, then it is always good to do X.
3. If it is always good to do X, then everyone has a moral duty to do X.
4. If X is a moral duty, then it is, or ought to be, a legal duty as well.
5. If X is a legal duty, then it is moral to force everyone to do X; willful failure to fulfill that legal duty can and should be punished.

Applied to masks, this becomes:

1. "Masks work" — wearing a mask reduces transmission of COVID-19 or some other disease.
2. If masks work to reduce transmission of COVID (or another disease), then it is always good to wear a mask.
3. If it is always good to wear a mask, then everyone has a moral duty to wear a mask.
4. If wearing a mask is a moral duty, then it is, or ought to be, a legal duty as well.
5. If wearing a mask is a legal duty, then it is moral to force everyone to wear one; people who refuse to wear masks can and should be punished.

Often, this chain of reasoning is coupled with other hidden, incorrect assumptions. Two of the most common are that failure to wear a mask can only result from the worst motives, and that the legal duty to wear a mask is so important that willful failure justifies deprivation of basic human rights and

---

253   Economist Antony Davies has made a strong argument that the "if it saves one life" slogan is foolish even on utilitarian grounds, and never consistently applied as a matter of policy:

> "If we really believed that any law is justified if it saves just one life, we would require all Americans to pass a mental health evaluation on a regular basis or be institutionalized (more than 38,000 Americans commit suicide annually). We would outlaw all motor vehicles (almost 35,000 Americans die in vehicle accidents annually). We would require all houses to be single-story structures (more than 26,000 die in falls annually). We would ban alcohol (almost 17,000 die annually from alcohol-related liver disease). We would require people to be certified as swimmers before allowing them into any large body of water (more than 3,500 die from drowning annually). We would prohibit women from getting pregnant unless they had no family history of birth complications (more than 900 American women die in childbirth annually)."

Antony Davies & James R. Harrigan, "The 'If It Saves Just One Life' Fallacy," Intercollegiate Studies Institute, April 29, 2020, Online: https://isi.org/intercollegiate-review/if-it-saves-just-one-life-fallacy/

exclusion from society. Rather like the way a virus infiltrates the body by hiding inside a cell while it reproduces, these unstated premises smuggled themselves into acceptance by hiding inside the space between statements everyone agreed with, and conclusions that required more than what was said.

This explains in large part why debates over the empirical evidence regarding masks are so heated. The above chain of reasoning never gets off the ground in the first place if "X" (in this case, masking) does not work. Many people hold strongly to premises 2, 3, 4, and 5, and it is much more comfortable (and often intellectually easier) to keep arguments in the external, non-moral, realm of physical science on premise 1.

But not one premise in this entire chain of reasoning is necessarily true. Even if masks worked, premise 2 would still be false. It would not *always* be good to wear a mask, because slowing possible disease transmission is not always the foremost good to pursue in any given situation. If you are wearing a mask to protect yourself from a disease, you may have any number of needs or desires which mask-wearing interferes with, or things that you rightly value more highly than not catching a particular disease. If you are wearing a mask to protect others, the odds are overwhelming at any given time that you have nothing which a mask would protect them *from* (even if your mask worked).

The premise that if something is good to do, then you have a duty to do it (premise 3), is likewise faulty from the get-go. Just because it would be *good* for you to do something does *not* mean you have a moral duty to do it. Even a good act which directly results in saving lives may not be a moral duty. For example, it would be good for you to become a doctor, or an emergency medical technician, and if you entered either of those occupations, you would undoubtedly save many lives. However, it would be absurd to conclude from this that everyone has a moral *duty* to become a doctor or an EMT. The same applies to other good occupations, like farming and construction. Everyone needs food to eat and a place to live, but this does not imply that you have a moral duty to go into either of these occupations. An individual may *possibly* have such a moral duty, but that isolated, individualized duty, *if* it exists, would depend on numerous other factors that do not apply to everyone else. The majority of specific goods that people have an affirmative moral duty to do are *individualized* to themselves. In the case of masks, those demanding that you wear a mask are very possibly mis-perceiving their *own* moral duty in the matter, and are even less likely to be accurately perceiving *yours*.

If the reasoning behind compulsory masking fails with regard to things that are directly life-saving, such as becoming a physician or EMT, and also fails with regard to things that are directly life-preserving, such as growing food, then it fails even more in regards to compulsory masking, with its benefits that are indirect at best (assuming they work), and fictional in reality.

The fourth premise in this faulty chain of reasoning is that if wearing a mask is a moral duty, then it is, or ought to be, a legal duty. Another way of saying this is, "What should not be should not be permitted."[254] The differences between what is moral and what is legal are notorious. Just as something being *legal* does not necessarily imply that it is *moral*, something being immoral does not necessarily imply that it should also be illegal. You no more have a presumptive moral obligation to defend everyone against possible infection with a particular virus than you have a presumptive moral obligation to defend everyone against possible mass shootings, much less a legal obligation. Even assuming masks work, refusing to wear a mask no more "imposes" a risk of infection on others than refusal to get firearms training and a concealed carry license "imposes" risk of a mass shooting by leaving you less prepared to deal with one of *those*. Likewise, you are not "imposing risk" on your friends and family when you give them a ride in your car after midnight, even though car accident fatalities are much more likely to occur in a car after midnight than in a pickup truck during daylight hours.[255] (Also, masks are not analogous to seatbelts. More on this later, and in Part 5.) The same reasoning behind compulsory masking, if applied consistently, would dictate any number of universalized "public safety" or "public health" measures that hardly anybody thinks are feasible or a good idea.

The fifth and final premise, that if wearing a mask is a legal duty, then it is a moral duty, and it is moral to force everyone to wear one, is just as flawed.[256] St. Augustine wrote, "a law that is not just is not a law."[257] St. Thomas Aquinas quotes and expands on Augustine, adding that unjust laws are more analogous to acts of violence, and thus impose no moral obligation.[258] Martin Luther King Jr. echoed and referenced both St. Augustine and Thomas Aquinas in 1963 when he wrote: "… there are two types of laws: just and unjust. I would be the first to advocate obeying just laws. One has not

---

254   William Lane Craig, "Consequentialism and the Problem of Evil," Reasonable Faith Question of the Week #448, November 15, 2015. Online at:
https://www.reasonablefaith.org/writings/question-answer/consequentialism-and-the-problem-of-evil

255   U.S. Department of Transportation, National Highway Traffic Safety Administration (NHTSA), "2020 Traffic Safety Facts Annual Report," DOT HS# 813141 (2022), in particular, *see* Table 3 on page 23, and Figure 21 on pg 120, Available Online: https://crashstats.nhtsa.dot.gov/Api/Public/ViewPublication/813375

256   The erroneous conflation of legal duty and moral duty to which even higher education provides no immunity, can be shown in a single sentence by one PhD author: "Where there is policy (or other) regarding the wearing of masks, the morally correct thing to do, in a deontological tradition, is to wear the mask."
Gillian R. Rosenberg, "The Moral Failing of Not Wearing a Mask," LinkedIn, October 21, 2020, Online:
https://www.linkedin.com/pulse/moral-failing-wearing-mask-dr-gillian-r-rosenberg/

257   Saint Augustine, *de libero arbitrio voluntatis* (*On the free choice of the will*), Book 1, Section 5, 1.5.11.33 (c. 388 A.D.). Anna S. Benjamin and L.H. Hackstaff (translators), The Library of Liberal Arts, Bobbs-Merrill (1964). Available online: https://archive.org/details/onfreechoiceofwi00augu/page/10/mode/2up?q=unjust

258   Saint Thomas Aquinas. *The Summa Theologica*, First Part of the Second Part, Treatise on Law, Question 96, Article 4, "On The Power of Human Law, Whether Human Law Binds a Man in Conscience," Objection 3. Catholic Way Publishing. Kindle Edition. Alternative translation available online:
http://www.sophia-project.org/uploads/1/3/9/5/13955288/aquinas_law.pdf

only a legal but a moral responsibility to obey just laws. Conversely, one has a moral responsibility to disobey unjust laws." Additionally, "law and order exist for the purpose of establishing justice."[259] People fight bitterly over whether particular laws are just and therefore valid precisely *because* they agree with St. Augustine, St. Thomas Aquinas, and Dr. King.

The moral assumption behind many of the arguments for compelling individuals to wear masks or to get vaccinated is that if there is something which an individual can do that reduces the risk of infecting others (and thus lightens the burden of the disease for individuals and the society), then they have a moral obligation to do it. This is nothing more than a disease-specific variation of the false premise that if doing X contributes to achieving some good, then you have a moral obligation to do X. If you become a healthcare provider, that would indisputably contribute to public health and lighten the burden of disease on society, but that does not at all imply you therefore have a moral *duty* to become a healthcare provider.

Doctors may have a moral "duty to treat" patients, but that does not imply that it is moral or legal to force doctors to provide care against their will, or in ways that violate their conscience. Even if, for the sake of argument, we assume that everyone has a moral duty to personally assist in preventing the transmission of any given infectious disease, it would not necessarily follow that everyone has a moral or legal duty to use any particular *means* of doing so, much less *every possible means* of doing so. Not only do slogans like "we do not have a right to infect others," fail to carry their burden to show the absolute open-ended moral duty to assist which they presume, but they also wrongly conflate moral and legal duties. In order to establish a legal duty sufficient to justify mandates, it is not enough to establish a moral duty, and even a *general* legal duty to assist does not necessarily imply a legal duty to use a *specific* means of rendering assistance (e.g. wearing a mask). Unless you assume that someone has a duty to assist using *every possible means*, alternative preventative measures could be used to fulfill that duty. This alone would preclude universal compulsory masking.

Every step of the most important chain of moral reasoning supporting compulsory masking fails miserably. Even if wearing a mask "works" and is a good thing for someone to do, that does not necessarily imply a moral duty to do so. Authority figures who violated essential individual liberties by forcing masks on everyone and the justifiers rationalizing those actions, never showed that a moral duty to wear masks existed, they simply asserted one by brute force. Moral duty does not imply legal duty, and even if every citizen in a community had a legal duty to take affirmative action to prevent disease transmission, it would *still* not follow that a *specific means* of preventing infection could be

---

259   King, Martin Luther, Jr. (16 April 1963). "Letter from Birmingham Jail." Stanford University, Martin Luther King, Jr. Papers Project (2004). Available online:
Lhttps://kinginstitute.stanford.edu/sites/mlk/files/letterfrombirmingham_wwcw_0.pdf

mandated. Compulsory masking rests on two pillars: the scientific argument that masks work, and the moral argument that if masks work, forcing them onto everyone is a moral imperative. Both pillars crumble when subjected to the slightest pressure. Compulsory masking was unjust and immoral from day-one.[260]

---

260   It might be tempting at this point to accuse me of constructing a straw-man argument and then demolishing it. To this, I have two main points of response. First, if I had encountered any similarly-robust attempt to link "masks work" to "it is moral to force everybody to wear one," I would have repeated, cited, and demolished *that* instead. No one clamoring for masks that I could find — including the so-called "experts" in ethics getting their articles published in journals like *JAMA* or the *NEJM* — was attempting to connect all of these dots. They simply committed one or more of the equivocation or conflation errors described, or started from an *a priori* utilitarian attempt to ground objective moral values and duties in human flourishing and the flourishing of conscious creatures. Thus, I articulated the most obvious and intuitive pathway to compulsory masking based on those premises. Second, anyone inclined to say that the chain of reasoning summarized above is a straw-man or caricature ought to have something better to substitute in its place. If someone wants to articulate a better, more plausible and intellectually-robust way to connect "masks work" with "everyone has to wear a mask" (and that is not simply a variant of the premises described above), they are welcome to put it forward — I would very much like to see it. If someone puts forward such an argument, they have already done more thinking about the underlying issues than 99.99% of those who forced masks onto everyone or were clamoring for third-party authorities to do so.

Key Takeaways

🔑 Science may be able to tell you what *will happen* if you do something, but science cannot tell you whether or not you *should* do something. Getting from the belief that masks mitigated COVID to the conclusion that it was moral to force everyone to wear masks required at least four additional hidden steps:

- o If masks work, then it is always good to wear a mask.
- o If it is always good for someone to wear a mask, then they have a moral duty to do so.
- o If wearing a mask is a moral duty, then it ought to be a legal duty.
- o If wearing a mask is a legal duty, then it is moral to force everyone to wear one, and those who resist can and should be punished.

🔑 These hidden steps are moral assumptions that science cannot provide answers to, and not one of them is actually true.

- o Even if masks work, that does not always make wearing a mask a good thing.
- o Just because it is good to do something does not necessarily mean there is a universal moral duty to do it.
- o Not every moral duty should be a legal duty.
- o Just because some requirement is a law or regulation does not automatically make it (or its *enforcement*) morally right.

🔑 To retain legitimacy, laws must reflect a higher moral reality, as moral luminaries from St. Augustine to St. Thomas Aquinas to Martin Luther King Jr. have consistently held. Instead of reflecting this higher moral reality, mask mandates violated it.

Flawed theories of causality, culpability, and consequentialism

Mask mandates have been frequently compared to seatbelt laws, but even some of the more thoughtful pro-mask advocates quickly (and rightly) repudiated this bad analogy.[261] The obvious disanalogy is that seatbelts only protect the wearer, while the primary justification put forward for forcing masks onto everyone is that masks supposedly protect everyone *from* the wearer — as though masks somehow behave almost like one-way valves. Further evidence of this rationale is found in the fact that any masks with *actual* one-way valves allowing free flow of exhaled air were outright banned in many places during the COVIDcrisis. In any event, arguments for masks based on seatbelt law analogies could never be expected to convince those who see both masks and seatbelt laws as just another manifestation of toxic risk aversion and overextension of paternalistic safety culture. One

---

261   A similar bad analogy compares mask mandates to the classic "No Shirt, No Shoes, No Service." (e.g. Christopher Dolan and Lourdes De Armas, "No Shoes, No Shirt, No Mask Means No Service," February 11, 2021, https://dolanlawfirm.com/2021/02/no-shoes-no-shirt-no-mask-no-service/ .) Such analogies attempt to legitimate mask mandates by likening requirements to wear a mask to requirements to wear a shirt and shoes. Ordinances against public indecency or public defecation are also sometimes cited as a precedent. The fatal disanalogy in these comparisons is that different body parts are simply not comparable in this way. Different body parts have very different functions and so have acquired entirely different social conventions. Just because it is considered impolite to walk around someone's home or business with dirty bare feet, or to dine bare-chested, does not mean that it should therefore also be considered rude or illegal to breathe with an uncovered face in public. It is silly to suggest (much less assert) that conventions associated with body parts involved in reproduction or the evacuation of liquid or solid wastes should be applied to another part designed for entirely different functions like communication and respiration. We do not respire through our abdomens, we do not ingest nutrients through our feet, and our butts do not have 80 muscles allowing 7,000 different configurations for the purpose of communication like our faces do. Different body parts with different functions necessitate different conventions or regulations, and one's face should not be treated like one's feet, chest, or genitals. Slogans like "No Shirt, No Shoes, No Masks, No Service," assumed that faces are just another body part which ought to be covered, and to the extent they did so, got it wrong.

For a critique of the seat-belt analogy in a vaccination context, *see also*: Iñigo de Miguel Beriain, "Coercive vaccination: using the 'seat belt analogy' is not effective." *Blog: Journal of Medical Ethics*, British Medical Journal, August 22, 2021, https://blogs.bmj.com/medical-ethics/2021/08/22/coercive-vaccination-using-the-seat-belt-analogy-is-not-effective/?int_source=trendmd&int_campaign=usage-042019&int_medium=cpc

might as well ban mountaineering, skydiving, and any number of far more dangerous activities that are perfectly within individuals' absolute rights of personal liberty.

Alternative driving-related analogies put forward to justify compulsory masking were no better. One author in the *Desert News* argued that refusing to wear a mask is more analogous to driving drunk than to refusing to wear a seatbelt:

> "When I refuse to wear a mask in a time of pandemic… I affirm my right to act or not act, but I do so in such a way as to refuse to accept any responsibility for the consequences that my actions (or inactions) may bring about. In the case of COVID-19, I consent to the very real possibility that my body may be used by an external biological agent so as to bring about potentially life-threatening consequences for which I likely will not be held accountable. In such cases, then, I do not truly act but rather am acted upon, as I consent to allow the virus to act through me, perhaps bringing about consequences which I did not directly intend. The virus is in the driver's seat, not me. This is why the seat belt analogy is perhaps less illuminating than the analogy of drunken or impaired driving. If I were to choose to drink to the point of intoxication and then get behind the wheel, I have chosen to cede my agency to an external force — the alcohol — which then works through me and may inflict harm upon others."[262]

This analogy fails in so many ways that it turns out to be worse than the seatbelt comparison. Absent force or fraud, intoxication is entirely under an individual's control. One cannot become intoxicated simply through contact with a drunk person. On the other hand, viral infections, especially airborne *respiratory* viral infections, are under no such control and cannot be precisely dosed and titrated like alcohol. Compulsory masking is more like assuming that all drivers on the road are drunk until proven otherwise, and mandating ignition breathalyzers for all vehicles. God help us if that attitude ever becomes widespread.

The seatbelt or drunk driver arguments for compulsory masking make use of a flawed understanding of causality. This flawed understanding of causality ignores the crucial moral insight inherent in pushback against "victim blaming." Accepting *risk* of a particular bad outcome by failing to take every possible precaution is in no way the same as *consenting* to that bad outcome. This is still true

---

262   David Laraway, "Why comparing seat belts to wearing a mask is unhelpful," *Deseret News*, July 17, 2020. Online: https://www.deseret.com/opinion/2020/7/17/21327346/guest-opinion-covid-19-seat-belt-analogy-wear-mask-mandate-public-health-pandemic-liberty-safety (last accessed, January 24, 2023)

even if the precaution in question has been shown to have some demonstrable preventative benefit. Failing to wear a mask no more constitutes "consent" to SARS-CoV-2 infection than failing to use mosquito repellent constitutes "consent" to contracting malaria. Similarly, crossing a busy intersection does not constitute "consent" to being hit by a car, wearing provocative clothing does not constitute "consent" to sexual assault, mountain climbing does not constitute "consent" to breaking a bone, and surfing without a life jacket does not constitute "consent" to drowning. Taking an action with a higher intrinsic risk, or refusing to take an action that might lower risk, is not morally equivalent to being complicit in facilitating the harm that occurs when a risk materializes.

People who favor compulsory masking may object that placing primary responsibility for a person's health onto that person is, itself, a form of "victim blaming." But this would make the definition of "victim blaming" so broad that blaming anyone for how they contracted an illness could be called "victim blaming," regardless of how they did so. In addition to his earlier insights on the psychology of masking, psychiatrist Robert Freudenthal also elaborated on this specific error:

> "Instructing people that they need to cover their faces in a certain way, and if they do not do this, then they are behaving irresponsibly and inviting danger, and therefore bear the responsibility if there are negative consequences if they do not wear masks, is analogous to the experience that some people have, particularly women, of being instructed to 'cover up,' with the message of 'If you don't wear certain clothes you are immoral, and are inviting tragedy.'"[263]

The critical moral insight in pushback against victim blaming is that being an important, even essential, part of a causal chain of events leading to a bad outcome does not necessarily confer moral culpability *for* that outcome.[264]

Moreover, any "outcome" is itself, part of an ongoing chain of events which no human can foresee the conclusion of. As philosopher William Lane Craig pointed out during a lecture on the problem of evil and suffering:

> "[O]ne of the decisive objections to utilitarianism (which is the ethical theory that says that we should do that action which is likely to bring about the greatest

---

263   Freudenthal, Robert, "The True Meaning of Masking," Brownstone Institute Articles, October 30, 2021. https://brownstone.org/articles/the-true-meaning-of-masking/ last accessed December 31, 2022.

264   This has been recognized for centuries by the legal doctrine of proximate cause and "but for" test in cases of civil or criminal liability.

good for the greatest number of people) is that we have no idea of the ultimate outcome of our actions. Some short-term good might lead in the long run to untold misery while some action that looks disastrous in the short-term may turn out to bring about the greatest good for humanity."[265]

Utilitarianism falls under the broader philosophical category of consequentialism, which is the view that an action's morality is determined by its consequences, and there is so much overlap between the two that, in casual usage, the words are often used almost interchangeably. Utilitarianism and consequentialism fail as systems of ethics because no human can ever know what the final outcome of their actions will be, and because some actions are still right or wrong, even if they bring about a net bad outcome or a greater good. Bringing a child into this world remains a good thing, even though you cannot know if that child will grow up to be the next Chairman Mao or Mother Teresa. Elsewhere, Dr. Craig points out: "Just because the consequences of some action are good doesn't make that action right or good. The end doesn't justify the means. When a martyr voluntarily gives up his life to save others, we admire the martyr, but we don't admire the people who killed him."[266]

An alternative to utilitarianism is what is called a deontological view. Dr. Craig explains:

> "By contrast, on what is called a deontological view, our decisions are to be guided by certain moral principles, which we can know to be true without having to look into the future to see the outcomes of our choices…. Now, of course, sometimes the application of these moral principles will require us to consider the consequences of our actions—for example, whom shall we treat first in the plague? — but that is still not consequentialism because the rightness or wrongness of those actions is not determined just by their consequences. Even if by the vicissitudes of history deliberately infecting someone with the disease should issue in some great benefit for mankind, it would still be morally wrong to commit such an atrocity…

265   William Lane Craig, "Excursus on Natural Theology (Part 32): The Problem of Evil and Suffering (3)," *Reasonable Faith*: Defenders Podcast Series 3 (June 29, 2016), Transcript online at: https://www.reasonablefaith.org/podcasts/defenders-podcast-series-3/s3-excursus-on-natural-theology/excursus-on-natural-theology-part-32 (last accessed January 28, 2023)

266   William Lane Craig, "A Muslim Asks about Jesus" Reasonable Faith Question of the Week #288, October 21, 2012. Online at: https://www.reasonablefaith.org/writings/question-answer/a-muslim-asks-about-jesus (last accessed January 31, 2023)

"The assumption of those who do not see this is that 'What should not be should not be permitted.' That principle is false. There can be genuinely evil acts, things which should not be, and yet we—and God—can be morally justified in allowing them to occur."[267]

Dr. Craig cautions, "The consequentialism to be most wary of, I think, would make 'achieve a greater good' both a necessary *and sufficient* condition for permitting some when-considered-by-itself evil."[268] While "consideration of consequences can be relevant factually to deciding which moral principle is applicable in a situation," this does not mean that bringing about a net greater good provides enough justification for committing active wrongs in pursuit of that good: "I don't think that the well-being of conscious creatures is a good guide to discovering our moral duties. Indeed, it seems to me that it leads naturally to eugenics."[269] Ethics also cannot be grounded in the recommendations of some committee of experts — even an *ideal* committee of medical experts opining on mask mandates: "Why think that if you could assemble a committee of perfectly rational human beings and they all would agree that you should do some action *A*, this constitutes a moral obligation for you to do *A*?"[270]

To say that we have an open-ended moral duty to wear a mask because doing so brings about the greater good of human flourishing or public health simply smuggles in consequentialism or utilitarianism using a different vocabulary. Moral obligations do not and *cannot* originate from committees of "experts" or panicked mobs; still less from panicked mobs of "experts."[271]

267   William Lane Craig, "Consequentialism and the Problem of Evil," Reasonable Faith Question of the Week #448, November 15, 2015. Online at:
https://www.reasonablefaith.org/writings/question-answer/consequentialism-and-the-problem-of-evil

268   William Lane Craig, "Is God a Consequentalist?" Reasonable Faith Question of the Week #333, September 02, 2013. Online at: https://www.reasonablefaith.org/writings/question-answer/is-god-a-consequentialist

269   William Lane Craig, "Sam Harris on Objective Moral Values and Duties" Reasonable Faith Question of the Week #208, April 11, 2011. Online at
https://www.reasonablefaith.org/writings/question-answer/sam-harris-on-objective-moral-values-and-duties

270   William Lane Craig, "Contemporary Moral Arguments" Reasonable Faith Question of the Week #116, July 05, 2009. Online at: https://www.reasonablefaith.org/writings/question-answer/contemporary-moral-arguments

271   Note: It occurs to me that one or more readers may attempt to rebut the arguments I make in this section simply by sending me a picture of Dr. Craig wearing a mask — which I am aware that he has done on numerous occasions. As a proponent of Divine Command Theory, Dr. Craig would no doubt agree with me that God places different moral duties on different people depending on their individual situations, and that it is entirely possible that one person may have a moral duty to wear a mask while another person right next to them has a moral duty to refuse to wear a mask. Despite his insanely high intelligence, Dr. Craig is exceptionally humble, and would be the first to acknowledge that masks only fall under the umbrella of his expertise insofar as they involve philosophical issues. It is my position that the infection control properties of masks (or rather, lack thereof), never did and never could give rise to a moral duty

Key Takeaways

🔑 Seatbelt, drunk driving, public decency, and other analogies used to try to justify compulsory masking all fail.

🔑 Compulsory masking was based on a flawed view of causality and culpability which treated failing to take preventative measures against a bad outcome as a form of consenting to that bad outcome. Accepting risk of a bad outcome is not the same as consenting to that bad outcome or facilitating it.

🔑 Just because a committee of experts insists that a mask mandate is moral or ethical does not make it so. An act can still be morally wrong even if it meets the requirements of a utilitarian ethical system and brings about the greatest good for the greatest number of people. Measures like mask mandates would *still* have been wrong even if masks worked.

---

for anyone to wear a mask for COVID. However, the pressures arising from other moral forces might very well make it one person's duty to wear a mask for other reasons. In any event, Dr. Craig would no doubt also agree with me that a picture of him wearing a mask does nothing at all to defeat his criticisms of utilitarianism and consequentialism as systems of ethics.

Even consequentialism argues against compulsory masking

Even approaching masks using a purely consequentialist ethical framework, one which refuses to admit any individual right to conscientious objection *and* assumes for the sake of argument that masks work, there is *still* a powerful pragmatic case for honoring (or at least not penalizing) people (including healthcare providers) who refuse masks or other medical interventions on conscientious grounds. On a purely practical policy level, "although granting conscience-based exemptions from legal obligations as a matter of course may wreak havoc on the state's ability to maintain order, the same can be said of a state that rejects claims of conscience altogether." [272]

Punishment in civil society is typically for the purposes of reform, deterrence, or retribution, but punishing someone for taking non-violent action or refusing to take an action according to the dictates of their conscience almost never serves these ends effectively. "Reform" is not served by punishing conscientious objectors, because it is impossible to coerce belief against someone's wishes, and punishment of conduct motivated by sincerely held conscientious belief is likely to result in hardening rather than changing those beliefs. For the subset of conscientious objectors who *can* be successfully coerced into changing their outward behavior in spite of their beliefs, the result is in many ways even worse. In his book on *Conflicts of Law and Morality*, legal scholar Kent Greenawalt points out: "Deterrence of those who lack the will to act on their convictions exacts a terrible price. Their feeling that they have yielded to compulsion and violated their most deeply held beliefs and principles may involve profound resentment and loss of self-respect."[273] Even if the coercion is successful, someone performing a task which have been forced into but which they believe to be

272   Nadia N. Sawicki, *The Hollow Promise of Freedom of Conscience*, 33 Cardozo L. Rev. 1389 2011-2012. Available online: https://papers.ssrn.com/sol3/papers.cfm?abstract_id=1666278

273   Kent Greenawalt, *Conflicts of Law and Morality*, (New York, Oxford: Oxford University Press, 1989), pp. 315-316 available online: https://archive.org/details/conflictsoflawmo0000gree

immoral is simply not going to do as good of a job as they would otherwise. For example, drafting pacifist Quakers or Mennonites into a combat unit does not increase that unit's combat potential. A prime historical example of this is how members of pacifist sects conscripted into Napoleonic armies avoided using their weapons to injure or kill.[274] This made it worse than useless to try to force them to violate their sincerely held religious beliefs by conscripting them in the first place.

Using retributive punishment or trying to make an example of people who adhere to their beliefs can also be expected to fail, because willingness to act against self-interest in order to do what one believes to be morally right is widely (and rightly) considered to be a praiseworthy character trait. That kind of strength of conviction is, by itself, persuasive to many observers. After all, someone willing to suffer for their beliefs is more likely to have a good reason for holding those beliefs. People intuitively recognize that acts grounded in conscience, and taken contrary to temporal self-interest in the face of social punishments, are far more likely to reflect external objective moral truth than acts taken merely on the basis of legalistic compliance with rules or external mandates.[275]

Cultivating the development of personal conscience and self-determination is a good in and of itself, not just a means to an end. As legal scholar Lynn Wardle commented on James Madison's famous *Memorial and Remonstrance*:

> "[I]f men are not loyal to their duty to their God and their conscience, it is
> folly to expect them to be loyal to mere legal rules, statutes, judicial orders,
> or professional duties. If you demand that a man betray his conscience, you
> have eliminated the only moral basis for his fidelity to the rule of law, and have
> destroyed the moral foundation for democracy."[276]

---

274   Peter Brock, "Conscientious Objection in Revolutionary France," *The Journal of the Friends Historical Society*, 1995. 57(2), page 177-178 and footnote 46 on page 182. https://journals.sas.ac.uk/fhs/article/view/3500 Most likely those conscripted pacifists who were unwilling to confront the military judicial and penal system avoided using their weapons to harm others by simply going through enough of the motions to make it nearly impossible to tell during the heat of battle that this is what they were doing.

275   An analogous pro-mask-wearing person would be someone who insists on wearing a mask even in the face of comparable pressure and legal mandates *against* wearing masks. However, widespread pressure to take off masks has not equaled the intensity of the pressure to put on masks, nor is it likely to ever do so. Therefore, any such individuals who hold to wearing masks with corresponding intensity are virtually impossible to identify with certainty.

276   Lynn D. Wardle, "Protection of Health-Care Providers' Rights of Conscience in American Law: Present, Past, and Future," 9 Ave Maria Law Review, Volume 9, Number 1, (2010), p. 8, Available online: https://papers.ssrn.com/sol3/papers.cfm?abstract_id=1898456

Punishing people for following the dictates of their conscience undermines the rule of law because doing so highlights the distinctions between what is legal and what is moral, and if the law does not serve justice, it loses legitimacy. Thus, even on a purely consequentialist or utilitarian view, there is a powerful argument to be made that individual conscience should be honored rather than penalized when it comes to masks and other medical interventions. This fact makes the consequentialist arguments underlying compulsory masking particularly ill-considered.

Key Takeaways

🔑 Even under a utilitarian or consequentialist ethic, there is a strong case to be made for respecting conscientious objection.

🔑 Punishing conscientious refusal to comply with compulsory masking and other mandated medical interventions tends to undermine both the specific goals and the overall legitimacy of those mandates.

The Founders' views of medical liberty in practice

Even most of those people who think that mask mandates are defensible will concede that there are some circumstances under which it is morally permissible to deliberately infect another person with a deadly disease. The classic example of this is the medical precursor of vaccination, variolation for smallpox, which involved controlled infections with live smallpox virus in an effort to deliberately instigate a more controlled form of the disease. Ironically, the same mentality behind trying to force everyone to wear a mask or get the COVID vaccine is the same mentality that banned people from inoculating themselves with smallpox on the grounds that doing so imposed risk on others. Of course, deliberately infecting oneself with a disease like smallpox, as in variolation (and some live attenuated vaccines today), is a far cry from simply refusing to take one or more general preventative measures to *avoid* infection with a respiratory virus. The fact that variolation was widely practiced in America in the 1700s (albeit with localized restrictions in some areas) is an indicator of the degree to which individual medical freedom was respected.

One of the most straightforward ways to ascertain what the authors of the Constitution believed about where the line between essential liberty and the public good should be drawn during times of public health crises is to look at how they handled epidemics in their own lifetimes. Much has been made of George Washington's smallpox inoculations of the Continental Army during the American Revolution, or the Vaccine Act of 1813 signed by James Madison, but what is not fully appreciated is that The Founders lived through *multiple* epidemics next to which COVID pales by comparison. Yet their responses to these worse threats were nowhere near as extreme as ours was to COVID.

The authors of the United States Constitution fought the Revolutionary War during a 7-year smallpox epidemic that was proportionately far more deadly than COVID-19, and this contributed to the fact that American combatants alone racked up battlefield and disease deaths totaling nearly 1% of the overall population. [277] This is in addition to the civilian deaths and sacrifices involved in attaining American independence. It is obvious that the Founders valued their God-given liberties far more highly than the cost of even a worst-case Spanish Flu or coronavirus pandemic. *Proportionately,* one of the worst epidemics in our nation's history occurred in 1793 during George Washington's first term as president. This happened not just anywhere, but in Philadelphia, which was the capital of the United States at that time. This was the Philadelphia Yellow Fever Epidemic of 1793. In a city of 50,000 residents, nearly 10% of the population — 4,000 to 5,000 people — died. [278] Scenes occurred in Philadelphia that were never paralleled during COVID, "Bodies were found lying in the streets. A man and wife were discovered dead in bed, their little child between them still sucking at its mother's breasts… Children were wandering in the streets, their parents dead at home." [279] Even more instructive for our COVID experience, doctors at the time thought that Yellow Fever could be passed directly from person to person or through contact with contaminated items. [280] Various towns enacted quarantine measures on ships and individuals who were actively sick, and many nearby cities and towns used their militia to actively deny entry to any people or trade goods coming out of Philadelphia. However, when we look in depth at the history of that event, what we do *not* see is even more telling. There were no bans on "mass gatherings" of less than a dozen people backed up by threats of fines or imprisonment. We do not see officials categorizing their citizens' livelihoods and employment as essential or nonessential and then forcibly suspending all pastimes they consider to be nonessential. People within Philadelphia during the Yellow Fever Epidemic had freedom of movement. They conducted business. They visited neighbors, both sick and well. *They attended church.* Pennsylvania as a whole was not "locked down," much less any other state, and there were *no mask mandates.* Citizens in Philadelphia actively, contagiously, suffering from Yellow Fever were still free to choose

277   Elizabeth A. Fenn, *Pox Americana: the great smallpox epidemic of 1775-82*, New York: Hill and Wang (2001). 2002 Kindle ed.

278   Griscom, J.H., *A history, chronological and circumstantial, of the visitations of yellow fever at New York.* 1858, New York: Hall, Clayton, p. 5. Available online: https://collections.nlm.nih.gov/ext/dw/101586195/PDF/101586195.pdf

279   John Harvey Powell, *Bring Out Your Dead : The Great Plague of Yellow Fever in Philadelphia in 1793,* Philadelphia: University of Pennsylvania Press (1949), p. 96, available online: https://archive.org/details/bringoutyourdead0000powe_v4j6/page/96/mode/2up.

280   *See generally:* John Harvey Powell, *Bring Out Your Dead : The Great Plague of Yellow Fever in Philadelphia in 1793,* Philadelphia: University of Pennsylvania Press (1949), e.g. pp. 86-87, available online: https://archive.org/details/bringoutyourdead0000powe_v4j6/page/86/mode/2up.

what treatments to accept or refuse, according to their individual preferences and personal beliefs.[281] Measures like those used during COVID appear to have never even been *considered*.

Despite killing 10% of the city's population, the Philadelphia Yellow Fever Epidemic lasted from just August to November of 1793, meaning it did its damage in a much shorter span of time than the initial COVIDcrisis lockdowns. [282]If a disease that killed 10% of the population of the national capital, and the largest city in the United States at the time, did not constitutionally justify locking

---

281   *See generally*: John Harvey Powell, *Bring Out Your Dead : The Great Plague of Yellow Fever in Philadelphia in 1793*, Philadelphia: University of Pennsylvania Press (1949), pp. 99, 272 available online: https://archive.org/details/bringoutyourdead0000powe_v4j6/page/98/mode/2up, https://archive.org/details/bringoutyourdead0000powe_v4j6/page/272/mode/2up.

282   This 3- to 4-month duration is typical of historic plagues. For example, the Black Death ran through communities in about 3-4 months, as it did in the English village of Walsham in west Suffolk:

> "As far as it is possible to judge, the pestilence struck Walsham close to Easter Day, which this year fell on April 12, was fading fast by early June, and had departed by late June. Not a single tenant death was reported at the Walsham manor court session of March 6, but no fewer than 103 were reported at the next session, which was held on June 15. That the pestilence was virtually over by this latter date is indicated by the proceedings of the next Walsham court of August 1, in which a mere five further deaths were registered. Fortunately, additional evidence of the duration of the epidemic in the parish is provided by the proceedings of the courts of High Hall, where eleven out of the total of fifteen tenant deaths suffered on the small submanor occurred before May 25. It was in April and May 1349 that the plague in England probably reached its peak, in terms of geographical spread and number of deaths."
>
> John Hatcher, *The Black Death: A Personal History*,
> Cambridge, MA: Da Capo Press (2008), page 136.

Each wave of the Spanish Flu was distinct. While Australia escaped in 1918, with each province hit with two 3-month waves in 1919, the Spanish Flu hit New Zealand hardest over just a 2- to 3-month period from the end of October through December of 1918:

New Zealand's experience of the great 'Spanish influenza' of 1918 was severe but relatively short. Mortality from influenza and pneumonia had dwindled to single figures well before Christmas and, despite many people's fears, another wave of flu did not eventuate in 1919.

Geoffrey W. Rice, *Black November: The 1918 Influenza pandemic in New Zealand*, (1988) Revised and enlarged second edition, Canterbury University Press: Christchurch, New Zealand (2005, ebook ed. 2016), page 508.

The point is that if COVID was really a plague on the order of the Black Death or Spanish Flu, the ordeal would have been over by the end of summer 2020, with distinct breaks between any annual waves. History-making plagues may *spread* over a period of years, but they rip through communities in relatively short periods of time despite all efforts to contain them. They do not drag on interminably for more than a year like the COVIDcrisis. Apart from the large number of suspiciously empty hospitals, on its own, this historic pattern should have been enough to cast doubt on claims from the latter half of 2020 onwards that COVID was truly an ongoing emergency justifying mandates. We refer to flu *season* for a good reason, because there is no sense agonizing over it all year long, and coronaviruses are *all* seasonal.

See also:

everything down and masking up, then how could it possibly have been appropriate in 2020 for *any* leadership to push this? My argument is not that the authors of the Constitution did something, therefore it was good. Rather, my argument is that the authors of the Constitution understood the meaning and practical intent of the document that they wrote. The authors of the Constitution suffered through many national emergencies (military, disease, natural disaster, you name it), all of which were proportionately far more dire than COVID-19, and they did not see those events as providing justification for suspending constitutional recognition and protections of civil liberties. Therefore, *we* had no business doing so in the case of COVID-19.

The mentality that gave rise to the COVID response was not unique to 2020. It has been present in a subset of humans at every point throughout history, and so the Constitution of the United States lacks an emergency clause not because of accident or omission, but by design and from conscious intent. In their approach to epidemic disease, America's Founders recognized far more latitude for individual conscience and discretion than did the modern medical tyrannies who used the common good as a rhetorical jackboot to force masks and other medical interventions onto and into everybody.

John M. Barry, *The Great Influenza: The Story of the Deadliest Pandemic in History*, Penguin Books: New York (2005, 3rd ed. and ebook 2018).

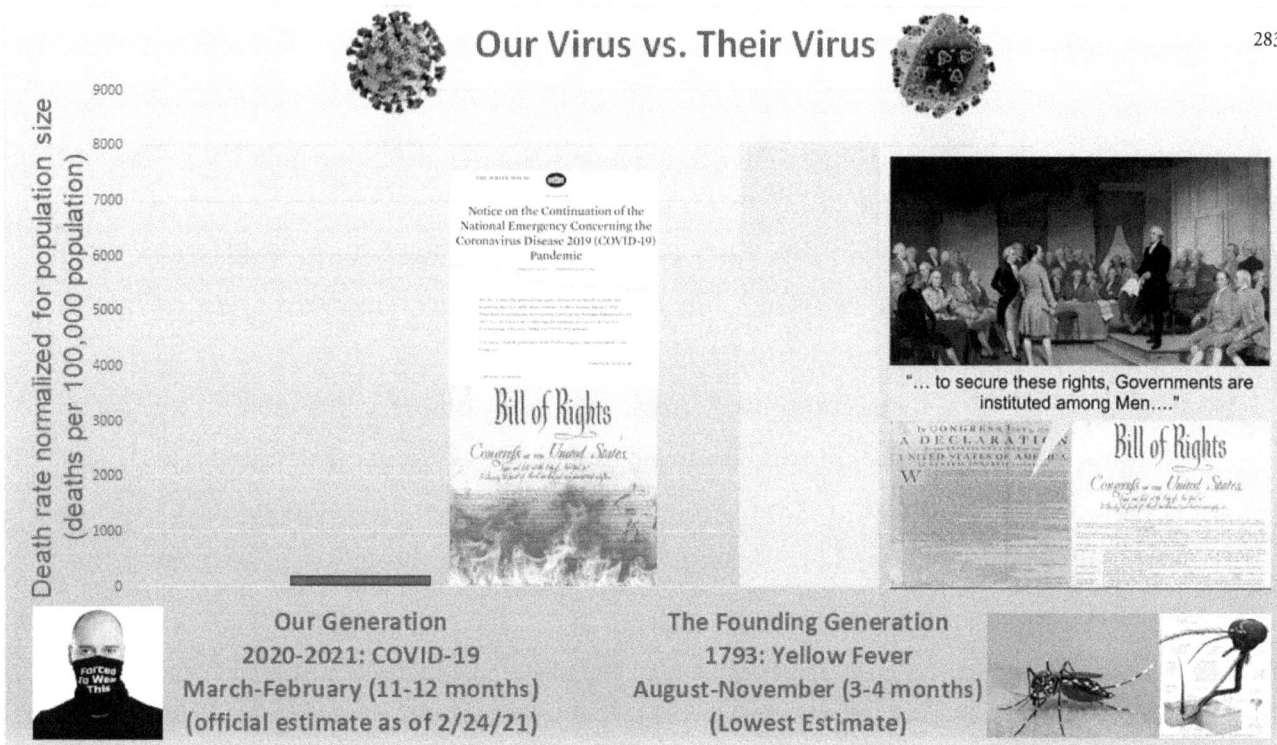

Our Virus vs. Their Virus

"... to secure these rights, Governments are instituted among Men...."

Our Generation	**The Founding Generation**
2020-2021: COVID-19	1793: Yellow Fever
March-February (11-12 months)	August-November (3-4 months)
(official estimate as of 2/24/21)	(Lowest Estimate)

"Our Virus vs. Their Virus" chart data sources:

John H. Griscom, *A history, chronological and circumstantial, of the visitations of yellow fever at New York*. 1858, New York: Hall, Clayton, page 5. https://archive.org/details/101586195.nlm.nih.gov

Pennsylvania Department of Health, *COVID-19 Data for Pennsylvania*, 2021, available from https://www.health.pa.gov/topics/disease/coronavirus/Pages/Cases.aspx

---

283   "Our Virus vs. Their Virus" chart data sources:

John H. Griscom, *A history, chronological and circumstantial, of the visitations of yellow fever at New York*. 1858, New York: Hall, Clayton, page 5. https://archive.org/details/101586195.nlm.nih.gov

Pennsylvania Department of Health, *COVID-19 Data for Pennsylvania*, 2021, available from https://www.health.pa.gov/topics/disease/coronavirus/Pages/Cases.aspx

Key Takeaways

🔑 The U.S. Constitution lacks an emergency exception to the Bill of Rights not by accident or omission, but by design and from conscious intent.

🔑 The authors of the U.S. Constitution lived through multiple episodes of disease that were proportionately far worse than COVID, yet they never even considered measures like compulsory masking.

🔑 At the time the U.S. Constitution was drafted, individual liberty in a medical context extended to deliberate self-infection with smallpox, one of the most dangerous and contagious diseases in history. Contrast these experiences with the response to COVID, where merely forgoing touted preventative measures like masking was banned and penalized.

Perverting the precautionary principle

Often, vague invocations of the precautionary principle are used to justify compulsory public health measures such as lockdown and masking. This perversion of the precautionary principle holds that if you can't prove you're not sick, or if you can't prove that not wearing a mask is safe, then you should wear a mask. However, a consistent understanding of the precautionary principle would also apply to every measure that was being made compulsory, including masks, lockdowns, and vaccines. Our knowledge is imperfect, and because we do not know how many lives proposed impositions may save, and we do not know how many lives they may take, we ought to leave individuals free to choose what risks they will accept and to make mutual accommodations — *especially* in situations like a plague (real or imagined).

For all we know, some new and far more deadly coronavirus will come along in 20 years, and everyone who got natural immunity to SARS-CoV-2 will be resistant, but those who avoided this virus or got the vaccine will have no such benefit. That is a wild scenario, and we should not plan on it, but that is part of the point; you cannot deprive people of their innate liberties based on a worst-case scenario that someone dreams up, because such catastrophizing scenarios will always exist. Freedom to choose is valuable enough that people ought not to be deprived of it, even if their use of that freedom entails that someone who is *possibly* infectious may *possibly* expose someone else to airborne particles which may *possibly* contain virions which may *possibly* be viable and may *possibly* cause a secondary infection which may *possibly* lead to illness which may *possibly* lead to hospitalization and *possibly* result in death. If you want nuanced incremental tradeoffs leading to a more optimal outcome, the best way to achieve that is by mutual accommodation at the individual level based on individual health and risk assessments, not by blanket mandates and coercion.

Another argument for compelling masking and other medical interventions involves invoking a version of the "Bystander at the Switch" ethical dilemma. In this dilemma, the reader is cast in the role of a bystander, watching a runaway train hurtling towards a large group of people standing on the tracks. There is a switch within arm's length which will send the train down a different track to crush a smaller group of people. Those pushing compulsory masking and other measures argue that everyone who might be in arm's reach of such a switch has a moral and legal duty to throw that switch, diverting the train onto the track holding the smaller group of people. On this view, anyone who tries to prevent the switch from being thrown becomes directly responsible for the death or potential death of everyone on the first track.

In real life, however, the number of people on each track is unknown because much of the track is out of sight (and sometimes being deliberately obscured). How the people currently on the track got there, as well as whether or not they remain there, is also outside the bystander's control. In real life, flipping the switch may involve pushing one group of people *onto* the track in order to divert the train's path when it hits them, but with no guarantee the train will actually change course (everyone harmed by COVID lockdowns or coerced mRNA injections falls into this category). The real point of the dilemma is to show that in real life, no person or government has the right to unilaterally arrogate such a choice to themselves.

There is a sense in which epidemics, pandemics, and public health can be likened to a battlefield or emergency room triage situation, where grim externally imposed urgent necessity compels people in leadership positions on the ground to allocate limited resources as best they can. But compulsory interventions like masks and vaccinations are entirely different. They are a deliberate self-directed seizure of fundamental choice from those who more rightfully possess this limited level of control over that aspect of their lives. One blogger answered the bystander at the switch dilemma in the context of COVID by applying a negative formulation of the Golden Rule: "Are you prepared to allow the government to throw you in front of the train to save someone else? If not, then demand that the government stop doing it to somebody else!" [284]

A rebuttal at this point might be to argue that *not* mandating masks would be to unjustifiably permit deadly infections in pursuit of a different "greater good" — that of individual liberty. It is true that declining to actively avoid something may, at the worst, *permit* something to occur, but that is not the same as *causing* that thing to occur. Conflating these two things is bad reasoning, and leads to destructive outcomes and misattribution of responsibility because it holds people responsible for things that they did not, in fact, cause. Thomas Sowell dryly observed that, "We seem to be getting closer and closer to a situation where nobody is responsible for what they did but we are all responsible for what somebody else did."[285]

---

284   Julius Ruechel, "Bystander at the Switch (updated): The Moral Case Against Mandatory Public Health Measures," Blog, February 1, 2022. Online: https://www.juliusruechel.com/2022/02/bystander-at-switch-updated-moral-case.html

285   Thomas Sowell, *The Thomas Sowell Reader*, New York: Basic Books (2011), p. 402, Kindle Edition

It is bad enough trying to play whack-a-mole with risk-tradeoffs when all the interactions involved are voluntary, but coercion adds another dimension entirely. When you refrain from interfering with people's free choices, *they* bear primary responsibility for all good and bad outcomes. But when you forcibly remove those choices, *you* take on primary responsibility, not just for the good outcomes, but for the bad ones as well. It is not some "noble sacrifice" to forcibly take such responsibility onto oneself against the will of those to whom such choices rightfully belong. It is the sort of hubris that has more in common with iconic evil overlords of fantasy literature (and with real-world history's worst totalitarians) than with any tragic hero. It is worse than the worst dereliction of social duty or criminal negligence which could be ascribed to refusal to wear a mask even if masks worked.

People who argue that they do not choose to accept the risk "imposed" on them by people in their community refusing to make use of one or more medical interventions, personal protections, or personal hygiene measures are badly mis-framing the argument. Objections like this draw a false equivalency between positive actions which *actually* impose risk (e.g. drinking and then driving) and non-actions. To accept such an objection in favor of compulsory masking, or any other measure, would be to accept a pathological degree of personal responsibility for the well-being of others. Moreover, the people making this objection are *already* at risk, and are using the existence of this risk to demand that they receive protection at the cost of harm to others. Who is being more selfish: the person who refuses to give assistance, or the person who demands that assistance, and forces others to provide it against their will?

Just as I would still be guilty of murder if I paid a hitman rather than doing the deed myself, outsourcing the imposition of medical interventions to a third-party, including government officials, would not make my actions in forcing such measures on others any less wrong. The same applies even if I ostensibly impose these medical interventions on *behalf* of a third party like my family. Third parties may in any case be totally unaware of (or even opposed) to my forcing such measures on others. A foresighted commentator in 2015 put it this way in the context of public health:

> "Never before have people widely asserted that they have the right to demand that everyone around them take all possible precautions at whatever cost to themselves to make this environment absolutely risk free. If, as the mandatory vaccination proponents contend, we can demand that everyone around us take every conceivable precaution against every communicable disease, what else can we demand of them?

"How about superbugs? What are we going to do about all those people who abuse antibiotics, ultimately leading to the creation of superbugs. Antibiotic-resistant bacteria are responsible for nearly 15,000 deaths in the US each year, far outstripping pre-vaccine deaths for measles, mumps and whooping cough combined. Can we not hold the irresponsible people who take antibiotics every time they have a minor infection accountable for this?

"Here's why not: because your right to protect 'public health' — whatever you think that may be given the interest-driven media hysteria of the moment — ends where my body begins. Herd immunity is not something anyone has a 'right' to. It is a positive externality, and like other such externalities it is not something you have a right to demand that your fellow human beings provide for you. More to the point, you do not have a right to demand that other parents impose risks on their children that they are not comfortable with, in order to protect your child or anyone else's children. Can there ever be a point where spreading a disease becomes 'assault'? Of course there can: A person who knows that they are infected with Ebola, for example, stepping into a crowded subway car and proceeding to cough all over the other passengers, could easily be considered guilty of assault.

"[N]ot being vaccinated does not equate to being infected with a disease, far less to knowingly infecting others. Failure to take every precaution against getting a disease is hardly 'assault'. Even in the case of a truly deadly illness like Ebola, there is no justification for forcing a particular method of prevention on those who have not contracted it, or forcing treatment on anyone who has. All that anyone has a right to do is demand that those people not infect others."[286]

Immunity acquired through natural means is every bit as beneficial of an externality as immunity acquired through artificial means. A state of non-immunity or sub-optimal protection cannot be legitimately equated with a state of active and transmissible disease. Rhetorical appeals to uncertainty, the precautionary principle, or other sophisms do nothing to defeat an individual's reasonable belief that they are healthy and noncontagious in the absence of signs, symptoms, or other definitive evidence to the contrary. The fact that such a belief may occasionally be mistaken is not grounds for forcing the

---

286   Bretigne Shaffer, "First They Came for the Anti-Vaxxers," LewRockwell.com, April 23, 2015. Online: https://www.lewrockwell.com/2015/04/bretigne-shaffer/first-they-came-for-the-anti-vaxxers/

general population to act as though it is *always* mistaken. In the matter of your own health and that of your children and other immediate family members, you have much more first-hand knowledge of your medical history and risk factors than even your family doctor, much less a public health official. You certainly care more, and you ultimately have a far stronger moral and legal claim over your own body than they do, because your body is your most fundamental property.

When a masked or unmasked person is simply going about their business, content to let others do the same, nothing has been "imposed" on anyone. Real imposition occurs when an authority makes it their business to put a mask on you, simply because someone else somewhere else in the community has a disease which they think a mask might retard. Even assuming that every argument in favor of wearing a mask is true, neglecting to put on a mask would remain, at worst, a sin of omission, while forcing one onto someone against their will is — and would remain — a sin of commission. At worst, it would be the difference between failing to help someone being stabbed, and doing the stabbing oneself. Both can be bad, but all other things being equal, sins of commission are worse. Attempts to justify compulsory masking based on claims of community self-defense fail because the right of self-defense does not permit pre-emptive strikes based on the assumption that every member of a community is an aggressor.

It still surprises many people to learn that in case law over the last few decades, the Supreme Court has ruled in multiple decisions that individuals do not possess a "legitimate claim of entitlement" to receive "protective services" from the police under the Fourteenth Amendment's Due Process Clause, even when the facts involved are "undeniably tragic."[287]

> "It forbids the State itself to deprive individuals of life, liberty, or property without 'due process of law,' but its language cannot fairly be extended to impose an affirmative obligation on the State to ensure that those interests do not come to harm through other means. Nor does history support such an expansive

---

287   *DeShaney v. Winnebago County*, 489 U.S. 189, 191 (1989), Available online at: https://www.loc.gov/item/usrep489189/

Castle Rock v. Gonzales, 545 U.S. 748, 756-757 (2005), Available online at: https://www.loc.gov/item/usrep545748/
For less technical, more popular-level summaries, *see also:*
Amanda Marcotte, "Uvalde shooting timeline exposes an ugly truth: The police have no legal duty to protect you," *Salon*, May 27, 2022, https://www.salon.com/2022/05/27/uvalde-timeline-exposes-an-ugly-truth-the-police-have-no-legal-duty-to-protect-you/
Ryan McMaken, "Police Have No Duty to Protect You, Federal Court Affirms Yet Again," MisesInstitute, December 20, 2018, online at: https://mises.org/power-market/police-have-no-duty-protect-you-federal-court-affirms-yet-again

reading of the constitutional text."[288]

If police agencies whose motto includes "to protect and serve" do not necessarily have a legal duty to protect me from direct harm by other individuals, there is no basis for myself or anyone else to claim that random neighbors have a legal duty to wear a mask to provide some theoretical measure of indirect protection from a pathogen.

During COVID, failing to "assist" by wearing a mask was rhetorically conflated with assault or manslaughter by means of a deadly bioweapon. This false equivalency is so outrageously bad that even panic-addled mental states cannot excuse it. In a nod to George Orwell's 1984, one particularly insightful internet blogger (*el gato malo*, who we met in Part 3) coined the term "riskcrime" to describe this mentality of "demanding a right to convict for that which exists only in potential and has never been acted upon."[289]

The mere existence of each one of us creates both risks and benefits for those around us, and everything we do creates both good and bad externalities. The fact that individuals do not always make choices which lead to optimal outcomes is no excuse for depriving them of those choices, much less for placing those choices in the hands of equally flawed and fallible politicians and public health officials. Policies such as banning indoor and outdoor gatherings or closing large public parks never had any calm, educated, or rational genesis. Constantly reminding people to stay 6 feet from each other at all times, and to avoid handshakes at all costs, had no more medical virtue than the practice of bloodletting. Does anyone seriously think it is a good idea to let experts who have spent 30 years obsessing over traffic fatalities dictate speed limits, or to let people with pathological anxiety or an unhealthy penchant for micromanagement have control over what you can and cannot do with your life? It is not the exercise of rights that should be regarded as conditional and carefully monitored, but the *abrogation* of those rights that needs these restraints.

In their own ways, the authors C. S. Lewis and J. R. R. Tolkien warned against the form of utilitarianism underlying compulsory masking, compulsory vaccination, and many other measures that compel *goods* of commission. In an essay against the humanitarian theory of punishment, Lewis wrote:

> "Of all tyrannies a tyranny sincerely exercised for the good of its victims may
> be the most oppressive. It may be better to live under robber barons than under

---

288   *DeShaney v. Winnebago County*, 489 U.S. 189, 195 (1989), Available online at: https://www.loc.gov/item/usrep489189/

289   el gato malo, "Due consideration of due process: tearing down Chesterton's defense against' tyranny," *Bad Cattitude* (blog), August 4, 2022. Online at: https://boriquagato.substack.com/p/due-consideration-of-due-process

omnipotent moral busybodies. The robber baron's cruelty may sometimes sleep, his cupidity may at some point be satiated; but those who torment us for our own good will torment us without end for they do so with the approval of their own conscience. They may be more likely to go to Heaven yet at the same time likelier to make a Hell of earth.[290]

Similarly, in a 1951 letter to his publisher, Tolkien explained:

"The Enemy in successive forms is always 'naturally' concerned with sheer Domination, and so the Lord of magic and machines; but the problem: that this frightful evil can and does arise from an apparently good root, the desire to benefit the world and others — speedily and according to the benefactor's own plans — is a recurrent motive."[291]

In a later draft of another letter to a fan, Tolkien described how even the wisest character in his *Lord of the Rings*, the wizard Gandalf, would be unable to rightly wield such a degree of power over others:

"Gandalf as Ring-Lord would have been far worse than Sauron. He would have remained 'righteous', but self-righteous. He would have continued to rule and order things for 'good', and the benefit of his subjects according to his wisdom (which was and would have remained great). Thus while Sauron multiplied evil, he left 'good' clearly distinguishable from it. Gandalf would have made good detestable and seem evil.[292]

In Tolkien's story, Gandalf's wisdom and humility are great enough that he consciously recognizes this truth, immediately rejects such power even when it is offered freely, and works to prevent others from seizing it. During COVID, public health officials and other authorities failed this same test miserably and repeatedly when it came to masks and other compulsory public health measures.

---

290   Clive Staples Lewis, *God in the dock: essays on theology and ethics*, Grand Rapids, Michigan: William B. Eerdmans Publishing Company (1970), p. 292. Available online: https://archive.org/details/godindockessayso0000lewi/page/292/mode/2up

291   John Ronald Ruel Tolkien, Humphrey Carpenter & Christopher Tolkien (editors), *The Letters of J.R.R. Tolkien*, Boston New York: Houghton Mifflin Harcourt (letter #131, p. 146). Kindle Edition.

292   John Ronald Ruel Tolkien, Humphrey Carpenter & Christopher Tolkien (editors), *The Letters of J.R.R. Tolkien*, Boston New York: Houghton Mifflin Harcourt (letter #246, pp. 332-333). Kindle Edition.

Tolkien's characterization of the evil that comes from this will to power in the name of doing good was prescient, and C. S. Lewis nailed it when he coined the phrase "humanitarian tyranny." This term accurately describes the profoundly wrong and unbelievably destructive measures forced on the citizens of virtually every country in the name of public health.

Key Takeaways

- Compulsory masking was based on a distorted, selective version of the precautionary principle, which used catastrophizing imaginative scenarios as the basis for policymaking.
- Using mandates to take away people's ability to come to mutual voluntary accommodations means becoming responsible for not just the desired good outcomes (which may or may not actually occur), but also becoming responsible for all the bad outcomes, too.
- The right to swing one's fist ends with someone else's face. Likewise, any "right" to public health *also* ends with someone else's face. Even if refusing to help someone by taking a particular precaution like wearing a mask is selfish, forcing someone to provide that help by making them wear a mask against their will is even *more* selfish.
- Having a good objective does not automatically make every measure taken in *pursuit* of that objective moral and right. Tyranny perpetrated with the best of motives is still tyranny.

Militarizing public health provides no justification

David Walsh, an ethicist and political scientist, was once asked to define when a state has the moral authority to override a citizen's otherwise-inviolable autonomy in cases of communicable disease. He responded: "when the number of deaths caused by a disease in a community outweighs the number of births." [293] Though I think even this criteria gives too much latitude to state authorities against individual rights, on a practical level, no modern disease, *including* COVID-19, meets this standard. Arguably not even epoch-defining plagues like the Black Death or Plague of Justinian met this standard.

A classic argument for compulsory military service is that "even a single soldier's refusal to fight in a war of great import is likely to result in an increased risk of harm to his fellow soldiers, as well as an increased risk to the safety of his countrymen." [294] During the COVIDcrisis, arguments like this were repurposed and turned against those who objected to compulsory masking, vaccination, social distancing, or any of the other mandated public health measures. However, attempts to militarize society in the name of some desirable end like public health fail the tests of reason, morality, and legality. Diseases simply do not pose the same type of existential threat to a society or nation as a war for national survival. Even if the threat from a plague *was* truly comparable to a threat from war, the same rights of conscientious objection would still apply.

America's Founders, in particular, were adamant on the primacy of conscience. In 1775, at the outset of the Revolution, George Washington wrote to then-loyal Colonel Benedict Arnold:

---

293   Evans G, Bostrom A, Johnston RB, Fisher BL, Stoto MA, Ed. Risk Communication and Vaccination: Summary of a Workshop. Pg. 12. Institute of Medicine Vaccine Safety Forum. Washington D.C. National Academy Press 1997. Quoted in Barbara Loe Fisher, "The Moral Right to Conscientious, Philosophical and Personal Belief Exemption to Vaccination," National Vaccine Information Center, (original published 1997, references updated and expanded in 2015). Available online at: https://www.nvic.org/vaccination-decisions/informed-consent

294   Nadia N. Sawicki, *The Hollow Promise of Freedom of Conscience*, 33 Cardozo L. Rev. 1389 2011-2012, p. 48. Available online: https://papers.ssrn.com/sol3/papers.cfm?abstract_id=1666278

"Check every Idea; & crush in its earliest Stage every Attempt to plunder even those who are known to be Enemies to our Cause… as far as lays in your Power you are to protect & support the free Exercise of the Religion of the Country & the undisturbed Enjoyment of the Rights of Conscience in religious Matters with your utmost Influence & Authority."[295]

In a 1792 essay on property, James Madison declared, "In a word, as a man is said to have a right to his property, he may be equally said to have a property in his rights," and that, "Conscience is the most sacred of all property[.]"[296] Thomas Jefferson wrote in 1809, "No provision in our constitution ought to be dearer to man, than that which protects the rights of conscience against the enterprises of the civil authority."[297]

The primacy of individual conscience even in matters of personal or national survival was, itself, a sincerely held religious belief for some Americans. A complaint and remonstrance from a group of these Americans, addressed to Pennsylvania's Berks County Committee, was read to the Pennsylvania General Assembly on the eve of the American Revolution in 1775. It protested the County Committee's resolution to fine all those who refused military service, as such mandates violated the:

"Sacred Right and Property to every Person inhabiting in this Province who Shall confess and acknowledge One Almighty God… whereby it is declared 'That no Person or Persons… Shall be in any Case molested or prejudiced in his or their Person and Estate, because of his or their conscientious Persuasion or practice, nor be compelled… to do or Suffer any Act or Thing contrary to their religious Persuasion. Therefore to wrest the Enjoyment of the Same from any Body must be Sacrilege [and e]xcite divine Vengeance, and must be void in Effect.'"
[298]

295   George Washington, "Instructions to Colonel Benedict Arnold, 14 September 1775," *Founders Online,* National Archives, https://founders.archives.gov/documents/Washington/03-01-02-0356 . [Original source: *The Papers of George Washington*, Revolutionary War Series, vol. 1, *16 June 1775–15 September 1775*, ed. Philander D. Chase. Charlottesville: University Press of Virginia, 1985, pp. 457–460.]

296   James Madison, "For the *National Gazette*, 27 March 1792," *Founders Online,* National Archives, https://founders.archives.gov/documents/Madison/01-14-02-0238 . [Original source: *The Papers of James Madison*, vol. 14, *6 April 1791–16 March 1793*, ed. Robert A. Rutland and Thomas A. Mason. Charlottesville: University Press of Virginia, 1983, pp. 266–268.]

297   Thomas Jefferson, "From Thomas Jefferson to Richard Douglas, 4 February 1809," The Thomas Jefferson Papers at the Library of Congress, Series 1: General Correspondence. 1651-1827, Microfilm Reel: 043, available online at: https://www.loc.gov/item/mtjbib019696/

298   Richard K. MacMaster et al., *Conscience in crisis: Mennonites and other peace churches in America, 1739-1789: interpretation and documents*, Scottsdale, PA: Herald Press (1979), pp. 256-258 Available Online:

These conscientious objectors cited the Pennsylvania Assembly's own recommendation from June 30, 1775:

> "The house taking into Consideration, that many of the good People of this Province, are conscientiously scrupulous of bearing Arms, do hereby earnestly recommend to the Associators for the Defence of their Country and others, that they bear a brotherly Regard towards this Class of their fellow Subjects and Country Men; and to these conscientious People, it is also recommended, that they cheerfully assist in Proportion to their Abilities[.]"[299]

The petitioners also referred[300] to the resolution of the Continental Congress from July 18, 1775:

> "As there are some people, who, from religious principles, cannot bear arms in any case, the Congress intended no violence to their consciences, but earnestly recommend it to them, to contribute liberally in this time of universal calamity, to the relief of their distressed brethren in the several colonies, and to do all other services to their oppressed Country, which they can consistently with their religious principles."[301]

---

https://archive.org/details/conscienceincris0000macm

299   Richard K. MacMaster et al., *Conscience in crisis: Mennonites and other peace churches in America, 1739-1789: interpretation and documents*, Scottsdale, PA: Herald Press (1979), pp. 256-258 Available Online: https://archive.org/details/conscienceincris0000macm

300   ("these Great Politicians nobly Scorn to violate Peoples Consciences")

301   Library of Congress, "Resolution of July 18, 1775," reprinted in *Journals of the Continental Congress 1774-1789*, Worthington Chauncey Ford ed., Washington Government Printing Office (1905), p. 189. Available online: https://archive.org/details/journalscontine07statgoog/page/n6/mode/2up

Quoted in McConnell, "The Origins and Historical Understanding of Free Exercise of Religion," *103 Harv L Rev 1410-1517 (May 1990)*. Available Online: https://chicagounbound.uchicago.edu/cgi/viewcontent.cgi?article=12614&context=journal_articles

McConnell says "The language as well as the substance of this policy is particularly significant, since it recognizes the superior claim of religious "conscience" over civil obligation, even at a time of "universal calamity," and leaves the appropriate accommodation to the judgment of the religious objectors."

Ellis M. West disputes McConnell's characterization, stating that Congress "was at best expressing only an opinion or hope, for the revolutionary army consisted of state militias, exemptions from which could only be granted **by** the state governments." West, Ellis M. "The Right to Religion-Based Exemptions in Early America: The Case of Conscientious Objectors to Conscription." *Journal of Law and Religion* 10, no. 2 (1993): 367-402. doi:10.2307/1051141 Available online: https://scholarship.richmond.edu/cgi/viewcontent.cgi?article=1132&context=polisci-faculty-publications

Unfortunately, as with the declaration that "all men are created equal," many implications of the right to freedom of religion which underpinned the American Revolution were also infringed, deferred, and/or imperfectly realized during the Founders' lifetimes. Under the pressure of war, even explicit constitutional protections for the rights of conscience were repeatedly violated at a local and sometimes even a state level. For example, even though Chapter 1, Article VIII of the Pennsylvania Constitution of 1776 included a specific prohibition that, "nor can any man who is conscientiously scrupulous of bearing arms be justly compelled thereto if he will pay such equivalent,"[302] members of pacifist sects such as Quakers and Mennonites nonetheless found themselves repeatedly subject to forms of persecution including confiscation of property, attempted conscription, and even denial of the right to vote because of their refusal to engage in violence. These burdens were imposed under the dogma that the right to claim the protections of society for one's rights to life, liberty, and property, depended on providing *protection* for society through personal military service or some equivalent.[303] Exactly how failing to provide society with protection from *outside* attack gave "society" the right to not only forego protecting one's rights from those outside attackers but to, *itself,* actively violate those same rights from within, was usually ignored.

Concerns over the lack of an explicit protective clause for religious conscience when it conflicted with other legal obligations became one of many objections raised in the ratification debates, and helped maintain the driving force behind the ultimate adoption of the Bill of Rights.[304] As one writer who adopted the pseudonym of an "Old Whig" put it,

---

Having personally read the entirety of this particular entry in the Journals of the Continental Congress (which includes allocation of money and detailed recommendations about how the state militias were to be formed), my own judgement is that McConnell's characterization is by far the more accurate.

302   "Saturday, September 28, 1776. Pennsylvania Constitution of 1776, Chapter 1, Article VIII." *Minutes of the Convention of 1776*, Pennsylvania Archives Volume X, (1896) p. 770 Available Online:
https://www.paconstitution.org/wp-content/uploads/2017/11/const-1776-pa-archives-vol10.pdf

303   Richard K. MacMaster et al., *Conscience in crisis: Mennonites and other peace churches in America, 1739-1789: interpretation and documents*, Scottsdale, PA: Herald Press (1979), p. 531, Available Online:
https://archive.org/details/conscienceincris0000macm

304   Total consensus about anything in the 18th century was as unobtainable and illusory as it is today. James Madison's initial proposal for Article II of the Bill of Rights included a provision that, "no one religiously scrupulous of bearing arms shall be compelled to render military service." At least one delegate, Judge Egbert Benson of New York, moved to strike this clause on the argument that, "It may be a religious persuasion, but it is no natural right, and therefore ought to be left to the discretion of the government"
Richard K. MacMaster et al., *Conscience in crisis: Mennonites and other peace churches in America, 1739-1789: interpretation and documents*, Scottsdale, PA: Herald Press (1979), p. 534, Available Online:
https://archive.org/details/conscienceincris0000macm

On the other hand, but still contending for deletion of this clause, was the possibility that including such a specific provision might imply that the Rights of Conscience *only* applied to the very narrow field of military service. If the right to religious freedom already involved a right to *live* according to one's religious beliefs — as many Americans

"'What is there in the new proposed constitution to prevent his [a conscientious objector's] being dragged like a Prussian soldier to the camp and there compelled to bear arms?' Without a Bill of Rights, the 'Old Whig' said, liberty of conscience and all other personal rights would depend 'on the will and the pleasure' of their rulers."[305]

Highlighting such violations, debates, and shortcomings does not disprove the existence of a legitimate natural right of individual conscience any more than pointing to the delay of the 13th Amendment and universal emancipation can disprove the unalienable rights recognized by the Declaration that "all men are created equal." If anything, these debates show that many, if not most, Americans were aware of this right and wrestling with its implications prior to, and throughout the birth of the Republic.[306]

asserted — then any explicit inclusion of such a provision became either redundant or a source of unnecessary confusion. Though Benson's initial motion narrowly failed, the clause was ultimately dropped, and the challenge of working out many details of what religious freedom and the rights of conscience required was ultimately left to the states and future generations.

305   Robert A. Rutland, *The Birth of the Bill of Rights 1776-1791* (New York, 1955), pp. 136-137. Available Online: https://archive.org/details/birthofbillofrig0000unse/page/136/mode/2up?q=%22Old+Whig%22

306   The Pennsylvania Militia Act of 1755 provides an example of the Quakers' practical efforts to tolerate, honor, and allow latitude for potentially competing demands of individual conscience immediately prior to the onset of the Seven Years' War:

> "The militia law passed in that year, however, was most unusual; it authorized a militia and set up rules for its organization, but enrollment in the militia was not required of anyone. The law basically legalized a voluntary militia and was designed to avoid the dilemma of having no militia or having one that drafted only non-pacifists. This is clear from the language of the law itself. First, it said that any law that compelled Quakers to bear arms would violate the Charter of Privileges, but that exempting only Quakers from military service "would be inconsistent and Partial." Then, it noted that "great Numbers of People of other Religious Denominations are come among us who are under no such Restraint, some of whom... Conscientiously think it their Duty to fight in defence of their Country..., and such have an Equal Right to Liberty of Conscience with others." Therefore, the statute concluded, "We do not think it reasonable that any should thro' a want of legal powers be in the least restrain'd from doing what they Judge it their Duty to do for their own Security and the publick Good."

Ellis M. West, "The Right to Religion-Based Exemptions in Early America: The Case of Conscientious Objectors to Conscription." *Journal of Law and Religion* 10, no. 2 (1993): 367-402. (at 387) doi:10.2307/1051141 . Available online: https://scholarship.richmond.edu/cgi/viewcontent.cgi?article=1132&context=polisci-faculty-publications

Benjamin Franklin (himself no Quaker) defended this law using a rhetorical dialogue published in the Pennsylvania Gazette. Franklin argued that this unusual militia law was the most equitable way to satisfy the requirements of conscience for all parties concerned. He praised the Quakers, who, 'being a Majority in the Assembly... They might indeed have made the Law compulsory on all others. But it seems they thought it more equitable and generous to leave to all as much Liberty as they enjoy themselves, and not lay even a seeming Hardship on others, which they themselves declined to bear."

Benjamin Franklin, "A Dialogue between X, Y, and Z, 18 December 1755," *Founders Online,* National Archives,

U.S. Supreme Court decisions like *Sherbert v. Verner*[307] and *Wisconsin v. Yoder*[308] have not only presumed and affirmed that the right to live according to one's religious beliefs is indeed entitled to Constitutional protection, but in *United States v. Seeger*[309] and *Welsh v. United States*[310], the court made clear that beliefs entitled to Constitutional protections are not limited to traditional definitions of religion, but can include even atheistic moral or ethical beliefs "about what is right and wrong" as long as "these beliefs be held with the strength of traditional religious convictions."[311]

For millions of Americans and others throughout the world, wearing a mask was and remains a violation of conscience. This makes wearing a mask, in the words of *Sherbert v. Verner*, a regulation to "compel affirmation of a repugnant belief."[312] This means that it takes more than simply proffering a plausible public health rationale for wearing a mask, because when rights protected by the First Amendment conflict with a State regulation, "It is basic that no showing merely of a rational relationship to some colorable state interest would suffice[.]"[313] Pointing out that masks are a public health measure does not defeat this argument, because it is

---

https://founders.archives.gov/documents/Franklin/01-06-02-0131 . [Original source: *The Papers of Benjamin Franklin*, vol. 6, *April 1, 1755, through September 30, 1756*, ed. Leonard W. Labaree. New Haven and London: Yale University Press, 1963, pp. 295–306.]

Regrettably, this militia law was overruled and declared invalid by the authorities in London eleven months after passage, and unfortunately for the Quakers, their spirit of generosity and adherence to principle was not fully reciprocated after they lost control of the Pennsylvania General Assembly.

It is also worth noting that in late 1775, the Quaker association known as the Philadelphia Meeting for Suffering "appealed to the Assembly for unconditional exemptions from military duty on the grounds that the 'liberty of conscience' guaranteed in the colony's Charter 'was not limited to the Acts of Public Worship only.'"
Quoted in Ellis M. West, "The Right to Religion-Based Exemptions in Early America: The Case of Conscientious Objectors to Conscription." *Journal of Law and Religion* 10, no. 2 (1993): 367-402. (at 389-390) doi:10.2307/1051141 . Available online:
https://scholarship.richmond.edu/cgi/viewcontent.cgi?article=1132&context=polisci-faculty-publications

307   *Sherbert v. Verner*, 374 U.S. 398 (1963), Available Online: https://supreme.justia.com/cases/federal/us/374/398/
308   *Wisconsin v. Yoder*, 406 U.S. 205 (1972), Available Online: https://supreme.justia.com/cases/federal/us/406/205/
309   *United States v. Seeger*, 380 U.S. 163 (1965), Available Online:
https://supreme.justia.com/cases/federal/us/380/163/
310   *Welsh v. United States*, 398 U.S. 333, (1970). Available Online:
https://supreme.justia.com/cases/federal/us/398/333/
311   *Welsh v. United States*, 398 U.S. 333, 340 (1970). Available Online:
https://supreme.justia.com/cases/federal/us/398/333/
312   *Sherbert v. Verner*, 374 U.S. 398, 402 (1963), Available Online:
https://supreme.justia.com/cases/federal/us/374/398/
313   *Sherbert v. Verner*, 374 U.S. 398, 406 (1963), Available Online:
https://supreme.justia.com/cases/federal/us/374/398/

As we shall see in Part 5, decisions like *Sherbert v. Verner* explicitly state that mere rational or reasonable basis scrutiny are insufficient for laws that infringe on fundamental dictates of conscience.

possible for a gesture or practice to simultaneously serve both symbolic and practical functions, and practical benefits do not automatically confer legitimacy in compelling symbolic speech.

Even taking militarization of public health as a legitimate option, doing so would still not justify compulsory masking, or would, at the very least, require conscience-based exemptions. Soldiers voluntarily enter a contractual obligation to forgo exercising a range of rights (including some aspects of personal security and self-preservation) under particular circumstances for a set duration of time. However, even for Soldiers, courts have repeatedly and emphatically ruled that "the cost of military service has never entailed the complete surrender of all "basic rights[.]""[314] One can certainly make a much stronger case that Soldiers have a presumptive moral or legal duty to assist each other than one can make for civilians, but even that would still fail to show that Soldiers have a presumptive moral or legal duty to make use of a particular *means* to provide such assistance. A paragon example of this distinction comes from the case of Medal of Honor winner Desmond Doss. A Seventh Day Adventist "conscientious cooperator" who refused to carry any weapon but was perfectly willing to act as a medic, Desmond Doss volunteered for service during World War 2. Accommodating Desmond Doss' sincerely held religious beliefs led to saving the lives of 75 Soldiers in *one day* on Okinawa. This does not include his life-saving services in other battles and campaigns on Guam and Leyte. Compare this to the 80 deaths from COVID in 2020 and 2021 throughout all the Armed Services *combined*.[315] Similarly, chaplains in the Armed Forces do not usually carry weapons. Citizen-civilians do not lack any rights that Citizen-Soldiers possess. If carrying a weapon in a World War is not an essential moral or legal duty for Soldiers, then wearing a mask cannot be an essential moral or legal duty for all citizens during a "war" against a disease. If there is a place in direct combat for a Soldier who refuses to carry a weapon, there must also be a place in direct patient care for even healthcare workers who refuse to wear a mask or get a vaccine.

The same principles which sustain a citizen's right to various degrees of conscientious objection also uphold a citizen's right to conscientiously abstain from making use of one or more medical interventions to combat a specific disease, no matter how effective such interventions may or may not be. A citizen who refuses to wear a mask or get a vaccination is, at the worst, analogous to a Soldier who refuses to make use of a particular weapon system in combat,

---

314   *Jaskirat Singh v. David Berger*, No. 22-5234 (D.C. Cir. 2022), Document #1978908, Filed: 12/23/2022. Available Online: https://law.justia.com/cases/federal/appellate-courts/cadc/22-5234/22-5234-2022-12-23.html

315   "Collectively, our armed forces have lost 80 lives to COVID-19 over the course of the pandemic. Defs.' App. 263, ECF No. 44-3." *U.S. Navy SEALs 1-26 v. Biden*, No. 22-10077 (5th Cir. 2022), Document: 66, Date Filed: 01/03/2022, available online:
https://storage.courtlistener.com/recap/gov.uscourts.txnd.355696/gov.uscourts.txnd.355696.66.0_5.pdf

or at the best analogous to the situation of Medal of Honor recipient Desmond Doss.[316]

I emphasize the story of Desmond Doss, not just because he was a true hero whose story single-handedly refutes broad-brush claims of selfishness on the part of conscientious objectors, but also because the grudging accommodation he received was partially due to what those who came before him were willing to suffer and endure for their beliefs. Both Desmond Doss' memory and precedent deserve to be honored. Those principles his case exemplifies should be consistently applied today, and those principles require accommodating anyone who refuses to wear a mask or get a vaccine, regardless of the pathogen involved. During the COVIDcrisis, the same officials who had a duty to reaffirm and vigorously defend those principles repeatedly threw them to the wayside based on nothing more than unquantified, unexemplified, dogmatic assertions of public health "necessity."[317] The Bill of Rights specifically protects freedom of religion rather than public health for good reason. Matters of religion are no less important than matters of public health. Coercion is wrong in matters of religion. Therefore, coercion is wrong in matters of public health.

---

316   As a side note, before being discharged from the Army, Desmond Doss developed tuberculosis while on campaign from the rigors he endured, and eventually had to have his left lung and five ribs removed. Too many people today would have forced him to breathe through an N95 with one lung in the name "not imposing risk" on others. *See also,*

Frances Doss & Jeremy Albret, *Desmond Doss: Conscientious Objector*, Pacific Press Publishing Association (2005).

Booton Herndon, *Redemption at Hacksaw Ridge: The Gripping True Story That Inspired the Movie*, Remnant Publications (2016).

317   For competing interpretations of the Founders' views of what the right to Freedom of Religion entailed in practice, *See also:*

Campbell, W.J., *Religious Neutrality in the Early Republic.* Regent University Law Review, 2012. **24**: p. 311.
https://scholarship.richmond.edu/cgi/viewcontent.cgi?httpsredir=1&article=2317&context=law-faculty-publications
https://www.regent.edu/acad/schlaw/student_life/studentorgs/lawreview/docs/issues/v24n2/03Campbellvol.24.2.pdf

McConnell, "The Origins and Historical Understanding of Free Exercise of Religion," *103 Harv L Rev 1410-1517 (May 1990).* Available Online:
https://chicagounbound.uchicago.edu/cgi/viewcontent.cgi?article=12614&context=journal_articles

Salmon, D.A., *Religious and Philosophical Exemptions from Vaccination Requirements and Lessons Learned from Conscientious Objectors from Conscription.* 2001. **116**(4): p. 289-295. https://dx.doi.org/10.1093/phr/116.4.289

Ellis M. West, "The Case Against a Right to Religion-Based Exemptions." *Notre Dame Journal of Law, Ethics & Public Policy*, 1990. **4**: p. 591. https://scholarship.law.nd.edu/cgi/viewcontent.cgi?article=1554&context=ndjlepp

Ellis M. West "The Right to Religion-Based Exemptions in Early America: The Case of Conscientious Objectors to Conscription." *Journal of Law and Religion* 10, no. 2 (1993): 367-402. doi:10.2307/1051141 . Available online:
https://scholarship.richmond.edu/cgi/viewcontent.cgi?article=1132&context=polisci-faculty-publications

Key Takeaways

🔑 Analogizing public health programs to a form of war only strengthens the moral and legal case for requiring conscience-based exemptions.

🔑 The same principles which uphold a citizen's right to various degrees of conscientious objection also uphold a citizen's right to conscientiously abstain from making use of one or more medical interventions to combat a specific disease, regardless of how effective such interventions may be.

🔑 There is a place in direct combat for Soldiers like Desmond Doss who refuse to carry a weapon, therefore there must also be a place — even in direct patient care — for healthcare workers who refuse to wear a mask or get a vaccine.

Would-be justifiers of compulsory masking

In *The Lucifer Effect*, Dr. Philip Zimbardo concisely describes how diffusion of responsibility and administrative evil can propagate within a hierarchy, organization, or society until it becomes pervasive. 'Senior Architects' design the broad policy and rely on 'Justifiers' to come up with new language and concepts to make immoral actions more palatable. This, in turn, allows 'Foremen' to supervise and give boots-on-the-ground orders to 'Technicians' and 'Grunts' who do the most visible hands-on enforcement and act as fall guys if necessary.[318] Compulsory masking found so many eager would-be 'Justifiers' that an exhaustive list and rebuttal would take several books. I have selected two individuals, in particular, to focus on as examples, because their work incorporates many of the issues that we have already covered, and because of their visibility and repeated publication in both popular-level media and prestigious medical journals such as the *Journal of the American Medical Association*, the *New England Journal of Medicine*, *Bioethics*, and the *Journal of Medical Ethics*.

Because they rest on the same core assumptions, arguments for compulsory masking and compulsory vaccination overlap so extensively that they are often interchangeable. Medical journals have been saturated with a one-sided presentation of these arguments for decades, and this editorial bias was on full display during the COVIDcrisis. Authors who published opinion (sometimes called "viewpoint" or "perspective") articles in journals pushing compulsory masking in 2020 also published pieces in 2021 supporting compulsory vaccination. For example, the *Journal of the American Medical Association* published an article by one author supporting mask mandates in August 2020, and another article by the same author supporting vaccine mandates in October 2021.[319]

---

318   Philip Zimbardo. *The Lucifer Effect*, New York: Random House Publishing Group (2007), Kindle Edition, p. 403

319   Lawrence O. Gostin, I Glenn Cohen, Jeffrey P. Koplan, "Universal Masking in the United States: The Role of Mandates, Health Education, and the CDC," *Journal of the American Medical Association (JAMA)*. Published online

In 2020, this authors' ethical case for mask mandates was just one sentence: "The ethical justification for face coverings is their utility in preventing transmission of serious disease to community members."[320] That's it. (Recall the hidden chain of faulty reasoning behind this which we covered earlier.) This author also used seatbelt and intoxicated driving laws as an analogy and legal precedent for laws mandating masks.

The same author's October 2021 article on vaccines, titled, "COVID-19 Vaccine Mandates — A Wider Freedom," went into slightly more depth. He spends on sentence summarizing the constrained and unconstrained visions of freedom which Thomas Sowell devoted an entire book to: "There are at least 2 types of freedom — freedom from personal restraint and a wider freedom to engage in daily life without significant risk of exposure to safety hazards." The author simply asserts as an *a priori* given that, "the public has the right to engage in daily social and economic life without fear of avoidable harms." Ignoring the fact that this newly discovered open-ended collective "right" to live without fear of avoidable harms is more a question of individual psychology than of science, ethics, or constitutionality, the author immediately wielded it to dismember the *actual* fundamental right of informed consent. He writes, "Certainly, competent adults have the right to bodily integrity and to make their own health care decisions. Yet, the right of informed consent has clear limits." What carefully reasoned, meticulously constructed argument did this author put forward to justify his conclusion that the limits of informed consent permit vaccine mandates? Nothing more than the same tired, predictable, rhetorical sleight-of-hand used to wrongly conflate inaction with assault, non-immunity with active disease, and to manufacture an open-ended universal duty to assist: "No one has the right to expose others to a potentially serious infectious disease."[321]

The author of the two articles discussed above is Lawrence O. Gostin, a distinguished Georgetown University law professor. Born in 1949,[322] Professor Gostin has sat on multiple Legal and Public Health editorial boards, including the *American Journal of Tropical Medicine and Public Health*, the *Journal*

August 10, 2020. https://dx.doi.org/10.1001/jama.2020.15271

Lawrence O. Gostin, "COVID-19 Vaccine Mandates—A Wider Freedom." JAMA Health Forum, 2021. **2**(10): p. e213852. https://dx.doi.org/10.1001/jamahealthforum.2021.3852

320   Lawrence O. Gostin, I Glenn Cohen, Jeffrey P. Koplan, "Universal Masking in the United States: The Role of Mandates, Health Education, and the CDC," Journal of the American Medical Association, JAMA. Published online August 10, 2020. https://dx.doi.org/10.1001/jama.2020.15271

321   Lawrence O. Gostin, "COVID-19 Vaccine Mandates—A Wider Freedom." JAMA Health Forum, 2021. **2**(10): p. e213852. https://dx.doi.org/10.1001/jamahealthforum.2021.3852

322   By the time COVID came on the scene, Professor Gostin's age put him squarely in the top risk brackets for COVID. In case anyone missed the innuendo, I am suggesting that Gostin's arguments for the vaccine mandates in the context of COVID have been at least partially driven by self-serving fear-based motivated reasoning. I am *not*, however, *rebutting* Gostin's arguments on this basis (that would be an *ad hominem* fallacy). Gostin's arguments in favor of mandating things like masks and vaccines are wrong no matter who makes them or how old they happen to be.

of Law and Medicine, and the *American Journal of Bioethics*.[323] His career demonstrates that a medical degree is not necessary to opine on public health in the pages of medical journals or to be elected to a lifetime membership of the National Academy of Medicine. Professor Gostin does not have a public health degree. This highlights the outrageous "Where's-your-public-health-degree?" double-standard applied to silence dissenting medical professionals and everyday conscientious layman-researchers.

Professor Gostin was also instrumental in helping the CDC draft the Model State Emergency Health Powers Act (MSEHPA) in 2001, following the anthrax scare.[324] At the time this Act was drafted, the Association of American Physicians and Surgeons (AAPS) warned that it included such broad emergency powers and unilateral authority to *declare* those emergencies in the first place that it "turns governors into dictators."[325] Between 2001 and 2020, the vast majority of states passed some variation of this Act. AAPS predictions were fully borne out during the COVIDcrisis, when many governors used the emergency authority conferred by the Act to award themselves powers that most kings would envy. Gostin concluded his 2021 *JAMA Health Forum* article by saying that "Using every tool — including mandates to achieve high vaccination coverage enhances freedom."[326] This was a first-rate Orwellian equivocation, substituting the individual, unalienable, freedoms recognized by the Bill of Rights with a collective "freedom" allowing the government to override informed consent in the name of the greater good.

Arguments for compulsory masking from authors *with* medical and public health degrees turn out to be no better. Matthew K. Wynia, MD, MPH, is the Director of the Center for Bioethics and Humanities at the University of Colorado and served as the Director of the American Medical Association's Institute for Ethics from 2000 to 2013. His positions on masks and vaccines during the COVIDcrisis provide an excellent example of the primitive consequentialism, selective critical thinking, paternalism, and ethical double-standards at work among too many people with prestigious letters after their names.

---

323   Lawrence O. Gostin, *Curriculum Vitae*, 2023, p. 4 available online: https://oneill.law.georgetown.edu/wp-content/uploads/2021/06/Gostin-Lawrence-CV.pdf

324   Lawrence O. Gostin, *The Model State Emergency Health Powers Act*, October 23, 2001, Prepared by The Center for Law and the Public's Health at Georgetown and Johns Hopkins Universities For the Centers for Disease Control and Prevention, Available online: https://biotech.law.lsu.edu/blaw/bt/MSEHPA.pdf

325   "Model Emergency Health Powers Act (MEHPA) Turns Governors Into Dictators," *AAPS Analysis*, Association of American Physicians and Surgeons, December 3, 2001. Available online: https://aapsonline.org/testimony/emerpower.htm

The AAPS analysis also accurately predicted many aspects of how the measure would be implemented, from the stick-and-carrot tying of the Act to HHS federal funding, to the conveniently-vague terms "substantial risk" and "significant number," as well as Governors being able to declare the "public health emergency" unilaterally.

326   Lawrence O. Gostin, "Vaccine Mandates—A Wider Freedom." *JAMA Health Forum*, 2021. **2**(10): p. e213852. https://dx.doi.org/10.1001/jamahealthforum.2021.3852

In the wake of the Supreme Court's 2022 ruling in *Dobbs v. Jackson Women's Health Organization*, Dr. Wynia published a perspective in the prestigious *New England Journal of Medicine* in which he made a strong case for collective civil disobedience among the medical profession when confronted with laws that threaten their patients' well-being.[327] In his 2022 *NEJM Perspective*, Dr. Wynia suggested that "professional 'conscience' protections should apply equally to physicians refusing to participate in abortions and those who are compelled by conscience to provide abortion care."[328] Starting with the American Medical Association's code of ethics' allowance that, "In exceptional circumstances of unjust laws, ethical responsibilities should supersede legal duties," Dr. Wynia approvingly cited St. Augustine and Martin Luther King, Jr., "an unjust law is no law at all." There is, said Dr. Wynia, a "moral responsibility to disobey unjust laws." Dr. Wynia vigorously defended the principle that "medically nuanced decisions are best left in the hands of individual patients and their physicians — not state lawmakers." He also argued that "professional civil disobedience poses little threat of anarchy." Dr. Wynia's articulation of these principles should be applauded, as well as his recognition that, "Professional civil disobedience would undoubtedly require tremendous courage."

One would naturally expect Dr. Wynia to respond similarly when citizens and healthcare providers apply the same principles that he defends in an abortion context to masks and vaccines mandated by executive or administrative fiat. Instead, he consistently advocated compulsory masking and vaccination, continues to do so, and shows no patience for those who disagree. He cheered the firing of a Florida ER doctor who offered mask exemptions for school children[329] and approvingly re-tweeted

---

327   Matthew Wynia, *Professional Civil Disobedience — Medical-Society Responsibilities after Dobbs.* New England Journal of Medicine, 2022. **387**(11): p. 959-961. https://dx.doi.org/10.1056/nejmp2210192

328   Matthew Wynia, *Professional Civil Disobedience — Medical-Society Responsibilities after Dobbs.* New England Journal of Medicine, 2022. **387**(11): p. 961. https://dx.doi.org/10.1056/nejmp2210192

329   Matthew Wynia, Twitter Post, *@MatthewWynia*, August 28, 2021, 4:57 p.m. https://twitter.com/MatthewWynia/status/1431722598245683201

Matthew Wynia
@MatthewWynia   ...

"...he offered [medical exemption] letters for $50 to parents who wanted their kids exempt from school mask mandates."

BTW, same thing would happen if he sold disability plate letters, sight unseen, for $50 via a Facebook group.
talkingpointsmemo.com/news/florida-e... via @TPM

> talkingpointsmemo.com
> Florida ER Doctor Booted From Hospital After Offering $5...
> A Florida emergency room doctor got the boot from the hospital he...

4:57 PM · Aug 28, 2021

Summer Concepcion, "Florida ER Doctor Booted From Hospital After Offering $50 Mask Opt-Out Letters to Parents," TalkingPointsMemo, August 26, 2021, online:

criticism of Florida Surgeon General Joseph Ladapo's refusal to wear a mask.[330] If his Twitter feed is anything to go by, Dr. Wynia vehemently denies conscience protections for doctors who refuse to participate in masking or who help their patients do so, no matter how highly credentialed those doctors may be. While American courts have rightly ruled multiple times that "religious beliefs need not be acceptable, logical, consistent, or comprehensible to others in order to merit First Amendment protection,"[331] in the context of COVID, Dr. Wynia asserted that "it is ethically justified for employers to question whether one's belief is sincere if it is inconsistently applied."[332] I invite readers to apply Dr. Wynia's proposed standard of judgment to his own work. Apparently, the impact on public health and human life from individual decisions about whether to wear a mask or get a vaccine completely lacks the medical and moral nuances involved in getting an abortion.

When former Undersecretary of State for Public Diplomacy and Public Affairs Karen Hughes published a July 2020 opinion piece in the *Washington Post* likening non-masking to "yelling 'fire!' in a packed theater or brandishing a loaded gun in a crowd," and saying that we should "label failure to wear a

---

https://talkingpointsmemo.com/news/florida-er-doctor-removal-medical-opt-out-letters-school-mask-mandates

Kim Bellware and Lateshia Beachum, "A Florida ER doctor offered $50 mask exemption letters for children. Then, his hospital found out." *The Washington Post*, August 26, 2021, online: https://www.washingtonpost.com/health/2021/08/26/brian-warden-masks/

330   Matthew Wynia, Twitter Post, *@MatthewWynia*, October 25, 2021, 11:27 a.m. https://twitter.com/MatthewWynia/status/1452658145856090114

Matthew Wynia
@MatthewWynia   ...

FL Senate President Wilton Simpson (R)Trilby, "criticized the refusal to put on a mask after being asked by Polsky...'it shouldn't take a cancer diagnosis for people to respect each other's level of comfort with social interactions during a pandemic.'

health.wusf.usf.edu
Florida's new surgeon general refuses mask and is told to l...
Senate President Wilton Simpson sent a memo to senators saying it was disappointing to learn of the incident at the ...

11:27 AM · Oct 25, 2021

Referencing Associated Press, "Florida's new surgeon general refuses mask and is told to leave meeting with state senator," *WUSF Public Media*, October 25, 2021, Online: https://health.wusf.usf.edu/health-news-flori-da/2021-10-25/florida-surgeon-general-refuses-mask-and-is-told-to-leave-meeting-with-state-senator

331   *Thomas v. Review Bd., Ind. Empl. Sec. Div.*, 450 U.S. 707 (1981). at 714. https://supreme.justia.com/cases/federal/us/450/707/

332   Matthew K. Wynia, Thomas D. Harter, Jason T. Eberl "Why A Universal COVID-19 Vaccine Mandate Is Ethical Today," *Health Affairs Blog*, November 3, 2021. DOI: 10.1377/hblog20211029.682797 Online: https://www.healthaffairs.org/do/10.1377/forefront.20211029.682797

mask as what it really is: an incredibly selfish act that puts other people's lives at risk," [333]Dr. Wynia approvingly commented, "Better late than never for this fundamental realization." [334]

"Fire in a crowded theater" is, at best, a misleadingly incomplete way of phrasing that famous qualification on free speech. The original 1919 court ruling this analogy was taken from (*Schenck v. United States*), actually said "falsely shouting fire in a theatre and causing a panic."[335] Being both a false statement and resulting in a dangerous and harmful panic is a *very* different thing altogether. Applied consistently, this standard would be much more likely to condemn causing a panic in a crowded community with apocalyptic modeling projections or symbolically shouting "Plague! Plague!" by wearing a mask.

In September 2021, after Floridian chiropractor Dan Busch came under criticism for writing mask exemptions, [336]Dr. Wynia unsympathetically opined, "you can't use state-sanctioned privileges to promote private beliefs. If you do, expect to lose your license." [337] In other words, if doctors' personal beliefs contradict the party line, they must still parrot the medical establishment's party line which they believe to be false – "or else."

---

333   Karen Hughes, "Opinion I've watched in alarm as my fellow Republicans shun masks. It's selfish." *The Washington Post*, July 1, 2020, Online: https://www.washingtonpost.com/opinions/2020/07/01/mask-wearing-isnt-political-test-its-moral-test-my-fellow-republicans-are-failing/

334   Matthew Wynia, Twitter Post, @*MatthewWynia*, July 5, 2020, 9:35 a.m., https://twitter.com/MatthewWynia/status/1279770931217367040

Matthew Wynia @MatthewWynia · Jul 5, 2020 · · ·
Better late than never for this fundamental realization: "...label failure to wear a mask as what it really is: an incredibly selfish act that puts other people's lives at risk." washingtonpost.com/opinions/2020/...
♡ 2

Schenck v. United States, 249 U.S. 47 (1919), available online:
https://tile.loc.gov/storage-services/service/ll/usrep/usrep249/usrep249047/usrep249047.pdf

335   *Schenck v. United States*, 249 U.S. 47 (1919), available online:
https://tile.loc.gov/storage-services/service/ll/usrep/usrep249/usrep249047/usrep249047.pdf

336   Katie Shepherd, "Hundreds lined up at a Florida chiropractor's office after he signed mask-exemption forms for students," *The Washington Post*, September 2, 2021, Online:
https://www.washingtonpost.com/nation/2021/09/02/florida-chiropractor-mask-mandate-exemption/

337   Matthew Wynia, Twitter Post, @*MatthewWynia*, September 2, 2021,
https://twitter.com/MatthewWynia/status/1433421426829135874

Matthew Wynia @MatthewWynia · Sep 2, 2021 · · ·
"I am not an anti-mask person or an anti-vax person, but I am a pro-freedom, pro-choice person."
Fine, but you can't use state-sanctioned privileges to promote private beliefs. If you do, expect to lose your license.
washingtonpost.com/nation/2021/09...

In April 2022, Judge Kathryn Kimball Mizelle struck down the CDC's Airline mask mandate for relying on an incorrect meaning of the word "sanitation" contained in the Public Health Service Act (PHSA), as well as for being arbitrary and capricious.[338] Despite Judge Kimball's careful, multi-page analysis citing authoritative dictionaries from when the Public Health Service Act was first passed, as well as the statute's historical usage which never came close to such sweeping mandates prior to 2020, Dr. Wynia dismissed her ruling as showing "zero understanding of how the word 'sanitation' was actually used in the 1940s, when the PHSA was passed."[339] The most charitable response to this is that Dr. Wynia showed zero understanding of Judge Mizelle's decision.

Dr. Wynia's views on masks and vaccines seem to involve generalizing his strong view of healthcare providers' moral and ethical "duty to care" into an open-ended *legal* duty assigned to the general public. When defending compulsory vaccination at length in a November 2021 article,[340] Dr. Wynia made use of the same deficient arguments we dealt with earlier. His final answer to the question of "What makes any public health mandate ethical, whether it's for mask-wearing, social distancing, or vaccination?" is the same as Lawrence Gostin's: non-action is "imposing risk." For both Dr. Wynia and Professor Gostin, the vague, unquantified "risk of one person harming many others, even inadvertently, provides ethical justification for limiting the choice to go unvaccinated during a pandemic."[341] In so many words, Dr. Wynia argues that essential liberties recognized by the Bill of Rights are subservient

---

338   *Health Freedom Defense Fund, Inc. et al v. Biden et al*, No. 8:2021cv01693 - Document 53 (M.D. Fla. 2022) Available online: https://law.justia.com/cases/federal/district-courts/florida/flmdce/8:2021cv01693/391798/53/

339   Matthew Wynia, Twitter Post, *@MatthewWynia*, April 24, 2022, 10:22 a.m. https://twitter.com/MatthewWynia/status/1518233935083294720

340   Matthew K. Wynia, Thomas D. Harter, Jason T. Eberl "Why A Universal COVID-19 Vaccine Mandate Is Ethical Today," *Health Affairs Blog*, November 3, 2021. DOI: 10.1377/hblog20211029.682797 Online: https://www.healthaffairs.org/do/10.1377/forefront.20211029.682797

341   Matthew K. Wynia, Thomas D. Harter, Jason T. Eberl "Why A Universal COVID-19 Vaccine Mandate Is Ethical Today," *Health Affairs Blog*, November 3, 2021. DOI: 10.1377/hblog20211029.682797 Online: https://www.healthaffairs.org/do/10.1377/forefront.20211029.682797

to public health, and "avoiding mandates to uphold the ideals of personal and religious freedom is not worth the risk to others that would ensue in the current environment," as though any non-zero risk of transmission is categorically sufficient to justify sweeping mandates and violations of essential liberties. Dr. Wynia also completely ruled out traditionally acquired natural immunity when he asserted that, "Herd immunity for COVID-19 will only occur through vaccination."[342] This is inexcusable coming from *anyone* with a MD or public health degree.

As a side note, Dr. Wynia also cited the Bangladesh mask study[343] and Alihsan et al.'s 2022 "systematic review" — which Wynia refers to as a meta-analysis[344] — as evidence in favor of mask-wearing. Neither of these citations is to his credit, and the Alihsan literature review is a particular embarrassment (as the detailed discussion near the end of the science section on mask efficacy shows).

---

342   Matthew K. Wynia, Thomas D. Harter, Jason T. Eberl "Why A Universal COVID-19 Vaccine Mandate Is Ethical Today," *Health Affairs Blog*, November 3, 2021. DOI: 10.1377/hblog20211029.682797 Online: https://www.healthaffairs.org/do/10.1377/forefront.20211029.682797

343   Abaluck, J., et al., *Impact of community masking on COVID-19: A cluster-randomized trial in Bangladesh.* Science, 2021. **0**(0): p. eabi9069. https://www.science.org/doi/abs/10.1126/science.abi9069
Matthew Wynia, Twitter Post, @*MatthewWynia*, September 3, 2021, 2:04 p.m.
https://twitter.com/MatthewWynia/status/1433853378937851904

> **Matthew Wynia**
> @MatthewWynia   · · ·
>
> RCT on masking in Bangladesh: an intervention led to "...tripling of mask usage" & "villages randomized to surgical masks (n = 200)" had reduced COVID by "11.2% overall (aPR = 0.89 [0.78,1.00]) and 34.7% among individuals 60+ (aPR = 0.65 [0.46, 0.85])." poverty-action.org/sites/default/...
>
> 2:04 PM · Sep 3, 2021

344   Alihsan, B., et al., *The Efficacy of Facemasks in the Prevention of COVID-19: A Systematic Review.* 2022, Cold Spring Harbor Laboratory. https://dx.doi.org/10.1101/2022.07.28.22278153
Matthew Wynia, Twitter Post, @*MatthewWynia*, August 1, 2022, 9:23 a.m.
https://twitter.com/MatthewWynia/status/1554095471383048192

> **Matthew Wynia**
> @MatthewWynia   · · ·
>
> Meta-analysis, in pre-print, and doesn't differentiate between high and low filtration masks, but still finds: "The probability of getting COVID-19 for mask wearers was 7% (97/1463, p=0.002), for non-mask wearers, probability was 52% (158/303, p=0.94)."
>
> **medRxiv**   medrxiv.org
>   THE PREPRINT SERVER FOR HEALTH SCIENCES   The Efficacy of Facemasks in the Prevention of COVID-19: ... Facemasks have become a symbol of disease prevention in the context of COVID-19; yet, there still exists a paucity of ...
>
> 9:23 AM · Aug 1, 2022
>
> **1** Retweet   **1** Quote Tweet

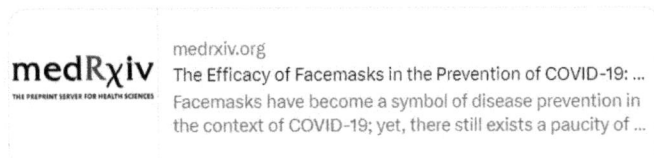

For Dr. Wynia and his co-authors, mask mandates are justified because refusal to wear a mask impinges on everyone else's liberty to "breathe clean air."[345] Meanwhile, Federal District Judge Susan Mollway dismissed a challenge to compulsory masking In Hawaii on the grounds that, "The 'right to breathe oxygen without restriction' is not a fundamental right,"[346] and Federal District Judge William Conley did the same in Wisconsin, stating that "Certainly, being denied air may be actionable, but being denied the 'cleanest air readily available' does not implicate a fundamental right."[347] In other words, according to this reasoning, mask mandates are *ethical* because *everyone* has an open-ended fundamental liberty to "breathe clean air," and mask mandates are *legal* because *you* have no fundamental right to breathe the "cleanest air readily available" or even to "breathe oxygen without restriction." Heads they win; tails you lose. Moreover, either line of reasoning, if applied consistently, could just as easily be used to strike down a mask mandate instead of upholding one. If there is no right to breathe the cleanest air readily available, the justification for forcing everyone to wear a filter of any kind evaporates.

Key Takeaways

🔑 Case studies of experts who advocated for compulsory masking serve to highlight many of the flawed premises we have dealt with in this section on mask philosophy.

🔑 Holding prestigious positions and having lots of letters after one's name does not guarantee either careful reasoning or even intellectual consistency when it comes to things like masks.

🔑 "The ignorance, prejudices, and groupthink of an educated elite are still ignorance, prejudice, and groupthink." (Thomas Sowell, *Intellectuals and Society*, 2011)

---

345   Matthew K. Wynia, Thomas D. Harter, Jason T. Eberl "Why A Universal COVID-19 Vaccine Mandate Is Ethical Today," *Health Affairs Blog*, November 3, 2021. DOI: 10.1377/hblog20211029.682797 Online: https://www.healthaffairs.org/do/10.1377/forefront.20211029.682797

346   *Denis v. Ige et al*, No. 1:2021cv00011 - Document 62 (D. Haw. 2021), p. 34, available online: https://law.justia.com/cases/federal/district-courts/hawaii/hidce/1:2021cv00011/152658/62/

347   *Joseph v. Becerra*, 22-cv-40-wmc, Document 21, p. 12 (W.D. Wis. Nov. 29, 2022), available online: https://law.justia.com/cases/federal/district-courts/wisconsin/wiwdc/3:2022cv00040/48728/21/

The Masking Mentality in case law: Jacobson, Buck v. Bell, and Korematsu

The poisonous precedent set by compulsory mask-wearing did not come into being out of nothing. It was a natural extension of the flawed underlying rationale and legal precedents that drove some of the worst injustices in the history of American jurisprudence.

In 1944, the Supreme court decision in *Korematsu vs. United States* upheld the mandated internment of over 100,000 Japanese Americans during World War 2, irrespective of actual innocence or guilt, on the belief that a minority of them were harboring anti-American sympathies and intentions that could get others killed. The 6-3 *Korematsu* decision permitted authorities to use an emergency as an excuse for taking preemptive action using the convenient and highly abusable proxy criteria of race to predict social harm and death. In his incisive dissent, Justice Frank Murphy pointed out that the court majority's decision unilaterally deprived more than 100,000 Japanese-Americans of their Fifth Amendment rights as well as "the constitutional rights to live and work where they will, to establish a home where they choose and to move about freely." Moreover, "In excommunicating them without benefit of hearings, this order also deprives them of all their constitutional rights to procedural due process."[348] During the COVIDcrisis, the same pathogenic reasoning underlying the racist anti-Japanese WWII mandate was applied to the population as a whole, merely substituting obsessive PCR testing, case counts, and arbitrary CDC "COVID-19 Community Levels" in place of race. Legally penalizing or constraining Americans, especially those who refused to wear a facemask or refused vaccination, on the tissue-thin rationale that they might *possibly* be harboring an infectious virus, was a medicalized version of the *Korematsu* travesty. The difference is that there was even *less* excuse for such actions in 2020 than during World War II.

---

348   *Korematsu v. United States*, 323 U.S. 214 (1944), (Justice Murphy, dissenting), p. 335, Available online: https://supreme.justia.com/cases/federal/us/323/214/

Korematsu earned its place of dishonor alongside the pre-Civil War *Dred Scott* decision, but its other ideological progenitor, *Jacobson v. Massachusetts* (1905), has been cited *ad nauseum* by fans of lockdown measures, mask mandates and compulsory vaccination. In *Jacobson v. Massachusetts*, the Supreme Court ruled that it was within a state's constitutional police power to fine a Swedish pastor, Henning Jacobson, for refusing a smallpox vaccine.[349]

In Part 5, we will discuss the legal issues with *Jacobson v. Massachusetts* at length. For now, leaving aside the obvious point that COVID-19 was never even *remotely* comparable to historical smallpox on *any* metric, when we look at the history of the Supreme Court case of *Jacobson v. Massachusetts*, and how it has been applied since 1905, it is easy to see the great evil that can be traced directly to affirming a definition of individual liberty and autonomy that is too narrow. Mandates were not needed to eliminate smallpox late in the 20th century, and there is no evidence that fining Henning Jacobson for refusing a vaccine prevented even a single case of smallpox in Massachusetts, but there is *irrefutable* evidence that doing so led directly to another of the most shameful episodes in American history.

Jacobson v. Massachusetts was, in its turn, cited just 22 years later to support *Buck v. Bell* (1927), a eugenicist ruling that resulted in the forcible sterilization of tens of thousands of poor and minority American women.[350] Supreme Court Justice Oliver Wendell Holmes Jr. delivered the *Buck v. Bell* decision:

> "We have seen more than once that the public welfare may call upon the best
> citizens for their lives. It would be strange if it could not call upon those who
> already sap the strength of the State for these lesser sacrifices... The principle that

---

349   *Jacobson v. Massachusetts*, 197 U.S. 11, 31 (1905), available online: https://supreme.justia.com/cases/federal/us/197/11/

350   *Buck v. Bell*, 274 U.S. 200 (1927), Available online: https://supreme.justia.com/cases/federal/us/274/200/

    The literature on the compulsory or uninformed sterilization of thousands of American women following *Buck v. Bell* — especially poor and minority American women — is extensive. Journalist Adam Cohen provides an excellent overview in the epilogue of his book on *Buck v. Bell* and the eugenics movement in America (*Imbeciles: The Supreme Court, American Eugenics, and the Sterilization of Carrie Buck*, Penguin, 2017), and a short 30-minute interview on the subject was published online by NPR in 2016 ("The Supreme Court Ruling That Led To 70,000 Forced Sterilizations," *NPR Health News*, March 7, 2016, available online: https://www.npr.org/sections/health-shots/2016/03/07/469478098/the-supreme-court-ruling-that-led-to-70-000-forced-sterilizations)

    The specific 70,000 victim number comes from William Edward Leuchtenburg describing the aftermath of *Buck v. Bell*: "Over the next generation some seventy thousand persons in the United States were sterilized by state order." *The Supreme Court reborn : the constitutional revolution in the age of Roosevelt*, New York: Oxford University Press (1995), p. 15, Available online: https://archive.org/details/supremecourtrebo0000leuc/page/14/mode/2up

See also Jane Lawrence, "The Indian health service and the sterilization of Native American women," *American Indian Quarterly* 24, no. 3 (2000): 400-419, Available online: https://www.jstor.org/stable/1185911

sustains compulsory vaccination is broad enough to cover cutting the Fallopian tubes. *Jacobson v. Massachusetts*, 197 U. S. 11."[351]

The lawyer Irving Whitehead filed a legal brief to the Supreme Court supporting Carrie Buck's right to *not* be sterilized against her will. He cited multiple precedent cases where compulsory sterilization was ruled unconstitutional, and accurately predicted the natural consequences of ruling against Carrie Buck based on "expert" opinions and general appeals to societal duty, predicting "A reign of doctors will be inaugurated and in the name of science new classes will be added, even races may be brought within the scope of such a regulation and the worst forms of tyranny practiced."[352] Irving Whitehead did not have to wait until mandates in 2020 to see the accuracy of his prediction. A 1937 Third Reich legal review of "Race Protection Laws of Other Countries" directly quoted *Buck v. Bell*, as did Nazi doctors' defense lawyers at Nuremberg.[353]

In the wake of *Buck v. Bell*, thousands of institutionalized minors were sterilized in Virginia alone, boys and girls alike.[354] Author Stephen Trombley obtained access to institutional medical board transcripts from a class action lawsuit mounted by former inmates, as well as statements provided by former inmates to the Richmond office of the ACLU. Trombley reproduced excerpts from these in his 1988 book, *The Right to Reproduce: A History of Coercive Sterilization*. Once victims were institutionalized, "consent" was easy to coerce, and when it could not be coerced, it was ignored. All of these reports are chilling, but one transcript from the case of a 16-year-old girl sterilized in 1964 is enough to show the institutional mockery that passed for "due process" once the inviolability of informed consent

---

351   It is worth noting that the sole dissenting Justice, Pierce Butler, dissented based on his devout Catholicism, at least according to Chief Justice Oliver Wendell Holmes, Jr., who remarked of Butler: "I wonder whether he will have the courage to vote with us in spite of his religion." Butler's dissenting vote in spite of peer pressure from the nation's highest justices will forever be to his credit.
Oliver Wendell Holmes Jr., *Quoted in* William Edward Leuchtenburg, *The Supreme Court reborn : the constitutional revolution in the age of Roosevelt*, New York: Oxford University Press (1995), p. 15, Available online: https://archive.org/details/supremecourtrebo0000leuc/page/14/mode/2up?q=religion&view=theater
352   Irving P. Whitehead, "Whitehead: Supreme Court Brief" (1926). *Buck v Bell Documents*, Paper 97, p. 18, Georgia State University College of Law Reading Room, Faculty Publications. Available online: http://readingroom.law.gsu.edu/buckvbell/97
353   "Translation of Hofmann Document 54, Hofmann Defense Exhibit 61: Extract from 'Information Service of the Racial-Political Office of the NSDAP Reich Administration,' 30 July 1937, Concerning Race Protection Laws of Other Countries," *Trials of War Criminals Before the Nuernberg Military Tribuanls Under Control Council law No. 10*, Volume IV (Nuernberg October 1946-April 1949), Washington D.C.: U.S. Government Printing Office. reprinted in "Nuremberg Documents" (2009). *Buck v Bell Documents*. 45, p. 4. Georgia State University College of Law Reading Room, Faculty Publications. Available Online: https://readingroom.law.gsu.edu/buckvbell/45
354   Ironically, at least in Virginia, Jim Crow discrimination had a silver lining in this one respect — hospital treatment refused on racial grounds often included refusing to provide sterilization procedures.

and individual liberty were bypassed in the name of public health and the general welfare.[355] The girl "was truant from school because she was a slow learner and did not like being teased by other children. She was collected by a social worker and told she was 'going to go on a long trip'." She was then involuntarily detained for more than three months and given a "diagnosis" of "Cultural Familial Mental Retardation, undetermined genetic mechanism present," before her hearing in front of the medical board that would decide whether she was to be allowed the "privilege" of someday getting a chance to have children.

The girl's parents were not permitted to attend the board in-person, which is a major red flag on its own. The girl's parents also refused to sign the papers consenting to have their daughter sterilized. The girl's mother wrote to the board, pleading on her daughter's behalf:

> "My husband and myself are not going to sign these papers... Send her home to me for good. I am not ruining her life.

> "Put yourself in my place[.] if your daughter or son was to be fix[ed] and they wanted to get married someday[,] and the girl or boy could not become a mother or father[,] would you want this to happen to your daughter or son[?] [W]ell I don't either[.] I want [my daughter] home for good[.] [S]he is a nice girl and a good girl. I am proud of her."[356]

The girl herself stated clearly, "I say if I do go home, ask the social worker to let me go home next year for good then I'm going to go home and do what my mama tells me... I do not want to get sterilized either." The "expert" board's chilling final ruling was issued the same day and carried out without hesitation, against the protests of the girl and her parents. No criminal, or even civil, conviction necessary. "The Board orders that sexual sterilization be performed on this patient no sooner than 30 days from this date." Even rapists do not deprive their victims of the physical capacity to have children.

Whether pro-life or pro-choice, everyone who values women's reproductive rights and essential liberties should be automatically *very* leery of *any* argument that cites *Jacobson v. Massachusetts* as a

---

355   Stephen Trombley, *The Right to Reproduce: A History of Coercive Sterilization*, Weidenfeld and Nicolson: London, 1988. pg. 245-248.

356   Stephen Trombley, *The Right to Reproduce: A History of Coercive Sterilization*, Weidenfeld and Nicolson: London, 1988. pg. 245-248. Trombley mentions his belief that the grammatical and punctuation errors in the transcript were due to illiteracy on the part of the transcribing staff member rather than the girl's mother, who "clearly... was able to read and write very well." He also notes that the doctor's "entire diagnosis was based on hearsay and assumption.... Her mother *is said to be* mentally retarded; as sister is *said to have been* a patient at another institution."

supporting precedent. 95 years after *Buck v. Bell*, another lawyer, Robert Barnes, eloquently described the resultant trajectory. This is not a slippery slope fallacy because the predictable downhill slide has already happened, and worse, because we failed to uproot and repudiate the flawed philosophy undergirding the *Jacobson* decision:

"Born amidst malaria and smallpox pandemics, the Constitution authorized no emergency exception to the liberties secured under it. The Founding Fathers understood the virus of concentrated power posed more of a threat than any biological virus ever could. The Ninth Amendment to the Constitution safeguarded all ancient rights and liberties, including the ancient tort of battery. United States Constitution, Amendment IX. The right against battery assured 'the right of every individual to the possession and control of his own person, free from all restraint or interference of others,' which would be 'sacred' and protected under the law. Union Pacific R. Co. v. Botsford, 141 U.S. 250, 251 (1891). The famed Justice Benjamin N. Cardozo defined the doctrine as the universal right of every person 'to determine what shall be done with his own body.' Schloendorff v. Society of New York Hospital, 105 N.E. 92, 93 (1914).

"This right to informed consent incorporates necessarily the right to refuse treatment: 'The forcible injection of medication into a nonconsenting person's body represents a substantial interference with that person's liberty.' Washington v. Harper, 494 U.S. 210, 229 (1990). The Nuremberg Code enshrines the right of informed consent as a matter of universal law, so widely recognized, that courts consider it a jus cogens legal principle enforceable everywhere. Abdullah v. Pfizer, Inc., 562 F.3d 163 (2d Cir. 2009). Based on these precepts, courts require clear and convincing evidence that a person poses an imminent, severe risk to others before those individuals may be subject to any forced medical care. O'Conner v. Donaldson, 422 U.S. 563 (1975); Addington v. Texas, 441 U.S. 418 (1978).

"We only deviated from this Informed Consent standard of medical care during the Eugenics Era, a diseased doctrine birthed in the medical academies of the United States at the turn of the last century, as a deformed outgrowth of the then in-vogue school of Social Darwinism. A trio of decisions carved out emergency exceptions to Constitutional liberties, including authorizing a criminal fine for not taking a vaccine (Jacobson v. Massachusetts, 197 U.S. 11 (1905)), forced sterilization of poor and politically unprotected populations (Buck v. Bell, 274 U.S. 200 (1927)), which relied exclusively on expanding Jacobson, and the decisions culminated in the kind of 'emergency exception' logic that led the Supreme Court to authorize forced detention camps based on race alone. Korematsu v. United States, 323 U.S. 214 (1944). This trilogy of infamy sees its corpses rise again as 'precedents,' seemingly permitting governments to reinstate Eugenics-Era logic across the legal landscape."[357]

Appeals to historical and legal precedent from other diseases like smallpox and tuberculosis to provide grounds for COVID-19 restrictions fail for at least two reasons, even on purely utilitarian grounds. First, smallpox and tuberculosis were demonstrably far more deadly than COVID-19. Even according to the CDC, as late as 1900, tuberculosis killed 194 out of every 100,000 Americans, and was barely edged out by pneumonia as the leading cause of death.[358] By contrast, even including all the deaths *with* COVID instead of *from* COVID, the CDC was only able to report less than half that number throughout all of 2020 (91.5 COVID deaths per 100,000).[359] Smallpox, with its 7% to 43% mortality rate, spent seven years ravaging North America during the American Revolution.[360] COVID-19 is not even in the same league as these other diseases, and so any restrictions placed on the general population to protect from COVID-19 should have been lighter than those used in the case of diseases like smallpox and tuberculosis, not heavier. Second, safeguards for individual rights when dealing with these more dangerous diseases were either not honored or totally ignored in the case of COVID-19. A short perusal of the CDC's *Tuberculosis Control Laws and Policies Handbook* makes this abundantly clear.[361] The guide even provides a handy checklist, including an entire section on "Safeguarding Rights."

---

357   *Children's Health Defense v. Food & Drug Administration*, No. 6: 22-CV-00093-ADA (W.D. Tex. Jan. 12, 2023). Document 1, Filed 01/24/22, pp. 38-40, *at* 133-135. Online: https://storage.courtlistener.com/recap/gov.uscourts.txwd.1159970/gov.uscourts.txwd.1159970.1.0.pdf

358   Cole, D, *Achievements in public health, 1900-1999: Control of infectious diseases*, Morbidity and Mortality Weekly Report (1999), **48**(29), 621. Available online: https://www.cdc.gov/mmwr/preview/mmwrhtml/mm4829a1.htm

359   Ahmad, F.B., et al., *Provisional mortality data—united states, 2020*. Morbidity and Mortality Weekly Report, 2021. **70**(14): p. 519. https://www.cdc.gov/mmwr/volumes/70/wr/mm7014e1.htm

360   Fenn, E.A., *Pox Americana : the great smallpox epidemic of 1775-82*. 2002 ebook ed. 2001, New York: Hill and Wang, p. 21

361   Centers for Disease Control and Prevention. *Tuberculosis control laws and policies: A handbook for public health and legal practitioners*. (2009). At the time of this writing in 2023, this handbook was last updated in 2009. Available online: https://www.cdc.gov/tb/programs/TBlawPolicyHandbook.pdf

Not one of the items on that list was fully honored in the case of COVID-19, and most have been ignored entirely.

Not a single one of the CDC's own articulated safeguards for individuals rights in the case of diseases like tuberculosis was fully honored in the case of COVID, and most were ignored entirely.

IV. Safeguarding Rights		
A. Due Process		
46. Do laws provide persons subject to TB control measures that restrict personal freedoms the following due process protections: Written notice? Written delineation of rights? Prior hearing or post-deprivation hearing? Access to representation?		
Right to culturally-appropriate translations for non-English speaking persons? Decision from an impartial tribunal or court of law? Right to present and cross-examine witnesses? Right to review the record? Right to appeal? Other specific due process guarantees?		
B. Confidentiality and Privacy		
47. Do laws limit the acquisition, use, or disclosure of identifiable data by public health authorities concerning the identities, health status, or health records of persons with suspected or confirmed TB?		
48. Do laws limit the sharing of identifiable data by public health authorities in other jurisdictions concerning the identities, health status, or health records of persons with suspected or confirmed TB?		
C. Anti-discrimination		
49. Do laws protect persons with suspected or confirmed TB from unwarranted discrimination via governmental or private sector entities?		
50. Do laws require that TB services be provided on a non-discriminatory basis?		
51. Do laws require that special provisions be made for persons with suspected or confirmed TB and have vision, hearing, or other impairments or disabilities?		
52. Do laws require that special provisions be made for persons with suspected or confirmed TB and are non-English speakers?		
D. Religious Exemption		
53. Do laws recognize exemptions from TB prevention and control measures based on individual religious or other beliefs?		
V. Special Populations		
54. Concerning minors in custody, inmates, or other wards, do laws require specific measures or protections for persons with suspected or confirmed TB?		
55. If so, is responsibility for TB control and prevention shared with non-public health authorities?		

CDC Tuberculosis Control Laws and Policies Handbook
Table 2. TB Control Law Checklist.
IV. Safeguarding Rights.

The Centers for Law and the Public's Health: A Collaborative at Johns Hopkins and Georgetown Universitites, *Tuberculosis Control Laws and Policies: A Handbook for Public Health and Legal Practitioners*, prepared for The Centers for Disease Control and Prevention (CDC), 2009, Table 2(IV), page 35, available online: https://www.cdc.gov/tb/programs/TBLawPolicyHandbook.pdf 362

---

362   The Centers for Law and the Public's Health: A Collaborative at Johns Hopkins and Georgetown Universities, *Tuberculosis Control Laws and Policies: A Handbook for Public Health and Legal Practitioners*, prepared for The Centers for Disease Control and Prevention (CDC), 2009, Table 2(IV), page 35, available online: https://www.cdc.gov/tb/programs/TBLawPolicyHandbook.pdf

Universal compulsory masking implicitly reverses the presumption of innocence that is supposed to be at the heart of our American legal system. It visually shifts the proper burden of proof from a potential accuser onto the accused. Just as we presume that a human is born free rather than a slave, or that someone is telling the truth until proven to be a liar, the proper default presumption is one of health rather than sickness. The mere presence of a virus in a community — even a new, contagious, or particularly dangerous virus — is not grounds for treating all apparently healthy people as active plague vectors. The presence of HIV in the gay community in the 1980s and 1990s did not justify treating all gay people as though they had active AIDS, and the presence of SARS-CoV-2 in the general community never justified treating all citizens as though they had transmissible-stage COVID-19. Courts have already ruled (multiple times) that policies discriminating on the basis of HIV infection cannot withstand scrutiny even in a military setting.[363] If this is true vis-à-vis an *actual* infection (bloodborne, incurable, possibly transmissible, with no vaccine and requiring daily medication and periodic blood testing), then discrimination is even less defensible on the basis of a *potential* infection which, while airborne, is imminently treatable, does not require daily medication or periodic blood testing, from which total recovery is the rule rather than an exception, and which never posed a substantial threat to over 99% of the population in the first place.

Lockdowns, mask-wearing and other COVIDcrisis mandates subjected Americans' constitutionally recognized natural rights and liberties to the most sustained invasion and occupation since *Korematsu*, and these violations of Natural Law were not unique to the United States. The public benefit of reaffirming and preserving essential Constitutional rights vastly outweighs any harm to a "collective" interest in public health from any disease. Individuals are not collective property, and individual medical decisions must never be collectivized. For all the invective aimed at "selfish" individual medical choices, the "collective" cares far less about the results of an individuals' medical decision than individuals themselves do, and experts purporting to speak for the collective have far less detailed information on which to base any such decisions. We each have the right to influence disease progression in our communities by persuasion, and by the example of how we handle our own personal health. We each have *no* right, either singly or as a group, to unilaterally determine what diseases other people do or do not catch, even if doing so would indirectly benefit us. Trying to force such determinations ostensibly on behalf of a third party like children or the elderly is no better.

Avoiding a risk and acting with caution may sometimes overlap with responsible behavior, but it does not *equate* to responsible behavior, and doing so with *every* risk would be both irresponsible and

---

363   *Roe v. Department of Defense* 947 F.3d 207 (4th Circuit, 2020) No. 19-1410. Available online: https://law.justia.com/cases/federal/appellate-courts/ca4/19-1410/19-1410-2020-01-10.html
Harrison v. Austin, E.D. Va. 1:18-cv-00641-LMB-IDD, Document 307, Filed 04/06/22 Available online: https://www.lambdalegal.org/sites/default/files/legal-docs/downloads/harrison_roe_msj_opinion_4-6-2022.pdf

cowardly. Politicians and legislators do not stand *in loco parentis* for the citizens who elected them, nor can they confer this power on any public health officials they appoint. Compulsory masking revealed that paternalistic attitudes justly condemned in other contexts are alive and well, not just in the medical community, but throughout the political classes, the laptop professional caste, and an all-too-large segment of the general citizenry.

The principle that sustains compulsory masking and compulsory vaccination is the principle that says your unalienable rights are contingent on the extent to which someone (or many someones) can derive benefit from violating them. Your body is only your own to the extent the communal organism does not "need" it. By contrast, the principle that *forbids* compulsory masking and compulsory vaccination is the same principle that forbids performing involuntary medical interventions, including sterilizations and experiments, in the name of the common good. In the wake of the Nuremberg trials, Pope Pius XII articulated the basis for this principle in 1952, "It must be noted that, in his personal being, man is not finally ordered to usefulness to society. On the contrary, the community exists for men."[364] Individuals do not relate to their community the way that cells relate to their bodies, entirely subordinated to the good of the whole. Forcing every citizen to wear a smart watch or smart collar to monitor vitals and record biometric data would no doubt be immensely helpful to public health. It would also be an immense, absolutely unacceptable violation of individual unalienable rights, even if it were imposed in the name of public safety and public health during a war for national survival. The same holds true of compulsory masking for COVID or any other disease.

Key Takeaways

- Compulsory masking was a medicalized extension of the reasoning and belief systems dominant in some of the most infamous legal rulings in American history, including *Buck v. Bell* and *Korematsu*.
- Responses to COVID, including compulsory masking, ignored or ran roughshod over previously established safeguards for individual rights in the context of historically worse communicable respiratory diseases like tuberculosis.
- Individuals in a community are not like cells in a human body. Individuals have rights that cannot be overruled by any amount of communal benefit from violating them. The right to forgo making use of particular medical interventions, or participating in particular medical experiments, are among these.

---

364   Pope Pius XII (1952), "The Moral Limits of Medical Research and Treatment — Pope Pius XII," *The Linacre Quarterly*: Vol. 19: No. 4, Article 2, p. 104, available online: https://epublications.marquette.edu/lnq/vol19/iss4/2

Part 4 Conclusion: when compliance becomes complicity

> *"They conform to a particular requirement and in so doing they themselves perpetuate that requirement."*
>
> – Václav Havel, *The Power of the Powerless*, 1978[365]

Mask-wearing, like many other practices over the course of human history (including flag salutes) is a visual slogan.[366] Our affected mutual indifference to one another's masks was and remains a fragile illusion. In reality, by dutifully wearing our masks when ordered to do so, we reaffirm to one another

---

365   Václav Havel, The Power of the Powerless. International Journal of Politics October, 1978. https://www.nonviolent-conflict.org/resource/the-power-of-the-powerless/

366   One could write a book solely about visual slogans and expressive speech, but a good single 20th century example is how portraits of Chairman Mao became visual slogans during the Chinese Cultural Revolution. (Frank Dikötter, *The Cultural Revolution, A People's History 1962-1976*, Bloomsbury Publishing: New York, 2017.)

> "On 27 August, for instance, a directive from Beijing demanded that only portraits of the Chairman be displayed on public occasions. But nobody wanted to fall behind in the cult of the leader. As the range of objects condemned as feudal or bourgeois expanded, ordinary people increasingly turned to the only politically safe commodities available — Mao photos, Mao badges, Mao posters and Mao books." (p. 97)

> "In the capital miniature shrines went on sale, with three leaves in the form of a triptych. The centre had a portrait of the Chairman, while the outer panels carried quotations. These cult objects took

not only our knowledge of the rules of the "game," but pressure each other to accept those rules. As Václav Havel put it, we "confirm thereby the power that requires the slogans in the first place. Quite simply, each helps the other to be obedient. Both are objects in a system of control, but at the same time they are its subjects as well. They are both victims of the system and its instruments."[367]

What Václav Havel described as a "system" includes what we think of as the permeating culture or social atmosphere, "What we understand by the system is not, therefore, a social order imposed by one group upon another, but rather something which permeates the entire society and is a factor in shaping it, something which may seem impossible to grasp or define."[368] Every person who wore a mask (myself included) helped to legitimize the practice, and contributed, along with thousands of other rituals, to enhancing the panoramic atmosphere of "pandemic emergency" used to justify and perpetuate the accompanying self-destructive and immoral Public Health™ measures. As Havel put it:

"This panorama, of course, has a subliminal meaning as well: it reminds people where they are living and what is expected of them. It tells them what everyone else is doing, and indicates to them what they must do as well, if they don't want to be excluded, to fall into isolation, alienate themselves from society, break the rules of the game, and risk the loss of their peace and tranquility and security."[369]

"[T]heir slogans are mutually dependent: both were displayed with some awareness of the general panorama and, we might say, under its diktat. Both, however, assist in the creation of that panorama, and therefore they assist in the creation of that diktat as well."[370]

---

the place of the old family altar at home, as people met the gaze of the Chairman the moment they woke up and reported back to him in the evening." (p. 168-169)

"Each year National Day was celebrated with processions in all major cities, as hundreds of thousands of workers, peasants and students marched in serried ranks, shouting slogans, waving red flags and holding aloft portraits of the Chairman." (p. 245)

367   Václav Havel, *The Power of the Powerless*. International Journal of Politics October, 1978, p. 15. Available Online: https://www.nonviolent-conflict.org/resource/the-power-of-the-powerless/

368   Václav Havel, *The Power of the Powerless*. International Journal of Politics October, 1978, p. 15. Available Online: https://www.nonviolent-conflict.org/resource/the-power-of-the-powerless/

369   Václav Havel, *The Power of the Powerless*. International Journal of Politics October, 1978, pp. 14-15. Available Online: https://www.nonviolent-conflict.org/resource/the-power-of-the-powerless/

370   Václav Havel, *The Power of the Powerless*. International Journal of Politics October, 1978, p. 15. Available Online: https://www.nonviolent-conflict.org/resource/the-power-of-the-powerless/

This kind of complicity provides another excellent example of how and why moral duties and legal duties *do not and should not* completely overlap in the way mask mandates assume. It also explains why there is, and ought to be, a high bar for a moral duty to surmount before it can become a legal duty. This kind of complicity involves shades of grey. For individuals like myself, who recognized compulsory masking for what it was from the get-go, and who know (or, like the "experts," *claimed* to know) so much about masks, it includes moral culpability. But just as moral duties vary individually, such culpability does not apply to everyone equally. The point at which compliance becomes complicity and culpability varies from person to person, as does the extent of one's moral duty to push back. There is no doubt that individual motivations at every level of responsibility for implementing and enforcing coerced masking covered the whole spectrum of motivation from opportunism or altruism to the fear, arrogance, and self-righteousness wrongly ascribed to people who refused to comply. Regardless of the motivation or intent behind coerced masking, however, the net effect was indisputably to get a disturbingly large segment of the population to, in Havel's words, "participate in the common responsibility for it, so they may be pulled into and ensnared by it, like Faust by Mephistopheles."[371]

Compulsory masking was sold and pushed using malformed philosophies and pathological, boundary-violating notions of "social responsibility." Because human nature does not change, the mentality underlying compulsory masking will always seek to manifest in various forms. Unless or until enough people develop the courage to, as Havel put it, "live within the truth," we will be left to live within the alternative, "accepting the given rules of the game… thus making it possible for the game to go on, for it to exist in the first place."[372]

---

371   Václav Havel, *The Power of the Powerless.* International Journal of Politics October, 1978, p. 16. Available Online: https://www.nonviolent-conflict.org/resource/the-power-of-the-powerless/

372   Václav Havel, *The Power of the Powerless.* International Journal of Politics October, 1978, p. 10. Available Online: https://www.nonviolent-conflict.org/resource/the-power-of-the-powerless/

Part 5:

Under Color of Law — Legal Duels Over Compulsory Masking

> "We can have intellectual individualism and the rich cultural diversities that we owe to exceptional minds only at the price of occasional eccentricity and abnormal attitudes. When they are so harmless to others or to the State as those we deal with here, the price is not too great. But freedom to differ is not limited to things that do not matter much. That would be a mere shadow of freedom."
>
> — United States Supreme Court, *West Virginia Board of Education v. Barnette*, 1943.[373]

During the COVIDcrisis, a small cadre of lawyers demonstrated why doctors of law are every bit as essential as medical doctors. I am not a lawyer and what follows is not legal advice. Take this section on the legal arguments surrounding masking as you would take editorial material from a journalist, a blogger, or a friend who did a *lot* of homework. Sources are cited in the footnotes. An entire book could be written on this topic alone. This section is intended to be a broad overview of the legal aspects involved with compulsory masking (especially the constitutional issues) and a jumping off point for your own researches. Do not spend money or file a lawsuit based purely on what you read here. Consulting a lawyer and — most importantly — doing your own research are essential prerequisites

---

373   *West Virginia Board of Education v. Barnette et al.*, 319 U.S. 624, 641-642 (1943), available online: https://www.loc.gov/item/usrep319624/

for success pushing back on masking through any judicial or administrative route. Quite apart from the specific legal issues involving masks, administrative and courtroom procedures are complicated and difficult to navigate in-and-of themselves. At the time of this writing, mask-related litigation is still very active, and much remains unresolved.

When mask mandates hit the scene, they were just one out of many types of unprecedented and invasive mandates. These mandates were issued, changed, repealed, and re-issued with modifications so quickly that it was practically a full-time job just to stay up-to-date on what was "allowed" and what was not. In the face of such a policy-making frenzy, imposed in an atmosphere of widespread uncertainty and fear, plenty of people, even among those inclined to push back against COVIDcrisis overreach, simply acquiesced to mask mandates as a comparatively minor nuisance. For many, that is truly all masks ever were. For others, fighting mask mandates was a priority, but still had to be put on the back burner. Business owners facing financial ruin through mandated restrictions on their operations and parents whose children were now at home 24/7 had more pressing concerns. Still, many Americans recognized the deeper threat that mask mandates presented and fought back as best they could — whether in the court of public opinion or in the courts of law. Even the unsuccessful lawsuits were helpful, functioning like artillery rounds that serve to hone in on a target by trial and error. As is the case with mask science, upon close examination, the laws, regulations, rulings and arguments cited to sustain compulsory masking turn out to be far weaker than advertised.

Just who, precisely, is putting these mandates out, and how are they doing it?

The legal powers invoked to mandate masks varied depending on the mandate, involving a labyrinthine maze of statutes, emergency powers, and administrative authorizations. Though House Speaker Nancy Pelosi implemented a masking policy in the House of Representatives as a "matter of decorum,"[374] to its credit, the U.S. Congress never passed a mandatory masking law, despite shrill howls for such unconstitutional legislation from many quarters. It was mostly the executive branches of State and Federal government — State Governors, Federal agencies, and local municipalities — that mandated masks using interlocking disaster declarations and their respective emergency powers, playing an administrative shell game of diffuse-the-responsibility.

Official emergencies give executives and other authorities broad powers to bypass normal processes and constitutional safeguards, allowing them to enact orders that infringe rights normally sacrosanct. Naturally, then, the first step in mandating masks was a chain reaction of public health emergency and disaster declarations that started well before most people were even aware of SARS-CoV-2 or the administrative avalanche bearing down on them.

On January 30, 2020, the World Health Organization declared that COVID-19 constituted a public health emergency of international concern.[375] This third-party declaration provided political cover and psychological impetus for United States Health and Human Services Secretary Alex M. Azar to declare a public health emergency for the entire United States pursuant to the Public Health

374   167 Congressional Record H40–41 (daily ed. Jan. 4, 2021) (announcement by the Speaker Pro Tempore) https://www.congress.gov/congressional-record/volume-167/issue-2/daily-digest

375   Emergency Committee on the Novel Coronavirus, "Novel Coronavirus (2019-nCoV) Situation Report — 10" World Health Organization, January 30, 2020, https://www.who.int/docs/default-source/coronaviruse/situation-reports/20200130-sitrep-10-ncov.pdf?sfvrsn=d0b2e480_2

Service Act (PHSA) the next day.[376] Four days later, on February 4, 2020, the HHS Secretary put out a *separate* public health emergency declaration under a subsection of the PHSA, known as the Public Readiness and Emergency Preparedness Act (PREP Act, for short).[377] This *second* public health emergency declaration, based on his authority under section 564 of the Food, Drug, and Cosmetic Act,[378] set the stage for immunity from liability regarding the use of multiple medical countermeasures to COVID-19, including masks.[379] This HHS public health emergency declaration was followed by an Emergency Use Authorization declaration, which took effect on March 27, 2020.[380]

376   January 31, 2020, invoking Section 319 of the Public Health Service (PHS) Act, 42 U.S.C. 247d
Department of Health and Human Services, Office of the Secretary, "Declaration Under the Public Readiness and Emergency Preparedness Act for Medical Countermeasures Against COVID-19," February 4, 2020, notice published March 17, 2020,
https://www.federalregister.gov/documents/2020/03/17/2020-05484/declaration-under-the-public-readiness-and-emergency-preparedness-act-for-medical-countermeasures
https://www.govinfo.gov/content/pkg/FR-2020-03-17/pdf/2020-05484.pdf

377   Office of the Secretary, Department of Health and Human Services, "Determination of Public Health Emergency," determination and declaration February 4, 2020, published in The Federal Register, Vol. 85, No. 26, Friday, February 7, 2020,
https://www.federalregister.gov/documents/2020/02/07/2020-02496/determination-of-public-health-emergency
https://www.govinfo.gov/content/pkg/FR-2020-02-07/pdf/2020-02496.pdf
The PREP act was enacted on December 30, 2005, as Public Law 109-148, Division C, Section 2. It was an amendment to the Public Health Service (PHS) Act, which added Sections 319F-3 and 319F-4 (https://www.govinfo.gov/content/pkg/PLAW-109publ148/pdf/PLAW-109publ148.pdf).
These sections are codified under
Title 42, U.S.C. §247d-6d (https://www.govinfo.gov/content/pkg/USCODE-2021-title42/pdf/USCODE-2021-title42-chap6A-subchapII-partB-sec247d-6d.pdf), and
Title 42, U.S.C. §247d-6e (https://www.govinfo.gov/content/pkg/USCODE-2021-title42/pdf/USCODE-2021-title42-chap6A-subchapII-partB-sec247d-6e.pdf)

378   Section 564(b)(1)(C) of the Federal, Food, Drug, and Cosmetic Act (Title 21 U.S.C. 360bbb-3)
https://uscode.house.gov/view.xhtml?req=(title:21%20section:360bbb%20edition:prelim)

> "The Secretary may make a declaration that the circumstances exist justifying the authorization under this subsection for a product on the basis of… a determination by the Secretary that there is a public health emergency, or a significant potential for a public health emergency, that affects, or has a significant potential to affect, national security or the health and security of United States citizens living abroad, and that involves a biological, chemical, radiological, or nuclear agent or agents, or a disease or condition that may be attributable to such agent or agents…"

379   Department of Health and Human Services, Office of the Secretary, "Declaration Under the Public Readiness and Emergency Preparedness Act for Medical Countermeasures Against COVID-19,"declaration effective February 4, 2020, published in the Federal Register, Vol. 85, No. 52, Tuesday, March 17, 2020,
https://www.federalregister.gov/documents/2020/03/17/2020-05484/declaration-under-the-public-readiness-and-emergency-preparedness-act-for-medical-countermeasures
https://www.govinfo.gov/content/pkg/FR-2020-03-17/pdf/2020-05484.pdf

380   Office of the Secretary, Department of Health and Human Services, "Emergency Use Authorization Declaration," determination February 4, 2020, declaration effective March 27, 2020, published in The Federal Register Vol. 85, No. 63, Wednesday, April 1, 2020,

Federal-level emergency and disaster declarations provided additional political cover for State Governors already empowered to declare emergencies through their own state laws and emergency health powers acts (known by various names in each state). Each State had its own separate but similar process. Giving just two examples from Texas and Massachusetts, in Texas, Governor Greg Abbott issued a disaster proclamation on March 13, certifying based on a mere 30 confirmed cases (not deaths, but *cases*) that "COVID-19 poses an imminent threat of disaster," and renewed the disaster declaration monthly thereafter.[381] When he mandated masks in July, Governor Abbot cited to Texas Department of State Health Services (DSHS) commissioner Dr. John Hellerstedt's determination that COVID-19 counted as a public health disaster within the meaning of the Texas Health and Safety Code.[382] In Massachusetts, Governor Charles Baker declared an emergency three days earlier on March 10, 2020, and made masks mandatory on May 6th citing only a vague reference to the CDC's advice to wear cloth face coverings.[383] Other states followed their own version of this pattern.

Most of these executive orders were easy to locate and were neither lengthy nor complicated, and contained numerous broad exceptions that undermined their stated rationale of infection control. Governor Abbot's July 2020 mask mandate, for example, was just 5 pages long and made exceptions for medical conditions, eating and drinking, voting, security identification, religious worship, giving a speech to an audience, and exercising while more than 6 feet away from other people. Massachusetts Governor Baker's mask mandate outright said, "All persons are strongly discouraged from using medical-grade masks to meet the requirements of this Order," — on the rationale that healthcare workers needed medical-grade masks more.

In practice, mask-wearing often went far beyond what was required by the mandates, as everyone wanted to signal that they were part of the "solution," and feared being seen as part of the problem. Too many people who were enforcing theser orders simply treated them as applying to every person

https://www.federalregister.gov/documents/2020/04/01/2020-06905/emergency-use-authorization-declaration
https://www.govinfo.gov/content/pkg/FR-2020-04-01/pdf/2020-06905.pdf
381   Governor Abbott cited authority from Texas Government Code Section 418.014.
Governor Greg Abbott, Disaster proclamation, March 13, 2020, Executive Department, Austin, Texas, available online:
https://gov.texas.gov/uploads/files/press/DISASTER_covid19_disaster_proclamation_IMAGE_03-13-2020.pdf
382   Specifically, within the meaning of Chapter 81 of the Texas Health and Safety Code.
Governor Greg Abbot, Executive Order GA 29 "Relating to the use of face coverings during the COVID-19 disaster," July 2, 2020, Executive Department, Austin, Texas, Available online:
https://gov.texas.gov/uploads/files/press/EO-GA-29-use-of-face-coverings-during-COVID-19-IMAGE-07-02-2020.pdf
383   Charles D. Baker, Executive Order 591 "Governor's Declaration of Emergency," March 10, 2020, Boston, Massachusetts,
https://www.mass.gov/doc/governors-declaration-of-emergency-march-10-2020-aka-executive-order-591/download
Charles D. Baker, COVID-19 Order No. 31 "Order Requiring Face Coverings in Public Places Where Social Distancing is Not Possible," May 1, 2020, Boston, Massachusetts,
https://www.mass.gov/doc/may-1-2020-masks-and-face-coverings/download

engaged in all activities in a public setting, ignoring exceptions or construing those exceptions as narrowly as possible. Children and adolescents who were never at any real risk from COVID were frequently forced to wear facemasks while playing and exercising outdoors by adults who could have and should have known better.[384]

Many local districts issued their own mask mandates weeks or even months before their state governors did so. In El Paso, Texas, for example, where I lived at the time, Mayor Dee Margo and Health Authority Dr. Hector Ocaranza mandated masks on April 24, 2020.[385] Under Dee Margo's mayoral successor, Oscar Leeser, El Paso was zealously suing to keep those mask mandates in place more than a year later, months after Governor Abbott finally banned mask mandates in July 2021.[386] At an even more local level, school boards, businesses, and private institutions implemented too many mask mandates to count.

Foremost among the Federal agencies pushing for compulsory masking was the CDC. When asked to provide medical justification for compulsory masking, public health authorities universally gestured in the general direction of CDC "recommendations" and publications.[387] CDC Director Robert Redfield co-authored a July 2020 editorial in the *Journal of the American Medical Association* calling

---

384   I repeatedly observed this in El Paso, Texas, throughout 2020, lasting late into 2021. It was both frustrating and heartbreaking to watch children walking home from school wearing masks (sometimes even when alone).

385   Dee Margo, Mayor of the City of El Paso, "Second Amendment to Local Emergency Directive," April 24, 2020, available online https://www.epstrong.org/documents/covid19/2020.04.23%20-%20Second%20Amendment%20to%20Local%20Emergency%20Directive.pdf?1587690490

386   Aaron A. Bedoya, "Will there be a mask mandate in El Paso County on Wednesday? Here's what we know," *El Paso Times*, August 18, 2021, Available Online: https://www.elpasotimes.com/story/news/health/2021/08/17/mask-mandate-el-paso-texas-abbott-what-we-know-for-wednesday/8162823002/

Governor Greg Abbot, Executive Order GA 38 "Relating to the continued response to the COVID-19 disaster," July 29, 2021, Executive Department, Austin, Texas, Available online: https://gov.texas.gov/uploads/files/press/EO-GA-38_continued_response_to_the_COVID-19_disaster_IMAGE_07-29-2021.pdf

Texas Attorney General Ken Paxton, COVID-19: List of Government Entities Unlawfully Imposing Mask Mandates, last updated October 5, 2021, Online: https://www.texasattorneygeneral.gov/covid-governmental-entity-compliance (lists El Paso City and El Paso Community College still imposing mask mandates)

387   Especially the CDC's MMWR publications and the Science Brief: *Community Use of Masks to Control the Spread of SARS-CoV-2 | CDC.* December 6, 2021. https://www.cdc.gov/coronavirus/2019-ncov/science/science-briefs/masking-science-sars-cov2.html?CDC_AA_refVal=https%3A%2F%2Fwww.cdc.gov%2Fcoronavirus%2F2019-ncov%2Fmore%2Fmasking-science-sars-cov2.html#print

Previous versions from November 20, 2020, and May 7, 2021, can be found from the wayback machine:

https://web.archive.org/web/20210329185803/https://www.cdc.gov/coronavirus/2019-ncov/science/science-briefs/masking-science-sars-cov2.html?CDC_AA_refVal=https%3A%2F%2Fwww.cdc.gov%2Fcoronavirus%2F2019-ncov%2Fmore%2Fmasking-science-sars-cov2.html

https://web.archive.org/web/20210507231754/https://www.cdc.gov/coronavirus/2019-ncov/science/science-briefs/masking-science-sars-cov2.html?CDC_AA_refVal=https%3A%2F%2Fwww.cdc.gov%2Fcoronavirus%2F2019-ncov%2Fmore%2Fmasking-science-sars-cov2.html

for universal masking.[388] In a September 2020 Senate hearing, Dr. Redfield famously declared while brandishing a surgical mask: "these masks are *the* most important, powerful, public health tool we have," and "I might even go so far as to say that this facemask is more guaranteed to protect me against COVID than when I take a COVID vaccine."[389] If Dr. Redfield's successor, Dr. Rochelle Walensky, was not a true fanatic for masks, then she was at least truly committed to faking it. As late as January 11, 2022, she was testifying, double-masked, before the Senate Health Committee (next to a conspicuously unmasked Dr. Anthony Fauci).[390] It is no exaggeration to say that the CDC served as the most important lynchpin for the policy of compulsory masking.

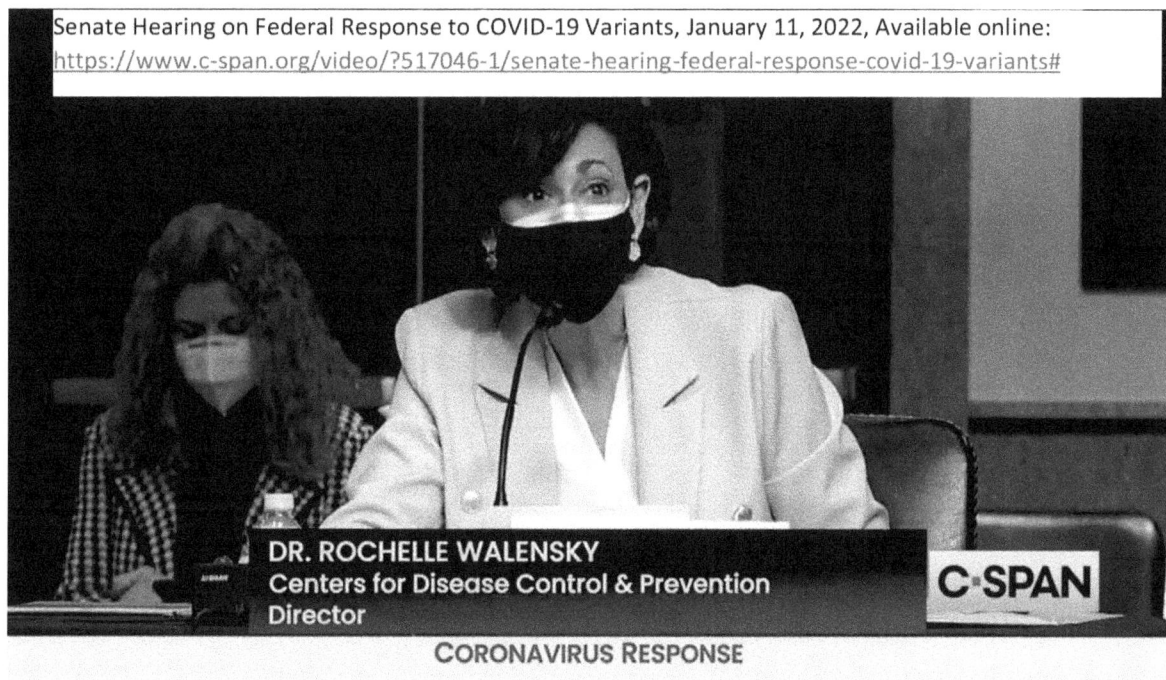

The Department of Defense, which presided over one of the healthiest subpopulations of the United States, and one of the least-vulnerable to COVID, was nevertheless the fastest to jump on the mask mandate bandwagon. On April 5, Secretary of Defense Mark Esper issued an order mandating the use of cloth face coverings for "all individuals on DoD property, installations, and facilities."[391] By

388   Brooks, J.T., J.C. Butler, and R.R. Redfield, "Universal Masking to Prevent SARS-CoV-2 Transmission—The Time Is Now," *JAMA*, 2020. **324**(7): p. 635. https://dx.doi.org/10.1001/jama.2020.13107

389   Dr. Robert Redfield Testimony, Hearing entitled, "Department of Health and Human Services Cornoavirus Response: A Review of Efforts to Date and Next Steps," United States Senate Committee on Appropriations Subcommittee Hearing, September 16, 2020, 10:00 a.m., G50 Dirksen Senate Office Building, Senator Roy Blunt (R-Missouri) presiding, **timestamp 1:31:09-1:31:44**, available online: https://www.appropriations.senate.gov/hearings/review-of-coronavirus-response-efforts

390   Senate Hearing on Federal Response to COVID-19 Variants, January 11, 2022, Available online: https://www.c-span.org/video/?517046-1/senate-hearing-federal-response-covid-19-variants#

391   Mark T. Esper, memorandum, April 5, 2020, "SUBJECT: Department of Defense Guidance on the Use of Cloth Face Coverings," available online:

then, plenty of local hospital, clinic, and medical unit commanders had already been effectively mandating masks since March.[392]

Citing CDC recommendations, other Federal executive agencies like the Food and Drug Administration (FDA) enthusiastically facilitated mandates by providing carefully lawyered Emergency Use Authorizations (EUAs) for facemasks (including cloth facemasks), as well as for other paraphernalia like goggles and face shields.[393] The FDA also put out enforcement policy guidance recommendations regarding masks as medical devices.[394] Notably, the FDA's EUA even for surgical masks only mentions protection from "respiratory droplets and large particles," *not aerosols*, despite the fact that on the ground then and later, virus-containing *aerosols* were cited as the reason for forcing people to wear cloth and surgical masks. This was just one of many internal contradictions in masking doctrine.

The Occupational Safety and Health Administration (OSHA) already had detailed regulations for certain workplace personal protective equipment (PPE) in healthcare settings.[395] These regulations were further extended in the name of controlling COVID while referring and deferring to the CDC.[396] On June 21, 2021, OSHA adopted a Healthcare Emergency Temporary Standard (ETS), and promised to "vigorously enforce" the general duty clause and general standards of personal

---

https://media.defense.gov/2020/Apr/05/2002275059/-1/-1/1/DOD-GUIDANCE-ON-THE-USE-OF-CLOTH-FACE-COVERINGS.PDF?source=GovDelivery

392   Author's personal experience.

393   The FDA issued a facemask EUA letter on April 24th, 2020, on mask use by the general public as "source control" (revising its first mask EUA from April 18th). It issued an EUA regarding surgical masks on August 5th.

FDA Chief Scientist Denise M. Hinton, Emergency Use Authorization Letter, April 24, 2020, U.S. Food and Drug Administration, Available online: https://www.fda.gov/media/137121/download

FDA Chief Scientist Denise M. Hinton, Emergency Use Authorization Letter, August 5, 2020, U.S. Food and Drug Administration, p. 3 Available online: https://www.fda.gov/media/140894/download

394   The FDA's enforcement policy specifically says that "Face masks, face shields, and respirators are devices when they are intended for a medical purpose, such as prevention of infectious disease transmission (including uses related to COVID-19)."

Food and Drug Administration (FDA), "Enforcement Policy for Face Masks, Barrier Face Coverings, Face Shields, Surgical Masks, and Respirators During the Coronavirus Disease (COVID-19) Public Health Emergency," April 2020 (revised September 2021), April 2020:

https://web.archive.org/web/20200522021154/https://www.fda.gov/media/136449/download;

September 2021: https://web.archive.org/web/20230110120628/https://www.fda.gov/media/136449/download

395   OSHA Title 29 Code of Federal Regulations 1910 Subpart I (29 CFR 1910 subpart I).

https://www.ecfr.gov/current/title-29/subtitle-B/chapter-XVII/part-1910/subpart-I

396   "CDC provides information about face coverings as one type of mask among other types of masks. OSHA differentiates face coverings from the term "mask" and from respirators that meet OSHA's Respiratory Protection Standard."

United States Department of Labor, Occupational Safety and Health Administration, COVID-19 Frequently Asked Questions, available online: https://www.osha.gov/coronavirus/faqs#cloth-face-coverings

protective equipment and respiratory protection standards with regard to COVID-19.[397] In spite of all these EUAs, cloth face coverings never met the criteria to serve as personal protective equipment (PPE) under OSHA standards.

---

397   United States Department of Labor, Occupational Safety and Health Administration, "COVID-19 Healthcare Emergency Temporary Standard (ETS)," last updated December 27, 2021 (as of February 20, 2023), available online: https://www.osha.gov/coronavirus/ets

COVID theater with a sharp point

Like other anti-COVID measures, masks were imposed on the unwilling using threats of harsh fines and even prison time, backed up by force. In the United States, Hawaiian Kaua'i County Mayor Derek Kawakami issued a particularly harsh mask mandate prescribing up to $5,000 fines and 1 year of prison time for violators.[398] Doctors exercising their rights of free speech to oppose masking could and did have their licenses threatened, investigated, or revoked if medical boards did not like their speech, as happened to Dr. Steven LaTulippe in Oregon.[399] In a bitter irony, these draconian measures were often justified on the rationale that dissident doctors' speech and teachings violated medical ethics. This, despite the fact that dissident doctors maintained their positions based on science and their sense of duty to speak the truth. Ordinary citizens fared no better. In Ohio, Alecia Kitts was tased and arrested for not wearing a mask outdoors at a middle school game.[400] Idaho church deacon Gabriel Rench was arrested in September 2020 along with fellow congregants Sean and Rachel Bohnet for not wearing masks while singing psalms at an outdoor protest worship service.[401] Citizens

398   Derek S.K. Kawakami, "Mayor's Emergency Rule #6 https://www.kauai.gov/Portals/0/Civil_Defense/EmergencyProclamations/Mayor%27s%20Emergency%20Rule%20%236%2020200413.pdf

399   To cite just two examples, Dr. Steven LaTulippe and Dr. Scott Jensen both argued that mask use is not effective to stop SARS-CoV-2. Dr. Steven LaTulippe had his license revoked by the Oregon Medical Board, and Dr. Scott Jensen had his medical license investigated 5 times by June 2022.

Brian Bakst, "In campaign, Jensen vents over medical board," *MPRnews*, June 17, 2022, online: https://www.mprnews.org/story/2022/06/17/jensen-vows-board-of-medical-practice-makeover

Steven LaTulippe, "COVID-19 and Medical Board Tyranny," Journal of American Physicians and Surgeons, 2022, Volume 27, Number 4. https://www.jpands.org/vol27no4/latulippe.pdf

Ralph Ellis, "Doctor Who Claimed Masks Hurt Health Loses License," WebMD, September 21, 2021 https://www.webmd.com/covid/news/20210920/doctor-who-claimed-masks-hurt-health-loses-license

400   She later was fined $200 for resisting arrest, $150 for criminal trespassing, and court costs.

Jesse Wharff, Phyllis Smith, and Zach Shrivers, "Update: Marietta woman pleads 'no contest' in mask refusal case," *WTAP News*, September 27, 2021, online:
https://www.wtap.com/2020/09/24/woman-refuses-to-wear-mask-is-tased-at-football-game-in-logan-ohio/

Kenneth Garger, "Woman Arrested, tased for not wearing a mask at middle school football game," *New York Post*, September 24, 2020, online: https://nypost.com/2020/09/24/woman-tased-arrested-for-not-wearing-mask-at-football-game/

401   *Rench et al v. Moscow et al*, Case 3:21-cv-00138-MCE, Document 40, Filed 02/01/23, available online:
https://moscowidaho.news/wp-content/uploads/2023/02/40-MemorandumAndOrder.pdf

throughout the country protesting mandatory masks at school board and city council meetings were denied entry or not allowed to speak *against* masks without first putting *on* masks.[402] In El Paso, Texas, *months* after Governor Abbott banned school mask mandates, 17-year-old Skyler Brown was arrested for "disrupting" an El Paso Independent School Board meeting by refusing to put on a face mask.[403] Even as late as October of 2022, when Gregory Hahn of North Carolina appeared in person to serve on jury duty, Judge Charles Gilchrist sentenced him to 24 hours in jail for contempt of court because Mr. Hahn refused to put on a mask.[404] "What the irony of this whole thing is, is that the judge was talking to me without a mask on," Hahn said in an interview. Even sitting congressmen, including Thomas Massie, Marjorie Taylor Greene, Ralph Norman, and Andrew Clyde, were fined for not wearing masks.[405]

Bad as mask hysteria was in the United States, other countries had it worse. In Mexico, a 30-year-old bricklayer named Giovanni López was arrested for not wearing a mask and died from police-inflicted blunt trauma to the head.[406] China's ruthless lockdown measures, including masks, could fill a book on their own, and India saw multiple incidents where brutal police beatings were precipitated either by not wearing a mask or by wearing a mask "improperly."[407] When German Judge Christian Dettmar

---

Caleb Parke, "Idaho man arrested for not wearing mask at outdoor worship service: 'Unbelieveable,'" *Fox News*, September 25, 2020, online:

https://www.foxnews.com/us/coronavirus-idaho-arrest-mask-outdoor-worship-service-laura-ingraham-gabriel-rench

402   Too many to list, and I have direct personal experience of this when I provided scientific information during the Las Cruces City Council's public comments (on August 9, 2021, as well as other times), and the Las Cruces Public School board (August 17, 2021).

403   Jim Parker and JC Navarrete, "Teen arrested for not wearing mask at EPISD board meeting won't face charges," *KVIA.com*, October 5, 2021, online: https://kvia.com/news/education/2021/10/04/teen-arrested-for-not-wearing-mask-at-episd-board-meeting-wont-face-charges/

404   Jenny Goldsberry, "North Carolina judge throws man in jail for 24 hours for not wearing mask in court," *Washington Examiner*, October 14, 2022, online:

https://www.washingtonexaminer.com/news/judge-throws-man-in-jail-not-wearing-mask

Emma Camp, "A North Carolina Man Was Jailed for Refusing To Wear a Mask in Court," *Reason*.com, October 18, 2022, online at: https://reason.com/2022/10/18/a-north-carolina-man-was-jailed-for-refusing-to-wear-a-mask-in-court/

405   *Massie v. Pelosi*, 590 F. Supp. 3d 196 (D.C. 2022). Available online:

https://scholar.google.com/scholar_case?case=735226917287193397&hl=en&as_sdt=4000003&scfhb=1

406   David Agren, "Death of man after face mask arrest shines light on Mexican police brutality," *The Guardian*, June 4, 2020, available online:

https://www.theguardian.com/world/2020/jun/05/mexican-arrested-for-not-wearing-face-mask-later-found-dead

407   Just four examples:

On June 5, 2020, Rajasthan policemen accosted Mukesh Kumar for not wearing a mask outdoors, leading to a heated argument and ending with viral video of Kumar being thrashed by 4-5 policemen, who at one point were kneeling on his neck. (Zee Media Bureau, edited by Ritesh K Srivastava, "Policemen thrash man in Rajasthan's Jodhpur for not wearing mask, video goes viral," *ZeeNews*, June 5, 2020, available online: https://zeenews.india.com/rajasthan/policemen-thrash-man-in-rajasthans-jodhpur-for-not-wearing-mask-video-goes-viral-2288202.html) courageously ruled that continuing to mask schoolchildren "would violate numerous rights of the children and their parents under the law, the constitution and international conventions,"[408] his office, private premises, and car were shortly afterwards searched, his cell phone was confiscated, and he had to retain an attorney to defend himself.[409] In August 2023, he received a suspended sentence of two years' imprisonment on criminal charges of incorrectly applying the law because of his ruling.[410]

On October 1, 2020, the Hindustan Times Correspondent in New Delhi reported that a Delhi Police head constable filed a complaint after being beaten by civil defense volunteers after an argument over him not wearing a mask turned violent. (HTCorrespondent, New Delhi, "Head constable thrashed in argument over face mask in Delhi," *Hindustan Times Delhi News*, October 1, 2020, https://www.hindustantimes.com/delhi-news/head-constable-thrashed-in-argument-over-face-mask-in-delhi/story-7kXm3rDSyRnREu1g9OnO3M.html)

On April 6, 2021, two constables from the Madhya Pradesh Indore police assaulted autorickshaw driver Krishna Kunjir for wearing his mask improperly. (Express News Service, "Indore: Police 'assault' autorickshaw driver for not wearing mask properly; cops say provoked," *TheIndian*Express, April 7, 2021, https://indianexpress.com/article/india/indore-police-autorickshaw-driver-assault-video-7262608/

On September 1, 2021, NDTV.com reported that villagers had to intervene to stop Jharkhand policemen from brutally beating Pawan Kumar Yadav, an Indian Army Jawan, for not wearing a mask. Several of the police on video inflicting the beating were also not wearing masks. (Manish Kumar, "Video: Army jawan Thrashed Mercilessly By Jharkhand Cops Over Mask," *NDTV*.com, September 1, 2021, https://www.ndtv.com/india-news/video-army-jawan-thrashed-mercilessly-by-jharkhand-cops-over-mask-2526723

408   [Translated From German] Weimar Local Court, Order dated 08.04.2021 (8 April 2021), Ref.: 9 F 148/21, https://2020news.de/wp-content/uploads/2021/05/ENGLISH-TRANSLATION-COMPLETE-DOCUMENT-Amtsgericht_Weimar_9_F_148_21_EAO_Beschluss_anonym_2021_04_08-en.pdf

409   "Search and Seizure at Home of Judge Who Rendered the Sensational Weimar Mask-Judgment," *2020 News*, April 26, 2021 https://2020news.de/en/house-warrant-executed-on-weimar-judge-for-political-reasons/

410   Yudi Sherman, "German judge who ruled against mandates sentenced to two years probation," *Frontline News*, August 24, 2023, online: https://frontline.news/post/german-judge-who-ruled-against-mandates-sentenced-to-two-years-probation

Jacobson v. Massachusetts

Humans are rational beings, but human rationality is often the most vigorously employed to justify decisions made on non-rational grounds. Long before the dawn of the American Republic, it was proverbial that you cannot reason people out of a position they did not reason themselves into.[411] This tendency is every bit as characteristic of people in highly educated professional and intellectual occupations as it is of people who make up the general population. As Thomas Sowell insightfully pointed out:

> "Although the talents and education of intellectuals would seem to enable them
> to be proficient at engaging in logically structured arguments, using empirical
> evidence to analyze contending ideas, many of their political or ideological views

---

411   For example, Federalist U.S. Congressman Fisher Ames said: "men are not to be reasoned out of an opinion that they have not reasoned themselves into." This phrasing is most likely an improved rephrase of Jonathan Swift's statement in a 1721 pamphlet that, "Reasoning will never make a Man correct an ill Opinion, which by Reasoning he never acquired."

Source of Jonathan Swift Quote: Jonathan Swift, A Person of Quality [pseudonym], "Letter to a Young Gentleman, Lately enter'd into Holy Orders", Second Edition, (Letter Dated January 9, 1720), Quote Page 27, Printed for J. Roberts at the Oxford Arms in Warwick Lane, London, 1721. Available online (Google Books Full View), pdf pg. 27: https://repository.monash.edu/items/show/65167#?c=0&m=0&s=0&cv=0

Source of Fisher Ames Quote: Lucius Junius Brutus [pseudonym], *Works of Fisher Ames*, Independent Chronicle, at Boston, October 12, 1786, p. 5; reprinted in *Works of Fisher Ames Compiled by a Number of His Friends*, Frances Ames (editor), Printed and Published by T. B. Wait & Company, Boston, Massachusetts, 1809. Available online (Google Books Full View), pdf pg. 56: https://books.googleusercontent.com/books/content?req=AKW5QacKR1CLqLhgabLN-TimBPK7McjTl-bSQtqsQDIZHYc6JWl8M1Ov7xe7ga4LHp-B6607BeP9jB2DmEPJ-sh3WScpKb3b9CR9pIRKx-2S38aGtoXDC4FsQ1sQSwlhIhDSmJaE0YsFdvNi0jWzw-c5TMS7RncRBjhx_z5lciwLHA7-jSjUHSEAf8fptEFYuD-hXbAeEZIJkaLevCYrAn6vdf5nJNVNEidGKFcLcdHLSWWI2PGkAim9EBL2xHeoe0eoIe_LjorG0T1

QuoteInvestigator.com (with special thanks to Daniel Gilligan and Stephen Goranson) provided a more detailed analysis of the origins of this aphorism (last accessed February 21, 2023): https://quoteinvestigator.com/2015/07/10/reason-out/#f+11618+1+4

are promoted by verbal virtuosity in *evading* both structured arguments and empirical evidence." [412]

Judges are people too. They possess every foible, fault, and potential virtue that comes from being human. Judges are not unbiased because *humans* are not unbiased. Judges are no less prone to fears, confirmation bias, or motivated reasoning than anyone else. They are no less vulnerable to propaganda, psychological manipulation, or social pressures than are doctors, dentists, scientists, politicians, and the general public. A judge's context and experience form the lens through which they read and interpret the law. The multi-layered judicial system, with its appellate courts, *assumes* that judges make mistakes. Just as doctors were subjected to professional pressures during the COVIDcrisis, so were judges. Additionally, the career path of becoming a judge tends to favor personality types that like rules and defer to authority figures more than average, and the lengthy timeline for progression in this occupation ensured that, compared to the general population, a higher proportion of judges fell into the age brackets most vulnerable to COVID. The COVIDcrisis threw all of this into stark relief, as courts across the United States employed great ingenuity and verbal virtuosity to undermine or dismiss legal challenges to lockdown measures — challenges which many judges found inherently repugnant for much deeper reasons.

For much of 2020 (and even into 2021 and 2022) the 1905 case of *Jacobson v. Massachusetts* proved to be one of the most convenient, powerful, and widely recycled additives making up the legal formulas used to scour away challenges to COVID mandates. *Jacobson v. Massachusetts* lurks behind almost every ruling that upheld compulsory masking. If courts did not cite it directly, they cited other cases which relied on it. Illinois Federal District Judge Sara Ellis provided a good summary of the standard COVIDcrisis interpretation of *Jacobson*:

> "Over a century ago in Jacobson, the Supreme Court developed a framework
> by which to evaluate a State's exercise of its emergency authority during a public
> health crisis… Jacobson explained that '[u]pon the principle of self-defense,
> of paramount necessity, a community has the right to protect itself against an
> epidemic of disease which threatens the safety of its members.'" [413]

---

412   Thomas Sowell, *Intellectuals and Society*, New York: Basic Books, 2011, Kindle Edition, p. 96. My own self-application of this observation prevented me from starting this book until December 2022, at which point, after observing so many even *more* foolish and *less*-informed opinions on these issues being given highly influential spotlights, I figured that at the very least my own good faith attempt couldn't do any worse after involving so much homework.

413   *Illinois Republican Party v. Pritzker*, 470 F. Supp. 3d 813, 820 (page 7), (Northern District of Illinois, 2020), available online: https://libertyjusticecenter.org/wp-content/uploads/016-IL-Republican-Party-Opinion-denying-Motion-for-TRO-and-PI.pdf and https://casetext.com/case/ill-republican-party-v-pritzker

"*Jacobson v. Commonwealth of Massachusetts* provides the applicable standard in cases challenging public health orders," ruled Judge Robert C. Chambers in West Virginia.[414] "Since the challenged orders are public health measures to address a disease outbreak, *Jacobson* provides the proper scope of review," said United States District Judge Catherine C. Blake in Maryland.[415] Judge Martin Feldman in Louisiana said: "The police power outlined in *Jacobson*... precludes this Court from 'second-guess[ing] the wisdom or efficacy' of measures taken by state officials in response to the COVID-19 pandemic."[416] In reviewing the role played by *Jacobson v. Massachusetts* during COVIDcrisis court battles, law professor Josh Blackman observed: "*Jacobson* or no *Jacobson*, courts still would have deferred to unprecedented lockdown measures. Yet, *Jacobson* proved to be a useful prop[.]"[417]

The precipitating issue in *Jacobson v. Massachusetts* was compulsory vaccination, which Swedish pastor Henning Jacobson challenged under the Fourteenth Amendment, and that is what the case is best known for. In *Jacobson*, the United States Supreme Court ruled that the Fourteenth Amendment did not prevent states from using their police power to enact a compulsory vaccination law with a modest one-time fine for refusing the medical procedure. Yet, over and over again during the COVIDcrisis, courts across the United States relied heavily on this 1905 Supreme Court decision as an indispensable tool for rejecting a far wider range of legal challenges to COVIDcrisis measures like lockdowns and compulsory masking. "[T]he Mask Mandates are emergency measures to which this court should apply the 'highly deferential standard' articulated in *Jacobson v. Commonwealth of Massachusetts*, 197 U.S. 11 (1905), rather than the usual constitutional framework," declared Judge Susan Mollway in Hawaii. During the 9 months from March through December 2020, alone, *Jacobson* was cited in court decisions more than 200 times.[418] Illinois District Judge Staci M. Yandle pointed out in November 2021 that her own reliance on *Jacobson* was hardly unique: "relying on *Jacobson*, courts across the country have consistently declined to enjoin state and local restrictions aimed at

---

414   *Stewart v. Justice* 502 F. Supp. 3d 1057, 1062 (S.D.W. Va. 2020), available online: https://casetext.com/case/stewart-v-justice-1

415   *Antietam Battlefield KOA v. Hogan*, 461 F. Supp. 3d 214, 228 (D. Md. 2020), available online: https://casetext.com/case/antietam-battlefield-koa-v-hogan

416   *4 Aces Enters., LLC v. Edwards*, 479 F. Supp. 3d 311, 323 (E.D. La. 2020), https://casetext.com/case/4-aces-enters-llc-v-edwards (also at https://www.leagle.com/decision/infdco20200818d71).

417   Josh Blackman, "The Irrepressible Myth of *Jacobson v. Massachusetts*," *Buffalo Law Review*, Vol. 70 (January 2022): 131-270, p. 268, available online: https://papers.ssrn.com/sol3/papers.cfm?abstract_id=3906452

418   Josh Blackman, "The Irrepressible Myth of *Jacobson v. Massachusetts*," *Buffalo Law Review*, Vol. 70 (January 2022): 131-270, p. 227, available online: https://papers.ssrn.com/sol3/papers.cfm?abstract_id=3906452

Minnesota Voters All. v. Walz, 492 F. Supp. 3d 822, 837 (D. Minn. 2020), available online: https://case-law.vlex.com/vid/minn-voters-alliance-v-891222192

Denis v. Ige et al, No. 1:2021cv00011 - Document 62 (D. Haw. 2021), pg. 15, available online: https://law.justia.com/cases/federal/district-courts/hawaii/hidce/1:2021cv00011/152658/62/

protecting the public against the spread of COVID-19."[419] One article in *The National Law Review* summarizing the results of legal challenges to compulsory masking up through September 2021 wryly noted: "Often, the only question the courts ponder in these cases is whether to apply the Supreme Court's tiers of scrutiny analysis or the older framework articulated in *Jacobson v. Massachusetts*."[420]

In Part 4 of this book, we saw how the cancerous legacy of *Jacobson v. Massachusetts* metastasized over the course of later rulings. Even if taken at face value on its own terms, however, the original 1905 decision, when taken as a whole, does not support the extreme interpretations put forward to justify its widespread COVIDcrisis application. The *Jacobson* case is usually cited as an automatic checkmate that an individual's rights over their body are not absolute, but it actually makes an even more effective argument that governmental power to override individual liberty does not become absolute *even during a public health emergency*. Setting aside its own closing statement that "We now decide only that the statute covers the present case,"[421] which limited its future broad usage, the *Jacobson* decision actually affirmed *multiple* limitations on the scope of state power which future courts had to construe as narrowly as possible in order to wield *Jacobson* the way that they did.

In *Jacobson*, the majority opinion delivered by Justice Harlan explicitly affirmed that state police powers are not unlimited, citing previous Supreme Court decisions where the court had invalidated sanitary laws which had "invaded the domain of Federal authority and violated rights secured by the Constitution."[422] The court also affirmed that the Massachusetts vaccination law would not have been valid if it was primarily for some purpose other than public health (i.e. pretextual), and that a reading

---

419   *H's Bar, LLC v. Berg*, No. 20-CV-1134-SMY, 2020 WL 6827964, at *4 (Southern District of Illinois Nov. 21, 2020), available online: https://www.ilsd.uscourts.gov/opinions/ilsd_live.3.20.cv.1134.4679379.0.pdf

420   Brandon O. Mouland & Laura Lashley, "Masks Up, Pens Down: (Still) Litigating Mask Mandates in 2021," *The National Law Review*, September 24, 2021, Online:
https://www.natlawreview.com/print/article/masks-pens-down-still-litigating-mask_mandates-2021

421   *Jacobson v. Massachusetts*, 197 U.S. 11, 31 (1905), p. 39, available online:
https://supreme.justia.com/cases/federal/us/197/11/

422   Specifically (p. 28), Justice Harlan cited *Railroad Company v. Husen*, 95 U.S. 465, 471-473 (1877), available online: https://supreme.justia.com/cases/federal/us/95/465/). In his excellent analysis of *Jacobson v. Massachusetts*, Professor Blackman argues that the "rights secured by the Constitution" mentioned by Justice Harlan in referencing the *Railroad* decision "did not concern the violation of individual rights." However, I believe Professor Blackman's interpretation is mistaken on this point. In *Railroad*, the Missouri law which the Supreme Court struck down *also* unacceptably burdened *individual* liberties, even though the Supreme Court chose to invalidate the law using a different, more accessible, line of attack.

Professor Blackman specifically highlights elsewhere in his analysis (p. 138) that during this period of the Supreme Court's history, "the Court treated economic property rights in the same fashion as it treated personal liberty." This was because one's "personality" — the combination of one's physical body, personal liberties, legal rights, personal privileges, intellectual achievements, reputation, etc, - was, itself, considered a form of property, as in Justice Warren and Justice Brandeis' joint essay on "The Right of Privacy": "The right of property in its widest sense, including all possession, including all rights and privileges, and hence embracing the right to an inviolate personality."

of the Massachusetts law which excluded reasonable medical exemptions could justify intervention to protect the individual. Moreover, while the police power of the State of Massachusetts might be able to extend to imposing a one-time modest *fine* on an individual for refusing vaccination, it did *not* extend to vaccinating that individual by force. Perpetual quarantine or exclusion from civil society seems not to have crossed anyone's mind.

When *Jacobson v. Massachusetts* was decided in 1905, the First Amendment had not yet been "incorporated" to apply against the states via the Fourteenth Amendment. In fact, the Supreme court had yet to strike down *any* law based on the incorporation doctrine even by the time of *Buck v. Bell*

The Supreme Court decision in *Railroad* specifically mentions that the "right of steamboat owners and railroad companies to transport such property through the state is loaded by the law with onerous liabilities because of their agency in the transportation" (p. 95 U.S. 471). The effect of the Missouri law struck down in *Railroad* was to "discriminate between the property of citizens of one state and that of citizens of other states."

> "Regarding the statutes as mere police regulations, intended to protect domestic cattle against infectious disease, those courts have refused to inquire whether the prohibition did not extend beyond the danger to be apprehended, and whether, therefore, the statutes were not something more than exertions of police power. That inquiry, they have said, was for the legislature, and not for the courts. With this we cannot concur. The police power of a state cannot obstruct foreign commerce or interstate commerce beyond the necessity for its exercise, and under color of it objects not within its scope cannot be secured at the expense of the protection afforded by the federal Constitution."

In deciding *Railroad v. Husen*, the Supreme Court cited to two earlier cases (*Henderson v. Mayor of City of New York*, 92 U.S. 259 (1875) and *Chy Lung v. Freeman*, 92 U.S. 275 (1875)), where the "right of a state to protect herself" which can "only arise from vital necessity" had been "carried beyond the scope of that necessity," resulting in burdens on the individual liberties of so many excess persons that the laws in question were held to be invalid because they were overextended (i.e. they were not narrowly tailored).

Applying the terms of *Railroad v. Husen* to masks: modern universal mask mandates intended to protect against the small class of "persons afflicted by contagious or infectious diseases" should be held unconstitutional because they are so "far-reaching" that they affect citizens "not of any class which the state could lawfully exclude" (i.e. every healthy person in the state).

It is also worth noting that, going back even farther, the language of *Chy Lung v. Freeman*, the case cited to support *Railroad v. Husen*, makes clear that the violations of individual rights and the violations of the commerce clause could be inextricably linked, and that the presumption of validity does not necessarily lie with the state:

> "If the right of the states to pass statutes to protect themselves in regard to the criminal, the pauper, and the diseased foreigner landing within their borders exists at all, it is limited to such laws as are absolutely necessary for that purpose, and this mere police regulation cannot extend so far as to prevent or obstruct other classes of persons from the right to hold personal and commercial intercourse with the people of the United States."

Chy Lung v. Freeman, 92 U.S. 275 (1875)
https://supreme.justia.com/cases/federal/us/92/275/

in 1927.[423] The Supreme Court transcript of the *Jacobson* trial reveals that Massachusetts' mandatory vaccination law made it an outlier among the other states of the Union: "three-quarters of the states of the Union have not entered upon the policy of enforcing vaccination upon their inhabitants by legal penalty. Not one of the states undertakes forcible vaccination of its inhabitants, while the states of Utah and West Virginia expressly provide that no such compulsion shall be used."[424] Moreover, despite courts citing *Jacobson* to dismiss Free Exercise and other First Amendment challenges to COVIDcrisis mandates, the *Jacobson* decision provided no such governing precedent because Henning Jacobson never raised *any* defenses based in the First Amendment.

Finally, and so important that it is worth repeating, the Massachusetts law at issue in *Jacobson* imposed only a modest penalty for refusing vaccination– a one-time fine of just $5 (a little over $150 in 2023 dollars).[425] This fact was part of the reason that Jacobson's conviction under the compulsory vaccination law was upheld. As law professor Josh Blackman writes:

> "The five-dollar penalty operated in a similar fashion as nominal damages. The amount was large enough to prosecute in court, but not large enough to actually punish offenders. And a one-time fee of five dollars was unlikely to deter opponents of vaccination. Because there was no possibility for jail time, an offender could simply pay the penalty and move on with his life."[426]

The Massachusetts vaccination law Henning Jacobson was charged under was thus, in effect, a low-cost form of public health commutation fee — analogous to paying a fee to avoid a military draft. Damningly for anyone citing *Jacobson* to uphold draconian public health measures and societal exclusion for the unmasked and unvaccinated, an unvaccinated person who paid the Massachusetts

---

423   William Edward Leuchtenburg, *The Supreme Court reborn : The Constitutional Revolution in the Age of Roosevelt*, New York: Oxford University Press (1995), p. 23, Available online: https://archive.org/details/supremecourtrebo0000leuc/page/14/mode/2up?q=religion&view=theater

424   *Henning Jacobson*, Plaintiff in Error, vs. *The Commonwealth of Massachusetts*, 197 U.S. 11 (1905) (No. 70-175) Transcript of Record at page 7 (pdf page 33), https://perma.cc/EUK9-TX4V

425   The U.S. Bureau of Labor Statistics provides an inflation calculator that goes back to 1913 and provides a helpful back-of-the-envelope benchmark (https://www.bls.gov/data/inflation_calculator.htm). The number obtained by using this calculator is also consistent with Professor Josh Blackman's research. Josh Blackman, "The Irrepressible Myth of Jacobson v. Massachusetts," *Buffalo Law Review*, Vol. 70 (January 2022): 131-270, p. 181, available online: https://papers.ssrn.com/sol3/papers.cfm?abstract_id=3906452

426   Josh Blackman, "The Irrepressible Myth of *Jacobson v. Massachusetts*," *Buffalo Law Review*, Vol. 70 (January 2022): 131-270, p. 185, available online: https://papers.ssrn.com/sol3/papers.cfm?abstract_id=3906452

fine was then fully "in compliance" with the law *and* remained free to spread smallpox.[427] Unvaccinated Pastor Henning Jacobson remained completely free to engage in large social and political gatherings during the years his case worked its way up through the courts, and did so across multiple states, including when he served as part of a delegation to the St. Louis World's Fair of 1904.[428]

Since COVID-19 never came close to approaching the same league as Smallpox even according to the most apocalyptic divinations, a truly consistent application of the *Jacobson* standard in the case of COVID would have resulted in even *smaller* one-time opt-out fees for both masks and vaccines and *no* societal restrictions. Instead, the unmasked and unvaccinated faced vitriolic ostracization and demonization by pundits,[429] politicians, and public health officials, including but not limited

---

427 Henning Jacobson actually highlighted this internal contradiction in his submissions to the Supreme Court: "The utmost that the law undertakes to do is to provide a penalty for its violation. After I have paid the penalty I am as much a menace to the community as I was before, and if the effect of the law be that with one payment I may continue to be a menace to the public health, as was held in England (Pilcher v. Stafford, 4 Vest & S. 775), the law is too absurd to justify its existence. If I may be repeatedly fined for recalcitrancy… then the argument made by the Supreme Court of Massachusetts as to the insignificance of the fine of $5 is fallacious and unjust. The fine of $5 may be increased to $100 by the costs of prosecution, and if, as has occurred in Massachusetts, repeated prosecutions have been brought against the same individual, the accumulation of fines and taxes may become an insupportable burden and result in the imprisonment of the offender."
Henning Jacobson, Plaintiff in Error, vs. *The Commonwealth of Massachusetts*, 197 U.S. 11 (1905) (No. 70-175) Transcript of Record at page 17 (pdf page 43), https://perma.cc/EUK9-TX4V

428 Josh Blackman, "The Irrepressible Myth of *Jacobson v. Massachusetts*," *Buffalo Law Review*, Vol. 70 (January 2022): 131-270, p. 164, available online: https://papers.ssrn.com/sol3/papers.cfm?abstract_id=3906452

429 One of the more well-written and less-profane screeds is quoted at length below:
"Ziggy" [pseudonym], "A Message For People Who Refuse To Wear A Face Covering," June 30, 2020, *travelingformiles.com*, online: https://travelingformiles.com/a-message-for-people-who-refuse-to-wear-a-face-covering/
"Your attempts to use the freedom and liberty that others died to preserve to justify your own selfishness is insulting to their memories, and the fact that you're so willing to trample upon those memories marks you out as a truly hateful human being and one that's deserving of nothing but society's contempt.

"Refusing to wear a mask or a face covering isn't a badge of honor, it's a badge that marks you out as a thoughtless individual with very little understanding of what the words liberty and freedom really mean, what the real costs and sacrifices the protection of those ideals have been, or what the responsibilities of living in a free society really are.

"I'm sure you've convinced yourself that your ridiculous refusal to wear a face covering makes you a "patriot" and "defender of civil liberties", but you're really just an idiot parroting words you don't understand and who couldn't be further from embodying the ideals of freedom and liberty.

to job loss,[430] fines of hundreds or thousands of dollars, and (in a few cases) deprivation of further fundamental rights through jail time. Mask and vaccine mandates during the COVIDcrisis went far, *far* beyond anything resembling a proportionate extension of the *Jacobson* decision.[431]

In his dissent from the majority that was upholding Japanese-Americans' World War II internment without trial or conviction in *Korematsu v. United States*, Justice Robert Jackson warned that when the Court validates a bad principle, "The principle then lies about like a loaded weapon ready for the hand of any authority that can bring forward a plausible claim of an urgent need."[432] In his own analysis, Professor Blackman applies this quote to the creeping, over-extensive "myth of *Jacobson v. Massachusetts*." During the COVIDcrisis, politicians, pundits, public health officials and judges picked up *Jacobson* and applied it to everything they could get away with in a constitutional Reign of Terror. In so doing, they often misused it not only on *unenumerated* rights, but on *every other right* explicitly recognized and protected by the Constitution as well. "Traditional doctrine does not control during a pandemic; *Jacobson* does," United States District Judge Martin Feldman bluntly ruled in Louisiana.[433] New Mexico District Judge James Browning cited *Jacobson* to restrict churchgoers' enumerated First

---

"Allow me to sum things up this way: In the last hundred years, members of your family or, at the very least members of your friends' families, have almost certainly given their lives so that others may live theirs. Now here you are refusing to do something as simple and painless as covering your nose and mouth so the same may happen again. Aren't you embarrassed to be the person you are?"

430   Paige Smith and Robert Iafolla, "Refusing to Wear a Mask at Work Could Get You Fired," May 20,2020, *Bloomberg Law*, online at: https://news.bloomberglaw.com/daily-labor-report/refusing-to-wear-a-mask-at-work-could-get-you-fired

431   As a side note, I agree with Henning Jacobson and his supporters that even this degree of infringement on individual bodily autonomy is unacceptable. On a practical level, though, my experiences during the COVIDcrisis have taught me that I more than likely would have simply paid the $5 and got on with my life. However, my COVID experiences have also taught me that doing so would have been a bad call in the long run, and a particular dereliction of duty in my own case.

As a further side note, 19th century state courts generally upheld school vaccination laws on the rationale that school attendance was a privilege, rather than a right, and that states could therefore set conditions for the receipt of this privilege (*see* Josh Blackman, p. 153). The 20th and 21st century arrangement of turning school into a compulsory "right" while retaining conditions like mandatory vaccination is an administrative catch-22 designed to have the cake and eat it too. Indeed, the first overextension of Jacobson came not with *Buck v. Bell* in 1927, but five years earlier, in 1922, when Jacobson was cited to uphold an even more draconian vaccination law that San Antonio public health officials used to bar an unvaccinated girl from attending not just public, but also *private* school (*Zucht v. King*, 260 U.S. 174 (1922), Available online: https://supreme.justia.com/cases/federal/us/260/174/).

432   *Korematsu v. United States*, 323 U.S. 214 (1944), (Justice Jackson, dissenting), p. 323, Available online: https://supreme.justia.com/cases/federal/us/323/214/

433   To his credit, the Judge Feldman's decision makes clear that the government's lockdown measures survived challenge solely because of the recent Chief Justice Roberts shadow docket concurrence and the resulting 5th Circuit opinion: "The bar owners cannot overcome the deference due state officials during this pandemic. It is Jacobson and Abbott's high bar that requires this result. Make no mistake: The Governor's victory does not mean his proclamations are sound policy; nor does it mean the proclamations are sufficiently solicitous of the interests of Louisiana small-business

Amendment rights of assembly *and* free exercise of religion.[434] Judge Sara Ellis did the same in Illinois, declaring that during a crisis where *Jacobson v. Massachusetts* is implicated, constitutional claims have "a less than negligible chance of prevailing."[435] Even within the 5th Circuit, which in many ways was a bastion for embattled federal employees and military service members fighting unlawful vaccine mandates, two judges (Stuart Kyle Duncan and Jennifer Walker Elrod) insisted in April 2020 that *Jacobson* created a pandemic exception to the Bill of Rights: "*Jacobson* instructs that *all constitutional rights* may be reasonably restricted to combat a public health emergency… *Jacobson* governs a state's emergency restriction of *any individual right*" (emphasis added).[436]

Fortunately for Americans, other judges rightly and vocally disagreed, and even more privately questioned the scientific, moral, or legal legitimacy of various COVIDcrisis mandates. Judge Daniel Collins of the 9th Circuit wrote:

> "Nothing in Jacobson supports the view that an emergency displaces normal constitutional standards… Jacobson says nothing about what standards would apply to a claim that an emergency measure violates some other, enumerated constitutional right; on the contrary, Jacobson explicitly states that other constitutional limitations may continue to constrain government conduct."[437]

---

owners, like the plaintiffs here; it means quite simply that the proclamations are constitutional." Even so, this ruling was one of many pandemic black pills to swallow.

4 Aces Enters., LLC v. Edwards, 479 F. Supp. 3d 311, 323 (E.D. La. 2020), https://casetext.com/case/4-aces-enters-llc-v-edwards (also at https://www.leagle.com/decision/infdco20200818d71).

434   *Legacy Church, Inc. v. Kunkel*, 472 F. Supp. 3d 926,(D.N.M. 2020), Available online:
https://casetext.com/case/legacy-church-inc-v-kunkel-1
https://www.justsecurity.org/wp-content/uploads/2020/04/LegacyChurch.Browning.opinion.pdf

435   It is worth noting that Judge Ellis' ruling was also premised on the patently false assumption that: "There is no cure, vaccine, or effective treatment for COVID-19." Even in 2020, multiple effective treatment existed, and recovery (i.e. "cure") was the overwhelming norm. Additionally, the only "scientific" sources that Judge Ellis cited to were from the CDC.

Illinois Republican Party v. Pritzker, 470 F. Supp. 3d 813, 818 (pages 2 & 8), (Northern District of Illinois, 2020), available online: https://libertyjusticecenter.org/wp-content/uploads/2020/06/016-IL-Republican-Party-Opinion-denying-Motion-for-TRO-and-PI.pdf and https://casetext.com/case/ill-republican-party-v-pritzker

436   *In re Abbott*, 954 F.3d at 786 and 778 n 1, available online: https://casetext.com/case/in-re-abbott-1994.

437   To his immense credit, Judge Collins also wrote regarding California Governor Gavin Newsom's anti-COVID measures: "Even the most ardent proponent of a broad reading of Jacobson must pause at the astonishing breadth of this assertion of government power over the citizenry, which in terms of its scope, intrusiveness, and duration is without parallel in our constitutional tradition."

South Bay United Pentecostal Church v. Newsom, 959 F.3d 938, 942 (9th Cir. 2020) (Collins, J., dissenting), No. 20-555333, May 22, 2020, Available online: https://casetext.com/case/s-bay-united-pentecostal-church-v-newsom

Judge Lance Walker, for the District of Maine, quoted and affirmed a dissent by Supreme Court Justice Alito: "[I]t is a mistake to take language in *Jacobson* as the last word on what the Constitution allows public officials to do during the COVID–19 pandemic. Language in *Jacobson* must be read in context[.]"[438] In the same dissent, Justice Alito rightly pointed out that, "a public health emergency does not give Governors and other public officials *carte blanche* to disregard the Constitution for as long as the medical problem persists."[439] In September 2020, while coronaphobia[440] was still very much at its peak, Judge William S. Stickman IV issued a lionhearted ruling against Pennsylvania Governor Thomas Wolf's lockdown measures despite *Jacobson*. Judge Stickman ruled:

> "[A]n extraordinarily deferential standard based on Jacobson is not appropriate…
> The Constitution cannot accept the concept of a 'new normal' where the basic
> liberties of the people can be subordinated to open-ended emergency mitigation
> measures. Rather, the Constitution sets certain lines that may not be crossed,
> even in an emergency."[441]

In November 2020, a majority of the Supreme Court indicated that they, too, disagreed with the overextended reading of *Jacobson v. Massachusetts*. The Supreme Court's holding that month in *Roman Catholic Diocese of Brooklyn v. Cuomo* was a critical turning point for embattled unalienable rights everywhere.[442] In that decision, the Court used traditional First Amendment jurisprudence to grant relief against the severe and discriminatory restrictions New York Governor Cuomo placed on houses of worship in the name of public health. In addition to the majority's now-famous pronouncement

438   *Savage v. Mills*, 478 F. Supp. 3d 16, 26 (D. Me. 2020), available online: https://casetext.com/case/rick-savage-two-bros-llc-v-mills quoting *Calvary Chapel Dayton Valley v. Sisolak*, 140 S. Ct. at 2608 (2020) (Alito, J., dissenting), available online: https://casetext.com/case/calvary-chapel-dayton-valley-v-sisolak.

439   *Calvary Chapel Dayton Valley v. Sisolak*, 140 S. Ct. at 2608 (2020) (Alito, J., dissenting), available online: https://casetext.com/case/calvary-chapel-dayton-valley-v-sisolak.

440   Though it would not surprise me if multiple people independently coined this term, the earliest use of which I am aware at the time of this writing, and to whom I ascribe credit for my own use, was Jeffrey A. Tucker ("The Downfall of the Gurus,", *The Epoch Times*, January 23, 2023, Online: https://www.theepochtimes.com/the-downfall-of-the-gurus_5004053.html).

441   *County of Butler v. Wolf*, 486 F. Supp. 3d 883 (W.D. Pa. 2020), pp. 17 & 66; available online: https://cases.justia.com/federal/district-courts/pennsylvania/pawdce/2:2020cv00677/266888/79/0.pdf?ts=1600184121

        Depressingly, Judge Stickman's decision was stayed by the higher court and later vacated as moot. However, importantly and encouragingly, the final vacatur was due to mootness because the Pennsylvania Assembly restricted Governor Wolf's powers, rather than any flaws in Judge Stickman's constitutional analysis. (County of Butler v. Governor of Pennsylvania, No. 20-2936 (3d Cir. 2021), available online: https://law.justia.com/cases/federal/appellate-courts/ca3/20-2936/20-2936-2021-08-10.html

442   *Roman Catholic Diocese of Brooklyn v. Cuomo*, 141 S. Ct. 63, 208 L. Ed. 2d 206, 592 U.S. (2020), available online: https://www.supremecourt.gov/opinions/20pdf/20a87_4g15.pdf.

that "even in a pandemic, the Constitution cannot be put away and forgotten,"[443] Justice Neil Gorsuch wrote a concurrence which directly addressed the widespread erroneous readings of *Jacobson v. Massachusetts*:

> "Jacobson hardly supports cutting the Constitution loose during a pandemic. That decision involved an entirely different mode of analysis, an entirely different right, and an entirely different kind of restriction...

> "Put differently, Jacobson didn't seek to depart from normal legal rules during a pandemic, and it supplies no precedent for doing so...

> "Even if judges may impose emergency restrictions on rights that some of them have found hiding in the Constitution's penumbras, it does not follow that the same fate should befall the textually explicit right to religious exercise...

> "In fact, Jacobson explained that the challenged law survived only because it did not 'contravene the Constitution of the United States' or 'infringe any right granted or secured by that instrument.'"[444]

Like a painfully slow turning of the tide, challenges to COVIDcrisis overreach after November 2020 began to make gradual headway as lower courts followed the Supreme Court's lead in *Roman Catholic Diocese*. Even so, progress was incremental, and multiple courts stubbornly clung to *Jacobson* well into 2021 and 2022. District Judge William Young countered: "... a chorus of scholars, Justices, and courts argue that *Jacobson*'s standard is improper, particularly when applied to First Amendment challenges... The Supreme Court, however, has not yet ruled on whether the tiers of scrutiny overrule *Jacobson* despite recent opportunity to do so."[445] In March 2021, five months after the Supreme Court's decision, Judge Cynthia Bashant dismissed a substantive due process claim against masks by saying, "Because *Jacobson* remains good law, the Court will apply it to Plaintiff's substantive due process claim."[446] "Although *Jacobson* is more than a century old, recent case law shows that it is still

---

443   *Roman Catholic Diocese of Brooklyn v. Cuomo*, 141 S. Ct. 63, 208 L. Ed. 2d 206, 592 U.S. (2020), p. 5, available online: https://www.supremecourt.gov/opinions/20pdf/20a87_4g15.pdf

444   *Roman Catholic Diocese of Brooklyn v. Cuomo*, 141 S. Ct. 63, 208 L. Ed. 2d 206, 592 U.S. (2020), (Gorsuch, N., concurring, pp. 3-5), available online: https://www.supremecourt.gov/opinions/20pdf/20a87_4g15.pdf

445   *Delaney v. Baker*, 511 F. Supp. 3d 55, *72 (D. Mass. 2021), available online: https://casetext.com/case/delaney-v-baker-2

446   *Forbes v. County of San Diego*, Case No. 20-cv-00998-BAS-JLB, at *7 (S.D. Cal. Mar. 4, 2021), available online: https://casetext.com/case/forbes-v-cnty-of-san-diego

good law," insisted Judge Robert Chambers, three months after the Supreme Court's ruling.[447] After being reminded of Justice Gorsuch's concurrence and the U.S. Supreme Court's admonition that: "we may not shelter in place when the Constitution is under attack[,]" Judge Chambers proceeded to do that very thing, upholding a mask mandate with the snippy retort that: "This Court has not sheltered in place; it simply finds no invasion of fundamental constitutional rights here, even under Plaintiff's preferred standard."[448] Likewise, Ninth Circuit Judge Jennifer Dorsey cited *Jacobson* to uphold a mask mandate in late December 2021.[449]

The mentality behind the erroneous *Jacobson* standard continues to subtlety infect court decisions.[450] Many courts that did not quote *Jacobson* directly still used it *indirectly* by citing to court rulings which relied on it.[451] Even if *Jacobson* had been good law, the Supreme Court still specifically referenced *Jacobson* prior to the COVIDcrisis, saying that "a State's interest in the protection of life falls short of justifying any plenary override of individual liberty claims."[452] Regardless of how many courts refused to recognize it, the fact remains that even under the *Jacobson* standard, compulsory masking was and *is* "a plain, palpable invasion" of rights secured by fundamental law. The first major legal victory against compulsory masking would have to wait until 2021, and it would come in state, rather than federal, court. In the following sections, we will review constitutional and other challenges brought

---

447   *Stewart et al. v. Justice et al.* No. 3:2020cv00611 - Document 45 (S.D.W. Va. 2021), p. 4 & 5, available online: https://law.justia.com/cases/federal/district-courts/west-virginia/wvsdce/3:2020cv00611/230251/45/

448   *Stewart et al. v. Justice et al.* No. 3:2020cv00611 - Document 45 (S.D.W. Va. 2021), p. 6, available online: https://law.justia.com/cases/federal/district-courts/west-virginia/wvsdce/3:2020cv00611/230251/45/

449   *Branch-Noto v. Sisolak*, 576 F. Supp. 3d 790, 799-800 (D. Nev. 2021), available online: https://casetext.com/case/branch-noto-v-sisolak-1;
2:21-cv-01507 JAD-DJA, p. 10-11 (D. Nev. Dec. 22, 2021) https://ag.nv.gov/uploadedFiles/agnvgov/Content/News/PR/PR_Docs/2021/DOC%2038%20Order%20Denying%20Motion%20for%20Preliminary%20Injunction,%20Granting%20in%20Part%20Motions%20to%20Dismiss%20and%20Closing%20Case.pdf

450   For example, in *Stewart et al. v. Justice et al.* (February 9, 2021), Judge Robert Chambers explicitly rejected Judge Gorsuch's analysis of *Jacobson* and applied the *Jacobson* standard anyway (*Stewart et al. v. Justice et al.* No. 3:2020cv00611 - Document 45 (S.D.W. Va. 2021), p. 6, available online: https://law.justia.com/cases/federal/district-courts/west-virginia/wvsdce/3:2020cv00611/230251/45/)
        Professor Blackman cites several such examples on page 267 (pdf. Pg 137) of his analysis, and the erroneous *Jacobson* standard still cropped up as an obstacle in multiple masking cases well into 2021 (for example, *Denis v. Ige et al*, No. 1:2021cv00011 - Document 62 (D. Haw. 2021), available online: https://law.justia.com/cases/federal/district-courts/hawaii/hidce/1:2021cv00011/152658/62/).

451   A good example of such indirect referencing and usage is how, even in August 2022, a Washington State panel of judges in *Sehmel v. Shah* cited to the 2020 *Stewart v. Justice* decision which relied on *Jacobson* (*Sehmel v. Shah*, 514 P.3d 1238 (Wash. Ct. App. 2022), at 16, available online: https://caselaw.findlaw.com/wa-court-of-appeals/2182709.html)

452   *Planned Parenthood of Southeastern Pennsylvania v. Casey*, 505 U.S. 833, 857 (1992), available online: https://tile.loc.gov/storage-services/service/ll/usrep/usrep505/usrep505833/usrep505833.pdf referencing *Jacobson v. Massachusetts*, 197 U.S. 11, 24-30 (1905)

against compulsory masking, and explore the reasons why some legal challenges succeeded where others failed.

Challenging Mask Mandates

It is one thing to know or sense that one's rights are being violated, but it is much harder to explain precisely *which* rights are being violated and *how*. *Articulating* that explanation within a legal system to obtain relief is at a still higher level of difficultly. Compulsory masking is no exception to this rule, and as we have already seen from the psychology and philosophy involved, most legal challenges to compulsory masking started with severe hidden handicaps.

What specific natural rights and legally protected interests does compulsory masking violate? *Whose* rights and interests are they? How are these natural rights and legal interests recognized and protected under local, state, or federal law? What kind of judicial scrutiny do these violations have to survive, and what standard of review do judges apply when adjudicating them? At the time of this writing, in the late summer of 2023, precise answers to these questions are still being hammered out. The direction that process takes, as well as the outcome, are by no means set in stone. In a sense, it will never be completely resolved, because the issues underlying compulsory masking are as old as mankind. This makes it worth the time and effort to look at some of the mask lawfare to-date.

Challenges to mask mandates fall into one of two broad categories: rights-based constitutional claims, and procedure-based or administrative claims. If it is not sneaking in a *Jacobson-v.-Massachusetts*-type standard, under current jurisprudence a court will subject the challenged mask mandate to either rational basis scrutiny, strict scrutiny, or some form of intermediate scrutiny, depending on the type of challenge and what criteria that challenge meets. Where rational basis scrutiny is applied, mask mandates are generally upheld. Where strict scrutiny is applied, mask mandates are generally struck down.

Legal challenges to mask mandates are inevitably met with motions to dismiss. For the purposes of deciding whether or not to dismiss a lawsuit, judges are supposed to assume that well-pleaded allegations of fact in an initial complaint are true, and draw all reasoned inferences in favor of the plaintiff. "Even if it seems 'almost a certainty to the court that the facts alleged cannot be proved to support the legal claim,' the claim may not be dismissed so long as the complaint states a claim."[453] When it came to lawsuits challenging compulsory masking, however, this presumption was effectively flipped to favor the mandates. Before the political tide started turning against masks and the CDC's general malfeasance during the COVIDcrisis became widely known, only a handful of allegations challenging compulsory masking managed to qualify as "well-pleaded" enough to survive a motion to dismiss (pure coincidence, no doubt).

"Rational basis" and "strict scrutiny"

For all the loud insistence that mask mandates were just "following the science," compulsory masking was always driven far more by fear, social pressure, and a pathological obsession for control than by anything else. Mask mandates proved effectively immune to direct challenge by *any* volume of science because the standard of review applied to mask mandates was generally "rational basis" scrutiny. In the words of the Supreme Court, this standard of scrutiny is "the most relaxed and tolerant form of judicial scrutiny[.]"[454] Rational basis scrutiny starts by presuming the authority's actions are justified and legal, and the burden is placed on the challenger to prove otherwise. A plaintiff must prove (or at least convincingly allege), largely in advance, that no conceivable legitimate governmental purpose could serve as a basis for the mandates he is challenging. The rational basis test is so easy for a defendant authority to satisfy that it is just one level removed from being a rubber-stamp approval, and it was the go-to torpedo used for sinking legal challenges to compulsory masking that could not be dismissed using technicalities or by referring to *Jacobson v. Massachusetts*.

Under rational basis scrutiny, mask mandates are "accorded a strong presumption of validity" from the moment of inception.[455] Not just a *nominal* presumption, mind you, but a *strong* presumption. All rational basis scrutiny requires of mask mandates in order to be valid is: 1) that they promote a "*legitimate* governmental *purpose*," and 2) that there be a "rational relationship between" the

---

453   This legal standard is so commonly known that there is not much point even citing references for it, as a basic search of some variant of this wording will turn up many cases from every circuit, but the specific quote I have used is taken from the reference below.
Clark v. Amoco Production Company, 794 F.2d 967, 970 (5th Cir. 1986), quoting *Boudeloche v. Grow Chemical Coatings Corporation*, 728 F.2d 759, 762 (5th Cir. 1984), available online
https://law.justia.com/cases/federal/appellate-courts/F2/794/967/231003/
See also *Jakupovic v. Curran*, 850 F.3d 898, 902 (7th Cir. 2017), available online:
https://casetext.com/case/jakupovic-v-curran-3
454   *City of Dallas v. Stanglin*, 490 U.S. 19, 20 (1989), available online:
https://tile.loc.gov/storage-services/service/ll/usrep/usrep490/usrep490019/usrep490019.pdf
455   *Denis v. Ige et al*, No. 1:2021cv00011 - Document 62 (D. Haw. 2021), pg. 24
(https://law.justia.com/cases/federal/district-courts/hawaii/hidce/1:2021cv00011/152658/62/)
quoting *Heller v. Doe by Doe*, 509 U.S. 312, 319 (1993) (https://supreme.justia.com/cases/federal/us/509/312/)

mandates and that purpose. Regardless of any fundamental philosophical disputes about the role of government, when masks are examined using "rational basis" scrutiny, the "legitimate governmental purpose" criterion is always satisfied because decreasing the spread of *any* disease is always considered a legitimate governmental purpose.

As for the second half of the rational basis test, the working definition of "rational" in "rational relationship" is so broad that it includes judges' and legislators' intuition, regardless of how irrational, uninformed, or tainted by mass hysteria that intuition may be. As a practical matter, in an "emergency" where more than half the population is scared witless, anything that might theoretically-possibly-maybe "work" will pass the rational basis test. By the same token, reliance on a public health expert or institution is *also* always considered rational for purposes of passing the rational basis test. In other words, an "expert" *opinion* that masks work is the only thing necessary to pass muster under "rational basis" scrutiny unless the judge decides to go out of their way to give this standard of review some genuine teeth. In practice, judges treated the CDC as a source that "cannot reasonably be questioned."[456] CDC "recommendations" and statements in favor of masking functioned as *ex cathedra* pronouncements of public health doctrine from an institutional *pontifex maximus*. Heresies or minority beliefs that contradicted this masking dogma were given short shrift, no matter how well-argued and supported. When challenged, authorities and judges who privately supported mask mandates or who were unwilling to contravene them for other reasons simply gestured in the general direction of the CDC's recommendations to justify leaving everyone's masks firmly in place and punish any "self-centered" "anti-masker" heretics.

The bottom line is that under a "rational basis" standard of review, numbers, evidence, and actual science were — and remain — optional depending on the judge's whim. One district court judge was very blunt about this when applying the rational basis standard to dismiss a legal challenge to masks. The authorities issuing the mask mandate had "'no obligation to produce evidence to sustain the rationality' of the Mask Mandates," she declared, and when it comes to containing a pandemic, "the government's choice of means 'is not subject to courtroom fact-finding and may be based on rational speculation unsupported by evidence or empirical data.'"[457] In other words, under rational basis review, authorities could and did get away with imposing mask mandates based on pure speculation, and the balance of the scientific evidence was only one step up from irrelevant. Like the *Jacobson v. Massachusetts* standard, the onus was then thrown on the Plaintiff to show that there was "no real or substantial relation," between the mandate and its objective, and this was effectively impossible

---

456   *Denis v. Ige et al*, No. 1:2021cv00011 - Document 62 (D. Haw. 2021), pg. 5 n 3, available online: https://law.justia.com/cases/federal/district-courts/hawaii/hidce/1:2021cv00011/152658/62/

457   *Denis v. Ige et al*, No. 1:2021cv00011 - Document 62 (D. Haw. 2021), p. 25, available online: https://law.justia.com/cases/federal/district-courts/hawaii/hidce/1:2021cv00011/152658/62/

to do because scared and biased courts simply either deferred to the public health authorities or declared that "rational" speculation was good enough even when faced with a mountain of contrary evidence. When the numbers, evidence, and actual science contradicted pre-existing biases that favored masking, the uncooperative evidence and science were simply dismissed or ignored. When these could *not* be ignored, courts simply said that the most such evidence could show was "that there was a debate among experts concerning the effectiveness of face masks. That does not mean it was irrational for state and local officials to follow the CDC, rather than [Plaintiff's] chosen experts."[458] Mask mandates were imposed in the name of The Science™, but upheld because they were *allowed* to be illogical and unscientific.[459]

In order for a mask mandate to fail the rational basis test (or its administrative counterpart), it must be irrational or "arbitrary and capricious." Most mask mandates actually *were* arbitrary and capricious and irrational, but showing that to be the case took both Herculean effort and great good luck in reaching a judge who would grant a genuinely fair hearing on the subject. In order to survive a motion to dismiss, the initial complaint not only needed to allege that the mandate in question was irrational, but explain in enough excruciating detail *why* that was so. As at least one lawsuit proved,[460] this is not impossible, but we will look more at the heroic lawsuit which showed mask mandates can fail even rational basis-type scrutiny later on.

Strict scrutiny, by contrast, is the "most demanding test known to constitutional law." It is rare that government actions fail rational basis scrutiny, and it is almost as rare for government actions to pass true strict scrutiny. As one judge put it, "strict scrutiny leaves few survivors."[461] It should come as no

---

458   *Denis v. Ige et al*, No. 1:2021cv00011 - Document 62 (D. Haw. 2021), pp. 27-28, available online: https://law.justia.com/cases/federal/district-courts/hawaii/hidce/1:2021cv00011/152658/62/

459   *See* also the U.S. Supreme Court's summary of rational basis scrutiny:

"'The problems of government are practical ones and may justify, if they do not require, rough accommodations - illogical, it may be, and unscientific.'"

            Metropolis Theatre Co. v. City of Chicago, 228 U. S. 61, 69-70

Quoted in *City of Dallas v. Stanglin*, 490 U.S. 19, 27 (1989), available online: https://tile.loc.gov/storage-services/service/ll/usrep/usrep490/usrep490019/usrep490019.pdf

460   *Health Freedom Defense Fund, Inc. et al v. Biden et al*, No. 8:2021cv01693 - Document 53 (M.D. Fla. 2022), Available online: https://www.courtlistener.com/docket/60052717/health-freedom-defense-fund-inc-v-biden/ and https://law.justia.com/cases/federal/district-courts/florida/flmdce/8:2021cv01693/391798/53/

    HFDF v. Biden was hugely important for several reasons: 1) it was a defeat for a mask mandate even under the rubber-stamp rational basis standard; 2) it undermined all other mask mandates by going straight to the root of the problem, challenging the rationality of the main authority that everyone else pointed to in order to pass rational basis scrutiny for their mask mandates; and 3) it challenged the CDC's agency overreach to issue mandates under its own authority.

461   *Los Angeles v. Alameda Books, Inc.*, 535 U.S. 425, 455 (2002). Available online: https://supreme.justia.com/cases/federal/us/535/425/

surprise that governmental defendants *hate* being subjected to strict scrutiny and will do everything they can to evade it.

Under strict scrutiny, the burden of proof is not on the plaintiff, but on the *defending authority* to justify their policy. Moreover, to survive strict scrutiny, a mandate must actually further a *"compelling governmental interest."* The merely *"legitimate* governmental *purpose,"* required in rational basis scrutiny is nowhere near enough. A *compelling* governmental *interest* is something that the government *must* do as a legal duty, rather than something discretionary which it merely *wants* to do or can perhaps *justify* doing. For example, providing national defense is a compelling governmental interest, whereas providing education past a certain level may be a *legitimate* governmental *purpose*, but does not rise to the level of a *compelling* governmental *interest*.

In the case of COVID-19, the Supreme Court has already ruled that: "Stemming the spread of COVID–19 is unquestionably a compelling interest,"[462] but this concession still does not give mask mandates a free pass under strict scrutiny — far from it. Under strict scrutiny, a policy must not only serve a compelling governmental interest, but it must be *"substantially* related"[463] to furthering the compelling interest, rather than just *"reasonably* related" or *"rationally* related" as in rational basis scrutiny. Under strict scrutiny, vague appeals to fear-saturated intuition, hand-waving in the direction of experts with conflicts of interest, and magic words like "health and safety," or "risk of transmission" are not enough to meet the burden of proof to show that something is *"substantially related"* to its objective. Under strict scrutiny, it actually matters on an evidential level how well the policy *works*. To meet the evidential requirements for a compelling interest test, courts have ruled that a "generalized statement of interests, unsupported by specific and reliable evidence, is not sufficient[.]"[464] Additionally, unlike the more permissive rational basis scrutiny, the compelling interest test bars defending authorities from coming up with "after-the-fact explanations"[465] to do some borderline-

---

462   *Roman Catholic Diocese of Brooklyn v. Cuomo*, 141 S. Ct. 63, 208 L. Ed. 2d 206 *67 U.S. (2020), available online: https://www.supremecourt.gov/opinions/20pdf/20a87_4g15.pdf.

463   "To meet the burden of justification, a State must show "at least that the [challenged] classification serves 'important governmental objectives and that the discriminatory means employed' are 'substantially related to the achievement of those objectives.'" *United States v. Virginia*, 518 U.S. 515, 516 (1996) (https://supreme.justia.com/cases/federal/us/518/515/), quoting *Wengler v. Druggists Mutual Insurance. Company.,* 446 U. S. 142, 150 (https://supreme.justia.com/cases/federal/us/446/142/)

464   *Davila v. Gladden*, No. 13-10739, p. 13 (11th Cir. 2015) available online: https://law.justia.com/cases/federal/appellate-courts/ca11/13-10739/13-10739-2015-01-09.html

465   "the compelling interest test bars the government from invoking 'after-the-fact explanations' to justify actions that have burdened religion... as the Supreme Court explained in the free-exercise context, "[g]overnment 'justification[s]' for interfering with First Amendment rights 'must be genuine, not hypothesized or invented post hoc in response to litigation.'"

Haight v. Thompson 763 F.3d 554 (6th Cir. 2014) (https://casetext.com/case/haight-v-thompson-1)

cheating by retconning their rationale. As the U.S. Supreme Court has pointed out, for purposes of strict scrutiny, "The justification must be genuine, not hypothesized or invented *post hoc* in response to litigation. And it must not rely on overbroad generalizations[.]"[466] Mask mandates fail *all* of these criteria when closely interrogated. Mask mandates were defended on the grounds that the state has a "compelling interest" in slowing the spread of COVID, but the real, unstated, far-less-legitimate interest was in creating artificial solidarity and uniformity, showing that everyone was doing their part while displaying a not-so-subtle reminder of coercion and abnormality.

Even under strict scrutiny, however, the scientific evidence regarding masks has relevance only insofar as it affects governmental defendants' ability to satisfy *part* of their total burden of proof. Those lawsuits that successfully challenged compulsory masking took pains to make clear to the courts that the issue being litigated was *not* about whether or not masks "worked." Even *if* the balance of empirical evidence could be shown to favor a challenged mask mandate, or if the mandate could somehow be shown to have "some effect" in achieving the stated goal, this *still* may not be enough to survive strict scrutiny and justify burdening individual liberties, because strict scrutiny *also* requires that the policy be "narrowly tailored" and use the "least restrictive means" of furthering the compelling interest. A policy is not "narrowly tailored" if it is excessive or overbroad, and the Supreme Court has consistently affirmed that: "The least-restrictive-means standard is exceptionally demanding[.]"[467] As the U.S. Supreme Court said in *NAACP v. Alabama*, a "governmental purpose to control or prevent activities constitutionally subject to state regulation may not be achieved by means which sweep unnecessarily broadly and thereby invade the area of protected freedoms."[468] This also applies to a governmental purpose to control or prevent the spread of infectious disease.

It is easy to see why strict scrutiny is lethal to COVIDcrisis mandates, including compulsory masking. Quarantining individuals who are sick and symptomatic is narrowly tailored (and historically effective). Universal masking, on the other hand, was always grossly overbroad by definition. Forcing everyone to wear a medical device regardless of their individual medical history, immune status, and other risk factors is self-evidently *neither* narrowly tailored *nor* the least-restrictive means available. At any given time during the COVIDcrisis, the vast majority of the population being forcibly masked was either uninfected or recovered, and either way was incapable of transmitting SARS-CoV-2. According to the CDC's own numbers, this non-infectious overwhelming majority was virtually always over 98% of the population and never dropped below 94%, with the largest recorded relative "peak" in

---

466   *United States v. Virginia*, 518 U.S. 515, 516 (1996) (https://supreme.justia.com/cases/federal/us/518/515/),

467   *Burwell v. Hobby Lobby Stores*, Inc., 573 U.S. 682 (2014), available online:
https://supreme.justia.com/cases/federal/us/573/682/

468   *NAACP v. Alabama*, 377 U.S. 288, 307 (1964), available online:
https://tile.loc.gov/storage-services/service/ll/usrep/usrep377/usrep377288/usrep377288.pdf

COVID infections coming during January 2022, by which point a large segment of the population being forced to wear masks had already acquired natural immunity.[469]

Historically, the government has a compelling interest in preventing crime, but anti-mask laws designed to further the government's compelling interest in crime prevention were unconstitutionally overbroad if they also banned expressive mask wearing. As a 1992 analysis by attorney Stephen J. Simoni, published in the *Fordham Law Review*, noted:

> "Anti-mask laws that explicitly permit only certain expressive mask-wearing
> unnecessarily ban First Amendment exercise, as individuals engaged in political

---

469   I qualify the CDC's recorded peaks because the CDC routinely failed to adjust or normalize its recorded infection rate based on the number of tests administered or the population being tested.

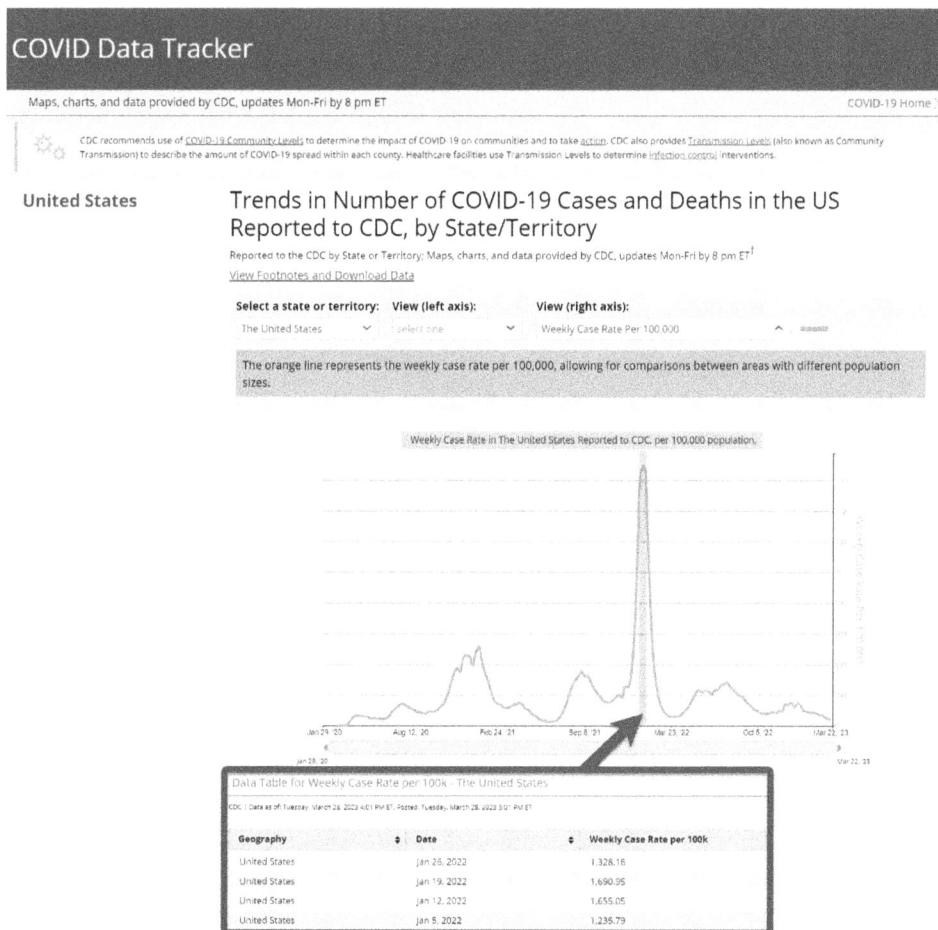

Throughout the vast majority of the COVIDcrisis, mask mandates required more than 98% of the population to use a preventative measure that was not necessary because they either could not get or could not pass on SARS-CoV-2 infection. Even during the CDC's recorded infection peak of January 2022, more than 94% of the population remained unable to spread the virus.

Data Source: https://covid.cdc.gov/covid-data-tracker/#trends_weeklycases_totalcasesper100k_00

"CDC COVID Data Tracker," https://covid.cdc.gov/covid-data-tracker/#trends_weeklycases_totalcasesper100k_00,

expression do not generally pose the danger that individuals intending to commit crime do. Therefore, such laws fail the narrowly tailored requirement."[470]

Likewise, mask *mandates* also unnecessarily ban First Amendment exercise because individuals engaged in political expression overwhelmingly do not pose the danger to public health that infected individuals do. Under these circumstances, mask *mandates* fail the narrow tailoring requirement as badly as mask *bans*. Analogously, even though the state has a compelling interest in preventing crime, speculative, even "rational" fears about masked individuals' higher potential to commit crimes have often been insufficient to overcome First Amendment protections for expressive mask-wearing.[471] The same reasoning applies with equal force to speculative fears about *un*masked individuals' potential to spread disease. When it comes to face coverings, mandates and bans are *both* wrong. True liberty resides in the middle ground, where people decide on an individual basis whether to put on or take off a face covering.

Grounds (or excuses) for dismissal — "standing"

In many cases, just getting to any kind of scrutiny was an achievement in its own right. During the COVIDcrisis, many judges saw challenges to compulsory masking as inherently offensive, dangerous, or both, and found multiple technical ways to simply dismiss them outright without really investigating the actual merits. Judges possess a host of subtle discretionary tools that can nudge a case towards a particular outcome: they can decide what evidence to admit or not admit; they can rule the plaintiff does not have standing to sue in the first place; they can admit the plaintiff has standing to sue but declare the case "unripe" for adjudication; and they can defer to bodies like

---

470 Stephen J. Simoni, "Who Goes There?" — Proposing a Model Anti-Mask Act, 61 *Fordham Law Review*, 241, 256 (1992), available online: https://ir.lawnet.fordham.edu/cgi/viewcontent.cgi?article=3009&context=flr

471 For just three examples:

  Fear that masked individuals demonstrating in front of an embassy were planning or more likely to storm a foreign embassy (*Ghafari v. Municipal Court*, 150 Cal. Rptr. 813, 816 (Cal. Ct. App. 1978), https://casetext.com/case/gates-v-municipal-court).

  Fear that agitators and terrorists with non-expressive purposes may use masked protests as cover, or that masks may cause wearers to lose their inhibitions and therefore have a greater tendency to disrupt the peace (*Aryan v. MacKey*, 462 F. Supp. 90 (N.D. Tex. 1978), available online: https://law.justia.com/cases/federal/district-courts/FSupp/462/90/2142290/).

  When a Georgia anti-mask law was challenged under the First Amendment in *State v. Miller*, "it was saved only by implementing a narrowing construction of the statutory text that circumscribed the conduct subject to criminal penalty."

Gates v. Khokhar, 884 F.3d 1290, 1299 (11th Cir. 2018), *cert. denied*, 139 S. Ct. 807 (2019), 31a (Williams, J., dissenting) available online: https://www.supremecourt.gov/DocketPDF/18/18-511/66921/20181015162248885_Gates%20v.%20Khokhar%20Appendix.pdf

the CDC without meaningful analysis. In addition, many of the litigants challenging compulsory masking were *pro se* — meaning they filed suit on their own without attorney assistance (usually for financial reasons) — and made errors that gave judges easy outs, such as bringing a challenge under federal law to a state court, or a state law challenge to federal court.[472] (Whether a mask mandate had to be challenged under federal law or state and local law depended on whether a federal or a state authority was imposing the mandate.)

The requirement that a plaintiff have "standing" was a common tool used to cull lawsuits challenging mask mandates. In order to have "standing" to sue on masks, plaintiffs needed to allege an injury to one or more of their own *personal* legally protected interests, that injury had to be fairly traceable to the conduct of the defendant they were suing, *and* it had to be something a favorable judicial decision could redress. The requirement for standing (as well as the fine distinction between who was *issuing* the mandate being challenged and who was *enforcing* it) led to multiple lawsuits against mask mandates being dismissed on what were essentially pedantic technicalities that refused to acknowledge the full weight and threats — both implicit and explicit — behind many of the mask mandates.[473]

For example, in Minnesota, Judge Patrick Schiltz ruled that: "because plaintiffs have not clearly alleged that they intend to enter public indoor spaces without face coverings, they likely lack standing to pursue their challenges to the validity of [the mask mandate] under the Elections Clause and First Amendment."[474] In Massachusetts, the *generalized* threat of a fine for noncompliance was not deemed sufficient injury to establish standing. A plaintiff who fell under Massachusetts Governor Baker's mask mandate had her case dismissed because she failed to allege "that she has *personally* been forced to wear a mask or to require her employees to wear a mask on any occasion. She thus has not established that she suffered any concrete and particularized injury with respect to the mask requirement."[475] The dismissal of a Louisiana plaintiff's challenge to mask mandates focused on the idea that the requested redress was too "speculative" to confer standing because an injunction against the *Governor's* mask

---

472   This sort of error is a surefire way to taint the rest of one's lawsuit in the eyes of a presiding judge.

473   For an example typical of such tendentious hairsplitting in a related context, see Judge Robert C. Chambers' dismissal of a Fourteenth Amendment Claim against lockdown orders by affirming the right of *inter*state travel (which he could hardly deny, as it has been repeatedly upheld by the Supreme Court) while denying that *intra*state travel is also a fundamental right protected by the Fourteenth Amendment.
(*Stewart v. Justice* 502 F. Supp. 3d 1057, 1068 (S.D.W. Va. 2020), available online:
https://casetext.com/case/stewart-v-justice-1)

474   *Minnesota Voters All. v. Walz,* 492 F. Supp. 3d 822, page 26 (D. Minn. 2020), available online:
https://law.justia.com/cases/federal/district-courts/minnesota/mndce/0:2020cv01688/189156/51/

475   *Bechade v. Baker* CIVIL ACTION NO. 20-11122-RGS (D. Mass. Sep. 23, 2020)
https://casetext.com/case/bechade-v-baker

mandate would not necessarily prevent *private* actors from imposing their own mask mandates in the vein of "No Shirt, No Shoes, No Service."[476]

In Pennsylvania, a challenge to Governor Wolf's mask mandate foundered on a combination of all three of the above excuses. "Plaintiffs here have neither sufficiently established an intention to engage in the proscribed conduct, nor have they persuaded us that they face a credible threat of future enforcement by the Defendants," ruled Judge John Jones III in *Parker v. Wolf*.

> "An injunction here will simply not alleviate the injuries Plaintiffs claim to suffer when they wear a mask—those injuries will almost certainly persist even in the absence of state-level enforcement of the Mask Mandate.... While businesses who no longer face a threat of state enforcement may (subject to local orders) allow Plaintiffs to enter their establishment without a mask, it is too speculative to conclude this would actually happen."[477]

In plain English, the mandate's enforcement by the authorities in *Governor Wolf's* administration was *not* credible, but the threat of enforcement by *other — local —* authorities was *so* credible that any injunctive relief against the *Governor's* mask mandate would be too inadequate or speculative to justify. Who knew that the standard for judging threat credibility could be so flexible? These categorical all-or-nothing applications of standing doctrine also ignored the fact that "standing" does not require perfect relief; as long as the requested redress will at least *lessen* the injury, standing can be established.[478] Just because an injury cannot be redressed *entirely* is no excuse for refusing to acknowledge that an injury has occurred in the first place. The plaintiffs in *Parker v. Wolf* vigorously appealed, but delay was another common judicial avoidance tactic. By the time the appellate court issued its ruling, Governor Wolf's mask mandate had been rescinded and the Court of Appeals simply dismissed the case as being moot — on the grounds that the plaintiffs had not shown a "reasonable expectation" that the statewide mask mandate was likely to be reinstated. Never mind the fact that Governor Wolf's administration continued to claim the authority to impose a mask mandate again

---

476   *Cangelosi v. Edwards* , No. 20-cv-1991, 2020 WL 6449111 (E.D. La. Nov. 3, 2020), https://casetext.com/case/cangelosi-v-edwards

477   *Parker* v. *Wolf,* 506 F. Supp. 3d 271, 288-290, (M.D. Pa. 2020), available online: https://casetext.com/case/parker-v-wolf

478   For example, see *Delaney v. Baker*, 511 F. Supp. 3d 55, *68 (D. Mass. 2021), available online: https://casetext.com/case/delaney-v-baker-2, " See *Antilles Cement Corporation v. Fortuno*, 670 F.3d 310, 318 (1st Cir. 2012) (holding that the plaintiff had standing where it demonstrated that if the laws at issue were preempted, the plaintiff would have significantly greater business opportunities)."

at any time, and never mind the experiences of 2020 and 2021, the threat of future mask mandates simply wasn't *credible* enough.[479]

---

479   *Parker* v. *Wolf,* No. 20-3518, Document 41, (United States Court of Appeals for the 3rd Circuit, Nov. 23, 2021), available online:

https://www.americanfreedomlawcenter.org/wp-content/uploads/2020/09/Opinion-affirming-district-court.pdf

Constitutional claims against masks

So, which of the many constitutional and procedural claims against compulsory masking got to rational basis scrutiny, how were some able to reach strict scrutiny, and what were the results? Mask mandates implicate a wide range of constitutional and procedural rights under federal, state, and local law. Mask mandates violate all three of the absolute rights of individuals listed by William Blackstone that we examined in Part 4: personal security, personal liberty, and personal property. They violate personal security and personal property by interrupting people's enjoyment of their most fundamental and intimate property — their own bodies. Mask mandates violated personal liberty when people's movements in public and access to places of public accommodation were conditioned on restraining their face behind a piece of snug cloth or other material.

Depending on the individuals and mandates involved, compulsory masking can involve the First Amendment, the Fourteenth Amendment, the Nineth and Tenth Amendments, and potentially even the Fourth and Fifth Amendments. This does not include any procedural statutes that compulsory masking potentially violates. Some of these rights have already been upheld against masks in several critical cases. Others, courts have so far failed or refused to recognize (either willfully or because the claims involved were inadequately formulated and pled). A few claims against masks have not yet really been tested, and some were simply never ruled on because courts used other claims brought by the plaintiffs to strike down the mask mandates being challenged. Many constitutional claims against masks have never been fully developed and articulated simply because the few times they were alleged or pleaded, the pleadings were based on a layman understanding of the terms involved, or composed in a way that made professional jurists discount them almost immediately.[480] We will look at these in turn, starting with the most important – the First Amendment.

---

480   Thus highlighting, in the process, how wide the gap between the common understanding of laws and the actual practice of law has become in America.

The First Amendment — Free Speech

> *"Congress shall make no law respecting an establishment of religion, or prohibiting the free exercise thereof; or abridging the freedom of speech, or of the press; or the right of the people peaceably to assemble, and to petition the government for a redress of grievances."*
>
> — United States Constitution, Amendment I

Background and intermediate *"O'Brien"* scrutiny

> *"Freedom of discussion, if it would fulfill its historic function in this nation, must embrace all issues about which information is needed or appropriate to enable the members of society to cope with the exigencies of their period... The health of the present generation and of those as yet unborn may depend on these matters[.]"*
>
> -United States Supreme Court, *Thornhill v. Alabama*, 1940[481]

The most intuitive and common constitutional challenges to compulsory masking were made under the First Amendment and the Fourteenth Amendments on grounds of free speech, free exercise, privacy, and due process. As the United States Supreme Court has repeatedly recognized, conduct and speech are not mutually exclusive. Conduct can *be* a form of speech, though this does not mean that any

---

[481]   *Thornhill v. Alabama*, 310 U. S. 88, 102-103 (1940), available online: https://tile.loc.gov/storage-services/service/ll/usrep/usrep310/usrep310088/usrep310088.pdf

and all conduct is expressive enough to count as "speech" protected under the First Amendment.[482] The U.S. Supreme Court put it like this: "It is possible to find some kernel of expression in almost every activity a person undertakes — for example, walking down the street or meeting one's friends at a shopping mall — but such a kernel is not sufficient to bring the activity within the protection of the First Amendment."[483] According to the U.S. Supreme Court, an act of expressive conduct falls within the scope of the First Amendment if "[a]n intent to convey a particularized message was present, and in the surrounding circumstances the likelihood was great that the message would be understood by those who viewed it."[484]

Symbolic speech and expressive conduct protected under the First Amendment include both actions and *non*-actions. Symbolic speech jurisprudence prior to 2020 has gradually refined the legal understanding of when conduct becomes speech that brings the First Amendment into play. Cases involving expressive conduct from the state level all the way up to the U.S. Supreme Court involve a wide range of specific precedents, including (but not limited to): picketing and demonstrating;[485] parading;[486] dancing (including erotic dancing);[487] conducting a "sit-in" protest;[488] flying a particular

---

482   "We cannot accept the view that an apparently limitless variety of conduct can be labeled 'speech' whenever the person engaging in the conduct intends thereby to express an idea."
United States v. O'Brien, 391 U.S. 367, 376 (1968), available online:
https://tile.loc.gov/storage-services/service/ll/usrep/usrep391/usrep391367/usrep391367.pdf
        "… we have rejected "the view that an apparently limitless variety of conduct can be labeled 'speech' whenever the person engaging in the conduct intends thereby to express an idea,"
Texas v. Johnson, 491 U.S. 397, 404 (1989),
https://tile.loc.gov/storage-services/service/ll/usrep/usrep491/usrep491397/usrep491397.pdf
483   *City of Dallas v. Stanglin*, 490 U.S. 19, 25 (1989), available online:
https://tile.loc.gov/storage-services/service/ll/usrep/usrep490/usrep490019/usrep490019.pdf
484   *Spence v. Washington*, 418 U.S. 405, 410-411 (1974), available online:
https://tile.loc.gov/storage-services/service/ll/usrep/usrep418/usrep418405/usrep418405.pdf
485   *Thornhill v. Alabama*, 310 U. S. 88 (1940), available online:
https://tile.loc.gov/storage-services/service/ll/usrep/usrep310/usrep310088/usrep310088.pdf
Edwards v. South Carolina, 372 U.S. 229 (1963), available online:
https://tile.loc.gov/storage-services/service/ll/usrep/usrep372/usrep372229/usrep372229.pdf
486   *Hurley v. Irish-American Gay, Lesbian and Bisexual Group of Boston, Inc.*, 515 U. S. 557, 566 (1995), available online https://supreme.justia.com/cases/federal/us/515/557/case.pdf
487   *Attwood v. Purcell*, 402 F. Supp. 231 (1975), available online:
https://law.justia.com/cases/federal/district-courts/FSupp/402/231/1413216/
488   *Garner v. Louisiana*, 368 U.S. 157 (1961), available online: https://supreme.justia.com/cases/federal/us/368/157/

flag;[489] saluting or refusing to salute a flag;[490] altering or defacing a flag;[491] burning a flag;[492] contributing money to your own political campaign;[493] wearing particular articles of clothing such as armbands[494] or "Freedom Buttons";[495] sharing food in the park;[496] baking a customized wedding cake;[497] building an expressive website;[498] and — yes — wearing or not wearing masks.[499] These historical court rulings in the context of symbolic speech are crucial for understanding why masking and non-masking indisputably involve fundamental First Amendment freedoms.

---

489   *Stromberg v. California*, 283 U.S. 359 (1931), available online:
https://supreme.justia.com/cases/federal/us/283/359/

490   *West Virginia Board of Education v. Barnette et al.*, 319 U.S. 624, 63 S. Ct. 1178, 87 L. Ed. 1628 (1943), available online: https://www.loc.gov/item/usrep319624/

491   *Spence v. Washington*, 418 U.S. 405, (1974), available online:
https://tile.loc.gov/storage-services/service/ll/usrep/usrep418/usrep418405/usrep418405.pdf

492   *Texas v. Johnson*, 491 U.S. 397 (1989),
https://tile.loc.gov/storage-services/service/ll/usrep/usrep491/usrep491397/usrep491397.pdf

493   *Citizens United v. Federal Election Commission*, 558 U.S. 310 (2010), available online:
https://tile.loc.gov/storage-services/service/ll/usrep/usrep558/usrep558310/usrep558310.pdf

494   *Tinker v. Des Moines Independent Community School District*, 393 U.S. 503, (1969), available online:
https://www.loc.gov/item/usrep393503/

495   *Burnside v. Byars*, 363 F.2d 744 (5th Cir. 1966), available online:
https://law.justia.com/cases/federal/appellate-courts/F2/363/744/264045/

496   *Fort Lauderdale Food Not Bombs v. City of Fort Lauderdale*, 901 F.3d 1235 (11th Cir. 2018), available online:
https://casetext.com/case/fort-lauderdale-food-not-bombs-v-city-of-fort-lauderdale

497   *Masterpiece Cakeshop v. Colorado Civil Rights Commission,* 138 S. Ct. 1719 (2018) (Thomas, J., concurring in part and concurring in the judgment), available online: https://www.supremecourt.gov/opinions/17pdf/16-111_new2_22p3.pdf or https://casetext.com/case/masterpiece-cakeshop-ltd-v-colo-civil-rights-commn-3

498   303 Creative LLC v. Elenis, No. 21-476 (U.S. June. 30, 2023), available online:
https://www.supremecourt.gov/opinions/22pdf/21-476_c185.pdf

499   *Schumann v. New York*, 270 F. Supp. 730 (S.D.N.Y. 1967), available online:
https://law.justia.com/cases/federal/district-courts/FSupp/270/730/1609402/

Ghafari v. Municipal Court, 87 Cal.App.3d 255, 150 Cal.Rptr. 813, 815 (1978), available online:
https://casetext.com/case/ghafari-v-municipal-court

Aryan v. MacKey, 462 F. Supp. 90 (N.D. Tex. 1978), available online:
https://law.justia.com/cases/federal/district-courts/FSupp/462/90/2142290/

Robinson v. State, 393 So.2d 1076 (Fla. 1980), available online:
https://law.justia.com/cases/florida/supreme-court/1980/58232-0.html

State v. Miller, 260 Ga. 669, 671 n 2 (Ga. 1990), available online:
https://law.justia.com/cases/georgia/supreme-court/1990/s90a1172-1.html

Hernandez v. Superintendent, 800 F. Supp. 1344 (E.D.Va. 1992), available online:
https://law.justia.com/cases/federal/district-courts/FSupp/800/1344/1393592/

Daniels v. State, 264 Ga. 460, 448 S.E.2d 185 (1994), available online:
https://law.justia.com/cases/georgia/supreme-court/1994/s94g0362-1.html

Ryan v. County of DuPage, 45 F.3d 1090 (7th Cir. 1995), available online: https://casetext.com/case/ryan-v-county-of-dupage

Gates v. Khokhar, 884 F.3d 1290 (11th Cir. 2018), *cert. denied,* 139 S. Ct. 807 (2019), available online: https://www.supremecourt.gov/DocketPDF/18/18-511/66921/20181015162248885_Gates%20v.%20Khokhar%20Appendix.pdf

In a particularly important sequence of cases beginning in the 1960s, the Supreme Court ruled that defacing or burning an American flag constituted symbolic speech protected by the First Amendment,[500] whereas publicly burning a draft card did *not*.[501] What was the difference? Three things. First, burning a flag is more "inherently expressive" than burning a draft card (flags are inherently symbolic objects — that is their purpose[502] — whereas this is not true of draft cards). Second, any power to make laws against "desecration of a venerated object" like the American flag are nowhere near as directly and closely related to explicitly granted constitutional powers as are draft card laws passed pursuant to Congress' constitutional power to raise and support armies. Finally, even though "the flag as readily signifies this Nation as does the combination of letters found in 'America,'"[503] the state's interest in preserving the flag as a symbol of national unity or a form of national intellectual property is nowhere near as strong as the state's compelling interest in national defense or an individual's First Amendment liberty interests.

This sequence of cases was important because it gave rise to what is known as the "*O'Brien* test" — named after the 1968 *United States v. O'Brien* Supreme Court decision involving draft card burning described above. *O'Brien* scrutiny is a type of intermediate scrutiny that falls between rational basis and strict scrutiny, and one which has been applied to masks both before and during the COVIDcrisis. Under the "*O'Brien* standard," a governmental regulation burdening free speech must pass four criteria: 1) it must be "within the constitutional power of the Government"; 2) it must be that the regulation "furthers an important or substantial governmental interest" which is 3) "unrelated to the suppression of free expression"; *and* 4) "the restriction on alleged First Amendment freedoms is no greater than is essential."[504] If a law or mandate which burdens speech fails this test, then it is unconstitutional.

The more lenient *O'Brien* scrutiny *may* be used when "speech and nonspeech elements are combined in the same course of conduct," and the restrictions being challenged are an "incidental" side effect of measures pursuant to "an important governmental interest" which is "unrelated to the suppression of

500   *Spence v. Washington*, 418 U.S. 405 (1974), available online:
https://tile.loc.gov/storage-services/service/ll/usrep/usrep418/usrep418405/usrep418405.pdf
Texas v. Johnson, 491 U.S. 397 (1989),
https://tile.loc.gov/storage-services/service/ll/usrep/usrep491/usrep491397/usrep491397.pdf
501   *United States v. O'Brien*, 391 U.S. 367 (1968), available online:
https://tile.loc.gov/storage-services/service/ll/usrep/usrep391/usrep391367/usrep391367.pdf
502   Flags are a type of emblem, alongside talismans, totems, sigils, crests, insignias, trademarks, etc.
503   *Texas v. Johnson*, 491 U.S. 397, 405 (1989),
https://tile.loc.gov/storage-services/service/ll/usrep/usrep491/usrep491397/usrep491397.pdf
504   *United States v. O'Brien*, 391 U.S. 367, 377 (1968), available online:
https://tile.loc.gov/storage-services/service/ll/usrep/usrep391/usrep391367/usrep391367.pdf

expression."[505] Restrictions on "pure speech" *require* strict scrutiny, but restrictions on symbolic speech or "expressive conduct" which would fail *strict* scrutiny can often sail through under *O'Brien* scrutiny. Judges typically apply *O'Brien* scrutiny when, in their opinion, the symbolic speech in question is not so "inherently expressive" that the presence of "speech" in the conduct is indisputable. This wiggle room, combined with motivated reasoning, allowed mask mandates to evade strict scrutiny and survive First Amendment challenges that should have been fatal. In the hands of several judges during the COVIDcrisis, *O'Brien* scrutiny was hardly better than rational basis scrutiny. Yet, when applied impartially and objectively, *O'Brien* scrutiny has historically vindicated First Amendment rights in a number of important cases.

Unlike destroying a draft card, the "inevitable effect" of wearing masks is potent communication, but when reading COVIDcrisis cases challenging compulsory masking under the First Amendment, one gets the impression that judges would have preferred to catch COVID themselves rather than acknowledge the speech inherent in mask-wearing and non-mask-wearing. Based on such cases, a rather one-sided September 2021 summary in *The National Law Review* concluded: "Courts have roundly rejected claims that masks unlawfully regulate speech or expression. Mask mandates regulate conduct, not speech, and therefore do not implicate the Free Speech Clause at all."[506]

Some Judges, perhaps not 100% convinced Free Speech claims against masking were truly without merit, expended a great deal of effort to dismiss First Amendment challenges on other grounds such as standing. Judge John Jones III in the Pennsylvania *Parker v. Wolf* case described earlier may be one of these. The point of Judge Jones' ruling in that case was that even *assuming* masks implicate Free Speech, the plaintiffs *still* failed to bring a justiciable claim because they supposedly could not meet the "particularity" and "redressability" requirements needed for standing. But as a prelude to dismissal, Judge Jones rightly conceded that, assuming for the sake of argument "a mask does indeed constitute compelled speech," then "we initially find that Plaintiffs have stated a concrete, *de facto*

505
> "Texas has not asserted an interest in support of Johnson's conviction that is unrelated to the suppression of expression and would therefore permit application of the test set forth in *United States v. O'Brien*, 391 U. S. 367, whereby an important governmental interest in regulating nonspeech can justify incidental limitations on First Amendment freedoms when speech and nonspeech elements are combined in the same course of conduct."
>
> *Texas v. Johnson*, 491 U.S. 397, 397 (1989), https://tile.loc.gov/storage-services/service/ll/usrep/usrep491/usrep491397/usrep491397.pdf

506   Brandon O. Mouland & Laura Lashley, "Masks Up, Pens Down: (Still) Litigating Mask Mandates in 2021," *The National Law Review*, September 24, 2021, Online: https://www.natlawreview.com/print/article/masks-pens-down-still-litigating-mask-mandates-2021

injury," because: "While the harm associated with compelled speech might be intangible, the Supreme Court has held that such a First Amendment injury may 'nevertheless be concrete.'"[507]

Even when a court or governmental defendant would concede that masks do indeed involve symbolic speech, they would still argue that mask mandates were justified under *O'Brien* scrutiny because the *purpose* of the mask mandates was to further a compelling governmental interest unrelated to speech, and the "incidental" restrictions placed on First Amendment freedoms by mask mandates were "no greater than is essential" or "no greater than necessary." A good example of this comes from a March 2022 ruling. A group of Republican congressmen led by Kentucky Representative Thomas Massie sued House Speaker Nancy Pelosi over the House of Representatives' internal mask mandate. Both parties acknowledged the symbolic speech inherent in masking or non-masking, but District Judge Reggie Walton dismissed the suit anyway, using the *O'Brien* criteria. Judge Walton asserted that the mask mandate: 1) "is within the constitutional power of the [g]overnment"; 2) "the governmental interest is unrelated to the suppression of free expression"; and 3) "the incidental restriction on alleged First Amendment freedoms is no greater than is essential."[508] As we shall see, Judge Walton — and every other judge who similarly dismissed First Amendment challenges to compulsory masking — got it wrong.

Clothing can be speech

As we saw in in our examination of the psychology behind masking, mask-wearing is such a powerful expressive act that not only is it symbolic speech, but it is an *unusually potent and persuasive* form of symbolic speech. Even though wearing or not wearing a mask lacks the precision of spoken or written words, it still conveys a "particularized message," with a great likelihood that the message will

---

507   It is also worth noting that Judge John Jones III's dismissal of *Parker v. Wolf* has been referenced as an example of a Free Speech claim against masks being dismissed because masks do not implicate free speech (for example, the *National Law Review* article from September 2021 does this), but that is not really what Judge Jones' ruling says. Judge Jones actually *avoided* ruling on the merits of the plaintiffs' Free Speech claim against masks and instead dismissed the case on the standing grounds described earlier. I personally suspect that Judge Jones took such great care to show that the plaintiffs did not have standing because he was not completely confident their Free Speech claim would fail on the merits.
Parker v. *Wolf*, 506 F. Supp. 3d 271, 287-290, (M.D. Pa. 2020), available online: https://casetext.com/case/parker-v-wolf
508   *Massie v. Pelosi*, 590 F. Supp. 3d 196 (D.C. 2022), *20. Available online:
https://scholar.google.com/scholar_case?case=735226917287193397&hl=en&as_sdt=4000003&scfhb=1
Judge Walton's ruling in *Massie v. Pelosi* also took for granted the very issues being disputed. It assumed that compulsory masking is within the power of the government. It assumed masks work and thereby further the interest of public health. It assumed that infection control was the primary motivation involved; and it absurdly assumed that masking every Congressman in the House of Representatives (including those who had already recovered from COVID) was no greater than essential.

be understood by those who view it. This was especially true in the context of the COVIDcrisis. The sheer volume of lawsuits filed challenging masks on Free Speech grounds is, on its own, a powerful attestation to widespread intuitive awareness of the fact that mask-wearing or non-wearing is, indeed, a form of speech. Both supporters and detractors of compulsory masking recognized this, and many "experts" and authorities were remarkably frank about the non-medical ulterior communicative motive behind mask mandates.

Some articles of clothing, like patches on uniforms, are *purely* symbolic, and the utility of other articles of clothing is often vastly outweighed by their symbolism. In ancient times, wearing rough sackcloth was a sign of distress, grief, penitence, or humility. Political party members show their affiliation by wearing iconic hats, pins, and shirts. Judges and priests wear special garments indicating their legal and religious roles and status, as they have throughout human history. Mennonites and Amish have distinctive forms of dress that make them immediately identifiable. Quakers have historically refused to wear certain common symbols based on what such things communicate. Articles of clothing worn on or about the head and face are especially prone to symbolic communication, from the metal collars that designated enslaved thralls, to Roman use of the *pileus* felt cap for signifying manumission and liberty.[509] Traditional veils, a bishop's miter, and the artistic full-face costumes and armor that appear in cultures everywhere from ancient Greece to Feudal Japan all attest to the special communicative significance of the head and face.[510]

Historically, some feminists used public unveiling as a symbolic declaration of emancipation, as when Egyptian Feminist Union founder Huda Sharawi caused a commotion in 1923 Cairo by removing her veil (some sources report that she also threw it into the Mediterranean).[511] Conversely, in Oman, where female slaves were not allowed to wear veils, some female ex-slaves put *on* veils as a symbolic statement of their newly improved legal status after slavery was legally abolished in 1970.[512] Debates

---

509   Harry Thurston Peck, *Harper's dictionary of classical literature and antiquities*, 1896, New York: Harper and brothers, pp. 1003-1004, available online: https://archive.org/details/cu31924027019482/page/1004/mode/2up?q=pileus *See* the last paragraph in the entry on *Manumissio* — the legal act by which slaves were freed.

510   To say nothing of what can be communicated by the astounding variety of decorative and communicative piercings and jewelry placed in ears, lips, nose, tongue and even teeth.

511   Yedida Kalfon Stillman and Norman A. Stillman, *Arab Dress, A Short History: From the Dawn of Islam to Modern Times*, rev. 2nd ed., (Boston: Brill, 2003), p. 155, available online: https://archive.org/details/ArabDress.FromTheDawnOfIslamToModernTimesByYedidaKalfonStillman

Also related in Also related in Albert Habib Hourani, "The Vanishing Veil: A Challenge to the Old Order," UNESCO Courier (No. 11, 1955), p. 35, 37 available online: https://unesdoc.unesco.org/ark:/48223/pf0000069477

512   Lloyd Llewellyn-Jones, *Aphrodite's Tortoise: The Veiled Woman of Ancient Greece*, (Llandysul, Wales: Gomer Press, 2003), p. 141.

Suzanne Miers, *Slavery in the Twentieth Century: The Evolution of a Global Problem*, (United Kingdom: AltaMira Press, 2003), p. 347

over *burqa*-wearing highlight the inescapable symbolic speech and fundamental rights involved in face coverings. French President Nicolas Sarkozy called *burqas* "a sign of subservience" and German feminist Alice Schwarzer argued that such veiling is dehumanizing.[513] France banned face coverings in 2011, threatening stiff fines for violators, and proclaiming: "The Republic lives with its face uncovered, in all public places: public roads, public transport, shops and shopping centers, schools, post office, *hospitals…*" (emphasis added).[514] The French Republic lived with its face uncovered for all of 9 years until it *mandated* face coverings in 2020 — *without* bothering to repeal the earlier ban.[515] This contradiction reveals the real driving mentality underlying both measures: a pathological desire to control sources of collective anxiety by imposing artificial uniformity.

Masks are directly expressive, and intuitive recognition of this fact prompted people to wear masks in an exaggerated or hyperbolic manner as a means of speech when engaging in protest long before COVID. Protestors against human trafficking wore cloth masks on a 2017 march in London to represent the silence of victims,[516] and pro-choice protestors donned medical masks to convey the idea that they were being censored following the U.S. Supreme Court's 1991 *Rust v. Sullivan* decision.[517]

---

513   "The Burqa Debate: Are Women's Rights Really the Issue?" *Spiegel International*, June 24, 2010, available online: https://www.spiegel.de/international/europe/the-burqa-debate-are-women-s-rights-really-the-issue-a-702668.html

514   "La République se vit à visage découvert," March 8, 2011, http://archives.gouvernement.fr/fillon_version2/gouvernement/la-republique-se-vit-a-visage-decouvert.html

515   Jason Silverstein, "France will still ban Islamic face coverings even after making masks mandatory," May 12, 2020, CBS News *World*, https://www.cbsnews.com/news/france-burqa-ban-islamic-face-coverings-masks-mandatory/

516   Photo captioned: "People march against modern slavery in London on 14 October 2017, wearing masks that represent the silence of victims of exploitation."

Kate Hodal, Annie Kelly, and Harriet Sherwood, "True scale of UK slavery likely to involve 'tens of thousands' of victims," *The Guardian*, October 17, 2017, online: https://www.theguardian.com/global-development/2017/oct/17/true-scale-of-uk-slavery-tens-of-thousands-of-victims-kevin-hyland

517   For just two examples, see pictures accompanying:

Tara Block, "How Women Have Protested Through History," *POPSUGAR*, August 18, 2014, online: https://www.popsugar.com/love/photo-gallery/22387726/image/22387752/Pro-Choice-Rally-US-1991

Scott Lemieux, "Rust v. Sullivan (1991)," *The Free Speech Encyclopedia*, original article published 2009), online: https://www.mtsu.edu:8443/first-amendment/article/316/rust-v-sullivan (last accessed March 30, 2023)

Medical masks are inherently symbolic the same way that white lab coats are inherently symbolic, but to an even greater degree. In the context of the COVIDcrisis, masks were more akin to a military uniform or an emblem of ideological allegiance like the tricolor cockade of the French Revolution. Less than a year after the United States had finished ratifying the Bill of Rights in December 1791 to protect against all such forms of ideological tyranny, the French revolutionary Legislative Assembly *mandated* that all men residing or traveling in France wear the national form of this decorative disc in July 1792. In fact, the French Legislative Assembly went even further, and declared that the wearing of "Any cockade other than that with the three national colors, is a sign of rebellion."[518] During the COVIDcrisis, refusal to wear a mask was likewise widely interpreted and treated as a sign of rebellion. Based on how they dismissed challenges to compulsory masking, no doubt most of the COVIDcrisis judges would have called such bald-faced tyranny by the French Legislative Assembly a "trivial

Also see Drea Maier, "Demonstrating Differences; NOW Rallies Against 'Gag,'" N.Y. Newsday, July 7, 1991, at 17, cited in Stephen J. Simoni, "Who Goes There?" — Proposing a Model Anti-Mask Act, 61 *Fordham Law Review*, 241, 248 n 43 (1992), available online: https://ir.lawnet.fordham.edu/cgi/viewcontent.cgi?article=3009&context=flr

518   French Legislative Assembly, Mobilization for War, 5 July 1792:

> "16. Any man residing or travelling in France is required to wear the national cockade. Ambassadors and accredited agents of foreign powers are excepted from this provision.
>
> 17. …Any cockade other than that with the three national colors, is a sign of rebellion."
>
> [Translated from the original French using Google machine translation.]
>
> "XVI. Tout homme résidant ou voyageant en France , est tenu de porter la cocarde nationale. Sont exceptés de la présente disposition les ambassadeurs et agens accrédités des puissances étrangères.
>
> XVII. … Toute cocarde autre que celle aux trois couleurs nationales , est un signe de rebellion."]
>
> > Imprimerie nationale, *Journal des débats et des décrets, ou Récrit de ce qui s'est passé aux séances de l'assemblée nationale depuis le 17 juin 1789, jusqu'au premier septembre de la même année*, Volume 31, pages 166-167, available online:
> >
> > https://www.google.com/books/edition/Journal_des_d%C3%A9bats_et_des_d%C3%A9crets_ou_R/gjlEAAAAcAAJ?hl=en (pdf pages 170-173)

It is also worth noting that the tricolor cockade was mandatory in Paris from even earlier than this. Many people were harassed or directly punished for refusing to wear the French national tricolor cockade. At least one Quaker reported being arrested for this very offense:

> "It has pleased the lord to suffer us to fall under divers tryals, which in our weak state, we have found painful & grievious…. I was arrested at Paris because I had not the National Cockade, & signified my reasons for noncompliance, before the Judges of the Peace, & since that, before Petition Mayor of Paris, who had me set at liberty…"
>
> > - Letter from Jean Marsillac, 10 July 1792, Library ref. MS VOL 314/70, quoted in "The French Revolution: Quakers and cockades," June 15, 2015, Blog by *Library of the Society of Friends*, available online: https://quakerstrongrooms.org/2015/06/15/the-french-revolution-quakers-and-cockades/, last accessed March 16, 2023).

See also: Peter Brock, "Conscientious Objection in Revolutionary France," *The Journal of the Friends Historical Society*, 1995. 57(2), page 173 and footnote 28 on page 181. https://journals.sas.ac.uk/fhs/article/view/3500

The 18th century French Revolution, as well as the 20th century Russian Revolution and Chinese Communist takeover, provide numerous examples of how articles of clothing originated with — or quickly acquired — virulent political symbolism.

imposition on an individual's freedom."[519] But like being forced to wear a French Revolutionary Cockade or any other ideological badge, there was nothing "trivial" about being forced to wear a mask. The fact that virtually everyone was forced to do it just made the violation of fundamental liberties that much worse for being so widespread.

The presence or absence of a mask indisputably sends a message, especially when the act is performed by a minority in the context of a declared emergency mask mandate. Even if masks were successfully shown to have a practical function in controlling viruses, that would do nothing to diminish the compelled speech inherent in COVIDcrisis masking and non-masking. As we shall see, masks are so close to what the Supreme Court has called "pure speech," that they involve "direct, primary First Amendment rights[.]"[520] The surprise should not be that so many people perceived the compelled symbolic speech in compulsory masking, but that anyone managed to claim masks were *not* symbolic speech while keeping a straight face.

Steel-manning the arguments for dismissing First Amendment challenges to masking

Despite the extensive historical and legal precedents pointing to masks as a form of speech, when compulsory masking was repeatedly challenged under the Free Speech Clause of the First Amendment, courts stubbornly refused to acknowledge the reality that masks are symbolic speech.[521] "Mask Mandates regulate conduct, not speech, and do not implicate the Free Speech Clause at all... They require specific conduct: wearing a mask in public. That conduct does not include a significant expressive element," insisted Hawaii District Judge Susan Mollway, dismissing a mask lawsuit in mid-2021; "individuals can still engage in expressive activity. They just have to wear masks while they do."[522]

---

519   *Stewart et al. v. Justice et al.* No. 3:2020cv00611 - Document 45 (S.D.W. Va. 2021), p. 10, available online: https://law.justia.com/cases/federal/district-courts/west-virginia/wvsdce/3:2020cv00611/230251/45/.

520   *Tinker v. Des Moines Independent Community School District*, 393 U.S. 503, 508 (1969), available online: https://www.loc.gov/item/usrep393503/

521   Many of these challenges (e.g. *Denis v. Ige et al, No. 1:2021cv00011 - Document 62 (D. Haw. 2021)* or *Joseph v. Becerra*, No. 22-cv-40-wmc (W.D. Wis. Nov. 29, 2022)) were filed *pro se* by concerned citizens with no legal experience working in their free time who could not afford lawyers, and even professionally drafted claims that masks were symbolic speech implicating the First Amendment often lacked citations to references or authorities which supported this allegation. For example, the amended complaint in *Parker v. Wolf* (506 F. Supp. 3d 271, 288-290, (M.D. Pa. 2020) is like this (*see* page 18, https://www.americanfreedomlawcenter.org/wp-content/uploads/2020/09/First-Amended-Complaint-Parker-v-Wolf-Filed.pdf).

522   In this case the judge pointed out that the plaintiff had not alleged that the masks burdened his ability to speak.

Those challenges to compulsory masking that were not dismissed on technical grounds were shut down using a broadly two-pronged constitutional and pragmatic approach. Sometimes a judge dismissing on technical grounds would employ this formula anyway, for good measure. On the constitutional side of the equation, this approach would start by playing up COVID-19 as a national emergency of the highest order, then entirely denying or heavily downplaying the First Amendment implications of compulsory masking. After all, the First Amendment can't apply if masks are not speech. Even *if* masking and non-masking *are* speech under the First Amendment, went this reasoning, mask mandates can only warrant rational basis scrutiny, or at the *very* most, intermediate *O'Brien* scrutiny, both of which they easily survive. On the pragmatic side of the equation, the invariable starting point finding would be that masks work well enough to qualify as basic, essential, commonsense safety measures (after all, the CDC *said so*). The clincher would be that masks have no significant side effects and leave "ample alternative channels" for communication of any objections to mask-wearing. Plaintiffs challenging mask mandates were somehow never *quite* able to carry their burden of proof in these courts.

In Florida, Magistrate Judge Jared Strauss recommended dismissal of a Free Speech claim against mask wearing because:

> "[N]either wearing or not wearing a mask is inherently expressive. In the context of COVID-19, wearing a mask does not evince an intent to send a message of subservience to authority — or any message at all... Thus, wearing a mask is not 'inherently expressive' such that the challenged mask mandates constitute compelled speech subject to First Amendment analysis."[523]

"[T]he act of refusing to wear a face covering does not carry a meaning that is 'overwhelmingly apparent' such that it is protected speech," ruled Federal District Judge Robert Chambers in West Virginia. "Here, although Plaintiffs feel that refusing to wear a face covering expresses 'nonconformity with unconstitutional and un-American laws...' that meaning is not 'overwhelmingly apparent.'"[524] But, said Judge Chambers, "even if refusing to wear a mask is protected speech, the Mask Mandate withstands intermediate scrutiny as a content-neutral time, place, and manner restriction... it is a trivial imposition on an individual's freedom outweighed by the reasonableness of such precautions

---

Denis v. Ige et al, No. 1:2021cv00011 - Document 62 (D. Haw. 2021), pp. 29, available online: https://law.justia.com/cases/federal/district-courts/hawaii/hidce/1:2021cv00011/152658/62/

523   *Zinman v. Nova Southeastern University*, No. 21-CIV-60723-RAR, Document 81, p. 25-26 (S.D. Fla. Aug. 30, 2021), https://www.govinfo.gov/content/pkg/USCOURTS-flsd-0_21-cv-60723/pdf/USCOURTS-flsd-0_21-cv-60723-1.pdf

524   *Stewart v. Justice* 502 F. Supp. 3d 1057, 1066 (S.D.W. Va. 2020), available online: https://casetext.com/case/stewart-v-justice-1

during a pandemic."[525] For support, Judge Chambers quoted from the Supreme Court's 2006 decision in *Rumsfeld v. Forum for Academic and Institutional Rights, Inc*: "'[t]he fact that such explanatory speech is necessary is strong evidence that the conduct at issue here is not so inherently expressive that it warrants protection' as symbolic speech."[526]

Judge Chambers' 2020 use of this quote was, itself, later directly cited to strike down *another* 2022 Free Speech challenge to compulsory masking at the appellate level in Washington State.[527] Using almost identical reasoning, the Washington State court claimed the First Amendment was never really involved:

> "[T]he mask mandate does not implicate speech, therefore, we do not address whether the mask mandate survives strict scrutiny or compels speech."[528]

> "The expression must be 'overwhelmingly apparent' and not simply a kernel of expression… The fact that "'explanatory speech is necessary is strong evidence that the conduct at issue … is not so inherently expressive that it warrants protection" as symbolic speech.'"[529]

---

525   *Stewart et al. v. Justice et al.* No. 3:2020cv00611 - Document 45 (S.D.W. Va. 2021), p. 10, available online: https://law.justia.com/cases/federal/district-courts/west-virginia/wvsdce/3:2020cv00611/230251/45/.

526   *Stewart v. Justice* 502 F. Supp. 3d 1057, 1066 (S.D.W. Va. 2020), available online: https://casetext.com/case/stewart-v-justice-1 quoting *Rumsfeld v. Forum for Academic & Institutional Rights, Incorporated*, 547 U.S. 47, 66 (2006), https://tile.loc.gov/storage-services/service/ll/usrep/usrep547/usrep547046/usrep547046.pdf It is also worth noting that Judge Chambers' ruling here used the *Jacobson* standard, and was delivered on November 24, just one day before the Supreme Court's ruling in *Roman Catholic Diocese of Brooklyn v. Cuomo*.

527   *Sehmel v. Shah*, 514 P.3d 1238 (Wash. Ct. App. 2022), at 16, available online: https://caselaw.findlaw.com/wa-court-of-appeals/2182709.html
    "The fact that " 'explanatory speech is necessary is strong evidence that the conduct at issue … is not so inherently expressive that it warrants protection' as symbolic speech." Stewart v. Justice, 502 F. Supp. 3d 1057, 1066 (S.D. W.Va. 2020)" (quoting Rumsfeld v. Forum for Acad. & Inst. Rights, Inc., 547 U.S. 47, 66, 126 S. Ct. 1297, 164 L.Ed.2d 156 (2006)).

528   *Sehmel v. Shah*, 514 P.3d 1238 (Wash. Ct. App. 2022), at 2. available online: https://caselaw.findlaw.com/wa-court-of-appeals/2182709.html

529   *Sehmel v. Shah*, 514 P.3d 1238 (Wash. Ct. App. 2022), at 16. available online: https://caselaw.findlaw.com/wa-court-of-appeals/2182709.html
parroting the analysis in *Stewart v. Justice* 502 F. Supp. 3d 1057, 1066 (S.D.W. Va. 2020), available online: https://casetext.com/case/stewart-v-justice-1

"[T]he act of wearing or not wearing a mask is not expressive conduct, and thus does not invoke the protections of the First Amendment."[530]

According to the Washington Court of Appeals, possible alternative interpretations of masking and non-masking somehow prevented these gestures from rising to the level of protected speech: "the choice to wear a mask is not expressive conduct because 'there are several non-political reasons why one may not be wearing a mask at any given moment'... 'failing to wear a face covering would likely be viewed as inadvertent or unintentional, and not as an expression of disagreement with the Governor.'"[531]

> "While an individual may choose to wear, or not wear, a mask as a way to make a political statement, the subjective intent of the person engaging in the conduct is not determinative... In addition to the person's intent to communicate a message, that message must also be 'overwhelmingly apparent' and understandable by the person viewing the message. Here, there is a host of reasons why a person may not be wearing a mask. Therefore, not wearing a mask is not 'overwhelmingly apparent' as communicating a political message."[532]

When the appellant plaintiffs argued that if the judges could not affirm masks that masks are speech, this finding of fact should be left up to a jury to determine (as provided for by the Seventh Amendment[533]), the Washington Court disagreed and refused to permit a jury trial.

Minnesota Federal District Judge Patrick Schiltz derisively described a First Amendment claim against compulsory masking as "meritless" and having "no chance of success." Compulsory masking, "either does not implicate the First Amendment at all or, at most, has an incidental and trivial impact on First Amendment freedoms."[534] Minnesota Governor Tim Walz' mask mandate, he said, "is

---

530   *Sehmel v. Shah*, 514 P.3d 1238 (Wash. Ct. App. 2022), at 19. available online: https://caselaw.findlaw.com/wa-court-of-appeals/2182709.html

531   *Sehmel v. Shah*, 514 P.3d 1238 (Wash. Ct. App. 2022), at 17. available online: https://caselaw.findlaw.com/wa-court-of-appeals/2182709.html quoting *Stewart v. Justice* 502 F. Supp. 3d 1057, 1066 (S.D.W. Va. 2020), available online: https://casetext.com/case/stewart-v-justice-1 and *Antietam Battlefield KOA v. Hogan*, 461 F. Supp. 3d 214, 236 (D. Md. 2020)

532   (internal citations omitted) *Sehmel v. Shah*, 514 P.3d 1238 (Wash. Ct. App. 2022), at 18. available online: https://caselaw.findlaw.com/wa-court-of-appeals/2182709.html

533   "In suits at common law, where the value in controversy shall exceed twenty dollars, the right of trial by jury shall be preserved, and no fact tried by a jury, shall be otherwise reexamined in any court of the United States, than according to the rules of the common law." U.S. Constitution, Amendment VII.

534   *Minnesota Voters All. v. Walz*, 492 F. Supp. 3d 822, page 31 (D. Minn. 2020), available online: https://law.justia.com/cases/federal/district-courts/minnesota/mndce/0:2020cv01688/189156/51/

clearly constitutional, whether analyzed under *O'Brien* or *Jacobson*."[535] "To merit First Amendment protection," explained Judge Schiltz, "the conduct regulated by the challenged law must be 'inherently expressive.'"[536] But, he said, wearing or not wearing a face covering "is not inherently expressive."[537] Why? Because "an observer would have no idea why someone is not wearing a face covering. Absent explanation, the observer would not know whether the person is exempt… or simply forgot to bring a face covering, or is trying to convey a political message. That fact takes the conduct outside of the First Amendment protection[.]"[538] After all, Judge Schiltz went on: "the incidental restriction on alleged First Amendment freedoms is no greater than is essential[.]"[539] He concluded: "plaintiffs are free to express their opinions about [the mask mandate] in every conceivable way *except* by violating its provisions and putting at risk the lives and health of their fellow citizens… In short, plaintiffs have no chance of success on their claim that [the mask mandate], standing alone, violates the First Amendment."[540]

---

535   *Minnesota Voters All. v. Walz*, 492 F. Supp. 3d 822, page 30 (D. Minn. 2020), available online: https://law.justia.com/cases/federal/district-courts/minnesota/mndce/0:2020cv01688/189156/51/

536   *Minnesota Voters All. v. Walz*, 492 F. Supp. 3d 822, page 29 (D. Minn. 2020), available online: https://law.justia.com/cases/federal/district-courts/minnesota/mndce/0:2020cv01688/189156/51/

537   *Minnesota Voters All. v. Walz*, 492 F. Supp. 3d 822, page 29 (D. Minn. 2020), available online: https://law.justia.com/cases/federal/district-courts/minnesota/mndce/0:2020cv01688/189156/51/

538   *Minnesota Voters All. v. Walz*, 492 F. Supp. 3d 822, page 30 (D. Minn. 2020), available online: https://law.justia.com/cases/federal/district-courts/minnesota/mndce/0:2020cv01688/189156/51/

539   *Minnesota Voters All. v. Walz*, 492 F. Supp. 3d 822, page 30 (D. Minn. 2020), available online: https://law.justia.com/cases/federal/district-courts/minnesota/mndce/0:2020cv01688/189156/51/

540   *Minnesota Voters Alliance v. Walz*, 492 F. Supp. 3d 822, 837-838 (D. Minn. 2020), (https://law.justia.com/cases/federal/district-courts/minnesota/mndce/0:2020cv01688/189156/51/) quoting *United States v. O'Brien*, 391 U.S. 367, 377 (1968), available online: https://tile.loc.gov/storage-services/service/ll/usrep/usrep391/usrep391367/usrep391367.pdf

This was one of many 2020 orders that improperly depended on *Jacobson v. Massachusetts*. At least Judge Schiltz' ruling had the excuse of being issued in October 2020, one month before the Supreme Court pulled the rug out from under COVIDcrisis rulings based on *Jacobson v. Massachusetts*, though this was no doubt cold comfort to the plaintiffs Judge Schiltz dismissed without leave to amend.

As an aside, the Minnesota plaintiffs before Judge Schiltz also advanced a First Amendment argument that Governor Walz' mask mandate chilled their ability to engage in political activities protected by the First Amendment, including voting, campaigning, and associating by creating a contradiction in the law. An earlier state statute made it a misdemeanor for a person to conceal their identity "in a public place by means of a robe, mask, or other disguise." The plaintiffs argued that Governor Walz' mask mandate placed them in an impossible situation where they committed a misdemeanor by *wearing* a mask, but also committed a misdemeanor by *not* wearing a mask. This was unsuccessful. Courts universally ruled that wearing a mask to prevent COVID did not run afoul of laws against concealing one's identity by using a mask. Despite this, the pervasiveness of temporary exceptions written into mask mandates for when "the mask is required to be lowered briefly for identification or security purposes" (Department of Defense Office of the Undersecretary of Defense for Personnel and Readiness, "Consolidated Department of Defense Coronavirus Disease 2019 Force Health Protection Guidance — Revision 5," March 24, 2023,

The preferred default was to find that mask mandates either did not involve the First Amendment, or were subject only to rational basis scrutiny, which they easily passed. As Judge John Kastrenakes put it: "this Court finds that no constitutional right is infringed by the Mask Ordinance's mandate to wear a facial covering, and that the requirement to wear such a covering has a clear rational basis based on the protection of public health."[541] However, when complaints specifically alleged "an intent to convey a particularized message" by not wearing a mask, or raised another constitutional argument that was harder to ignore, such as that the mask mandates were *not*, in fact, "neutral and of general applicability," a few courts took it in hand to address intermediate *O'Brien* scrutiny.[542]

Even assuming that mask-wearing is expressive conduct or symbolic speech, these courts would argue, freedom of speech is not absolute, and "The government generally has a freer hand in restricting expressive conduct than it has in restricting the written or spoken word."[543] Mask mandates: "can only be construed as a content-neutral time, place, and manner restriction that is subject to intermediate scrutiny,"[544] said Judge Robert Chambers. Judge Patrick Schiltz said that mask mandates: "easily pass muster" under the four intermediate scrutiny *O'Brien* tests of 1) being within the constitutional

---

https://media.defense.gov/2023/Mar/28/2003187831/-1/-1/1/CONSOLIDATED-DEPARTMENT-OF-DEFENSE-CORONAVIRUS-DISEASE-2019-FORCE-HEALTH-PROTECTION-GUIDANCE-REVISION-5.PDF), or "to remove a face covering to verify an identity for lawful purposes" (*Snell v. Walz*, No. A21-0626, p. 5 (Minn. Feb. 8, 2023), available online: https://mn.gov/law-library-stat/archive/supct/2023/OPA210626-020823.pdf), undermined not only the stated medical rationale for mask mandates, but also any governmental defendant arguments that COVIDcrisis masks "neither conceal one's identity nor constitute a 'disguise'" (*Minnesota Voters Alliance v. Walz*, 492 F. Supp. 3d 822, 837-838 (D. Minn. 2020), https://law.justia.com/cases/federal/district-courts/minnesota/mndce/0:2020cv01688/189156/51/).

541   *Machovec v. Palm Beach County*, No. 2020CA006920AXX — Filing #110806335, 27 July 2020 (15th Judicial Circuit, Florida), p. 7, available online: https://clearinghouse.net/doc/109346/ & https://www.floridacivilrights.org/wp-content/uploads/2020/07/Order-Denying-Temporary-Injunction.pdf
Machovec v. Palm Beach County, 310 So. 3d 941 (Fla. Dist. Ct. App. 2021), https://casetext.com/case/machovec-v-palm-beach-cnty
c.f. *Denis v. Ige et al*, No. 1:2021cv00011 - Document 62 (D. Haw. 2021), pg. 24-29, available online: https://law.justia.com/cases/federal/district-courts/hawaii/hidce/1:2021cv00011/152658/62/
Massie v. Pelosi, 590 F. Supp. 3d 196, *226 (D.C. 2022). Available online: https://scholar.google.com/scholar_case?case=735226917287193397&hl=en&as_sdt=4000003&scfhb=1
Delaney v. Baker, 511 F. Supp. 3d 55, *73-74 (D. Mass. 2021), available online: https://casetext.com/case/delaney-v-baker-2
542   Even those courts which addressed *O'Brien* scrutiny had often already concluded on other grounds that masks should be upheld by the time they got around to doing so.
543   *Massie v. Pelosi*, 590 F. Supp. 3d 196, *225 (D.C. 2022). Available online: https://scholar.google.com/scholar_case?case=735226917287193397&hl=en&as_sdt=4000003&scfhb=1
544   *Stewart v. Justice* 502 F. Supp. 3d 1057, 1066 (S.D.W. Va. 2020), available online: https://casetext.com/case/stewart-v-justice-1

power of the government, 2) furthering an important or substantial government interest, 3) being unrelated to the suppression of speech, and 4) being no greater than essential.[545]

Mask mandates generally breezed through the first three *O'Brien* criteria. As part of its general police powers, a state government "has the constitutional authority to enact measures to protect the health and safety of its citizens."[546] Thus, to most courts there was no question that, assuming mask mandates do not unacceptably infringe any rights, they are "within the constitutional power of the government."[547] In that same vein, the substantial governmental interest supposedly furthered by mask mandates, that of controlling the spread of a highly contagious and potentially deadly disease in the name of public health, "is unrelated to the suppression of free expression."[548] In practice, any semi-plausible rationale for compulsory masking presented by a governmental defendant, no matter how pretextual it may be in reality, will satisfy the third *O'Brien* criteria of needing to be unrelated to the suppression of speech. As the Georgia Supreme Court put it in a 1990 case on masks: "under settled principles of constitutional law the court will not strike down an otherwise constitutional statute on the basis of an alleged illicit legislative motive."[549] With so much benefit of the doubt, there was never a realistic chance that mask mandates would be found to *target* speech.

Discussions of the first three *O'Brien* criteria in court rulings were usually framed by a formulaic recitation of the death toll from COVID, an assertion that there was no cure or effective treatment, a mention of transmission by "droplets" or "aerosols," a citation to the CDC, and an appeal to scientific consensus. This framing set judges up to vault the biggest hurdle in intermediate *O'Brien* scrutiny — the requirement that universal compulsory masking be "no greater than essential"; in other words, "narrowly tailored."

In the hysteria-charged atmosphere of the early COVIDcrisis, judges were under great internal and external pressure to uphold mask mandates. Once things calmed down, they found themselves under

---

545   *Minnesota Voters All. v. Walz*, 492 F. Supp. 3d 822, pp. 28 and 30-31 (D. Minn. 2020), available online: https://law.justia.com/cases/federal/district-courts/minnesota/mndce/0:2020cv01688/189156/51/
Massie v. Pelosi, 590 F. Supp. 3d 196, *226-228 (D.C. 2022). Available online: https://scholar.google.com/scholar_case?case=735226917287193397&hl=en&as_sdt=4000003&scfhb=1
546   *Minnesota Voters All. v. Walz*, 492 F. Supp. 3d 822, page 30 (D. Minn. 2020), available online: https://law.justia.com/cases/federal/district-courts/minnesota/mndce/0:2020cv01688/189156/51/
547   *Massie v. Pelosi*, 590 F. Supp. 3d 196, *226 (D.C. 2022). Available online: https://scholar.google.com/scholar_case?case=735226917287193397&hl=en&as_sdt=4000003&scfhb=1
548   *Massie v. Pelosi*, 590 F. Supp. 3d 196, *227 (D.C. 2022). Available online: https://scholar.google.com/scholar_case?case=735226917287193397&hl=en&as_sdt=4000003&scfhb=1
Minnesota Voters All. v. Walz, 492 F. Supp. 3d 822, page 30 (D. Minn. 2020), available online: https://law.justia.com/cases/federal/district-courts/minnesota/mndce/0:2020cv01688/189156/51/
549   *State v. Miller*, 260 Ga. 669, 672 n 3 (Ga. 1990), available online: https://law.justia.com/cases/georgia/supreme-court/1990/s90a1172-1.html

pressure to appear consistent and avoid admitting their own errors (or calling out their colleagues' errors) that were nearly as strong. This makes it less surprising, but still every bit as appalling, that courts during the COVIDcrisis repeatedly found any limitations on First Amendment freedoms from universal compulsory masking to be "incidental," "no greater than is essential," and "outweighed by the strong governmental interests behind the policies."[550] Judge Robert Chambers in West Virginia provided one of the best succinct summaries of the combined reasoning that got mask mandates past intermediate *O'Brien* scrutiny:

> "[E]ven if refusing to wear a mask is protected speech, the Mask Mandate withstands intermediate scrutiny as a content-neutral time, place, and manner restriction: being required to wear a mask in conformity with the Governor's Order may be an inconvenience or annoyance, but it is a trivial imposition on an individual's freedom outweighed by the reasonableness of such precautions during a pandemic…

> "The Mask Mandate passes this test: (1) it is a justifiable exercise of the state's police power under the United States Constitution; (2) it furthers a compelling interest by slowing the spread of the coronavirus and saving lives; (3) that interest is unrelated to the alleged restrictions on Plaintiffs' speech; and (4) it imposes no greater restriction than is required because it is limited by age, activity, and place."[551]

In a separate opinion earlier in the same case, Judge Chambers also wrote:

> "The Court also finds that [the mask mandate] is narrowly tailored. Although the state-wide order is undeniably broad, it is no less narrowly tailored to its purpose. The order is geographically broad and indefinite, but so too is the virus. The order applies to all individuals regardless of their apparent health, but the evidence shows that asymptomatic individuals are nonetheless contagious."[552]

---

550   *Massie v. Pelosi*, 590 F. Supp. 3d 196, *227, *231 (D.C. 2022). Available online: https://scholar.google.com/scholar_case?case=735226917287193397&hl=en&as_sdt=4000003&scfhb=1

551   *Stewart et al. v. Justice et al.* No. 3:2020cv00611 - Document 45 (S.D.W. Va. 2021), pp. 10-11, available online: https://law.justia.com/cases/federal/district-courts/west-virginia/wvsdce/3:2020cv00611/230251/45/

552   *Stewart v. Justice* 502 F. Supp. 3d 1057, 1067 (S.D.W. Va. 2020), available online: https://casetext.com/case/stewart-v-justice-1

Another factor that courts repeatedly cited to show that universal masking was "no greater than essential" — and therefore not violative of fundamental human or constitutional rights — was that mask mandates left open "ample alternative channels" for any desired communications inhibited by compulsory masking.[553] As Federal Judge Susan Mollway said in Hawaii: "People wearing masks can still speak… the Mask Mandates do not impose more than an incidental burden on speech. To the contrary, individuals can still engage in expressive activity. They just have to wear masks while they do."[554] In his dismissal of Republican congressmen's legal challenge to compulsory masking, Judge Reggie B. Walton said there were "myriad means" of communicating the congressmen's message opposing masks, "including wearing masks or other clothing containing the messages they wanted to convey, or making speeches from the House Chamber or elsewhere on the subject."[555]

The real overriding reason mask mandates survived "scrutiny," was that COVID-19 was "a public health crisis." "The Pandemic." A plague. An *emergency*. Individual liberties, no matter how inalienable, simply *had* to take a back seat (especially in light of *Jacobson v. Massachusetts*). A Florida court stated this attitude the most directly: "the spread of infectious or contagious diseases or other potential public calamity, presents an exigent circumstance before which all private rights must immediately give way under the government's police power."[556] In every decision, the assumption that masks "work" was omnipresent — an unquestionable finding of fact which no judge or jury of ignorant laypeople could be qualified or permitted to evaluate for themselves. After all, the CDC *said so*.

- "Contrary to the plaintiffs' assertion regarding the existence of 'recent scientific findings that the use of face coverings has no appreciable effect on slowing or halting the spread of COVID-19…' the consensus within the scientific community is clear that masks — and, in particular, well-fitting protective masks — are effective in slowing the spread of the COVID-19 virus[.]" *Massie v. Pelosi*, 590 F. Supp. 3d 196, *228-229 (D.C. 2022)[557]

553   *Massie v. Pelosi*, 590 F. Supp. 3d 196, *225 (D.C. 2022). Available online: https://scholar.google.com/scholar_case?case=735226917287193397&hl=en&as_sdt=4000003&scfhb=1

554   *Denis v. Ige et al*, No. 1:2021cv00011 - Document 62 (D. Haw. 2021), pg. 30-31, available online: https://law.justia.com/cases/federal/district-courts/hawaii/hidce/1:2021cv00011/152658/62/

555   *Massie v. Pelosi*, 590 F. Supp. 3d 196, *225 (D.C. 2022). Available online: https://scholar.google.com/scholar_case?case=735226917287193397&hl=en&as_sdt=4000003&scfhb=1

556   *Machovec v. Palm Beach County*, 310 So. 3d 941 (Fla. Dist. Ct. App. 2021), available online: https://casetext.com/case/machovec-v-palm-beach-cnty quoting Davis v. City of S. Bay, 433 So. 2d 1364, 1366 (Fla. 4th DCA 1983), available online: https://casetext.com/case/davis-v-city-of-south-bay

557   *Massie v. Pelosi*, 590 F. Supp. 3d 196, *228-229 (D.C. 2022). Available online: https://scholar.google.com/scholar_case?case=735226917287193397&hl=en&as_sdt=4000003&scfhb=1

- "...face coverings and other social distancing restrictions are effective at slowing the spread of the virus." *Stewart v. Justice* 502 F. Supp. 3d 1057, 1065 (S.D.W. Va. 2020)[558]

- [The West Virginia mask mandate] "serves to slow the spread of COVID-19 by requiring individuals to wear masks indoors." *Stewart v. Justice* 502 F. Supp. 3d 1057, 1066 (S.D.W. Va. 2020)[559]

- "The evidence for the efficacy of wearing masks in preventing the spread of COVID-19 is indisputable and is in accordance with CDC guidelines." *Stewart v. Justice.* No. 3:2020cv00611 - Document 45, p.7 (S.D.W. Va. 2021)[560]

- "Plaintiffs do not deny the existence of COVID-19, or that it is a dangerous disease, or that it is easily spread (including by people who do not know that they are infected), or that face coverings slow its spread and thus save lives." *Minnesota Voters Alliance v. Walz,* 492 F. Supp. 3d 822, p. 4 (D. Minn. 2020)[561]

- "For example, the CDC stated that 'face coverings are one of the most powerful weapons we have to slow and stop the spread of the virus — particularly when used universally within a community setting.'" *Sehmel v. Shah*, 514 P.3d 1238, at 8 (Wash. Ct. App. 2022)[562]

- "It has been proven that the wearing of masks can slow the transmission of the spread of the coronavirus." *Delaney v. Baker*, 511 F. Supp. 3d 55, *67 (D. Mass. 2021)[563]

- "Additionally, requirements for face coverings also reduce the chance that respiratory droplets containing the virus will infect others." *Antietam Battlefield KOA v. Hogan*, 461 F. Supp. 3d 214, 229 (D. Md. 2020)[564]

COVIDcrisis court rulings on masks and mask mandates present readers with one long, slogan-infected litany after another: "we do not have a constitutional or protected right to infect others,"[565]

558   *Stewart v. Justice* 502 F. Supp. 3d 1057, 1065 (S.D.W. Va. 2020), available online:
https://casetext.com/case/stewart-v-justice-1

559   *Stewart v. Justice* 502 F. Supp. 3d 1057, 1066 (S.D.W. Va. 2020), available online:
https://casetext.com/case/stewart-v-justice-1

560   *Stewart et al. v. Justice et al.* No. 3:2020cv00611 - Document 45, p. 7 (S.D.W. Va. 2021), available online:
https://law.justia.com/cases/federal/district-courts/west-virginia/wvsdce/3:2020cv00611/230251/45/

561   *Minnesota Voters All. v. Walz,* 492 F. Supp. 3d 822, p. 4 (D. Minn. 2020), available online:
https://law.justia.com/cases/federal/district-courts/minnesota/mndce/0:2020cv01688/189156/51/

562   *Sehmel v. Shah*, 514 P.3d 1238, at 8 (Wash. Ct. App. 2022), available online:
https://caselaw.findlaw.com/wa-court-of-appeals/2182709.html

563   *Delaney v. Baker*, 511 F. Supp. 3d 55, *67 (D. Mass. 2021), available online:
https://casetext.com/case/delaney-v-baker-2

564   *Antietam Battlefield KOA v. Hogan*, 461 F. Supp. 3d 214, 229 (D. Md. 2020), available online:
https://casetext.com/case/antietam-battlefield-koa-v-hogan

565   *Machovec v. Palm Beach County*, No. 2020CA006920AXX — Filing #110806335, 27 July 2020, page 12
(15th Judicial Circuit, Florida), available online: https://clearinghouse.net/doc/109346/ &
https://www.floridacivilrights.org/wp-content/uploads/2020/07/Order-Denying-Temporary-Injunction.pdf

and one "cannot plausibly claim that anyone has died as a result of having had to wear a mask."[566] Wearing or not wearing a mask is conduct, ran the arguments. It cannot be speech protected under the First Amendment. There are multiple non-communicative alternative reasons to wear or not wear masks, so it is not "inherently expressive," or it has only a "kernel of expression." Explanatory speech is necessary, and so the meaning of masking and non-masking is not "overwhelmingly apparent." Universal masking is a basic, commonsense safety requirement like seatbelts and safety helmets. Masks are a minor, trivial inconvenience. Any injury masks might cause is speculative. Masks do not inflict harm, especially any kind of irreparable harm. Opponents of masks failed to carry their burden of proof or show adequate threat of irreparable, redressable harm to establish standing.[567] A mask mandate in church did not constitute a redressable injury sufficient to establish standing against a governor's mask mandate because it was the *archdiocese*, not the *governor* imposing the church's mask mandate.[568] Mask mandates in public locations might constitute a redressable injury sufficient to establish standing, but the mask mandate being challenged survives every type of applicable scrutiny.[569]

On and on it went. Under no circumstances would COVIDcrisis jurisprudence subject masks to strict scrutiny on the basis of the First Amendment Free Speech clause. According to these courts, under the Free Speech clause of the First Amendment, universal masking warranted either no scrutiny, rational basis scrutiny, or — at *most* — remarkably deferential intermediate scrutiny. Such was the majority consensus. But the majority consensus was in error, and every court that dismissed or ruled against a Free Speech challenge to compulsory masking got the law in this area badly wrong.

The COVIDcrisis is far from the first time in American history that courts have stubbornly refused to uphold citizens' rights until the political atmosphere made it safe to do so. Rulings that dismissed challenges to compulsory masking reasoned in circles, used brute force denial, or often completely ignored the central allegations in the First Amendment claims they dismissed. Virtually every premise in such decisions remains erroneous: they presume that masks work; they nonsensically presume that

---

566   *Denis v. Ige et al*, No. 1:2021cv00011 - Document 62 (D. Haw. 2021), p. 28, available online: https://law.justia.com/cases/federal/district-courts/hawaii/hidce/1:2021cv00011/152658/62/

567   *Minnesota Voters All. v. Walz*, 492 F. Supp. 3d 822, page 32 (D. Minn. 2020), available online: https://law.justia.com/cases/federal/district-courts/minnesota/mndce/0:2020cv01688/189156/51/

568   *Delaney v. Baker*, 511 F. Supp. 3d 55, *70 (D. Mass. 2021), available online: https://casetext.com/case/delaney-v-baker-2

       The court said there was "no evidence" to support plaintiff Delaney's allegation that his archdiocese (the Archdiocese of Boston) would not have implemented a mask mandate absent Massachusetts' Governor Baker's order, therefore it was only speculative that injunctive relief against the governor's mandate would relieve the plaintiff of the parish-level mask mandate. "Delaney has not proven that, but for Governor Baker's orders, the Archdiocese would institute rules that would remedy his wearing a mask inside[.]"

569   *Delaney v. Baker*, 511 F. Supp. 3d 55, *71 (D. Mass. 2021), available online: https://casetext.com/case/delaney-v-baker-2

masking *everyone* — including the healthy, the least vulnerable, and the already-recovered — is both reasonable *and* "no greater than is essential"; they deny the incontrovertible communicative effect that mask-wearing or non-wearing has; and they ultimately forced the plaintiffs to symbolically affirm masking even while trying to verbally argue against it. Judges who said plaintiffs had no chance of success challenging masks on a Free Speech basis were correct — at least insofar as they were describing their own courts. What they were *not* correct about was the actual law and legal precedents involved.

Masks are speech

Mask speech prior to COVID

The fact of the matter is that going into 2020, there was already extensive case law precedent holding that mask-wearing *can be* inherently expressive. Mask-speech cases have involved everything from cloth masks and dust masks to V for Vendetta and KKK masks. Masks have been used to communicate everything from threats and intimidation to "protesting against politicians, banks, and financial institutions."[570] Even if mask-wearing or non-wearing were not *inherently* expressive when judged in isolation and out of context, the critical context of the COVIDcrisis *made* it expressive. Eating and drinking or walking from point A to point B are not "inherently expressive" either, but case law precedents decisively show that both of these can be symbolic speech protected under the First Amendment in the right "surrounding circumstances."

In their 1890 joint essay, "The Right to Privacy," U.S. Supreme Court Justices Samuel Warren and Louis Brandeis explained how multiple earlier Supreme Court Rulings assumed and protected both a broad individual right to privacy and to the control of one's speech, regardless of the form that speech may take:

> "The common law secures to each individual the right of determining, ordinarily, to what extent his thoughts, sentiments, and emotions shall be communicated to others. Under our system of government, he can never be compelled to express them (except when upon the witness-stand)… The existence of this right does not depend upon the particular method of expression adopted. It is immaterial whether it be by word, or by signs, in painting, by sculpture, or in music."[571]

---

570   *Gates v. Khokhar*, 884 F.3d 1290, 1299 (11th Cir. 2018), *cert. denied,* 139 S. Ct. 807 (2019), available online: https://www.supremecourt.gov/DocketPDF/18/18-511/66921/20181015162248885_Gates%20v.%20Khokhar%20Appendix.pdf

571   Samuel D. Warren and Louis D. Brandeis, "The Right to Privacy," December 15, 1890, Harvard Law Review Vol. IV, No. 5 (197) available online: https://www.jstor.org/stable/1321160

Deliberately hiding or showing one's face to others has been understood to communicate a message for millennia. This form of communication is as old as the Bible (see the appendix for a few dozen examples). In the United States, this was so widely understood that Stephen J. Simoni's review of court cases and laws involving mask-wearing through 1992, published in the *Fordham Law Review*, concluded:

> "Virtually all existing anti-mask laws violate the United States Constitution. Because mask-wearing can itself serve an expressive function, as well as enable speakers who need anonymity to express their beliefs publicly, anti-mask laws both directly violate and inhibit the First Amendment's free speech and free assembly guarantees. Furthermore, as many instances of mask-wearing are completely innocent, anti-mask laws likely violate the Fourteenth Amendment's substantive due process guarantee."[572]

The review noted, "demonstrators often use masks only to express ideas and not to conceal their identities."[573] The review included a proposed model anti-mask law with an explicit umbrella exemption for expressive mask-wearing "speech" protected by the First Amendment, including "expressive mask-wearing when the wearers do not desire their resulting anonymity."[574] There would be no point including a provision for mask-speech if, as many courts asserted during the COVIDcrisis, mask-wearing is not or cannot *be* speech protected by the First Amendment. Another 2012 mask law analysis published in the *Hastings International & Comparative Law Review* pointed out that *anti-mask laws* failed the First Amendment neutrality requirement for similar reasons: "The enumerated lists of exceptions in general anti-mask laws belie such content neutrality. For example, the inclusion of exemptions for masquerades, but not for political protests, implicitly discriminates between entertaining and persuasive content."[575] The truth is that mask-wearing *can* be — and often *is* — speech, and if *wearing* a mask can be speech, *refusing* to wear a mask can *also* be speech, especially when the wearers and non-wearers are in the minority or acting contrary to a mandate. The more charged the surrounding circumstances become, the more likely masking and non-masking are to

---

572   Stephen J. Simoni, *"Who Goes There?" — Proposing a Model Anti-Mask Act*, 61 *Fordham Law Review*, 241, 273 (1992), available online: https://ir.lawnet.fordham.edu/cgi/viewcontent.cgi?article=3009&context=flr

573   Stephen J. Simoni, *"Who Goes There?" — Proposing a Model Anti-Mask Act*, 61 *Fordham Law Review*, 241, 250 n 61 (1992), available online: https://ir.lawnet.fordham.edu/cgi/viewcontent.cgi?article=3009&context=flr

574   Section 100(b)(1)
Stephen J. Simoni, *"Who Goes There?" — Proposing a Model Anti-Mask Act*, 61 *Fordham Law Review*, 241, 267-268 (1992), available online: https://ir.lawnet.fordham.edu/cgi/viewcontent.cgi?article=3009&context=flr

575   Evan Darwin Winet, "Face-Veil Bans and Anti-Mask Laws: State Interests and the Right to Cover the Face," Hastings International & Comparative Law Review, Vol. 35, 217, 235-236 (2012), available online: https://repository.uchastings.edu/hastings_international_comparative_law_review/vol35/iss1/8/

be a form of speech. In the context of the COVIDcrisis, because universal mask *mandates* effectively restricted expressive unmasking based on the content (whether political, medical, and/or religious), they did not serve the state interest in reducing the spread of COVID-19 in the least restrictive manner needed to pass scrutiny under the First Amendment.

Just 10 years after the U.S. Supreme Court's 1968 expressive conduct decision in *United States v. O'Brien*, the *O'Brien* standard was applied directly to masks in the 1978 Texas case of *Aryan v. MacKey*.[576] Texas Tech University granted a permit to students to hold a protest march with the proviso that they were not allowed to wear masks while doing so. Students who wanted to wear masks as a means of preserving their anonymity *and* as a form of symbolic speech sued for violation of their First Amendment rights. In his ruling, Judge Patrick Higginbotham recognized "two distinct functions which the masks serve. The first is noncommunicative… The second is communicative: the masks have become symbols of protests… Serious First Amendment questions arise, however, when there is such a nexus between anonymity and speech that a bar on the first is tantamount to a prohibition on the second."[577] Judge Higgenbotham's insight did not suddenly become untrue in the context of SARS-CoV-2 simply because public health was substituted for anonymity. If anything, even *more* serious First Amendment questions arise when there is such a nexus between public health and speech that a mandate imposed ostensibly to serve public health becomes a content-based bar on speech or an outright form of compelled speech.

Judge Higginbotham also rightly pointed out in his ruling that the burden of proof was not on the plaintiff students whose rights were being violated, but on the defendant university authorities to prove that their regulation on masks did not violate the First Amendment. During the COVIDcrisis, many courts wrongly placed this burden of proof on the plaintiffs who sought to challenge mask regulations. As in the COVIDcrisis, the potential harm the university authorities sought to avoid in 1978 was related to masks' non-communicative functions, and the *O'Brien* standard required the university authorities "to demonstrate a substantial causal relationship between the restriction and the interest the restriction is supposed to further."[578] In other words, under *O'Brien* scrutiny, evidence makes a difference. It matters how well masks work.

Texas Tech University lost in court. It was not even able to get to the final "no greater than essential" part of the *O'Brien* test, because it could not show that its rationale for forbidding masks was based

576   *Aryan v. MacKey*, 462 F. Supp. 90 (N.D. Tex. 1978),
https://law.justia.com/cases/federal/district-courts/FSupp/462/90/2142290/

577   *Aryan v. MacKey*, 462 F. Supp. 90, *92 (N.D. Tex. 1978),
https://law.justia.com/cases/federal/district-courts/FSupp/462/90/2142290/

578   *Aryan v. MacKey*, 462 F. Supp. 90, *93 (N.D. Tex. 1978),
https://law.justia.com/cases/federal/district-courts/FSupp/462/90/2142290/

on anything other than speculation. In upholding the students' First Amendment rights, Judge Higginbotham quoted another 1969 Supreme Court decision on symbolic speech, *Tinker v. Des Moines Independent Community School District*: "in our system, undifferentiated fear or apprehension of disturbance Is not enough to overcome the right to freedom of expression."[579] Yet time and again during the COVIDcrisis, courts allowed "studies" and institutional pronouncements which were no better than elaborate speculation to satisfy this criterion despite the mountain of contrary evidence.

The symbolic speech inherent in masks came up in court again in 1994, when an Illinois man, Timothy Ryan, Jr., was concerned about possible airborne Legionnaires' Disease in the ventilation system of a courthouse. Ryan insisted on wearing a mask into the courthouse, was briefly arrested by a Sherriff's deputy for refusing to remove it, and promptly sued, charging violations of the First and Fourth Amendments. The Seventh Circuit appellate court which reviewed this case and issued its final ruling in 1995 was headed by Judge Richard Posner, the most cited legal scholar in the United States during the 20th century.[580] The panel's unanimous ruling held that the rule *against* mask-wearing in the courthouse, which led to Ryan's arrest, was reasonable and constitutional *because* of what mask-wearing communicates. The panel took it for granted that mask-wearing *is* communicative and that a mask can be a symbol. "The wearing of a mask inside a courthouse implies intimidation... it would be the height of irresponsibility to allow masked people into courthouses."[581] Even if the wearer did not verbally explain *why* he was wearing a mask, Judge Posner said, "most people would assume that he had a respiratory disease, or allergies, or an impaired immune system, or some other condition… One assumes when one sees such people that they have a problem peculiar to themselves, that they are not in normal health."[582] The court also did not dispute that a mask-wearer could very conceivably "cause a panic" among the unmasked if he verbally told them "that the air was not fit to breathe unless one was wearing a protective mask."[583] During the COVIDcrisis, many people strenuously — and very reasonably — objected to being forced to broadcast propositions like these at everyone who saw

579   *Aryan v. MacKey*, 462 F. Supp. 90, *93 (N.D. Tex. 1978), https://law.justia.com/cases/federal/district-courts/FSupp/462/90/2142290/ quoting *Tinker v. Des Moines Independent Community School District*, 393 U.S. 503, 508 (1969), available online: https://www.loc.gov/item/usrep393503/

580   Judge Posner came in first with 7,981 citations to his credit, nearly doubling that of the 4,488 citations from second place Ronald Dworkin.
Shapiro, Fred R. (2000). "The Most-Cited Legal Scholars". *Journal of Legal Studies*. **29** (1): 409–426. https://www.jstor.org/stable/10.1086/468080

581   *Ryan v. County of DuPage*, 45 F.3d 1090, 1092 (7th Cir. 1995), available online: https://casetext.com/case/ryan-v-county-of-dupage

582   *Ryan v. County of DuPage*, 45 F.3d 1090, 1095 (7th Cir. 1995), available online: https://casetext.com/case/ryan-v-county-of-dupage

583   *Ryan v. County of DuPage*, 45 F.3d 1090, 1094-1095 (7th Cir. 1995), available online: https://casetext.com/case/ryan-v-county-of-dupage

them, and Judge Posner's hypothetical was proven correct: mask-wearing accompanied by words that the air was not fit to breathe unless one was wearing a protective mask *did* cause widespread panic more contagious than any virus.

Ryan's lawsuit defending his mask-wearing failed in part for the same reason that some of the COVIDcrisis lawsuits against masks did: he failed to allege that his primary purpose in wearing a mask was to engage in symbolic speech:

> "Ryan did not say that he was wearing it to protest the poor quality of the air in the courthouse; he said he was wearing it to save his throat and lungs from the bad air. The mask was not being used as a symbol; it was being used for its original purpose as a medical appliance."[584]

Ryan was not allowed to invoke the First Amendment after the fact because his own stated primary purpose in wearing a mask was not communicative. At the same time, Judge Posner and the rest of the panel held that the rule *against* wearing masks under which Ryan was arrested was constitutionally valid *because* the menacing messages sent by wearing a mask interfered with "security and decorum" in the courthouse — "in which rational reflection and disinterested judgment will not be disrupted by intimations of violence, or by the subtler hints of coercion[.]"[585] Subtle "hints of coercion," might be involved in mask-wearing? You don't say! I, for one, am shocked.

Judge Posner's ruling was not charting entirely new territory. In 1990, a case involving intentional symbolic speech using a KKK mask went up to the Georgia Supreme Court in *State v. Miller*. It was simply taken without controversy for purposes of analysis that "wearing a mask was conduct 'sufficiently imbued with elements of communication' to implicate the First Amendment."[586] In that case the Supreme Court of Georgia held that Georgia's *anti*-mask statute was constitutional because the "implied threats and intimidation," associated with KKK mask-wearing "had helped to create a

---

584   *Ryan v. County of DuPage*, 45 F.3d 1090, 1094 (7th Cir. 1995), available online: https://casetext.com/case/ryan-v-county-of-dupage

585   *Ryan v. County of DuPage*, 45 F.3d 1090, 1095 (7th Cir. 1995), available online: https://casetext.com/case/ryan-v-county-of-dupage

586   *State v. Miller*, 260 Ga. 669, 671 n 2 (Ga. 1990), available online: https://law.justia.com/cases/georgia/supreme-court/1990/s90a1172-1.html

     Technically, the court "assumed[d] without deciding that… wearing a mask was conduct sufficiently imbued with elements of communication to implicate the First Amendment." Just four years later, the same court said: "In Miller, we held that the [Anti-Mask] Act does not unconstitutionally infringe upon any expressive conduct protected by the First Amendment which may be implicated." (Daniels v. State, 264 Ga. 460, *462, 448 S.E.2d 185 (1994), available online: https://law.justia.com/cases/georgia/supreme-court/1994/s94g0362-1.html)

climate of fear that prevented Georgia citizens from exercising their civil rights."[587] In order to uphold Georgia's anti-mask law, however, the Georgia court had to read an unwritten degree of specificity into the statute. The Court held that a conviction under the anti-mask statute required an intent or reasonable expectation on the part of the mask-wearer that their mask would provoke "apprehension of intimidation, threats, or impending violence."[588] This narrowing judicial interpretation was needed to ensure that: "To the extent that the statute does proscribe the communicative aspect of mask-wearing conduct, its restriction is limited to threats and intimidation, which is not protected expression under the First Amendment."[589] In other words, Georgia's anti-mask law only survived a First Amendment constitutional challenge because the Georgia Supreme Court made it narrowly tailored enough that "It would be absurd to interpret the statute to prevent non-threatening political mask-wearing."[590] In Florida, a similar Anti-Mask Act was struck down as unconstitutional before it even went into effect.[591]

In the *State v. Miller* decision, both the Georgia Supreme Court majority and the dissent found it uncontroversial that wearing a mask included symbolic speech. The majority simply held that the speech communicated by wearing a mask did not fall under First Amendment protection because, as the dissenting justice complained, it was "irrebuttably presumed to be symbolic speech that gives rise to apprehension of intimidation, threats or impending violence." The dissenting justice also highlighted the problems of consistency in upholding a ban on symbolic speech using Klan masks when symbolic speech through wearing Nazi uniforms had already been upheld by the United States Supreme Court in a 1978 ruling.[592]

By 2018, at least in Georgia, the only real question remaining was whether police officers could escape civil liability for wrongfully arresting mask-wearers who refused to remove their masks but did not *intend* "to threaten or intimidate." This was decided in the 2018 case *Gates v. Khokhar*, when a man

---

587   *State v. Miller*, 260 Ga. 669, 672 (Ga. 1990), available online:
https://law.justia.com/cases/georgia/supreme-court/1990/s90a1172-1.html
588   *State v. Miller*, 260 Ga. 669, 676-677 (Ga. 1990), available online:
https://law.justia.com/cases/georgia/supreme-court/1990/s90a1172-1.html
589   *State v. Miller*, 260 Ga. 669, 673 (Ga. 1990), available online:
https://law.justia.com/cases/georgia/supreme-court/1990/s90a1172-1.html
590   *State v. Miller*, 260 Ga. 669, 676 (Ga. 1990), available online:
https://law.justia.com/cases/georgia/supreme-court/1990/s90a1172-1.html
591   *Robinson v. State*, 393 So.2d 1076 (Fla. 1980),
https://law.justia.com/cases/florida/supreme-court/1980/58232-0.html
592   *State v. Miller*, 260 Ga. 669, 680 (Ga. 1990), available online:
https://law.justia.com/cases/georgia/supreme-court/1990/s90a1172-1.html summarizing Collin v. Smith, 578 F2d 1197, 1202 (?th Cir. 1978) (cert. denied, 439 U.S. 916 (99 SC 291, 58 LE2d 264) (1978),
https://law.justia.com/cases/federal/appellate-courts/F2/578/1197/448646/

sued under the First Amendment after being improperly arrested for wearing a V for Vendetta mask during a protest. To reach its verdict, the Georgia Supreme Court referred back to its previous cases from the 1990s.[593] While the majority held that the arresting police officers had enough arguable probable cause to escape civil liability, both the majority and the dissent took it as a settled fact that wearing a mask was a form of speech.[594] The dissenting justice put this most clearly: "There can be no doubt that the order to remove the masks was directed at what would be constitutionally protected expression."[595] The dissent reminded the majority that "the *Miller* court acknowledged this right in finding that engaging in non-threatening political mask-wearing was clearly protected conduct[.]"[596]

Whether the reader is more inclined to side with the majority ruling or the dissent in these Georgia cases, the most important takeaway is that while the specific *nature* and *content* of the speech inherent in masking was disputed, *nobody* disputed that wearing a mask *was* speech which implicated the First Amendment. In all these cases, even the acknowledged compelling state interest in preventing crime and preserving the public peace did not allow "rational basis" dismissal of Free Speech challenges where masks were concerned. This case law history completely forecloses the legitimacy of trite assertions by COVIDcrisis courts that compulsory masking does not implicate the First Amendment. Fights over COVID masks did not break out because people were confused about whether or not masking or non-masking sent a message. Fights over masks broke out because people *got* the messages and treated them like fighting words. The presence or absence of a mask cannot *both* be provocative enough to repeatedly serve as potential fighting words *and* fail to qualify as expressive conduct protected under the First Amendment. Mask proponents cannot have this argument both ways. Court rulings during the COVIDcrisis which found that masking does not constitute compelled speech flew in the face of copious scientific literature, case law, strong pronouncements from both sides of the political aisle, and universal experience.

---

593   *Daniels v. State*, 264 Ga. 460, 448 S.E.2d 185 (1994), available online: https://law.justia.com/cases/georgia/supreme-court/1994/s94g0362-1.html

594   *Gates v. Khokhar*, 884 F.3d 1290, 1299 (11th Cir. 2018), *cert. denied,* 139 S. Ct. 807 (2019), available online: https://www.supremecourt.gov/DocketPDF/18/18-511/66921/20181015162248885_Gates%20v.%20Khokhar%20Appendix.pdf

595   *Gates v. Khokhar*, 884 F.3d 1290, 1299 (11th Cir. 2018), *cert. denied,* 139 S. Ct. 807 (2019), 37a (Williams, J., dissenting), available online: https://www.supremecourt.gov/DocketPDF/18/18-511/66921/20181015162248885_Gates%20v.%20Khokhar%20Appendix.pdf

596   *Gates v. Khokhar*, 884 F.3d 1290, 1299 (11th Cir. 2018), *cert. denied,* 139 S. Ct. 807 (2019), 37a n 6 (Williams, J., dissenting), available online: https://www.supremecourt.gov/DocketPDF/18/18-511/66921/20181015162248885_Gates%20v.%20Khokhar%20Appendix.pdf

Mask-wearing *is* symbolic speech — specifically, "intimidation, violence, and implied threats."[597] Whether the actual or implied threat is of violence communicated by a Klan mask, of protest communicated by a V for Vendetta mask, or of disease communicated by a medical mask, it is indisputable that mask-wearing can and *does* communicate actual or implied threats. The only question is how believable viewers find those threats to be, and how legally actionable they are. The United States Supreme Court has already held that First Amendment protections for speech extend even to the extreme of displaying and parading in actual Nazi uniforms and armbands.[598] Yet Klan masks and protest masks communicate such plain, palpable threats of violence and intimidation that even the First Amendment recoils or permits arguable probable cause for arrests. This being the case, then regardless of whatever non-communicative function masks might ostensibly serve, forcing *anyone* to use their body as a billboard for the plain, palpable threats of infection and illness communicated by medical masks must, at the very least, warrant not just strict scrutiny but the *strictest* scrutiny. Moreover, as in the 2021 case of *Forbes v. County of San Diego,* when compulsory masking was being challenged under grounds other than free speech, court filings defending mask use outright *admitted* the communicative effect of masking and cited this as a *reason* to mandate masks: "in issuing the Mask Rules, the State explained that: '… The use of face coverings by everyone… can also reinforce physical distancing by signaling the need to remain apart.'"[599]

Getting around the precedents

How did COVIDcrisis courts purport to dispatch all of these precedents? Mostly, they tried to do so by brute force denial. They also subtly equivocated and misapplied two descriptors from precedent cases: "inherently expressive," and "overwhelmingly apparent." These two descriptors appear over and over again in COVIDcrisis court rulings upholding mask mandates. According to judges upholding mask mandates, masking and non-masking must be "inherently expressive," with an "overwhelmingly apparent" meaning before the First Amendment can apply.

The phrase "inherently expressive" comes from the 2006 Supreme Court case of *Rumsfeld v. Forum for Academic & Institutional Rights, Incorporated.* In this case, the Supreme Court ruled that denying

---

597   *Gates v. Khokhar,* 884 F.3d 1290, 1299 (11th Cir. 2018), *cert. denied,* 139 S. Ct. 807 (2019), available online: https://www.supremecourt.gov/DocketPDF/18/18-511/66921/20181015162248885_Gates%20v.%20Khokhar%20Appendix.pdf

598   Despite the fact that the First Amendment protects people displaying the swastika on their person while wearing Nazi uniforms in public parades.
National Socialist Party of America v. Village of Skokie, 432 U.S. 43 (1977), available online: https://tile.loc.gov/storage-services/service/ll/usrep/usrep432/usrep432043/usrep432043.pdf

599   *Forbes v. County of San Diego,* Case No. 20-cv-00998-BAS-JLB, at *7 (S.D. Cal. Mar. 4, 2021), available online: https://casetext.com/case/forbes-v-cnty-of-san-diego

military recruiters the same access provided to recruiters from other organizations was not expressive conduct protected under the First Amendment. Law schools discriminating against military recruiters argued that excluding military recruiters was a form of expressive conduct protected under the First Amendment, but the Supreme Court rejected this, responding that:

> "[W]e have extended First Amendment protection only to conduct that is inherently expressive…

> "[L]aw schools 'expressed' their disagreement with the military by treating military recruiters differently from other recruiters. But these actions were expressive only because the law schools accompanied their conduct with speech explaining it…

> "The expressive component of a law school's actions is not created by the conduct itself but by the speech that accompanies it. The fact that such explanatory speech is necessary is strong evidence that the conduct at issue here is not so inherently expressive that it warrants protection under O'Brien."[600]

The phrase "overwhelmingly apparent" comes from the famous 1989 flag-burning case of *Texas v. Johnson*, where the Supreme Court ruled that Gregory Lee Johnson's burning of an American flag qualified as protected speech under the First Amendment because "the expressive, overtly political nature of the conduct was both intentional and overwhelmingly apparent."[601] Some courts have gone so far as to say that this 1989 case created a new "*Johnson* test" for symbolic speech and expressive conduct:

> "To determine 'whether particular conduct possesses sufficient communicative elements to bring the First Amendment into play,' the two-part Johnson test asks: (1) "whether '[a]n intent to convey a particularized message was present,'" and (2) whether "the likelihood was great that the message would be understood by those who viewed it."[602]

600   *Rumsfeld v. Forum for Academic & Institutional Rights, Incorporated*, 547 U.S. 47, 66 (2006), https://tile.loc.gov/storage-services/service/ll/usrep/usrep547/usrep547046/usrep547046.pdf
601   *Texas v. Johnson*, 491 U.S. 397, 397 (1989), https://tile.loc.gov/storage-services/service/ll/usrep/usrep491/usrep491397/usrep491397.pdf
602   *Zinman v. Nova Southeastern University*, No. 21-CIV-60723-RAR, Document 81, p. 25 (S.D. Fla. Aug. 30, 2021), https://www.govinfo.gov/content/pkg/USCOURTS-flsd-0_21-cv-60723/pdf/USCOURTS-flsd-0_21-

When quoting these two cases to uphold mask mandates, judges would rattle off possible *non-communicative* interpretations of masking and non-masking which opposed those of the plaintiffs. The existence of these alternative interpretations, the judges would say, meant that any message being communicated by masking or non-masking was not "overwhelmingly apparent" enough to qualify as "inherently communicative" for First Amendment purposes. A good example of how this line of reasoning usually ran comes from Judge Jared M. Strauss in *Zinman v. Nova Southeastern University*:

> "While Plaintiff may interpret mask wearing as subservience or akin to worship of idols, that message would not be 'overwhelmingly apparent' to one observing him or anyone else wearing a mask… Thus, wearing a mask is not 'inherently expressive' such that the challenged mask mandates constitute compelled speech subject to First Amendment analysis.

> "Similarly, the act of not wearing a mask is not inherently expressive under Johnson. Even positing Plaintiff's intent to convey a message of protest against authority by not wearing a mask, he cannot satisfy the second prong of the Johnson test because there is no great likelihood that such a message would be understood by one observing his conduct. There are myriad reasons why someone may not be wearing [a] mask — the person may qualify for a medical exemption, may be apathetic towards or unconcerned about COVID, may have simply forgotten their mask, or, indeed, may be attempting to send a political or religious message as Plaintiff contends. However, there is no way for an observer to know the reason why without additional explanation."[603]

This cookbook argument may appear solid at first glance, but falls apart on cross-examination. Simply looking closer at both the 1989 *Johnson* case and the 2006 *Rumsfeld* case is enough to do this, but more extensive reading of the Supreme Court's rulings in the area of symbolic speech and expressive conduct makes it "overwhelmingly apparent" that COVID-panicked courts got it badly wrong on masks, misapplying the law and moving the goalposts for what counts as symbolic speech and expressive conduct protected under the First Amendment.

---

cv-60723-1.pdf quoting *Burns v. Town of Palm Beach*, 999 F.3d 1317 at 1336 (11th Cir. 2021), available online: https://casetext.com/case/burns-v-town-of-palm-beach-1

603   *Zinman v. Nova Southeastern University*, No. 21-CIV-60723-RAR, Document 81, p. 25 (S.D. Fla. Aug. 30, 2021), https://www.govinfo.gov/content/pkg/USCOURTS-flsd-0_21-cv-60723/pdf/USCOURTS-flsd-0_21-cv-60723-1.pdf

First, the Supreme Court never said in either the 1989 *Johnson* or 2006 *Rumsfeld* case that the message communicated by expressive conduct or symbolic speech *must* be "overwhelmingly apparent" to qualify as "inherently expressive" under the First Amendment. To turn this phrase into a requirement confuses (and conflates) necessity with sufficiency. Sufficiency and necessity are two very different things, and any judge or lawyer worth their salt is well aware of the distinction. For example, being 50 years old is *sufficient* to serve as a United States senator, but it is only *necessary* to be 30 years old in order to do so.[604] Communicating a message that is "overwhelmingly apparent" to the vast majority of one's audience is — as the Supreme Court ruled in 1989 — obviously *sufficient* to invoke First Amendment protections of expressive conduct, but all that is *necessary* according to the Supreme Court precedents used to *decide* the 1989 *Johnson* flag-burning case is that the activity be "sufficiently imbued with elements of communication to fall within the scope of the First and Fourteenth Amendments."[605]

The demand that the specific message inherent in masking or non-masking be "overwhelmingly apparent" has more in common with the minority dissent in *Texas v. Johnson* than with the majority ruling. "Far from being a case of 'one picture being worth a thousand words,'" complained the dissenting justices, "flag burning is the equivalent of an inarticulate grunt or roar that, it seems fair to say, is most likely to be indulged in not to express any particular idea, but to antagonize others."[606] What the dissenting minority in *Texas v. Johnson* missed with regard to flags, and what judges upholding mask mandates missed (or refused to recognize) during the COVIDcrisis, is that speech — in any form — does not need to be eloquent, articulate, or immune to misunderstanding in order to be comprehensible, effective, and entitled to protection under the First Amendment. Speech which is the most genuine often tends to be more rudimentary.

Judges who upheld mask mandates claimed that possible alternative interpretations of the motivation behind masking and non-masking meant that neither act could qualify as "inherently expressive" symbolic speech because the message was not "overwhelmingly apparent." But the Supreme Court has already issued multiple rulings on symbolic speech and expressive conduct that go against this argument, starting with the landmark 1943 case, *West Virginia v. Barnette et al.* During World War II, members of the Jehovah's Witnesses refused to salute the American flag on religious and free speech grounds, and were punished for doing so, until finally vindicated by the Supreme Court. At the height of worldwide war — a far greater national emergency than COVID ever was — the Supreme Court held that compelling even a *momentary* flag salute was unconstitutional coercion of symbolic

---

604   U.S. Constitution, Article I, section 3, clause 3.

605   *Spence v. Washington*, 418 U.S. 405, 409 (1974), available online:
https://tile.loc.gov/storage-services/service/ll/usrep/usrep418/usrep418405/usrep418405.pdf

606   *Texas v. Johnson*, 491 U.S. 397, 432 (1989),
https://tile.loc.gov/storage-services/service/ll/usrep/usrep491/usrep491397/usrep491397.pdf

speech.[607] Justice Robert H. Jackson (who later served as prosecutor during the Nuremburg Trials) delivered the Supreme Court's ruling. It was a ringing affirmation that echoes just as strongly today: ""If there is any fixed star in our constitutional constellation, it is that no official, high or petty, can prescribe what shall be orthodox in politics, nationalism, religion, or other matters of opinion or force citizens to confess by word or act their faith therein."[608]

West Virginia Board of Education v. Barnette is especially relevant to masking for at least three reasons. First, it took place in the middle of a national emergency — World War II. Second, it was another case of symbolic speech through *non*-action. Third, it was another case where only a minority of those who observed the non-action would have understood the message being sent. Every objection raised against masking or non-masking being a form of speech during COVID was also applicable to the Jehovah's witnesses who objected to saluting the United States flag during World War II. The exact reason why a Jehovah's Witness was refusing to salute the American Flag might be "overwhelmingly apparent" to members of the same minority sect, but not to most other observers or viewers without additional explanation. Nevertheless, saluting a flag, like wearing a mask, *is* a form of speech, and the First Amendment protects the right to *refrain* from speaking as powerfully as it protects the right to speak. During the COVIDcrisis, wearing a mask was a *perpetual* salute, every bit as expressive as holding a sign with written words. Compulsory masking forced wearers to act as instruments for fostering public adherence to a specific ideological point of view, symbolically shouting: "Unclean! Unclean!" in a crowded community, creating or exacerbating panic. It was an overt attempt to prescribe what shall be orthodox in medicine and politics.

Just because some potential subset (or even the majority) of recipients or observers fail to recognize or acknowledge a message does not mean that no message has been sent. Speech protected under the First Amendment does not have to be articulate, eloquent, or comprehensible to every viewer.

607 *West Virginia Board of Education v. Barnette et al.*, 319 U.S. 624, 63 S. Ct. 1178, 87 L. Ed. 1628 (1943), available online: https://www.loc.gov/item/usrep319624/
"To sustain the compulsory flag salute we are required to say that a Bill of Rights which guards the individual's right to speak his own mind, left it open to public authorities to compel him to utter what is not in his mind." The Court's reasoning in *West Virginia* also applies to masks because mask-wearing is highly communicative symbolic speech and expressive conduct.
608 *West Virginia Board of Education v. Barnette et al.*, 319 U.S. 624, 63 S. Ct. 1178, 87 L. Ed. 1628 (1943), at 642, available online: https://www.loc.gov/item/usrep319624/
 Note: Professor Lawrence Gostin, whose strong published support for compulsory masking we discussed earlier, made a revealing response to Justice Jackson in 1999 that starkly illustrates the mentality given full rein during the COVIDcrisis. Gostin declared that Justice Jackson's "statement is as false as it is eloquent. All health agencies must regulate social meaning to promote health."
Lawrence O. Gostin, Scott C. Burris, and Zita Lazzarini, "The Law and the Public's Health: A Study of Infectious Disease Law in the United States," *Columbia Law Review*, Vol. 99, No. 59, p. 93, (January 1999), Available at SSRN: https://ssrn.com/abstract=139923

Whether or not speech of some type takes place is independent of the size of the receptive audience. Ancient Christians communicated their believing status to one another using the simple pictogram of a fish. That this would have been lost on most contemporary pagan viewers was by design, and does nothing to defeat the fact that symbolic speech occurred when such iconography was employed. In *West Virginia Board of Education v. Barnette*, even though the vast majority of Americans did not share the Jehovah's Witnesses perception of saluting the flag as an act of worship sufficient to violate the commandments found in the Book of Exodus, the fact that members of the minority sect did so was sufficient to implicate First Amendment protections of both speech and religious practice. In the context of COVID-19, wearing or not wearing a mask clearly created a signal and sent a message — the very *definition* of symbolic speech.

The failure of a portion of onlookers to accurately receive an intended message from masking or non-masking does not prove that no speech has taken place. If you say something and a deaf person does not hear you, or if a deaf person signs to you and you do not look at their hands or recognize the meaning of their gestures, speech has *still* taken place. The same applies to the speech inherent in masking and non-masking. To most people refusing to wear a mask, not wearing a mask communicated, "I am being open with you. I am not a threat to you." The fact that many people read such an act as, "I do not respect you," or "I do not care if you get sick," was a mistaken interpretation arising from their own internal filter. Misunderstandings or disagreements about what has been said can only happen if speech has, in fact, taken place. The fact that many people disliked the message communicated by masking or non-masking, or that people disagreed about the details or virtues of the message, just provides further proof that a message was being sent in the first place. Heated disagreement over precisely *what* meaning attaches to wearing or not wearing a mask merely serves to underscore that mask-wearing or non-wearing in the context of COVID was, in fact, an act of symbolic speech, and as such, implicated the First Amendment.

Published scientific literature since 2020 recognized the speech in masking and non-masking. A viewpoint article in the *Journal of General Internal Medicine* remarked:

> "Others refuse to wear a mask for political reasons, seeing it as a symbolic statement. Republicans are less likely to wear masks than Democrats, with less than 50% of Republicans wearing masks compared with over 75% of Democrats. For some, mask-wearing can be viewed as a sign of weakness and shame, particularly for men. Sixty-seven percent of women reported that they wore masks outside compared to just 56% of men."[609]

---

609   Eliyahu. Y. Lehmann and Lisa S. Lehmann, "Responding to Patients Who Refuse to Wear Masks During the Covid-19 Pandemic." *Journal of General Internal Medicine*, 2020. https://dx.doi.org/10.1007/s11606-020-06323-x

Another commentary published in the *International Journal for Environmental Research and Public Health* observed: "… mask adherence may, in part, be a function of whether or not mask wearing is perceived to be normative among one's political party."[610]

Mask-wearing indisputably implicates the free speech clause of the First Amendment. In the words of *West Virginia v. Barnette*, it "invades the sphere of intellect and spirit which it is the purpose of the First Amendment to our Constitution to reserve from all official control."[611] Court rulings that argued non-masking was somehow not protected expression because a reasonable observer would simply think the mask wearer was complying with public health law, or that the mask-wearer could distance themselves from the message communicated by masks through some sort of disclaimer, were directly addressed and demolished two years before COVID by Supreme Court Justice Clarence Thomas in his 2018 *Masterpiece Cakeshop v. Colorado Civil Rights Commission* concurrence (joined by Justice Neil Gorsuch). In *Masterpiece Cakeshop,* the Supreme Court reversed the Colorado Court of Appeals, which had ruled against baker Jack Phillips' refusal to bake a cake for a same-sex wedding. Phillips asserted his free speech and free exercise rights under the First Amendment, which the Colorado Court of Appeals denied using the same reasoning employed during COVID to disparage and deny the symbolic speech in masking and non-masking. In his concurrence, Justice Thomas quickly exposed and refuted the flawed general arguments used by the lower court:

> "The Colorado Court of Appeals was wrong to conclude that Phillips' conduct was not expressive because a reasonable observer would think he is merely complying with Colorado's public-accommodations law. This argument would justify any law that compelled protected speech. And, this Court has never accepted it. From the beginning, this Court's compelled-speech precedents have rejected arguments that 'would resolve every issue of power in favor of those in authority'…

> "The Colorado Court of Appeals also erred by suggesting that Phillips could simply post a disclaimer, disassociating Masterpiece from any support for same-sex marriage. Again, this argument would justify any law compelling speech.

---

610   Scheid, J.L., et al., *Commentary: Physiological and Psychological Impact of Face Mask Usage during the COVID-19 Pandemic.* Int J Environ Res Public Health, 2020. **17**(18). https://pubmed.ncbi.nlm.nih.gov/32932652/

611   *West Virginia Board of Education v. Barnette et al.*, 319 U.S. 624, 642 (1943), available online: https://www.loc.gov/item/usrep319624/

And again, this Court has rejected it. We have described similar arguments as 'beg[ging] the core question.'"[612]

Both of these flawed general arguments refuted by Justice Thomas were repeatedly and routinely employed by courts to justify compulsory masking, and both were indefensibly wrong, especially coming just two or three years after the Supreme Court's *Masterpiece Cakeshop* Decision.

Moreover, the U.S. Supreme Court has *already* said that changes to apparel far more subtle than the sudden appearance of masks are entitled to first Amendment Protection. In the landmark 1969 Free Speech case *Tinker v. Des Moines Independent Community School District*, the same nine United States Supreme Court Justices who issued the *O'Brien* decision just one year earlier held that "the wearing of armbands" was: "closely akin to 'pure speech' which, we have repeatedly held, is entitled to comprehensive protection under the First Amendment."[613] This decision applies directly to masks in the context of COVID in at least two ways. First, like the Supreme Court's decision in *Barnette*, *Tinker* was also decided in the context of that most archetypal national emergency — a war. Second, armbands are far more subtle articles of clothing and elements of a person's appearance than are masks. If armbands qualify as nearly "pure speech" under the First Amendment, then masks — being far more expressive — are even closer to pure speech. If mere *armbands* can, as the Supreme Court has said, implicate "direct, primary First Amendment rights akin to 'pure speech,'"[614] then masking and non-masking cannot avoid doing so too. Judicial decisions which denied this obvious truth were only able to do so using procedural technicalities or mental gymnastics that robbed their conclusions of legitimacy.

Despite this, courts during the COVIDcrisis resolutely disregarded the relevance of *Tinker* and its sibling pre-COVID precedents in order to uphold mask mandates. "*Tinker* provides no support for the plaintiffs' position," insisted Judge Reggie B. Walton, as he dismissed the suit against compulsory masking led by Congressman Thomas Massie. "'Releasing respiratory fluids during exhalation' that 'carry virus and transmit infection' without any sort of preventative mechanism in the middle of a deadly pandemic… is not comparable to wearing an armband on top of one's clothing."[615] But all Judge Walton did here was to couple brute-force denial with an appeal to the nonspeech component

612   *Masterpiece Cakeshop v. Colorado Civil Rights Commission,* 138 S. Ct. 1719, 1744-1745 (2018), (Thomas, J., concurring in part and concurring in the judgment) available online: https://www.supremecourt.gov/opinions/17pdf/16-111_new2_22p3.pdf or https://casetext.com/case/masterpiece-cakeshop-ltd-v-colo-civil-rights-commn-3

613   *Tinker v. Des Moines Independent Community School District,* 393 U.S. 503, 505–06 (1969), available online: https://www.loc.gov/item/usrep393503/

614   *Tinker v. Des Moines Independent Community School District,* 393 U.S. 503, 508 (1969), available online: https://www.loc.gov/item/usrep393503/

615   *Massie v. Pelosi,* 590 F. Supp. 3d 196, 230 (D.C. 2022), *22. Available online: https://scholar.google.com/scholar_case?case=735226917287193397&hl=en&as_sdt=4000003&scfhb=1

of masking. His prejudicial reframing made no actual attempt to rebut the point that masks and armbands are both symbolic speech. The objection Judge Walton put forward was simply an indirect attempt to imply that the assumed infection control benefits of masks automatically overruled any First Amendment protections without having to formally make or defend such an argument.

Ultimately, every legal argument which tries to suggest that masking and non-masking are not speech protected under the First Amendment fails miserably. The Supreme Court has repeatedly and consistently recognized that "the Constitution looks beyond written or spoken words as mediums of expression."[616] According to the U.S. Supreme Court, conduct can be a form of speech that falls within the scope of the First Amendment if: "An intent to convey a particularized message was present, and in the surrounding circumstances the likelihood was great that the message would be understood by those who viewed it."[617] In 1974, the Supreme Court specifically said that a symbolic act does not even *need* to convey a "particularized message" to fall under the protection of the First Amendment: "a narrow, succinctly articulable message is not a condition of constitutional protection, which if confined to expressions conveying a 'particularized message,' would never reach the unquestionably shielded painting of Jackson Pollock, music of Arnold Schöenberg, or Jabberwocky verse of Lewis Carroll."[618] In other words, conveying a "particularized message" is *sufficient* to invoke the First Amendment, but it is not a *necessary* condition for doing so.

The Supreme Court has ruled *multiple* times that the First Amendment can be — and very often *is* — invoked by expressive conduct of a far less expressive or specific nature than masking or non-masking. In its 1995 ruling in *Hurley v. Irish-American Gay, Lesbian and Bisexual Group of Boston*, the Supreme Court further undermined the reasoning used to dismiss First Amendment challenges to compulsory masking when it explained why the First Amendment applied to the diffuse and imprecise expressive conduct and symbolic speech present in parades:

> "[A] private speaker does not forfeit constitutional protection simply by
> combining multifarious voices, or by failing to edit their themes to isolate an
> exact message as the exclusive subject matter of the speech. Nor, under our

---

616   *Hurley v. Irish-American Gay, Lesbian and Bisexual Group of Boston, Inc.*, 515 U. S. 557, 569 (1995), available online https://supreme.justia.com/cases/federal/us/515/557/case.pdf

617   *Spence v. Washington*, 418 U.S. 405, 410-411 (1974), available online:
https://tile.loc.gov/storage-services/service/ll/usrep/usrep418/usrep418405/usrep418405.pdf

618   *Hurley v. Irish-American Gay, Lesbian and Bisexual Group of Boston, Inc.*, 515 U. S. 557, 569 (1995), available online https://supreme.justia.com/cases/federal/us/515/557/case.pdf (c.f. *Spence v. Washington*, 418 U. S. 405, 411 (1974) (*per curiam*))

precedent, does First Amendment protection require a speaker to generate, as an original matter, each item featured in the communication." (emphasis added)[619]

Nevertheless, Judge Jared Strauss in Florida argued that the Supreme Court's ruling in *Hurley* could not apply to masking because "The inherent purpose of… holding a parade is to convey a particular message… The same can hardly be said about wearing or not wearing a mask."[620] But this "refutation" is just a variant way of saying "neither wearing or not wearing a mask is inherently expressive."[621] Apart from being obviously false in the case of people who are wearing or not wearing masks with the intent to convey a particular message, it is nothing more than a statement of disbelief which does nothing to address the Supreme Court's actual reasoning that still applies every bit as much to masks as to parades.

The power of context

Even if masking and non-masking were not "inherently expressive," both would *still* qualify as symbolic speech and expressive conduct protected under the First Amendment because *context* — "the surrounding circumstances" — is the crucial factor in determining what constitutes symbolic or expressive speech under the First Amendment. Judges who dismissed Free Speech challenges to compulsory masking either ignored, discounted, or prejudicially misread this all-important component of symbolic speech. Their decisions erroneously assumed that because masks may not be symbolic speech in some contexts, such as a hospital operating room, that they therefore cannot constitute symbolic speech in other contexts.

Meaning is extremely context-dependent. Gestures which are completely innocuous in one context may have great import in others. Eating a piece of bread, taking a sip of wine, or immersing oneself in water are not "inherently expressive" either, but when done in the company of others in a place of worship, these acts become sacramental professions of religious faith, with the nuanced manner in which they are performed even communicating a celebrant's denominational branch. Small circular bands are not "inherently expressive" either, but when made of a precious metal and worn on the fourth digit of the hand, they communicate one's marital status to all observers. Context and detail make

---

619   *Hurley v. Irish-American Gay, Lesbian and Bisexual Group of Boston, Inc.*, 515 U. S. 557, 569-570 (1995), available online https://supreme.justia.com/cases/federal/us/515/557/case.pdf

620   *Zinman v. Nova Southeastern University*, No. 21-CIV-60723-RAR, Document 81, p. 26 footnote 13 (S.D. Fla. Aug. 30, 2021), https://www.govinfo.gov/content/pkg/USCOURTS-flsd-0_21-cv-60723/pdf/USCOURTS-flsd-0_21-cv-60723-1.pdf

621   *Zinman v. Nova Southeastern University*, No. 21-CIV-60723-RAR, Document 81, p. 25-26 (S.D. Fla. Aug. 30, 2021), https://www.govinfo.gov/content/pkg/USCOURTS-flsd-0_21-cv-60723/pdf/USCOURTS-flsd-0_21-cv-60723-1.pdf

the difference between a keyring and a wedding band, between taking a bath and getting baptized, between having a snack or taking communion. It is the difference between walking or driving from point A to point B and participating in a march to protest against COVID lockdowns. Trying to disprove the symbolic speech of such acts by reducing them to their mechanical components is simply a cheap rhetorical trick of the sort one learns in high school. Treating a mask as just another medical device is like treating a wedding band as just another ring of metal, or treating a communion wafer as just another cracker.

Deliberately eschewing public health or hygiene measures to send a message is as old as the Bible. Jesus Himself engaged in an act of symbolic speech when He declined to participate in the common hygiene practice of handwashing before eating at a dinner party (which most likely involved communal dishes). The point He made in doing this was considered highly offensive by many present.[622] It should come as no surprise, therefore, that many professing to be Jesus' followers immediately and intuitively recognized the symbolic speech inherent in wearing or not wearing a mask in the context of COVID, and acted accordingly. Evangelicals, for example, were less likely than the general population to wear masks throughout the events surrounding COVID, and Evangelicals as a group were harshly criticized for this trend.[623]

Prior to the COVID hysteria, courts like the 11th Circuit in Florida used the previous Supreme Court decisions in this area to determine that under the right conditions, even food-sharing can be speech protected under the First Amendment. As the 11th Circuit observed in a 2018 ruling on symbolic speech and expressive conduct: "History may have been quite different had the Boston Tea Party been viewed as mere dislike for a certain brew and not a political protest against the taxation of the American colonies without representation."[624] In that 2018 ruling, the 11th Circuit held a particular instance of food-sharing in a public park to be expressive conduct protected by the First Amendment based on five contextual factors. These five contextual cues built a winning case that the actions in question met the Supreme Court's criteria: "in the surrounding circumstances the likelihood was great that the message would be understood by those who viewed it."[625] First, the event was readily

622   Luke 11:37-53

623   Relevant Staff [pseud.], "Study: Young White Evangelicals Are Way Less Likely to Wear a Mask," *RELEVANT Magazine*, June 23, 2020, online:
https://relevantmagazine.com/culture/study-young-white-evangelicals-are-way-less-likely-to-wear-a-mask/
Ryan P. Burge, "White Evangelicals' Coronavirus Concerns Are Fading Faster," *Christianity Today*, June 23, 2020, online:
https://www.christianitytoday.com/news/2020/june/evangelicals-covid-19-less-worry-social-distancing-masks.html

624   *Fort Lauderdale Food Not Bombs v. City of Fort Lauderdale*, 901 F.3d 1235 at 1241 (11th Cir. 2018), available online: https://casetext.com/case/fort-lauderdale-food-not-bombs-v-city-of-fort-lauderdale

625   *Fort Lauderdale Food Not Bombs v. City of Fort Lauderdale*, 901 F.3d 1235 at 1240 (11th Cir. 2018), available online: https://casetext.com/case/fort-lauderdale-food-not-bombs-v-city-of-fort-lauderdale quoting *Spence v.*

distinguishable from normal conduct — the food sharing event in question was clearly different from relatives and friends simply eating together at the park. Second, the symbolic acts or events were open to the public –the food sharing event was open for participation to everyone. Third, the symbolic actions took place in a traditional public forum — a city park. Fourth, the issue involved was one of public concern: "the record demonstrates without dispute that the treatment of the City's homeless population is an issue of concern in the community." Fifth, the significance of the act had a historical basis; in the case of food sharing, this dates back millennia. As the court noted: "Jesus shared meals with tax collectors and sinners to demonstrate that they were not outcasts in his eyes."[626]

The 11th Circuit applied the same set of criteria again in 2021[627] to determine that the architecture of a particular home did *not* constitute expressive speech protected under the First Amendment. First, the architecture of the home was not so obviously symbolic that the building was readily distinguishable from other large custom-built homes in a communicative sense. Second, the home was not readily viewable by the public — in fact, much of it was carefully "shielded from public view with a limestone wall, louvered gate, heavy landscaping, and substantial vegetation." Third, private residences are not traditional public forums "historically associated with the exercise of First Amendment rights." Fourth, residential architecture — and the homeowner's design in particular — had not been a matter of widespread public concern or discussion. Fifth, the court found that "there was no evidence that residential architecture, specifically, has a historical association with communicative elements that would put a reasonable observer on notice of a message from [the homeowner's] house."[628]

Washington, 418 U.S. 405, 410-411 (1974), available online: https://tile.loc.gov/storage-services/service/ll/usrep/usrep418/usrep418405/usrep418405.pdf

626   *Fort Lauderdale Food Not Bombs v. City of Fort Lauderdale*, 901 F.3d 1235 at 1243 (11th Cir. 2018), available online: https://casetext.com/case/fort-lauderdale-food-not-bombs-v-city-of-fort-lauderdale

627   *Burns v. Town of Palm Beach*, 999 F.3d 1317 at 1336 (11th Cir. 2021), available online: https://casetext.com/case/burns-v-town-of-palm-beach-1

628   *Burns v. Town of Palm Beach*, 999 F.3d 1317 at 1345 (11th Cir. 2021), available online: https://casetext.com/case/burns-v-town-of-palm-beach-1

Personally, my own historical readings lead me to strongly suspect that the court was at least partially mistaken on this point, though a genuine rebuttal would be the subject for another book and is outside my personal area of expertise and interest. Judge Stanley Marcus wrote a vigorous dissent, rightly criticizing the other judges' refusal to look at architectural evidence outside the record submitted by the parties in the suit, and his objections were weighty enough that the majority felt the need to respond at length to several of them. For example, in rebutting Judge Marcus' criticisms of their finding regarding the fifth criteria, the majority outright said that Thomas Jefferson's residence of Monticello would currently pass all five criteria (though few of the reasons they give could be argued to have applied when Jefferson was actually *living* there). Bottom line, architecture can certainly be a form of art (Judge Marcus cites as an example the historic Farnsworth House Outside of Chicago), and the First Amendment protects art as a form of expressive speech. I don't see any good reason to *a priori* exclude *residential* architecture from this protection. The majority says their ruling does not do this, but Judge Marcus' dissent raises legitimate concerns. In any case, the ruling here does not detract from the case against masks.

The 11th Circuit's five criteria are clues to assist analysis, so failing one or two of them does not preclude First Amendment involvement, but the more of them that are met, the stronger the First Amendment connection becomes. In the context of COVID, wearing or refusing to wear a mask easily meets every one of these five criteria, plus one more. First, non-masking during a public health emergency is easily distinguishable from non-masking under normal circumstances, especially in areas where masks are mandated or strongly recommended. Similarly, because *non*-masking is (and remains) the norm for the majority of humans throughout history, widespread masking broadcasts a visual signal that things are not normal, and so it is likewise readily distinguishable from normal conduct. Second, masking and non-masking are open to all in both a viewing and a participatory sense. Going about in either a masked or an unmasked state is readily seen as an implicit endorsement via social modeling for others to do likewise. This artificial "consensus" and social proof were a major motive behind mask mandates. Third, masking and the conspicuous abstention from wearing a mask are both by their very nature public actions that take place in traditional public forums and other locations open to the public. This was especially apparent in places like government buildings, school board meetings, and town council meetings where people (including myself) were repeatedly refused entry without a mask — thus inhibiting and often outright prohibiting their speech as a direct result of their scruples against wearing masks. Fourth, the activity or non-activity involved in masking undoubtedly implicates issues of public concern: public health, public policy regarding a particular source of risk, and the extent of individual liberties (to list just three). In the case of masks, this concern grew to the level of public and political polarization. Fifth, deliberately hiding or showing one's face has incontrovertibly been understood to convey a message over the millennia — from Biblical times to the present (more on this in the appendix). Sixth, the inescapable communicative properties of mask-wearing had already been acknowledged and affirmed by multiple scientific studies and court rulings prior to the COVIDcrisis. The moment someone puts on or takes off a mask with the intent of sending a message, doing so becomes a form of speech protected under the First Amendment.

When upholding mask mandates, judges would claim that serving a significant governmental interest and leaving "ample alternative channels" for communication meant that compulsory masking qualified as a permissible time, place, or manner" regulation subject only to intermediate scrutiny. However, as the Supreme Court has pointed out, "For a time, place, or manner regulation to be valid, it must be neutral as to the content of the speech to be regulated."[629] This is a condition which compulsory masking by its nature simply cannot meet. Mask-wearing cannot be content-neutral because it is, *itself*, content.

---

629   *Pacific Gas and Electric Company v. Public Utilities Commission of California*, 475 U.S. 1 at 20 (1986), https://tile.loc.gov/storage-services/service/ll/usrep/usrep475/usrep475001/usrep475001.pdf

Restrictions on speech do not automatically become constitutional just because speakers have a viable alternative. The Supreme court said in 1939 and again in 1974, "one is not to have the exercise of his liberty of expression in appropriate places abridged on the plea that it may be exercised in some other place."[630] As Justice Alito put it in 2020, "Respecting some First Amendment rights is not a shield for violating others."[631] Public places are an appropriate place to show one's face and exercise one's First Amendment right of free speech. Up until 2020, it was *hiding* one's face in public that was viewed as aberrant behavior. Also, in those few cases where Courts have permitted abridgements of Free Speech, the vast majority were cases where the legislative branch of government, not the executive branch, imposed the restriction. Mask mandates were almost entirely creatures of executive action, passed in violation of multiple procedural safeguards.

Straightforward, unbiased application of the First Amendment's Free Speech Clause and Supreme Court precedents should have been enough to show at a glance that compulsory masking was completely unconstitutional, but there are three *additional* elements which made compulsory masking one of the worst First Amendment violations in American history, and an imposition which should not have survived *any* level of scrutiny. First, compulsory masking affected every opposing speaker's intended message; second, compulsory masking violated every American's right to *receive* speech; and third, compulsory masking inhibited or prohibited wearers' use of one of their most primal and fundamental forms of pure speech — their facial expressions.

Compulsory masking alters objecting wearers' speech

The Supreme Court's 2006 *Rumsfeld* opinion specifically referenced the first of these three additional violations, and a thorough reading of this decision and its cited precedents reveals that this case should have been referenced to *enjoin* mask mandates rather than uphold them. When, in 2006, the Supreme Court said, "we have extended First Amendment protection only to conduct that is inherently expressive,"[632] it was not laying out some new, stricter and more subjective test for what counts as symbolic speech or expressive conduct. Rather, it was affirming its previous judgements in this area, including more subtle instances of symbolic speech and expressive conduct like those in *Tinker*, *West Virginia*, and *Hurley* — and there are more where those came from.

---

630   *Spence v. Washington*, 418 U.S. 405, 411 (1974), available online: h
ttps://tile.loc.gov/storage-services/service/ll/usrep/usrep418/usrep418405/usrep418405.pdf
quoting *Schneider v. State*, 308 U. S. 147, 163 (1939),
https://tile.loc.gov/storage-services/service/ll/usrep/usrep308/usrep308147/usrep308147.pdf
631   *Calvary Chapel Dayton Valley v. Sisolak*, 140 S. Ct. at 2608 (2020) (Alito, J., dissenting), available online:
https://casetext.com/case/calvary-chapel-dayton-valley-v-sisolak
632   *Rumsfeld v. Forum for Academic & Institutional Rights, Incorporated*, 547 U.S. 47, 66 (2006),
https://tile.loc.gov/storage-services/service/ll/usrep/usrep547/usrep547046/usrep547046.pdf

In 1986, three years before its flag-burning decision in *Texas v. Johnson*, the Supreme Court applied strict scrutiny to a state regulation that burdened a corporation's speech because of how the regulation altered the speakers' own message. In *Pacific Gas and Electric Company v. Public Utilities Commission of California*, the Supreme Court ruled that citizens cannot be forced to "use their private property as a 'mobile billboard' for the State's ideological message."[633] The Court ruled that the coerced insertion of a leaflet in the envelope of a private utility's bill and newsletter constituted an unconstitutional violation of Free Speech because the utility "may be forced either to appear to agree with [the intruding leaflet] or to respond."[634] The utility, the Court said, "'might well conclude' that, under these circumstances, 'the safe course is to avoid controversy,' thereby reducing the free flow of information and ideas."[635] It is unconstitutional, ruled the Supreme Court to "require speakers to affirm in one breath that which they deny in the next."[636] Mask mandates forced opponents of compulsory masking to affirm by their expressive conduct that which they denied with their words, and we all know that it is *actions* which speak the loudest.

Companies derive their free speech rights from the rights of the citizens that own them, and there is no First Amendment Free Speech right which a corporation possesses that an individual citizen does not also possess. There is a common thread throughout the Supreme Court's First Amendment rulings, "all speech inherently involves choices of what to say and what to leave unsaid."[637] It is axiomatic that: "A system which secures the right to proselytize religious, political, and ideological causes must also guarantee the concomitant right to decline to foster such concepts."[638] In fact, the Supreme Court listed "ordinary people engaged in unsophisticated expression" alongside the press and professional publishers when describing the primary beneficiaries of this right to decide "what not to say."[639]

---

633   *Pacific Gas and Electric Company v. Public Utilities Commission of California*, 475 U.S. 1 at 17 (1986), https://tile.loc.gov/storage-services/service/ll/usrep/usrep475/usrep475001/usrep475001.pdf

634   *Pacific Gas and Electric Company v. Public Utilities Commission of California*, 475 U.S. 1 at 15 (1986), https://tile.loc.gov/storage-services/service/ll/usrep/usrep475/usrep475001/usrep475001.pdf

635   *Pacific Gas and Electric Company v. Public Utilities Commission of California*, 475 U.S. 1 at 15 (1986), https://tile.loc.gov/storage-services/service/ll/usrep/usrep475/usrep475001/usrep475001.pdf
quoting *Miami Herald Publishing Co. v. Tornillo*, 418 U. S. 241, 257 (1974), https://tile.loc.gov/storage-services/service/ll/usrep/usrep418/usrep418241/usrep418241.pdf .

636   *Pacific Gas and Electric Company v. Public Utilities Commission of California*, 475 U.S. 1 at 16 (1986), https://supreme.justia.com/cases/federal/us/475/1/

637   *Pacific Gas and Electric Company v. Public Utilities Commission of California*, 475 U.S. 1 at 11 (1986), https://tile.loc.gov/storage-services/service/ll/usrep/usrep475/usrep475001/usrep475001.pdf

638   *Wooley v. Maynard*, 430 U.S. 704, 714 (1977), available online: https://tile.loc.gov/storage-services/service/ll/usrep/usrep430/usrep430705/usrep430705.pdf
reiterated in *Pacific Gas and Electric Company v. Public Utilities Commission of California*, 475 U.S. 1 at 16 n 13 (1986), https://supreme.justia.com/cases/federal/us/475/1/

639   *Hurley v. Irish-American Gay, Lesbian and Bisexual Group of Boston, Inc.*, 515 U. S. 557, 573-574 (1995), https://supreme.justia.com/cases/federal/us/515/557/case.pdf

The fact that a mask mandate "does not, on its face, target speech or discriminate on the basis of its content,"[640] is not enough to dismiss Free Speech claims against compulsory masking, because it is the effect of a law's *application* that determines whether the First Amendment is being violated. The Supreme Court ruled in *Hurley* that when private parade organizers were forced to include a particular group, "the state courts' application of the statute produced an order essentially requiring petitioners to alter the expressive content of their parade," which "had the effect of declaring the sponsors' speech itself to be the public accommodation."[641]

As in *Hurley* and *Pacific Gas and Electric*, compulsory masking uses one of a person's most intimate pieces of private property – their face – as a place of public accommodation. Compulsory masking makes people's faces a mobile billboard for the ideological messages that COVID is a deadly threat, that they are personally susceptible either to contracting or to spreading it, that compulsory masking is warranted, and that the government may control their body to the extent someone else may derive benefit. Whenever a masked person speaks out against compulsory masking, they are forced to contend with the fact that they are simultaneously helping to disseminate hostile views, and, in the words of the Supreme Court, "might well conclude" that "the safe course is to avoid controversy." Many people *did* reach this conclusion. Compelling someone to wear a mask while they argue against masking impermissibly requires them to associate with speech with which they disagree. At the very least, a person wearing a mask against their objections is forced either to appear to agree with mask-wearing or to respond.[642] In The words of the Supreme Court, "That kind of forced response is antithetical to the free discussion that the First Amendment seeks to foster."[643]

On its own, forcing someone to include an unwanted message within or alongside their own — as in *Hurley* or *Pacific Gas and Electric Company* — would be enough to make mask mandates presumptively unconstitutional and require strict scrutiny. But compulsory masking goes *beyond* even *these* unconstitutional extremities. In their application, mask mandates have the effect of forcing wearers to *physically superimpose* a message with which many vehemently disagree *over* their own message in a way that obstructs and undermines it. Compulsory masking does not merely *alter* opponents' speech, it *suppresses* opposing speakers' messages and *substitutes* the masker's message. This

---

640   *Hurley v. Irish-American Gay, Lesbian and Bisexual Group of Boston, Inc.*, 515 U. S. 557, 572 (1995), available online https://supreme.justia.com/cases/federal/us/515/557/case.pdf

641   *Hurley v. Irish-American Gay, Lesbian and Bisexual Group of Boston, Inc.*, 515 U. S. 557, 572-573 (1995), https://supreme.justia.com/cases/federal/us/515/557/case.pdf

642   c.f. *Pacific Gas and Electric Company v. Public Utilities Commission of California*, 475 U.S. 1 at 14-15 (1986), https://tile.loc.gov/storage-services/service/ll/usrep/usrep475/usrep475001/usrep475001.pdf

643   *Pacific Gas and Electric Company v. Public Utilities Commission of California*, 475 U.S. 1 at 16 (1986), https://tile.loc.gov/storage-services/service/ll/usrep/usrep475/usrep475001/usrep475001.pdf

is *worse* than the violations of free speech the Supreme Court struck down in *Hurley* and *Pacific Gas and Electric Company*.

Mask mandates had the effect of appropriating all citizens' speech during the COVIDcrisis to create and proclaim an artificial orthodoxy — actions explicitly repudiated by the Supreme Court on multiple occasions. Compulsory masking was thus presumptively unconstitutional from inception. Regardless of the aim behind it, universal compulsory masking inevitably entails compelled speech. When the choice of whether to wear or not wear a mask is made based on the message masking or non-masking conveys, "it boils down to the choice of a speaker not to propound a particular point of view, and that choice is presumed to lie beyond the government's power to control."[644]

The Supreme Court's 1995 *Hurley* decision was cited in its 2006 *Rumsfeld* opinion. Referring to its 1995 *Hurley* decision, the Court said in *Rumsfeld* that: "This Court has found compelled-speech violations where the complaining speaker's own message was affected by the speech it was forced to accommodate."[645] Judges who wanted to uphold compulsory masking would ignore this crucial reference while quoting the 2006 *Rumsfeld* decision. It is disingenuous to cite the 2006 *Rumsfeld* decision to uphold masks, saying "The fact that such explanatory speech is necessary is strong evidence that the conduct at issue here is not so inherently expressive that it warrants protection,"[646] while ignoring the Supreme Court's crucial reference to its 1995 precedent. Having the complaining speaker's own message affected by the speech it was forced to accommodate is *exactly* the situation produced by compulsory masking. Rulings and arguments which upheld compulsory masking thus used sophistry to ignore half of the chief complaint. The complaints challenging compulsory masking were not simply that mandated mask-wearing *inhibits or prevents* expressive activity (though it indisputably does this by hiding facial expressions and increasing the effort required to speak or be understood). The First Amendment complaints challenging compulsory invariably charged that forced masking invasively affects speakers' own messages by *compelling* expression of a message that many found to be false, objectionable, or repugnant for multiple valid reasons. In addition to being

644   *Hurley v. Irish-American Gay, Lesbian and Bisexual Group of Boston, Inc.*, 515 U. S. 557, 575 (1995), https://supreme.justia.com/cases/federal/us/515/557/case.pdf

645   *Rumsfeld v. Forum for Academic & Institutional Rights, Incorporated*, 547 U.S. 47, 49 (2006), https://tile.loc.gov/storage-services/service/ll/usrep/usrep547/usrep547046/usrep547046.pdf referencing *Hurley v. Irish-American Gay, Lesbian and Bisexual Group of Boston, Inc.*, 515 U. S. 557, 566 (1995), https://supreme.justia.com/cases/federal/us/515/557/case.pdf

646   *Rumsfeld v. Forum for Academic & Institutional Rights, Incorporated*, 547 U.S. 47, 66 (2006), https://tile.loc.gov/storage-services/service/ll/usrep/usrep547/usrep547046/usrep547046.pdf (cleaned up) *Sehmel v. Shah*, 514 P.3d 1238 (Wash. Ct. App. 2022), at 16, available online: https://caselaw.findlaw.com/wa-court-of-appeals/2182709.html
Stewart v. Justice 502 F. Supp. 3d 1057, 1066 (S.D.W. Va. 2020), available online: https://casetext.com/case/stewart-v-justice-1

a form of compelled speech in-and-of itself, compulsory masking visually overwrote every opposed speaker's intended contrary message in direct violation of the First Amendment.

The objection that mask mandates are simply a use of the state's power to impose reasonable restrictions on the time, place, and manner in which views are presented does not hold water. In no way, shape or form is it a reasonable time, place, and manner restriction to categorically forbid the presentation of a particular view in all public places in a way that forces people to alter and undermine their own speech by overwriting it with a message they find repugnant.

Another combination justification and objection raised to defend overriding people's speech is that mask-wearers "can expressly disavow any connection with the message" sent by masks. This is what Judge Reggie B. Walton was getting at when he said objecting Congressmen had "myriad means" of getting their message opposing masks across, "including wearing masks or other clothing containing the messages they wanted to convey, or making speeches from the House Chamber or elsewhere on the subject."[647] We have already seen how Justice Clarence Thomas officially refuted this, but even if Justice Thomas had not done so, all justifications like Judge Walton's fail in the case of masks because actions speak louder than words, making the disavowals Judge Walton proposed either worthless or hypocritical. Even if a mask-wearer is able to convince viewers that they do not believe in the message sent by wearing a mask, their message is still forcibly attenuated by the presence of the mask on their face. After all, the very fact that they are still wearing a mask is visible evidence telling observers that their convictions about masks are not strong enough to make them risk the penalties for not wearing one.

Mask mandates were an especially pernicious form of suppressed and compelled speech because they "produce[d] their restrictive results at the precise time when public interest in the matters discussed would naturally be at its height."[648] The right to communicate through facial expressions, gestures, and dress has, in the words of the U.S. Supreme Court, "from ancient times, been a part of the privileges, immunities, rights, and liberties of citizens." The fundamental individual rights of free speech, autonomy, and privacy are not privileges conferred by society or the government. They are deeply rooted in American heritage, and founded upon historical notions and federal constitutional expressions of ordered liberty. Declaring a public health emergency does not magically exempt a person's body from the protections of the First Amendment. Your face is not a place of public accommodation, neither is your speech, and no authority *ever* has a right to use *either one* to

---

647   *Massie v. Pelosi*, 590 F. Supp. 3d 196, *225 (D.C. 2022). Available online: https://scholar.google.com/scholar_case?case=735226917287193397&hl=en&as_sdt=4000003&scfhb=1

648   *Bridges v. California*, 314 U.S. 252, 268 (1941), available online: https://tile.loc.gov/storage-services/service/ll/usrep/usrep314/usrep314252/usrep314252.pdf

propound their preferred message against your will. Violation of the First Amendment rights not to speak and not to associate with particular repugnant speech, as compulsory masking indisputably does, triggers strict scrutiny.

Compulsory masking violates observers' First Amendment right to receive speech

It is foundational Supreme Court jurisprudence that "the protection of the Bill of Rights goes beyond the specific guarantees to protect from congressional abridgment those equally fundamental personal rights necessary to make the express guarantees fully meaningful."[649] This means that, among other things, the Free Speech clause of the First Amendment does not merely protect the rights to speak and refrain from speaking, but also the right to *receive* speech (even in cases when that speech is lacking in social worth[650]).

In 1965, in *Lamont v. Postmaster General*, the Supreme Court declared a postal regulation unconstitutional because it inhibited and limited the "unfettered exercise of the addressee's First Amendment rights" to receive written speech. The postal regulation struck down had inhibited addressees' First Amendment rights to receive speech by detaining foreign mail tagged as "communist political propaganda" until the addressees affirmatively opted-in to receiving this mail, even though it had already been addressed to them by the sender ("the addressee in order to receive his mail must request in writing that it be delivered").[651] In striking down this postal service regulation, the Court also quoted a 1921 opinion by Justice Oliver Wendell Holmes, Jr.: "the use of the mails is almost as much a part of free speech as the right to use our tongues[.]"[652] This is true, and even *more* a part of free speech than the use of the mails is the use of our faces and facial expressions.

As it did during the COVIDcrisis with regard to masks, the Government argued in 1965 that it was not, in fact, violating any rights: "only inconvenience and not an abridgment is involved." Supreme Court Justice Brennan responded that "inhibition as well as prohibition against the exercise of precious

---

649   *Lamont v. Postmaster General*, 381 U.S. 301, 308 (1965), (Brennan, J., concurring) available online: https://tile.loc.gov/storage-services/service/ll/usrep/usrep381/usrep381301/usrep381301.pdf

650   *Stanley v. Georgia*, 394 U.S. 557 (1969), available online: https://tile.loc.gov/storage-services/service/ll/usrep/usrep394/usrep394557/usrep394557.pdf

651   "These might be troublesome cases if the addressees predicated their claim for relief upon the First Amendment rights of the senders."
Lamont v. Postmaster General, 381 U.S. 301, 307 (1965), available online: https://tile.loc.gov/storage-services/service/ll/usrep/usrep381/usrep381301/usrep381301.pdf

652   *Lamont v. Postmaster General*, 381 U.S. 301, 305 (1965), available online: https://tile.loc.gov/storage-services/service/ll/usrep/usrep381/usrep381301/usrep381301.pdf quoting Milwaukee Social Democratic Publishing Company v. Burleson, 255 U.S. 407, 437 (1921) (Holmes, J., dissenting), available online: https://tile.loc.gov/storage-services/service/ll/usrep/usrep255/usrep255407/usrep255407.pdf

First Amendment rights is a power denied to the government... we cannot sustain an intrusion on First Amendment rights on the ground that the intrusion is only a minor one."[653] Justice Brennan then quoted a decision issued by the Supreme Court 80 years before, in 1886. It is a reminder that should never have been forgotten during the COVIDcrisis:

> "It may be that it is the obnoxious thing in its mildest and least repulsive form; but illegitimate and unconstitutional practices get their first footing in that way, namely, by silent approaches and slight deviations from legal modes of procedure. This can only be obviated by adhering to the rule that constitutional provisions for the security of person and property should be liberally construed. A close and literal construction deprives them of half their efficacy, and leads to gradual depreciation of the right, as if it consisted more in sound than in substance. It is the duty of courts to be watchful for the constitutional rights of the citizen, and against any stealthy encroachments thereon."[654]

The Supreme Court's 1965 *Lamont* decision fed into in another 1976 case greatly affecting public health, *Virginia State Board of Pharmacy v. Virginia Citizens Consumer Council,* which challenged bans on the advertising of prescription drug prices. When *recipients'* First Amendment rights to receive speech were invoked, the "ample alternative channels" reasoning so frequently used by courts to dismiss challenges to masking during the COVIDcrisis was not good enough. In *Virginia State Board of Pharmacy,* the Supreme Court said:

> "We are aware of no general principle that freedom of speech may be abridged when the speaker's listeners could come by his message by some other means, such as seeking him out and asking him what it is. Nor have we recognized any such limitation on the independent right of the listener to receive the information sought to be communicated."[655]

---

653   *Lamont v. Postmaster General,* 381 U.S. 301, 309 (1965), (Brennan, J., concurring) available online: https://tile.loc.gov/storage-services/service/ll/usrep/usrep381/usrep381301/usrep381301.pdf

654   *Boyd v. United States,* 116 U.S. 616, 635 (1886), available online: https://tile.loc.gov/storage-services/service/ll/usrep/usrep116/usrep116616/usrep116616.pdf quoted in *Lamont v. Postmaster General,* 381 U.S. 301, 309-310 (1965), (Brennan, J., concurring) available online: https://tile.loc.gov/storage-services/service/ll/usrep/usrep381/usrep381301/usrep381301.pdf

655   "Certainly, the recipients of the political publications on Lamont could have gone abroad and thereafter disseminated them themselves." *Virginia State Board of Pharmacy v. Virginia Citizens Consumer Council, Inc.,* 425 U.S. 748, 757-758 n 15 (1976), available online: https://tile.loc.gov/storage-services/service/ll/usrep/usrep425/usrep425748/usrep425748.pdf

Here, again, the Supreme Court has repeatedly stated: "where a speaker exists... the protection afforded is to the communication, to its source and to its recipients both... this Court has referred to a First Amendment right to 'receive information and ideas,' and that freedom of speech 'necessarily protects the right to receive.'"[656]

These cases apply to masks because the right to engage in speech through expressive conduct also includes the right to *receive* speech through expressive conduct, and mask mandates outrageously violated this right. When describing the return to true normalcy after the Philadelphia Yellow Fever Epidemic of 1793, John Adams' son, Thomas, wrote: "The idea of danger is dissipated in a moment when we perceive thousands walking in perfect security about their customary business, & no ill consequences ensuing from it."[657] As we saw in Part 3, on mask psychology, a 2021 article in the *International Journal of Environmental Research and Public Health* aptly described the importance of the messaging involved in this type of expressive conduct:

> "A minority can just ignore the collective panic and continue to live their normal lives, because they are free to do so. Such a minority can be an example and a wake-up call to those that do succumb to the collective hysteria or are close to doing so... Having this example, the anxiety of observers may fall. Observers may follow the example, and the group of hysterics shrinks. It is one of the core characteristics of decentralized systems that they allow for competition, error detection, and correction. If the people that ultimately become role models for others through their interaction become ill and die, the panic would be confirmed."[658]

Mask mandates were a virulent attack on just this sort of expressive conduct. They violated not just everyone's right to *make* these expressive communications, but also everyone's right to *receive* them. In so doing, they extended and exacerbated the duration of the COVIDcrisis. This violation was especially egregious and unjustifiable in the case of dissenting healthcare workers and others

---

656   *Virginia State Board of Pharmacy v. Virginia Citizens Consumer Council, Inc.*, 425 U.S. 748, 756-757 (1976), available online: https://tile.loc.gov/storage-services/service/ll/usrep/usrep425/usrep425748/usrep425748.pdf

657   Thomas Boylston Adams to Abigail Adams, 24 November 1793," *Founders Online,* National Archives, https://founders.archives.gov/documents/Adams/04-09-02-0263. [Original source: *The Adams Papers*, Adams Family Correspondence, vol. 9, *January 1790–December 1793*, ed. C. James Taylor, Margaret A. Hogan, Karen N. Barzilay, Gregg L. Lint, Hobson Woodward, Mary T. Claffey, Robert F. Karachuk, and Sara B. Sikes. Cambridge, MA: Harvard University Press, 2009, pp. 455–457.]

658   Bagus, P., J.A. Peña-Ramos, and A. Sánchez-Bayón, "COVID-19 and the Political Economy of Mass Hysteria." *International Journal of Environmental Research and Public Health*, 2021. 18(4): p. 1376. https://dx.doi.org/10.3390/ijerph18041376

perceived to have special knowledge about masks, SARS-CoV-2, and infection control in general. When doctors, dentists, and other healthcare providers widely perceived as having special expertise in the field wore masks, it constituted an even more powerful endorsement of the practice. "Silence" by conformity and compliance from healthcare workers or anyone else who either knew better (or had reservations) was at the very least a form of enforced tacit consent, recognized in the old Latin legal maxim: "he who is silent is taken to agree in cases where he should and could speak."[659] In controversial matters of public concern, and especially during a public health emergency, citizens are entitled to receive all available information that their fellows wish to communicate, including the information communicated by both masking and non-masking.

After reviewing the arguments put forward to justify banning the advertisement of prescription drug prices, the Supreme Court concluded: "on close inspection it is seen that the State's protectiveness of its citizens rests in large measure on the advantages of their being kept in ignorance."[660] Does that sound at all familiar? The Court went on to say:

> "There is, of course, an alternative to this highly paternalistic approach. That alternative is to assume that this information is not in itself harmful, that people will perceive their own best interests if only they are well enough informed, and that the best means to that end is to open the channels of communication rather than to close them."[661]

This was and is just as true during a declared public health emergency. The solution was not — and will never be — imposing a coerced artificial orthodoxy in politics and science. Rather, the solution protected by the First Amendment is more information, more debate, more speech: vocal; written; symbolic; and expressive. As the Supreme Court ruled decades ago, "It is precisely this kind of choice, between the dangers of suppressing information, and the dangers of its misuse if it is freely available, that the First Amendment makes for us."[662]

---

659   *qui tacet consentire videtur, ubi loqui debuit ac potuit*

660   *Virginia State Board of Pharmacy v. Virginia Citizens Consumer Council, Inc.*, 425 U.S. 748, 769 (1976), available online: https://tile.loc.gov/storage-services/service/ll/usrep/usrep425/usrep425748/usrep425748.pdf

661   *Virginia State Board of Pharmacy v. Virginia Citizens Consumer Council, Inc.*, 425 U.S. 748, 770 (1976), available online: https://tile.loc.gov/storage-services/service/ll/usrep/usrep425/usrep425748/usrep425748.pdf

662   *Virginia State Board of Pharmacy v. Virginia Citizens Consumer Council, Inc.*, 425 U.S. 748, 770 (1976), available online: https://tile.loc.gov/storage-services/service/ll/usrep/usrep425/usrep425748/usrep425748.pdf

Facial expressions are pure speech

> *"Your face, my thane, is as a book where men / May read strange matters."*
>
> — William Shakespeare, *Macbeth*, 1.5.73-74.

> *"And in thy face strange motions have appear'd, / Such as we see when men restrain their breath / On some great sudden hest. O, what portents are these? / Some heavy business hath my lord in hand, / And I must know it, else he loves me not."*
>
> — William Shakespeare, *Henry IV*, 2.3.62-67

> *"'You do not ask me or tell me much that concerns yourself, Frodo,' said Gildor. 'But I already know a little, and I can read more in your face and in the thought behind your questions. You are leaving the Shire, and yet you doubt that you will find what you seek, or accomplish what you intend, or that you will ever return. Is not that so?'*
>
> *'It is,' said Frodo."*
>
> — J. R. R. Tolkien, *The Lord of The Rings: The Fellowship of the Ring*,
> Chapter 3: "Three is Company"

But Digory was more interested in the faces, and indeed these were well worth looking at. The people sat in their stone chairs on each side of the room and the floor was left free down the middle. You could walk down and look at the faces in turn.

"They were nice people, I think," said Digory.

Polly nodded. All the faces they could see were certainly nice. Both the men and women looked kind and wise, and they seemed to come of a handsome race. But after the children had gone a few steps down the room they came to faces that looked a little different. These were very solemn faces. You felt you would have to mind your P's and Q's, if you ever met living people who looked like that. When they had gone a little further, they found themselves among faces they didn't like: this was about the middle of the room. The faces here looked very strong and proud and happy, but they looked cruel. A little further on they looked crueller. Further on again, they were still cruel but they no longer looked happy. They were even despairing faces: as if the people they belonged to had done dreadful things and also suffered dreadful things.

— C. S. Lewis, *The Chronicles of Narnia: The Magician's Nephew,*
Chapter 4: "The Bell and the Hammer"

Masks are far closer to pure speech than any other article of clothing, but the *faces* and the *facial expressions* that masks overwrite *are* pure speech. Facial expressions are the most precise, articulate, and unfiltered manifestation of body language. They have as much claim to being pure speech as the written word. If anything, facial expressions may possess an even stronger claim to that title. What they lack in precision, they more than make up in speed, intensity, nuance, and candor. Facial expressions, like vocalizations, are among humanity's most primordial means of communication — in constant use thousands of years before the first alphabets or hieroglyphics were developed. Human infants use facial expressions to communicate with their parents long before they can use even the most basic words. If primate communication is anything to go by, facial expressions pre-date formal language itself.

Recognition of this fact is baked into our use of language at such a basic level that we rarely even notice it. It is right there in the term: facial *expressions*.[663] To de*face* something is to disfigure it, especially its most visually important parts. When something is ef*faced*, it is erased. If you are self-ef*faci*ng, you make yourself less noticeable. We dare someone to, "Say that to my face." We invite people to collaborate, to "get together and hash this out face-to-face" because that provides for the best communication. We try to "save face" and avoid "losing face." We may *say* one thing, but if our *face* tells a different story, our words will not be believed. If someone says "it's fine" with a smile, we know things are all good. If they say "it's fine" with a scowl or compressed lips, we know things are very much *not* fine.

A thing's *face* denotes not just its most important outward appearance and presence, but what that appearance *communicates*. We prefer to be able to take things (and people) "at face value." A document's face expresses its most manifest, direct, or obvious sense. Statutes, editorials, and court records express things *on their face*. We have the right to *face* our accusers in court. To do something *in facie* means to do it in the open, where people see it. Plaintiffs bring *prima facie* evidence to support their complaints. If a question of federal law "affirmatively appears on the *face* of the record,"[664] a federal court may have jurisdiction to review it. A statute expresses something, or speaks "on its *face*."[665] It can be "judged

---

663   Note: lawyer Jeff Childers' formulation deserves credit for being the first time I found the seeds of this argument in print, and is too good to not quote:

"The same thing a person says with a smile means something completely different when said with a scowl or even with a neutral expression. The Mask Mandate materially impairs Justin's verbal and non-verbal communications—his facial and verbal expressions—preventing him from speaking with a degree of normalcy and expressing his thoughts, feelings, and personal identity."

Green v. Alachua County, 323 So. 3d 246, Appellant's Initial Brief, Filing #109556371 E-Filed 06/29/2020 (Fla. Dist. Ct. of App., 2021), p. 33, go to:

https://1dca.flcourts.gov/Oral-Arguments/Briefs-and-Petitions-for-Cases-Scheduled-for-Oral-Argument/20-1661
and search for "case 20-1661"

664   "[T]his Court acquires no jurisdiction to review the judgment of a state court of last resort on a writ of error, unless it affirmatively appears on the **face** of the record that a federal question constituting an appropriate ground for such review was presented in and expressly or necessarily decided by such state court."
Whitney v. California, 274 U.S. 357, 360 (1927), available online:

https://tile.loc.gov/storage-services/service/ll/usrep/usrep274/usrep274357/usrep274357.pdf

665   "On its **face** the editorial merely expressed exulting approval of the verdict, a completed action of the court, and there is nothing in the record to give it additional significance."
Bridges v. California, 314 U.S. 252, 297 (1941), available online:

https://tile.loc.gov/storage-services/service/ll/usrep/usrep314/usrep314252/usrep314252.pdf

upon its *face*."[666] After "testing the section on its *face*,"[667] it may be "invalid on its *face*,"[668] because "the inevitable effect of a statute on its *face* may render it unconstitutional."[669] In determining this, Judges may sometimes need to go "behind the *face* of the statute or of the complaint"[670] to get at its full meaning and implications.

Facial expressions are pure speech. They are pure speech more surely than are those symbols on a page that we train our brains to interpret as letters. Through dedicated effort and lifelong practice, we train our brains to recognize and translate specific designs on paper as letters of the alphabet, but infants and toddlers learn to recognize, differentiate, and read faces and facial expressions far earlier than they develop the capacity to sound out letters or even speak full sentences. Alphabetic illiteracy is no barrier to reading facial expressions. The trope of individuals skilled with letters and numbers but painfully deficient at reading obvious facial and social cues is pervasive in film and literature for a reason, as is the trope of artists who capture in portraits what cannot be put into words.

Our faces, and how we use them, communicate what words do not, will not, or cannot. Bowing down with one's face to the ground communicates special obeisance. Men were reading faces long before they could read books, scrolls, or cuneiform tablets. Once men developed writing, they wrote chapters and whole books on how to read and interpret people's facial expressions to get at more truth than can be had from each other's written or spoken words alone.[671] Subtle cues from lip compression to

---

666   "The statute must be judged upon its **face**."
Thornhill v. Alabama, 310 U. S. 88, 88 (1940), available online:
https://tile.loc.gov/storage-services/service/ll/usrep/usrep310/usrep310088/usrep310088.pdf

667   "There is a further reason for testing the section on its **face**."
Thornhill v. Alabama, 310 U. S. 88, 97 (1940), available online:
https://tile.loc.gov/storage-services/service/ll/usrep/usrep310/usrep310088/usrep310088.pdf

668   "The statute is invalid on its **face**."
Thornhill v. Alabama, 310 U. S. 88, 89 (1940), available online:
https://tile.loc.gov/storage-services/service/ll/usrep/usrep310/usrep310088/usrep310088.pdf

669   "…the inevitable effect of a statute on its **face** may render it unconstitutional."
United States v. O'Brien, 391 U.S. 367, 390 (1968), available online:
https://tile.loc.gov/storage-services/service/ll/usrep/usrep391/usrep391367/usrep391367.pdf

670   "…there is no occasion to go behind the **face** of the statute or of the complaint…"
Thornhill v. Alabama, 310 U. S. 88, 96 (1940), available online:
https://tile.loc.gov/storage-services/service/ll/usrep/usrep310/usrep310088/usrep310088.pdf

671   c.f.
Mac Fulfer, *The Power of Face Reading: A simple illustrated guide to understanding our universal language*, (Las Vegas, NV: Global Insight Communications, 1994). Reprint, 2019.
Paul Ekman and Wallace V. Friesen, *Unmasking the Face: A guide to recognizing emotions from facial expressions* (Los Altos, CA: Major Books, 2003). Reprint, Kindle edition, 2009.
Mark G. Frank, *Understanding Nonverbal Communication* Course Guidebook, (Buffalo, NY: The Teaching Company, 2016), pp. 29-33

how facial expressions form and disappear communicate a great deal.[672] We see examples of this of this in every medium from literature to film — fiction and non-fiction. An observer might not know exactly *why* someone's face is happy, sad, or contorted in pain, but the lack of additional detailed information does not negate the fact that the face and facial expressions already communicated details about an internal frame of mind.

Judges trying to shoehorn facial expressions into failing intermediate *O'Brien* scrutiny argued: "The fact that "'explanatory speech is necessary is strong evidence that the conduct at issue ... is not so inherently expressive that it warrants protection' as symbolic speech."[673] But facial expressions communicate *more* than mere words, and they modify every spoken word. One could just as accurately claim that the fact that explanatory *facial expressions* are necessary is strong evidence that the *words* at issue are "not so inherently expressive" they warrant protection as speech. In both cases, where words and facial expressions are used to communicate, the fact that additional explanation may be needed means nothing more than when you need to give a friend more detail after telling them "I am sad," or "I am angry." Whether your voice or your face communicated the message, your friend still received it, and "speech" implicating the First Amendment still took place. Moreover, this pure speech also implicates the Fourteenth Amendment right to privacy, as Supreme Court Justice Warren and Justice Brandeis recognized in their 1890 essay on "The Right to Privacy":

> "If, then, the decisions indicate a general right to privacy for thoughts, emotions, and sensations, these should receive the same protection, whether expressed in writing, or in conduct, in conversation, in attitudes, or in facial expression."[674] (emphasis added)

If anything, facial expressions are needed to refine the message of words on paper, as evinced by the notorious misunderstandings of tone that can happen when reading letters and email. Similarly, when talking with people on the phone, we often deliberately use our facial expressions to enhance

---

672   "Genuine facial expressions fade. False facial expressions will suddenly go away."
Chase Hughes, *Six-Minute X-Ray: Rapid Behavior Profiling* (Milton, DE: Evergreen Press, Kindle Edition, 2020), p. 56.
673   *Sehmel v. Shah*, 514 P.3d 1238 (Wash. Ct. App. 2022), at 16, available online:
https://caselaw.findlaw.com/wa-court-of-appeals/2182709.html
The fact that " 'explanatory speech is necessary is strong evidence that the conduct at issue ... is not so inherently expressive that it warrants protection' as symbolic speech." Stewart v. Justice, 502 F. Supp. 3d 1057, 1066 (S.D. W.Va. 2020) (quoting Rumsfeld v. Forum for Acad. & Inst. Rights, Inc., 547 U.S. 47, 66, 126 S. Ct. 1297, 164 L.Ed.2d 156 (2006)).
674   Samuel D. Warren and Louis D. Brandeis, "The Right to Privacy," December 15, 1890, Harvard Law Review Vol. IV, No. 5 (206) available online: https://www.jstor.org/stable/1321160
Justices Brandeis and Warren next consider and discard the possible objection that involuntary expressions do not warrant the same protections as deliberate expressions: "If the test of deliberateness of the act be adopted, much casual correspondence which is now accorded full protection would be excluded[.]"

vocal communication even when we know the other person cannot see us. The rise of stylized facial expressions in emojis and emoticons as part of written messages is tacit recognition of this fact and a shorthand attempt to compensate for it. Moreover, facial expressions are often more honest because they are less voluntary and harder to fake. They are as natural and appropriate in any pubic space as any written text and spoken words.[675] *Any* restriction on the pure speech of facial expressions demands strict scrutiny in every sense of the term.

Facial expressions are pure speech. Freedom of speech, recognized in the First Amendment, is every human's unalienable birthright. The unalienable First Amendment right to *refrain* from speaking secures the right to wear or not wear a mask, and the unalienable First Amendment right to *speak* categorically prohibits compulsory masking, regardless of the emergency. Even assuming masks to be 100% effective with no other side effects, and even assuming some worst-case respiratory variant of smallpox or the Black Death, compulsory masking would *still* not have been justified or constitutional. The *fact* that more narrowly tailored measures which have stood the test of millennia were available — like temporary quarantine limited to known infections — only made compulsory masking that much more inexcusable.

Dealing with objections and counterarguments

But "freedom of speech is not absolute"!

On the morning of March 2, 1961, one-hundred and eighty-seven black high school and college students met at Zion Baptist Church in Columbia, South Carolina. From there, around noon, they walked in smaller separate groups to the South Carolina State House grounds while the legislature was in session. There, they found 30 law enforcement officers already waiting, and peacefully protested segregation while walking in an "orderly way" single file or two abreast carrying signs for about 45 minutes. They did not even disrupt or impede traffic within the statehouse grounds. The United States Supreme Court summarized what happened next:

> "When told by police officials that they must disperse within 15 minutes on pain
> of arrest, they failed to do so and sang patriotic and religious songs after one of
> their leaders had delivered a 'religious harangue.' There was no violence or threat

---

675   c.f. "streets are natural and proper places for the dissemination of information and opinion." *Schneider v. State*, 308 U.S. 147, 163 (1939), available online:
https://tile.loc.gov/storage-services/service/ll/usrep/usrep308/usrep308147/usrep308147.pdf
quoted in *Thornhill v. Alabama*, 310 U. S. 88, 105-106 (1940), available online:
https://tile.loc.gov/storage-services/service/ll/usrep/usrep310/usrep310088/usrep310088.pdf

of violence on their part or on the part of any member of the crowd watching them; but petitioners were arrested and convicted of the common-law crime of breach of the peace, which the State Supreme Court said 'is not susceptible of exact definition.'"[676]

The best the arresting and convicting authorities could do was to characterize the students' conduct as "boisterous," "loud," and "flamboyant" while stamping their feet and clapping their hands, but they nevertheless imposed penalties ranging from a $10 fine or five days in jail to a $100 fine or 30 days in jail (in 2023 dollars these fines would be $100 to $1,000, respectively).[677] Almost two years later, in February 1963, the Supreme Court reversed the conviction, decisively ruling that:

> "South Carolina infringed their rights of free speech, free assembly and freedom to petition for a redress of grievances-rights guaranteed by the First Amendment and protected by the Fourteenth Amendment from invasion by the States… The circumstances in this case reflect an exercise of these basic constitutional rights in their most pristine and classic form."[678]

I bet you'll never guess how Supreme Court Justice Clark began his dissent: "The priceless character of First Amendment freedoms cannot be gainsaid, but it does not follow that they are absolutes immune from necessary state action reasonably designed for the protection of society."[679] Some variation of this platitude has been raised as an objection at virtually every vindication of individual liberties in America's history. During the COVIDcrisis, it was parroted ceaselessly as some sort of irrefutable checkmate when citizens objected to the unprecedented invasions of their rights.

"Freedom of speech is not absolute."

"Your liberties are not absolute."

And so on, *ad nauseum.*

---

676   *Edwards v. South Carolina*, 372 U.S. 229, 229 (1963), available online:
https://tile.loc.gov/storage-services/service/ll/usrep/usrep372/usrep372229/usrep372229.pdf
677   *Edwards v. South Carolina*, 372 U.S. 229, 233-234 (1963), available online:
https://tile.loc.gov/storage-services/service/ll/usrep/usrep372/usrep372229/usrep372229.pdf
678   *Edwards v. South Carolina*, 372 U.S. 229, 229 & 235 (1963), available online:
https://tile.loc.gov/storage-services/service/ll/usrep/usrep372/usrep372229/usrep372229.pdf
679   *Edwards v. South Carolina*, 372 U.S. 229, 239 (1963), available online:
https://tile.loc.gov/storage-services/service/ll/usrep/usrep372/usrep372229/usrep372229.pdf

"Freedom of speech is not absolute" has been the trite starting excuse for every violation of individuals' unalienable First Amendment rights in America's history, and a universal objection when those same rights are vindicated.

When the Supreme Court ruled in 1941 that judges cannot punish journalists for out-of-court publications opining on pending cases, Justice Frankfurter and three others dissented: "… freedom of public expression alone assures the unfolding of truth, it is indispensable to the democratic process. But even that freedom is not an absolute and is not predetermined."[680] When the Supreme Court ruled in 1974 that hanging an American flag upside down with a peace symbol on it was protected symbolic speech, which the owner could not be arrested for, how did Justice Rehnquist begin his dissenting opinion? "[T]he right of free speech is not absolute at all times and under all circumstances."[681] When, in 1989, the Court affirmed Americans' right to symbolic speech by flag burning in *Texas v. Johnson*, the dissenting objection was the same: "the Court insists that the Texas statute prohibiting the public burning of the American flag infringes on respondent Johnson's freedom of expression. Such freedom, of course, is not absolute."[682]

At the best, the phrase "freedom of speech is not absolute" can be used as an introduction, but, like saying, "that's just your opinion" backed up by nothing else, more often it was used as a vapid dodge or a subtle attempt to foist the burden of proof back onto the person who objected to having their rights stolen by fiat. Like saying, "burden of proof is on you," it concealed lazy thinking, and allowed those with the real burdens of proof to sit back and play the part of sniping skeptics. There is no

---

680   *Bridges v. California*, 314 U.S. 252, 293 (1941) (Frankfurter, J., dissenting), available online: https://tile.loc.gov/storage-services/service/ll/usrep/usrep314/usrep314252/usrep314252.pdf

The dissenting justices also quoted the Roman orator Cicero's maxim that "the safety of the people is the supreme law," as though the human rights record of a civilization which raised crucifixion to an art form provides any kind of model for Americans to emulate. This feels like a cheap shot, both at Rome and the dissenting U.S. Supreme Court justices, but we all just spent the last three years living through just one of the many destinations such safety-reasoning leads to.

Two years later in Martin v. City of Struthers, the Supreme Court ruled that it was an unconstitutional infringement on First Amendment rights to enforce a municipal ordinance "forbidding any person to knock on doors, ring doorbells, or otherwise summon to the door the occupants of any residence for the purpose of distributing to them handbills or circulars." What was the go-to dissenting argument? Justice Frankfurter argued: "neither the First nor the Fourteenth Amendment is to be treated by judges as though it were a mathematical abstraction, an absolute having no relation to the lives of men."
Martin v. City of Struthers, 319 U.S. 141, 152 (1943) (Frankfurter, J., dissenting), available online: https://tile.loc.gov/storage-services/service/ll/usrep/usrep319/usrep319141/usrep319141.pdf

681   *Spence v. Washington*, 418 U.S. 405, 417 (1974) (Rehnquist, J., dissenting), available online: https://tile.loc.gov/storage-services/service/ll/usrep/usrep418/usrep418405/usrep418405.pdf

682   *Texas v. Johnson*, 491 U.S. 397, 430 (1989) (Rehnquist, J., dissenting)
https://tile.loc.gov/storage-services/service/ll/usrep/usrep491/usrep491397/usrep491397.pdf

good reason why the specifically enumerated right to the freedom of speech *ever* automatically takes a backseat to a limited, unenumerated, "right" to public health. Rather, the presumption is in favor of individual liberty. Indirect, marginal public health benefits, even if proven, cannot justify direct and massive violations of essential individual liberties. This goes double for when the benefits are projections based on theory and computer models, as was the case with SARS-CoV-2. The burden is on the infringer to justify their actions, and hand-waiving, use of magic words like "safety," or COVIDcrisis sloganeering was not — and never will be — good enough. In the words of the Supreme Court:

> "[T]he First Amendment does not speak equivocally. It prohibits any law 'abridging the freedom of speech, or of the press.' It must be taken as a command of the broadest scope that explicit language, read in the context of a liberty-loving society, will allow."[683]

Few things are absolute; some narrowly defined speech is not protected, including statements "which by their very utterance inflict injury."[684] This includes defamation & libel, threats & intimidation, certain types of obscenity, incitement, and other "utterances inimical to the public welfare, tending to incite to crime, disturb the public peace, or endanger the foundations of organized government and threaten its overthrow by unlawful means."[685] Masking and non-masking do not fit into any of these categories. They do not even come close. Yet mask mandates treated merely going about one's business with one's face uncovered as "constitutionally unprotected communications 'which by their very utterance inflict injury.'"[686]

---

683   *Bridges v. California*, 314 U.S. 252, 263 (1941), available online:
https://tile.loc.gov/storage-services/service/ll/usrep/usrep314/usrep314252/usrep314252.pdf

684   *Virginia State Board of Pharmacy v. Virginia Citizens Consumer Council, Inc.*, 425 U.S. 748, 775-776 (1976), (Stewart, J., concurring), available online:
https://tile.loc.gov/storage-services/service/ll/usrep/usrep425/usrep425748/usrep425748.pdf quoting *Chaplinsky v. New Hampshire*, 315 U. S. 568, 572 (1942), available online:
https://tile.loc.gov/storage-services/service/ll/usrep/usrep315/usrep315568/usrep315568.pdf

> "There are certain well-defined and narrowly limited classes of speech, the prevention and punishment of which have never been thought to raise any Constitutional problem These include the lewd and obscene, the profane, the libelous, and the insulting or "fighting" words-those which by their very utterance inflict injury or tend to incite an immediate breach of the peace."

685   *Whitney v. California*, 274 U.S. 357, 371 (1927), available online:
https://tile.loc.gov/storage-services/service/ll/usrep/usrep274/usrep274357/usrep274357.pdf

686   *Virginia State Board of Pharmacy v. Virginia Citizens Consumer Council, Inc.*, 425 U.S. 748, 775-776 (1976), (Stewart, J., concurring), available online:
https://tile.loc.gov/storage-services/service/ll/usrep/usrep425/usrep425748/usrep425748.pdf quoting *Chaplinsky v. New Hampshire*, 315 U. S. 568, 572 (1942), available online:
https://tile.loc.gov/storage-services/service/ll/usrep/usrep315/usrep315568/usrep315568.pdf

Justice Oliver Wendel Holmes Jr.'s famous example of "falsely shouting fire in a theatre and causing a panic,"[687] was oversimplified and overused during the COVIDcrisis. In 1927, just two weeks after he handed down the worst verdict of his career in *Buck v. Bell*, Justice Holmes joined with Justice Brandeis to write a concurring opinion for the First Amendment case of *Whitney v. California* that was in many ways antithetical to the principles underlying *Buck v. Bell*. Justices Holmes and Brandeis reiterated that even dangerous speech is not outside the protection of the First Amendment:

> "[E]ven advocacy of violation, however reprehensible morally, is not a justification for denying free speech where the advocacy falls short of incitement and there is nothing to indicate that the advocacy would be immediately acted on. The wide difference between advocacy and incitement, between preparation and attempt, between assembling and conspiracy, must be borne in mind."[688]

While Justices Holmes and Brandeis said that "only an emergency can justify repression,"[689] they heavily qualified this apparent concession, making clear that the starting presumption is in favor of individual liberty: "a State is, ordinarily, denied the power to prohibit dissemination of social, economic and political doctrine which a vast majority of its citizens believes to be false and fraught with evil consequence."[690] They went on to say:

> "Those who won our independence believed that the final end of the State was to make men free to develop their faculties; and that in its government the deliberative forces should prevail over the arbitrary. They valued liberty both as an end and as a means. They believed liberty to be the secret of happiness and courage to be the secret of liberty. They believed that freedom to think as you will and to speak as you think are means indispensable to the discovery and spread of political truth; that without free speech and assembly discussion would be futile; that with them, discussion affords ordinarily adequate protection against the dissemination of noxious doctrine; that the greatest menace to freedom is an

---

687   *Schenck v. United States*, 249 U.S. 47 (1919), available online:
https://tile.loc.gov/storage-services/service/ll/usrep/usrep249/usrep249047/usrep249047.pdf
688   *Whitney v. California*, 274 U.S. 357, 376 (1927) (Brandeis and Holmes, J. J., concurring), available online:
https://tile.loc.gov/storage-services/service/ll/usrep/usrep274/usrep274357/usrep274357.pdf
689   "Only an emergency can justify repression."
Whitney v. California, 274 U.S. 357, 377 (1927) (Brandeis and Holmes, J. J., concurring), available online:
https://tile.loc.gov/storage-services/service/ll/usrep/usrep274/usrep274357/usrep274357.pdf
690   *Whitney v. California*, 274 U.S. 357, 374 (1927) (Brandeis and Holmes, J. J., concurring), available online:
https://tile.loc.gov/storage-services/service/ll/usrep/usrep274/usrep274357/usrep274357.pdf

inert people; that public discussion is a political duty; and that this should be a fundamental principle of the American government.[691]

The main point, in their words, is that "Fear of serious injury cannot alone justify suppression of free speech and assembly."[692] Consistent application of this principle would almost entirely rule out "rational basis" scrutiny for measures that infringe citizens' First Amendment rights. Justices Holmes and Brandeis emphasized that even when a First Amendment violation serves a compelling state interest and is effective at producing the desired results, it may *still* be unconstitutional: "A police measure may be unconstitutional merely because the remedy, although effective as means of protection, is unduly harsh or oppressive."[693] Despite the State's compelling interest in preventing crime, "The fact that speech is *likely* to result in some violence or in destruction of property is not enough to justify its suppression" (emphasis added).[694] The implications for mask mandates are obvious. Even if failing to wear a mask had been *likely* to spread COVID, that would still not have been enough to justify suppression of the speech inherent in masking and non-masking and facial expressions. Even "dangerous" speech that makes others uncomfortable is protected by the First Amendment.

Masking and non-masking are, at their core, more ideological communication than anything else. First Amendment protections are at their most robust for this sort of speech, and over it the state correspondingly has far less latitude to impose regulations, even — or especially — during a public health crisis. Even mistaken and erroneous speech is protected. As the U.S. Supreme Court said in its 1976 *Virginia State Board of Pharmacy* ruling on the right to receive speech:

> "Ideological expression, be it oral, literary, pictorial, or theatrical, is integrally related to the exposition of thought-thought that may shape our concepts of the whole universe of man. Although such expression may convey factual information relevant to social and individual decisionmaking, it is protected by the Constitution, whether or not it contains factual representations and even if it includes inaccurate assertions of fact."[695]

---

691   *Whitney v. California*, 274 U.S. 357, 375 (1927) (Brandeis and Holmes, J. J., concurring), available online: https://tile.loc.gov/storage-services/service/ll/usrep/usrep274/usrep274357/usrep274357.pdf

692   *Whitney v. California*, 274 U.S. 357, 376 (1927) (Brandeis and Holmes, J. J., concurring), available online: https://tile.loc.gov/storage-services/service/ll/usrep/usrep274/usrep274357/usrep274357.pdf

693   *Whitney v. California*, 274 U.S. 357, 377 (1927) (Brandeis and Holmes, J. J., concurring), available online: https://tile.loc.gov/storage-services/service/ll/usrep/usrep274/usrep274357/usrep274357.pdf

694   *Whitney v. California*, 274 U.S. 357, 378 (1927) (Brandeis and Holmes, J. J., concurring), available online: https://tile.loc.gov/storage-services/service/ll/usrep/usrep274/usrep274357/usrep274357.pdf

695   *Virginia State Board of Pharmacy v. Virginia Citizens Consumer Council, Inc.*, 425 U.S. 748, 779 (1976), (Stewart, J., concurring), available online:

While the court recognized that "the Constitution does not provide absolute protection for false factual statements that cause private injury," (this is why there is a First Amendment exception for slander and libel) it also recognized that "factual errors are inevitable in free debate, and the imposition of liability for erroneous factual assertions can 'dampen the vigor and limit the variety of public debate by inducing self-censorship.'"[696] The Court stressed that "it is a cardinal principle of the First Amendment that 'government has no power to restrict expression because of its message, its ideas, its subject matter, or its content[.]'"[697] In the Court's own words, this is true of speech: "be it oral, literary, *pictorial*, or *theatrical*."[698] This is worth restating for emphasis because masking and non-masking are both pictorial and theatrical forms of speech.

But COVID is "a clear and present danger"!

In an effort to delineate and clarify when speech is and is not protected, the U.S. Supreme Court has gradually articulated and refined a number of descriptors. In its 1963 *Edwards v. South Carolina* ruling that upheld the right of minority citizens to peacefully protest against unjust majority laws, the Court accepted that:

> "Speech is often provocative and challenging. It may strike at prejudices and preconceptions and have profound unsettling effects as it presses for acceptance of an idea. That is why freedom of speech... is... protected against censorship or punishment, unless shown likely to produce a clear and present danger of a serious substantive evil that rises far above public inconvenience, annoyance, or unrest... There is no room under our Constitution for a more restrictive view. For the alternative would lead to standardization of ideas either by legislatures, courts, or dominant political or community groups." (emphasis added)[699]

---

https://tile.loc.gov/storage-services/service/ll/usrep/usrep425/usrep425748/usrep425748.pdf

696   (internal quotation marks omitted) *Virginia State Board of Pharmacy v. Virginia Citizens Consumer Council, Inc.*, 425 U.S. 748, 777 (1976), available online:

https://tile.loc.gov/storage-services/service/ll/usrep/usrep425/usrep425748/usrep425748.pdf

697   *Virginia State Board of Pharmacy v. Virginia Citizens Consumer Council, Inc.*, 425 U.S. 748, 776 (1976), (Stewart, J., concurring), available online:

https://tile.loc.gov/storage-services/service/ll/usrep/usrep425/usrep425748/usrep425748.pdf

quoting *Chaplinsky v. New Hampshire*, 315 U. S. 568, 572 (1942), available online:

698   *Virginia State Board of Pharmacy v. Virginia Citizens Consumer Council, Inc.*, 425 U.S. 748, 779 (1976), (Stewart, J., concurring), available online:

https://tile.loc.gov/storage-services/service/ll/usrep/usrep425/usrep425748/usrep425748.pdf

699   *Edwards v. South Carolina*, 372 U.S. 229, 237 (1963), available online:

https://tile.loc.gov/storage-services/service/ll/usrep/usrep372/usrep372229/usrep372229.pdf

Compulsory masking was an extreme standardization of public health ideas. The "clear and present danger" exception referenced by the Court in 1963 originated in the same 1919 case as the famous "fire in a theatre" analogy: *Schenck v. United States*,[700] and is one of these clarifying tests for unprotected speech. In 1941, the Supreme Court explained:

> "[W]e have also suggested that 'clear and present danger' is an appropriate guide in determining the constitutionality of restrictions upon expression where the substantive evil sought to be prevented by the restriction is 'destruction of life or property, or invasion of the right of privacy.'"[701]

This standard invokes two of the most basic, longstanding, and universal legal justifications and defenses for both individual and governmental actions that would otherwise be impermissible: self-defense and (more broadly) necessity as the lesser of two evils. Laws challenged on grounds of Free Speech are often defended on this basis: that limitations on Free Speech *may* be permissible when, as the Supreme Court put it, "some speakers will be destroyed in the absence of the challenged law."[702] In the context of free speech, what the debate on mask efficacy gets at is the implicit assumption that *if* masks "work" well enough, *then* limitations on free speech applied during COVID were permissible. The formulaic recitation of cases and deaths from COVID, mentions of "aerosols," citations to the CDC, and appeals to scientific consensus and CDC guidelines in every dismissal of a challenge to compulsory masking were an indirect way of trying to evoke this exception without having to defend it on an even playing field. They were attempts to avoid admitting that mask mandates were impermissibly prescribing "what shall be orthodox" in science, medicine, and public health, and dealing with the constitutional fact that even dangerous speech is protected.

Claims that compulsory masking is an essential public health measure are, in effect, claims of necessity. In the words of the 9th Circuit, a necessity defense must demonstrate "that a direct causal relationship be reasonably anticipated to exist between the defender's action and the avoidance of harm."[703]

quoting *Terminiello v. Chicago*, 337 U.S. 1, 4-5 (1949), available online:
https://tile.loc.gov/storage-services/service/ll/usrep/usrep337/usrep337001/usrep337001.pdf

700   *Schenck v. United States*, 249 U.S. 47, 52 (1919), available online:
https://tile.loc.gov/storage-services/service/ll/usrep/usrep249/usrep249047/usrep249047.pdf

701   *Bridges v. California*, 314 U.S. 252, 255 (1941), available online:
https://tile.loc.gov/storage-services/service/ll/usrep/usrep314/usrep314252/usrep314252.pdf quoting *Thornhill v. Alabama*, 310 U. S. 88, 105 (1940), available online:
https://tile.loc.gov/storage-services/service/ll/usrep/usrep310/usrep310088/usrep310088.pdf

702   *Hurley v. Irish-American Gay, Lesbian and Bisexual Group of Boston, Inc.*, 515 U. S. 557, 577,
https://supreme.justia.com/cases/federal/us/515/557/case.pdf

703   *United States v. Simpson*, 460 F.2d 515, 518 (9th Cir. 1972), available online:
https://casetext.com/case/united-states-v-simpson-20

Compulsory masking assumes that the imposing authorities are faced with an unavoidable choice of evils and that compulsory masking is the lesser evil — for which there are no viable alternatives. It assumes that the harm compulsory masking prevents is imminent, severe, and that there is a *direct* causal connection between compulsory masking of *every* individual and the severity of harm from COVID. This was why mask efficacy was a dogma which could never be disputed. It was an indispensable pillar supporting the implicit claims of self-defense and necessity that compulsory masking rested on.

This is a major underlying reason why, in practice, the judicial deck was stacked in favor of the dogma that masks work, and why no scrutiny or rational basis scrutiny was preferred. Just because there is a *debate* among experts about something does not necessarily make something irrational, and so even when disputed, compulsory masking still passed rational basis scrutiny. This put plaintiffs challenging masks in an impossible catch-22. When plaintiffs conceded for the sake of argument that masks either work or might work, this was used as grounds to rule against them as well, because making this concession merely reinforced the supposed "rational" or "reasonable" relationship between mask-use and viral spread.[704] But challenging the efficacy of masks outright was a surefire way to start out on the wrong foot. In cases where plaintiffs disputed the efficacy of masks and provided scientific sources, their sources were dismissed on various pretextual grounds, such as that the plaintiffs' references predated the pandemic or did not involve SARS-CoV-2 transmission.[705] Such brush-offs just demonstrated the motivated reasoning and outrageous double-standards being applied. New science may help *explain* old findings, but good science does not become bad science with the passage of time, and authorities like the CDC that pushed masks did so primarily using studies that did not involve SARS-CoV-2 either.

Typically, these dismissals followed a basic pattern. They would start with a recitation of the total number of purported deaths *with* COVID, ignoring the actual mortality rate of SARS-CoV-2, which was comparable to that of other coronaviruses or a bad flu.[706] They would also usually say something

---

704   *Minnesota Voters All. v. Walz*, 492 F. Supp. 3d 822, 837-838 (D. Minn. 2020), p. 2, available online: https://case-law.vlex.com/vid/minn-voters-alliance-v-891222192

705   *Denis v. Ige et al*, No. 1:2021cv00011 - Document 62 (D. Haw. 2021), pg. 27 n 15, available online: https://law.justia.com/cases/federal/district-courts/hawaii/hidce/1:2021cv00011/152658/62/

706   Yanis Roussel et al., SARS-CoV-2: fear versus data. Int J Antimicrob Agents, 2020. 55(5): p. 105947. https://www.ncbi.nlm.nih.gov/pmc/articles/PMC7102597/pdf/main.pdf

John P. A. Ioannidis, "Reconciling estimates of global spread and infection fatality rates of COVID-19: An overview of systematic evaluations." *European Journal of Clinical Investigation*, 2021. https://dx.doi.org/10.1111/eci.13554

along the lines that "there is currently no cure and no vaccine,"[707] thus sidestepping the fact that for more than 99% of those infected, no "cure" was needed, as well as the fact that there were effective treatment protocols for COVID-19 available in March 2020. Mention of natural immunity in these court decisions ranges from rare to non-existent. Playing up the risk from COVID as much as possible, these court decisions would then assert that masks are effective and essential, that universal compulsory masking is a rational, proportionate response to respiratory infections, and conclude by deciding that the plaintiff was not entitled to relief (i.e., they should stop complaining and wear a mask). Clearly, according to these courts, the threat from any unmasked individual outweighed any alleged violations of those individuals' First Amendment rights.

But pointing out the deadliness of a particular disease does not automatically make every ostensible prevention measure rational or justified. In order for necessity or "clear and present danger" justifications to be valid in the case of masks, proponents also needed to demonstrate by clear and convincing evidence, at a minimum, that masks work, that alternative means like temporary quarantine were not viable (i.e., that universal compulsory masking is narrowly tailored), *and* that any unmasked individuals (including those who already recovered from COVID) are not just a speculative *theoretical* risk, but an *actual* imminent and *severe* threat to life. A fatal defect in every implied or explicit necessity defense for public health masking is that even *if* masks had been *proven* to work, any benefits would, *at best*, still be as indirect as the benefits from denying wartime accommodation for conscientious objectors, and "Congress and courts have long held that granting combat exemptions for conscientious objectors is not likely to seriously hamper the state's defense efforts."[708]

Moreover, claims of necessity require a *balancing* of public interests, but rulings which upheld mask mandates simply categorically (and unjustifiably) presumed that the state interest in public health outweighed everything else. In reality, the public interest favors *non*-compulsion. Upholding fundamental individual liberties is *always* in the public interest. In the words of the 6th Circuit, "it is

---

707   c.f. "There is no cure, vaccine, or effective treatment for COVID-19."

Illinois Republican Party v. Pritzker, 470 F. Supp. 3d 813, 818 (page 7), (Northern District of Illinois, 2020), available online: https://libertyjusticecenter.org/wp-content/uploads/016-IL-Republican-Party-Opinion-denying-Motion-for-TRO-and-PI.pdf and https://casetext.com/case/ill-republican-party-v-pritzker

"There is currently no cure and no vaccine."

Minnesota Voters All. v. Walz, 492 F. Supp. 3d 822, p. 3 (D. Minn. 2020), available online:
https://law.justia.com/cases/federal/district-courts/minnesota/mndce/0:2020cv01688/189156/51/

"[T]here currently is no known cure, no effective treatment, and no vaccine."

Machovec v. Palm Beach County, No. 2020CA006920AXX — Filing #110806335, 27 July 2020, page 2 (15th Judicial Circuit, Florida), available online: https://clearinghouse.net/doc/109346/ &
https://www.floridacivilrights.org/wp-content/uploads/2020/07/Order-Denying-Temporary-Injunction.pdf

708   Nadia N. Sawicki, *The Hollow Promise of Freedom of Conscience*, 33 *Cardozo Law Review* 1389 2011-2012, p. 57. Available online: https://papers.ssrn.com/sol3/papers.cfm?abstract_id=1666278

always in the public interest to prevent the violation of a party's constitutional rights."[709] This includes the First Amendment. In the words of the 7th Circuit, "injunctions protecting First Amendment freedoms are always in the public interest."[710] Judge Kurt D. Engelhardt of the 5th Circuit articulated this foundational principle especially well in the COVID-19 vaccine context: "The public interest is also served by maintaining our constitutional structure and maintaining the liberty of individuals to make intensely personal decisions according to their own convictions—even, or perhaps particularly, when those decisions frustrate government officials."[711] This applies no less to masks. Additionally, the public benefits from the security provided by a civil service and governmental administration which gives individual liberties the broadest construction possible. Powerful and highly effective federal, state, and local governments are worse than useless without an even *more* powerful and highly effective Bill of Rights.[712]

Even taking public health prophylaxis as a compelling state interest, that only makes it analogous to other compelling state interests like crime-fighting and prevention, because the same basic justification applies, that "some speakers will be destroyed in the absence of the challenged law."[713] The compelling

---

709   *G & V Lounge, Inc. v. Mich. Liquor Control Comm'n*, 23 F.3d 1071, 1079 (6th Cir.1994), available online: https://casetext.com/case/g-v-lounge-v-michigan-liquor-control-comn#p1079

This 6th Circuit quotation has been cited approvingly in multiple other circuits, including the 10th and the 5th Circuits: *Awad v. Ziriax*, 670 F.3d 1111, 1132 (10th Cir.2012), available online: https://casetext.com/case/awad-v-ziriax#p1132 *Jackson Women's Health Org. v. Currier*, 760 F.3d 448, 470 n. 9 (5th Cir.2014), available online: https://casetext.com/case/jackson-womens-health-org-v-currier-3#p470

710   *Christian Legal Soc'y v. Walker*, 453 F.3d 853, 859 (7th Cir.2006), available online: https://casetext.com/case/christian-legal-society-v-walker#p859

Quoted with approval in multiple other Circuits, including the 5th Circuit
Texans for Free Enter. v. Tex. Ethics Comm'n, 732 F.3d 535, 539 (5th Cir. 2013), available online: https://casetext.com/case/texans-for-free-enter-v-tex-ethics-commn

711   *BST Holdings, LLC v. Occupational Safety and Health Administration*, No. 21-60845 (5th Cir. 2021), Document: 00516091902, p. 20, filed November 12, 2021, available online: https://law.justia.com/cases/federal/appellate-courts/ca5/21-60845/21-60845-2021-11-12.html & https://int.nyt.com/data/documenttools/5th-cir/90be1e20702cd0de/full.pdf

712   While this wording is my own, and represents a substantial development rather than a direct quote, I first heard the nucleus of this point made by lawyer Robert Barnes in a February 15, 2022 podcast:

"I care about our constitutional republic. I don't care about our military prowess — our military prowess means nothing if we have no Constitutional rights in the first place; it's not worth the paper it's printed on, it's not worth anything, it ain't worth the paper it's printed on if we're now a military state."

Robert Barnes, "Ep. 100: Biden-Gate; SBF Bail; Elon Musk Victory; 2020 Election Stuff & MORE! Viva and Barnes LIVE!" *vivafrei* YouTube channel, Timestamp 1:41:23-1:41:34 (streamed on February 15, 2023) Available Online: https://www.youtube.com/watch?v=zhzOj3671n4 (last accessed May 9, 2023)

713   *Hurley v. Irish-American Gay, Lesbian and Bisexual Group of Boston, Inc.*, 515 U. S. 557, 577, https://supreme.justia.com/cases/federal/us/515/557/case.pdf

state interest in crime prevention does not outweigh everything else, especially when any benefits are abstract and indirect and the measures in question involve concrete and immediate infringements on liberties explicitly recognized by the Bill of Rights. The court rulings affirming this are too numerous to even begin listing. In the same manner, any compelling state interest in public health must also yield to individual liberties. The real burden of proof was always on the authorities,[714] but the cases that dismissed challenges to masks ignored this.

In 1927, Justices Holmes and Brandeis recognized that the "clear and present danger" standard was still especially vague:

> "This Court has not yet fixed the standard by which to determine when a danger shall be deemed clear; how remote the danger may be and yet be deemed present; and what degree of evil shall be deemed sufficiently substantial to justify resort to abridgement of free speech and assembly as the means of protection."[715]

At first glance, Justice Holmes' and Justice Brandeis' 1927 concurrence may seem to suggest that reasonable basis scrutiny is enough:

> "To justify suppression of free speech there must be reasonable ground to fear that serious evil will result if free speech is practiced. There must be reasonable ground to believe that the danger apprehended is imminent. There must be reasonable ground to believe that the evil to be prevented is a serious one."[716]

However, the Supreme Court made clear in a number of later cases any "emergency" exceptions for basic liberties are much narrower than Justice Holmes' and Justice Brandeis' language in 1927 might seem to suggest.

The Court clarified in 1941, that:

---

714 "When the Government restricts speech, *the Government bears the burden* of proving the constitutionality of its actions." *United States v. Playboy Entertainment Group, Inc.,* 529 U. S. 803, 816 (2000), available online: https://tile.loc.gov/storage-services/service/ll/usrep/usrep529/usrep529803/usrep529803.pdf
Quoted with emphasis in *Watchtower Bible & (and) Tract Society of New York, Inc., et al. v. Village of Stratton et al.,* 536 U.S. 150, 170 (2002), available online:
https://tile.loc.gov/storage-services/service/ll/usrep/usrep536/usrep536150/usrep536150.pdf
715 *Whitney v. California,* 274 U.S. 357, 374 (1927) (Brandeis and Holmes, J. J., concurring), available online: https://tile.loc.gov/storage-services/service/ll/usrep/usrep274/usrep274357/usrep274357.pdf
716 *Whitney v. California,* 274 U.S. 357, 376 (1927) (Brandeis and Holmes, J. J., concurring), available online: https://tile.loc.gov/storage-services/service/ll/usrep/usrep274/usrep274357/usrep274357.pdf

"The 'clear and present danger' cases, decided by this Court, indicate that the substantive evil likely to result must be extremely serious and the degree of imminence extremely high before utterances can be punished... The First Amendment's prohibition of 'any law abridging the freedom of speech or of the press' must be given the broadest scope that can be countenanced in an orderly society."[717]

Unlike the reasonable basis test, "The clear and present danger doctrine requires a weighing of the evidence."[718] Even successfully showing that the speech in question may cause the feared harm is not enough: "In accordance with what we have said on the 'clear and present danger' cases, neither 'inherent tendency' nor 'reasonable tendency' is enough to justify a restriction of free expression."[719] Mask mandates imposed a *de facto* "reasonable tendency" test, but even assuming that non-masking could be shown to have a "reasonable tendency" to increase COVID transmission, the Supreme Court has already said, "The reasonable tendency test is so vague and indefinite that it 'is repugnant to the guarantee of liberty contained in the Fourteenth Amendment.'"[720]

The Supreme Court has also said what risks do *not* meet the "clear and present danger" standard, and its rulings make clear that "reasonable basis" scrutiny or "intermediate" scrutiny are *far* too deferential. In 1940, the Supreme Court ruled that "no clear and present danger of destruction of life or property, or invasion of the right of privacy, or breach of the peace can be thought to be inherent in the activities of every person who approaches the premises of an employer and publicizes the facts of a labor dispute involving the latter."[721] By the same token, no "clear and present danger" of communicable disease can reasonably be assumed to exist in every member of the population who approaches within 6 feet of another person when no signs and symptoms of disease are present, regardless of whatever communicable diseases may be extant in the community as a whole.

In 1941 (incidentally, the day after the Japanese attack on Pearl Harbor) the Supreme Court clarified this even further in *Bridges v. California*. The Court said, "the likelihood, however great, that a

---

717   *Bridges v. California*, 314 U.S. 252, 252 (1941), available online:
https://tile.loc.gov/storage-services/service/ll/usrep/usrep314/usrep314252/usrep314252.pdf
718   *Bridges v. California*, 314 U.S. 252, 255 (1941), available online:
https://tile.loc.gov/storage-services/service/ll/usrep/usrep314/usrep314252/usrep314252.pdf
719   *Bridges v. California*, 314 U.S. 252, 273 (1941), available online:
https://tile.loc.gov/storage-services/service/ll/usrep/usrep314/usrep314252/usrep314252.pdf
720   *Bridges v. California*, 314 U.S. 252, 255 (1941), available online:
https://tile.loc.gov/storage-services/service/ll/usrep/usrep314/usrep314252/usrep314252.pdf
721   *Thornhill v. Alabama*, 310 U. S. 88, 105 (1940), available online:
https://tile.loc.gov/storage-services/service/ll/usrep/usrep310/usrep310088/usrep310088.pdf

substantive evil will result cannot alone justify a restriction upon freedom of speech or the press."[722] In the same case, the Court noted: "Restatement of the phrase 'clear and present danger' in other terms has been infrequent. Compare, however: '. . . the test to be applied . . . is not the remote or possible effect.'"[723] During COVID, the entire justification for compulsory masking was *based* on the remote or possible effect of masks on viral transmission. A *possible* increase in the chances of transmitting a disease which one might *possibly* have, with an average infection fatality rate of less than 0.15%,[724] cannot *possibly* meet this standard. Any benefit from universal masking would be, at *best*, a "remote or possible effect." If marginally altering one's risk of contracting a single communicable disease could justify such invasive abridgements of fundamental rights, then no infringement of basic rights would be off the table.

Freedom of speech may not be absolute, but the government's constitutional license to infringe freedom of speech and other essential rights is *less* absolute still. Even assuming every benefit ascribed to masks during the COVIDcrisis had been accurate, compelling their use would *still* be blatantly unconstitutional. Mask mandates irrationally treated every set of lungs as a "clear and present danger," even though no generalized "clear and present danger" of communicable disease can reasonably be assumed to exist where no signs and symptoms are present.

Seatbelts and safety helmets are a bad legal analogy

The argument that masks are just a basic commonsense safety device akin to seatbelts and safety helmets also fails. In his initial appellant brief that would ultimately lead to the first major direct

---

722   *Bridges v. California*, 314 U.S. 252, 255 (1941), available online:
https://tile.loc.gov/storage-services/service/ll/usrep/usrep314/usrep314252/usrep314252.pdf

723   *Bridges v. California*, 314 U.S. 252, 262 (1941), available online:
https://tile.loc.gov/storage-services/service/ll/usrep/usrep314/usrep314252/usrep314252.pdf quoting *Schaefer v. United States*, 251 U. S. 466, 486 (1920) (Brandeis, J., dissenting), available online:
https://tile.loc.gov/storage-services/service/ll/usrep/usrep251/usrep251466/usrep251466.pdf

724   As highlighted elsewhere in this book, by January 2021, Dr. Ioannidis of Stanford University had already compiled meta-analytic data showing that the true infection average fatality rate of SARS-CoV-2 was 0.15% or less — in other words, a bad flu. For those following the studies included in his meta-analysis, this broad conclusion was glaringly apparent from the start of the COVIDcrisis (it was clear from the beginning with the data coming from the *Diamond Princess* cruise ship outbreak). Dr. Ioannidis was also one of the authors in a 2022 age-stratified meta-analysis that found the median infection fatality rates of SARS-CoV-2 to be far lower even than this (0.035% for ages 0-59, and 0.095% for ages 0-69 years old).
John P. A. Ioannidis, "Reconciling estimates of global spread and infection fatality rates of COVID-19: An overview of systematic evaluations." *European Journal of Clinical Investigation.* 2021 May;51(5):e13554. doi: 10.1111/eci.13554. Epub 2021 Apr 9. PMID: 33768536; PMCID: PMC8250317 https://www.ncbi.nlm.nih.gov/pmc/articles/PMC8250317/
Pezzullo, A.M., et al., *Age-stratified infection fatality rate of COVID-19 in the non-elderly informed from pre-vaccination national seroprevalence studies.* 2022, Cold Spring Harbor Laboratory. https://dx.doi.org/10.1101/2022.10.11.22280963

legal setback for compulsory masking, attorney Jeff Childers identified at least nine legally relevant distinctions between masks and helmets or seatbelts.[725]

- Unlike masks, seatbelts and safety helmets have never been considered a form of symbolic speech, and do not "occupy a unique role historically and fundamentally, particularly as related to speech."

- Unlike masks, seatbelts and most safety helmets do not interfere with non-verbal speech like facial expressions.

- Unlike masks, seatbelts and most safety helmets do not obscure or interfere with the long-recognized protectable interest a person has in their likeness.[726]

- Unlike masks, seatbelts and safety helmets have historically been regarded positively by most people.

- Unlike masks, seatbelts and safety helmets lack "a long, storied history of symbolizing threats, intimidation, and violence." Masks have a historical association with criminality and disease that helmets and seatbelts completely lack.[727]

- Unlike masks, state legislatures have never attempted to outlaw seatbelts or safety helmets.

- Unlike masks, no states have legislated automatic increases in criminal penalties just because someone wears a seatbelt or safety helmet while committing a crime.[728]

- Unlike masks, no one ever made the use of seatbelts and helmets a requirement to obtain essential goods and services.

- Unlike with masks, "there is no broad scientific controversy about whether seatbelts and helmets 'actually work.'"

---

725   *Green v. Alachua County*, 323 So. 3d 246, Appellant's Initial Brief, Filing #109556371 E-Filed 06/29/2020 (Fla. Dist. Ct. of App., 2021), go to: https://1dca.flcourts.gov/Oral-Arguments/Briefs-and-Petitions-for-Cases-Scheduled-for-Oral-Argument/20-1661 and search for "case 20-1661"

726   "There is a 'long recognized … common law right of privacy which includes protection against appropriation, for the defendant's advantage, of the plaintiff's name or likeness.' Abdul-Jabbar v. Gen. Motors Corp., 85 F. 3d 407 (9th Cir. 1996) (internal cites, ellipses, and brackets omitted, emphasis added)."
Green v. Alachua County, 323 So. 3d 246, Appellant's Initial Brief, Filing #109556371, page 41, E-Filed 06/29/2020 (Fla. Dist. Ct. of App., 2021), go to: https://1dca.flcourts.gov/Oral-Arguments/Briefs-and-Petitions-for-Cases-Scheduled-for-Oral-Argument/20-1661 and search for "case 20-1661"

727   *State v. Miller*, 260 Ga. 669, 671 (Ga. 1990), available online:
https://law.justia.com/cases/georgia/supreme-court/1990/s90a1172-1.html

728   "See § 775.0845, Florida Statutes (reclassifying crimes to the next higher felony or misdemeanor if citizen was wearing and mask while committing crime)."
Green v. Alachua County, 323 So. 3d 246, Appellant's Initial Brief, Filing #109556371, page 30, E-Filed 06/29/2020 (Fla. Dist. Ct. of App., 2021), go to: https://1dca.flcourts.gov/Oral-Arguments/Briefs-and-Petitions-for-Cases-Scheduled-for-Oral-Argument/20-1661 and search for "case 20-1661"

If one insists on making an automobile analogy, compulsory masking was most akin to mandating that all drivers keep their emergency lights perpetually flashing, combined with the invasive intimacy of forcibly installing an ignition breathalyzer in every car (including ignition breathalyzers for people that don't drink at all).

There is at least one more legal disanalogy between masking laws and seatbelt and helmet laws. Seatbelt and helmet laws are a subcategory of motor vehicle safety statutes, and motor vehicle safety statutes are typically subject to strict liability. Prior to 2020, the Georgia Anti-Mask Act cases we reviewed earlier[729] explicitly rejected this sort of interpretation in the case of legislation applying to masks. The legal basis for banning masks is, if anything, far stronger than the legal basis for mandating masks. Georgia's Anti-Mask Act was far more narrowly tailored than any COVIDcrisis mask mandate, and it was passed by the legislature according to well-established conventional procedures rather than emergency executive fiat. Despite this, without substantial pruning by Georgia's Supreme Court, Georgia's Anti-Mask Act was *still* unable to survive *less-than-strict* scrutiny from a rudimentary constitutional challenge. The Supreme Court of Georgia found that the statute would be overbroad and unconstitutional if applied with "no reference to intent, such as some motor vehicle safety statutes... Although no intent or fault requirement was included by the legislature in the Anti-Mask Act, the need for a narrowing construction… prohibited a strict liability interpretation."[730]

In a 2018 ruling, the Eleventh Circuit Court of appeals reaffirmed that: "the Georgia mask statute must be read in light of the limitations placed on it by the Georgia Supreme Court,"[731] and Judge Kathleen Williams specifically emphasized that it was "the limiting construction that saved the statute from constitutional infirmity."[732] At the very least, therefore, Judges in the Eleventh Circuit who dismissed First Amendment challenges to compulsory masking by referencing seatbelt and helmet laws erred in doing so.[733]

729   *State v. Miller*, 260 Ga. 669 (Ga. 1990), available online:
https://law.justia.com/cases/georgia/supreme-court/1990/s90a1172-1.html

730   *Daniels v. State*, 264 Ga. 460, *464 n 4; 448 S.E.2d 185 (1994), available online:
https://law.justia.com/cases/georgia/supreme-court/1994/s94g0362-1.html

731   *Gates v. Khokhar*, 884 F.3d 1290, 1299 (11th Cir. 2018), *cert. denied*, 139 S. Ct. 807 (2019), 15a, available online: https://www.supremecourt.gov/DocketPDF/18/18-511/66921/20181015162248885_Gates%20v.%20Khokhar%20Appendix.pdf

732   *Gates v. Khokhar*, 884 F.3d 1290, 1299 (11th Cir. 2018), *cert. denied*, 139 S. Ct. 807 (2019), 34a (Williams, J., dissenting) available online: https://www.supremecourt.gov/DocketPDF/18/18-511/66921/20181015162248885_Gates%20v.%20Khokhar%20Appendix.pdf

733   *Machovec v. Palm Beach County*, No. 2020CA006920AXX — Filing #110806335, 27 July 2020 (15th Judicial Circuit, Florida), available online: https://clearinghouse.net/doc/109346/ & https://www.floridacivilrights.org/wp-content/uploads/2020/07/Order-Denying-Temporary-Injunction.pdf
Machovec v. Palm Beach County, 310 So. 3d 941 (Fla. Dist. Ct. App. 2021), https://casetext.com/case/machovec-v-palm-beach-cnty

Mask mandates fail intermediate "O'Brien" scrutiny

But what about those times when government *can* legitimately infringe on symbolic speech and expressive conduct? What about those times when the governmental interest is compelling, or when speech is "not so inherently expressive" that it qualifies for full First Amendment protection, and is instead subject to intermediate *O'Brien* scrutiny? After all, the Supreme Court has said: "No one supposes, for example, that a city need permit a man with a communicable disease to distribute leaflets on the street or to homes, or that the First Amendment prohibits a state from preventing the distribution of leaflets in a church against the will of the church authorities."[734]

This quote comes from the 1943 case *Martin v. City of Struthers*. During World War 2, the City of Struthers, Ohio, passed a municipal ordinance "forbidding any person to knock on doors, ring doorbells, or otherwise summon to the door the occupants of any residence for the purpose of distributing to them handbills or circulars." When challenged under the First Amendment, one of the defenses raised by the city was its compelling governmental interest in preventing crime. The Supreme Court responded that: "While door to door distributers of literature may be either a nuisance or a blind for criminal activities, they may also be useful members of society engaged in the dissemination of ideas in accordance with the best tradition of free discussion."[735] The non-zero chance that a door-to-door canvasser might be a criminal did not justify infringing all door-to-door canvassers' First Amendment rights, especially when individual property owners could simply exclude solicitors. Likewise, the non-zero chance that a door-to-door canvasser might have COVID or any other communicable disease does not justify infringements on First Amendment rights any more than does the non-zero chance that they might be a criminal. The Supreme Court could just as easily (and just as accurately) have said, "While door-to-door distributers of literature may have a communicable disease, they may also be perfectly healthy members of society engaged in the dissemination of ideas in accordance with the best tradition of free discussion." Indeed, the Supreme Court reaffirmed its 1943 decision in the 2002 case of *Watchtower Bible & (and) Tract Society of New York v. Village of Stratton*, despite the documented existence of crimes — including murder — using door-to-door sales or canvassing as a cover.[736] Public safety and public health do not automatically trump the First Amendment.

---

734   *Martin v. City of Struthers*, 319 U.S. 141, 143 (1943), available online:
https://tile.loc.gov/storage-services/service/ll/usrep/usrep319/usrep319141/usrep319141.pdf

735   *Martin v. City of Struthers*, 319 U.S. 141, 145 (1943), available online:
https://tile.loc.gov/storage-services/service/ll/usrep/usrep319/usrep319141/usrep319141.pdf

736   *Watchtower Bible & (and) Tract Society of New York, Inc., et al. v. Village of Stratton et al.*, 536 U.S. 150, 177 (2002), available online: https://tile.loc.gov/storage-services/service/ll/usrep/usrep536/usrep536150/usrep536150.pdf

Intermediate *O'Brien* scrutiny deals less with what *constitutes* speech than it does with when the government is allowed to infringe on symbolic speech and expressive conduct. Even if compulsory masking somehow did not warrant strict scrutiny, compulsory masking still fails the *O'Brien* standard in multiple ways. Under the *O'Brien* standard, a governmental regulation burdening free speech must: 1) be "within the constitutional power of the Government"; 2) further an important or substantial governmental interest; 3) be "unrelated to the suppression of free expression"; and 4) be "no greater than is essential."[737]

Public health emergency powers are typically defined and inherently *limited* by the statutes granting those powers, and the power of magistrates to reduce the spread of disease is not as "broad and sweeping" as the power of Congress to raise and support armies invoked in the *O'Brien* decision to penalize burning draft cards.[738] Furthermore, the requirement that "the restriction on alleged First Amendment freedoms is no greater than is essential,"[739] means that any measure infringing the First Amendment rights of individuals must be "narrowly tailored" and not "overbroad." If there is "alternative means that would more precisely and narrowly"[740] reduce the spread of COVID than universal masking (such as, for example, *quarantine of the symptomatic* or any number of other nonpharmaceutical interventions), then compulsory masking is greater than essential and thus fails intermediate *O'Brien* scrutiny. But the restriction on First Amendment freedoms from mask mandates *was* (and remains) far greater than is essential to further any interest in stemming the spread of COVID. In other words, universal masking is not, in fact, "narrowly tailored." The restriction on First Amendment freedoms from mask mandates is much greater than is essential, and mask mandates fail intermediate scrutiny because of this.

When the Supreme Court ruled that Gregory Johnson's expressive flag-burning was constitutionally protected in 1989, two state interests were proffered to justify the suppression of expression in laws prohibiting desecration of the United States flag: 1) preventing breaches of the peace ("imminent lawless action"), and 2) preserving the flag as a symbol of nationhood and national unity. The strongest of these two was the State's compelling interest in preventing crime and breaches of the peace.[741]

---

737   *United States v. O'Brien*, 391 U.S. 367, 377 (1968), available online:
https://tile.loc.gov/storage-services/service/ll/usrep/usrep391/usrep391367/usrep391367.pdf
738   *United States v. O'Brien*, 391 U.S. 367, 368 (1968), available online:
https://tile.loc.gov/storage-services/service/ll/usrep/usrep391/usrep391367/usrep391367.pdf
739   *United States v. O'Brien*, 391 U.S. 367, 377 (1968), available online:
https://tile.loc.gov/storage-services/service/ll/usrep/usrep391/usrep391367/usrep391367.pdf
740   *United States v. O'Brien*, 391 U.S. 367, 381 (1968), available online:
https://tile.loc.gov/storage-services/service/ll/usrep/usrep391/usrep391367/usrep391367.pdf
741   "Few individuals would argue that preventing crime is not an important or substantial government interest. Accordingly, all anti-mask laws satisfy this component of the O'Brien test."
"The prevention of crime, which is the stated objective of anti-mask laws, is certainly a compelling government interest."

However, since no breach of the peace occurred as a result of the flag burning, this interest was not implicated. Also, it is established constitutional law that breaches of the peace caused by lack of self-control in offended hearers do not override speakers' rights to Free Speech. By extension, the State's interest in preventing the spread of COVID should not be implicated in the case of any particular individual refusing to wear a mask unless COVID is actually transmitted (and even then, there would need to be some good evidence that "but for failure to wear a mask, COVID transmission would not have occurred").

Additionally, when reading court rulings upholding mask mandates, one finds two universal refrains: that masks are a reasonable measure which *may* help, and that their primary function is to protect *others* from the *wearer*.[742] There is a huge gap between *possibly* helping vs. being *necessary* or *essential*. The word "necessary" does not include anything and everything that might possibly-maybe-conceivably-theoretically help. It is completely false to say that SARS-CoV-2 would be or could be controlled *but for* individual failure to wear masks, yet this is precisely the implication from carelessly applying words like "necessary" and "essential" to masks and mask mandates. Showing that masks *may* help or *possibly* help mitigate disease spread would not come anywhere *close* to being the same as showing that masks are *necessary* or *essential* to mitigate disease spread. Universal compulsory masking self-evidently goes beyond what is "necessary" or "essential" because it includes many people who have no chance of transmitting COVID whatsoever (either because they do not have it or are immune), and for whom it would therefore be irrational to wear a mask even if masks could be shown to "work." *Coercing* irrational behavior only compounds the inherent wrong.

Mask mandates commit a basic fallacy of composition, assuming that what is true of the whole is true of every part. But what is true of the whole is *not* necessarily true of every part (and vice versa). For example, just because a book as a whole is interesting does not necessarily mean that every individual chapter is also interesting, and just because one sheet of paper in a book is lightweight does not mean that the book as a whole is lightweight. If a disease is present in the community as a whole because one or two out of every 100 people have it, that does not make it rational or proportionate to act like every individual in the community is infectious, and still less does it give anybody the right to *force* everybody else to act this way.

Stephen J. Simoni, "Who Goes There?" — Proposing a Model Anti-Mask Act, 61 *Fordham Law Review*, 241, 250 and 257 (1992), available online: https://ir.lawnet.fordham.edu/cgi/viewcontent.cgi?article=3009&context=flr

742   e.g. "requiring individuals to cover their nose and mouth while out in public is intended to prevent the transmission from the wearer of the facial covering to others (with a secondary benefit being protection of the mask wearer)."

Machovec v. Palm Beach County, 310 So. 3d 941, 946 (Fla. Dist. Ct. App. 2021)

Unlike other conditions which may be yes/no propositions, such as anonymity, one's chance of being infected or infecting others is always a matter of degree; a matter of degree that can be affected by multiple interchangeable prevention methods. The fact that masks cannot contribute to an individual's inherent immunity means that masking is exchangeable and interchangeable with other common nonpharmaceutical infection control methods such as social distancing, handwashing, and quarantining. This interchangeability means that even *if* masks "work," the State can always achieve the same effect on public health "through means that would not violate… First Amendment rights."[743]

Finally, as the U.S. Supreme Court has pointed out in other contexts, just because the state has a compelling interest in achieving some aggregate result, "does not mean that it has one 'in each marginal percentage point by which' it achieves this abstract interest."[744] In other words, just because the government may have a compelling interest in stemming the spread of COVID-19 *in the nation as a whole*, it does not therefore follow that the government *also* has a compelling interest in enforcing measures which create a marginal shift in any particular *individual's* chances of catching or transmitting a virus. Even if mask-wearing were decisively shown to affect an individual's chances of catching or transmitting any given virus, it would still be just one out of many measures which produce (at best) a marginal effect on risk to an individual who is *himself* a marginal fraction of a marginal percentage point in the government's aggregate compelling public health interest.

The compelling interest of public health does not carry all the way to broad compulsory masking, and compulsory masking should presumptively fail even rational basis scrutiny when challenged under any article of the Bill of Rights because no incremental interest the State can possibly have in the theoretical benefits from forcing one individual to wear a mask can possibly outweigh the real, severe, and immediate harms caused by the violative mandates necessary to accomplish this. The mask mandates were *never* "narrowly tailored," and repeatedly saying so did not make it any less false.[745] Universal compulsory mask-wearing, by its very nature, *cannot* be narrowly tailored enough to qualify as "no greater than is essential."

---

743 *Pacific Gas and Electric Company v. Public Utilities Commission of California*, 475 U.S. 1 at 19 (1986), https://tile.loc.gov/storage-services/service/ll/usrep/usrep475/usrep475001/usrep475001.pdf

744 *Doster v. Kendall*, 54 F.4th 398, 42 (6th Cir. 2022), available online: https://casetext.com/case/doster-v-kendall-6?sort=relevance&p=1&type=case&resultsNav=false referencing Brown v. Entertainment Merchants Association, 564 U.S. 786, 803-804 n.9 (2011) (https://tile.loc.gov/storage-services/service/ll/usrep/usrep564/usrep564786/usrep564786.pdf)

745 The Florida District Court of Appeal's comment in Jeff Childers' case is equally applicable here: "Despite this extraordinarily broad language, the trial court characterized the mandate as one applicable 'in limited circumstances.'" *Green v. Alachua County*, 323 So. 3d 246, page 11 note 4 (Fla. Dist. Ct. App. 2021), available online: https://law.justia.com/cases/florida/first-district-court-of-appeal/2021/20-1661.html

Alternative channels are worthless when the primary message is already being overwritten

A final argument purporting to justify mask mandates under intermediate scrutiny, and which was regularly trotted out during the COVIDcrisis, is that mask mandates are constitutional because people who wish to criticize compulsory masking have "ample alternative channels" to get their message across. "People wearing masks can still speak," said Federal Judge Susan Mollway; "the Mask Mandates do not impose more than an incidental burden on speech. To the contrary, individuals can still engage in expressive activity. They just have to wear masks while they do."[746] Judge Reggie B. Walton said that Congressmen opposing mask mandates had "myriad means" of getting their message across while still wearing masks, "including wearing masks or other clothing containing the messages they wanted to convey, or making speeches from the House Chamber or elsewhere on the subject."[747] "[A]s the Governor rightly states, the [mask] order does not prevent Plaintiffs from expressing their disagreement with the face covering requirement in other ways," said Judge Robert Chambers.[748]

This particular argument was put forward to satisfy a criterion of intermediate scrutiny nascent in Justice Harlan's 1968 *O'Brien* concurrence. After quoting the four main intermediate scrutiny criteria listed by the majority, Justice Harlan went on to say:

> "I wish to make explicit my understanding that this passage does not foreclose consideration of First Amendment claims in those rare instances when an 'incidental' restriction upon expression, imposed by a regulation which furthers an 'important or substantial' governmental interest and satisfies the Court's other criteria, in practice has the effect of entirely preventing a 'speaker' from reaching a significant audience with whom he could not otherwise lawfully communicate."[749]

In other words, Justice Harlan was saying that it was possible for a restriction on speech to still violate the First Amendment even if it met the other four *O'Brien* criteria. He was alluding to earlier cases

---

746   *Denis v. Ige et al*, No. 1:2021cv00011 - Document 62 (D. Haw. 2021), pg. 30-31, available online: https://law.justia.com/cases/federal/district-courts/hawaii/hidce/1:2021cv00011/152658/62/

747   *Massie v. Pelosi*, 590 F. Supp. 3d 196, *225 (D.C. 2022). Available online: https://scholar.google.com/scholar_case?case=735226917287193397&hl=en&as_sdt=4000003&scfhb=1

748   *Stewart v. Justice* 502 F. Supp. 3d 1057, 1067 (S.D.W. Va. 2020), available online: https://casetext.com/case/stewart-v-justice-1

749   *United States v. O'Brien*, 391 U.S. 367, 388-389 (1968) (Harlan, J., concurring), available online: https://tile.loc.gov/storage-services/service/ll/usrep/usrep391/usrep391367/usrep391367.pdf

like *Martin v. City of Struthers* in 1943,[750] where the Supreme Court gave greater weight to First Amendment claims against restrictions on speech when other equally effective alternative means of conveying their message were not available to the speakers. This concern of Justice Harlan gradually developed into an additional intermediate scrutiny criterion on top of the original four: "The final requirement of intermediate scrutiny is that a regulation leave open ample alternatives for expression," said Chief Justice Rehnquist in 2002.[751]

Mask bans and mask mandates inherently fail this final *O'Brien* criterion. By their nature, mask bans and mandates cannot avoid also banning or mandating more than just the "non-communicative aspect" of masking and non-masking, and alternatives left open need to be legitimately comparable to whatever speech has been closed off in terms of effectiveness and audience reach, as well as uncontaminated by accompanying compelled and coerced speech. Wearing a mask with an anti-mask slogan or delivering a speech against compulsory masking *while wearing a mask* does not meet this requirement. Furthermore, the "ample alternative channels" argument goes both ways. If the *State* has "ample alternative channels" to achieve a public heath objective "through means that would not violate… First Amendment rights,"[752] then universal masking is not narrowly tailored enough to survive intermediate scrutiny.

A restriction on speech does not become constitutional just because speakers have a viable alternative means of communicating their message. As the Supreme Court said, "one is not to have the exercise of his liberty of expression in appropriate places abridged on the plea that it may be exercised in some other place."[753] Yet that is exactly what compulsory masking did: public places and the company of others are a "natural and proper"[754] place to show one's face for purposes of communication, but this was prohibited in traditional public forums; other public places; private businesses; and even houses of worship — all on the fear-enhanced speculation that a pathogen carried by *someone, somewhere* in

---

750   *Martin v. City of Struthers*, 319 U.S. 141 (1943), available online:
https://tile.loc.gov/storage-services/service/ll/usrep/usrep319/usrep319141/usrep319141.pdf

751   *Watchtower Bible & (and) Tract Society of New York, Inc., et al. v. Village of Stratton et al.*, 536 U.S. 150, 180 (2002) (Rehnquist, C.J., dissenting), available online:
https://tile.loc.gov/storage-services/service/ll/usrep/usrep536/usrep536150/usrep536150.pdf

752   *Pacific Gas and Electric Company v. Public Utilities Commission of California*, 475 U.S. 1 at 19 (1986),
https://tile.loc.gov/storage-services/service/ll/usrep/usrep475/usrep475001/usrep475001.pdf

753   *Spence v. Washington*, 418 U.S. 405, 411 (1974), available online:
https://tile.loc.gov/storage-services/service/ll/usrep/usrep418/usrep418405/usrep418405.pdf

754   *Schneider v. State*, 308 U.S. 147, 163 (1939), available online:
https://tile.loc.gov/storage-services/service/ll/usrep/usrep308/usrep308147/usrep308147.pdf
quoted in *Thornhill v. Alabama*, 310 U. S. 88, 106 (1940), available online:
https://tile.loc.gov/storage-services/service/ll/usrep/usrep310/usrep310088/usrep310088.pdf

the community had a non-zero chance of lethal transmission by persons with no signs or symptoms, *and* that masks *might* marginally decrease that chance.

Mask mandates, and the COVIDcrisis court rulings that upheld them, rest on the presumption that the governmental interest in public health extends in an open-ended, effectively limitless manner, to every conceivable risk factor and mitigation measure. Many immediate, direct, concrete harms traceable to the exercise of individual liberties are not enough to justify infringement of First Amendment rights. Appeals to theoretical, future, indirect harms, as in compulsory masking, are even *less* able to justify concrete, immediate, direct violations of essential liberties. Pointing to any potential aggregate benefit is not sufficient. The Bill of Rights does not turn on a numbers game.[755]

Free speech conclusion — the unique and important nature of a person's face

In his book, *Six Minute X-Ray: Rapid Behavior Profiling*, body language expert Chase Hughes devotes a full chapter to the face, and makes it clear that his brief overview is just the tip of this iceberg:

> "A person typically glances at the face 11 times per minute in conversations. The most impactful researcher in facial movement science was Dr. Paul Eckman. Eckman traveled to the depths of jungles to seek out tribes who had never been exposed to outside human contact to verify that facial movements and facial expressions are universal. We truly are born with the same facial expressions and nonverbal communication strategies, and Dr. Eckman proved it. His groundbreaking book, Unmasking the Face, paved the way for modern researchers in behavior science."[756]

It should have come as no surprise to anyone that mandates directly regulating a body part as unique and important as the face would constitute an imposition of the most invasive sort. As attorney Jeff Childers articulated the issue:

---

755   This point was inspired by "The "general applicability test doesn't turn on [a] numbers game." *Air Force Officer v. Austin*, Civil Action 5:22-cv-00009-TES, p. 28 (M.D. Ga. Feb. 15, 2022), available online: https://law.justia.com/cases/federal/district-courts/georgia/gamdce/5:2022cv00009/123364/51/ quoting *Dr. A v. Hochul*, 142 S. Ct. 552, 556 (2021) (Gorsuch, J., dissenting), available online: https://www.supremecourt.gov/opinions/21pdf/21a145_gfbi.pdf

756   Chase Hughes, *Six-Minute X-Ray: Rapid Behavior Profiling* (Milton, DE: Evergreen Press, Kindle Edition, 2020), p. 52.

"Any ordinance requiring… [someone] to cover his face also impairs his face's 'unique role in communication,' his 'personal identity,' and his ability to 'express [his] thought and emotion.' That, after all, is just what the term 'facial expressions' means… A masked face is not at all 'normal,' but a 'degree of normalcy is expected for effective verbal and non-verbal communication.'"[757]

In its *Amicus* Brief supporting Health Freedom Defense Fund's challenge to the CDC's travel mask mandate, the Association of American Physicians and Surgeons stressed: "The ability to see another's demeanor while he is speaking is often as important as the content of what he says."[758]

According to the American Medical Association's *Guides to the Evaluation of Permanent Impairment*:

"The face plays a unique role in communication. No other part of the body serves as specific a function for personal identity and for the expression of thought and emotion. Facial expressions are an integral part of normal living postures. A degree of normalcy is needed for effective verbal and nonverbal communication. Facial anatomy contributes to identity, expression, normal functioning, and appearance of the forehead, cheeks, eyes, eyelids, eyebrows, lips, mouth, nose, chin, and neck. The face is such a prominent feature that it plays a critical role in the individual's physical, psychological, and emotional makeup. Facial disfigurement can affect all of these components and can result in social, vocational, and even psychiatric harm.[759]

This characterization of the face has been in official use for more than 30 years, and has been employed to reach decisions in court cases.[760]

---

757   *Green v. Alachua County*, 323 So. 3d 246, Appellant's Initial Brief, Filing #109556371 E-Filed 06/29/2020 (Fla. Dist. Ct. of App., 2021), p. 33, go to: https://1dca.flcourts.gov/Oral-Arguments/Briefs-and-Petitions-for-Cases-Scheduled-for-Oral-Argument/20-1661 and search for "case 20-1661"

758   *Amicus Curiae* Brief of the Association of American Physicians and Surgeons in Support of Plaintiffs, p. 11, *Health Freedom Defense Fund, Inc. et al v. Biden et al*, No. 22-11287 (11th Circuit Court of Appeals, Aug. 5, 2022), available online: https://healthfreedomdefense.org/wp-content/uploads/2022/08/Amicus-Brief-of-American-Assoc-of-Physicians-and-Surgeons.pdf

759   (emphasis added) Robert. D. Rondinelli, ed., *AMA Guides to the Evaluation of Permanent Impairment*, 6th ed. (Chicago, IL: American Medical Association, 2008), p. 260-261

760   It was present in the 3rd edition of the *AMA Guides* (revised 1990) and quoted in *Gonzales v. Advanced Component Sys.*, 949 P.2d 569, 572 (Colo. 1997). I first picked up this particular trail when Jeff Childers brought it up in his appellate brief for *Green v. Alachua County*.
The relevant passage from the Third Edition of the AMA Guide is located on page 171 and is available online: https://archive.org/details/guidestoevaluati00enge/page/170/mode/2up?q=communication

From ancient times, every person has possessed a fundamental property right in every part of their body. Legal recognition of the special status and value held by structures integral to the face - the eyes and teeth - goes back to the Code of Hammurabi[761] and the Law of Moses. The Law of Moses arguably recognized that even slaves retained ownership rights over the most important structures of their faces by providing that slaves who had an eye ruined or tooth broken by their owners were to be set free in compensation.[762] Legal recognition of the special nature of the face as a whole goes back at least as far as the Code of Justinian in the 6th century AD:

> "If someone has been sentenced to the mines because of the gravity of the crimes of which he or she has been convicted, he or she shall not at all be tattooed on the face, since the penalty imposed by conviction can be set forth in a single inscription either on the hands or the legs, in order that the face which has been designed to resemble the beauty of heaven, shall not at all be disfigured."[763]

In her survey of historical attitudes towards facial disfigurements, published in the *Transactions of the Royal Historical Society*, scholar Patricia Skinner noted: "in the early Irish law code Bretha Dein Checht, the shame of the public scar on the face exposed its victim to public ridicule — hence, the law states, a fine has to be paid for every public assembly the victim has to endure with facial disfigurement."[764] Additionally, a person's likeness, which is primarily — if not wholly — determined by their face, is a protectable property interest in most modern legal systems.

Depressingly, deliberate facial disfigurement as a particularly barbaric form of punishment for theft, adultery, administrative corruption, and political offenses is even more widely attested in the historical record than protection. An article published in the journal *Antiquity* in 2020 noted:

The 2001 Fifth Edition of the AMA Guide has this passage on page 255, and is available online: https://archive.org/details/guidestoevaluati00cocc/page/254/mode/2up?q=communication

761 c.f. The Code of Hammurabi, §196, §198-201, available online: https://www.ehammurabi.com/ Note: while §202-205 could be referring to the face, striking someone's cheek could also be used as a more general expression referring to assault and battery. That being said, if taken literally, this would be special legal recognition of the face going back to the earliest recorded law codes.

762 Exodus 21:26-27 (NIV): "An owner who hits a male or female slave in the eye and destroys it must let the slave go free to compensate for the eye. And an owner who knocks out the tooth of a male or female slave must let the slave go free to compensate for the tooth." see also Leviticus 21:18, where facial deformities or mutilations are one of the disqualification criteria for the priesthood.

763 Fred H. Blume & Bruce W. Frier, *The Codex Of Justinian: A New Annotated Translation With Parallel Latin And Greek Text*, 9.47.17 Vol. 3, p. 2427 (Cambridge, UK: 2016), available online: https://archive.org/details/fred-h.-blume-bruce-w.-frier-the-codex-of-justinian-a-new-annotated-translation-

764 Patricia Skinner, *'Better Off Dead Than Disfigured'? The Challenges of Facial Injury in the Pre-Modern Past.* Transactions of the Royal Historical Society, 2016. Vol. **26**: p. 39. https://dx.doi.org/10.1017/s0080440116000037

"[M]utilation has commonly been applied to the limb extremities and the head, but the face, especially serves as a brutally obvious medium for marking out certain individuals from others. This is because the human face is not only defined by the morphology of underlying bone, muscle and fat, but soft tissues, such as those forming the ears, nose, and lips, all serve to make each face distinct, facilitating social communication and allowing the expression of emotions."[765]

A person's face is the visible physical locus of their personality, and this deliberate targeting is *itself* an indicator of the special importance historically attached to what at least one highly published surgeon has called "the most noble and expressive part of the human body."[766] Compulsory masking may not *permanently* disfigure the face, but it indisputably infringes on that part of our bodies which most houses and expresses our identity, and which possesses a unique historical and legal status because of this role.

The private forum of your face and facial expressions is not exempt from the protections of the First Amendment. Saying that the First Amendment protects a man's speech but does not protect his mouth from being forcibly closed would be the worst kind of absurdity. Areas of the face forcibly covered by masks include the lips and teeth needed to properly form words, as is the nose that helps give the voice its resonant quality. It is worse than absurd to argue that the First Amendment protects speech but somehow permits a man's facial expressions to be forcibly overwritten by a mask that inherently communicates a message he opposes.

The facial expressions forcibly occluded by masks are pure speech, and there are multiple factors that decisively tip mask mandates into the realm of unconstitutional infringement on Free Speech, even under lesser forms of scrutiny than the strict scrutiny which compulsory masking unquestionably requires. The overwhelming majority of people compelled to wear masks were unable to transmit COVID either by reason of not being infected or because they had already recovered. This is self-evidently greater than is essential. People desiring to engage in expressive conduct and symbolic speech by not masking had their message "affected by the speech it was forced to accommodate."[767] Any "alternative channels" available to communicate opposition to masking were comparatively worthless in light of the potent communicative effective of mask-wearing. Verbal messages against masks were

---

765   Garrard Cole et al., "Summary justice or the King's will? The first case of formal facial mutilation from Anglo-Saxon England." *Antiquity*, 2020. **94**(377): p. 1263-1277. https://dx.doi.org/10.15184/aqy.2020.176 https://pure.hud.ac.uk/ws/files/19973933/Antiquity_paper_peer_review_copy.pdf

766   Giorgio Sperati, "Amputation of the nose throughout history," *Acta Otorhinolaryngol Ital*, 2009. **29**(1): p. 44-50, available online: https://www.actaitalica.it/issues/2009/1-09/Sperati.pdf

767   *Hurley v. Irish-American Gay, Lesbian and Bisexual Group of Boston, Inc.*, 515 U. S. 557, 566 (1995), available online https://supreme.justia.com/cases/federal/us/515/557/case.pdf

effectively overridden or at least rendered hypocritical and suspect by the symbolic message people were forced to convey by wearing one, and this effect was especially pernicious in the case of doctors and other healthcare providers who opposed compulsory masking. Additionally, compulsory masking violated every potential viewer's right to *receive* speech through facial expressions, symbolic speech, and expressive conduct.

Anything that artificially increases the effort required for speech is an inhibition. Speech cannot occur without airflow past the vocal chords, so anything that increases resistance to such airflow — e.g. a mask — is, *by definition,* a mechanical inhibition of vocal speech. Additionally, as we have already seen in the science section, published studies from 2020 and 2021 described voice disorders as a known side effect of mask-wearing.[768] Exceptions and exemptions from mask-wearing for public speaking and artistic vocal performances at the height of the pandemic implicitly conceded this fact.[769] Imposing a masking requirement as a prerequisite for public comment before legislatures, city councils, and school boards grossly inhibited citizens' exercise of their First Amendment rights to petition the government. The "minor" vocal effects of masks on speech *might* be justifiable under a rational basis standard, but the social modeling and social proof inherent in mask-wearing forces everyone who wants to argue against masks to symbolically contradict themselves.

The COVIDcrisis mask mandates were not at all narrowly tailored. Generations from now, the inherent madness of compulsory masking is going to become proverbial, and a cultural byword for tyrannical overreaction. Mask mandates do not and cannot qualify as being "necessary" because the government always has less intrusive alternatives available to accomplish the same goal. Attempts to minimize the violation inherent in mask mandates by saying some version of "the facial covering requirement is not permanent"[770] completely ignored the essential point which cannot be overstated. As the U.S. Supreme Court has said, "The loss of First Amendment freedoms, for even minimal periods of time,

---

768   Claudia Heider et al., *Prevalence of Voice Disorders in Healthcare Workers in the Universal Masking COVID -19 Era.* The Laryngoscope, 2021. **131**(4). https://dx.doi.org/10.1002/lary.29172

Łukasz Matusiak, et al., *Inconveniences due to the use of face masks during the COVID -19 pandemic: A survey study of 876 young people.* Dermatologic Therapy, 2020. **33**(4). https://dx.doi.org/10.1111/dth.13567

769   For example, Minnesota Governor Tim Walz' face covering mandate provided exceptions for "[w]hen testifying, speaking, or performing[.]"

Tim Walz, "Emergency Executive Order 20-81 Requiring Minnesotans to Wear a Face Covering in Certain Settings to Prevent the Spread of COVID-19," *State of Minnesota Executive* Department, July 22, 2020, available online: https://mn.gov/governor/assets/EO%2020-81%20Final%20Filed_tcm1055-441323.pdf

cited in: *Snell v. Walz*, No. A21-0626, p. 5 (Minn. Feb. 8, 2023), available online: https://mn.gov/law-library-stat/archive/supct/2023/OPA210626-020823.pdf

770   *Green v. Alachua County*, 323 So. 3d 246, *260 (Fla. Dist. Ct. App. 2021). Available online: https://law.justia.com/cases/florida/first-district-court-of-appeal/2021/20-1661.html

unquestionably constitutes irreparable injury."[771] How *long* mask mandates lasted was secondary compared to the fact that they were ever imposed in the first place! Other grounds used to dismiss challenges to compulsory masking were equally insulting legalisms. For example:

> "Here, the Plaintiff fails to allege a reasonable probability that a real injury will occur unless the temporary injunction is issued. The wearing of a face covering in public under the limited circumstances contained in the emergency order will not, in any way, alter the Plaintiff's physical person or result in permanent disfigurement."[772]

This, of course, is a textbook example of equivocation about what a "real injury" consists of. Physical or other tangible injuries are important and valid types of injury, but intangible injuries like deprivation of fundamental liberties, including freedom of speech, are still legally concrete and require injunctive relief. To limit *"real injury"* to "physical injury" is among the worst sort of judicial reasoning. You don't really believe in freedom of speech unless you are willing to respect it for those opinions you know are wrong and dangerous. Masking is speech. Non-masking is speech. Mandating either one was and remains a "plain, palpable invasion of rights secured by the fundamental law," and every court that ruled otherwise during the COVIDcrisis got it badly wrong to the extent they did so.

In fairness to the courts which found that masking did not implicate the First Amendment, no plaintiff alleged symbolic speech and expressive conduct through masking or non-masking as thoroughly or as well-researched as this book. But in fairness to the plaintiffs, based on their COVIDcrisis rulings, these courts would no doubt have insisted that throwing British tea into Boston harbor was not "inherently communicative" either. The burden of proof should never have been placed on the plaintiffs to the extent it was, initial complaints and responses to motions to dismiss are not generally permitted to be book-length,[773] and even on a charitable evaluation, multiple courts' refusal to acknowledge and uphold the First Amendment implications inherent in compulsory masking resonates with the worst episodes of jurisprudence in American history — from *Dred Scott* to *Buck v. Bell*, to *Korematsu*. When a statute (even more, an executive mandate) infringes on a fundamental right, the burden of proof is on the *State*, not the individual whose rights are being are infringed. Judges who upheld mask

---

771   *Elrod v. Burns*, 427 U.S. 347, 373 (1976),
https://tile.loc.gov/storage-services/service/ll/usrep/usrep427/usrep427347/usrep427347.pdf

772   (emphasis supplied) *Green v. Alachua County*, 323 So. 3d 246, *260 (Fla. Dist. Ct. App. 2021). Available online: https://law.justia.com/cases/florida/first-district-court-of-appeal/2021/20-1661.html

773   Many plaintiffs made what should have — and likely would have — been adequate allegations in normal times, and if I personally did not have the benefit of reading the courts' unsettling rulings in their cases, I would have expected far more minimal pleadings to be adequate.

mandates were not asleep at the wheel during the COVIDcrisis. Rather, they were wide awake and vigorously scrubbing their hands in the time-dishonored tradition of Pontius Pilate.

I take the position that it is absolutely beyond *any* legitimate use of an authority's power to compel a citizen to communicate a message which that citizen knows or believes to be false. At the *very least*, the strictest scrutiny must be met first. It is worth repeating: "The loss of First Amendment freedoms, for even minimal periods of time, unquestionably constitutes irreparable injury."[774] An *ongoing* constitutional violation of the magnitude and duration of compulsory masking certainly does this. Compulsory masking is "beyond all question, a plain, palpable invasion of rights secured by the fundamental law." Masking and non-masking is an incredibly potent form of symbolic speech and expressive conduct far closer to pure speech than numerous things the U.S. Supreme Court has already ruled are protected by the First Amendment. Compulsory masking affects opposing wearers' speech by forcing them to associate with speech they find repugnant, and overwrites their own message with that of the maskers. Compulsory masking violates every viewer's right to receive speech contradicting the message communicated by masking. Finally, the facial expressions censored by masks are pure speech in-and-of themselves. Any *one* of these things would require that compulsory masking be subject to strict scrutiny which it cannot survive. In combination, they raise compulsory masking to one of the worst, most invasive, most widespread violations of First Amendment rights in American history.

Regardless of what court rulings drafted in an atmosphere of artificially enhanced terror may have asserted based on motivated reasoning and a cherry-picked selection of case law, compulsory masking indisputably warrants not just strict scrutiny but *the strictest* scrutiny under the Free Speech Clause of the First Amendment, and every court that denied this made a clear error of law. The conclusion that mask-wearing or non-wearing did not implicate speech could only be reached through willful ignorance, motivated reasoning, and contemptible strawman argumentation.

---

774   *Elrod v. Burns*, 427 U.S. 347, 373 (1976), available online: https://tile.loc.gov/storage-services/service/ll/usrep/usrep427/usrep427347/usrep427347.pdf

The First Amendment — Free Exercise

"*It is the duty of every man to render to the Creator such homage and such only as he believes to be acceptable to him. This duty is precedent, both in order of time and in degree of obligation, to the claims of Civil Society. Before any man can be considered as a member of Civil Society, he must be considered as a subject of the Governour of the Universe: And if a member of Civil Society, who enters into any subordinate Association, must always do it with a reservation of his duty to the General Authority; much more must every man who becomes a member of any particular Civil Society, do it with a saving of his allegiance to the Universal Sovereign.*"

James Madison, 1785[775]

---

775   James Madison, "Memorial and Remonstrance against Religious Assessments, [ca. 20 June] 1785," *Founders Online,* National Archives, https://founders.archives.gov/documents/Madison/01-08-02-0163. [Original source: *The Papers of James Madison*, vol. 8, *10 March 1784–28 March 1786*, ed. Robert A. Rutland and William M. E. Rachal. Chicago: The University of Chicago Press, 1973, pp. 295–306.]
The Supreme Court also cited to James Madison's "Memorial and Remonstrance" in *Wisconsin v. Yoder*, 406 U.S. 205, 218-219 n 9 (1972), https://tile.loc.gov/storage-services/service/ll/usrep/usrep406/usrep406205/usrep406205.pdf

> *"COVID-19 is not a blank check for a State to discriminate against religious people, religious organizations, and religious services. There are certain constitutional red lines that a State may not cross even in a crisis. Those red lines include racial discrimination, religious discrimination, and content-based suppression of speech."*
>
> — U.S. Supreme Court Justice Brett Kavanaugh, *Calvary Chapel Dayton Valley v. Sisolak*, 2020[776]

The First Amendment provides legal protection, not just for beliefs, but also for actions (or non-actions) stemming *from* those beliefs. "Congress shall make no law respecting an establishment of religion, or prohibiting the free exercise thereof…" Freedom of religion, including the free *exercise* of religion, is the first fundamental freedom listed in the Bill of Rights. It comes even before freedom of speech, and with good reason. Historically, it is one of the unalienable rights most frequently violated, and also the one which men have been willing to suffer the most to uphold and defend. In his concurring opinion upholding First Amendment protection for religious exercise in the landmark 1963 Free Exercise case *Sherbert v. Verner*, U.S. Supreme Court Justice Potter Stewart wrote: "I am convinced that no liberty is more essential to the continued vitality of the free society which our Constitution guarantees than is the religious liberty protected by the Free Exercise Clause explicit in the First Amendment and imbedded in the Fourteenth."[777] Justice Stewart's words were prescient. The right of religious freedom enshrined in the First Amendment was one of the very few strong enough to directly take on the overreach of COVIDcrisis mandates supercharged by panic.

It may seem inconceivable that wearing a mask could violate anyone's sincerely held religious beliefs during a declared public health emergency. Yet for thousands of Americans, wearing a mask was and remains a major violation of sincerely held religious beliefs, depending on the specific context and type of mask — including medical masks in the name of stopping a respiratory virus. Just as sincerely held religious beliefs may dictate *wearing* a face-covering in certain contexts, they are equally, perhaps *more* likely to *forbid* doing so in other contexts. One (specifically Christian set) out of many possible religious beliefs that preclude mask-wearing in the name of COVID is detailed in the appendix.

---

776   *Calvary Chapel Dayton Valley v. Sisolak*, 140 S. Ct. at 2615-2616 (2020) (Kavanaugh, J., dissenting), available online: https://casetext.com/case/calvary-chapel-dayton-valley-v-sisolak

777   *Sherbert v. Verner*, 374 U.S. 398, 413 (1963) (Stewart, J., concurring), available online: https://tile.loc.gov/storage-services/service/ll/usrep/usrep374/usrep374398/usrep374398.pdf

During the COVIDcrisis, sincerely held religious beliefs drove thousands of Americans to refuse and actively oppose compulsory masking. Many people sensed that wearing a mask was edging into quasi-sacramental performative ritual territory, and at least one lawsuit alleged that in this sense compulsory masking amounted to an establishment of Scientism and Safety as secular religions in violation of the First Amendment.[778] The belief that wearing a mask was a repugnant ritual with religious significance could certainly feed into individual sincerely held religious beliefs regarding masks, but any claims that compulsory masking violated the Establishment Clause of the First Amendment fail as a standalone matter of law for at least two reasons. First, the legal threshold for something to count as an establishment of religion is high enough that even more overtly religious acts like opening Congressional sessions with prayer do not qualify, and secondly because secular ideologies like Scientism are not recognized as religions for legal purposes.

For First Amendment purposes, religious beliefs are not limited to tenets and practices specifically required or prohibited by an individual's faith. It is enough that the primary motivation be religious in origin. The Supreme Court has consistently ruled that, in order to be covered by the First Amendment, a religious belief "need not be confined in either source or content to traditional or parochial concepts of religion."[779] A religious belief does not have to fall under a historically established religious tradition, as long as it is "a sincere and meaningful belief occupying in the life of its possessor a place parallel to that filled by the God of those admittedly qualified for the exemption."[780] In one 1970 case, *Welsh v. United States*, the agnostic plaintiff went so far as to cross out "religious training" as a reason for opposing participation in war on his Selective Service statement, but the Court ruled that his moral and ethical beliefs in opposition to all war still qualified for a religious exemption because they were "beliefs about what is right and wrong and these beliefs are held with the strength of traditional religious convictions."[781]

It is also not necessary for other individuals from the same faith to agree in order for religious beliefs to receive First Amendment protection. As the Supreme Court (very properly) said when confronted with an intrafaith disagreement: "it is not within the judicial function and judicial competence to inquire whether the petitioner or his fellow worker more correctly perceived the commands of their

---

778   *Joseph v. Becerra*, 22-cv-40-wmc, Document 21, p. 6-7 (W.D. Wis. Nov. 29, 2022), available online: https://law.justia.com/cases/federal/district-courts/wisconsin/wiwdc/3:2022cv00040/48728/21/

779   *Welsh v. United States*, 398 U.S. 333, 339 (1970), available online: https://tile.loc.gov/storage-services/service/ll/usrep/usrep398/usrep398333/usrep398333.pdf

780   *United States v. Seeger*, 380 U.S. 163, 166, 176 (1965), available online: https://tile.loc.gov/storage-services/service/ll/usrep/usrep380/usrep380163/usrep380163.pdf

781   *Welsh v. United States*, 398 U.S. 333, 334 (1970), available online: https://tile.loc.gov/storage-services/service/ll/usrep/usrep398/usrep398333/usrep398333.pdf

common faith. Courts are not arbiters of scriptural interpretation."[782] This means that people from the same religion can have disagreements about whether or not to wear masks, and both positions receive First Amendment protection as long as they are sincerely held.

Religious exercise under the First Amendment includes "any exercise of religion, whether or not compelled by, or central to, a system of religious belief."[783] This definition, in the words of the Supreme Court, "provide[s] very broad protection for religious liberty," and applies to "'the performance of (or abstention from) physical acts' that are 'engaged in for religious reasons.'"[784] There is a legal distinction between the First Amendment's absolute protections for religious *beliefs* vs. its strong protections for the *exercise* of those beliefs. Obviously, no amount of sincerity would confer First Amendment protections on free exercise which extended to Aztec- or Canaanite-style human sacrifice. But still, in the words of the Supreme Court, "only those interests of the highest order and those not otherwise served can overbalance legitimate claims to the free exercise of religion."[785] Under the First Amendment, laws that burden religious exercise must be *both* neutral *and* generally applicable *unless* they are narrowly tailored. Otherwise, the law must satisfy strict scrutiny. COVID mask mandates were neither neutral, nor generally applicable, nor narrowly tailored.

Mask mandates are neither neutral, nor generally applicable, nor narrowly tailored

Since mask mandates could not survive true strict scrutiny, when faced with First Amendment Free Exercise challenges, advocates and imposing authorities had to insist that the mandates were "neutral and generally applicable" and thus not subject to strict scrutiny in the first place. Magistrate Judge Jared M. Strauss dismissed a Free Exercise challenge to compulsory masking in Florida by saying:

> "[R]ational basis review applies because the mask mandates are neutral and
> generally applicable. They are neutral because they do not have the object
> of 'infring[ing] upon or restrict[ing] practices because of their religious
> motivation.'… They are generally applicable because the government has not 'in

---

782   *Thomas v. Review Board of the Indiana Employent security Division et al.*, 450 U.S. 707, 716 (1981).

783   42 U.S.C. § 2000cc-5(7)

784   *Burwell v. Hobby Lobby Stores, Inc.*, 573 U.S. 682, 710 (2014),
https://supreme.justia.com/cases/federal/us/573/13-354/case.pdf
quoting *Employment Division*, Department of Human Resources *v. Smith*, 494 U.S. 872, 877 (1990),
https://tile.loc.gov/storage-services/service/ll/usrep/usrep494/usrep494872/usrep494872.pdf

785   *Wisconsin v. Yoder*, 406 U.S. 205, 215 (1972),
https://tile.loc.gov/storage-services/service/ll/usrep/usrep406/usrep406205/usrep406205.pdf

a selective manner impose[d] burdens only on conduct motivated by religious belief.'"[786]

In Massachusetts, District Judge William G. Young ruled that "Governor Baker's mask mandate in public places is 'neutral and of general applicability.'"[787] Judge Young's ruling on the neutrality and general applicability of mask mandates was then cited approvingly by Judge Susan Mollway in Hawaii.[788] A 6th Circuit panel of three judges acknowledged the multiple exceptions to the Michigan Department of Health and Human Services masking orders, but ruled that the orders were neutral and generally applicable anyway. The masking orders, said the 6th Circuit Panel:

> "[A]re not so riddled with secular exceptions as to fail to be neutral and generally applicable. The exceptions to the MDHHS Orders were narrow and discrete. First, many of the exceptions, such as medical intolerance to mask use, eating and drinking, swimming, or receiving a medical treatment during which a mask cannot be worn, are 'inherently incompatible with' wearing a mask."[789]

---

786   *Zinman v. Nova Southeastern University*, No. 21-CIV-60723-RAR, Document 81, p. 23 (S.D. Fla. Aug. 30, 2021), https://www.govinfo.gov/content/pkg/USCOURTS-flsd-0_21-cv-60723/pdf/USCOURTS-flsd-0_21-cv-60723-1.pdf

787   *Delaney v. Baker*, 511 F. Supp. 3d 55, *72 (D. Mass. 2021), available online: https://casetext.com/case/delaney-v-baker-2

788   *Denis v. Ige et al*, No. 1:2021cv00011 - Document 62 (D. Haw. 2021), pg. 22, available online: https://law.justia.com/cases/federal/district-courts/hawaii/hidce/1:2021cv00011/152658/62/

789   *Resurrection School v. Hertel*, 11 F.4th 437, 458 (6th Cir. 2021), available online: https://casetext.com/case/resurrection-sch-v-hertel

A later *en banc* hearing by 17 judges of the 6th Circuit declared the First Amendment challenge moot because by then the mask mandates being challenged had been dropped, so the plaintiffs were still ultimately denied meaningful relief, despite Judge John K. Bush's excellent dissenting opinion which tore apart the smaller panel's bad reasoning. Judge Bush showed in detail where the lower panel erred: "the underlying premise from which the panel majority reasoned is that there can exist some arbitrarily large number of exemptions disparately favoring secular conduct but that pose no First Amendment concern, at least until the exemptions can be deemed to 'riddle' the challenged law." As Judge Bush pointed out, this premise directly contradicts the Supreme Court's decision in *Tandon v. Newsom*, where treating *any* comparable secular activity more favorably than religious exercise triggers strict scrutiny under the First Amendment. Judge Bush directly addressed the panel's comment quoted above:

> "Take first, for instance, the panel majority's rationalization of the exemptions for eating, drinking, swimming, and medical treatments — said to be 'inherently incompatible with wearing a mask.' The apparent implication of this comment is that those activities are physically impossible while wearing a mask and thus are 'inherently incompatible,' while simultaneous masking and religious instruction is *physically* possible and thus 'compatible.' Yet the problems with this argument are legion."

Judge Bush went on to explain that just because it may be possible to *physically* do two things (like wear a mask and receive religious instruction) at the same time does not necessarily make those things "compatible," in the fuller sense needed, especially since the plaintiffs had consistently asserted that simultaneous masking and proper religious instruc-

But no matter how dogmatically judges, pundits, and any other authorities insisted that mask mandates were "neutral and generally applicable," no mask mandate truly met this standard as laid out by the U.S. Supreme Court:

> "First, government regulations are not neutral and generally applicable, and therefore trigger strict scrutiny under the Free Exercise Clause, whenever they treat any comparable secular activity more favorably than religious exercise."[790]

Though directly targeting religious conduct certainly does make a law non-neutral, just because a law does not explicitly target religious conduct does not always mean that the law is actually neutral in practice. Supreme Court Justice Alito explained that a "law cannot be regarded as protecting an interest of the highest order... when it leaves appreciable damage to that supposedly vital interest unprohibited."[791] In other words, if some particular conduct (such as not wearing a mask) undermines a claimed governmental interest, a law is not neutral if it permits that particular conduct on the basis of a secular reason while at the same time *forbidding* that particular conduct on the basis of a religious reason. A mask mandate that makes an exception for singing unmasked in a chorus while recording in a Hollywood sound studio, but refuses to let churchgoers take off their masks to sing, is not neutral even though it does not *directly* target religiously motivated non-masking. Allowing casinos to operate at 50% capacity but not churches is another example of this sort of non-neutrality, as is granting medical accommodations for not wearing a mask, but refusing religious accommodations. Non-neutrality also raises separate-but-related issues of disparate enforcement which implicate the Fourteenth Amendment. As another example, Justice Alito pointed to how mask mandates were eased to permit drinking in Las Vegas casinos, but not to permit worship in churches: "Casinos are

---

tion was both physically and spiritually impossible. Judge Bush concluded that "The panel majority's conclusion that masking and religious instruction are 'compatible' after all seems predicated on nothing more than a judicial reappraisal of what Resurrection School's religious scruples do and do not permit"

Resurrection School v. Hertel, 35 F.4th 524, 540-541 (2022), available online: https://scholar.google.com/scholar_case?case=5344911665347716699&q=resurrection+school+v+hertel&hl=en&as_sdt=4000003

790   *Tandon v. Newsom*, 141 S.Ct. 1294, 1296 (2021),
https://www.supremecourt.gov/opinions/20pdf/20a151_4g15.pdf

citing *Roman Catholic Diocese of Brooklyn v. Cuomo*, 141 S. Ct. 63, 67-68 (2020), available online:
https://www.supremecourt.gov/opinions/20pdf/20a87_4g15.pdf

791   *Calvary Chapel Dayton Valley v. Sisolak*, 140 S. Ct. at 2608 (2020) (Alito, J., dissenting), available online:
https://casetext.com/case/calvary-chapel-dayton-valley-v-sisolak

quoting *Church of the Lukumi Babalu Aye, Inc., v. City of Hialeah*, 508 U.S. at 547,113 S. Ct. 2217, available online:
https://tile.loc.gov/storage-services/service/ll/usrep/usrep508/usrep508520/usrep508520.pdf

permitted to serve alcohol, which is well known to induce risk taking, and drinking generally requires at least the temporary removal of masks."[792]

First Amendment free exercise analysis also does not require that the activities compared be identical. A mask mandate in all public schools does not mean that imposing a mask mandate on religious schools survives the First Amendment when gyms, piercing studios, and office buildings have lesser or no mask mandates. Regardless of whether one person is unmasked for a medical reason, another unmasked for a religious reason, and still others unmasked for a performance, identification, or some other practical reason, one thing remains the same — they're *all* unmasked. All such unmasked individuals pose an identical supposed hazard to the claimed governmental interest in public health and infection control, and favoring one of the secular categories over a religious rationale is a violation of the First Amendment which requires strict scrutiny. As Justice Kavanaugh wrote in *Calvary Chapel Dayton Valley v. Sisolak* (2020), "it is not enough for the government to point out that other secular organizations or individuals are also treated *unfavorably*. The point 'is not whether one or a few secular analogs are regulated. The question is whether a single secular analog is *not* regulated.'"[793] If a law treats *any* comparable secular activity more favorably than religious exercise, it must satisfy strict scrutiny.[794]

In a parallel vein, according to the Supreme Court, "A law is not generally applicable if it invites the government to consider the particular reasons for a person's conduct by providing a mechanism for individualized exemptions."[795] In other words, a mandate which allows for exemptions is not truly "generally applicable," and if secular exemptions are permitted — including medical and administrative exemptions — religious exemptions must also be permitted. "In *Employment Div., Dept, of Human Resources of Ore. v. Smith…* for example, the Court explained that 'where the State has in place a system of individual exemptions, it may not refuse to extend that system to cases of religious hardship *without compelling reason.*'"[796] Granting medical and administrative exemptions to mask-wearing while refusing to grant religious exemptions undermines any claim to neutrality and general applicability.

---

792   *Calvary Chapel Dayton Valley v. Sisolak*, 140 S. Ct. at 2605 (Alito, J., dissenting), available online: https://casetext.com/case/calvary-chapel-dayton-valley-v-sisolak

793   *Calvary Chapel Dayton Valley v. Sisolak*, 140 S. Ct. at 2613 (2020) (Kavanaugh, J., dissenting), available online: https://casetext.com/case/calvary-chapel-dayton-valley-v-sisolak

794   *Tandon v. Newsom*, 141 S.Ct. 1294, 1296 (2021), https://www.supremecourt.gov/opinions/20pdf/20a151_4g15.pdf

795   *Fulton v. City of Philadelphia*, --- U.S. ----, 141 S. Ct. 1868, 1877 (2021), available online: https://www.supremecourt.gov/opinions/20pdf/19-123_g3bi.pdf

796   (internal quotation marks omitted; emphasis added by the Court)
Calvary Chapel Dayton Valley v. Sisolak, 140 S. Ct. at 2612 (2020) (Kavanaugh, J., dissenting), available online: https://casetext.com/case/calvary-chapel-dayton-valley-v-sisolak
quoting *Employment Division, Department of Human Resources of Oregon v. Smith*, 494 U.S. 872, 876, 884 (1990), available online: https://tile.loc.gov/storage-services/service/ll/usrep/usrep494/usrep494872/usrep494872.pdf

Since every mask mandate explicitly or implicitly allowed multiple secular exemptions — including medical exemptions — any refusal to also grant religious exemptions to mask-wearing violated the Free Exercise clause of the First Amendment.

The various branches of the military learned in lower courts that with regard to COVID vaccines, "any" comparable secular activity does indeed include medical and administrative exemptions. Judge Matthew McFarland in the 6th Circuit explained: "the Air Force's COVID-19 vaccination mandate is not generally applicable because it allows for medical and administrative exemptions as well as religious exemptions."[797] A 6th Circuit panel opinion delivered by Judge Eric Murphy later in the same case likewise took the Air Force to task:

> "[T]he Air Force appears to freely grant medical and administrative exemptions from its vaccine mandate. These exemptions 'produc[e] substantial harm' to the health and readiness interests that the Air Force claims to be compelling… the Air Force says that it must reject religious exemptions because those working in 'close physical contact' can spread COVID-19… But the Air Force has allowed medical or administrative exemptions even when these exemptions undercut that interest… Perhaps most striking, the Surgeon General denied a religious exemption for Major Corvi… because her assignment 'require[d] intermittent to frequent contact with others.' In the same month, she received a medical exemption for her pregnancy."[798]

Judge Reed O'Connor in the 5th Circuit addressed the same issues in the Navy:

> "The Navy's mandate is not neutral and generally applicable. First, by accepting individual applications for exemptions, the law invites an individualized assessment of the reasons why a servicemember is not vaccinated…. Second, the 'comparable secular activity' includes refusing the vaccine for medical reasons or participation in a clinical trial… an influx of religious accommodation requests is not a valid reason to deny First Amendment rights. No matter how small the

---

797   *Doster v. Kendall*, 596 F. Supp. 3d 995, 1018 (S.D. Ohio 2022), available online: https://casetext.com/case/doster-v-kendall?q=doster%20v%20kendall&sort=relevance&p=1&type=case&tab=keyword&jxs=&resultsNav=false

798   (citations omitted) *Doster v. Kendall*, 54 F.4th 398, 423 (6th Cir. 2022), available online: https://casetext.com/case/doster-v-kendall-6?sort=relevance&p=1&type=case&resultsNav=false
Later, the 6th Circuit noted: "That the Department has granted only a comparative handful of religious exemptions, while granting thousands of medical and administrative ones, is itself at this stage of the case significant proof of discrimination." *Doster v. Kendall*, 48 F.4th 608, 614 (6th Cir. 2022), available online: https://casetext.com/case/doster-v-kendall-5?sort=relevance&p=1&type=case&resultsNav=false

number of secular exemptions by comparison, any favorable treatment—in this case, deployability without medical disqualification—defeats neutrality. For these reasons, the mandate triggers strict scrutiny under the First Amendment."[799]

Another 5th Circuit case involving the Army highlighted one more way that secular medical and administrative exceptions can be given illegitimate non-neutral preference over religious exceptions: by allowing lower-level authorities to approve one but not the other.

> "[T]he procedure employed by the Army for reviewing exemption requests itself undermines the claim that the Army considers religious exemptions on an individualized basis. Temporary medical exemptions to the COVID-19 vaccine mandate can be directly granted by the applicant's healthcare provider… an applicant's regional health commanding general can approve a permanent medical exemption, with any appeals to be considered by the Surgeon General. In contrast, a religious exemption must be approved by the Surgeon General, and the ASA (M&RA) [Assistant Secretary of the Army (Manpower and Reserve Affairs)] reviews any appeals. Thus, as a matter of structure, the Army has opted to make the process for obtaining religious exemptions less personal and local than the process for medical exemptions."[800]

This reasoning applies just as much to mask mandates as it does to vaccine mandates. The mere *existence* of exceptions undermines a claim to be "generally applicable," and that claim becomes progressively weaker as the exceptions pile up. The Supreme Court of Louisiana quoted an especially good summary by a panel of judges in the Federal 6th Circuit:

> "As a rule of thumb, the more exceptions to a prohibition, the less likely it will count as a generally applicable, non-discriminatory law. At some point, an exception-ridden policy takes on the appearance and reality of a system of individualized exemptions, the antithesis of a neutral and generally applicable

---

799   *US Navy SEALs 1-26 v. Biden*, 4:21-cv-01236, Document 66, p.21-22 (N.D. Tex.) Available online: https://storage.courtlistener.com/recap/gov.uscourts.txnd.355696/gov.uscourts.txnd.355696.66.0_5.pdf additional case documents available: https://www.courtlistener.com/docket/60824061/us-navy-seals-1-26-v-biden/

800   (citations omitted) *Schelske et al v. Austin et al*, No. 6:2022cv00049 - Document 78, p. 51-52 (N.D. Tex. 2022), available online: https://law.justia.com/cases/federal/district-courts/texas/txndce/6:2022cv00049/368347/78/ additional case documents available: https://www.courtlistener.com/docket/65397073/schelske-v-austin/.
After reviewing the evidence, Judge James Hendrix found that "In fact, the plaintiffs 'are statistically far more likely—in fact, almost guaranteed—to be denied a religious [exemption] than to contract a severe case of COVID-19.'" (p. 61)

policy and just the kind of state action that must run the gauntlet of strict scrutiny."[801]

Both as-written and in-practice, mask mandates were riddled with unexplained secular exceptions to masking while disallowing religiously motivated non-masking, making them *neither* neutral *nor* generally applicable.[802] Additionally, the cumulative effect of exceptions progressively dilutes the purported infection control rationale behind imposing a mandate in the first place.

A prime example of a mask mandate so riddled with exceptions that it loses general applicability was the series of mask mandates put out by the United States Department of Defense. Despite supervising one of the healthiest and least-vulnerable sub-populations in the United States, the DoD pushed masking harder and further than any other governmental body with the possible exception of the CDC. Review of social media postings from late March 2020 reveals that service members were already wearing face coverings (even outdoors) before Secretary of Defense Mark Esper issued orders on April 5 that, "to the extent practical, all individuals on DoD property, installations, and facilities will wear cloth face coverings when they cannot maintain six feet of social distance in public areas or work centers."[803] On February 4, 2021, Secretary of Defense Lloyd Austin went even further, turning CDC masking "guidelines" into automatic orders (along with everything that entails) for military personnel:

> "Effective immediately, all individuals on military installations, as defined below, and all individuals performing official duties on behalf of the Department from

---

801   *State of Louisiana v. Mark Anthony Spell*, 339 So. 3d 1125, 1136 (La. 2022), available online: https://casetext.com/case/state-v-spell-11919?sort=relevance&p=1&type=case quoting
Roberts v. Neace, 958 F.3d 409, 413-414 (6th Cir. 2020), available online: https://casetext.com/case/roberts-v-neace-1#p413

802   Some courts finding that mask mandates are "neutral and of general applicability" while others find them to be narrowly tailored enough to pass intermediate scrutiny is what we politely call "inconsistent judicial review." Trying to point out that courts in different states were ruling on mask mandates with subtle differences does nothing to defeat this fact. Comparing mask mandates in Hawaii, Minnesota, Pennsylvania, Texas, West Virginia, and Florida may be like comparing Fuji Apples, Gala apples, and Granny Smith apples, but it is hardly comparing apples to oranges.

803   Secretary of Defense Mark Esper, Memorandum, "Department of Defense Guidance on the Use of Cloth Face Coverings," April 5, 2020, available online: https://media.defense.gov/2020/Apr/05/2002275059/-1/-1/1/DOD-GUIDANCE-ON-THE-USE-OF-CLOTH-FACE-COVERINGS.PDF

Just vising any official social media page and setting the post filter to March or April 2020 is illuminating. Masks were already being effectively mandated by lower commands even before SECDEF Esper put out his April 5th mandate.

In-person religious services on military installations were also suspended earlier than comparable-risk secular activities, as revealed by March 19 Facebook posts at Fort Bliss, which announced the suspension of all public religious services while Morale, Welfare, and Recreation amenities like restaurants, the golf club, and bingo hall remained open in some in-person capacity.

any location other than the individual's home, including outdoor shared spaces, will wear masks in accordance with the most current CDC guidelines."[804]

SECDEF Austin's memorandum incorporated exceptions for secular activities: "For brief periods of time when eating and drinking"; "briefly for identification or security purposes"; and "to reasonably accommodate an individual with a disability." The consolidated DoD COVID-19 Force Health Protection Guidance documents put out by the Office of Under Secretary of Defense Gilbert Cisneros listed even more exceptions, to "accommodate participation in a religious service," "when clear or unrestricted visualization of verbal communication is required for safe and effective operations," "when environmental conditions are such that mask wearing presents a health and safety hazard," and "*case-by-case exceptions*" (*emphasis added*).[805] If the previous host of exemptions were not enough, this last provision for *case-by-case exceptions* based on executive discretion, on its own, renders the mask mandate not "generally applicable" and therefore subject to strict scrutiny, because it creates a mechanism for granting individualized exceptions. As the U.S. Supreme Court said, "The creation of a formal mechanism for granting exceptions renders a policy not generally applicable, regardless whether any exceptions have been given, because it "invite[s]" the government to decide which reasons for not complying with the policy are worthy of solicitude[.]"[806] It is also worth noting that, while the DOD guidance explicitly permits religious and case-by-case exceptions to masking, all formal religious accommodation requests for masking exemptions were treated the same way as religious-based requests for vaccine exemptions: pre-determined blanket disapproval. This, despite the broad granting of situational and administrative exemptions and the exception-riddled rules which belied any assertions that masking was so mission-critical that no exemptions could be granted.

Mask mandates failed the test of general applicability because each one had multiple exceptions, and allowed executive discretion to grant modifications and exemptions. Mask mandates failed the test of neutrality because they included secular exceptions while denying religious ones. Finally, mask

804   Secretary of Defense Lloyd Austin, Memorandum, "SUBJECT: Use of Masks and Other Public Health Measures," February 4, 2021, available online: https://media.defense.gov/2021/Feb/04/2002576265/-1/-1/1/DOD-ANNOUNCES-USE-OF-MASKS-AND-OTHER-PUBLIC-HEALTH-MEASURES.PDF

805   Department of Defense Office of the Undersecretary of Defense for Personnel and Readiness, "Consolidated Department of Defense Coronavirus Disease 2019 Force Health Protection Guidance," Section 5.3. *Masks*, 4 April 2022, Section https://media.defense.gov/2022/Apr/06/2002971407/-1/-1/1/CONSOLIDATED-DEPARTMENT-OF-DEFENSE-CORONAVIRUS-DISEASE-2019-FORCE-HEALTH-PROTECTION-GUIDANCE.PDF
Department of Defense Office of the Undersecretary of Defense for Personnel and Readiness, "Consolidated Department of Defense Coronavirus Disease 2019 Force Health Protection Guidance — Revision 5," Section 5.3. *Masks*, March 24, 2023, https://media.defense.gov/2023/Mar/28/2003187831/-1/-1/1/CONSOLIDATED-DEPARTMENT-OF-DEFENSE-CORONAVIRUS-DISEASE-2019-FORCE-HEALTH-PROTECTION-GUIDANCE-REVISION-5.PDF

806   *Fulton v. City of Philadelphia*, --- U.S. ----, 141 S. Ct. 1868, 1879 (2021), available online: https://www.supremecourt.gov/opinions/20pdf/19-123_g3bi.pdf

mandates were not narrowly tailored, because in addition to being *under*inclusive by exempting categories of activity seen to have comparable risk, mask mandates were also *over*inclusive because they swept up the overwhelming majority of people who would not have needed to wear a mask even if masks worked. Treating everyone the same, regardless of their individual medical health, risk factors, and immune status, *can* be neutral and generally applicable, but by definition, it is *not* narrowly tailored. *True* narrowly tailored, time-tested alternatives (e.g., quarantine of the symptomatic) were always readily available. Universal compulsory masking only really ever addressed one additional, *theoretical* risk: lethal spread of COVID from *asymptomatic* or *presymptomatic* individuals. The majority of people covered by mask mandates did not have COVID and were at no meaningful *risk* from COVID. This, too, prevented mask mandates from qualifying as narrowly tailored. COVID mask mandates were neither neutral, nor generally applicable, nor narrowly tailored, and thus required strict scrutiny under the Free Exercise Clause of the First Amendment.

The Religious Freedom Restoration Act (RFRA)

A critical offshoot of the First Amendment's Free Exercise clause is the Religious Freedom Restoration Act of 1993, usually referred to simply as "RFRA." Current First Amendment jurisprudence allows laws that infringe on freedom of religious exercise to pass under rational basis scrutiny if they are neutral and generally applicable. (This is why too many courts during the COVIDcrisis twisted the definition of "neutral and generally applicable" to apply it to mask mandates.) RFRA, on the other hand, subjects even neutral and generally applicable laws to strict scrutiny if they substantially burden the exercise of religion.

Congress passed RFRA in response to a series of late 20th century decisions by the Supreme Court which undermined (or, arguably, neutered) the First Amendment's Free Exercise clause in certain areas, especially correctional and military contexts. The explicit purpose of RFRA was to restore and guarantee protections for the exercise of religion that were applied via the First Amendment in landmark Supreme Court cases like *Sherbert v. Verner* in 1963 and *Wisconsin v. Yoder* in 1972, but which were not applied in later rulings like the Court's 1986 5-4 decision in *Goldman v. Weinberger* (which allowed minutiae of military uniform regulations to take precedence over the First Amendment's Free Exercise Clause).[807] RFRA applies to all components of the Federal Government, including

---

807   In *Goldman*, the U.S. Supreme Court upheld an Air Force uniform regulation that had the effect of preventing an officer who was also an ordained Orthodox Jewish rabbi from wearing a yarmulke, despite the fact that this prohibition demonstrably burdened the officer's religion.
Goldman v. Weinberger, 475 U.S. 503 (1986), available online:
https://tile.loc.gov/storage-services/service/ll/usrep/usrep475/usrep475503/usrep475503.pdf

military and correctional institutions, but does not apply to State or local actors[808] (that being said, at the time of this writing, about half of states have passed laws substantially similar to RFRA, and even for those states that do not have RFRA-like statutes, the First Amendment still applies).

In the statutes own words, a person whose religious practices are burdened in violation of The Religious Freedom Restoration Act, "may assert that violation as a claim or defense in a judicial proceeding and obtain appropriate relief."[809] RFRA specifically says that the Federal Government may not substantially burden a person's exercise of religion, "even if the burden results from a rule of general applicability."[810] The statute makes just one exception under two conditions:

> "Government may substantially burden a person's exercise of religion only if it demonstrates that application of the burden to the person —
>
> > (1) is in furtherance of a compelling governmental interest; and
> >
> > (2) is the least restrictive means of furthering that compelling governmental interest."[811]

Under RFRA, a claimant has to show two things: a sincerely held religious belief, and a substantial burden on that belief. Once that is done, the burden of proof is on the government to satisfy strict scrutiny and show that its infringement 1) furthers a compelling governmental interest by 2) the least restrictive means.

A "substantial burden" that would activate RFRA's protections includes when a person is "forced to choose between following the tenets of their religion and receiving a governmental benefit," (as in the 1963 landmark case *Sherbert v. Verner*) or when they are "coerced to act contrary to their religious beliefs by threat of civil or criminal sanctions" (as in the 1972 landmark case *Wisconsin v. Yoder*).[812]

---

808   RFRA defines "government" to include any "branch, department, agency, instrumentality, and official (or other person acting under color of law) of the United States. 42 U.S.C. § 2000bb-2(1)
("In *City of Boerne v. Flores*, however, the Supreme Court held that to the extent RFRA applied to states or munici-palities, Congress had exceeded its powers and the law was unconstitutional.")
Denis v. Ige et al, No. 1:2021cv00011 - Document 62 (D. Haw. 2021), pg. 22, available online:
https://law.justia.com/cases/federal/district-courts/hawaii/hidce/1:2021cv00011/152658/62/
809   42 U.S.C. § 2000bb-1(c)
810   United States Code, 2006 Edition, Supplement 3, Title 42 - THE PUBLIC HEALTH AND WELFARE CHAPTER 21B - RELIGIOUS FREEDOM RESTORATION 42 U.S.C.] § 2000bb-1(a)
811   42 U.S.C. § 2000bb-1(b)
812   *Navajo Nation v. U.S. Forest Service*, 535 F.3d 1058, 1069–70 (9th Cir. 2008), available online:
https://www.courtlistener.com/opinion/1324254/navajo-nation-v-us-forest-service/
Referencing *Sherbert v. Verner* and *Wisconsin v. Yoder*, respectively.

Thus, if someone has a sincerely held religious belief against wearing a mask (including or especially for COVID), any requirement to wear a mask is a "substantial burden" if that requirement is backed by threats of fines, job loss, denial of services like education or public transportation, denial of access to places of public accommodation, or other penalties. In the words of the 6ᵗʰ Circuit, "This intangible injury (the coerced violation of religious beliefs) is irreparable even when the coercion comes from such lesser forms of pressure as the threatened loss of a civilian job or the loss of the ability to play a college sport."[813] Additionally, as Illinois Judge Raylene Grischow ruled in 2022:

> "To demonstrate irreparable injury, the moving party need not show an injury that is beyond repair or compensation in damages, but rather need show only transgressions of a continuing nature… There is no adequate remedy at law because the loss of the continuous sacrifice of legal rights cannot be cured retroactively once the issues are decided on the merits." [814]

More than this, the compelling interest that the Government has to show in order to satisfy RFRA is not merely a generalized one as in lesser forms of scrutiny. In 2006, the U.S. Supreme Court made clear that "RFRA requires the Government to demonstrate that the compelling interest test is satisfied through application of the challenged law 'to the person'—the particular claimant whose sincere exercise of religion is being substantially burdened."[815] To satisfy the requirements of RFRA, the Government needs to show it has a compelling interest in applying its mandate to the *specific individual*. This is important, because as we covered before, even when the government has a *generalized* compelling interest, the default is that "the government does not have a compelling interest in each marginal percentage point by which its goals are advanced."[816] Even in prison or military contexts,

---

813   *Doster v. Kendall*, No. 22-3702, p. 35 (6th Cir. 2022), available online:
https://law.justia.com/cases/federal/appellate-courts/ca6/22-3702/22-3702-2022-11-29.html
referencing *Dahl v. Board of Trustees of Western Michigan University*, 15 F.4th 728, 730, 736 (6th Cir. 2021) (per curiam) (https://casetext.com/case/dahl-v-bd-of-trs-of-w-mich-univ); see TransUnion LLC v. Ramirez, 141 S. Ct. 2190, 2204 (2021); *Elrod*, 427 U.S. at 373 (plurality opinion).

814   *Austin et al. v. The Board of Education of Community Unit School District #300 et al.*, No. 2021-CH-500002, 21-CH-500003, 21-CH-500005 & 21-CH-500007, Temporary Restraining Order, p. 25 (Circuit Court of Sangamon County, Illinois, Feb. 04, 2022), available online:
https://media.nbcchicago.com/2022/02/Temporary-Restraining-Order.pdf or
https://www.scribd.com/document/557478467/Sangamon-County-Illinois-Mask-mandate-Ruling

815   *Gonzales v. O Centro Espirita Beneficente Uniao Do Vegetal*, 546 U.S. 418, 430-431 (2006), available online:
https://tile.loc.gov/storage-services/service/ll/usrep/usrep546/usrep546418/usrep546418.pdf

816   *Brown v. Entertainment Merchants Association*, 564 U.S. 786, 803-804 n.9 (2011) available online:
https://tile.loc.gov/storage-services/service/ll/usrep/usrep564/usrep564786/usrep564786.pdf

"RFRA 'squarely' places the burden on the government to demonstrate a compelling interest achieved through the least restrictive means… RFRA's focus on 'the burden to the person' demands more than dismissive, encompassing, and inflexible generalizations about the government's interest and about the absence of a less restrictive alternative. ('Broadly formulated interests' and 'generalized statement[s]' will not suffice). Instead, the government must proffer 'specific and reliable evidence' (not formulaic commands, policies, and conclusions)" (emphasis added).[817]

What might an individualized "to the person" analysis need to include in the case of an individual with religious objections to masking for COVID? To start, it would require far more than a reasonable basis showing that masks actually work. The U.S. Supreme Court has already ruled that in the context of individuals with *active Tuberculosis*, an individualized inquiry requires looking at "(a) the nature of the risk… (b) the duration of the risk… (c) the severity of the risk… and (d) the probabilities the disease will be transmitted and will cause varying degrees of harm."[818] A "to the person" analysis under RFRA would be even more stringent. At the very least, a "to the person" analysis of masking for COVID would require showing by "clear and convincing" evidence that masks work, and this showing would just be the *first* step. The analysis would need to fit the benefits of masks into the objecting individual's specific circumstances, looking at the disease in question, its lethality (stratified by age), how prevalent it is in the individual's community and workplace, the relevant transmission rates, the individual's age, overall health and fitness, occupation, and infection history. The analysis would then need to use all this information in some back-of-the-envelope show-your-work math to arrive at an *individualized* disease risk estimate with and without masks. In order to satisfy the "least restrictive means" requirement, the analysis would also need to evaluate and discard alternative methods (pharmaceutical and non-pharmaceutical), which could be used singly or combined to achieve the same effect without violating any sincerely held religious beliefs. Finally, due process would require that the individual in question be given a formal opportunity to interrogate, critique, and rebut any and every part of this analysis, or at least produce their own competing analysis.

---

817   (internal citations omitted from quote and expanded below)
Navy SEAL 1 v. Austin, 586 F. Supp. 3d 1180, 1198-1199 (M.D. Fla. 2022), available online:
https://storage.courtlistener.com/recap/gov.uscourts.flmd.395057/gov.uscourts.flmd.395057.111.0_4.pdf
quoting
Davila v. Gladden, No. 13-10739, p. 13 (11th Cir. 2015) available online:
https://law.justia.com/cases/federal/appellate-courts/ca11/13-10739/13-10739-2015-01-09.html
818   *School Board of Nassau County, Florida, et al. v. Arline*, 480 U.S. 273, 274 (1987), available online:
https://tile.loc.gov/storage-services/service/ll/usrep/usrep480/usrep480273/usrep480273.pdf

In sum, a "to the person" analysis that satisfies RFRA would need to show that the individual in question posed a quantified, imminent, substantial health risk to themselves and others without a mask, and that there were no viable alternatives available to achieve a comparable risk reduction. This threshold is hard enough to meet for anyone that is *not sick*, and basically impossible for someone that already had the disease in question and recovered. No COVIDcrisis mask mandate came anywhere close to meeting the "to the person" standard required in RFRA.

Objections to religious exceptions

Academic debates over of freedom of religion and the First Amendment go into detail discussing the tension between religious accommodations and the principle of neutrality contained in the Establishment Clause of the First Amendment, which prohibits the government from enacting any law "respecting an establishment of religion." Critics list the practical difficulties involved in adjudicating religious claims, and how the existence of a right to religious accommodation encourages false claims or perceptions of unfairness. Some have gone so far as to argue against the very existence of a generalized constitutional right of religious accommodation.[819] At the everyday ground level, though, three objections to religious exemption claims predominate, and none of them are strong. A vague but common objection is that adding religious accommodations on top of secular exemptions would somehow exceed some arbitrary amount of what counts as (or feels like) "too many" total exceptions. However, even in the more constraining military context, courts have ruled that "an influx of religious accommodation requests is not a valid reason to deny First Amendment rights."[820] The related objection based on fear of potential abuse by insincere applicants bringing false claims is one of the reasons why sincerity of belief, rather than orthodoxy or denominational size, is the legal test in religious free exercise cases. The legal presumption of innocence allows some guilty criminals to evade courtroom justice, but that does not justify punishing the innocent by presuming guilt. In the same way, the fact that a number of insincere people may manage to take advantage of religious accommodations does not justify denying accommodations to the sincere. There are multiple ways of reasonably verifying sincerity for legal purposes, and the sincerity standard both protects minorities and makes exploitative gamesmanship more difficult.

---

819   Ellis M. West, "The Case Against a Right to Religion-Based Exemptions." Notre Dame Journal of Law, Ethics & Public Policy, 1990. 4: p. 591. https://scholarship.law.nd.edu/cgi/viewcontent.cgi?article=1554&context=ndjlepp Ellis M. West, "The Right to Religion-Based Exemptions in Early America: The Case of Conscientious Objectors to Conscription." Journal of Law and Religion 10, no. 2 (1993): 367-402. doi:10.2307/1051141 . Available online: https://scholarship.richmond.edu/cgi/viewcontent.cgi?article=1132&context=polisci-faculty-publications

820   *US Navy SEALs 1-26 v. Biden*, 4:21-cv-01236, Document 66, p.21-22 (N.D. Tex.) Available online: https://storage.courtlistener.com/recap/gov.uscourts.txnd.355696/gov.uscourts.txnd.355696.66.0_5.pdf additional case documents available: https://www.courtlistener.com/docket/60824061/us-navy-seals-1-26-v-biden/

One of the strongest objections is that permitting religious exemptions begins to "make the professed doctrines of religious belief superior to the law of the land, and in effect to permit every citizen to become a law unto himself."[821] This objection was put forward by the State of Louisiana while prosecuting Pastor Mark Anthony Spell for violating Louisiana Governor Jon Bel Edwards' lockdown orders by holding in-person worship services in 2020. The Louisiana Supreme Court responded directly:

> "The state nevertheless maintains the orders are neutral and generally applicable and that any holding to the contrary would allow the defendant to 'become a law unto himself' through 'professed doctrines of religious belief…' The state's argument misses a crucial point. The defendant does not argue he is 'a law unto himself' or the executive orders violate 'his own standards.' What defendant seeks--and what our Constitution ensures--is that his religious activities be treated no differently than comparable secular activities. Disparate treatment implicates the Free Exercise Clause… We interpret Pastor Spell's request not as one for special treatment, but for equal treatment." [822]

Compared to those COVID crisis court rulings which sought to evade these foundational principles using any and every means available, the Louisiana Supreme Court's finale reads like an unequivocal blast of clean air:

> "We reject any contention that early in a crisis, the Constitution's protection of fundamental rights must always yield to the needs of the state to respond to the crisis. A public health emergency does not relegate the First Amendment to a proposition or allow violations thereof to be judged on a sliding scale of constitutionality. The infringement of the fundamental right of the free exercise of religion, whether in times of crisis or calm, must always be strictly scrutinized by our courts."[823]

---

821   *Reynolds v. United States*, 98 U.S. 145, 167 (1878); available online: https://tile.loc.gov/storage-services/service/ll/usrep/usrep098/usrep098145/usrep098145.pdf

822   *State of Louisiana v. Mark Anthony Spell*, 339 So. 3d 1125, 1136 (La. 2022), available online: https://casetext.com/case/state-v-spell-11919?sort=relevance&p=1&type=case

823   *State of Louisiana v. Mark Anthony Spell*, 339 So. 3d 1125, 1139 (La. 2022), available online: https://casetext.com/case/state-v-spell-11919?sort=relevance&p=1&type=case

The Fourth and Fifth Amendments — self-incrimination, due process, and takings

"No person… shall be compelled in any criminal case to be a witness against himself, nor be deprived of life, liberty, or property, without due process of law; nor shall private property be taken for public use, without just compensation."

— United States Constitution, Amendment V

When groping for the right legal terms to describe how their fundamental liberties were being violated by compulsory masking, many Americans intuitively gravitated towards the Fourth and Fifth Amendments because of their well-known protections for the right to privacy, the right to remain silent, the right to due process, and the presumption of innocence. This intuitive association with amendments designed to protect citizens accused of wrongdoing was only reinforced by the language used to characterize refusal to wear a mask. Failure or refusal to wear a mask was constantly described by authorities in ways that implied it was a form of criminal negligence. Loaded terms like "reckless," "dangerous," and "negligent," were routinely mixed with other epithets like "selfish," "anti-social," and "freedumb," then venomously flung at people who resisted wearing masks. The overtly punitive nature of the mask mandates, threatening hefty fines and jail time, reinforced this perception.

Realistic legal minds knew that actual criminal negligence requires a standard of proof which refusal to wear a mask was unlikely to satisfy, including "reckless disregard" for the consequences of one's conduct, a "heedless indifference" to the rights and safety of others, and "reasonable foresight" that injury would probably result.[824] Such a standard would be especially hard to meet against someone

---

824   *Daniels v. State*, 264 Ga. 460, 464 (1994), available online: https://law.justia.com/cases/georgia/supreme-court/1994/s94g0362-1.html

refusing to wear a mask as a form of conscientious objection, especially if they had knowledge of their own health, prior recovery from COVID, or a recent negative test. Conscientious inaction and refusal to wear a mask which is meant to be a statement or demonstration that one is non-threating is also tough to construe as a form of reckless endangerment, much less any kind of overt threat. Additionally, as courts like the Georgia Supreme Court have ruled multiple times, "It is a long-standing rule that criminal statutes must be strictly construed against the state and liberally in favor of the accused."[825] Nevertheless, legal standards took a backseat to politics when it came to masks, especially at the rhetorical and street levels.

Many people forced to wear masks very reasonably saw mask-wearing as a form of symbolic self-accusation of sickness and disease in violation of the right to remain silent and not self-incriminate. Compulsory masking felt like a public health Bill of Attainder — a law pronouncing someone guilty of some wrongdoing, nullifying their civil rights without a trial, and expressly forbidden by Article I, Section 9 of the Constitution. Mask mandates assumed that everyone was infectious until proven healthy, the opposite in principle of "innocent until proven guilty," and the fact that mask mandates were most often imposed by executive fiat rather than the legislative process only exacerbated this impression.

In spite of this intuitive association, however, direct challenges to masking under the Fourth and Fifth Amendments couched in these terms got nowhere because of the highly specific nature of the protections involved. The Supreme Court has affirmed that "the principal object of the Fourth Amendment is the protection of privacy, rather than property,"[826] but the Fourth Amendment's privacy protections apply in specific criminal, search and seizure, and evidentiary contexts, and are not as broad as the privacy protections in the Fourteenth Amendment. As for the Fifth Amendment, when one frustrated military veteran sued after having been denied medical care at the Veterans Administration clinic and threatened with police action for refusing to wear a mask, alleging that he was "treated, handled, and regarded as a direct threat to himself or others, and presumed to be guilty," the presiding judge responded brusquely: "Plaintiff does not explain by whom he is being 'presumed guilty' or how, but a presumption of innocence has been recognized only in the context of a criminal trial." The judge also felt the need to add (for some reason), "The fact that plaintiff's refusal to wear a mask is a cause for suspicion is neither surprising nor actionable."[827]

---

825   *State v. Miller*, 260 Ga. 669, 677 (Ga. 1990) (Hunt, J., concurring specially), available online: https://law.justia.com/cases/georgia/supreme-court/1990/s90a1172-1.html

826   *Warden v. Hayden*, 387 U.S. 294, 304 (1967), available online: https://tile.loc.gov/storage-services/service/ll/usrep/usrep387/usrep387294/usrep387294.pdf

827   *Joseph v. Becerra*, 22-cv-40-wmc, Document 21, p. 10 (W.D. Wis. Nov. 29, 2022), available online: https://law.justia.com/cases/federal/district-courts/wisconsin/wiwdc/3:2022cv00040/48728/21/

"Innocent until proven guilty" used in the more colloquial sense *can* have legal weight in other non-criminal contexts, but these broader applications of the principle arise from legal sources other than the Fifth Amendment. In an employment context, for example, attorney Jeff Childers has pointed out:

> "There is a small but well-developed set of case law holding that employers can't just assume that a healthy employee is sick simply because they are in some at-risk category. Consider the case of an employer treating a healthy gay person as having AIDS and requiring them to work from home or something. That's not fair. It turns out that kind of differential treatment violates the Americans with Disabilities Act."[828]

Discrimination or denial of services based on the presence or absence of a mask is, at bottom, discrimination based on either presumed medical status or ideological status.

The Fifth Amendment is not solely concerned with criminal justice, however. The Fifth Amendment's takings clause provides various degrees of protection against direct confiscations or regulations restricting the use or enjoyment of private property, depending on the nature of the property and the regulation in question. Takings can also be indirect, as when the government forces citizens to purchase something. At least one initial complaint challenging compulsory masking included an argument that forcing citizens to purchase masks using their own money under threat of criminal sanction was just this sort of indirect taking[829] (this challenge ultimately succeeded, but not based on that particular argument).

Even more importantly in the context of masks, a fundamental component of property ownership protected by the Fifth Amendment is the right to exclude, which is violated by forcing something into or onto someone's property — i.e. *taking* access. Even invasions as indirect as overflying private property can be this kind of "taking" under the Fifth Amendment.[830] For example, in 2021, the Supreme Court ruled that a California regulation forcing agricultural employers to "allow union organizers onto their property for up to three hours per day, 120 days per year" was an unconstitutional

---

828   Jeff Childers, "A New World," *Coffee and Covid* (Blog), July 30, 2022, online:
https://www.coffeeandcovid.com/p/-coffee-and-covid-saturday-july-30
829   *Green v. Alachua County*, 323 So. 3d 246, Amended Complaint, Filing # 107300778 E-Filed 05/11/2020 (Fla. Dist. Ct. of App., 2021), p. 14,
https://www.fl-counties.com/sites/default/files/2020-07/5.11.20%20Green%20Amended%20Complaint_0.pdf
830   *Cedar Point Nursery v. Hassid et al.*, 141 S. Ct. 2063, 2075 (2021), available online:
https://www.supremecourt.gov/opinions/20pdf/20-107_ihdj.pdf

per se taking under the Fifth and Fourteenth Amendments.[831] The Court explained: "The fact that a right to take access is exercised only from time to time does not make it any less a physical taking."[832]

A person's mind and body are their most intimate and fundamental property. It stands to reason, therefore, that any regulation forcing persons to allow something such as a mask (or a therapeutic of any sort) onto (or into) their body in the name of a public benefit like public health would constitute a *per se* physical taking of the most invasive sort under the Fifth Amendment. COVIDcrisis mask mandates forced people all over the world to allow masks onto their intimate property — their faces — for hours at a time, 365 days each year, for multiple years. To hold that public health authority extends to universal compulsory mask mandates effectively "takes" a legally continuous right of access to one of the most intimate parts of every American's most fundamental property — their face and airways — for purposes of public benefit.

James Madison declared, "as a man is said to have a right to his property, he may be equally said to have a property in his rights."[833] Rights are a form of property, and the COVIDcrisis global deprivation of multiple rights in the name of public health was certainly a taking in derogation of the spirit, if not the letter, of the Fifth Amendment. At the very least, mask mandates were indisputably "a direct and immediate interference" with people's "enjoyment and use" of their faces. Ultimately, though, any potential challenges to compulsory masking based on the Fourth and Fifth Amendments have yet to be fully developed or litigated, and did not play a major role in COVIDcrisis litigation.

---

831   *Cedar Point Nursery v. Hassid et al.*, 141 S. Ct. 2063, 2069 (2021), available online: https://www.supremecourt.gov/opinions/20pdf/20-107_ihdj.pdf

832   *Cedar Point Nursery v. Hassid et al.*, 141 S. Ct. 2063, 2075 (2021), available online: https://www.supremecourt.gov/opinions/20pdf/20-107_ihdj.pdf

833   James Madison, "For the *National Gazette*, 27 March 1792," *Founders Online*, National Archives, https://founders.archives.gov/documents/Madison/01-14-02-0238 . [Original source: *The Papers of James Madison*, vol. 14, *6 April 1791–16 March 1793*, ed. Robert A. Rutland and Thomas A. Mason. Charlottesville: University Press of Virginia, 1983, pp. 266–268.]

Thirteenth Amendment challenges to masks

As an aside, when property is involuntarily taken or its use is restricted by authorities, "[t]he Court has often described the property interest taken as a servitude or an easement."[834] This concept of a taking as a "servitude" combined with an understanding of one's body as one's most fundamental property led at least two plaintiffs to challenge compulsory masking under the 13th Amendment.[835] In Wisconsin, Mark Joseph argued that, at least constructively, "by having to wear a mask, he is being forced to provide a medical service to others in violation of the Thirteenth Amendment's prohibition on involuntary servitude."[836] In Oregon, a parent argued on behalf of her child that tying school receipt of federal funds to the enforcement of mask mandates was an indirect way of placing children in servitude for money.[837]

Unfortunately for these challenges, while the word "servitude" can be accurately used in a *general* sense to refer to being controlled by a person or thing, as well as to deprivation or lack of liberty (and both of these definitions certainly applied to compulsory masking), the *legal* definition of servitude for Thirteenth Amendment purposes is much more specific. In legal terms, servitude *can* refer to enslavement, but it also includes legal obligations laid on properties and estates, as well as the right of requiring someone to do something which they are obligated to perform, as in the case of a contract. In dismissing Thirteenth Amendment challenges to masking, judges acidly remarked that such claims were "utterly without merit,"[838] and that "plaintiff's argument that masking is 'a sign of slavery' 'both trivializes the horrors of slavery and fundamentally misconstrues the nature of the Mask Mandates.'"[839]

We saw in the psychology section of this book how physical props can be powerful psychological cues, getting people to internalize assigned situational rules and roles, and even inducing them to do things they later regret. While Thirteenth Amendment claims against masks fail as a matter of law, their existence provides further evidence that many Americans strongly perceived masks as a

---

834   *Cedar Point Nursery v. Hassid et al.*, 141 S. Ct. 2063, 2073 (2021), available online: https://www.supremecourt.gov/opinions/20pdf/20-107_ihdj.pdf

835   *Joseph v. Becerra*, 22-cv-40-wmc, Document 21, p. 13 (W.D. Wis. Nov. 29, 2022), available online: https://law.justia.com/cases/federal/district-courts/wisconsin/wiwdc/3:2022cv00040/48728/21/

Gunter v. North Wasco County School District Board of Education, 577 F. Supp. 3d 1141 (D. Or. 2021), available online: https://casetext.com/case/gunter-v-n-wasco-cnty-sch-dist-bd-of-educ#p1151

836   *Joseph v. Becerra*, 22-cv-40-wmc, Document 21, p. 13 (W.D. Wis. Nov. 29, 2022), available online: https://law.justia.com/cases/federal/district-courts/wisconsin/wiwdc/3:2022cv00040/48728/21/

837   *Gunter v. North Wasco County School District Board of Education*, 577 F. Supp. 3d 1141, 1151 n 6 (D. Or. 2021), available online: https://casetext.com/case/gunter-v-n-wasco-cnty-sch-dist-bd-of-educ#p1151

838   *Gunter v. North Wasco County School District Board of Education*, 577 F. Supp. 3d 1141, 1151 n 6 (D. Or. 2021), available online: https://casetext.com/case/gunter-v-n-wasco-cnty-sch-dist-bd-of-educ#p1151

839   *Joseph v. Becerra*, 22-cv-40-wmc, Document 21, p. 14 (W.D. Wis. Nov. 29, 2022), available online: https://law.justia.com/cases/federal/district-courts/wisconsin/wiwdc/3:2022cv00040/48728/21/

contemporary form of slave-collar, which reinforces First Amendment claims that press the symbolic speech inherent in masking and non-masking. Additionally, the dismissing judges' curt rejections of Thirteenth Amendment challenges to masking failed to even acknowledge the long and well-documented history of distinctive articles of clothing signifying servile legal status, as well as the rhetoric of the decade prior to 2020 when politicians were trying to make *non*-masking compulsory, and which routinely characterized face coverings as signs of subservience. As noted earlier in this book, expressive mask-wearing was associated with protests of human trafficking prior to 2020, and forced masking was inflicted on prisoners in Guantanamo Bay as part of procedures designed to make them more psychologically submissive.

Implicitly limiting the characterization of slavery to its practice in the 19th Century American Antebellum South, as these judges did, also does a disservice to contemporary victims of human trafficking coerced using more subtle methods, and ignored plaintiffs' underlying point that slaves by definition cannot say "no" to mask-wearing imposed on them in the name of public health or any other rationale. The historical record also contains indisputable examples where dehumanizing and punitive masking was indeed "a sign of slavery."[840] For example, in his 1856 book, *Sketches of Life in Brazil*, Thomas Ewbank remarked:

> "It is said slaves in masks are not so often encountered in the streets as formerly, because of a growing public feeling against them. I met but three or four, and in each case the sufferer was a female. The mask is the reputed ordinary punishment

---

840   c.f.

Jean Baptiste Debret, "Masque de fer blanc que l'on fait porter aux nègres", ["White Iron Mask that One Makes Negro Wear"], watercolor, Rio de Janeiro, *Slavery Images: A Visual Record of the African Slave Trade and Slave Life in the Early African Diaspora*, public domain image accessed May 17, 2023, online:
http://www.slaveryimages.org/s/slaveryimages/item/2996
"Untitled Image (Metal Face Mask)", *Slavery Images: A Visual Record of the African Slave Trade and Slave Life in the Early African Diaspora*, accessed May 17, 2023, http://www.slaveryimages.org/s/slaveryimages/item/2592
From Thomas Ewbank, Life in Brazil (New York, 1856), p. 437, public domain image, available online:
https://archive.org/details/lifeinbrazil00ewba/page/436/mode/2up
"Iron Mask and Collar for Punishing Slaves, Brazil, 1817-1818", *Slavery Images: A Visual Record of the African Slave Trade and Slave Life in the Early African Diaspora*, accessed May 17, 2023,
http://www.slaveryimages.org/s/slaveryimages/item/1299
     It is also worth noting that least one former slave, Fountain Hughes, who was 12 or 13 years old when the American Civil War began, was interviewed about his experiences for the Library of Congress in 1949. What stood out in his memory at age 101 were not the stereotypical whips and chains, but the constant monitoring, control of his movements, and restrictions on education.
Hermond Norwood and Fountain Hughes. *Interview with Fountain Hughes, Baltimore, Maryland*. Baltimore, Maryland, June 11, 1949. Audio Recording: https://www.loc.gov/item/afc1950037_afs09990a/ , pdf transcript:
https://tile.loc.gov/storage-services/service/afc/afc1950037/afc1950037_afs09990/afc1950037_afs09990a.pdf

and preventative of drunkenness… Minute holes are punched to admit air to the nostrils, and similar ones in front of the eyes."[841]

While the masks described by Thomas Ewbank were made of metal, a slave collar remains a symbol of servility regardless of whether it is made of cloth, leather, iron, gold, or melt-blown polypropylene. The connection between masks and slavery is similarly baked into the Spanish language and history. The Spanish word *bozal* refers to a muzzle, a covering for an animal's mouth, or to restrain from speech.[842] It also refers to the state of being wild, untamed, green, and inexperienced. In Spanish New World colonies, *bozales* were recently enslaved Africans, as compared to Hispanicized *ladinos* — whether free or slave.[843] At the very least, given both the history and psychology involved, COVIDcrisis concerns about masks being a sign of servility ought not to have been dismissed so casually by courts who were confronted with this particular challenge.

---

841   Thomas Ewbank, *Life in Brazil* (New York, 1856), p. 437, available online: https://archive.org/details/lifeinbrazil00ewba/page/436/mode/2up

842   Salvatore Ramondino, ed. "bozal," *The New World Spanish/English English/Spanish Dictionary* (New York: New American Library, 1968), pp. 92 & 910, available online: https://archive.org/details/newworldspanishe0000unse_x0m3/page/910/mode/2up?q=bozal

843   City University of New York Dominican Studies Institute, "Ladinos and Bozales," *First Blacks in the* Americas, Online: http://firstblacks.org/en/summaries/arrival-02-ladinos-and-bozales/, last accessed: August 13, 2023.

    I verified this connection by reviewing several independent sources, but I want to make sure I acknowledge my debt to Roberto Strongman for first bringing this final historical-linguistic connection between masks and slavery to my awareness via his article: "The Mask of Your Enslavement: The Image, History, and Meaning of Escrava Anastácia," November 4, 2021, *The Brownstone Institute*, online https://brownstone.org/articles/mask-enslavement-history-meaning-escrava-anastacia/

The Ninth and Fourteenth Amendments

> "The enumeration in the Constitution, of certain rights, shall not be construed to deny or disparage others retained by the people."
>
> — United States Constitution, Amendment IX

> "No state shall make or enforce any law which shall abridge the privileges or immunities of citizens of the United States; nor shall any state deprive any person of life, liberty, or property, without due process of law; nor deny to any person within its jurisdiction the equal protection of the laws."
>
> — United States Constitution, Amendment XIV, Section 1

William Penn's hat, unenumerated rights, the right to be let alone, and the right to privacy

The Constitution of the United States protects those fundamental rights that make up the minor day-to-day matters of liberty — the minutiae and *leviculae* — just as much as it does the more obvious or momentous rights. An exchange during the original Congressional debates that took place when drafting the Bill of Rights reveals how the more "trivial" rights involving things like articles of apparel can be intimately and inextricably connected to the more momentous and well-known ones, and how petty tyranny undermines the more prominent enumerated rights by chipping away at the at the unenumerated ones.

During the debates over how the First Amendment was to be worded, Theodore Sedgwick, one of the delegates from Massachusetts, objected to the explicit inclusion of the right of assembly: "it is derogatory to the dignity of the House to descend to such minutiae." Sedgwick argued that going to such a level of detail was like declaring: "that a man should have a right to wear his hat if he pleased," and made a motion that the clause protecting the right of assembly be struck. Representative John Page of Virginia responded: "The gentleman from Massachusetts objects to the clause, because the right is of so trivial a nature. He supposes it no more essential than whether a man has a right to wear his hat or not; but let me observe to him that such rights have been opposed, and a man has been obliged to pull off his hat when he appeared before the face of authority; people have also been prevented from assembling together on their lawful occasions[.]"[844]

John Page was referring to a trial in England from a century earlier which involved Pennsylvania's Quaker Founder William Penn. This trial was part of the founding generation's communal memory, and involved violations of many rights which America's Founders were anxious to preserve and protect. Having previously been banned from preaching inside any building, William Penn delivered an outdoor sermon on a street corner in London. He was immediately arrested and thrown into jail, then hauled before the London Court of Sessions on charges that he "did unlawfully and tumultuously assemble" and "did take upon himself to preach and speak… to the great disturbance of [the king's] peace." Penn entered the courtroom bareheaded, at which the mayor, who was the head magistrate, took offense and ordered that Penn's hat be put back on again. Penn, now wearing his hat, was brought to the bar. But William Penn's religious beliefs dictated that he refuse to remove his hat in the customary gesture of respect. At Penn's non-action, the mayor *again* took offense and fined him 40 marks for contempt — roughly the amount a skilled tradesman would earn in a year at the time. After hearing the case, the jury found Penn guilty of the non-crime of "speaking in Gracechurch street." The mayor was furious, and threatened the *jury* with punishment if they did not convict. He even kept the jury imprisoned overnight without fire, food, drink, or even a chamber pot. To its everlasting credit, the jury, led by its foreman, Edward Bushell, refused to cave to either threats or duress, and the next day returned an even more direct: "Not Guilty." In a rage, the magistrate had the twelve *jurors* imprisoned for refusing to obey the court's order to convict, and also refused to

---

844   Irving Brandt, *The Bill of Rights: Its Origins and Meaning*, (New York: Bobs-Merrill, 1965), p. 54-55, available online: https://archive.org/details/billofrightsitso00bran

release William Penn, citing the fine for not removing his hat, which the magistrate had ordered put on in the first place.[845]

After America's first Congressional delegates were reminded of William Penn's experiences, the freedom of assembly was explicitly included in the First Amendment, and the Ninth Amendment was passed to help protect those other more "trivial" fundamental rights not specifically enumerated in the first eight Amendments. Later, protection for personal liberty was further reinforced by the Due Process Clause of the Fourteenth Amendment. The presumption starts in favor of individual liberty, not governmental power. If wearing or not wearing a *hat* implicates fundamental rights, even in the controlled setting of a courtroom, then wearing or not wearing a mask certainly does as well.

In 1970, the First Circuit Court of appeals applied the history of First Amendment debates and William Penn's trial to the seemingly trivial matter of school hairstyle mandates. The Court recognized that:

> "What appears superficially as a dispute over which side has the burden of persuasion is, however, a very fundamental dispute over the extent to which the Constitution protects such uniquely personal aspects of one's life[.]"[846]

When confronted with the same issue in 2001, the Alaskan Supreme Court held "control of personal appearance" to be "a fundamental liberty right"[847] Hairstyles have been a historical target of busybody tyranny the world over, and mask mandates are far more analogous to hairstyle regulations than they are to seatbelt and helmet laws. Requiring plaintiffs who were challenging compulsory masking to show why they should not have to wear a mask represented a fundamental cheat, reversing the proper burden of proof.

In overturning a hair-length mandate, a 7th Circuit panel took judicial notice of the U.S. Supreme Court's recognition for "the special right of an individual to control his physical person."[848] The 7th Circuit referenced Chinese and Russian history to explain:

---

845   Irving Brandt, *The Bill of Rights: Its Origins and Meaning*, (New York: Bobs-Merrill, 1965), p. 53-67, available online: https://archive.org/details/billofrightsitso00bran

846   *Richards v. Thurston*, 424 F.2d 1281, *1283 (1st Cir. 1970), https://casetext.com/case/richards-v-thurston

847   *Sampson v. State*, 31 P.3d 88, 93 (Alaska 2001), available online: https://casetext.com/case/sampson-v-state-96

848   "More recently the Supreme Court has recognized that the special right of an individual to control his physical person weighs heavily against arbitrary state intrusions. *Breithaupt v. Abram*, 352 U.S. 432, 439, 77 S.Ct. 408, 1 L.Ed.2d 448 (1957); *Rochin v. California*, 342 U.S. 165, 172, 72 S.Ct. 205, 96 L.Ed. 183 (1952)."
Crews v. Cloncs, 432 F.2d 1259, 1264 (7th Cir. 1970), (available online: https://casetext.com/case/crews-v-cloncs

"[T]he freedom guaranteed by the Bill of Rights could be significantly diluted through subtle forms of state repression and tyranny. History contains many examples of regimes which have attacked and silenced their opponents by requiring conformity of hairstyle or dress. Thus, following the Manchus' invasion of China in 1644, in order to consolidate its power over Chinese citizens, the founders of the Ching dynasty transformed the appearance of the country by compelling the male population to shave the front of the head, and to wear the hair in a queue. Official dress was altered from the Chinese style… to the skullcap and Tartar gown with high collar and fastening at the side. In the first year or two after the conquest… the opposition to these measures was intense, and thousands of people from all walks of life chose to die rather than adapt the marks of servitude. Another example includes official prohibition of beards during the reign of Peter the Great.[849]

Defenders of mask mandates might argue, as Alachua County in Florida did, that controlling citizens' hairstyles is more invasive than "'merely' covering their faces,"[850] but that is simply wrong. While removing a mask is certainly faster than regrowing hair, mask mandates are inherently more invasive than any rules on hair or beards. One's face is a more intimate area than the crown of one's head, and no hairstyle increases dead air space and resistance to airflow when breathing like masks do. Moreover, masks impose a far greater interference with citizens' abilities to express themselves by overwriting the pure speech of facial expressions with a superimposed message.[851] In any case, masking *does* directly implicate hairstyles, at least for men. According to the theory used to try to justify mask mandates, a good seal around the mask is essential to maximize a mask's effectiveness.

---

849   (internal citations omitted) *Crews v. Cloncs*, 432 F.2d 1259, 1264, 1264 n 7 (7th Cir. 1970), (available online: https://casetext.com/case/crews-v-cloncs) quoting Henry McAleavy, *The Modern History of China* (New York: Praeger, 1967), (available online: https://archive.org/details/modernhistoryofc0000unse),
and referencing Will and Ariel Durant, *The Story of Civilization VIII: The Age of Louis XIV*, (New York: Simon and Schuster, 1963), p. 398, (available online: https://archive.org/details/TheStoryOfCivilizationcomplete/Durant Will - The story of civilization 8/page/n455)

> "Soon an edict went throughout Russia that all laymen were to shave their chins; mustaches might remain. The beard had been almost a religious symbol in Russa; it had been worn by the Prophets and the Apostles; and the reigning Patriarch, Adrian, only eight years before, had condemned the shaving of the beard as irreligious and heretical. Peter accepted the challenge: beardlessness was to be a sign of modernity, of willingness to enter into Western civilization."

850   *Green v. Alachua County*, 323 So. 3d 246, Appellant's Initial Brief, Filing #109556371 E-Filed 06/29/2020 (Fla. Dist. Ct. of App., 2021), p. 39, go to: https://1dca.flcourts.gov/Oral-Arguments/Briefs-and-Petitions-for-Cases-Scheduled-for-Oral-Argument/20-1661 and search for "case 20-1661"

851   Hair length on its own may not be enough to implicate the First Amendment (see *Richards v. Thurston*), but this ruling does not preclude hair length, color, cut, and style doing so when combined.

Gaps greatly decrease the efficacy of masks. Beards create *many* such gaps. If bearded "expert" mask advocates like Dr. Peter Hotez and CDC director Robert Redfield truly believed that COVID-19 was dangerous, and that masks were so essential as to justify mandates, they would have shaved their beards off back in March of 2020 and pushed for bans on facial hair. Instead, they kept their own facial hair throughout the COVIDcrisis, and there was no serious talk of emergency facial hair bans on grounds of public health. Perhaps at least a few decision-makers or their legal advisors were aware that bans on facial hair had already been ruled either unconstitutional or automatically suspect (c.f. the 3rd Circuit: "we held unconstitutional the Newark Police Department's policy that officers could not have facial hair").[852]

In a sense, the mandates of the COVIDcrisis were, at their roots, nothing new. Using the pretexts of safety, security, or other generally desirable ends — including the government's compelling interest in preventing crime — busybodies, busybullies,[853] and anxiety-ridden authorities habitually attempt to take away or criminalize fundamental individual liberties not explicitly listed in the Bill of Rights, but which are every bit as basic and essential as the enumerated rights. Nineteenth and twentieth-century bans on interracial marriage were a glaring and long-lasting violation of this type, but violations of the more "trivial" rights are just as indefensible and even more insidious. U.S. Supreme Court Justice Potter Stewart used a memorable turn of phrase to describe this sort of encroachment "in a manner made the more pernicious by its very subtlety."[854]

One such attempt by the Florida city of Jacksonville to ban *multiple* "trivial" rights had to be struck down by the Supreme Court in 1972. In *Papachristou v. City of Jacksonville*, the city had passed an ordinance permitting arrest and conviction of people engaged in many usually innocent pastimes, including "common night walkers," "persons wandering or strolling around from place to place without any lawful purpose or object," and "habitual loafers." Being a habitual nightwalker and nightjogger myself, I find the thought of such an ordinance particularly grating. The Supreme Court very properly struck down the Jacksonville ordinance for violating the Fourteenth Amendment's due process guarantee by being so vague and generalized that it cast too wide a net when seeking to prevent crime:

---

852   *Fulton v. City of Philadelphia.*, 922 F.3d 140, 156, (3d Cir. 2019), available online: https://www.leagle.com/decision/infco20190422056

853   I am indebted to whoever coined this delightful term, which I first encountered in Antony Davies and James Harrigan's book *Cooperation and Coercion: How Busybodies Became Busybullies and What That Means for Economics and Politics* (Intercollegiate Studies Institute, 2020).

854   *Stanley v. Georgia*, 394 U.S. 557, 570 (1969), (Stewart, J., concurring), available online: https://tile.loc.gov/storage-services/service/ll/usrep/usrep394/usrep394557/usrep394557.pdf

"The Jacksonville ordinance makes criminal activities which by modern standards are normally innocent... these activities are historically part of the amenities of life as we have known them. They are not mentioned in the Constitution or in the Bill of Rights. These unwritten amenities have been in part responsible for giving our people the feeling of independence and self-confidence, the feeling of creativity. These amenities have dignified the right of dissent and have honored the right to be nonconformists and the right to defy submissiveness. They have encouraged lives of high spirits rather than hushed, suffocating silence."[855]

The Supreme Court reasoned that "A direction by a legislature to the police to arrest all 'suspicious' persons would not pass constitutional muster."[856] Similarly, going about unmasked has historically been innocent — including throughout the worst plagues of history — and a direction by a legislature (never mind an executive order) to officials to mask all "asymptomatic" persons does not pass constitutional muster, either. Compulsory masking in the name of the Spanish Flu and COVID *criminalized* the right to be nonconformist and right to defy submissiveness. Compulsory masking *mandated* lives of anxious spirits and suffocating silence.

The point is that not being explicitly enumerated in the first eight amendments to the Constitution does not magically render a fundamental liberty non-existent. The burden of proof is not on the individual whose rights are being violated, but on the authority to show why its interest is compelling enough and its infringement mild enough to survive scrutiny. Often during the COVIDcrisis, rights were successfully disparaged, ignored, and violated because they were not enumerated. Instead, this reasoning should have been applied to emergency powers. If specific rights are presumed to be questionable or suspect simply because they are unenumerated or fall under the broad headings of the Ninth and Fourteenth Amendments, courts ought to have been even *more* reticent to uphold unenumerated emergency powers. COVIDcrisis mandates all too often gained traction not because of any inherent intellectual substance or integrity, but because of widespread, deliberately cultivated, and totally unnecessary abject fear. Face covering mandates used altruism, civic-mindedness, and infection control to mask what was more often, at bottom, political expediency, cowardice, and intellectual laziness.

---

855   *Papachristou v City of Jacksonville*, 405 U.S. 156, 163-164 (1972), available online:
https://tile.loc.gov/storage-services/service/ll/usrep/usrep405/usrep405156/usrep405156.pdf
856   *Papachristou v City of Jacksonville*, 405 U.S. 156, 169 (1972), available online:
https://tile.loc.gov/storage-services/service/ll/usrep/usrep405/usrep405156/usrep405156.pdf

Why Ninth Amendment challenges to masking failed

It is a funny thing about constitutional interpretation: generalized clauses which can be interpreted to give the government more power, such as the commerce clause or the "necessary and proper" clause in Article I, Section 8,[857] are consistently interpreted as broadly as possible. Constitutional clauses which protect individual rights using similarly generalized wording, on the other hand, have a much more uphill struggle, and are often only recognized after being repeatedly and egregiously violated, as demonstrated by the history of efforts to uphold basic unenumerated individual liberties like the right not to be forcibly sterilized. Admittedly, there have been exceptions to this trend where the Courts or other authorities have affirmed or implied "rights" that do not exist (usually at the expense of *actual* rights),[858] but this general pattern holds true. Even so, as covered in our discussion of the First Amendment right to *receive* speech, to its credit the United States Supreme Court has repeatedly recognized that specific guarantees in the Bill of Rights have zones of control that protect "equally fundamental personal rights necessary to make the express guarantees fully meaningful,"[859] — even when those rights are not explicitly stated in the text of the relevant amendment.

The Ninth Amendment recognizes that not all fundamental, inalienable rights and liberties are explicitly listed in the previous eight amendments. It was included in the Bill of Rights to provide formal recognition and reassurance that the explicit listing of a few specific individual rights in the first eight Amendments would not be interpreted to mean that the rest were unprotected[860] (though this is exactly what happened on a massive scale with compulsory masking and other mandates during the

---

857   Giving Congress the power "to make all Laws which shall be necessary and proper for carrying into Execution the foregoing Powers, and all other Powers vested by this Constitution in the Government of the United States, or any Department or Officer thereof[.]"

858   No doubt many examples of COVIDcrisis mandates upholding some "right" not to get an infection (as opposed to the narrower right to not be intentionally infected against one's will) at the expense of multiple genuine rights will spring to reader's minds, but for a general discussion of this topic, I refer readers to Thomas Sowell's excellent book *Intellectuals and Society* (New York: Basic Books, 2011), especially Chapter 8's section on "The Rhetoric of 'Rights.'"

859   *Lamont v. Postmaster General*, 381 U.S. 301, 308 (1965), (Brennan, J., concurring) available online: https://tile.loc.gov/storage-services/service/ll/usrep/usrep381/usrep381301/usrep381301.pdf

860   This was a concern discussed by Alexander Hamilton in *Federalist* 84:

> "I go further, and affirm that bills of rights, in the sense and to the extent in which they are contended for, are not only unnecessary in the proposed Constitution, but would even be dangerous. They would contain various exceptions to powers not granted; and, on this very account, would afford a colorable pretext to claim more than were granted. For why declare that things shall not be done which there is no power to do? Why, for instance, should it be said that the liberty of the press shall not be restrained, when no power is given by which restrictions may be imposed?"

The Federalist No. 84, [28 May 1788]," *Founders Online*, National Archives, https://founders.archives.gov/documents/Hamilton/01-04-02-0247 [Original source: *The Papers of Alexander*

COVIDcrisis). The text of the Ninth Amendment is simple and straightforward: "The enumeration in the Constitution, of certain rights, shall not be construed to deny or disparage others retained by the people." The Ninth Amendment's history, purpose, and simplicity made it an intuitive tool for *pro se* litigants attempting to challenge mask mandates.

Unfortunately, the Ninth Amendment has been functionally relegated to a position as one of the Fourteenth Amendment's sidekicks. Claims brought against compulsory masking directly under the Ninth Amendment (if they were not removed in amended complaints[861]) were dismissed for the simple reason that, under modern jurisprudence, the Ninth Amendment is treated as "a rule of interpretation rather than a source of rights."[862] According to the courts, it "has never been recognized as independently securing any constitutional right, for purposes of pursuing a civil rights claim."[863]

The most prominent 20th century Supreme Court Decision to heavily feature the Ninth Amendment was *Griswold v. Connecticut* in 1965.[864] In *Griswold*, the Supreme Court referred to the Ninth Amendment to strengthen its inference that the right to marital privacy falls under the general, protected, individual right to privacy, and includes the use of contraception. The Court's majority opinion held (using rather cumbersome wording) that "specific guarantees in the Bill of Rights have

---

Hamilton, vol. 4, *January 1787–May 1788*, ed. Harold C. Syrett. New York: Columbia University Press, 1962, pp. 702–714.

861   Mark Joseph's initial complaint in *Joseph v. Becerra* included a Ninth Amendment claim which was removed in the amended complaint.

862   *Froehlich v. State of Wisconsin Department of Corrections*, 196 F.3d 800, 801 (7th Cir. 1999), available online: https://casetext.com/case/froehlich-v-state-of-wis-dept-corr quoted by Judge Susan Mollway in *Denis v. Ige et al*, No. 1:2021cv00011 - Document 62 (D. Haw. 2021), pg. 33, available online: https://law.justia.com/cases/federal/district-courts/hawaii/hidce/1:2021cv00011/152658/62/available onlin

        Judge Dorsey dismissed a similar claim in Nevada on the same grounds. "The notion that the Ninth Amdnement supplies a basis to enjoin a school mask mandate deserves only short shrift." *Branch-Noto v. Sisolak*, 576 F. Supp. 3d 790, 803 (D. Nev. 2021), available online: https://casetext.com/case/branch-noto-v-sisolak-1; 2:21-cv-01507 JAD-DJA, p. 17 (D. Nev. Dec. 22, 2021) https://ag.nv.gov/uploadedFiles/agnvgov/Content/News/PR/PR_Docs/2021/DOC%2038%20Order%20Denying%20Motion%20for%20Preliminary%20Injunction,%20Granting%20in%20Part%20Motions%20to%20Dismiss%20and%20Closing%20Case.pdf

863   *Strandberg v. City of Helena*, 791 F.2d 744, 748 (9th Cir. 1986), available online: https://casetext.com/case/strandberg-v-city-of-helena quoted by Judge Susan Mollway in *Denis v. Ige et al*, No. 1:2021cv00011 - Document 62 (D. Haw. 2021), pg. 33, available online: https://law.justia.com/cases/federal/district-courts/hawaii/hidce/1:2021cv00011/152658/62/

864   *Griswold v. Connecticut*, 381 U.S. 479 (1965), available online: https://tile.loc.gov/storage-services/service/ll/usrep/usrep381/usrep381479/usrep381479.pdf

penumbras, formed by emanations from those guarantees that help give them life and substance."[865] In the words of the majority, "Without those peripheral rights, the specific rights would be less secure."[866]

Justice Arthur Goldberg put it more simply (with multiple references) in his concurrence: "This Court… has never held that the Bill of Rights or the Fourteenth Amendment protects only those rights that the Constitution specifically mentions by name."[867] In his concurrence, Justice Goldberg spent several pages laying out how "The language and history of the Ninth Amendment reveal that the Framers of the Constitution believed that there are additional fundamental rights, protected from governmental infringement, which exist alongside those fundamental rights specifically mentioned in the first eight constitutional amendments."[868] Justice Goldberg clearly defined at least one example of a violation of the Ninth Amendment, though he did so mostly by restating the Amendment's generalized wording in a way that applied to the specific example of marital privacy:

> "To hold that a right so basic and fundamental and so deep-rooted in our society as the right of privacy in marriage may be infringed because that right is not guaranteed in so many words by the first eight amendments to the Constitution is to ignore the Ninth Amendment and give it no effect whatsoever. Moreover, a judicial construction that this fundamental right is not protected by the Constitution because it is not mentioned in explicit terms by one of the first eight amendments or elsewhere in the Constitution would violate the Ninth Amendment"[869] (emphasis added).

However, Justice Goldberg carefully stopped short of saying that any rights could be grounded directly in the Ninth Amendment:

> "I do not take the position… that the entire Bill of Rights is incorporated in the Fourteenth Amendment… Nor do I mean to state that the Ninth Amendment

---

865   *Griswold v. Connecticut*, 381 U.S. 479, 484 (1965), available online:
https://tile.loc.gov/storage-services/service/ll/usrep/usrep381/usrep381479/usrep381479.pdf
866   *Griswold v. Connecticut*, 381 U.S. 479, 482-483 (1965), available online:
https://tile.loc.gov/storage-services/service/ll/usrep/usrep381/usrep381479/usrep381479.pdf
867   *Griswold v. Connecticut*, 381 U.S. 479, 486-487 n 1 (1965) (Goldberg, J., concurring), available online:
https://tile.loc.gov/storage-services/service/ll/usrep/usrep381/usrep381479/usrep381479.pdf
868   *Griswold v. Connecticut*, 381 U.S. 479 (1965), available online:
https://tile.loc.gov/storage-services/service/ll/usrep/usrep381/usrep381479/usrep381479.pdf
869   *Griswold v. Connecticut*, 381 U.S. 479, 491 (1965) (Goldberg, J., concurring), available online:
https://tile.loc.gov/storage-services/service/ll/usrep/usrep381/usrep381479/usrep381479.pdf

constitutes an independent source of rights protected from infringement by either the States or the Federal Government."[870]

This landmark case thus left the Ninth Amendment in a position where it does not act as a source of rights, and only provides protection from a single highly specific line of attack: the argument that something is not a protected fundamental right because it is not explicitly mentioned in the Constitution.

Personally, I hold to the very view which Justice Goldberg took pains to avoid affirming: "that the Ninth Amendment constitutes an independent source of rights protected from infringement by either the States or the Federal Government." I believe that the Ninth Amendment does *exactly* that, at least when combined with the Tenth Amendment's reservation of powers to the states and *to the people*. The context of the Bill of Rights and the discussions surrounding ratification of the United States Constitution took great care to emphasize the limited nature of government. As James Madison said in *Federalist No. 45*, "The powers delegated by the proposed constitution to the federal government, are few and defined. Those which are to remain in the state governments are numerous and indefinite."[871] The 10th Amendment recognizes and protects these limitations with regard to both the states *and* the people: "The powers not delegated to the United States by the Constitution, nor prohibited by it to the states, are reserved to the States respectively, or *to the people*" (emphasis added). The careful wording clearly indicates that "the States" and "the people" are separate and distinct categories composed of individuals acting in their own personal spheres (individual states and individual people).

The Ninth Amendment was drafted and proposed by James Madison, and he stated during the Congressional debates surrounding the Bill of Rights that he did this with the express intent of guarding against any implication "that those rights which were not singled out, were intended to be assigned into the hands of the General Government, and were consequently insecure."[872] Protection is implicit in any formal recognition, and it does not make sense to affirm that other amendments actively protect rights which extend beyond their precise wording, but deny this in the case of the one amendment which includes recognition of unlisted rights *within its own wording*. It seems highly

---

870   *Griswold v. Connecticut*, 381 U.S. 479, 492 (1965) (Goldberg, J., concurring), available online:
https://tile.loc.gov/storage-services/service/ll/usrep/usrep381/usrep381479/usrep381479.pdf
871   The Federalist Number 45, [26 January] 1788," *Founders Online*, National Archives,
https://founders.archives.gov/documents/Madison/01-10-02-0254. [Original source: *The Papers of James Madison*,
vol. 10, *27 May 1787–3 March 1788*, ed. Robert A. Rutland, Charles F. Hobson, William M. E. Rachal, and
Frederika J. Teute. Chicago: The University of Chicago Press, 1977, pp. 428–432.
872   James Madison in *Annals of the Congress of the United States, Volume I*: "The Debates and Proceedings in
the Congress of the United States, Comprising (With Volume II) the Period from March 3, 1789 to March 3, 1891,
Inclusive," (Washington: Gales & Seaton, 1834), p. 439 Available online:
https://archive.org/details/annalsofcongress01unit/page/n111/mode/2up

implausible (bordering on irrational) to me that the Amendment which explicitly recognizes the existence of unenumerated rights would not or should not also constitute a source of protection for those unenumerated rights in the same way that each of first eight Amendments does for the group of rights it recognizes. Even though the practical questions of what criteria a court can or should use to determine whether an unenumerated right is fundamental and protected are legitimate and challenging, that same difficulty has not prevented courts from doing this with regard to unenumerated fundamental rights protected by the Fourteenth Amendment. I personally cannot help seeing current judicial interpretation of the Ninth Amendment as unconstitutionally disparaging and denying rights that are "not enumerated in the Constitution" simply *because* they are not enumerated, violating the spirit (if not the letter) of the Ninth Amendment in the process.

Regardless of my own personal views, though, as a matter of practicality, bringing any civil rights claim grounded solely in the Ninth Amendment will simply have the effect of making the plaintiff look bad and making the presiding judge take the suit as a whole less seriously, especially if the judge in question happens to be negatively predisposed to begin with (as was the case for lawsuits challenging mask mandates). Under its current judicial construction, the Ninth Amendment provides a positive interpretative lens that helps ensure the rest of the Bill of Rights reaches its full potential, and especially strengthens the Fourteenth Amendment. On its own, however, the Ninth Amendment provides only the extremely narrow protection for fundamental rights of preventing them from being disparaged and infringed *solely* on the basis that they are not enumerated elsewhere in the Constitution. Perhaps that is good enough. I certainly hope so, and automatically invalidating the argument that something cannot be a fundamental right because it is not mentioned in the Constitution is no small thing. Like James Madison said, "the most minute provisions become important when they tend to obviate the necessity or the pretext for gradual and unobserved usurpations of power."[873] At this point, any broader role that may have been originally intended for the Ninth Amendment has gradually been folded into the Fourteenth Amendment. This does not mean that the Ninth Amendment is worthless and cannot play a useful (even important) part in a successful lawsuit, but it currently functions in a support role rather than on the front line, and on its own, the Ninth Amendment was not even a speed bump for mask mandates during the COVIDcrisis.

---

873   *The Federalist* Number 42, [22 January 1788]," *Founders Online,* National Archives, https://founders.archives.gov/documents/Madison/01-10-02-0244. [Original source: *The Papers of James Madison,* vol. 10, *27 May 1787–3 March 1788,* ed. Robert A. Rutland, Charles F. Hobson, William M. E. Rachal, and Frederika J. Teute. Chicago: The University of Chicago Press, 1977, pp. 403–409.

The Fourteenth Amendment

It is the Fourteenth Amendment that secures those unenumerated rights referred to by the Ninth Amendment which are not protected elsewhere in the Constitution. For example, while the First Amendment protects "freedom of *expressive* association," the Fourteenth Amendment protects "freedom of *intimate* association," which consists of "choices to enter into and maintain certain intimate human relationships."[874] Already by 1923 — over one-hundred years ago — when it was striking down an unconstitutional Nebraska law that banned the teaching of German before the eighth grade,[875] the Supreme Court had a substantial list of both enumerated and unenumerated rights protected by the Fourteenth Amendment:

> "While this Court has not attempted to define with exactness the liberty thus guaranteed, the term has received much consideration and some of the included things have been definitely stated. Without doubt, it denotes not merely freedom from bodily restraint but also the right of the individual to contract, to engage in any of the common occupations of life, to acquire useful knowledge, to marry, establish a home and bring up children, to worship God according to the dictates of his own conscience, and generally to enjoy those privileges long recognized at common law as essential to the orderly pursuit of happiness by free men."[876]

Since 1923, the list of recognized fundamental rights has only grown. Miscegenation laws banning interracial marriage were struck down for violating the Fourteenth Amendment,[877] as was the Connecticut law banning marital use of contraceptives mentioned during our discussion of the Ninth Amendment.[878] Fourteenth Amendment protections include the previously mentioned "freedom from bodily restraint,"[879] as well as "the liberty of the parents and guardians to direct the upbringing

---

874   *Roberts v. United States Jaycees*, 468 U.S. 609, 617 (1984), available online:
https://tile.loc.gov/storage-services/service/ll/usrep/usrep468/usrep468609/usrep468609.pdf
875   *Meyer v. Nebraska*, 262 U.S. 390, 399 (1923) available online:
https://tile.loc.gov/storage-services/service/ll/usrep/usrep262/usrep262390/usrep262390.pdf
876   *Meyer v. Nebraska*, 262 U.S. 390, 399 (1923) available online:
https://tile.loc.gov/storage-services/service/ll/usrep/usrep262/usrep262390/usrep262390.pdf
877   *Loving v. Virginia*, 388 U.S. 1 (1967), available online:
https://tile.loc.gov/storage-services/service/ll/usrep/usrep388/usrep388001/usrep388001.pdf
878   *Griswold v. Connecticut*, 381 U.S. 479 (1965), available online:
https://tile.loc.gov/storage-services/service/ll/usrep/usrep381/usrep381479/usrep381479.pdf
879   *Meyer v. Nebraska*, 262 U.S. 390, 399 (1923) available online:
https://tile.loc.gov/storage-services/service/ll/usrep/usrep262/usrep262390/usrep262390.pdf

of the children"[880] (which entails the right to decide *where* to educate one's children and *what* that education consists of).[881] Desegregation, interstate travel, the generalized constitutional right to personal liberty, the freedom to enter into contracts, aspects of voting rights, and more — if you can think of a protected fundamental liberty that does not easily fit into the original Bill of Rights, odds are that the Fourteenth Amendment is involved in protecting it, and it is through the Fourteenth Amendment that the majority of provisions contained in the Bill of Rights are applied ("incorporated") to protect individual liberties against the States.

Intuitively, one would think that something like masking, which so clearly affects the manner in which people associate, would implicate both expressive *and* intimate association, but at least one Federal District court ruled that mask mandates do not infringe on *any* freedom of association by analogizing statutes against public nudity[882] or requiring shirts and shoes in certain locations.[883] Other COVID-related mandates certainly infringed on these two fundamental rights of association, but the *mask* component of the mandates was theoretically a "minor restriction on the way they occur,"[884] which allowed anyone to freely assemble and associate so long as they were wearing a mask. To date, no one has brought a successful claim that mask mandates interfere with any freedom of association, although I have not yet seen a full exploration or articulation of this claim, either.

At least two portions of the Fourteenth Amendment are directly relevant to compulsory masking: the Due Process Clause and the Equal Protection clause. These are contained in Section 1 of the Fourteenth Amendment, which reads, in part: "No state shall make or enforce any law which shall abridge the privileges or immunities of citizens of the United States; nor shall any state deprive any person of life, liberty, or property, without due process of law; nor deny to any person within its jurisdiction the equal protection of the laws."

---

880   *Pierce v. Society of Sisters*, 268 U.S. 510 (1925), available online:
https://tile.loc.gov/storage-services/service/ll/usrep/usrep268/usrep268510/usrep268510.pdf

881   *Meyer v. Nebraska*, 262 U.S. 390 (1923) available online:
https://tile.loc.gov/storage-services/service/ll/usrep/usrep262/usrep262390/usrep262390.pdf

882   *Denis v. Ige et al*, No. 1:2021cv00011 - Document 62 (D. Haw. 2021), pg. 32, available online:
https://law.justia.com/cases/federal/district-courts/hawaii/hidce/1:2021cv00011/152658/62/

883   *Branch-Noto v. Sisolak*, 576 F. Supp. 3d 790, 797 (D. Nev. 2021), available online:
https://casetext.com/case/branch-noto-v-sisolak-1
2:21-cv-01507 JAD-DJA, p. 7 (D. Nev. Dec. 22, 2021) https://ag.nv.gov/uploadedFiles/agnvgov/Content/News/PR/PR_Docs/2021/DOC%2038%20Order%20Denying%20Motion%20for%20Preliminary%20Injunction,%20Granting%20in%20Part%20Motions%20to%20Dismiss%20and%20Closing%20Case.pdf

884   *Denis v. Ige et al*, No. 1:2021cv00011 - Document 62 (D. Haw. 2021), pg. 33, available online:
https://law.justia.com/cases/federal/district-courts/hawaii/hidce/1:2021cv00011/152658/62/

Equal Protection

The Equal Protection Clause of the Fourteenth Amendment is essentially what the name implies — a clause enforcing equal protection and neutral enforcement of the law. As the U.S. Supreme Court put it in 1942, "The guaranty of 'equal protection of the laws is a pledge of the protection of equal laws.'"[885] The Equal Protection Clause does not, on its own, guarantee any specific rights, but instead protects against disparate (i.e., discriminatory) treatment in the exercise of rights guaranteed elsewhere in the Constitution. When the government treats one individual or group (a "class") of individuals disparately compared to a similarly situated individual or group, then that disparate treatment is unconstitutional if it has no rational basis, targets a suspect class, or burdens a fundamental right. As you probably guessed based on that description, equal protection violations are only subject to rational basis scrutiny unless the disparate treatment targets a suspect class of persons or burdens a fundamental right. If the disparate treatment targets a suspect class or burdens a fundamental right, then heightened scrutiny is applied.[886] Suspect classes requiring heightened scrutiny include race, gender, religion, national origin, immigration status, and wedlock status, but not things like age, wealth, height, weight, occupation, and criminal history. (Some state-level laws include more categories of suspect classes.) Selective enforcement of facially neutral mandates could also give rise to Fourteenth Amendment claims under the Equal Protection Clause (a notorious mask-related example of this is the selective enforcement of mask mandates against parents protesting mask mandates for their children at school board meetings but not against George Floyd protestors).[887]

The Equal Protection Clause can be decisive in crucial cases involving fundamental rights. In 1942, for example, the Supreme Court applied strict scrutiny and unanimously held unconstitutional an Oklahoma law permitting punitive compulsory sterilization for certain repeat felonies including the

---

885   *Skinner v. Oklahoma*, 316 U.S. 535, 541 (1942), available online:
https://tile.loc.gov/storage-services/service/ll/usrep/usrep316/usrep316535/usrep316535.pdf

886   "Strict scrutiny applies to any state law or regulation that differently 'classifies by race, alienage, or national origin' or significantly burdens a class's exercise of a fundamental right, such as the right to interstate travel."
Branch-Noto v. Sisolak, 576 F. Supp. 3d 790, 804 (D. Nev. 2021), available online:
https://casetext.com/case/branch-noto-v-sisolak-1
2:21-cv-01507 JAD-DJA, p. 18 (D. Nev. Dec. 22, 2021) https://ag.nv.gov/uploadedFiles/agnvgov/Content/News/PR/PR_Docs/2021/DOC%2038%20Order%20Denying%20Motion%20for%20Preliminary%20Injunction,%20Granting%20in%20Part%20Motions%20to%20Dismiss%20and%20Closing%20Case.pdf

887   "Plaintiffs argue that by not enforcing the Order against protestors following the death of George Floyd, the Governor is favoring that speech over Plaintiffs' political speech." *Illinois Republican Party v. Pritzker*, 470 F. Supp. 3d 813, 820 (page 6), (Northern District of Illinois, 2020), available online: https://libertyjusticecenter.org/wp-content/uploads/2020/06/016-IL-Republican-Party-Opinion-denying-Motion-for-TRO-and-PI.pdf and
https://casetext.com/case/ill-republican-party-v-pritzker
The plaintiffs were correct in this allegation. Unfortunately, presiding Judge Sara Ellis was equally determined not to recognize this valid claim.

theft of more than $20 worth of property (which was defined in the Oklahoma law as "grand larceny"). The petitioner sentenced to sterilization in Oklahoma, Jack T. Skinner, had been convicted three times over a period of 9 years: once of stealing chickens and twice of committing robbery with the aid of a firearm. The Supreme Court held that Mr. Skinner's forced sterilization was an unconstitutional violation of the Equal Protection Clause of the Fourteenth Amendment because similarly harmful criminal offenses like embezzlement of the same amount were excepted from this particular penalty. The Court applied strict scrutiny because the penalty involved irreparable harm to the individual and because *not* being forcibly sterilized ("a right which is basic to the perpetuation of a race — the right to have offspring") was a fundamental right.[888]

Rather like a First Amendment neutrality test, forcing one group of people to wear masks but not another group of similarly situated people suggests the presence of a Fourteenth Amendment equal protection violation. However, the First Amendment neutrality test has the fundamental nature of the right built in, whereas the Fourteenth Amendment equal protection test does not. Whether an Equal Protection Clause claim is effective depends on how intellectually even-handed the judge is, the reason being cited for the unequal treatment, how blatantly unequal the treatment in question is, and (crucially) how self-evident or explicitly well-established in legislation or previous case law the suspect class or fundamental right involved are.

As it turned out, when lockdowns, compulsory masking, and other COVIDcrisis orders were challenged under the Fourteenth Amendment's Equal Protection Clause, most Judges had no difficulty denying that *any* fundamental right was implicated. Schools were among the last holdouts maintaining mask mandates, making an equal protection claim seem a natural route when students posing no real risk from COVID were not required to wear masks in any other place they congregated. As a summer 2021 lawsuit brought by parents in New Jersey trying to protect their children from being forcibly masked in school put it:

---

888   *Skinner v. Oklahoma*, 316 U.S. 535, 536 (1942), available online: https://tile.loc.gov/storage-services/service/ll/usrep/usrep316/usrep316535/usrep316535.pdf

   While its outcome was a helpful precedent for fundamental unenumerated individual liberties throughout America, because the *Skinner* case only involved sterilization as a *criminal* punishment, involuntary sterilization through *civil* actions (especially of the mentally disabled and mentally ill) was only somewhat curtailed rather than eliminated after this Supreme Court ruling.

   It is worth noting that in his concurrence, Chief Justice Stone argued that the Fourteenth Amendment's Due Process Clause was a more appropriate vehicle for the final verdict, because striking down the Oklahoma law based on equal protection implied that such an extreme invasion of personal liberty as sterilization could not be given the usual presumption of constitutionality even when passed by a legislature. However, I absolutely disagree with Justice Stone's repellant affirmation of the *Buck v. Bell* verdict, that "Undoubtedly a state may, after appropriate inquiry, constitutionally interfere with the personal liberty of the individual to prevent the transmission by inheritance of his socially injurious tendencies." Supreme Court decisions, like individual humans, are a mix of good and bad.

"Adults and children may sit in close packed rows at an indoor stadium, at a movie theater, in restaurants, in houses of worship, may dance in large groups at weddings or other gatherings, may mass at political meetings or protests, may sit close together at public lectures or at indoor theaters or playhouses, may mix freely in libraries and a host of other locations but schoolchildren are still subject to the full discipline of masking and other restrictive measures."[889]

But this lawsuit, as with multiple others filed by parents under the Equal Protection Clause, was capped at rational basis scrutiny and then dismissed. This was for two reasons. First, as Judge Kevin McNulty informed the New Jersey parents: "School children do not constitute a suspect class."[890] Second, the U.S. Supreme Court has not accepted education as a "fundamental right" which triggers heightened Fourteenth Amendment scrutiny. In the Supreme Court's own words: "Nor have we accepted the proposition that education is a 'fundamental right,' like equality of the franchise, which should trigger strict scrutiny when government interferes with an individual's access to it."[891] Thus, "Because the mask mandate does not target a suspect class or burden a fundamental right, it receives not strict scrutiny but rational-basis review,"[892] said Judge McNulty as he dismissed the parents' equal protection claim. For some reason, Judge McNulty also felt the need to remonstrate: "when faced by a common catastrophe like a pandemic, we must all make some sacrifices to protect ourselves and our more vulnerable neighbors."[893] Parents in Georgia fared no better: "Rational basis is the proper standard of review for the mask mandate. The mandate neither discriminates against a protected class nor infringes a fundamental right," said Judge Thomas Thrash in the Atlanta Division.[894] When two parents in Nevada brought an Equal Protection challenge against their children's school mask mandates because Governor Sisolak's Emergency Executive Directives exempted counties under 100,000 population, Judge Jennifer Dorsey dismissed the parents' claim on rational basis analysis because population density is not a suspect classification requiring heightened scrutiny unless it has a disparate impact on an actual suspect class.[895] After sticking in the knife, Judge Dorsey then

---

889   *Stepien v. Murphy*, Civ. 21-CV-13271 (KM) (JSA), Document 2: Initial Complaint, *83 p. 17 (D.N.J. July 9, 2021), https://8e81c69c-af5d-4ce4-a485-3addf6586c5d.filesusr.com/ugd/30ba77_51ca9e41938c45f4a19abf-324de5282f.pdf

890   *Stepien v. Murphy*, 574 F. Supp. 3d 229, 238 (D.N.J. 2021), https://casetext.com/case/stepien-v-murphy-1

891   *Kadrmas v. Dickinson Public Schools*, 487 U.S. 450, 458 (1988), available online: https://tile.loc.gov/storage-services/service/ll/usrep/usrep487/usrep487450/usrep487450.pdf

892   *Stepien v. Murphy*, 574 F. Supp. 3d 229, 238 (D.N.J. 2021), https://casetext.com/case/stepien-v-murphy-1

893   *Stepien v. Murphy*, 574 F. Supp. 3d 229, 233 (D.N.J. 2021), https://casetext.com/case/stepien-v-murphy-1

894   *W. S. v. Ragsdale*, 540 F. Supp. 3d 1215, 1218-1219 (N.D. Ga. 2021), https://casetext.com/case/ws-v-ragsdale#p1218

895   Judge Dorsey cited Supreme Court precedent on this particular issue: "The Fourteenth Amendment does not prohibit legislation merely because it is… limited in its application to a particular geographical or political subdivision of the state." (*Fort Smith Light and Traction Co. v. Board of Improvement of Paving District Number 16 of City of Fort*

twisted it by penalizing the parents' lack of legal expertise and awarding the school district $57,021 in attorney fees against the parents, saying "plaintiffs' claims were legally frivolous."[896] This award was particularly unjust, because the recommendations for school mask mandates which Judge Dorsey ostensibly subjected to rational basis scrutiny largely came from the CDC, and by then the CDC's Airline Mask Mandate had already been ruled "arbitrary and capricious."[897] I want to highlight that Judge Dorsey also cited Jehn et al. and Budzyn et al.'s contrived piece-of-trash CDC observational studies that we covered in the science section: characterizing the studies as "overwhelming, recent evidence that mask-wearing in public schools reduces the spread of COVID-19."[898]

Even when the schools involved were private religious schools, courts insisted on using only rational basis scrutiny to evaluate (and dismiss) equal protection claims against compulsory masking. Michigan Department of Health and Human Services Director Robert Gordon issued mask mandates in 2020 which favored multiple activities over religiously motivated conduct and education by allowing people to remove masks for identification, communicating with the deaf or hard of hearing, giving a speech, voting at a polling place, or "receiving a service for which temporary removal of the face covering is necessary."[899] Writing in an incisive and thorough dissent at the appellate level against mooting a legal challenge to this mandate, Judge John K. Bush highlighted that exempted activities under the mandate included: "dining at a restaurant, dining with friends at a private gathering, receiving a haircut, tattoo, or massage; sessions in a tanning booth; or the installation of a nose-ring."[900] Yet when a religious private school, Resurrection School, argued that Director Gordon's masking orders

---

Smith, Arkansas, 274 U.S. 387, 391 (1927)). However, it is worth noting that applying legislation solely to a particular geographical area *can* be an equal protection violation if it disproportionately impacts a protected class like race. *Branch-Noto v. Sisolak*, 576 F. Supp. 3d 790, 805 n 91 (D. Nev. 2021), available online: https://casetext.com/case/branch-noto-v-sisolak-1

2:21-cv-01507 JAD-DJA, page 19 note 91 (D. Nev. Dec. 22, 2021) https://ag.nv.gov/uploadedFiles/agnvgov/Content/News/PR/PR_Docs/2021/DOC%2038%20Order%20Denying%20Motion%20for%20Preliminary%20Injunction,%20Granting%20in%20Part%20Motions%20to%20Dismiss%20and%20Closing%20Case.pdf

896   *Branch-Noto v. Sisolak*, 2:21-cv-01507-JAD-DJA (D. Nev. Jul. 26, 2022), available online: https://casetext.com/case/branch-noto-v-sisolak-2

897   Judge Kathryn Kimball Mizelle's ruling against the CDC was handed down on April 18, 2022, while Judge Dorsey's award of attorneys' fees against the Nevada parents was put out more than three months later on July 26th. *Health Freedom Defense Fund v. Biden*, 599 F. Supp. 3d 1144 (M.D. Fla. 2022), available online: https://healthfreedomdefense.org/wp-content/uploads/2022/04/DE-53-ORDER-GRANTING-SJ-TO-PLAINTIFFS.pdf

898   *Branch-Noto v. Sisolak*, 576 F. Supp. 3d 790, 801 n 65-66 (D. Nev. 2021), available online: https://casetext.com/case/branch-noto-v-sisolak-1

2:21-cv-01507 JAD-DJA, page 13 note 65-66 (D. Nev. Dec. 22, 2021) https://ag.nv.gov/uploadedFiles/agnvgov/Content/News/PR/PR_Docs/2021/DOC%2038%20Order%20Denying%20Motion%20for%20Preliminary%20Injunction,%20Granting%20in%20Part%20Motions%20to%20Dismiss%20and%20Closing%20Case.pdf

899   *Resurrection School v. Hertel*, 35 F.4th 524, 527 (6th Cir. 2022), https://casetext.com/case/resurrection-sch-v-hertel-2

900   *Resurrection School v. Hertel*, 35 F.4th 524, 535 (6th Cir. 2022), https://casetext.com/case/resurrection-sch-v-hertel-2

"provide exceptions for other activity and conduct that is similar in its impact and effects, but not for Plaintiff's constitutionally protected activities,"[901] district Judge Paul Maloney coolly discounted the comparison by saying in so many words that the requirements of *equal* protection could somehow be met by the universal application of *un*equal exceptions:

> "Plaintiffs argue that Director Gordon's orders violate the Equal Protection clause because individuals may remove their face covering in some circumstances. Plaintiffs' argument here is conclusory and unpersuasive: There is nothing in the face-mask requirement that treats similarly situated groups of individuals differently. Every person over the age of five must wear a mask when outside the home, and the exceptions apply universally."[902]

While the Equal Protection Clause of the Fourteenth Amendment blunted authoritarian overreach in multiple areas during the COVIDcrisis, compulsory masking was not one of them. Equal protection claims that should have been viable, in practice, met with failure against compulsory masking and other overreaching measures (for example, restaurants that served alcohol were often given preferential treatment in lockdown orders over bars that served food, as though there was any meaningful difference in viral transmission risk[903]). I am not aware of any cases where a direct equal protection claim against compulsory masking was successful (though I will be delighted if I am later proved wrong about this). On the other hand, equal protection claims against compulsory masking often suffered from the same incomplete development and explication that other claims against masking did, and an especially well-pled equal protection claim[904] backed up by the right factual circumstances could still find success against compulsory masking in the future if put before an even-handed judge. As mentioned before, when it came to masks, getting to *any* sort of scrutiny was an achievement, and

---

901   *Resurrection School v. Gordon*, No. 1:20-CV-01016, Initial complaint, at 173, p. 28, filed 10/22/2020 (W.D. Mich. 2020) https://storage.courtlistener.com/recap/gov.uscourts.miwd.99403/gov.uscourts.miwd.99403.1.0.pdf other case documents can be accessed at: https://www.courtlistener.com/docket/18573245/resurrection-school-v-gordon/

902   *Resurrection School v. Gordon*, 507 F. Supp. 3d 897, 902 (W.D. Mich. 2020), https://casetext.com/case/resurrection-sch-v-gordon

903   One group of bar owners sued over this, and the result was as predictably depressing as many COVID-cases. Judge Martin Feldman found that the bar owners had brought a valid equal protection claim, but that the governmental infringements were justified ("but barely so") under the *Jacobson v. Massachusetts* standard. *4 Aces Enterprise v. Edwards*, 479 F. Supp. 3d 311, 327 (E.D. La. 2020), available online: https://casetext.com/case/4-aces-enters-llc-v-edwards

904   Though his case was ultimately successful on other grounds, lawyer Jeff Childers' equal protection claim from *Green v. Alachua County* is an example of one of those better-pleaded equal protection claims against compulsory masking which could have been expanded upon but should have succeeded as-is. *Green v. Alachua County*, Case No. 2020-CA-001249, Filing # 107300778, (Fla. 8th Cir. Ct., May 11, 2020), available online: https://www.fl-counties.com/sites/default/files/2020-07/5.11.20%20Green%20Amended%20Complaint_0.pdf

equal protection claims at least *did* routinely get to rational basis scrutiny. As we shall see later, mask mandates failed even rational basis-type scrutiny when finally brought before an open-minded judge.

Due Process

> "And as Dershowitz agreed: it can't just be public health…: 'Oh, you might be able to give COVID to someone.' Well, that's never been our standard for forcing anything on anybody, for doing any quarantine. It has to be clear and convincing evidence that unless you take the vaccine, you pose a substantial risk of severe harm, and again Dershowitz — big pro-vaccine guy — ultimately conceded that's the appropriate legal standard in this context."
>
> — Attorney Robert Barnes describing his public vaccine and quarantine debate with attorney Alan Dershowitz[905]

Absent the involvement of other fundamental individual rights like freedom of speech or exercise of religion, the general right to wear or not wear a mask is housed in the Due Process Clause of the Fourteenth Amendment, which guarantees two distinct but intertwined types of due process: procedural due process and substantive due process. The concept of "due process" recognizes that final outcome is not the only thing that matters — the process by which outcomes are achieved matters every bit as much.

Procedural due process concerns the steps and procedures that the government has to go through before it can lawfully deprive a citizen of life, liberty, or property. In a nutshell, according to the U.S. Supreme Court, "Procedural due process imposes constraints on governmental decisions which deprive individuals of 'liberty' or 'property' interests within the meaning of the Due Process Clause of the Fifth or Fourteenth Amendment."[906] The classic example of procedural due process is the trial by jury that is protected by the Sixth and Seventh Amendments, but administrative processes and procedures have also been ruled to satisfy this requirement, depending on the situation and interest at stake.

---

905   Robert Barnes, "Ep. 144: Biden-Gate; SBF Bail; Elon Musk Victory; 2020 Election Stuff & MORE! Viva and Barnes LIVE!" *vivafrei* Rumble channel, Timestamp 1:20:45-1:21:36 (streamed on January 15, 2023 6:00 p.m. EST) Available Online: https://rumble.com/v25p47q-ep.-144-biden-gate-sbf-bail-elon-musk-victory-2020-election-stuff-and-more-.html (last accessed January 18, 2023)
Original Debate: "Dersh vs. Barnes — Forced Vaccinations; Mask Mandates; HIPAA Violations," Viva Frei YouTube channel, (streamed live January 12, 2021), available online: https://www.youtube.com/watch?v=_30V5x0SD9g (last accessed June 27, 2023)
906   *Mathews v. Eldridge*, 424 U.S 319, 332, available online:
https://tile.loc.gov/storage-services/service/ll/usrep/usrep424/usrep424319/usrep424319.pdf

At a minimum, procedural due process includes notice, an opportunity to be heard, and impartial adjudication prior to any infringement or deprivation of liberty. Also, the hearing and adjudication must be meaningful. If the outcome is predetermined and the hearing is a mere rubber stamp formality, the requirements of due process are not satisfied.[907] Some Supreme Court rulings have implied that procedural due process also includes a general right to be free from official stigmatization.[908]

Substantive due process protects both the enumerated rights incorporated against the states and those unenumerated fundamental rights which the Ninth Amendment refers to, including the general right to liberty. In 1954, the U.S. Supreme Court stated, "Although the Court has not assumed to define 'liberty' with any great precision, that term is not confined to mere freedom from bodily restraint. Liberty under law extends to the full range of conduct which the individual is free to pursue, and it cannot be restricted except for a proper governmental objective."[909] This includes common law rights and privileges. Substantive due process rights default to rational basis scrutiny, but the more fundamental the right and the greater the infringement on it, the more scrutiny substantive due process requires, as when infringements on the right to marital privacy and the right to not be sterilized required strict scrutiny.

What *legally* counts as a fundamental right for purposes of due process, however, is much narrower than the popular concept of what counts as fundamental. For example, under the Fourteenth Amendment, police with the motto "to protect and serve" have no duty to protect.[910] As one court put it, "To succeed on a due process claim, whether procedural or substantive, a plaintiff must demonstrate that he or she possessed 'a constitutionally cognizable life, liberty, or property interest.'"[911] During the COVIDcrisis, courts gave the least-effective interpretations possible to claims that were trying to articulate the negative right to not have external artificial burdens unilaterally imposed on one's breathing. According to COVIDcrisis judges presented with challenges to compulsory masking, "The

---

907   *McCarthy v. Madigan*, 503 U.S. 140, 148 (1992), available online:
https://tile.loc.gov/storage-services/service/ll/usrep/usrep503/usrep503140/usrep503140.pdf

908   *See generally*: *Goss v. Lopez*, 419 U.S. 565 (1975), available online:
https://tile.loc.gov/storage-services/service/ll/usrep/usrep419/usrep419565/usrep419565.pdf and
Board of Regents v. Roth, 408 U.S. 564 (1972), available online:
https://tile.loc.gov/storage-services/service/ll/usrep/usrep408/usrep408564/usrep408564.pdf

909   *Bolling v. Sharpe*, 347 U.S. 497, 499-500 (1954), available online:
https://tile.loc.gov/storage-services/service/ll/usrep/usrep347/usrep347497/usrep347497.pdf

910   *Castle Rock v. Gonzales*, 545 U.S. 748 (2005), available online:
https://tile.loc.gov/storage-services/service/ll/usrep/usrep545/usrep545748/usrep545748.pdf

911   *Stewart v. Justice* 502 F. Supp. 3d 1057, 1067 (S.D.W. Va. 2020), available online:
https://casetext.com/case/stewart-v-justice-1 quoting *Sansotta v. Town of Nags Head*,
724 F.3d 533, 540 (4th Cir. 2013), available online: https://casetext.com/case/sansotta-v-town-of-nags-head-3

'right to breathe oxygen without restriction' is not a fundamental right."[912] One judge dismissed a *pro se* challenge to compulsory masking by saying: "Certainly, being denied air may be actionable, but being denied the 'cleanest air readily available' does not implicate a fundamental right.[913] One would think this kind of reasoning would also rule out forcing masks on people in the name of "source control," but that sort of consistency was too much to expect. The key for a Fourteenth Amendment substantive due process claim to receive anything more than rational basis scrutiny requires that the rights asserted be "deeply rooted in this Nation's history and tradition," and "implicit in the concept of ordered liberty."[914]

Substantive Due process says that even if you are demonstrably mentally or physically ill, the government cannot simply deprive you of liberty by involuntary institutionalization (also known as "involuntary commitment") or forcibly treat you against your will. Procedural due process says that even in those cases where involuntary institutionalization or compulsory treatment can *possibly* be justified, the state still has a burden of proof that it must meet before it can exercise its police power in the name of public health. In general, there are three levels of proof that procedural due process may require, depending on the infringement, and all of them require far more evidence than the "rational basis" standard used to uphold mask mandates. Involuntary commitment for psychiatric treatment, for example, is such an extreme deprivation of multiple fundamental liberties, and so close to imprisonment, that the state must show by a "clear and convincing" standard of evidence[915] in civil judicial proceedings that the mentally ill person is enough of a substantial, imminent danger to themselves or others to justify the intervention of state police power. Especially important in a public heath context, in 1980 the Supreme Court said that even in the case of convicted felons, the "medical nature of the inquiry ... does not justify dispensing with due process requirements."[916]

---

912   *Denis v. Ige et al*, No. 1:2021cv00011 - Document 62 (D. Haw. 2021), pg. 34, available online: https://law.justia.com/cases/federal/district-courts/hawaii/hidce/1:2021cv00011/152658/62/

913   *Joseph v. Becerra*, 22-cv-40-wmc, Document 21, p. 11 (W.D. Wis. Nov. 29, 2022), available online: https://law.justia.com/cases/federal/district-courts/wisconsin/wiwdc/3:2022cv00040/48728/21/

914   *Washington v. Glucksberg*, 521 U.S. 702, 721 (1997), available online: https://tile.loc.gov/storage-services/service/ll/usrep/usrep521/usrep521702/usrep521702.pdf quoting *Moore v. City of East Cleveland*, 431 U.S. 494, 503 (1977) (https://tile.loc.gov/storage-services/service/ll/usrep/usrep431/usrep431494/usrep431494.pdf) and *Palko v. Connecticut*, 302 U.S. 319, 325 (1937) (https://tile.loc.gov/storage-services/service/ll/usrep/usrep302/usrep302319/usrep302319.pdf)

915   *Addington v. Texas*, 441 U.S. 418 (1979), available online: https://tile.loc.gov/storage-services/service/ll/usrep/usrep441/usrep441418/usrep441418.pdf *O'Connor v. Donaldson*, 422 U.S. 563, 575 (1975), available online: https://tile.loc.gov/storage-services/service/ll/usrep/usrep422/usrep422563/usrep422563.pdf

916   *Vitek v. Jones*, 445 U.S. 480, 495 (1980), available online: https://tile.loc.gov/storage-services/service/ll/usrep/usrep445/usrep445480/usrep445480.pdf

"Clear and convincing" evidence is one tier down from the "beyond a reasonable doubt" standard procedural due process requires for criminal conviction. It is a tier above the simple "preponderance of the evidence" (i.e., 51% or better) standard for most civil proceedings, and a "preponderance of the evidence" itself is *far* above a mere "rational basis." Compulsory masking violated both substantive and procedural due process in different ways, and it was one of the most pernicious of many such infringements during the COVIDcrisis.[917]

The right to privacy; the right to be let alone

The Fourteenth Amendment also covers those facets of the right to privacy not protected elsewhere in the Bill of Rights. There is a longstanding common law right to be free from arbitrary government action — to be let alone. This foundational right undergirds the right of privacy in the same way that the foundational right of self-defense undergirds the Second Amendment. In his joint 1890 essay with U.S. Supreme Court Justice Samuel Warren, "The Right to Privacy," Supreme Court Justice Louis Brandeis elaborated on the deeply fundamental nature of this right to be let alone:

> "[T]he protection afforded to thoughts, sentiments, and emotions, expressed
> through the medium of writing or of the arts, so far as it consists in preventing
> publication, is merely an instance of the enforcement of the more general right of
> the individual to be let alone. It is like the right not to be assaulted or beaten, the
> right not to be imprisoned, the right not to be maliciously prosecuted, the right
> not to be defamed[.]"[918]

---

917   Another example of COVIDcrisis infringements on due process is how badly the liberty "to engage in any of the common occupations of life" (*Meyer v. Nebraska,* 262 U.S. 390, 399 (1923)) protected by the Fourteenth Amendment was especially disregarded and brutalized, while at the same time, the definition of "reasonable regulations" was repeatedly stretched past the breaking point by subjecting businesses and customers to all manner of arbitrary and ritualistic guidelines that in normal times would have been red flags for pathological anxiety and obsessive-compulsive disorder. A prime example of the sort of Pharisaic hairsplitting that courts have engaged in which undermined fundamental Fourteenth Amendment rights is summarized in Judge Martin Feldman's ruling in *4 Aces Enterprise v. Edwards.* Courts have rejected "a fundamental 'right to do business'" while at the same time ruling that the property interests in the *profits* of a business and the liberty interest in *operating* that business *do* "rise to the level of protectible interests." Notably, this rejection of a fundamental right to do business in one of "the common occupations of life" (which the Supreme Court has said the Fourteenth Amendment specifically protects) was based on cases which, themselves, ignored multiple other, previous cases that *affirmed* this precise right. In the end, even though Judge Feldman ultimately conceded that the plaintiffs had a valid substantive due process claim under the Fourteenth Amendment to challenge COVID mandates strangling their businesses, he then decided against the plaintiffs based on *Jacobson v. Massachusetts.*
4 Aces Enterprise v. Edwards, 479 F. Supp. 3d 311, 326 (E.D. La. 2020), available online: https://casetext.com/case/4-aces-enters-llc-v-edwards
918   Samuel D. Warren and Louis D. Brandeis, "The Right to Privacy," December 15, 1890, *Harvard Law Review* Vol. IV, No. 5 (205) available online: https://www.jstor.org/stable/1321160

According to Justice Brandeis and Justice Warren, this right to privacy — to be let alone — falls under "the more general right to the immunity of the person,—the right to one's personality."[919] Keep in mind that in this context, "personality" does not simply refer to a person's character traits, mannerisms, and other characteristics, but also includes a person's physical, mental, and social qualities. The right to one's personality is part of "The right of property in its widest sense, including all possession, including all rights and privileges[.]"[920] One of the conclusions that Justice Warren and Justice Brandeis drew was that: "the protection of society must come mainly through a recognition of the rights of the individual. Each man is responsible for his own acts and omissions only."[921]

Many Supreme Court rulings do not make sense in the absence of a broad and robust common law right to be let alone, and some reference this right directly. Just a year after Justice Warren and Justice Brandeis wrote their essay, the Supreme Court best summarized this fundamental common law right in an 1891 ruling against a mandated medical examination, which the Supreme Court quoted to reach a decision in 1990:[922]

> "No right is held more sacred or is more carefully guarded by the common law than the right of every individual to the possession and control of his own person, free from all restraint or interference of others unless by clear and unquestionable authority of law. As well said by Judge Cooley: 'The right to one's person may be said to be a right of complete immunity; to be let alone'" (emphasis added).[923]

In a 1928 case, Justice Brandeis wrote a dissenting opinion which referred to: "as against the Government, the right to be let alone — the most comprehensive of rights and the right most

---

Justices Brandeis and Warren next consider and discard the possible objection that involuntary expressions do not warrant the same protections as deliberate expressions: "If the test of deliberateness of the act be adopted, much casual correspondence which is now accorded full protection would be excluded[.]"

919   Samuel D. Warren and Louis D. Brandeis, "The Right to Privacy," December 15, 1890, *Harvard Law Review* Vol. IV, No. 5 (207) available online: https://www.jstor.org/stable/1321160

920   Samuel D. Warren and Louis D. Brandeis, "The Right to Privacy," December 15, 1890, *Harvard Law Review* Vol. IV, No. 5 (209) available online: https://www.jstor.org/stable/1321160

921   Samuel D. Warren and Louis D. Brandeis, "The Right to Privacy," December 15, 1890, *Harvard Law Review* Vol. IV, No. 5 (220) available online: https://www.jstor.org/stable/1321160

922   *Cruzan v. Director, Missouri Department of Health*, 497 U.S. 261, 269 (1990), available online: https://tile.loc.gov/storage-services/service/ll/usrep/usrep497/usrep497261/usrep497261.pdf

923   Union Pacific Railway Co. v. Botsford, 141 U.S. 250, 251 (1891) available online: https://tile.loc.gov/storage-services/service/ll/usrep/usrep141/usrep141250/usrep141250.pdf

As a further side note, this 1891 decision is one of the many United States Supreme Court rulings that quotes the work of English jurist William Blackstone.

valued by civilized men."[924] In his opinion, Justice Brandeis intertwined the right to be let alone with the right to privacy, because privacy is an indispensable component of the right to be let alone. In an important 1969 decision concerning both privacy and the First Amendment right to receive speech, the Supreme Court majority quoted Justice Brandeis' 1928 opinion *at length* after declaring: "also fundamental is the right to be free, except in very limited circumstances, from unwanted governmental intrusions into one's privacy."[925] That same year, in *Tinker v. Des Moines Independent Community School District*, which we discussed previously in the First Amendment symbolic speech context, the Supreme Court explicitly pointed out that, whatever the emotional impact produced by their message, some students wearing conspicuous armbands for the purpose of symbolic speech about the controversial Vietnam War produced no "interference, actual or nascent, with the schools' work or of collision with the rights of other students to be secure and to be let alone."[926] In this same decision, the Supreme Court said: "Our whole constitutional heritage rebels at the thought of giving government the power to control men's minds."[927]

The right to either wear or refuse to wear a mask is part of an individual's right to privacy — *to be let alone*. Forcing citizens to cover their faces with cloth or some other material is a major infringement of the common law right to be let alone, and we have seen in our review of the science and psychology of masking how mask mandates insidiously grasped at the power to shift wearers' beliefs in a particular direction. The COVIDcrisis mindset tended to disparage the right to be let alone and the right to privacy as being "trivial" or somehow weaker or less absolute than explicit rights like the right to free speech, but this perspective was false and completely ignored a major point that objectors to compulsory masking were making. Both the right to be let alone and the right to privacy have objective existence; both are broad and robust; both are deeply rooted in America's heritage and are founded upon historical notions and federal constitutional expressions of ordered liberty; both are recognized under the Nineth and Fourteenth Amendments of the United States Constitution (and protected under the Fourteenth); both of them entail the broad right to *not* wear a mask; and both were flagrantly and repeatedly violated by mask mandates.

Despite multiple courts' refusal to acknowledge or remedy these widespread violations of fundamental individual liberties, the real burden of proof was *always* on the authorities trying to mandate masks,

---

924   *Olmstead v. United States*, 277 U.S. 438, 478, 48 S. Ct. 564, 572, 72 L. Ed. 944 (1928) (Brandeis, J., dissenting), https://tile.loc.gov/storage-services/service/ll/usrep/usrep277/usrep277438/usrep277438.pdf

925   *Stanley v. Georgia*, 394 U.S. 557, 564 (1969), available online:
https://tile.loc.gov/storage-services/service/ll/usrep/usrep394/usrep394557/usrep394557.pdf

926   *Tinker v. Des Moines Independent Community School District*, 393 U.S. 503, 508 (1969), available online:
https://www.loc.gov/item/usrep393503/

927   *Stanley v. Georgia*, 394 U.S. 557, 565 (1969), available online:
https://tile.loc.gov/storage-services/service/ll/usrep/usrep394/usrep394557/usrep394557.pdf

and those authorities *never* truly met this burden. Compulsory masking is unquestionably a form of bodily restraint, and has been used in various forms as a dehumanizing punishment for hundreds, if not thousands, of years. No clear and present danger of destruction of life or property can be reasonably thought to be inherent in the activities of every person who goes about unmasked, even when the Black Death or Spanish Flu happen to be passing through a community. Substantive due process forbade compulsory universal masking. Procedural due process required, at a *minimum*, notice and an opportunity to respond and receive a meaningful hearing with impartial adjudication requiring "clear and convincing" evidence that every individual being forced to wear a mask was a legitimate, imminent, and substantial clear and present danger — a standard of proof which no one pushing universal compulsory masking ever *has* or ever *will* be able to truly meet.

Some courts have ruled that "when the State must act quickly or predeprivation process is impractical, and meaningful post-deprivation process is available, then due process is still satisfied."[928] But in the case of compulsory masking, no meaningful post-deprivation process was *ever* available. People across the globe were forced to wear masks with zero pre-deprivation due process — no hearing, no real discussion, and the only arbitration was by decisionmakers infected with political motivation and mass panic. During the COVIDcrisis, citizens were forced to sue for their most basic fundamental rights in hostile courts that were using all their judicial ingenuity to dismiss or rule against the claims being brought. And, as with violations of other individual liberties during the COVIDcrisis, multiple courts were just as unwilling to recognize violations of individual due process rights. Those who sought meaningful post-deprivation due process were far more likely to be reprimanded and penalized for "selfishly" making the attempt in the first place.

Green v. Alachua County

The U.S. Supreme Court explicitly referenced the connected rights to privacy and to be let alone in 1967: "the protection of a person's general right to privacy — his right to be let alone by other people — is, like the protection of his property and of his very life, left largely to the law of the individual States."[929] This reference to the right to be let alone being largely left to the individual States is important. The first major court win against compulsory masking in the United States came in state, rather than federal, court. This critical breakthrough lawsuit against masks was brought in Florida by attorney Jeff Childers on behalf of his client Justin Green in *Green v. Alachua County*.[930]

---

928   *Bowlby v. City of Aberdeen*, 681 F.3d at 200 n 1 (5th Cir. 2012), available online: https://www.casemine.com/judgement/us/5914f286add7b0493497f5df

929   *Katz v. United States*, 389 U.S. 347, 350-51, 88 S. Ct. 507, 511, 19 L. Ed. 2d 576 (1967), https://tile.loc.gov/storage-services/service/ll/usrep/usrep389/usrep389347/usrep389347.pdf

930   *Green v. Alachua County*, 323 So. 3d 246 (Fla. Dist. Ct. App. 2021). Available online: https://law.justia.com/cases/florida/first-district-court-of-appeal/2021/20-1661.html

In 1980 the citizens of the State of Florida made a move that ultimately benefited everyone else in the United States, when they explicitly enshrined the common law right to privacy and right to be let alone into Article 1, Section 23 of their state constitution: "Every natural person has the right to be let alone and free from governmental intrusion into the person's private life except as otherwise provided herein."[931] Still, when he took Justin Green as a client to challenge compulsory masking, Floridian lawyer Seldon Jeffrey ("Jeff") Childers knew he was facing an uphill fight, especially in an area as self-brainwashed when it came to masking as Alachua County. An experienced litigator with a flair for witty, consistently optimistic commentary,[932] Childers focused with laser-like precision on getting the Alachua County mask mandate to the strict scrutiny he knew it warranted and could not survive. He anticipated the usual adverse results at the district court level and focused his efforts on setting up the inevitable appeal. Childers would later say of his appellant brief for this case: "I worked harder on that brief than just about anything in my career."[933]

Though he also put forward solid claims based on the county's lack of administrative authority to issue mask mandates, the takings clause of the Fifth Amendment, equal protection under the Fourteenth Amendment, and the due process clause of the Fourteenth Amendment, Childers based his primary challenge in the right to privacy protected by the Fourteenth Amendment and Florida Constitution.[934] As anticipated, despite the explicit wording of Florida's Constitution, Childers' lawsuit met with the usual predetermined denial from Circuit Judge Donna Keim who not only completely denied that masks had anything to do with speech and personal medical decisions, but said "there is no broad legal or constitutional 'right to be let alone' by government."[935] "Not so," responded Childers in his appellate brief. "The Florida Constitution does provide for a 'right to be let alone... and, the constitutional right of privacy *is* broad... Our Supreme Court said that the right to be let alone from the government is not only broad but also strong, as strong as it can possibly be."[936] Childers directly

---

931   Florida State Constitution, Article 1, Section 23, Available online: https://www.flsenate.gov/Laws/Constitution

932   As evinced by the prolific entries in his blog, Coffee and Covid: https://www.coffeeandcovid.com/

933   Jeff Childers, "SWITCHEROO," *Coffee and Covid* (Blog), March 4, 2023, online: https://www.coffeeandcovid.com/p/switcheroo-saturday-march-4-2023

Note: All the appellant briefs leading up to the oral arguments can be downloaded directly from the Florida First District Court of Appeal website: https://1dca.flcourts.gov/Oral-Arguments/Briefs-for-Appeals-Scheduled-for-Oral-Argument/20-1661

934   *Green v. Alachua County*, Case No. 2020-CA-001249, Filing # 107300778, (Fla. 8th Cir. Ct., May 11, 2020), available online: https://www.fl-counties.com/sites/default/files/2020-07/5.11.20%20Green%20Amended%20Complaint_0.pdf

935   *Green v. Alachua County*, Case No. 2020-CA-001249, Filing # 107958230, page 3 (Fla. 8th Cir. Ct., May 26, 2020), available online: https://alachuacounty.us/Depts/Communications/Documents/ADACompliant/OrderDenyingMotionforPreliminaryInjuction-Green-v-AlachuaCounty.pdf

936   *Green v. Alachua County*, 323 So. 3d 246, Appellant's Initial Brief, Filing #109556371, pages 24-25, E-Filed 06/29/2020 (Fla. Dist. Ct. of App., 2021), go to: https://1dca.flcourts.gov/Oral-Arguments/Briefs-and-Petitions-for-Cases-Scheduled-for-Oral-Argument/20-1661 and search for "case 20-1661"

quoted the Florida Supreme Court: "[T]he drafters of the privacy amendment rejected the use of the words 'unreasonable' or 'unwarranted' before the phrase 'governmental intrusion' in order to make the privacy right as strong as possible."[937] He then quoted the Florida Supreme Court again when it specifically affirmed: "The concept of privacy or right to be let alone is deeply rooted in our heritage and is founded upon historical notions and federal constitutional expressions of ordered liberty."[938]

Jeff Childers additionally dealt with another frequent objection that was raised by the Trial Court — the argument that privacy rights do not apply in public locations where masks were being required. "The Trial Court stated — without qualification — that: 'There is no recognized right to a reasonable expectation to privacy in a public location[.]'"[939] It is true that some limited aspects of privacy rights do not apply in public settings (for example, one generally cannot forbid having one's picture taken when out in public except in areas where there is a reasonable expectation of privacy such as a changing room). However, as Childers pointed out, even under the U.S. Constitution's Fourteenth Amendment right to privacy — which is narrower than Florida's — the U.S. Supreme Court has held that, as when someone has a private conversation in a public telephone booth: "what he seeks to preserve as private, even in an area accessible to the public, may be constitutionally protected."[940] Americans do not lose all their privacy rights under either the U.S. Constitution or their state constitutions in public locations — especially not in Florida. Childers highlighted the fact that the Florida Supreme Court had already found that a right to privacy exists in public locations, and overturned unconstitutional laws accordingly. The Florida Supreme Court had previously found that even juveniles had fundamental rights to privacy which were unconstitutionally violated by curfew laws.[941] Jeff Childers also pointed out that the Florida Supreme Court had already ruled a statute *against* mask wearing was unconstitutionally overbroad, and thus a violation of due process.[942]

Jeff Childers' appellate brief in *Green v. Alachua County* is essential reading for anyone seeking to challenge compulsory masking.

937   *Gainesville Woman Care, LLC v. State*, 210 So. 3d 1243, 1252 (Fla. 2017), available online: https://clearinghouse.net/doc/134902/

938   *Winfield v. Div. of Pari-Mutuel Wagering, Dept. of Bus. Regulation*, 477 So. 2d 544, *546 (Fla. 1985), https://law.justia.com/cases/florida/supreme-court/1985/64793-0.html

939   *Green v. Alachua County*, 323 So. 3d 246, Appellant's Initial Brief, Filing #109556371, page 25, E-Filed 06/29/2020 (Fla. Dist. Ct. of App., 2021), go to: https://1dca.flcourts.gov/Oral-Arguments/Briefs-and-Petitions-for-Cases-Scheduled-for-Oral-Argument/20-1661 and search for "case 20-1661"

940   *Katz v. United States*, 389 U.S. 347, 350-51 (1967), https://supreme.justia.com/cases/federal/us/389/347/

941   Childers references *Warden v. Hayden*, 387 U.S. 294 (1967) (https://supreme.justia.com/cases/federal/us/387/294/), where the Supreme Court observed that "the principal object of the Fourth Amendment is the protection of privacy, rather than property."

942   *Green v. Alachua County*, 323 So. 3d 246, Appellant's Initial Brief, Filing #109556371, page 24-25, E-Filed 06/29/2020 (Fla. Dist. Ct. of App., 2021), go to: https://1dca.flcourts.gov/Oral-Arguments/Briefs-and-Petitions-for-Cases-Scheduled-for-Oral-Argument/20-1661 and search for "case 20-1661"

To their lasting credit, Judges Adam Tannenbaum and Robert Long of the Florida First District Court of Appeal upheld the fundamental individual rights involved: "The right to be let alone by government does exist in Florida, as part of a right of privacy that our supreme court has declared to be fundamental… the supreme court has construed this fundamental right to be so broad as to include the complete freedom of a person to control his own body."[943] Judges Tannenbaum and Long, the panel majority, explained:

> "[T]he supreme court has explained repeatedly that within the right to be let alone is 'a fundamental right to the sole control of his or her person.'…. This right ostensibly covers 'an individual's control over or the autonomy of the intimacies of personal identity' and a 'physical and psychological zone within which an individual has the right to be free from intrusion or coercion, whether by government or by society at large'…

> "As defined by the supreme court, article I, section 23's guarantee of bodily and personal inviolability—which we are asked to follow—must include the inviolability of something so intimate as one's own face. A person then reasonably can expect to be free from governmental coercion regarding what he puts on it… ('Under a free government, at least, the free citizen's first and greatest right, which underlies all others [is] the right to the inviolability of his person; in other words, the right to himself[.]')"[944]

The third panel Judge, Joseph Lewis, dissented, objecting that "the majority does not explain how requiring a person to wear a mask when interacting with the public during a pandemic is the equivalent of controlling the person's body."[945] Judge Lewis' objection makes some sense if controlling someone else's body was limited to direct physical control, but such a narrow definition would only cover the worst and most outrageous forms of controlling someone else's body. The majority may not have articulated this point in detail because of how basic it was. Controlling a person's body is not limited to using another person as a literal puppet or physically restraining them. When using another

---

Citing *Robinson v. State*, 393 So. 2d 1076 (Fla. 1980), available online:
https://law.justia.com/cases/florida/supreme-court/1980/58232-0.html

943   *Green v. Alachua County*, 323 So. 3d 246, 251 (Fla. Dist. Ct. App. 2021). Available online:
https://law.justia.com/cases/florida/first-district-court-of-appeal/2021/20-1661.html

944   *Green v. Alachua County*, 323 So. 3d 246, 253 (Fla. Dist. Ct. App. 2021). Available online:
https://law.justia.com/cases/florida/first-district-court-of-appeal/2021/20-1661.html

945   *Green v. Alachua County*, 323 So. 3d 246, 264 (Fla. Dist. Ct. App. 2021). Available online:
https://law.justia.com/cases/florida/first-district-court-of-appeal/2021/20-1661.html

person's body is unilateral or involuntary, as in involuntary servitude, a military draft, compulsory flag saluting, compulsory sterilization, compulsory blood or organ donation, or forcing someone else to make use of a medical intervention, compulsion and coercion become a manner of controlling the person's body, regardless of how delicately or subtly the control is exercised. It is one thing to *prevent* someone from taking a particular action, but forcing them to *perform* an affirmative action goes far beyond that. In this sense, forcing someone to perform an affirmative act, especially when it is done involuntarily for another's indirect benefit, as in the rationale for mandating masks, is a form of controlling the person's body. Refusing to wear a mask is a non-action that does nothing to control anyone else's body, but forcing someone to put on a mask against their will indisputably constitutes a manner of controlling their body, and a gross invasion of the "physical and psychological zone within which an individual has the right to be free from intrusion or coercion."

In June 2021, more than a year after Jeff Childers filed the initial complaint, the Court of Appeal ruled: "the single question that the trial court must answer is the likelihood that the mask mandate would survive strict scrutiny."[946] The Alachua County mask mandate had, of course, *zero* chance of surviving strict scrutiny and was immediately struck down. *Green v. Alachua County* was a critical turning point against compulsory masking, and more successes were on the horizon.

Masks are medical interventions, using medical devices, involving medical choices

One thing that everyone agrees on is that masks are medical devices.[947] The FDA's facemask Emergency Use Authorization letter on mask use by the general public as "source control" also included cloth masks as a subset.[948] The line between medical treatments and medical devices gets pretty blurry at times, especially in the case of medical devices like removable prostheses that are *also* a form of treatment (think dentures and prosthetic limbs, eyes, ears, and noses, etc). Many, if not most, medical treatments *use* medical devices, and most medical devices are used during medical treatments. Common descriptions of different medical devices refer to them as "treatments." For example, the Mayo Clinic describes a ventricular assist device as "a device that helps pump blood from the lower chambers of the heart to the rest of the body. It's a treatment for a weakened heart or heart failure."[949] The similarities

---

946   *Green v. Alachua County*, 323 So. 3d 246, 254 (Fla. Dist. Ct. App. 2021). Available online: https://law.justia.com/cases/florida/first-district-court-of-appeal/2021/20-1661.html

947   Medical masks/Surgical masks are regulated under 21 CFR 878.4040 as Class II medical devices (https://www.ecfr.gov/current/title-21/chapter-I/subchapter-H/part-878/subpart-E/section-878.4040) *See also*, ASTM2100-23.

948   FDA Chief Scientist Denise M. Hinton, Emergency Use Authorization Letter, April 24, 2020, U.S. Food and Drug Administration, Available online: https://www.fda.gov/media/137121/download

949   Mayo Clinic, "Patient Care & Health Information, Tests and Procedures: Ventricular assist device (VAD)," online https://www.mayoclinic.org/tests-procedures/ventricular-assist-device/about/pac-20384529 (last accessed May 4, 2023).

between masks and prostheses or other removable treatment-devices such as ventricular assist devices, insulin pumps, or feeding tubes, naturally led some plaintiffs and their attorneys to make the case that masks also fit into this category — that forcing someone to wear a medical device like a mask is a form of compulsory medical treatment. We have also seen in the psychology section of this book how compulsory masking functioned as a form of behavior modification, and behavior modification *is* a form of therapy — i.e., medical treatment.

Judges, however, rejected the argument of masks as medical treatment because the purpose of masking was to *prevent* transmission rather than to actually *treat* any disease or disability. When faced with any claims against compulsory masking, especially Fourteenth Amendment claims grounded in due process or the general right to liberty, the standard pro-mask response was to totally ignore every non-physical aspect of mask-wearing (and many of the physical aspects), in order to conclude that compulsory masking is a minor, permissible infringement because masks are not compulsory medical *treatment*[950] (as though that fact somehow made forcing them onto people's faces more defensible). This reasoning underlay rulings like that of California Judge Cynthia Bashant:

> "The requirement that an individual wear a mask in public within six feet of persons from other households during a pandemic is a far cry from compulsory vaccination, mandatory behavior modification treatment in a mental hospital, and other comparable intrusions into personal autonomy."[951]

The quality of judicial reasoning went downhill still further after that. One appellate court asserted: "Requiring facial coverings in public settings is akin to the State's prohibiting individuals from smoking in enclosed indoor workplaces."[952] It is true that a preventative, protective intent underlay both mask mandates and smoking bans, but the analogy was superficial at best, and this sort of reasoning was typical. Because someone somewhere in a community had a particular virus, everyone's normal breathing suddenly became comparable to smoking; a process that everyone must perform every few seconds to avoid dying could be regulated like an unhealthy recreation that many people go their entire lives without partaking in; and broadly *mandating* an activity somehow became equivalent to *prohibiting* one.

---

950   *Forbes v. County of San Diego*, Case No. 20-cv-00998-BAS-JLB, at *13-14 (S.D. Cal. Mar. 4, 2021), available online: https://casetext.com/case/forbes-v-cnty-of-san-diego
Machovec v. Palm Beach County, 310 So. 3d 941, 945-946 (Fla. Dist. Ct. App. 2021),
https://casetext.com/case/machovec-v-palm-beach-cnty

951   *Forbes v. County of San Diego*, Case No. 20-cv-00998-BAS-JLB, at *13 (S.D. Cal. Mar. 4, 2021), available online: https://casetext.com/case/forbes-v-cnty-of-san-diego

952   *Machovec v. Palm Beach County*, 310 So. 3d 941, 946 (Fla. Dist. Ct. App. 2021), available online: https://casetext.com/case/machovec-v-palm-beach-cnty

The argument that compulsory masking is somehow justifiable because masks are not medical treatment, however, is a total red herring. Just because a violation of individual liberty is not as egregious or outrageous as it can possibly be does not make it acceptable. Pointing out that compulsory masking is not medical treatment *per se* does nothing to prove that it is acceptable under the Fourteenth Amendment guarantee of due process or any other amendment. Making use of any medical device is a medical *choice*. Receiving medical treatments and using medical devices are *both* medical *choices*. Masks are medical devices, and thus compulsory masking constitutes a coerced medical *intervention*, and this is what arguments that compulsory masking is a form of medical treatment were getting at.

The unalienable individual right — the moral and legal *requirement* for informed consent in medical treatment — is just *one* manifestation of the generalized "right of every individual to the possession and control of his own person." As we have seen, protective Due Process Clause precedent goes back to Supreme Court decisions ruling against compulsory medical examinations held in private,[953] which are far less invasive than forcibly covering another person's nose and mouth with *anything* in public. It is easiest to simply use the U.S. Supreme Court's own words to explain the link here:

> "Under the common law of torts, the right to refuse any medical treatment emerged from the doctrines of trespass and battery, which were applied to unauthorized touchings by a physician."[954]

> "At common law, even the touching of one person by another without consent and without legal justification was a battery."[955]

Forcibly covering someone's nose and mouth against their will, whether by indirect coercion or direct force, clearly falls under this common law umbrella, even if the person can still breathe. Actual physical *harm* is not a requirement for battery to have occurred. Any unauthorized forced touching can qualify, even in a medical context.[956] As U.S. Supreme Court Justice Brennan wrote in 1990,

---

953   *Union Pacific Railway Co. v. Botsford*, 141 U.S. 250, 251 (1891) available online: https://tile.loc.gov/storage-services/service/ll/usrep/usrep141/usrep141250/usrep141250.pdf
As a further side note, this 1891 decision is one of the many United States Supreme Court rulings that quotes the work of English jurist William Blackstone.

954   *Mills v. Rogers*, 457 U. S. 291, 294, n. 4 (1982), available online: https://tile.loc.gov/storage-services/service/ll/usrep/usrep457/usrep457291/usrep457291.pdf

955   *Cruzan v. Director, Missouri Department of Health*, 497 U.S. 261, 269 (1990), available online: https://tile.loc.gov/storage-services/service/ll/usrep/usrep497/usrep497261/usrep497261.pdf

956   There is a concept of implied consent that provides doctors with legal protection within certain contexts such as their clinics, but this by no means extends to every public area.

> "The right to be free from medical attention without consent, to determine what shall be done with one's own body, is deeply rooted in this Nation's traditions, as the majority acknowledges… This right has long been 'firmly entrenched in American tort law' and is securely grounded in the earliest common law."[957]

Pointing out that masks are not medical *treatment* is rhetorical misdirection that ultimately does nothing to justify compulsory masking. Masks are a medical *intervention*, using medical *devices*, involving medical *choices*. They are a highly visible physical embodiment of unwanted medical *attention* and coerced medical *decision-making*, and a particularly invasive form of limited bodily restraint.

Masks are a modified quarantine

What the courtroom debates over masks as a form of medical treatment *did* help to do was to clarify by process of elimination what compulsory masking *actually* is for the purposes of Fourteenth Amendment due process analysis. The preventative and isolating nature of compulsory masking is an important clue. *Masking is a modified quarantine*, and this has major implications for both substantive and procedural due process. Used in its general sense, the word "quarantine" refers to any measure that separates, isolates, or excludes, especially for prevention of disease, and includes the more technical uses of both "isolation" and "quarantine." Nineteenth and Twentieth Century legal rulings often use the term quarantine to refer to isolation for a disease, and it is not wrong to do so. However, in the most precise usage, isolation refers to separation because of a known active infection, whereas quarantine involves individuals who are only *possibly* infected. According to the CDC's 2023 definitions, "Isolation separates sick people with a quarantinable communicable disease from people who are not sick. Quarantine separates and restricts the movement of people who were exposed to a contagious disease to see if they become sick."[958] As we shall see in more detail later, when defending its infamous travel mask mandate, the CDC claimed that its authority to impose masks came from 42 U.S.C. §264 — "Quarantine and Inspection," and categorized masks under quarantine in its basic website organization (https://www.cdc.gov/quarantine/masks/mask-travel-guidance.html).[959]

---

957   *Cruzan v. Director, Missouri Department of Health*, 497 U.S. 261, 305 (1990) (Brennan, J., dissenting), available online: https://tile.loc.gov/storage-services/service/ll/usrep/usrep497/usrep497261/usrep497261.pdf

958   Centers for Disease Control and Prevention (CDC) website: "Quarantine and Isolation," (accessed June 17, 2023), https://www.cdc.gov/quarantine/aboutlawsregulationsquarantineisolation.html

959   *https://www.cdc.gov/quarantine/masks/mask-travel-guidance.html*.
42 U.S.C. §264(a),
Title 42 - THE PUBLIC HEALTH AND WELFARE
CHAPTER 6A - PUBLIC HEALTH SERVICE
SUBCHAPTER II - GENERAL POWERS AND DUTIES
Part G - Quarantine and Inspection
From the U.S. Government Publishing Office, www.gpo.gov

Quarantine and isolation are both drastic infringements on individual liberty, but if anything, quarantine impinges on the right to liberty *more* than isolation, because even though isolation is more stringent, it is based upon definitive knowledge of infection and ends upon recovery, whereas quarantine is based upon mere suspicion and theoretical possibility. As a 2007 legal review summarized:

> "The more sweeping the proposed isolation or quarantine in terms of number and geographic scope, the more factual support, such as clinical evidence or expert testimony, will be required to illustrate the necessity and warrant the liberty intrusion."[960]

In many ways, quarantine is akin to a type of pre-trial detention, but without even being accused of a crime. Even when historical inquiry is limited to the United States, the history of quarantine is replete with examples where quarantine power was misused and abused in ways that were more moralizing, prejudiced, and punitive than could be justified by any purely protective purpose.

Very early in U.S. History, the U.S. Supreme Court specifically listed quarantine as being among the police powers of the states.[961] After the Fourteenth Amendment was passed, the Supreme Court broadly declared in 1885 that "The Fourteenth Amendment of the Constitution does not impair the police power of a state."[962] In 1902, in an immigration context, The Supreme Court held that state quarantine power could even be used to exclude *healthy* persons from a quarantined area, rejecting a

---

available online: https://www.govinfo.gov/content/pkg/USCODE-2011-title42/html/USCODE-2011-title42-chap6A-subchapII-partG.htm

CDC, "Requirement for Persons to Wear Masks While on Conveyances and at Transportation Hubs," 86 Fed. Reg. 8025, available online: https://www.federalregister.gov/documents/2021/02/03/2021-02340/requirement-for-persons-to-wear-masks-while-on-conveyances-and-at-transportation-hubs or https://healthfreedomdefense.org/wp-content/uploads/2022/01/DE-1-2_Exhibit-B.pdf

960   Michelle A. Daubert, "Pandemic Fears and Contemporary Quarantine: Protecting Liberty through a Continuum of Due Process Rights," *Buffalo Law Review*, Vol. 54, No. 4, p. 1299 (2007). Available at: https://digitalcommons.law.buffalo.edu/buffalolawreview/vol54/iss4/6

961   "…that immense mass of legislation which embraces everything within the territory of a State not surrendered to the General Government; all which can be most advantageously exercised by the States themselves. Inspection laws, quarantine laws, health laws of every description, as well as laws for regulating the internal commerce of a State, and those which respect turnpike roads, ferries, &c., are component parts of this mass. No direct general power over these objects is granted to Congress, and, consequently, they remain subject to State legislation." *Gibbons v. Ogden*, 22 US. (9 Wheat.) 1, 203 (1824), available online: https://tile.loc.gov/storage-services/service/ll/usrep/usrep022/usrep022001/usrep022001.pdf

962   *Barbier v. Connolly*, 113 U.S. 27 (1884), available online: https://tile.loc.gov/storage-services/service/ll/usrep/usrep113/usrep113027/usrep113027.pdf

Fourteenth Amendment Due Process challenge in order to do so.[963] However, the actual analysis the Supreme Court used to dismiss the Fourteenth Amendment claim in this particular case was superficial and conclusory at best. The 1902 Court majority reasoned in a single three-sentence paragraph that because "the regulation was lawfully adopted and enforced," any invalidation on due process grounds would "strip the government, whether state or national, of all power to enact regulations protecting the health and safety of the people."[964] Such reasoning would invalidate virtually every other Supreme Court Decision issued before or since, which has relied on the Due Process Clause to overturn an unconstitutional state law.

Many lower courts took these Supreme Court rulings in combination with 1905's *Jacobson v. Massachusetts* as a virtual *carte blanche* to dismiss Fourteenth Amendment challenges against a wide variety of laws and ordinances — including (or especially) in the area of quarantine and public health. In a ruling that would have overjoyed COVIDcrisis politicians and public health officials seeking to squelch freedom of speech, association, privacy, and every other inconvenient individual right imposing limits on their ability to act in the name of public health, the Supreme Court of Washington State declared without qualification in 1918 that: "the public health is the highest law; and whenever a police regulation, is reasonably demonstrated to be a promoter of public health, all constitutionally guaranteed rights must give way, to be sacrificed without compensation to the owner."[965]

During World War I, the rationale of protecting military efficiency and public health from venereal disease was used as an excuse to forcibly examine and quarantine thousands of women (prostitutes and otherwise) arrested on charges of prostitution or vagrancy, more often than not denying or depriving them of their fundamental rights of due process while doing so. After all, according to officials, prostitution was "antisocial conduct," venereal disease was a "social pathology," and Soldiers infected with venereal disease cost taxpayer money for treatment, so what better way to combat this threat in the name of public health than with the involuntary detention of any unmarried woman (and some married women) suspected of having a venereal disease? As Allan Brandt noted in his

---

963   *Compagnie Francaise de Navigation a Vapeur v. Louisiana. State Board of Health*, 186 U.S. 380, 393 (1902), available online: https://tile.loc.gov/storage-services/service/ll/usrep/usrep186/usrep186380/usrep186380.pdf

964   Justice Brown entered an excellent dissent in this case, with which Justice Harlan concurred, arguing that Louisiana's use of the quarantine power was so broad and indefinite that it clearly exceeded the "what is absolutely necessary for its self-protection" limitation on state use of quarantine police power established in previous cases. Specifically, Justice Brown quoted *Railroad Company v. Husen*, 95 U.S. 465, 471-473 (1877), available online: https://supreme.justia.com/cases/federal/us/95/465/ (this case and its own antecedents are discussed at length in an earlier footnote in this book)

965   *State* ex rel. *McBride v. Superior Court*, 103 Wash. 409, 419 (1918), available online: https://cite.case.law/wash/103/409/

It is worth noting that some of the most draconian COVIDcrisis measures and worst state court rulings — including on masks — also came out of Washington State.

History of Venereal Disease in the United States, "Vice raids fell most heavily on working-class women, the unemployed, and the unescorted."[966]

During World War I, the Army took stringent punitive measures against Soldiers who caught a venereal disease, including assignment to labor battalions and special shaming quarantine camps. ("A sign that hung over the camp at Gievre announced: 'This is a Venereal Camp. These Men are Helping the Hun.'"[967]) Additionally, whether from prejudiced diagnosis, generally poorer pre-enlistment healthcare, recruiting efforts that focused on more affected communities, or some combination thereof, black soldiers in the American Expeditionary Force were diagnosed with venereal diseases at much higher rates than average, and were subjected to quarantine procedures that were unequally harsh.[968]

Directing public health authority against particularly stigmatized groups and diseases led to some truly outrageous and unconstitutional court rulings even outside of wartime. Unfortunately, the excessive judicial deference to public health authorities during the COVIDcrisis was not new. Historically, many courts have been *far* too deferential to public health authorities. In 1919, women with venereal disease in Kansas were forcibly quarantined at an industrial farm, while their male counterparts were confined in a part of the *state penitentiary* which had been turned over to public health officials[969] — in all but name, imprisonment without trial for having a sexually transmitted disease. When some of the men in Kansas challenged their detention under the Fourteenth Amendment, the Kansas Supreme Court simply declared that no evidence of disease was required beyond the say-so of a public health official. It reads like something straight out of the COVIDcrisis:

> "[I]t is stated that the petitioners are not diseased. The question is one of fact, determinable by practically infallible scientific methods. The city health officer was authorized to ascertain the fact. He has certified to the existence of disease, and, in the absence of a charge of bad faith, or conduct equivalent to bad faith, on his part, his finding is conclusive[.]"[970]

---

966   Allan M. Brandt, *No Magic Bullet: A Social History of Venereal Disease in the United States Since 1880*, p. 86 (New York: Oxford University Press, 1985), available online:
https://archive.org/details/nomagicbulletsoc0000bran/page/86/mode/2up

967   Allan M. Brandt, *No Magic Bullet: A Social History of Venereal Disease in the United States Since 1880*, p. 116 (New York: Oxford University Press, 1985), available online:
https://archive.org/details/nomagicbulletsoc0000bran/page/116/mode/2up?view=theater

968   Allan M. Brandt, *No Magic Bullet: A Social History of Venereal Disease in the United States Since 1880*, p. 116 (New York: Oxford University Press, 1985), available online:
https://archive.org/details/nomagicbulletsoc0000bran/page/116/mode/2up?view=theater

969   *In re McGee*, 105 Kan. 574, 576 (1919), https://cite.case.law/kan/105/574/

970   *In re McGee*, 105 Kan. 574, 580 (1919), https://cite.case.law/kan/105/574/

When the men objected that even if they *were* diseased and isolation *was* necessary, they could obtain proper treatment in an isolated place in the city of Topeka, Kansas, rather than part of the *state penitentiary*, the Court responded glibly: "the public health authorities are not obliged to take chances."[971]

While courts in some states mustered the courage to step in and put a brake on the worst of these violations, courts in other states helped grease the wheels by denying that such violations were at all offensive to the U.S. Constitution. In 1922, Montana Supreme Court Chief Justice Brantly baldly asserted in a solo ("At Chambers") ruling that when it came to public health

> "The provisions of the Fourteenth Amendment to the Constitution of the United
> States… to the effect that no person may be deprived of his liberty without due
> process of law, have no application to the case of one detained in quarantine
> because affected with a dangerous communicable disease."[972]

This particular ruling was all the more outrageous because the "dangerous communicable disease" in question was gonorrhea — certainly not life-threatening and quite straightforward for any truly concerned members of the community to avoid. In a separate ruling that same year, the Supreme Court of Ohio revealed the usually unstated reasoning behind targeting STDs in the name of public health, as well as the general judicial recalcitrance against upholding the constitutional rights of the individuals involved:

> "The protection of the health and lives of the public is paramount, and those
> who by conduct and association contract such disease as makes them a menace to
> the health and morals of the community must submit to such regulation as will
> protect the public."[973]

The history of these campaigns reveals that public health officials at the time were prone to the same sort of irrationalities, biases, self-contradictions, and pretextual thinking as during the COVIDcrisis:[974]

---

971   *In re McGee*, 105 Kan. 574, 580-581 (1919), https://cite.case.law/kan/105/574/

972   *In re Caselli*, 62 Mont. 201, 204 P. 364 (Montana, Jan. 24, 1922), available online:
https://cite.case.law/mont/62/201/

973   *Ex parte Company*, 106 Ohio St. 50, 57 (1922), available online:
https://cite.case.law/ohio-st/106/50/
That same year, the Ohio Supreme Court said that "All known prostitutes and persons associating with them shall be considered as reasonably suspected of having a venereal disease." (*53)

974   Allan M. Brandt, *No Magic Bullet: A Social History of Venereal Disease in the United States Since 1880*, p. 84-92 (New York: Oxford University Press, 1985), available online:

"Most detention houses and quarantine hospitals where the women received treatment did not permit visitors. Public health officials justified this regulation on the premise that the women were under quarantine; this, despite the well-known fact that venereal diseases could not be communicated through the air."[975]

This coercive, condescending medico-moral paternalism provided fertile intellectual soil for the "social hygiene" lead up to *Buck v. Bell* within the next decade. Not to mention that the diseases in question were already (understandably) quite stigmatizing. Such draconian measures only provided sufferers with an even stronger incentive to do everything they could to evade detection, including foregoing treatment by physicians who might be required to report cases to public health authorities.

As late as the mid-twentieth century, some courts were still articulating the doctrine that even though the Constitution is the highest law of the United States, and state constitutions are the highest laws of each state, somehow local public health exceptionalism supersedes the rights guaranteed by both. In 1944, the Illinois Supreme Court ruled: "constitutional guaranties must yield to the enforcement of the statutes and ordinances designed to promote the public health."[976] The California Second District Court of Appeals put it just as bluntly in 1966: "even drastic measures for the elimination of disease, whether in human beings, crops or cattle, in a general way are not affected by constitutional provisions, either of the state or national government."[977] In 1977, citing *Jacobson v. Massachusetts* (of course), the 5th Circuit ruled in a case out of the Eastern District of Texas that "A state should not be required to provide the procedural safeguards of a criminal trial when imposing a quarantine to protect the public against a highly communicable disease."[978]

One of the most egregious cases occurred in 1948, when the California Third District Court of Appeal went so far as to uphold an involuntary quarantine of women suspected of carrying unspecified venereal disease even after medical examinations by independent doctors came back negative for infection.[979] Moreover, the women in question were forcibly "quarantined" *in an overcrowded jail* with prisoners

---

https://archive.org/details/nomagicbulletsoc0000bran/page/84/mode/2up

975   Allan M. Brandt, *No Magic Bullet: A Social History of Venereal Disease in the United States Since 1880,* p. 89 (New York: Oxford University Press, 1985), available online:

https://archive.org/details/nomagicbulletsoc0000bran/page/88/mode/2up?view=theater

976   *People* ex rel. *Baker v. Strautz*, 386 Ill. 360, 365 (Illinois, 1944), available online:

https://cite.case.law/ill/386/360/

977   *In re Halko*, 246 Cal. App. 2d 553, 566 (1966), available online:

https://www.courtlistener.com/opinion/2186709/in-re-halko/

978   *Morales v. Turman*, 562 F.2d 993, 998 (5th Cir. 1977), available online:

https://casetext.com/case/morales-v-turman-5

979   In re *Martin*, 83 Cal.App.2d 164, 174-175 (Cal. Ct. App. 1948) (Adams, P.J., dissenting), available online:

https://casetext.com/case/in-re-martin-60

accused of crimes, at double the prescribed cell occupancy and *four to a bed*.[980] The fact that *any* court was willing to permit this highlights the *de facto* quasi-criminal nature of public health quarantine, which (without admitting it) illicitly stripped targets of due process protections they possessed under the Fourth, Fifth, and Sixth Amendments when accused of far more serious crimes than being sick. As during the COVIDcrisis, the fact that public health officials and law enforcement repeatedly got away with such self-evidently demented measures as forcibly crowding *more* people into *smaller* spaces in the name of preventing disease transmission, reveals the irrationality, undue deference to authority figures, and pathological disregard for basic individual rights that repeatedly manifest like a chronic infection in the context of public health quarantine measures.

Even so, as during the COVIDcrisis, there were judges who issued rulings that pointed in a far more constitutional direction. In the 1948 California case described above, Presiding Justice Adams wrote in a lengthy and scathing dissent that "a mere suspicion, unsupported by facts giving rise to reasonable or probable cause, will afford no justification at all for depriving persons of their liberty and subjecting them to virtual imprisonment under a purported order of quarantine."[981] In his dissent, Justice Adams cited more than half a dozen previous cases demonstrating that the burden of proof in quarantine cases did not lie with the person being quarantined to show that they were not infected or infectious, but rather lay with the health officer to show probable cause of an infectious threat to public health. Justice Adams also affirmed that traditional rules of evidence and presumption of innocence should apply, calling much of the evidence purporting to justify the women's quarantine "hearsay of the rankest kind."[982] He additionally cited other cases where petitioners seeking relief against involuntary quarantine used habeas corpus to secure their release.[983] Justice Adams concluded:

> "Even a prostitute is entitled to the protection of those fundamental principles of
> liberty which are the basis of our civil and political institutions; and the zeal of

---

980   In re *Martin*, 83 Cal.App.2d 164, 170-171 (Cal. Ct. App. 1948) (Adams, P.J., dissenting), available online: https://casetext.com/case/in-re-martin-60

981   In re *Martin*, 83 Cal.App.2d 164, 176 (Cal. Ct. App. 1948) (Adams, P.J., dissenting), available online: https://casetext.com/case/in-re-martin-60

982   In re *Martin*, 83 Cal.App.2d 164, 175 (Cal. Ct. App. 1948) (Adams, P.J., dissenting), available online: https://casetext.com/case/in-re-martin-60

983   In re *Martin*, 83 Cal.App.2d 164, 182 (Cal. Ct. App. 1948), available online: https://casetext.com/case/in-re-martin-60

Specifically, Judge Adams cited *In re Johnson*, 40 Cal.App. 242 (Cal. Ct. App. 1919), available online: https://casetext.com/case/in-re-johnson-205; *In re Travers*, 48 Cal.App. 764 (Cal. Ct. App. 1920), available online: https://casetext.com/case/in-re-travers-1#p768; and *Wragg v. Griffin*, 185 Iowa 243 (1919), available online: https://cite.case.law/iowa/185/243/

an officer ought never to be permitted to transcend the constitutional rights of an accused."[984]

Even the majority in this case conceded that, at a minimum, health officers had to have probable cause to believe an individual was infectious before they could impose an involuntary quarantine.[985] The terminology used here is telling, because police officers require probable cause, rather than mere reasonable suspicion, in order to make an arrest, and the justices clearly used this as a frame of reference when referring to public health quarantine.

Judges like Justice Adams were not always in the minority. The Ninth Circuit struck down an openly race-based involuntary quarantine in San Francisco on equal protection grounds in 1900.[986] In 1919, The Iowa Supreme Court asked rhetorically whether, even assuming involuntary quarantine is

> "[A] wise and valid exercise of the police power for the general good… does it follow that a person not known to be so diseased, and (so far as here appears) showing no visible evidence, sign, or symptom of such disease, may be subjected to arrest, imprisonment, and violation of his person, for no better reason than that he is 'suspected' by someone whose identity is not revealed, to be diseased, and to satisfy the board of health or some of its officers whether there is, in fact, any ground for such suspicion?"[987]

The Iowa Supreme Court answered this question with a resounding negative: *mere suspicion* of disease was nowhere near enough to justify minimizing "the ordinary rights of a person which law and the usages of civilized life regard as sacred[.]"[988] At the very least, said the Court, the use of any such invasive powers against a person can *only* be granted *by the state legislature*, which must *do so expressly*.

---

984   In re *Martin*, 83 Cal.App.2d 164, 182 (Cal. Ct. App. 1948), available online: https://casetext.com/case/in-re-martin-60

985   The judges used both the phrases "probable cause" and "reasonable cause" because the exact wording of "reasonable suspicion" needing to achieve "probable cause" justifying an arrest was still being refined.
In re *Martin*, 83 Cal.App.2d 164 (Cal. Ct. App. 1948), available online: https://casetext.com/case/in-re-martin-60

986   *Jew Ho v. Williamson*, Jew Ho, 103 F. 10 (9th Circuit, 1900), available online: https://casetext.com/case/jew-ho-v-williamson
Note: The court in *Jew Ho* never explicitly invoked the Equal Protection Clause of the Fourteenth Amendment, but it invoked the Fourteenth Amendment generally, the reasons it gave for invalidating the quarantine were consistent with the Fourteenth Amendment's Equal Protection Clause, and it favorably quoted an earlier ruling by the Supreme Court (*Yick Wo v. Hopkins*, 118 U.S. 356, 6 Sup.Ct. 1064, 30 L.Ed. 220), which similarly cited the Fourteenth Amendment to strike down a violation of equal protection.

987   *Wragg v. Griffin*, 185 Iowa 243, 247-248 (1919), available online: https://cite.case.law/iowa/185/243/

988   *Wragg v. Griffin*, 185 Iowa 243, 252 (1919), available online: https://cite.case.law/iowa/185/243/

On top of all that, the Iowa Supreme Court drew the analogy that was studiously avoided by courts faced with Fifth Amendment claims during the COVIDcrisis: likening a man suspected of a crime to a man suspected of infection.

> "Even when charged with the gravest of crimes, he cannot be compelled to give evidence against himself, nor can the state compel him to submit to a medical or surgical examination, the result of which may tend to convict him of a public offense; and if there be any good reason why the same objection is not available in a proceeding which may subject him to ignominious restraint and public ostracism, it is at least a safe and salutary proposition to hold that, before the courts will uphold such an exercise of power, it must be authorized by a clear and definite expression of the legislative will... [I]n our judgment, the restraint of the petitioner, not as a diseased person... but solely as a suspect, and for the avowed purpose... in search of evidence of a loathsome disease which may or may not exist, is a deprivation of his liberty without due process of law, and he is entitled to be set free."[989]

We saw earlier in this book how forcing a mask onto someone's face was historically an especially ignominious form of restraint, even when used against slaves and accused terrorists, and it is also indisputable that during the COVIDcrisis, masking publicly gave symbolic evidence of disease, ostracizing all wearers from one another.

In 1921, a three-judge panel in the California Second District Court of Appeal unanimously ordered that a woman held in an involuntary quarantine be released at once, holding that:

> "Paying just regard to the constitutional guaranties of the right to personal liberty and personal security, it must be asserted that more than a mere suspicion that an individual is afflicted with an isolable disease is necessary to give an officer 'reason to believe' that such person is so afflicted."[990]

Despite the wide variance in earlier rulings, over the latter half of the 20th Century, individual civil rights protections in the context of infectious disease quarantine law began to gradually accrue the components of a potential renaissance through the secondary effects of mental health and disability legislation and the resulting case law. Despite the fact that the U.S. Supreme Court has never ruled on

---

989   *Wragg v. Griffin*, 185 Iowa 243, 252 (1919), available online: https://cite.case.law/iowa/185/243/
990   ex parte *Shepard*, 51 Cal. App. 49, 51 (Cal. Ct. App. 1921), available online:
https://casetext.com/case/in-re-shepard-4

the specifics of what the Fourteenth Amendment Due Process Clause requires when individuals are involuntarily quarantined for a communicable disease, the Court *has* addressed specific due process requirements in other medical contexts like mental health. The U.S. Supreme Court made rulings with indirect but substantial effects on quarantine case law in the 1970s and again in 1987. Two important Supreme Court cases, *O'Connor v. Donaldson* (1975)[991] and *Addington v. Texas* (1979),[992] involved Fourteenth Amendment due process challenges to involuntary public health interventions for mental illness.

In *O'Connor v. Donaldson*, a mental patient, Kenneth Donaldson, was confined to a mental health institution against his will for fifteen *years* simply because he was diagnosed as mentally ill, with no findings that he was a danger to either himself or to others. Donaldson was not released until he finally got his case before a jury in civil proceedings, and to its credit, the jury awarded him damages. In *O'Connor*, the Supreme Court held that "A finding of 'mental illness' alone cannot justify a State's locking a person up against his will and keeping him indefinitely in simple custodial confinement."[993] To anyone who lived through the lockdowns and compulsory masking of the COVIDcrisis, the Supreme Court's description of Donaldson's confinement sounds uncomfortably familiar: "a simple regime of enforced custodial care, not a program designed to alleviate or cure his supposed illness."[994]

Four years later, in *Addington v. Texas*, the Supreme Court clarified that before someone with mental illness can be quarantined from society against their will, due process requires "clear and convincing" evidence in civil proceedings that they are dangerous to themselves or others.[995] The nation's highest court essentially used the same process of elimination familiar to anyone who has ever taken a multiple-choice test. Since it did not want to create any new standards for burden of proof, the U.S. Supreme Court was faced with three existing standards it could choose from. These were, in ascending order: a "preponderance of the evidence," "clear and convincing" evidence, and "beyond a reasonable doubt." The Court reasoned that given the uncertainties of diagnosis and the other existing administrative safeguards against indefinite civil confinement, "beyond a reasonable doubt" was too hard of a burden for the State to meet in non-criminal proceedings. But it also reasoned that the

---

991   *O'Connor v. Donaldson*, 422 U.S. 563 (1975), available online:
https://tile.loc.gov/storage-services/service/ll/usrep/usrep422/usrep422563/usrep422563.pdf
992   *Addington v. Texas*, 441 U.S. 418 (1979), available online:
https://tile.loc.gov/storage-services/service/ll/usrep/usrep441/usrep441418/usrep441418.pdf
993   *O'Connor v. Donaldson*, 422 U.S. 563, 575 (1975), available online:
https://tile.loc.gov/storage-services/service/ll/usrep/usrep422/usrep422563/usrep422563.pdf
994   *O'Connor v. Donaldson*, 422 U.S. 563, 569 (1975), available online:
https://tile.loc.gov/storage-services/service/ll/usrep/usrep422/usrep422563/usrep422563.pdf
995   *Addington v. Texas*, 441 U.S. 418, 433 (1979), available online:
https://tile.loc.gov/storage-services/service/ll/usrep/usrep441/usrep441418/usrep441418.pdf

"preponderance of the evidence" standard used in most civil proceedings did not adequately protect the critical individual rights involved. The only remaining choice was the "clear and convincing" standard of evidence — greater than a 51% "preponderance of the evidence," but less than a 99% "beyond a reasonable doubt."

The impact of the Supreme Court's rulings in *O'Connor* and *Addington* extended beyond the mental health context. Multiple lower courts have applied the Supreme Court's *Addington* ruling to isolation for airborne, respiratory infectious diseases like tuberculosis.[996] In 1980, the West Virginia Supreme Court recognized that the involuntary isolation procedures of the West Virginia Tuberculosis Control Act and West Virginia's involuntary commitment procedures for mental health imposed the same deprivations of individual liberty using the same fundamental rationale of public health and safety. Consequently, the state court ruled that the established due process requirements for mental illness must also be applied to corresponding measures taken in the name of controlling infectious disease. The court required the public health officials to either release the plaintiff they were holding in quarantine against his will, or show by "clear, cogent, and convincing evidence" in another hearing (a hearing where the plaintiff had to be provided the benefit of legal counsel) that he was *actually infected and the* "health menace to others" that they alleged.[997] The New York Supreme Court ruled likewise in a 1999 case based on identical reasoning.[998] The United States District Court for the Southern District of New York revisited the question yet again in 2003, with the same results:

> "The Supreme Court, in the civil commitment cases, has set constitutional standards that must be met before an individual can be detained. The central requirements set out by the Court are the right to a particularized assessment of an individual's danger to self or others and the right to less restrictive alternatives…

---

996   Wendy E. Parmet, "Quarantining the Law of Quarantine: Why Quarantine Law Does Not Reflect Contemporary Constitutional Law" (2018). Wake Forest Journal of Law & Policy, Vol. 9, No. 1, p. 11 (2019), Northeastern University School of Law Research Paper No. 342-2019, Available at SSRN: https://ssrn.com/abstract=3341375
Note: The Court has, on the other hand, issued a number of rulings on the quarantine of *goods*.

997   The West Virginia Supreme Court also ruled that the plaintiff had a right to adequate written notice detailing the grounds on which his commitment was sought, the right to counsel, the "right to be present, cross-examine, confront, and present witnesses," and "the right to a verbatim transcript of the proceeding for purposes of appeal."
Greene v. Edwards, 263 S.E.2d 661, 663 (W. Va. 1980), available online:
https://www.courtlistener.com/opinion/1323449/greene-v-edwards/

998   *Matter of Bradley v. Crowell*, 181 Misc. 2d 529, 530 (N.Y. Sup. Ct. 1999), available online:
https://casetext.com/case/matter-of-bradley-v-crowell

"Accordingly, the fact that an individual has active TB does not itself justify involuntary detention…"

"The Supreme Court in <u>Addington v. Texas</u> 441 U.S. 418, 428-29 (1979), held that to meet due process demands in a civil commitment proceeding the standard of proof used must be greater than a preponderance of the evidence."[999]

In 1987, the U.S. Supreme Court further contributed to the refinement of infectious disease case law when it ruled that a person who is handicapped because of a disease is not removed from existing legal protections against discrimination based on disability simply because that disease happens to be contagious.[1000] The Court made clear that disability protections "would not require a school board to place a teacher with active, contagious tuberculosis in a classroom with elementary schoolchildren."[1001] Rather, what disability protections *would* require was that "prejudice, stereotypes, or unfounded fear" be replaced with an *individualized assessment* and quantitative fact finding, including possible reasonable accommodations and analysis of the actual risks involved.[1002] This ruling was effectively codified into the Americans with Disabilities Act of 1990 (ADA).

In 1993, New Jersey Superior Court Judge Donald S. Goldman authored what was probably the most conscientious analysis in this series of cases to-date. He incorporated previous involuntary commitment case law, as well as the Supreme Court's 1987 ruling and the ADA. Judge Goldman concluded that a person with contagious tuberculosis *could* be involuntarily committed to hospital quarantine or isolation, *on the condition* that "the standards and procedures applicable to involuntary

---

999   *Best v. St. Vincent's Hospital*, 03 Cv. 0365 (RMB) (JCF) (S.D.N.Y. Jul. 2, 2003), available online: https://casetext.com/case/best-v-st-vincents-hospital

1000   "We do not agree with petitioners that, in defining a handicapped individual under § 504, the contagious effects of a disease can be meaningfully distinguished from the disease's physical effects on a claimant in a case such as this. Arline's contagiousness and her physical impairment each resulted from the same underlying condition, tuberculosis. It would be unfair to allow an employer to seize upon the distinction between the effects of a disease on others and the effects of a disease on a patient and use that distinction to justify discriminatory treatment."

School Board of Nassau County, Florida, et al. v. Arline, 480 U.S. 273, 282 (1987), available online: https://tile.loc.gov/storage-services/service/ll/usrep/usrep480/usrep480273/usrep480273.pdf

It is worth emphasizing that the Supreme Court avoided ruling on whether or not someone could be considered a "handicapped person" *solely* on the basis of their contagiousness.

School Board of Nassau County, Florida, et al. v. Arline, 480 U.S. 273, 282 n 7 (1987), available online: https://tile.loc.gov/storage-services/service/ll/usrep/usrep480/usrep480273/usrep480273.pdf

1001   *School Board of Nassau County, Florida, et al. v. Arline*, 480 U.S. 273, 287 n 16 (1987), available online: https://tile.loc.gov/storage-services/service/ll/usrep/usrep480/usrep480273/usrep480273.pdf

1002   *School Board of Nassau County, Florida, et al. v. Arline*, 480 U.S. 273, 287-288 (1987), available online: https://tile.loc.gov/storage-services/service/ll/usrep/usrep480/usrep480273/usrep480273.pdf

civil commitments must be followed[.]"[1003] He proceeded with special care in his legal analysis because he recognized that:

> "The claim of 'disease' in a domestic setting has the same kind of power as the claim of 'national security' in matters relating to foreign policy. Both claims are very powerful arguments for executive action. Both claims are among those least likely to be questioned by any other branch of government and therefore subject to abuse. The potential abuse is of special concern when the other interest involved is the confinement of a human being who has committed no crime except to be sick."[1004]

Judge Goldman concurred with the West Virginia and New York Courts that "The closest legal analogy is provided by court cases that have reviewed the constitutionality of state statutes permitting the involuntary commitment of mental patients on the basis that they have a disease that causes them to be dangerous."[1005] Importantly, "Illness alone cannot be the basis for confinement."[1006]

> "Accordingly, the ADA and its regulations require that a health officer seeking to infringe upon a diseased person's liberty by imposing detention, confinement, isolation or quarantine, must first establish, by clear and convincing evidence, that the person poses a significant risk of transmitting disease to others with serious consequences...

> "The court is obligated to scrutinize the evidence before determining whether the [government's] justifications reflect a well-informed judgment grounded in careful and open-minded weighing of the risks and the alternatives, or whether they are simply conclusory statements that are being used to justify reflexive reactions grounded in ignorance or capitulation to public prejudice... Thus proof that this specific person (and not similar persons) poses a significant risk

---

1003   *City of Newark v. J.S.*, 279 N.J. Super. 178, 184 (1993), available online: https://law.justia.com/cases/new-jersey/appellate-division-published/1993/279-n-j-super-178-0.html

1004   *City of Newark v. J.S.*, 279 N.J. Super. 178, 191 (1993), available online: https://law.justia.com/cases/new-jersey/appellate-division-published/1993/279-n-j-super-178-0.html

1005   *City of Newark v. J.S.*, 279 N.J. Super. 178, 200 (1993), available online: https://law.justia.com/cases/new-jersey/appellate-division-published/1993/279-n-j-super-178-0.html

1006   *City of Newark v. J.S.*, 279 N.J. Super. 178, 192 (1993), available online: https://law.justia.com/cases/new-jersey/appellate-division-published/1993/279-n-j-super-178-0.html

to others, a risk that may not be merely speculative, theoretical, remote or even 'elevated,' is required."[1007]

In his ruling, Judge Goldman only applied this analysis to those measures which the plaintiff in his case actually challenged — to involuntary hospitalization and quarantine. He did not extend it to any other measures imposed on the plaintiff. This is made clear by his casual remark that the plaintiff's "right to outdoor activities may have to be curtailed if he refuses to wear his mask."[1008] This blemish does not, however, change the overall substance of the ruling, and even Judge Goldman's offhand mask comment mitigates *against* universal compulsory masking because he was referring to a single individual who had been definitively diagnosed, was known to be infectious, and does not seem to have ever formally challenged whatever individualized masking requirements the hospital may or may not have been considering. Had the plaintiff chosen to challenge any masking requirements at the time Judge Goldman handed down his decision, there was already *ample* clear and convincing evidence (including Dr. Tunevall's studies conducted in hospital operating rooms which we reviewed in Part 1) showing that masks do not reduce even *bacterial* transmission. On top of this, the risk of respiratory disease transmission outdoors has *always* been negligible, and this fact was reaffirmed by universal experience during the COVIDcrisis.

The bottom line of Judge Goldman's 1993 ruling was that even in the context of an infectious, long-lasting, and historically deadly respiratory disease, "The decisive consideration where personal liberty is involved is that each individual's fate must be adjudged on the facts of his own case, not on the general characteristics of a 'class' to which he may be assigned."[1009] Consistently applied, this would completely preclude compulsory universal masking. A person who is not even infected cannot be subjected to *any* form of indefinite compulsory quarantine because there is *zero* risk of them transmitting a pathogen that is not even *present* in their body. Even if someone did have an infectious respiratory disease, compulsory masking could still not be imposed because the clear and convincing balance of scientific evidence regarding masks is that they have no statistically significant benefits when it comes to decreasing the transmission of pathogens.

The above cases are helpful comparisons, but they are not *perfect* analogies to masking for COVID because they involved individuals who were demonstrably infected with a contagious respiratory disease. One legal analysis in the *Buffalo Law Review* summed up the implications nicely in 2007:

1007   *City of Newark v. J.S.*, 279 N.J. Super. 178, 197-198 (1993), available online:
https://law.justia.com/cases/new-jersey/appellate-division-published/1993/279-n-j-super-178-0.html

1008   *City of Newark v. J.S.*, 279 N.J. Super. 178, 205 (1993), available online:
https://law.justia.com/cases/new-jersey/appellate-division-published/1993/279-n-j-super-178-0.html

1009   *City of Newark v. J.S.*, 279 N.J. Super. 178, 200 (1993), available online:
https://law.justia.com/cases/new-jersey/appellate-division-published/1993/279-n-j-super-178-0.html

"Obviously, a suspected case, contact, or carrier should be afforded greater procedural protection than a diagnosed case, contact or carrier."[1010]

The length of a quarantine is not supposed to exceed the period of incubation and communicability for the disease in question. Inflicting an indefinite involuntary modified quarantine like compulsory masking on whole age groups or *the entire population* of a state covering thousands of square miles simply because a few dozen out of every 100,000 have a particular communicable disease ought to have been unconscionable under *any* ethic — medical or otherwise. During the COVIDcrisis, the rules-lawyer method of getting around this basic commonsense limitation was to either ignore it or to declare that anyone in the area of quarantine automatically counted as having "been exposed," and then define the entire community as the "area of quarantine." Every time a new infection in the community-wide "area of quarantine" occurred, the quarantine timer restarted. Sweet, simple, self-perpetuating. No way out. Politicians and public health officials during the COVIDcrisis added the further innovation of denigrating and denying pre-existing or naturally acquired immunity, so even full recovery from COVID was not enough to end the masking-quarantine.

Unfortunately, legal challenges to compulsory quarantine are relatively rare. Quarantines tend to be imposed against a background of great fear, when public health officials are at their most powerful and courts are at their most deferential. Few plaintiffs have the legal savvy, time, or funding to initiate challenges. Health officials acting in their official capacities have qualified immunity, remedies like the right of habeas corpus do not provide for damages, and even when damages are theoretically attainable, they are almost unheard of in practice. The short duration of quarantines also makes legal challenges perpetually vulnerable to dismissal for being moot. There is, additionally, the chronic chicken-and-egg problem that constitutional challenges to quarantines are dismissed on the grounds that no clearly established constitutional rights have been violated, but the constitutional rights cannot be clearly established in the absence of previously successful challenges.

1010   Michelle A. Daubert, "Pandemic Fears and Contemporary Quarantine: Protecting Liberty through a Continuum of Due Process Rights," *Buffalo Law Review*, Vol. 54, No. 4, p. 1343 (2007). Available at: https://digitalcommons.law.buffalo.edu/buffalolawreview/vol54/iss4/6

> "[A]n analogy to civil commitments is not necessarily appropriate for quarantined individuals, who may or may not be ill with an infectious disease, because they may not be dangerous to others. Quarantine is further along on the continuum and requires greater procedural due process protections." (p. 1334)

> "[B]ecause the procedural protections in quarantine hearings ought to be greater than those in civil commitment hearings, the burden of proof to meet due process demands for quarantine ought to undoubtedly be greater than a preponderance of evidence, the lowest evidentiary standard that applies to civil commitments." (p. 1340)

One individual who found herself willing to pursue such a challenge was nurse Kaci Hickox. In 2014, Ms. Hickox worked as a medical team treatment leader for Doctors Without Borders, helping to care for Ebola victims in Sierra Leone. As a reward for a month of this selfless service, upon arrival in the United States she was immediately detained by New Jersey public health officials and held in an isolation tent with a portable toilet and no shower for 80 hours despite an absence of signs or symptoms.[1011] When she was finally released and driven straight to her home in the State of Maine by EMTs, nurse Hickox immediately had to fend off Maine's public health officials. Maine's Department of Health and Human Services tried to subject Ms. Hickox to onerous and restrictive requirements that everyone who lived through the COVIDcrisis is intimately familiar with. Among other things, they asked the district court to virtually eliminate Ms. Hickox's ability to travel or leave the immediate vicinity of her house, to deny her all access to congregate public places such as shopping malls, movie theaters, and her workplace, and to require her to maintain a 3-foot distance from others when in non-congregate public activities like walking or jogging in the park.[1012] The fact that a 3-foot distancing requirement for Ebola progressed to 6-foot distancing for COVID in just six years highlights the noxious tendency towards "safety creep" present in public health, as well as the lack of any real evidence behind many of the ritualistic and appearance-driven interventions.

Fortunately for Nurse Hickox, District Court Chief Judge Charles C. LaVerdiere had not succumbed to the popular hysteria surrounding Ebola. Judge LaVerdiere held a hearing without delay, and clearly worked long past normal business hours to review all the competing parties' submissions before the hearing. His final commonsense court order effectively freed Hickox, simply requiring "Direct Active Monitoring" and instructions to immediately notify public health authorities if any symptoms of Ebola appeared during the possible incubation period.[1013] In an example of what true public health transparency looks like (and which organizations like the CDC should have learned from), since Ms. Hickox waived her right to confidentiality, Judge LaVerdiere ordered "that all filings, orders, and hearings in this matter shall be open to the public."[1014]

Kaci Hickox filed Forth and Fourteenth Amendment claims as well as tort claims challenging the lawfulness of her involuntary quarantine in New Jersey. Her case, however, had the bad fortune to be

---

1011   Several forehead "temporal" thermometer readings suggested a slight temperature, but temporal thermometer readings are notoriously variable, and her more accurate oral thermometer temperatures taken at the same time were universally normal.

1012   *Temporary Order, Mayhew v. Hickox*, No. CV-2014-36 (Me. Dist. Ct., Fort Kent, Oct. 30, 2014), available online: https://www.scribd.com/doc/245115481/Mayhew-v-Hickox

1013   *Order Pending Hearing* at 3, *Mayhew v. Hickox*, No. CV-2014-36 (Me. Dist. Ct., Fort Kent, Oct. 31, 2014), available online: https://www.scribd.com/doc/245122439/Maine-Ebola-Order#

1014   *Order Pending Hearing* at 4, *Mayhew v. Hickox*, No. CV-2014-36 (Me. Dist. Ct., Fort Kent, Oct. 31, 2014), available online: https://www.scribd.com/doc/245122439/Maine-Ebola-Order#

assigned to the same Judge Kevin McNulty who would later uphold compulsory masking for New Jersey school children against a Fourteenth Amendment equal protection challenge in 2021.[1015] In nurse Hickox' case, Judge McNulty held that the previous civil commitment cases were not directly relevant enough to clearly establish the same due process rights in quarantine cases. He also held that following the CDC's cookie-cutter matrix for evaluating risk and taking nurse Hickox's temperature was enough to meet the requirements of "individualized assessment" for purposes of quarantine, and that the public health officials had probable cause to support Hickox's continued involuntary detention even after her blood work came back negative for Ebola.[1016]

As for the idea that *suspected* cases should be accorded more due process rights than *diagnosed* cases, Judge McNulty arrived at the opposite conclusion based on the rationale that quarantine is a shorter detention, and "Shorter detentions require 'less compelling' evidence of dangerousness."[1017] By that reasoning, shorter prison sentences could somehow justify a lower standard of evidence for criminal conviction, or shorter involuntary mental health commitments could justify less than the Supreme Court's "clear and convincing" evidence standard. Quarantine is indisputably a form of civil confinement, however temporary. As a *Nebraska Law Review* article put it: "The government may not intend for a quarantine to be punitive, but quarantine is nonetheless punitive."[1018] In fact, "Parolees in revocation hearings are afforded more procedural due process rights than a person facing quarantine, even though the former has been convicted of a crime."[1019] Kaci Hickox was far from the only one subjected to onerous quarantine without due process, and neither Kaci Hickox nor any of the more than 233 Americans[1020] subjected to quarantine or *de facto* quarantine in the name of Ebola ever contracted the disease. Kaci Hickox' experience was a warning, but not a warning about disease or the need for a stronger public health establishment. Just 4 years after her case concluded in 2016, the 2020 COVIDcrisis brought every questionable ruling and unresolved issue of quarantine back to the forefront with a vengeance.

1015   *Hickox v. Christie*, 205 F. Supp. 3d 579, 596 (D.N.J. 2016), available online: https://casetext.com/case/hickox-v-christie-1 or
https://www.politico.com/states/f/?id=00000157-0ad7-d870-a97f-4fd7c0b50001

1016   *Hickox v. Christie*, 205 F. Supp. 3d 579, 594, 596 (D.N.J. 2016), available online: https://casetext.com/case/hickox-v-christie-1 or https://www.politico.com/states/f/?id=00000157-0ad7-d870-a97f-4fd7c0b50001

1017   *Hickox v. Christie*, 205 F. Supp. 3d 579, 596 (D.N.J. 2016), available online: https://casetext.com/case/hickox-v-christie-1 or https://www.politico.com/states/f/?id=00000157-0ad7-d870-a97f-4fd7c0b50001

1018   Jennifer Jolly-Ryan, "Balancing Interests and Risk of Error: What Quarantine Process Is Due after Ebolamania," 96 *Nebraska Law Review*, Vol. 96, No. 1, 122 (2017) Available at: https://digitalcommons.unl.edu/nlr/vol96/iss1/4

1019   Jennifer Jolly-Ryan, "Balancing Interests and Risk of Error: What Quarantine Process Is Due after Ebolamania," 96 *Nebraska Law Review*, Vol. 96, No. 1, 124 (2017) Available at: https://digitalcommons.unl.edu/nlr/vol96/iss1/4

1020   American Civil Liberties Union & Yale Global Health Justice Partnership, "Fear, Politics, and Ebola: How Quarantines Hurt the Fight Against Ebola and Violate the Constituion," (December 2015), page 29, available online: https://www.aclu.org/sites/default/files/field_document/aclu-ebolareport.pdf

Austin et al. v. The Board of Education of Community Unit School District #300

Kaci Hickox's lawsuit contesting the due process violations of her quarantine failed, but as a result of the 2014 Ebola scare that led to Kaci Hickox' experiences in New Jersey, the Illinois Department of Public Health (IDPH), updated its regulatory definitions to distinguish "quarantine, modified" from "quarantine, isolated," and ensure due process protections for both. The definition of "quarantine, modified" which Illinoians had due process rights to object to, included procedures and devices "intended to limit disease transmission." The due process requirements for quarantine required that, among other things, "no person shall be ordered to be quarantined or isolated…. [e]xcept with the consent of the person… or upon the prior order of the court of competent jurisdiction."[1021] Procedures falling under the definition of quarantine included tests, vaccines, and "exclusion from school." Quarantine "devices," obviously, included masks "intended to limit disease transmission," as smart lawyers and parents who reviewed the official definitions quickly grasped.

In February 2022, eight months following the pivotal Florida ruling in *Green v. Alachua County* won by Jeff Childers, Illinois Judge Raylene Grischow issued what was not only one of the best mask-related rulings of the whole affair, but also one which stands among the best rulings in the long history of tension between quarantine and due process in general. Once again, as in so many other places, the instigating lawsuits were brought on behalf of hundreds of parents seeking to protect their children from abuse-by-compulsory-masking in schools. Judge Grischow's ruling consolidated not just one but *four* such lawsuits, the largest of which was spearheaded by attorney Thomas DeVore.[1022]

Of course, during the COVIDcrisis, it was too much to expect that public health authorities would play fair and keep to their own rules when those rules started to work against them. With the incoming wave of litigation hitting and about to hit, the Illinois Department of Public Health executed the classic COVIDcrisis public health play of shifting the goalposts. Almost a year and a half into "two weeks to flatten the curve," in September 2021 the IDPH deleted or changed its definitions of quarantine to shut down this avenue of recourse for due process, citing (what else?) the "emergency." Viable legal challenge was, no doubt, an "emergency" as far as the IDPH was concerned. The "emergency" was so urgent that the IDPH even decided to skip such trivial formalities as the legally required rulemaking

---

1021   *Austin et al. v. The Board of Education of Community Unit School District #300 et al.*, No. 2021-CH-500002, 21-CH-500003, 21-CH-500005 & 21-CH-500007, Temporary Restraining Order, p. 8 (Circuit Court of Sangamon County, Illinois, Feb. 04, 2022), available online:
https://media.nbcchicago.com/2022/02/Temporary-Restraining-Order.pdf or
https://www.scribd.com/document/557478467/Sangamon-County-Illinois-Mask mandate-Ruling
1022   *Austin et al. v. The Board of Education of Community Unit School District #300 et al.*, No. 2021-MR-91, Initial Complaint (Circuit Court of the Seventh Judicial Circuit, Macoupin County, Illinois, Oct. 20, 2021), available online:
https://s3.amazonaws.com/jnswire/jns-media/5d/fc/11625511/parents_v_school_districts_pritzker_10-20-21.pdf

process. Judge Grischow, however, was not among the many judges who let their state's public health agencies get away with such underhanded methods.

Strictly speaking, as in Jeff Childers' Florida lawsuit challenging masking under the right to privacy, the Illinois school masking case was decided on grounds of state law rather than the United States Constitution,[1023] but the same is true of many quarantine cases implicating and upholding constitutional rights, and the principles involved went straight to the heart of the due process guaranteed by the Fourteenth Amendment, as well as the separation of executive and legislative powers. Judge Grischow was unequivocal:

> "The Legislature has made it clear that citizens have individual due process rights, specifically the due process right to object to being subjected to quarantine, vaccination, or testing which is alleged to prevent the spread of an infectious disease. This Court finds that masks are also a device intended to limit the spread of an infectious disease, and as such, is a type of modified quarantine... Plaintiffs have a protectable interest to not be subjected to any mandates by the Governor, [Illinois State Board of Education] or the School Districts which interfere with the due process protections provided to Plaintiffs under the IDPH Act in regard to masks as a type of quarantine, as well as vaccination or testing."[1024]

A year and a half into "two weeks to flatten the curve" Judge Grischow did not find the government arguments that due process should not apply "because pandemic" to be persuasive: "Had our Legislature intended that the various due process provisions, as argued by the Defendants were not

---

1023   As Judge Grischow put it:
> "While Plaintiffs' filings contain constitutional due process language, their request for emergency relief is actually premised upon the statutory theory that the State Defendants do not have authority to require masking, close contact exclusion, vaccinations and/or testing in schools unless it is voluntary or an IDPH proceeding is initiated in compliance with Section 2 Procedures for each non-consenting student or teacher, resulting in court orders in compliance with Section 2 Procedures."

Austin et al. v. The Board of Education of Community Unit School District #300 et al., No. 2021-CH-500002, 21-CH-500003, 21-CH-500005 & 21-CH-500007, Temporary Restraining Order, p. 19-20 (Circuit Court of Sangamon County, Illinois, Feb. 04, 2022), available online:

https://media.nbcchicago.com/2022/02/Temporary-Restraining-Order.pdf or

https://www.scribd.com/document/557478467/Sangamon-County-Illinois-Mask mandate-Ruling

1024   *Austin et al. v. The Board of Education of Community Unit School District #300 et al.*, No. 2021-CH-500002, 21-CH-500003, 21-CH-500005 & 21-CH-500007, Temporary Restraining Order, p. 19 (Circuit Court of Sangamon County, Illinois, Feb. 04, 2022), available online:

https://media.nbcchicago.com/2022/02/Temporary-Restraining-Order.pdf or

https://www.scribd.com/document/557478467/Sangamon-County-Illinois-Mask mandate-Ruling

to apply, the Legislature would have specifically done so. The Legislature certainly has had time to make any amendments."[1025] Judge Grischow's decision also touched on the equal protection issues that other judges had refused to acknowledge: "When the Legislature created our laws, they did so knowing individuals have a fundamental right to due process when one's liberty and freedom is taken away by forcing them to do something not otherwise required of all other citizens."[1026]

Judge Grischow recognized that continued deprivation of due process through compulsory masking met the criteria for an irreparable injury. Her evaluation described millions of Americans and even more millions of people across the world:

> "'To demonstrate irreparable injury, the moving party need not show an injury that is beyond repair or compensation in damages, but rather need show only transgressions of a continuing nature'… The Court finds the Plaintiffs' legal rights to procedural and substantive due process are being sacrificed each and every day… especially when there has been zero evidence that those children are contagious or highly likely to spread a contagious disease. Due process of law is a guaranteed right to the Plaintiffs under the Illinois Constitution and has been specifically codified for circumstances such as these… There is no adequate remedy at law because the loss of the continuous sacrifice of legal rights cannot be cured retroactively once the issues are decided on the merits… where injuries are of a continuing nature, remedies at law are inadequate, and injunctions should be imposed…There is no remedy available after trial in this cause which would compensate these Plaintiffs for the harm caused them by being forced to accept the masking mandate, which this Court finds are, by definition, a type of quarantine, as well as the vaccination or testing policies, being lodged against

---

1025   *Austin et al. v. The Board of Education of Community Unit School District #300 et al.*, No. 2021-CH-500002, 21-CH-500003, 21-CH-500005 & 21-CH-500007, Temporary Restraining Order, p. 18 (Circuit Court of Sangamon County, Illinois, Feb. 04, 2022), available online:
https://media.nbcchicago.com/2022/02/Temporary-Restraining-Order.pdf or
https://www.scribd.com/document/557478467/Sangamon-County-Illinois-Mask mandate-Ruling

1026   *Austin et al. v. The Board of Education of Community Unit School District #300 et al.*, No. 2021-CH-500002, 21-CH-500003, 21-CH-500005 & 21-CH-500007, Temporary Restraining Order, p. 15 (Circuit Court of Sangamon County, Illinois, Feb. 04, 2022), available online:
https://media.nbcchicago.com/2022/02/Temporary-Restraining-Order.pdf or
https://www.scribd.com/document/557478467/Sangamon-County-Illinois-Mask mandate-Ruling

Plaintiffs at the whims and caprice of the Defendants, all without any procedural or substantive due process rights to object."[1027]

It is worth noting here that at least one Ohio court ruled against compulsory masking on the assumption that masks are *not* a form of quarantine. In the summer of 2020, Ashland County Health Commissioner Heather Reffett was threatening the food service licenses of restaurants that were unwilling to force masks onto their customers and employees. On July 15th, Commissioner Reffett summarily revoked the food service license of Cattlemans restaurant in Savannah, Ohio, claiming its unmasked employees were an immediate danger to public health, and insisting that "face coverings must be worn at all times unless exceptions apply."[1028] With the aid of attorney Maurice Thompson of The 1851 Center for Constitutional Law,[1029] Cattlemans restaurant owner Mandy Close sued for injunctive relief on grounds of due process.[1030] Though, as with the Illinois case, this Ohio case challenging masking was not (strictly speaking) a constitutional challenge, the resultant due process ruling by Judge Ronald P. Forsthoefel was one of the bright spots to come out of the early months of COVID panic. Judge Forsthoefel started by applying the sort of commonsense skeptical inquiry studiously avoided by many of his peers:

> "[I]f the Dine Safe Ohio Order recognizes exceptions to a blanket mask wearing
> rule (and as such would not consider the lack of wearing a mask an immediate
> danger to public health), then it begs the question as to whether the failure to
> wear a mask for any reason could ever constitute or serve as the basis for finding
> an immediate danger to the public health. The Court further finds that Plaintiffs

1027   *Austin et al. v. The Board of Education of Community Unit School District #300 et al.*, No. 2021-CH-500002, 21-CH-500003, 21-CH-500005 & 21-CH-500007, Temporary Restraining Order, p. 21-22 (Circuit Court of Sangamon County, Illinois, Feb. 04, 2022), available online: https://media.nbcchicago.com/2022/02/Temporary-Restraining-Order.pdf or https://www.scribd.com/document/557478467/Sangamon-County-Illinois-Mask mandate-Ruling

1028   *Cattlemans Inc v. Ashland County Health Department*, No. 20-CIV-099, Complaint for Declaratory Judgment and Immediate Injunctive Relief, p. 17 (Court of Common Pleas, Ashland County, Ohio General Division, July 22, 2020), available online: https://drive.google.com/file/d/17g7P-4PVcpYFdZWq-vE95ID-DWGs4zd42/view or https://www.ashlandcountycpcourt.org/eservices/searchresults.page?x=7tNsa33VRby-CZW0qre1f04Xhj*FegJOebMLJTAiH6AvFCu56GRG5225FvDcTCg-ixRB8-Ak2gdh3c944byRsCA

1029   1851 Center for Constitutional Law, "Ohio Court: Health Departments Cannot Suspend Licenses Over Masks," July 23, 2020, https://ohioconstitution.org/ohio-court-health-departments-cannot-suspend-licenses-over-masks/

1030   *Cattlemans Inc v. Ashland County Health Department*, No. 20-CIV-099, Complaint for Declaratory Judgment and Immediate Injunctive Relief (Court of Common Pleas, Ashland County, Ohio General Division, July 22, 2020), available online: https://drive.google.com/file/d/17g7P-4PVcpYFdZWq-vE95IDDWGs4zd42/view or https://www.ashlandcountycpcourt.org/eservices/searchresults.page?x=7tNsa33VRbyCZW0qre1f04Xhj*FegJOebM-LJTAiH6AvFCu56GRG5225FvDcTCg-ixRB8-Ak2gdh3c944byRsCA

have been denied a meaningful right to appeal... and are therefore being denied their civil liberties, including the right to earn a living and operate a commercial enterprise, without due process of law."[1031]

Apart from his refreshing affirmation of the "right to earn a living and operate a commercial enterprise" which was run roughshod over during the COVID crisis, Judge Forsthoefel discovered that the health authorities' actions had violated Mrs. Close's rights to procedural due process from start to finish. "No advance notice of violation, no order of corrective action, and no meaningful right to a hearing was afforded Plaintiffs prior to revocation of their food service license."[1032]

The progress of time did not improve the county health department's case, either. In his April 2021 judgment issued 9 months later after an extensive individualized assessment, Judge Forsthoefel found that "There was no meaningful evidence presented to this Court to support a finding that the license revocation (without a meaningful opportunity for hearing) was based on an immediate danger to public health[.]"[1033] Judge Forsthoefel's ruling also contains one of the earliest judicial acknowledgements that the totalitarian and unconstitutional measures imposed in the name of public health *did not work*: "the orders at issue in Erie County, Lake County, as well as the *Dine Safe Ohio Order* in this case, fail to accomplish anything scientifically demonstrable, or otherwise corroborated with empirical data, to prevent the spread of contagious or infectious disease even if that purpose were authorized[.]"[1034] Judge Forsthoefel found that the invasive compulsory masking promulgated by the health department was *not*, in fact, authorized by law, because the governing statute did not grant the Director of the Ohio Department of Health and its subsidiary health authorities the power to issue or enforce *any*

1031   *Cattlemans Inc v. Ashland County Health Department*, No. 20-CIV-099, Judgement Entry Temporary Restraining Order, p. 2-3 (Court of Common Pleas, Ashland County, Ohio General Division, July 23, 2020), available online: https://www.ashlandcountycpcourt.org/eservices/searchresults.page?x=7tNsa33VRbyCZW0qre1f04Xhj*FegJOebMLJTAiH6AvFCu56GRG5225FvDcTCg-ixRB8-Ak2gdh3c944byRsCA

1032   *Cattlemans Inc v. Ashland County Health Department*, No. 20-CIV-099, Judgement Entry Temporary Restraining Order, p. 2-3 (Court of Common Pleas, Ashland County, Ohio General Division, July 23, 2020), available online: https://www.ashlandcountycpcourt.org/eservices/searchresults.page?x=7tNsa33VRbyCZW0qre1f04Xhj*FegJOebMLJTAiH6AvFCu56GRG5225FvDcTCg-ixRB8-Ak2gdh3c944byRsCA

1033   *Cattlemans Inc v. Ashland County Health Department*, No. 20-CIV-104, Judgement Entry, p. 4 (Court of Common Pleas, Ashland County, Ohio General Division, April 6, 2021), available online: https://www.ashlandcountycpcourt.org/eservices/searchresults.page?x=7tNsa33VRbyCZW0qre1f04Xhj*FegJOebM-LJTAiH6AvFCu56GRG5234hcmENE9r3*jrzC*ic9K6kPLIxyJNs6A

1034   *Cattlemans Inc v. Ashland County Health Department*, No. 20-CIV-104, Judgement Entry, p. 6 (Court of Common Pleas, Ashland County, Ohio General Division, April 6, 2021), available online: https://www.ashlandcountycpcourt.org/eservices/searchresults.page?x=7tNsa33VRbyCZW0qre1f04Xhj*FegJOebM-LJTAiH6AvFCu56GRG5234hcmENE9r3*jrzC*ic9K6kPLIxyJNs6A

and every action to prevent the spread of infectious disease. Rather, "the Ohio Department of Health only has ultimate authority in matters of **quarantine and isolation**."[1035]

In other words, even taking the position that masks are *not* a form of quarantine, health authorities *still* do not have the authority to mandate masks unless they are expressly *given* that power by the state legislature, and even if those powers *were* to be granted *by the legislature*, statutory due process protections must *still* apply. However, because masks *are*, in truth, a form of modified quarantine, due process protections that are even more fundamental apply. Either way, due process was and remains a real obstacle to universal compulsory masking, even when the Fourteenth Amendment stays hidden in the background. Unless and until a dedicated medical freedom amendment to the US Constitution is passed, the general right to not wear a mask apart from the protections for speech, religion, or any other affected fundamental right will remain part of the general right to liberty housed within the Due Process Clause of the Fourteenth Amendment.

The right to due process puts the burden on public health officials to show by clear and convincing evidence, in an individualized analysis, that the person or persons they are seeking to forcibly quarantine pose a direct threat with significant risk to others' health. In summarizing the body of pre-COVID legal scholarship on quarantine, a 2018 article by professor Wendy Parmet, published in the *Wake Forest Journal of Law and Policy*, said:

> "The scholars who participated in this discussion disagreed about many points, but for the most part they agreed that the precedent that existed established that the quarantine power, although broad, is subject to significant constitutional restraints… At a minimum, these include the requirement that quarantine be imposed only when it is necessary for public health (or is the least-restrictive alternative) and only when it is accompanied by procedural due process protections, including notice, the right to a hearing before an independent decision-maker either before or shortly after confinement, the right to counsel, and the requirement that the state prove its case with clear and convincing evidence."[1036]

1035   *Cattlemans Inc v. Ashland County Health Department*, No. 20-CIV-104, Judgement Entry, p. 6 (Court of Common Pleas, Ashland County, Ohio General Division, April 6, 2021), available online: https://www.ashlandcountycpcourt.org/eservices/searchresults.page?x=7tNsa33VRbyCZW0qre1f04Xhj*FegJOebM-LJTAiH6AvFCu56GRG5234hcmENE9r3*jrzC*ic9K6kPLIxyJNs6A

1036   Wendy E. Parmet, "Quarantining the Law of Quarantine: Why Quarantine Law Does Not Reflect Contemporary Constitutional Law" (2018). *Wake Forest Journal of Law & Policy*, Vol. 9, No. 1, pp. 1-33 at pp. 4-5 (2019), Northeastern University School of Law Research Paper No. 342-2019, Available at SSRN: https://ssrn.com/abstract=3341375

As we saw in Part 4 of this book, prominent public health law professor Lawrence Gostin went full totalitarian for universal compulsory masking and vaccination during the COVIDcrisis. However, prior to authoring the Model State Emergency Health Powers Act (MSEHPA) in 2001 and his transition into a self-described sanitarian and collectivist,[1037] Professor Gostin's earlier scholarly work from the 1990s contained much excellent due process analysis of individual rights in the context of quarantine, as well as guidelines for the legalities of public health that ought to have been applied in the context of COVID, starting with the principle that "public health statutes should be based on provisions that apply equally to all communicable diseases."[1038] This would have meant not treating COVID as some special threat that warranted throwing all previously established public health practice, scientific evidence, and essential individual liberties out the window.

Prior to the COVIDcrisis, Professor Gostin stated: "The burden of proof should fall on the entity seeking to demonstrate significant risk."[1039] So, how do you determine when a risk is "significant"? Depending on who uses it, "significant" can be a real weasel word. In scientific studies, it is conventional to say that a finding is "statistically significant" when there is a less than 5% chance that the finding could have occurred by random luck,[1040] but the everyday and legal usage of the term "significant" is much more flexible, and is typically relative to some unstated gut-level standard of comparison. In the absence of hard numerical cutoffs somewhere in the standard, the final determination of whether or not a particular risk counts as "significant" says as much or more about the decision-maker's individual situation, internal psychology, priorities, fears, and sacred values as it does about the risk they are evaluating. Psychological studies have shown for decades just how manipulable our sense of which risks are "significant" can be when making decisions, and the COVIDcrisis weaponization of behavioral psychology illustrated this truth on a global scale. If some hazard is terrifying enough, a less than one-in-a-million risk can suddenly become "significant." If some objective is desirable enough, a virtually guaranteed risk can suddenly become insignificant. In theory, the flexibility of the "significant risk" standard helps decisionmakers tailor responses to new or rapidly changing circumstances, but when people with the power to decide what counts as "significant" get scared or develop conflicts

---

1037   Lawrence O. Gostin, "From a Civil Libertarian to a Sanitarian," *Journal of Law and Society*, Vol. 34, No. 4, pp 594-616, at 596 (2007), available online: https://www.yumpu.com/en/document/view/16918344/from-a-civil-libertarian-to-a-sanitarian-school-of-public-health-and-

1038   Lawrence O. Gostin, Scott C. Burris, and Zita Lazzarini, "The Law and the Public's Health: A Study of Infectious Disease Law in the United States," *Columbia Law Review*, Vol. 99, No. 59, p. 119, (January 1999), Available at SSRN: https://ssrn.com/abstract=139923

1039   Lawrence O. Gostin, "The Americans with Disabilities Act and the Corpus of Anti-Discrimination Law: A Force for Change in the Future of Public Health Regulation," *Health Matrix: The Journal of Law Medicine*, Vol. 3, Issue 1, p. 116 (1993), available online: https://scholarlycommons.law.case.edu/cgi/viewcontent.cgi?article=1305&context=healthmatrix

1040   Statistical significance is a measure of surprise. You start by assuming that the intervention you are studying does *not* work. It measures the likelihood of getting the observed difference between your control and intervention group by random chance if the intervention does nothing.

of interest, the standard becomes far more of a liability than an asset. The "significant risk" standard itself becomes a significant risk.

It is helpful to start with a process of elimination. All significant risks are non-zero risks, but significant risks make up only a small subset of non-zero risks. Significant risks are, by definition, less than 100% certain. As 1990s Professor Gostin says: "Significant risk is not a remote risk, possibly not even an 'elevated risk.' There must be a material, real, or substantial possibility that the disease can be transmitted."[1041] Risks that are remote, trivial, negligible, speculative, or theoretical are not significant. A slight or unlikely risk *can* still be significant, but most are not. Whether a slight risk is significant depends on how catastrophic the harm being risked is. The more treatable a disease is, the more transient, the harder it is to spread, the easier it is to detect and avoid, and the lower its infection fatality rate, the less significant is *any* risk associated with transmission. A 1% chance of a parachute not opening or of getting cancer from a one-time activity would be significant to most people, but a 10%, 20%, or even 30+% chance of catching a cold in any given year is not significant even though there is no "cure" for a cold other than to recover the traditional way. I would agree with 1990s Professor Gostin when he writes "Perambulation by people with the flu may be impolite, but it is not a major health threat[.]"[1042] This remains true even though cold and flu viruses can and do act as the final disease that carries many people away every year.[1043] I (and, I think, most other people) would agree with 1990s Professor Gostin that:

> "[S]ignificant risk must be determined on a case-by-case basis, and not
> under any type of blanket rule, generalization about a class of disabled
> persons, or assumptions about the nature of disease. This requires a fact-
> specific individualized inquiry resulting in a 'well-informed judgement
> grounded in a careful and open-minded weighing of risks and alternatives.'
> A specific determination must be made that the person is in fact a carrier of

1041   Lawrence O. Gostin, "The Americans with Disabilities Act and the Corpus of Anti-Discrimination Law: A Force for Change in the Future of Public Health Regulation," *Health Matrix: The Journal of Law Medicine*, Vol. 3, Issue 1, p. 116 (1993), available online: https://scholarlycommons.law.case.edu/cgi/viewcontent.cgi?article=1305&context=healthmatrix

1042   Lawrence O. Gostin, Scott C. Burris, and Zita Lazzarini, "The Law and the Public's Health: A Study of Infectious Disease Law in the United States," *Columbia Law Review*, Vol. 99, No. 59, p. 107, (January 1999), Available at SSRN: https://ssrn.com/abstract=139923

1043   As early as March 2020, researchers in France compared the mortality from SARS-CoV-2 to the mortality from other types of coronaviruses over the previous decade (2013 to 2019) and found no statistically significant difference between mortality rates in hospitalized patients.
Yanis Roussel, Audrey Giraud-Gatineau, Marie-Thérèse Jimeno, Jean-Marc Rolain, Christine Zandotti, Philippe Colson, Didier Raoult, "SARS-CoV-2: fear versus data," *International Journal of Antimicrobial Agents*, May 2020. **55**(5): p. 105947. Epub. March 19, 2020. Available online: https://www.ncbi.nlm.nih.gov/pmc/articles/PMC7102597/pdf/main.pdf

a communicable disease and that the disease is readily transmissible in the environment in which he or she will be situated."[1044]

1990s Professor Gostin also had some recommendations regarding coercive health measures. Specifically, "statutes authorizing coercive health measures should include due process protections, both to protect individuals from unjustified restrictions on their liberty and to prevent erroneous or careless fact- finding by health officials."[1045] He affirmed that some narrow emergency exceptions could be justified, but, "Once compulsory powers have been utilized, however, statutes should provide for an immediate hearing, a showing of current infectiousness, and evidence of the need for continued detention or isolation."[1046] Also, "Reformed law should condition the exercise of compulsory powers on a demonstrated public health need, such as when an individual poses a direct threat to others."[1047] Blink and you might miss it. "A showing of current infectiousness." "A direct threat." This in diametric opposition to how compulsory powers were used during the COVIDcrisis, where masks (and vaccines) were mandated based on their supposed *in*direct benefits, yet people who refused these interventions were treated as *direct* threats without *any* showing of individual infectiousness, and often despite repeated demonstrations of *non*-infectiousness.

1990s Professor Gostin was not part of some fringe minority. If politicians, judges, public health officials, and 2020s Professor Gostin had followed the guidance of 1990s Professor Gostin and those who were like-minded, universal compulsory masking would never have been seriously contemplated, much less enacted with the cultish fervor it ultimately was. Though compulsory masking was on the defensive by early 2022, it remained deeply entrenched in many places. Even if the powerful effect of mask-wearing on belief that we looked at in our review of mask psychology was non-existent, too many officials and politicized experts had already gone all-in on their public commitment to masks. If politicians and "experts" had stopped at strongly recommending masks while leaving everyone free to choose whether or not to wear one, and limited themselves to persuasion, debate, social pressure,

---

1044   Lawrence O. Gostin, "The Americans with Disabilities Act and the Corpus of Anti-Discrimination Law: A Force for Change in the Future of Public Health Regulation," *Health Matrix: The Journal of Law Medicine*, Vol. 3, Issue 1, p. 114-115 (1993), available online:
https://scholarlycommons.law.case.edu/cgi/viewcontent.cgi?article=1305&context=healthmatrix
1045   Lawrence O. Gostin, Scott C. Burris, and Zita Lazzarini, "The Law and the Public's Health: A Study of Infectious Disease Law in the United States," *Columbia Law Review*, Vol. 99, No. 59, p. 119, (January 1999), Available at SSRN: https://ssrn.com/abstract=139923
1046   Lawrence O. Gostin, Scott C. Burris, and Zita Lazzarini, "The Law and the Public's Health: A Study of Infectious Disease Law in the United States," *Columbia Law Review*, Vol. 99, No. 59, p. 123, (January 1999), Available at SSRN: https://ssrn.com/abstract=139923
1047   Lawrence O. Gostin, Scott C. Burris, and Zita Lazzarini, "The Law and the Public's Health: A Study of Infectious Disease Law in the United States," *Columbia Law Review*, Vol. 99, No. 59, p. 128, (January 1999), Available at SSRN: https://ssrn.com/abstract=139923

and moral criticism, backing out would have been much more feasible. But we all know that was not what happened. Masks were forced onto the willing and unwilling — men, women and children — using every available oppressive tool, from administrative penalties, to job loss, to civil and criminal penalties, to shameless public demonization of conscientious objectors.

Judge Grischow's February 2022 Temporary Restraining Order against masking Illinois school children was another major step in the direction of sanity and constitutional quarantine-related jurisprudence, and another nail in the coffin for COVID mask mandates. After the Board of Education lost at the district level, the appellate court dismissed the *Board's* appeal as moot.[1048] Less satisfyingly, once the mask mandate was no longer in place, the parents' claims were formally mooted, as well.[1049] The next major setback for compulsory masking would come just two months later, when the CDC airline mask mandate finally failed rational basis scrutiny by a different name.

1048   *Austin v. The Board of Education of Community Unit School District 300*, 2022 Ill. App. 4th 220090 (Ill. App. Ct. Feb. 17, 2022), available online: https://casetext.com/case/austin-v-the-bd-of-educ-of-cmty-unit-sch-dist

1049   *Austin et al. v. The Board of Education of Community Unit School District #300 et al.*, No. 2021-CH-500002, Order on Motion to Dismiss Second Amended Complaint (Circuit Court of Sangamon County, Illinois, Aug. 12, 2022), available online: https://thesouthlandjournal.com/wp-content/uploads/2022/08/doc20220812140722.pdf

"Arbitrary and capricious": ending the CDC's travel mask mandate

Everyone has a non-zero risk of getting hit by a car when crossing the street, but that does not make it rational to mandate that everyone has to wear a fluorescent safety vest at all times when outside their homes based on the speculation that doing so will achieve some marginal decrease in pedestrian risk (even apart from the risk tradeoff of creating that kind of widespread visual distraction). Even in a military context, simply showing (or positing) a non-zero risk of something — including infectious disease — is not on its own enough to pass even rational basis review. The D.C. Circuit Court struck down a longstanding military ban on the deployment of HIV-positive servicemembers based on rational basis review, saying, "It is irrational to categorically bar the deployment of every asymptomatic HIV-positive service member with an undetectable viral load who is otherwise fit to serve based on speculation about aberrant conduct."[1050] Becoming non-deployable is a career-ending designation for military servicemembers, yet otherwise-fit HIV-positive servicemembers were automatically designated as non-deployable until the D.C. Circuit Court took the Army and Air Force to task for their "flawed belief that any non-zero risk of HIV transmission to other service members is sufficient to justify a categorical bar of HIV-positive individuals from deployment."[1051]

Though HIV is blood borne and harder to transmit, the same reasoning applies just as strongly to SARS-CoV-2. SARS-CoV-2 is airborne and easier to transmit, but it is imminently treatable for the minority of the population that even needs treatment, it does *not* require daily medication or periodic

---

1050   *Harrison v. Austin*, 597 F. Supp. 3d 884, 910 (E.D. Va. 2022), available online:
https://casetext.com/case/harrison-v-austin-2?sort=relevance&p=1&type=case&resultsNav=false
1051   *Harrison v. Austin*, 597 F. Supp. 3d 884, 911 (E.D. Va. 2022), available online:
https://casetext.com/case/harrison-v-austin-2?sort=relevance&p=1&type=case&resultsNav=false

blood testing, and total recovery is the rule rather than an exception. If it is irrational to categorically bar the deployment of individuals who are *known* to have a particular chronic communicable infection, it is even more irrational to mask everyone based on speculation about *possible* vulnerability to (and transmission of) a particular *transient* infection. This is especially true in the case of people who already recovered from the disease in question.

One of the most aggravating features of the mask mandate regime (apart from its characteristic self-righteousness and dogmatic adherence to the pronouncements of chronically-yet-shamelessly-wrong institutional "experts") was its perpetual game of "hide the responsibility." Agencies like the CDC and expert bodies issued what they called "guidelines" or "recommendations" which were then turned into mandates by authorities and senior public health officials — mandates which were finally enforced by people who were either true believers in masks, afraid of savage *ad-hominem* attacks and accusations from the true believers in masks, or simply desperate not to lose their jobs like other victims of the response to COVID. Agencies like the CDC could claim to not be issuing any orders, politicians and other authorities could claim to just be following The Science™, and on-the-ground enforcers could console themselves (or take pride) in "just doing their jobs."

If there was one single, indispensable, cornerstone to the reign of compulsory masking (and COVIDcrisis terror in general) in America, it was the CDC. With its incessant, meticulous, and ever-changing "guidelines" and "recommendations," the CDC was like a bellowing administrative beast crouching at the center of a legal labyrinth. Most of the CDC's powers are conferred by the Public Health Service Act, enacted in 1944, and revised many times since.[1052] Most often, the CDC issues "guidelines" and "recommendations," and these are given force of law by other branches of government. For example, the Department of Defense gave the CDC's "guidelines" and "recommendations" regarding masking for COVID-19 force of law as orders in its "Consolidated Department of Defense Coronavirus Disease 2019 Force Health Protection Guidance": "masking of patients, visitors, and personnel working in DoD health care facilities (including military medical, dental, and veterinary treatment facilities) will occur in accordance with CDC guidelines."[1053]

Masks were forced on travelers throughout the majority of 2020 based on CDC guidelines and recommendations, but things went a step further, starting on January 21, 2021. Nine months into the COVIDcrisis, newly sworn-in President Biden issued Executive Order 13998, effectively making

---

1052   Title 42. USC Chapter 6A — Public Health Service. Available online:
https://uscode.house.gov/view.xhtml?path=/prelim@title42/chapter6A&edition=prelim
1053   Department of Defense Office of the Undersecretary of Defense for Personnel and Readiness, "Consolidated Department of Defense Coronavirus Disease 2019 Force Health Protection Guidance — Revision 5," 5.3.e., March 24, 2023, https://media.defense.gov/2023/Mar/28/2003187831/-1/-1/1/CONSOLIDATED-DEPARTMENT-OF-DEFENSE-CORONAVIRUS-DISEASE-2019-FORCE-HEALTH-PROTECTION-GUIDANCE-REVISION-5.PDF

the CDC's word into law by requiring masking in accordance with CDC guidelines in virtually all public transportation contexts, including airports, commercial aircraft, trains, train stations, busses, and even maritime vessels like ferries.[1054] Just over a week later, on February 3rd, the CDC's now-infamous travel mask mandate requiring masks in virtually all public conveyances and transportation hubs was published in the Federal Register.[1055] Apart from forcing masks onto travelers in virtually every travel context and conscripting transportation companies, operators, and their employees into the role of monitoring and enforcement whether they liked it or not, the CDC mask mandate went so far as to require "instructing persons that Federal law requires wearing a mask on the conveyance and failure to comply constitutes a violation of Federal law." In so many words, civil and criminal penalties were prominently on the table. For scientific justification, the CDC cited a handful of the same weak, cherry-picked sources referenced in its science briefs on masks that we examined in the first part of this book.[1056]

Mask mandates in America and across the globe cited to the CDC's pronouncements for legitimacy. A legal judgement against the CDC's *own* overreaching mask mandate would dramatically weaken the others. Multiple citizen-litigators (among whom Lucas Wall deserves special recognition for his leadership, hard work, and tenacity) filed suit to challenge the CDC's travel mask mandate.[1057] Numerous Congressmen led by Representative Thomas Massie and Senator Rand Paul also filed a joint suit.[1058] The most successful suit against the CDC's travel mask mandate, however, was filed in July 2021 by Health Freedom Defense Fund (HFDF), led by founder Leslie Manookian and attorneys

---

1054   "Executive Order on Promoting COVID-19 Safety in Domestic and International Travel," 86 Fed. Reg. 7205, available online: https://www.federalregister.gov/documents/2021/01/26/2021-01859/promoting-covid-19-safety-in-domestic-and-international-travel or https://healthfreedomdefense.org/wp-content/uploads/2022/01/DE-1-1_Exhibit-A.pdf

1055   CDC, "Requirement for Persons to Wear Masks While on Conveyances and at Transportation Hubs," 86 Fed. Reg. 8025, available online: https://www.federalregister.gov/documents/2021/02/03/2021-02340/requirement-for-persons-to-wear-masks-while-on-conveyances-and-at-transportation-hubs or https://healthfreedomdefense.org/wp-content/uploads/2022/01/DE-1-2_Exhibit-B.pdf

1056   CDC.gov. *The Science of Masking to Control COVID-19.* 2021 11/16/2020 [cited 2020]; Available from: https://www.cdc.gov/coronavirus/2019-ncov/downloads/science-of-masking-full.pdf.
CDC.gov, *Science Brief: Community Use of Masks to Control the Spread of SARS-CoV-2 | CDC.* 2021. https://www.cdc.gov/coronavirus/2019-ncov/science/science-briefs/masking-science-sars-cov2.html?CDC_AA_refVal=https%3A%2F%2Fwww.cdc.gov%2Fcoronavirus%2F2019-ncov%2Fmore%2Fmasking-science-sars-cov2.html#print

1057   *Wall v. Centers for Disease Control & Prevention* No. 6:21-cv-00975, (M.D. Florida, 2022) available online: https://www.courtlistener.com/docket/59968410/wall-v-centers-for-disease-control-prevention/

1058   *Massie v. Centers for Disease Control and Prevention* No. 1:22cv00031 (W.D. Kentucky, 2022), available online: https://www.courtlistener.com/docket/63157678/massie-v-centers-for-disease-control-and-prevention/

Brant Hadaway and George Wentz.[1059] On April 18, 2022, after nine months of legal fencing, HFDF received a ruling from Judge Kathryn Kimball Mizelle in the Middle District of Florida that had passengers everywhere breaking into spontaneous maskless celebration thousands of feet in the air.

Judge Mizelle vacated the CDC's mask mandate as exceeding the CDC's statutory authority and violating the procedures required for agency rulemaking under the Administrative Procedure Act (APA). Exceeding its statutory authority and violating the rulemaking procedural due process required by the APA was the CDC's COVIDcrisis *modus operandi*. For its travel mask mandate, the CDC seized on the first paragraph of Title 42 U.S.C. §264 — "Quarantine and Inspection,"[1060] to claim that it possessed the authority to force all travelers to wear masks pursuant to the President's executive order. The statutory paragraph in question reads:

> "The Surgeon General, with the approval of the Secretary, is authorized to make and enforce such regulations as in his judgment are necessary to prevent the introduction, transmission, or spread of communicable diseases from foreign countries into the States or possessions, or from one State or possession into any other State or possession. For purposes of carrying out and enforcing such regulations, the Surgeon General may provide for such inspection, fumigation, disinfection, sanitation, pest extermination, destruction of animals or articles found to be so infected or contaminated as to be sources of dangerous infection to human beings, and other measures, as in his judgment may be necessary."[1061]

1059   *Health Freedom Defense Fund, Inc. et al v. Biden et al*, No. 8:2021cv01693 (M.D. Fla. 2022), available online: Available online: https://www.courtlistener.com/docket/60052717/health-freedom-defense-fund-inc-v-biden/ and https://healthfreedomdefense.org/wp-content/uploads/2021/07/DE-1-COMPLAINT-AS-FILED.pdf

1060   42 U.S.C. §264(a), Title 42 - THE PUBLIC HEALTH AND WELFARE; CHAPTER 6A - PUBLIC HEALTH SERVICE; SUBCHAPTER II - GENERAL POWERS AND DUTIES; Part G - Quarantine and Inspection available online: https://www.govinfo.gov/content/pkg/USCODE-2011-title42/html/USCODE-2011-title42-chap6A-subchapII-partG.htm
CDC, "Requirement for Persons to Wear Masks While on Conveyances and at Transportation Hubs," 86 Fed. Reg. 8025, available online: https://www.federalregister.gov/documents/2021/02/03/2021-02340/requirement-for-persons-to-wear-masks-while-on-conveyances-and-at-transportation-hubs or https://healthfreedomdefense.org/wp-content/uploads/2022/01/DE-1-2_Exhibit-B.pdf

1061   42 U.S.C. §264(a), available online: https://www.govinfo.gov/content/pkg/USCODE-2011-title42/html/USCODE-2011-title42-chap6A-subchapII-partG.htm
Title 42 - The Public Health and Welfare, Chapter 6A — Public Health Service, Subchapter II — General Powers and Duties, Part G — Quarantine and Inspection, § 264 — Regulations to control communicable diseases, (a) promulgation and enforcement by Surgeon General.

The CDC unilaterally declared that forcibly masking all travelers fell under either "sanitation" or "other measures." Masking every traveler wasn't all the CDC claimed this paragraph gave it the power to do, either. Right at the start of the crisis, in March 2020, the CDC had used §264 to issue "no sail" (later "conditional sailing") orders that completely stopped cruise ships from operating for more than six months.[1062] In September 2020, the CDC cited §264 to issue a mandate barring landlords from evicting nonpaying tenants, "covering all residential properties nationwide and imposing criminal penalties on violators."[1063] Florida Judge Steven Merryday enjoined the CDC's unlawful cruise ship orders in June 2021, and an incredulous U.S. Supreme Court finally ruled decisively against the CDC's eviction moratorium two months later in August, saying:

> "[T]he CDC has imposed a nationwide moratorium on evictions in reliance on a decades-old statute that authorizes it to implement measures like fumigation and pest extermination. It strains credulity to believe that this statute grants the CDC the sweeping authority that it asserts."[1064]

Though both of these other unlawful CDC mandates were rightly struck down, they did considerable damage while in force, and the legal process to end them took long months. The same was true of the CDC's 15-month-long travel mask mandate.

Judge Mizelle's shutdown of the CDC's travel mask mandate came in two parts. First, she analyzed the statute cited by the CDC and determined that the statute did not actually say what the CDC claimed it said as far as powers granted. Second, she showed that even if the CDC *had* possessed the statutory authority to issue a mask mandate, the CDC's travel mask mandate was *still* invalid and should be vacated because the way it had been issued violated the rulemaking processes required under the Administrative Procedure Act.

Judge Mizelle quickly realized that whether masks fell under the term "sanitation," as the CDC argued, depended on whether "sanitation" in the statute was being used in the sense of cleaning something, or

---

1062   *Florida v. Becerra*, 544 F. Supp. 3d 1241, 1272 (M.D. Fla. 2021), available online: https://casetext.com/case/florida-v-becerra

1063   *Alabama Association of Realtors v. Department of Health and Human Services*, 594 U.S. ___ at 2 (2021), available online: https://www.supremecourt.gov/opinions/20pdf/21a23_ap6c.pdf or https://casetext.com/case/ala-assn-of-realtors-v-dept-of-health-human-servs

CDC, "Temporary Halt in Residential Evictions to Prevent the Further Spread of COVID-19," 85 Fed. Reg. 55292 (2020), available online: https://www.federalregister.gov/documents/2020/09/04/2020-19654/temporary-halt-in-residential-evictions-to-prevent-the-further-spread-of-covid-19

1064   Alabama Association of Realtors v. Department of Health and Human Services, 594 U.S. ___ at 2 (2021), available online: https://www.supremecourt.gov/opinions/20pdf/21a23_ap6c.pdf or https://casetext.com/case/ala-assn-of-realtors-v-dept-of-health-human-servs

in the sense of keeping something clean. Masks could not fit under the definition of actively cleaning something, but they could arguably fit under the umbrella of keeping something clean. The statute itself did not define which sense it was using the term "sanitation" in — whether cleaning, keeping clean, or both. The CDC, of course, argued for the broadest possible definition, but after a careful analysis, Judge Mizelle ruled otherwise.

Apart from the CDC's recent overextension and abuse of §264 that couldn't help but render any further newly discovered wide-ranging powers within the statute — including a mask mandate — highly suspect, Judge Mizelle identified four additional reasons why the narrower sense of sanitation as cleaning something was the most accurate statutory reading. First, there was the written context. The words accompanying "sanitation" in the statute — inspection, fumigation, disinfection, pest extermination, and destruction — all involved *identifying* and *removing* sources of disease rather than *maintaining* a non-diseased status. Second, a search of the *Corpus of Historical American English* for the 15-year period leading up to the original passage of the statute revealed that at the time the statute was passed, by far the most common use of "sanitation" referred to making something clean, whereas only about 5% of the documented uses of "sanitation" referred to *maintaining* cleanliness. Third, interpreting "sanitation" in the broadest possible sense of the term to include *anything* that promotes hygiene and prevents disease — as the government contended — would make the use of other terms in the statute (e.g. disinfection and fumigation) pointless and redundant, which violated basic rules about how to read and interpret statutes. If Congress had wanted to give the CDC power to do *anything* that promotes hygiene and prevents disease, it would have said so clearly. Fourth, the Federal Government's historical use of inspection and quarantine power was as a secondary adjunct to assist the states, and this was how §264 had previously been interpreted. The CDC's sudden, broad interpretation of "sanitation" to justify forcing masks onto all travelers both reversed these historical roles and extended the mandate to settings that had little (if any) effect on interstate disease spread (e.g., city buses and ride-sharing Ubers). These reasons also made it highly unlikely that a mask mandate would fit under "other measures" either.

Based on the overall context, then, Judge Mizelle ruled, "'sanitation' and 'other measures' refer to measures that clean something, not ones that keep something clean. Wearing a mask cleans nothing. At most, it traps virus droplets. But it neither 'sanitizes' the person wearing the mask nor 'sanitizes' the conveyance." And if Congress had not clearly delegated this power, the agency did not have it. The CDC's travel mask mandate was therefore *ultra vires* — beyond the scope of its legal power and authority.

Judge Mizelle went on to point out that the portion of the statute the CDC was invoking to justify its mandate referred to procedures applied to *property*, not procedures applied to *persons*. As-implemented, the requirements to wear a mask bore far more resemblance to a performance requirement for "conditional release" to avoid a more stringent quarantine — subjects covered elsewhere in the statute — than they did to "sanitation":

> "Anyone who refuses to comply with the condition of mask wearing is—in a sense—detained or partially quarantined by exclusion from a conveyance or transportation hub under authority of the Mask Mandate. They are forcibly removed from their airplane seats, denied boarding at the bus steps, and turned away at the train station doors—all on the suspicion that they will spread a disease. Indeed, the Mask Mandate enlists local governments, airport employees, flight attendants, and even ride-sharing drivers to enforce these removal measures. In short, their freedom of movement is curtailed in a way similar to detention and quarantine…

> "As a result, the Mask Mandate is best understood not as sanitation, but as an exercise of the CDC's power to conditionally release individuals to travel despite concerns that they may spread a communicable disease (and to detain or partially quarantine those who refuse)."[1065]

Why might the CDC prefer to anchor its mask mandate on "sanitation" rather than the more directly applicable quarantine and conditional release sections of the statute? Having read to this point, you likely already suspect the answer. Quarantine implicates individual rights to due process. As Judge Mizelle highlighted, the section of the statute *avoided* by the CDC allowed:

> "[D]etention of an individual traveling between States only if he is 'reasonably believed to be infected' and is actually found 'upon examination' to be infected. The Mask Mandate complies with neither of these subsections. It applies to all travelers regardless of their origins or destinations and makes no attempt to sort based on their health."[1066]

1065   *Health Freedom Defense Fund, Inc. et al v. Biden et al*, No. 8:2021cv01693 - Document 53, p. 24 (M.D. Fla. April 18, 2022), available online:
https://www.courtlistener.com/docket/60052717/health-freedom-defense-fund-inc-v-biden/ and https://healthfree-domdefense.org/wp-content/uploads/2022/04/DE-53-ORDER-GRANTING-SJ-TO-PLAINTIFFS.pdf

1066   *Health Freedom Defense Fund, Inc. et al v. Biden et al*, No. 8:2021cv01693 - Document 53, p. 24 (M.D. Fla. April 18, 2022), available online:
https://www.courtlistener.com/docket/60052717/health-freedom-defense-fund-inc-v-biden/ and https://healthfree-domdefense.org/wp-content/uploads/2022/04/DE-53-ORDER-GRANTING-SJ-TO-PLAINTIFFS.pdf

But Judge Mizelle was not done. Even if the CDC *had* possessed the statutory power to issue a mask mandate, the way it did so violated the rulemaking requirements of the Administrative Procedure Act, and this was, on its own, enough to void the mandate. Initially, the CDC tried to claim that its mask mandate was not a "rule" subject to the rulemaking procedures, even though it self-evidently was exactly that. The CDC then tried to claim it had "good cause" to skip the required public notice and comment period for rulemaking because COVID was such an urgent "emergency."[1067] That transparent excuse probably would have worked in March 2020, but by January 2021, the CDC had waited more than 9 months after COVID hit to issue its travel mask mandate, showing that the *real* "emergency" was far more likely a political desire to avoid giving the public notice and having to deal with the public's comments. This, alone, would have been enough to vacate the mask mandate or at least stay it pending the required procedures, but what was arguably the best part of Judge Mizelle's ruling came next, when she demonstrated for all to see what *real* rational basis scrutiny is supposed to look like.

Technically, what Judge Mizelle applied to the CDC's mask mandate is called arbitrary and capricious review, rather than rational basis scrutiny, but as we shall see, this is a distinction in terminology rather than core substance. Under the Administrative Procedure Act, agencies are required to engage in "reasoned decisionmaking." If an agency rule is "arbitrary and capricious," then the agency has *not* engaged in the required reasoned decisionmaking, and the agency's rule can and should be vacated. Technically, arbitrary and capricious review is distinct from rational basis scrutiny. Rational basis scrutiny is a test applied to laws, while arbitrary and capricious review is a test for the actions and rules of agencies like the CDC, especially under the Administrative Procedure Act, but this is a distinction without a real core difference. "Reasoned decisionmaking" requires a "rational connection between the facts found and the choice made."[1068] If something is arbitrary and capricious, it is by definition

---

1067   The exact wording used by the CDC was: "Considering the public health emergency caused by COVID-19, it would be impracticable and contrary to the public's health, and by extension the public's interest, to delay the issuance and effective date of this Order."
CDC, "Requirement for Persons to Wear Masks While on Conveyances and at Transportation Hubs," 86 Fed. Reg. 8025, available online: https://www.federalregister.gov/documents/2021/02/03/2021-02340/requirement-for-persons-to-wear-masks-while-on-conveyances-and-at-transportation-hubs or https://healthfreedomdefense.org/wp-content/uploads/2022/01/DE-1-2_Exhibit-B.pdf

1068   *Health Freedom Defense Fund, Inc. et al v. Biden et al*, No. 8:2021cv01693 - Document 53, p. 48 (M.D. Fla. 2022), available online: https://www.courtlistener.com/docket/60052717/health-freedom-defense-fund-inc-v-biden/ and https://healthfreedomdefense.org/wp-content/uploads/2022/04/DE-53-ORDER-GRANTING-SJ-TO-PLAINTIFFS.pdf (quoting *Motor Vehicle Manufacturers Association v. State Farm Mutual Automobile Insurance Company*, 463 U.S. 29, 43 (1983), available online: https://tile.loc.gov/storage-services/service/ll/usrep/usrep463/usrep463029/usrep463029.pdf; quoting *Burlington Truck Lines, Inc. v. United States*, 371 U.S. 156, 168 (1962), available online: https://tile.loc.gov/storage-services/service/ll/usrep/usrep371/usrep371156/usrep371156.pdf)

irrational, and irrational actions by definition lack a rational basis. If a mask mandate is arbitrary and capricious, it is irrational, and cannot hold up under true rational basis scrutiny.

Rules, orders, guidelines, and recommendations that are arbitrary and capricious have a number of telltale hallmarks, and the list the follows is not exhaustive. Any single hallmark may be enough to render the rule arbitrary and capricious, and the more that are present, the greater that likelihood becomes. Every mask mandate involved many of these at the same time, and the CDC's mask-related excretions were among the worst offenders in this regard. Every mask mandate that relied on the CDC's mask mandates, rules, orders, guidelines or recommendations took on their fatal deficiencies.

The most basic way to fail the test of reasoned decisionmaking is to fail to satisfactorily explain the reasoning behind the rules being imposed. In other words, "The articulated rationale must also be adequate to explain all major aspects of the decision."[1069] Failing to supply a requisite reasoned analysis and explanation is enough on its own to render an order arbitrary and capricious. Of course, if there is no real reasoning behind the rules, no reasoning can or will be laid out. Mask mandates as a whole provided little (usually no) explanation for their choices. Mask mandates — including the CDC's mask mandate — never truly explained why all masks, from cloth to N95 respirators, were sufficient to meet their requirements. They simply asserted or assumed that any mask was better than no mask, regardless of how it was worn, cared for, or disposed of. This was and remains a giant red flag to anyone knowledgeable about the topic.

When relevant data and evidence exist, adducing substantive evidence and engaging with objections, alternatives, and opposing evidence is an essential part of reasoned decisionmaking. Excluding viable alternatives or large bodies of evidence from consideration is arbitrary and capricious. It is also arbitrary and capricious to entirely fail to consider an important aspect of the problem which a rule, order, guideline, or recommendation is meant to address. Mask mandates failed to consider, arbitrarily rejected, and outright omitted any real explanation for rejecting obvious alternatives or reasonable accommodations such as random surveillance testing or engineering controls. Over one hundred years of established scientific evidence that refuted the presumed efficacy of masking was willfully ignored by the majority of policymakers not just in America but across the globe, and I have yet to find a single mask mandate which considers and addresses potential substitute or supplementary requirements to masking such as surveillance testing, HEPA filters, ventilation improvements, hand hygiene, temperature checks, or simply quarantining symptomatic individuals.

1069   *Health Freedom Defense Fund, Inc. et al v. Biden et al*, No. 8:2021cv01693 - Document 53, p. 48 (M.D. Fla. 2022), available online:
https://www.courtlistener.com/docket/60052717/health-freedom-defense-fund-inc-v-biden/ and https://healthfreedomdefense.org/wp-content/uploads/2022/04/DE-53-ORDER-GRANTING-SJ-TO-PLAINTIFFS.pdf

Failing to narrowly tailor interventions in the furtherance of a compelling interest is arbitrary and capricious. Mask mandates, as we have seen, were *never* narrowly tailored. Mask mandates ignored their own underlying rationale of infection control in favor of procedures and processes specific to one particular method of disease prevention (e.g., mandating mask-wearing but not handwashing). The broad scope of mask mandates was applied regardless of individual risk factors, making them overinclusive while at the same time being underinclusive through making exceptions for activities of comparable risk. They made no effort to explain why their purposes — preventing transmission of serious illness — allowed for the exceptions they included but not other potential exceptions they left out, especially when the exceptions they included cumulatively called into question the mandates' overall efficacy and indispensability for infection control. Age cutoffs ranging from 2 years old to 5 years old were never explained because the judgement was purely arbitrary gut-level reasoning or (at best) political science rather than medical science. Mask mandates never addressed why people in "low transmission risk" indoor settings suddenly became "high transmission risk" to one another simply by virtue of walking into particular buildings when the occupancy of those buildings was entirely composed of *similarly low-risk individuals*. A prime example of this was masking requirements in dental clinics. If the risk to public health from COVID could permit no unmasking, then by the nature of what dental treatment must always entail — generating copious aerosols from multiple unmasked patients' open mouths for hours on end, in close proximity, all day, every day — no routine dental treatment *whatsoever* could have been justified! The most outrageous example, however, was the dismissive, unsupported, anti-scientific and ultimately inexcusable rejection of natural immunity as a basis for exemption. In the same vein, widespread efforts to forcibly eliminate the non-intervention control group of conscientious objectors were not just unscientific but *anti*-scientific, apart from being a gross violation of essential individual liberties.

Failure to consider inescapable tradeoffs is arbitrary and capricious. Mask mandates failed to specify and quantify (or at the very least specify and exemplify) how the hypothesized effect of masks on public health outweighed the constitutionally fundamental rights that were violated in the process — including but not limited to freedom of speech, freedom of religion, and the rights to privacy and due process. Moreover, even a cursory utilitarian analysis shows that many of the CDC's own recommendations ostensibly aimed at combating COVID-19 did far greater damage to public health than universal *un*masking could ever have done even assuming masks worked.

The presence of a pretext, *post hoc* justification, or an explanation that runs counter to the evidence are all dead giveaways that a policy is arbitrary and capricious. As Judge Mizelle pointed out, under the APA, "the Court may only consider 'the grounds that the agency invoked when it took the action.'"[1070]

1070   *Health Freedom Defense Fund, Inc. et al v. Biden et al*, No. 8:2021cv01693 - Document 53, p. 52 (M.D. Fla. 2022), available online: https://www.courtlistener.com/docket/60052717/health-freedom-defense-fund-inc-v-biden/ and

Mask mandates were ideology pushed on a pretext of public health. Mask mandate rationales were often flagrantly pretextual — part of a broader program aimed at increasing fear of COVID and providing evidence that something effective was being done.

Lastly, an order not being the product of agency expertise is yet another sign of being arbitrary and capricious. Most mask mandates were issued by parties that had no expertise in infection control. They simply proffered a vague administrative excuse and then pointed to the CDC despite intense public debate and a trove of easily accessible scientific data on the strength and durability of natural immunity from COVID-19, as well as the overwhelming scientific evidence that masks do not work. They took for granted that appealing to CDC guidelines was "legitimate reliance."

Judge Mizelle laid out in her ruling how the CDC's travel mask mandate miserably failed the requirement for reasoned decisionmaking in multiple ways. Under the Administrative Procedure Act, agencies like the CDC are required to identify the considerations they found persuasive when making policy judgements. The CDC totally failed to do this in its travel mask mandate. The CDC provided zero explanation for rejecting possible alternatives to masking and for its system of exceptions. As described above, an obvious starting point was the total lack of explanation why homemade and medical-grade masks were both considered sufficient. Categorical exceptions for children under two years old and those with ADA-recognized disabilities went completely unexplained. As Judge Mizelle said, "The Mandate makes no effort to explain why its purposes — prevention of transmission and serious illness — allow for such exceptions. Nor why a two-year-old is less likely to transmit COVID-19 than a sixty-two-year-old."[1071] According to the CDC, identification, eating, drinking, and taking medication all permitted at least temporary mask removal, but if even the mere *possibility* of COVID were such a grave threat to public health, why could these activities be permitted at passengers' convenience? The CDC did not say, and the CDC's additional exceptions for persons who were "experiencing difficulty breathing" or who were "feeling winded" undermined its claims that masking does not significantly impede breathing.

After a thorough review, Judge Mizelle ruled: "Because 'our system does not permit agencies to act unlawfully even in pursuit of desirable ends,' the Court declares unlawful and vacates the Mask

---

https://healthfreedomdefense.org/wp-content/uploads/2022/04/DE-53-ORDER-GRANTING-SJ-TO-PLAINTIFFS.pdf

1071   *Health Freedom Defense Fund, Inc. et al v. Biden et al*, No. 8:2021cv01693 - Document 53, p. 49 (M.D. Fla. 2022), available online https://www.courtlistener.com/docket/60052717/health-freedom-defense-fund-inc-v-biden/ and https://healthfreedomdefense.org/wp-content/uploads/2022/04/DE-53-ORDER-GRANTING-SJ-TO-PLAINTIFFS.pdf

Mandate."[1072] Indoctrinated devotees and institutional supporters of compulsory masking greeted her ruling with horrified cries of judicial overreach and predictions of widespread COVID outbreaks caused by unmasked travelers. No such outbreaks, of course, ever materialized — as a brief filed by Airline workers two-and-a-half months later pointed out.[1073] This result was entirely predictable based on the scientific studies we reviewed in the first part of this book. The majority of Americans, especially those who had done any real research, responded with celebration, approval, and quiet relief. Judge Mizelle's ruling on the CDC travel mask mandate was another major milestone on the road to a real recovery.

One would think that if COVID was such a great threat to public health, and masks were *so* effective and essential, then the CDC would have appealed Judge Mizelle's decision and sought a stay of her injunction at the earliest possible date. One would *think* that. Yet, after Judge Mizelle's ruling, the CDC declined to seek a stay of the injunction, and ended up waiting nearly a month and a half before it actually filed its appeal in the 11th Circuit Court of Appeals.[1074] In fact, the CDC's delay in filing its appeal was longer than the notice and comment period which the CDC had tried to bypass by claiming COVID was too much of an emergency.

In the interim, on April 29, less than two weeks after Judge Mizelle's ruling, another Judge in the Middle District of Florida, Paul Byron, issued a ruling in Lucas Wall's lawsuit against the CDC's travel mask mandate.[1075] Judge Byron's ruling was in many ways the antithesis of Judge Mizelle's (which he had clearly read) and traveled the same well-worn rut of reasoning that dominated too many courts during the COVIDcrisis. First, Judge Byron denied that the CDC's mask mandate implicated any constitutionally protected due process liberty interest, or that it involved an infringement on the fundamental right to travel. ("A mere inconvenience caused by a reasonable government regulation

---

1072   *Health   Freedom   Defense   Fund,   Inc.   et   al   v.   Biden   et   al*, No. 8:2021cv01693 - Document 53, p. 58 (M.D. Fla. 2022), available online: https://www.courtlistener.com/docket/60052717/health-freedom-defense-fund-inc-v-biden/ and https://healthfreedomdefense.org/wp-content/uploads/2022/04/DE-53-ORDER-GRANTING-SJ-TO-PLAINTIFFS.pdf

1073   "There have been no reports of increased COVID-19 spread in the aviation sector as a result of the *vacatur* of the [Federal Travel Mask Mandate] in [*Health Freedom Defense Fund v. Biden*]."
Brief of *Amici Curiae* 313 Airline Workers in Support of Appellant Urging Reversal, *Wall v. Centers for Disease Control & Prevention*, No. 22-11532, p. 14 (11th Circuit Court of Appeals, July 5, 2022) Available online: https://lucas.travel/wp-content/uploads/2022/07/Amicus-Brief-of-Airline-Workers-FILED.pdf

1074   Opening Brief for Appellants, *Health Freedom Defense Fund, Inc. et al v. Biden et al*, No. 22-11287 (11th Circuit Court of Appeals, May 31, 2022), available online:
https://healthfreedomdefense.org/wp-content/uploads/2022/07/OPENING-BRIEF-OF-APPELLANTS.pdf

1075   *Wall v. Centers For Disease Control & Prevention*, 6:21-cv-00975 — Document 274, (Middle District of Florida, April 29, 2022), available online:
https://storage.courtlistener.com/recap/gov.uscourts.flmd.390847/gov.uscourts.flmd.390847.274.0_1.pdf
https://www.courtlistener.com/docket/59968410/wall-v-centers-for-disease-control-prevention/

is not enough to amount to a denial of this fundamental right."[1076]) To Judge Byron, "sanitation" in the statute the CDC cited *clearly* included keeping something clean — it was right there in the dictionary! — and the statutory language should be interpreted according to the CDC's broad reading. Mask mandates were a legitimate extension of the agency's quarantine and inspection authority which applied to precommunicable individuals and were "a matter of common sense."[1077] Courts should defer to the CDC's expertise. After all, the CDC had cited to "seven different studies that 'confirmed the benefit of universal masking in community level analyses.'"[1078] This "extensive scientific research" was *clearly* good enough to meet the requirements of reasoned decisionmaking. The CDC's failure to precisely follow administrative requirements like trying to call a rule an "order," or bypassing notice and comment, was either appropriate or "immaterial"[1079] when it came to an ongoing public health emergency like COVID which had taken so very many lives, and which was so very transmissible, continued to mutate, and from which no one was safe.

As an aside, it is not completely clear which seven studies Judge Byron was referring to in his ruling, but the most likely candidates from the CDC's travel mask mandate list of references are Ueki et al. (2020), Wang et al. (2020), Mitze et al. (2020), Gallaway et al. (2020), Lyu and Wehby (2020), Hatzius et al. 2021, and Karaivanov et al. (2020). We covered *all* of these in the science section. Four "intervention with trend analysis" studies (including the infamous Massachusetts General Brigham study), one surveillance study, one laboratory study, and one modeling study. Forcing masks onto millions of Americans based on these "studies" barely worthy of the name was a sick joke. The studies on the list included no control groups, none of the more than 20 randomized controlled trials available in January 2021 (probably because none of those RCTs supported mask efficacy), none of the multiple meta-analyses of these randomized controlled trials, and not even so much as a retrospective cohort

---

1076   *Wall v. Centers For Disease Control & Prevention*, 6:21-cv-00975 — Document 274, p. 6 n 9 (Middle District of Florida, April 29, 2022), available online:

https://storage.courtlistener.com/recap/gov.uscourts.flmd.390847/gov.uscourts.flmd.390847.274.0_1.pdf

https://www.courtlistener.com/docket/59968410/wall-v-centers-for-disease-control-prevention/

1077   *Wall v. Centers For Disease Control & Prevention*, 6:21-cv-00975 — Document 274, p. 17-18 (Middle District of Florida, April 29, 2022), available online:

https://storage.courtlistener.com/recap/gov.uscourts.flmd.390847/gov.uscourts.flmd.390847.274.0_1.pdf

https://www.courtlistener.com/docket/59968410/wall-v-centers-for-disease-control-prevention/

1078   *Wall v. Centers For Disease Control & Prevention*, 6:21-cv-00975 — Document 274, p. 20 (Middle District of Florida, April 29, 2022), available online:

https://storage.courtlistener.com/recap/gov.uscourts.flmd.390847/gov.uscourts.flmd.390847.274.0_1.pdf

https://www.courtlistener.com/docket/59968410/wall-v-centers-for-disease-control-prevention/

1079   *Wall v. Centers For Disease Control & Prevention*, 6:21-cv-00975 — Document 274, p. 2 n 2 (Middle District of Florida, April 29, 2022), available online:

https://storage.courtlistener.com/recap/gov.uscourts.flmd.390847/gov.uscourts.flmd.390847.274.0_1.pdf

https://www.courtlistener.com/docket/59968410/wall-v-centers-for-disease-control-prevention/

study with a meaningful control group or comprehensive literature review. When Judge Byron threw out his off-handed comment, he may have intended to highlight how strong the CDC's evidential basis for its travel mask mandate was, but all he really did was illustrate the indefensible weakness underlying the CDC's "reasoned decisionmaking" for anyone willing to do a modicum of digging, as well as its inexcusable cherry-picking of the literature and his own lack of due diligence in deferring to the agency. Meanwhile, he made Judge Mizelle's ruling look better by comparison. Unmentioned were any of the many *more* references that plaintiff Lucas Wall had submitted providing evidence that masks do not work. The CDC made fools out of many judges during the COVIDcrisis.

Fortunately for Americans, Judge Byron's refusal to grant relief in Lucas Wall's case could not override Judge Mizelle's injunction, and both cases proceeded to the appellate level. With these two conflicting rulings, and the CDC and Lucas Wall both appealing their district-level verdicts, the cases went up to the 11th Circuit Court of Appeals, where Judge Mizelle's injunction remained intact until the final ruling.

The lawsuits against the CDC's travel mask mandate attracted a great deal of attention, both in the form of news coverage and in the form of *amicus* briefs. *Amicus* briefs are legal briefs filed in a lawsuit by third parties to support of one side or another, or to inform the court about another facet of the case pertinent to its final decision. Even at the appellate level, only a tiny fraction of legal cases involve *amicus* briefs. *Amicus* briefers are not directly involved in the case, but may have an interest in the final outcome or take a strong position on the legal question at hand. Mask cases in general, and the cases challenging the CDC's travel mask mandate in particular, involved *multiple amicus* briefs on both sides. Industrial hygiene experts,[1080] disabled passengers,[1081] and a group of over three-hundred airline workers[1082] filed *amicus* briefs supporting Lucas Wall's lawsuit.[1083] In *HFDF v. Biden*, the American Medical Association filed a brief supporting the CDC. (As another side note, while it was better-written, the AMA's brief was even worse evidentially than the CDC's mask mandate or mask

---

1080   Brief of *Amici Curiae* 3 Industrial Hygiene Experts in Support of Appellant Urging Reversal, *Wall v. Centers for Disease Control & Prevention*, No. 22-11532 (11th Circuit Court of Appeals, July 5, 2022), available online: https://lucas.travel/wp-content/uploads/2022/07/Amicus-Brief-of-Industrial-Hygiene-Experts-FILED.pdf

1081   Brief of *Amici Curiae* 16 Disabled passengers in Support of Appellant Urging Reversal, *Wall v. Centers for Disease Control & Prevention*, No. 22-11532 (11th Circuit Court of Appeals, July 5, 2022), available online: https://affordablecareactlitigation.files.wordpress.com/2022/07/11c-wall-16-disabled-passengers-amicus-7-5.pdf

1082   Brief of *Amici Curiae* 313 Airline Workers in Support of Appellant Urging Reversal, *Wall v. Centers for Disease Control & Prevention*, No. 22-11532 (11th Circuit Court of Appeals, July 5, 2022), available online: https://lucas.travel/wp-content/uploads/2022/07/Amicus-Brief-of-Airline-Workers-FILED.pdf

1083   A group of dual citizens also filed an *amicus* brief supporting Mr. Wall's lawsuit.
Brief of *Amici Curiae* 3 Dual Citizens in Support of Appellant Urging Reversal, *Wall v. Centers for Disease Control & Prevention*, No. 22-11532 (11th Circuit Court of Appeals, July 5, 2022), available online: https://lucas.travel/wp-content/uploads/2022/07/Amicus-Brief-of-Dual-Citizens-FILED.pdf

science brief.[1084]) A large group of public health and public health law experts (including Wendy Parmet and Lawrence Gostin, who we met earlier) also signed on to a joint brief in support of the CDC's statutory reading.[1085] Supporting the Health Freedom Defense Fund in opposition to these were the industrial hygiene experts, disabled passengers, and airline worker *amici* who supported Lucas Wall, as well as Congressmen,[1086] the Association of American Physicians and Surgeons,[1087] America's Frontline Doctors,[1088] and the Liberty, Life, and Law Foundation.[1089] On top of this, twenty-three states and state attorneys general led by Florida filed a joint *amicus* brief supporting Health Freedom Defense Fund.[1090] This last was an indication of just how much the political tide had turned against compulsory masking, as well as the growing awareness among state leadership of the CDC's malfeasance and manipulative evidential presentations.

1084   Brief of American Medical Association as *Amicus Curiae* in Support of Defendants-Appellants, *Health Freedom Defense Fund, Inc. et al v. Biden et al*, No.22-11287 (11th Circuit Court of Appeals, June 7, 2022), available online: https://www.ama-assn.org/system/files/health-freedom-defense-fund-v-biden-ama-amicus.pdf

1085   Brief of *Amici Curiae* Public Health and Public Health Law Experts in Support of Defendants-Appellants and Reversal, *Health Freedom Defense Fund, Inc. et al v. Biden et al*, No.22-11287 (11th Circuit Court of Appeals, July 11, 2022), available online:

https://healthfreedomdefense.org/wp-content/uploads/2022/08/As-Filed-Amicus-of-Public-Health-Law-Experts.pdf

1086   Brief of Honorable Thomas Massie, et al. as *Amici Curiae* in Support of Health Freedom Defense Fund, *Health Freedom Defense Fund, Inc. et al v. Biden et al*, No. 22-11287 (11th Circuit Court of Appeals, November 21, 2022), available online: https://massie.house.gov/uploadedfiles/hffvsbiden.pdf

1087   *Amicus Curiae* Brief of the Association of American Physicians and Surgeons in Support of Plaintiffs, *Health Freedom Defense Fund, Inc. et al v. Biden et al*, No. 22-11287 (11th Circuit Court of Appeals, Aug. 5, 2022), available online: https://healthfreedomdefense.org/wp-content/uploads/2022/08/Amicus-Brief-of-American-Assoc-of-Physicians-and-Surgeons.pdf

1088   *Amicus Curiae* Brief of America's Frontline Doctors in Support of Plaintiffs-Appellees and Affirmance, *Health Freedom Defense Fund, Inc. et al v. Biden et al*, No. 22-11287 (11th Circuit Court of Appeals, Aug. 8, 2022), available online: https://res.cloudinary.com/aflds/image/upload/v1660068993/aflds/Health_Freedom_Defense_Fund_Inc_et_al_v_Biden_etc_et_al_Americas_Frontline_Doctors_Amicus_Motion_Brief_Stamped_8_8_2022_21_cv_1693_MDFL_22_11287_CA_11_759748977c.pdf

1089   Brief of *Amicus Curiae* Liberty, Life, and Law Foundation in Support of Plaintiffs-Appellees and Affirmance, *Health Freedom Defense Fund, Inc. et al v. Biden et al*, No.22-11287 (11th Circuit Court of Appeals, August 2, 2022), available online: https://affordablecareactlitigation.files.wordpress.com/2022/08/11c-hfdf-lllf-amicus-8-2.pdf

1090   *Amicus* Brief of the State of Florida and the States of Alabama, Alaska, Arizona, Arkansas, Georgia, Idaho, Indiana, Iowa, Kansas, Kentucky, Louisiana, Mississippi, Missouri, Montana, Nebraska, Ohio, Oklahoma, South Carolina, Texas, Utah, Virginia, and West Virginia, *Health Freedom Defense Fund, Inc. et al v. Biden et al*, No.22-11287 (11th Circuit Court of Appeals, August 8, 2022), available online:

https://healthfreedomdefense.org/wp-content/uploads/2022/08/AMICUS-BRIEF-OF-FLORIDA-22-STATES.pdf

602

States challenge masks: Louisiana v. Becerra

In an August 2020 Viewpoint article published in the Journal of the American Medical Association, Lawrence Gostin acknowledged that "It is not clear whether the CDC has the authority to mandate face coverings nationwide." He also warned that "a federal mandate might provoke political opposition to face coverings rooted in state sovereignty."[1091] Professor Gostin's prediction was accurate — many states rightly perceived the mask mandates implemented through executive agencies to be an attempt at bypassing the legislative branch of government and pre-empting state law altogether.

Prior to COVID, some states actually banned masking as a means of punishment. Texas' Education Code, for example, forbade "obstructing the student's airway, including placing an object in, on, or over the student's mouth or nose or placing a bag, cover, or mask over the student's face"[1092] as an abusive form of aversive behavior management. (This was ignored during COVID but is still very much on the books.) Many state governors like Florida's Ron DeSantis[1093] and Texas' Greg Abbott[1094] had already rescinded and largely banned mask mandates in the spring and summer of 2021. South Dakota's governor, Kristi Noem, never implemented a statewide mask mandate in the first place. By

---

1091   Lawrence O. Gostin, I Glenn Cohen, Jeffrey P. Koplan, "Universal Masking in the United States: The Role of Mandates, Health Education, and the CDC." *JAMA*. 2020;324(9):837–838. Published online August 10, 2020. https://dx.doi.org/10.1001/jama.2020.15271 Gostin also suggested that "a well-crafted use of federal spending powers would likely be constitutional." He might as well have added "hint, hint."

1092   Texas Public Law. Texas Education Code - Sec. 37.0023(7)(B) "Prohibited Aversive Techniques," (2019), available online: https://statutes.capitol.texas.gov/Docs/ED/htm/ED.37.htm
Credit to Allan Stevo, *Face Masks in One Lesson*, (Crafting 52, 2020), for directing me to this.

1093   Florida Governor Ron Desantis, "Office of the Governor Executive Order Number 21-102: Suspending All Remaining Local Government Mandates and Restrictions Based on the COVID-19 State of Emergency," May 3, 2021, available online: https://www.flgov.com/wp-content/uploads/orders/2021/EO_21-102.pdf

1094   Texas Governor Greg Abbott, "Executive Order GA 34: Relating to the opening of Texas in response to the COVID-19 disaster," March 2, 2021, Available online: https://open.texas.gov/uploads/files/organization/opentexas/EO-GA-34-opening-Texas-response-to-COVID-disaster-IMAGE-03-02-2021.pdf

the fall and winter of 2021, many state governments were actively joining the fight on behalf of their citizens, pushing back on compulsory masking with lawsuits of their own.

In November 2021, after a great deal of work by multiple state attorney generals, a Fifth Circuit Court of Appeals stayed enforcement of the Occupational Safety and Health Administration (OSHA) Emergency Temporary Standard vaccine mandate that (among other things) required employees to be vaccinated or take weekly COVID tests and wear a mask.[1095] Undeterred, other executive agencies and localities at the federal and state levels shoehorned their own mask and vaccine mandates onto whoever was unlucky enough to fall under their power. On November 30, 2021, the Office of Head Start (OHS), Administration of Children and Families (ACF), and the Department of Health and Human Services imposed the Head Start Mask Mandate on hundreds of thousands of toddlers, volunteers, and staff nationwide as an "administrative standard."[1096] Nevermind that the risk COVID posed to children was negligible, because children two to five years old could not be vaccinated, they would have to wear the masks. And, of course, by November 2021 the "emergency" was still so great that HHS decided to bypass notice and comment, while pointing to CDC guidance and a smattering of studies that appeared in the CDC's mask science brief (studies which we already looked at in detail in the science section of this book). Twenty-three states, with Louisiana as the lead plaintiff, filed suit in December to stop this latest abuse.[1097]

Nine months later, in September 2022, Judge Terry Doughty freed more than half a million of America's toddlers, issuing a permanent injunction against the Head Start Mask and Vaccine Mandate in *Louisiana v. Becerra*.[1098] Like the CDC, HHS exceeded its statutory authority. HHS' attempts to justify masks and vaccine mandates under program performance standards, administrative and financial management standards, or standards relating to the condition and location of facilities all crumpled like wet tissue paper on critical examination. After carefully reviewing the agency's rationale and legal defense, Judge Doughty said: "there is nothing… which would allow Agency Defendants

---

1095   *BST Holdings, LLC v. Occupational Safety and Health Administration*, No. 21-60845 17 F.4th 604 (5th Cir. November 12, 2021), available online: https://casetext.com/case/bst-holdings-v-occupational-safety-health-admin or https://scholar.google.com/scholar_case?case=12061140935871149182&q=Louisiana+v+Becerra&hl=en&as_sdt=4000006

1096   "Vaccine and Mask Requirements To Mitigate the Spread of COVID–19 in Head Start Programs," 86 Fed. Reg. 68052 (Nov. 30, 2021), available online: https://www.federalregister.gov/documents/2021/11/30/2021-25869/vaccine-and-mask-requirements-to-mitigate-the-spread-of-covid-19-in-head-start-programs

1097   *Louisiana v. Becerra*, 3:21-CV-04370, Document 1 (Western District of Louisiana, December 21, 2021), Available online: https://content.govdelivery.com/attachments/MTAG/2021/12/21/file_attachments/2029135/Head%20Start%20Complaint.pdf

1098   *Louisiana v. Becerra*, 3:21-CV-04370, Document 128 (Western District of Louisiana, September 21, 2022), available online: https://casetext.com/case/louisiana-v-becerra-2 or https://libertyjusticecenter.org/wp-content/uploads/2021/12/2022-9-21-Brick-v.-Biden.pdf

to make medical decisions for employees and volunteers, and/or to require two (2), three (3), and four (4), year-old students to wear masks the majority of the day."[1099]

When HHS passed its toddler mask mandate, public health officials in states like New York were either rushing into lockstep or had beaten it to the punch. On November 17, New York City's Health Commissioner signed an order masking children ages two to five indefinitely. But by this time, the spell had been gradually breaking even in strongholds of the masking mentality like New York. A group of parents found one brave lawyer willing to help them file a challenge.[1100] On April 6, 2022, nearly two weeks before Judge Mizelle ended the CDC's travel mask mandate, Judge Ralph Porzio declared the New York Health Commissioner's indefinite order to mask toddlers void and unenforceable for being "akin to a law" and a measure which "usurps the authority of the legislative branch" — not to mention failing rational basis-type scrutiny for being arbitrary and capricious:

> "In fact, there has been no demonstration or empirical data submitted by Respondents that unmasked children in daycares or schools outside of New York City face increased risk of severe COVID-19 infection... 'Capricious action in a legal sense is established when an administrative agency on identical facts decides differently...' The Respondents have done exactly this. They have made different decisions on identical sets of facts and statistics, such as granting exemptions to unvaccinated athletes and performing artists, or allowing unvaccinated children over the age of five to unmask. This 'toddler mask mandate,' that is slated to continue 'indefinitely' is not in accordance with the CDC guidance, nor is it rational or reasonable in light of the ever-changing data. This is the very definition of arbitrary and capricious."[1101]

1099   *Louisiana v. Becerra*, 3:21-CV-04370, Document 128, p.20 (Western District of Louisiana, September 21, 2022), available online: https://casetext.com/case/louisiana-v-becerra-2 or https://libertyjusticecenter.org/wp-content/uploads/2021/12/2022-9-21-Brick-v.-Biden.pdf

1100   *Goldenstein v. New York City Department of Health & Mental Hygiene*, Initial Petition, No. 2022-85057 (N.Y. App. Div. March 10, 2022), available online: https://www.docketalarm.com/cases/New_York_State_Richmond_County_Supreme_Court/85057---2022/ALEXANDRA_GOLDENSTEIN_et_al_v._NEW_YORK_CITY_DEPARTMENT_OF_HEALTH_AND_MENTAL_HYGIENE_et_al/1/

1101   *Goldenstein v. New York City Department of Health & Mental Hygiene*, Decision & Order, No. 2022-85057, pages 7-8 (N.Y. App. Div. April 6, 2022), available online: https://www.docketalarm.com/cases/New_York_State_Richmond_County_Supreme_Court/85057---2022/ALEXANDRA_GOLDENSTEIN_et_al_v._NEW_YORK_CITY_DEPARTMENT_OF_HEALTH_AND_MENTAL_HYGIENE_et_al/38/

Judge Porzio's opinion was non-published, meaning it was not binding precedent, but it was an indication of the undercurrents of liberty and a return to sanity gradually building below the surface even in places long-dominated by the masking mentality.

Technicalities vs. Fundamentals

Health Freedom Defense Fund v. Biden and *Louisiana v. Becerra* — the lawsuits that ended the CDC's travel mask mandate and HHS' Head Start Mask Mandate — exemplify what are primarily administrative or statutory rather than direct-constitutional challenges to masks. The provisions of the Bill of Rights stand in the background, but these cases were not so much won on whether masks *could* be mandated, but on whether masks could be mandated *by the specific authorities* that did so *in the specific way* that they did so. The distinction is subtle, but critical. On a pragmatic level, this more modest type of challenge had a much higher average chance of success.

One of many things that fear drove out of most people's minds in 2020 was that emergency powers are not open-ended. Even emergency powers are limited and definite. Emergency powers are specifically enumerated in the relevant statutes and charters that confer them. Emergency powers do not extend to every possible area which might conceivably be helpful to deal with an emergency. As several of the cases discussed in this section show, a close examination of the relevant statutes for a particular jurisdiction reveal that its public health authorities have limited granted powers, and these do not extend to things like confining *all* people to their homes, broadly forbidding travel, closing most businesses, and mandating masks.[1102]

The details of what powers agencies and executives have under states of emergency are usually confined to a specific list, often under some heading like "official authority" or "general powers and duties." A good example of this at a local level comes from Jeff Childers' appellant brief in *Green v. Alachua County*. Childers argued that the County did not show that it had authority to issue the mandates, because the legal authority conferred on the County through its charter or Florida statute did not include mask mandates. Alachua Charter 27.08 reads: "executive orders shall be limited to those necessary to eliminate or contain conditions that threaten the health, safety, or welfare of the citizens of the county." In his brief, Childers drew attention to how the Trial Court described Alachua

---

1102   Jeff Childers cited *Wisconsin Legislature v. Palm* (391 Wis. 2d 497, 507 (Wi. 2020), available online: https://casetext.com/case/wis-legislature-v-palm) as an example of this: "Secretary Palm had failed to work with the Wisconsin legislature to develop the emergency order as the emergency statute explicitly required." i.e. Secretary Palm did not follow the proper administrative procedure.

County's mask mandate as "<u>potentially</u> reducing the spread of COVID-19."[1103] Something which is "potentially" helpful cannot fall under emergency orders which are "*limited* to those *necessary* to eliminate or contain" a pathogen. As Childers put it: "'Shall be.' 'Limited.' 'Necessary.' These are not words of expanded powers. Just the opposite, they are explicit words of limitation."[1104]

1103   *Green v. Alachua County*, 323 So. 3d 246, Appellant's Initial Brief, Filing #109556371, page 22, E-Filed 06/29/2020 (Fla. Dist. Ct. of App., 2021), go to: <u>https://1dca.flcourts.gov/Oral-Arguments/Briefs-and-Petitions-for-Cases-Scheduled-for-Oral-Argument/20-1661</u> and search for "case 20-1661"

1104   *Green v. Alachua County*, 323 So. 3d 246, Appellant's Initial Brief, Filing #109556371, page 22-23, E-Filed 06/29/2020 (Fla. Dist. Ct. of App., 2021), go to: <u>https://1dca.flcourts.gov/Oral-Arguments/Briefs-and-Petitions-for-Cases-Scheduled-for-Oral-Argument/20-1661</u> and search for "case 20-1661"

Unbiased judiciary? You decide.

In 2020 and 2021, most judges simply took assertions by state and public health officials regarding masks' efficacy and necessity at face value, citing a need to avoid judicial overreach and not "second-guess" the experts, regardless how tissue-thin, *post hoc*, or outright manufactured the evidence underlying masks and other COVIDcrisis policies actually was. The *real* underlying attitudes behind many judicial decisions upholding mask mandates in 2020 and 2021 became clear when executive and legislative *bans* on mask mandates finally appeared. In 2020 and 2021, it was almost impossible to find a lower court willing to "second-guess" legislative and official "expertise" on mask *mandates*, but when the shoe was finally on the other foot, there were plenty of lower courts who not only found that plaintiffs suddenly had standing to sue against *bans* on mask mandates, but also did not hesitate to stay or enjoin those bans.

When the Arkansas legislature passed a bill in April 2021 banning mask mandates in state entities and public schools but allowing private businesses (and schools) to impose their own mandates,[1105] Judge Tim Fox was quick to grant an injunction against it, citing equal protection, saying that the legislature's ban on mask mandates: "discriminates, without a rational basis, between minors in public schools and minors in private schools."[1106] So, because *private* schools could still mandate masks, *public* schools must be able to do so as well. "Equal protection" twisted into equal abuse.

When Tennessee's governor Bill Lee issued an order in August 2021 that parents could opt out of masking their children in schools on an individual basis, parents of students with disabilities sued to allow schools to mandate masks on *other* parents' children, claiming that a *ban* on mask mandates

---

1105   Arkansas Senate Bill 590 (Act 1002), *An Act to End Mandatory Face Covering Requirements in the State of Arkansas*, passed April 29, 2021, https://legiscan.com/AR/bill/SB590/2021 https://www.arkleg.state.ar.us/Home/FTPDocument?path=%2FACTS%2F2021R%2FPublic%2FACT1002.pdf

1106   *McClane v. Arkansas*, 60CV-21-4692, *15, page 3 (Circuit Court of Pulaski County, 6th Division, Arkansas, Aug. 6, 2021), available online: https://repository.library.northeastern.edu/downloads/neu:4f16f9312?datastream_id=content

It is worth noting that Judge Fox cited the Arkansas Constitution rather than the Fourteenth Amendment, though the equal protection guaranteed by each is fundamentally the same. The Supreme Court of Arkansas finally removed the stay on procedural grounds in January 2023. *State v. McClane*, No. CV-22-254 (Ark. Jan. 26, 2023), https://www.scribd.com/document/622137900/State-v-McClane

forced them to choose between their children's health and safety and their children's education. As though forcing every other parent's child to wear a mask against their will could somehow be a "reasonable" accommodation for those students with disabilities who were supposedly at increased risk. Picking up the torch for masks, District Judge Ronnie Greer *ordered* the Knox County Board of Education to enforce the mask mandate that had been in effect from 2020, ruling that "failure to prospectively adopt a mask mandate — the alleged reasonable accommodation — *is* an injury, a concrete, actual, and ongoing injury[.]"[1107] By contrast, Judge Greer insisted that the Court "can identify no harm to others that would result from a mask mandate[.]"[1108]

In South Carolina, Judge Mary Geiger Lewis did the same thing, admonishing South Carolina's Governor and Attorney General that: "Federal Courts routinely enjoin state officials from enforcing unconstitutional state laws."[1109] According to Judge Lewis, the mere *threat* of COVID was "irreparable harm," but *actual* forced masking of children in violation of all the rights described in this book was no such thing. It takes a special kind of mental contortionist to justify giving schools the power to force masks onto students against their will in the name of individual rights, but Judge Mary Geiger Lewis somehow managed it:

> "[E]mergency powers exist against the backdrop of individual rights and liberties, including the right to be free from discrimination. Such rights are sacrosanct… This is not a close call. The General Assembly's COVID measures disallowing school districts from mandating masks… discriminates against children with disabilities."[1110]

So, *preventing* schools from forcing masks onto students was somehow unconstitutional, but *allowing* schools to mandate masks for students in the name of nondiscrimination was constitutionally kosher.[1111] The fact that the American Academy of Pediatrics filed *amicus* briefs in *support* of allowing schools to force masks onto children in this and other cases should tell readers all they need to know

---

1107   *S.B. v. Lee*, 566 F. Supp. 3d 835, 849-850 (E.D. Tenn., October 12, 2021), available online: https://casetext.com/case/sb-v-lee-1

1108   *S.B. v. Lee*, 566 F. Supp. 3d 835, 870 (E.D. Tenn., October 12, 2021), available online: https://casetext.com/case/sb-v-lee-1

1109   *Disability Rights South Carolina et al v. McMaster et al*, No. 3:2021cv02728 - Document 115, p. 5 (District of South Carolina, November 1, 2021), available online: https://law.justia.com/cases/federal/district-courts/south-carolina/scdce/3:2021cv02728/266541/115/

1110   *Disability Rights South Carolina et al v. McMaster et al*, No. 3:2021cv02728 - Document 80, p. 21-22 (District of South Carolina, September 28, 2021), available online: https://www.aclu.org/legal-document/disability-rights-south-carolina-v-mcmaster-order-granting-temporary-restraining

1111   *Disability Rights South Carolina et al v. McMaster et al*, No. 3:2021cv02728 - Document 115, p. 6 (D.S.C. 2021), available online: https://law.justia.com/cases/federal/district-courts/south-carolina/scdce/3:2021cv02728/266541/115/

about that particular body of organized medicine.[1112] The ACLU likewise supported this and other suits to maintain mask mandates on children.[1113]

Fortunately for most of South Carolina's schoolchildren, the governor's order was a reflection of how much the masking mentality was beginning to ebb. In January 2022, two out of three judges on a Fourth Circuit Court of Appeals panel ruled that the plaintiffs did not have standing to sue for the simple fact that even with Judge Lewis' temporary injunction in place, the "overwhelming majority" of South Carolina school districts *still* did not re-impose universal masking for students and staff, so a favorable ruling would not grant the relief the plaintiffs were seeking.[1114] Judge James Wynn wrote what was actually an excellent dissenting opinion. If Judge Wynn's reasoning had been applied consistently throughout 2020 (as it should have been) multiple cases *challenging* compulsory masking would not have been dismissed for lack of standing:

> "A plaintiff 'need not show that a favorable decision will relieve [their] every injury.' Rather, a plaintiff need only show that they 'personally would benefit in a tangible way from the court's intervention.' For that reason, 'removal of even one obstacle to the exercise of one's rights, even if other barriers remain, is sufficient to show redressability.'"[1115]

A parallel case in Iowa followed this same pattern,[1116] with a group of parents claiming that preventing mask mandates violated their disabled children's civil rights. The ACLU and American Academy of Pediatrics supported *this* effort to re-mask Iowa school children as well.[1117] It was, at bottom, the same argument heard throughout the COVIDcrisis: these people over here are more vulnerable than you

---

1112   *Disability Rights South Carolina et al v. McMaster et al*, No. 3:2021cv02728 - Document 115, p. 3 (D.S.C. 2021), available online: https://law.justia.com/cases/federal/district-courts/south-carolina/scdce/3:2021cv02728/266541/115/

1113   "Disability Rights Groups Challenge South Carolina Ban on School Mask Mandates," ACLU Press Releases, August 24, 2021, https://www.aclu.org/press-releases/disability-rights-groups-challenge-south-carolina-ban-school-mask-mandates

1114   *Disability Rights South Carolina v. McMaster*, No. 21-2070 (4th Cir. January 25, 2022), available online: https://law.justia.com/cases/federal/appellate-courts/ca4/21-2070/21-2070-2022-01-25.html

1115   *Disability Rights South Carolina v. McMaster*, No. 21-2070, p. 37 (4th Cir. January 25, 2022) (Wynn, J., dissenting), available online: https://law.justia.com/cases/federal/appellate-courts/ca4/21-2070/21-2070-2022-01-25.html

1116   Complaint for Declaratory and Injunctive Relief, *Arc of Iowa v. Reynolds*, Case No. 4:21-cv-00264, Document 1 (Southern District of Iowa September 3, 2021), available online https://thearc.org/wp-content/uploads/2021/09/The-Arc-of-Iowa-v.-Reynolds-complaint.pdf, main case documents: https://thearc.org/resource/the-arc-of-iowa-v-reynolds/

1117   Brief of *Amici Curiae* Iowa Chapter of American Academy of Pediatrics and American Academy of Pediatrics in Support of Plaintiffs' Motion for Preliminary Injunction, *Arc of Iowa v. Reynolds*, Case No. 4:21-cv-00264, Document 59 (Southern District of Iowa September 30, 2021), available online:

are to a disease you might theoretically carry, therefore *you* must wear a mask. The plaintiffs had no difficulty finding a pediatrician willing to opine that "[t]here's no evidence that wearing a mask has any negative impact whatsoever on that child's physical, social, or mental well-being."[1118] District Court Judge Robert Pratt quickly issued a temporary restraining order and then a preliminary injunction against Governor Kim Reynolds' and Education Director Ann Lebo's enforcement of the Iowa Legislature's *ban* on school mask mandates.[1119] Unlike in South Carolina, many school districts in Iowa rushed to re-mask, at least for a few more months. The masking mentality at its finest: executive mask mandates under questionable emergency powers were totally fine, but *bans* on mask mandates passed through the normal legislative process somehow violated basic civil rights.

Not all Judges were so quick to grant such injunctions against mask mandate bans, however. During the same time period, Florida Federal Judge Michael K. Moore ruled *against* blocking Governor DeSantis' ban on mask mandates, saying the parents who filed the suit had not exhausted their administrative remedies first.[1120] By this point, more than enough of the studies that we reviewed in detail in the science section of this book had been published to show decisively that masks made no difference to COVID-19 infections in schools or anywhere else. The speculated special threat posed by SARS-CoV-2 to individuals with disabilities was further undermined when a group of UK researchers published their findings of a nationwide prospective study of SARS-CoV-2 in immunocompromised pediatric patients recruited from 46 hospitals in November 2021. From March 2020 to March 2021, of the 1527 immunocompromised participants, only 38 cases of SARS-CoV-2 were diagnosed. Of those 38 cases, four were admitted to the hospital, but in *none* of the four cases was SARS-CoV-2 the cause of the hospital admission. The Authors concluded: "SARS-CoV-2 infections have occurred in immunocompromised children and young people with no increased risk of severe disease. No

---

https://thearc.org/wp-content/uploads/2021/10/American-Academy-of-Pediatrics-Amicus-Brief-IA.pdf, main case documents: https://thearc.org/resource/the-arc-of-iowa-v-reynolds/

1118   Order Granting Plaintiffs' Motion for a Temporary Restraining Order, *Arc of Iowa v. Reynolds*, Case No. 4:21-cv-00264, Document 32, p. 13 (Southern District of Iowa September 13, 2021), available online: https://thearc.org/wp-content/uploads/2021/09/Iowa-Order-Granting-Temporary-Restraining-Order.pdf, main case documents: https://thearc.org/resource/the-arc-of-iowa-v-reynolds/

1119   Order Granting Plaintiffs' Motion for a Preliminary Injunction, *Arc of Iowa v. Reynolds*, Case No. 4:21-cv-00264, Document 60 (Southern District of Iowa October 8, 2021), available online: https://thearc.org/wp-content/uploads/2021/10/Order-Granting-Motion-for-Preliminary-Injunction-IA.pdf, main case documents: https://thearc.org/resource/the-arc-of-iowa-v-reynolds/

1120   *Hayes v. DeSantis*, Case No. 1:21-cv-22863, Document 98 — Order Denying Preliminary Injunction, (September 15, 2021), available online: https://storage.courtlistener.com/recap/gov.uscourts.flsd.597646/gov.uscourts.flsd.597646.98.0_1.pdf, Additional case documents: https://www.courtlistener.com/docket/60114348/hayes-v-governor-ronald-dion-desantis/ Initial Complaint (August 6, 2021): https://www.justdigit.org/wp-content/uploads/2021/08/1-Complaint.pdf

children died."[1121] And yet Judges like those mentioned here continued to enable forced masking for school children in the name of The Science™ and a truly twisted concept of equality.

In May 2022, a month after Judge Mizelle stopped the CDC's travel mask mandate, the 8th Circuit Court of Appeals declared the attempt to re-mask students in Iowa to be moot.[1122] However, Judge Pratt still got the last word. In his November 2022 judgement, Judge Pratt declared in so many words, if a "school district concludes that requiring masks is a reasonable modification to protect the safety of the student making the accommodation request under federal disability law," then, according to Judge Pratt, the governor (and by extension, the legislature) "must permit the imposition of a mask mandate."[1123]

1121   Chappell, H., et al., *Immunocompromised children and young people are at no increased risk of severe COVID-19.* Journal of Infection, 2022. **84**(1): p. 31-39. https://dx.doi.org/10.1016/j.jinf.2021.11.005

1122   *Arc Iowa v. Reynolds*, 33 F.4th 1042 (8th Cir., May 16, 2022), available online: https://thearc.org/wp-content/uploads/2022/05/Eighth-Circuit-Per-Curiam-Opinion-IA.pdf or https://casetext.com/case/arc-iowa-v-reynolds-1; main case documents: https://thearc.org/resource/the-arc-of-iowa-v-reynolds/

1123   Order Granting in Part and Denying in Part Defendants' Motions to Dismiss and Granting Plaintiffs' Motion for Summary Judgment, *Arc of Iowa v. Reynolds*, Case No. 4:21-cv-00264, Document 114, p. 22-23 (Southern District of Iowa November 1, 2022), available online: https://thearc.org/wp-content/uploads/2022/11/Order-on-Motions-to-Dismiss-and-Summary-Judgment-IA.pdf, main case documents: https://thearc.org/resource/the-arc-of-iowa-v-reynolds/

The case is moot! ... Or is it?

If you happen to have a tyrannical or highly authoritarian bent (or even just the sort of anxious personality that copes with anxiety by trying to control other people as much as possible), and if you want to impose a mandate to get everyone else to "do the right thing" (for everyone's collective good, of course), part of Gaming the Courts 101 is getting a feel for whether the court is going to favor you, and then using that information to determine how you implement your mandates. If the court is clearly favorable, you can run any legal case to its natural conclusion and get a ruling that will give you a useful precedent. If you are not sure, or if the court is looking more and more skeptical, and plaintiffs challenging your mandate might (heaven forbid!) get a fair hearing and subject your mandate to fatal review, you can still maintain your power and long-term potential to try your agenda later by withdrawing or modifying the restrictions you imposed and then claiming the issue is now moot.

When a case is said to be "moot," it means that the controversy is no longer applicable, and the case is effectively ended because there is nothing left that needs to be decided. A ruling on the controversy would have no "practical effect." As an example, if two people claim ownership of the same property and sue each other, the case becomes moot if one of them suddenly decides (for whatever reason) to relinquish all claims to the property under dispute. In the case of mask mandates and legal challenges to those mask mandates, ending the mandate being challenged might "moot" the case if there is no "reasonable possibility" that the mask mandate would be re-issued. If you happen to be the party that issued the mask mandate, the odds will favor you on any mootness claims, because (if the mooting decisions in multiple mask mandate cases are anything to go by) courts tend to give governmental parties a stunningly generous benefit of the doubt when deciding whether mootness applies. I am not referring to the usual good faith required to do business or make for smooth administration. I am referring to benefit of the doubt that often equals or exceeds what most people would give close friends and family.

Additionally, if you happen to be a judge who (for whatever reason) *really* does not want to rule one way or another on a case, then drawing out the case until it becomes moot is an especially attractive and useful evasion tactic (when it is an option), because it is almost impossible to prove. After all,

615

judges are very busy, and giving both parties all the time they need to make their strongest case is only fair. If that means the case drags on until it becomes moot, that's too bad, but it was the only fair thing you could have done. The trajectories of many of the legal challenges to compulsory masking discussed in this book are consistent with (if not outright highly suggestive of) the hypothesis that multiple courts avoided ruling on the issue of masks by delaying proceedings until they could dismiss the case as moot (particularly in cases where plaintiffs made pleadings using unusually strong arguments or evidence that ought to have gotten their claims to strict scrutiny or some form of success). Then again, that could just be the way completely innocent timelines appear when viewed through a recently acquired lens of mistrust.

Of course, the mere fact that a case or controversy technically no longer exists does not mean that a court cannot still rule on the issue it if it wants to definitively resolve the dispute for future reference. If courts were unable to rule on controversies after the fact, then a large subset of cases in which the wrongdoing involved was too short in duration to be litigated would be excluded from the possibility of redress. Fair-minded judges long ago recognized the potential abuses inherent in the possibility of mootness. In response, they articulated a number of exceptions, including what are known as the "voluntary cessation" exception and the "capable of repetition, yet evading review" exception. These two classic exceptions to mootness target the same basic abuse: "a defendant continually mooting a challenge to his conduct by ceasing the challenged behavior to end the litigation, but then returning to the allegedly wrongful conduct after the litigation is dismissed as moot."[1124] U.S. Supreme Court Justice Neil Gorsuch alluded to this type of underhanded gamesmanship in *Roman Catholic Diocese v. Cuomo*:

> "The Governor has fought this case at every step of the way. To turn away religious leaders bringing meritorious claims just because the Governor decided to hit the 'off' switch in the shadow of our review would be, in my view, just another sacrifice of fundamental rights in the name of judicial modesty."[1125]

"Capable of repetition, yet evading review" is the classic exception to mootness, even (or especially) in COVID-related cases. As the United States Supreme Court has acknowledged, "even if the

---

1124   *Snell v. Walz*, No. A21-0626, p. 17-18 (Minn. Feb. 8, 2023), available online: https://mn.gov/law-library-stat/archive/supct/2023/OPA210626-020823.pdf

1125   *Roman Catholic Diocese of Brooklyn v. Cuomo*, 141 S. Ct. 63, 208 L. Ed. 2d 206, 592 U.S. (2020) (Gorsuch, J., concurring), available online: https://www.supremecourt.gov/opinions/20pdf/20a87_4g15.pdf.

government withdraws or modifies a COVID restriction in the course of litigation, that does not necessarily moot the case."[1126]

If the possibility of giving plaintiffs the minimal redress of some meaningful sense of closure does not provide enough motivation to carry a case to its definitive conclusion, the mere reasonable possibility that the controversy will recur provides sufficient legal justification. One would think that when authorities insisted that they maintained the power to mandate masks, that alone would have been enough to automatically create a "reasonable possibility" that mask mandates would recur. Once mask mandates were lifted, however, many courts suddenly upped the bar for what it takes for something to count as "reasonable possibility." Courts' gymnastic flexibility in defining what counts as a "reasonable possibility" when it came to masks working (no evidence needed) versus mask mandates being re-imposed (anything short of a direct confession of intent is insufficient) is astonishing. Astonishing, that is, until one realizes that the common thread is an exorbitant willingness on the part of too many judges to give the entities that mandated masks far more benefit of the doubt than they ever deserved. Then it simply becomes depressing.

Two challenges to compulsory masking were dismissed as moot by the Minnesota Supreme Court[1127] and 6th Circuit Court of Appeals.[1128] The courts said there was "no reasonable possibility" that Minnesota and Michigan's state governors or agencies would reimpose mask mandates comparable to those being challenged and which had just been lifted. The Minnesota Supreme Court acknowledged that "Governor Walz has not instituted another mask mandate or peacetime emergency in response to COVID-19, but he maintains that he retains that power."[1129] Despite the governor's insistence that he retained the power to mandate masks, and the experience of nearly three years during the COVIDcrisis, the Minnesota Supreme Court *still* ruled that mask mandates "are not, however, issues of statewide importance that require immediate resolution… Although the constitutional rights at issue are important ones, there is no serious harm or uncertainty that requires immediate resolution."[1130] Similarly, the 6th Circuit Court of Appeals said, "We are unlikely to see this mandate in a similar

---

1126   *Tandon v. Newsom*, 141 S. Ct. 1294, 1297 (2021), available online: https://www.supremecourt.gov/opinions/20pdf/20a151_4g15.pdf

1127   *Snell v. Walz*, No. A21-0626, (Minn. Feb. 8, 2023), available online: https://mn.gov/law-library-stat/archive/supct/2023/OPA210626-020823.pdf

1128   *Resurrection School v. Hertel*, 35 F.4th 524 (2022), available online: https://scholar.google.com/scholar_case?case=5344911665347166998&q=resurrection+school+v+hertel&hl=en&as_sdt=4000003

1129   *Snell v. Walz*, No. A21-0626, p. 8 (Minn. Feb. 8, 2023), available online: https://mn.gov/law-library-stat/archive/supct/2023/OPA210626-020823.pdf

1130   *Snell v. Walz*, No. A21-0626, p. 15-16 (Minn. Feb. 8, 2023), available online: https://mn.gov/law-library-stat/archive/supct/2023/OPA210626-020823.pdf

form again… we see no reasonable possibility that the State will impose a new mask mandate with roughly the same exceptions as the one originally at issue here."[1131]

What evidence could be so persuasive that it was able to convince high court majorities in these two cases that there was "no reasonable possibility" that mask mandates would recur? It certainly wasn't because the parties that imposed the mandates were willing to give any kind of *guarantee* of non-repetition. The state governors' offices and public health agencies argued strenuously that their mandates were constitutional and that they retained the power to impose them again. The mask mandates, they claimed, had been lifted for reasons unrelated to the litigation — specifically, "high vaccination rates, low case counts, new treatment options, and warmer weather," such that "COVID-19 conditions continued to improve."[1132] Nevermind the fact that even taking all of those reasons for lifting the mandates to be true, and assuming a lack of hidden motives, it would *still* do nothing to prevent *future* mask mandates the moment those trends reversed. These kinds of arguments *might* have held water after the mask mandates of the Spanish Flu, which were much shorter and a first-time occurrence, but in light of the COVIDcrisis mask mandates, which were often explicitly justified with reference to the Spanish Flu mask mandates, it was wishful thinking (at best) for any court to even *suggest* it was "entirely speculative"[1133] that there would be a recurrence of comparable mask mandates in the future. Yet courts across the country credulously did just that.

Health Freedom Defense Fund's lawsuit against the CDC's travel mask mandate suffered the same fate, dramatically weakening the precedent created by Judge Mizelle's excellent and much-needed April 2022 ruling.[1134] On April 10, 2023, President Biden signed a joint resolution of Congress that officially terminated the COVID-19 national emergency.[1135] The HHS Secretary's public health

---

1131   *Resurrection School v. Hertel*, 35 F.4th 524, 530 (6th Cir. 2022), https://casetext.com/case/resurrection-sch-v-hertel-2

1132   *Snell v. Walz*, No. A21-0626, p. 21-22 (Minn. Feb. 8, 2023), available online: https://mn.gov/law-library-stat/archive/supct/2023/OPA210626-020823.pdf quoting *Resurrection School v. Hertel*, 35 F.4th 524, *529 (2022), available online: https://scholar.google.com/scholar_case?case=5344911665534716699&q=resurrection+school+v+hertel&hl=en&as_sdt=4000003 Cases brought earlier in the COVIDcrisis were discounted because they were brought earlier in the crisis, when "the underlying emergency declarations were still in place."

1133   *Snell v. Walz*, No. A21-0626, p. 19 (Minn. Feb. 8, 2023), available online: https://mn.gov/law-library-stat/archive/supct/2023/OPA210626-020823.pdf

1134   *Health Freedom Defense Fund v. President of the United States*, No. 22-11287, (11th Cir. Jun. 22, 2023), available online: https://casetext.com/case/health-freedom-def-fund-v-president-of-the-united-states https://healthfreedomdefense.org/wp-content/uploads/2023/06/OPINION-DISMISSING-THE-CASE-AS-MOOT-00366714.pdf

1135   Act of April 10, 2023, Public Law No. 118-3, 137 Stat. 6 (2023), available online: https://www.congress.gov/118/plaws/publ3/PLAW-118publ3.pdf

emergency declaration expired on May 11.[1136] Consequently, a three-judge 11th Circuit panel ruled: "the Mandate has expired on its own terms… there is no longer any Mandate for us to set aside or uphold."[1137] The fact that the CDC continued to unapologetically defend the legality of its mandate made no difference. Indeed, this may have been part of the CDC's legal strategy once it started to lose, as one commentator observed, "The administration's real priority may have been to preserve the CDC's power. By stringing out the case until it became moot, they avoided a precedent that might have tied their hands in the future."[1138]

Many judges, to their lasting credit, recognized that the arguments for mooting challenges to mask mandates were dubious at best, and argued strenuously against doing so. Judge John K. Bush wrote a vigorous and cogent dissent against mooting Resurrection School's challenge to the compulsory masking of its students in Michigan: "Our collective experience with two years of on-again-off-again masking mandates demonstrates that there is at least a reasonable possibility this dispute could recur."[1139] Judge Bush emphasized that "the Supreme Court's own precedents instruct that a challenge to a rescinded policy is unlikely moot when the defendant mounts a vigorous defense of the policy's lawfulness." Not only did the Michigan Department of Health and Human Services continue to insist on the constitutionality of its mask mandates for school children, but "when asked at oral argument whether the state would commit not to reenact its earlier mandate, the state's counsel bluntly responded: 'Absolutely not.'"[1140] Unfortunately, Judges like John K. Bush were too often in the minority.

If one is not willing to stand on principle during a crisis, at a time when such stands are most needed, then at the very least, issues raised during the crisis should be definitively adjudicated once the panic has finally faded and the disastrous outcomes from coercive and unconstitutional policies are more visible and still at the forefront of public consciousness. Local governments across the United States

---

1136   End of the Federal COVID-19 Public Health Emergency (PHE) Declaration, Centers for Disease Control and Prevention (May 5, 2023), https://www.cdc.gov/coronavirus/2019-ncov/your-health/end-of-phe.html

1137   *Health Freedom Defense Fund v. President of the United States*, No. 22-11287, p.5 (11th Cir. Jun. 22, 2023), available online: https://casetext.com/case/health-freedom-def-fund-v-president-of-the-united-states https://healthfreedomdefense.org/wp-content/uploads/2023/06/OPINION-DISMISSING-THE-CASE-AS-MOOT-00366714.pdf

1138   Ilya Somin, "Appellate Court Dismisses Case Challenging CDC Transportation Mask mandate Because it has Become Moot," *reason.com: The Volokh Conspiracy*, June 23, 2023, available online: https://reason.com/volokh/2023/06/23/appellate-court-dismisses-case-challenging-cdc-airline-mask mandate-because-it-has-become-moot/

1139   *Resurrection School v. Hertel*, 35 F.4th 524, 532-533 (6th Cir. 2022) (Bush, J., dissenting), https://casetext.com/case/resurrection-sch-v-hertel-2

1140   *Resurrection School v. Hertel*, 35 F.4th 524, 532 (6th Cir. 2022) (Readler, J., dissenting), https://casetext.com/case/resurrection-sch-v-hertel-2

and many other countries repeatedly imposed — and then fought to maintain or re-impose — plenty of their own mask mandates even after statewide mandates were lifted (I am sad to say that my home city of El Paso, Texas, was among them). Even among those governors who lifted their mandates voluntarily, few formally relinquished or denied that they still retained the power to issue mask mandates in the future. Fewer still supported legislative safeguards against repeat COVID mandates.

For many courts, future prevention of COVID-style mandates and current psychological closure for the plaintiffs was not enough of a "practical effect" to justify ruling on such a politically hot issue, especially one that was rapidly receding into the rearview mirror. In the name of judicial modesty, multiple rulings against state governor COVIDcrisis mask mandates which would have gone a long way towards making local governments and future governors think twice about trying an encore performance to the mask-charade wound up marred by mootness. The ending of effective challenges to mask mandates before they could reach the binding precedent finish line ensured that many successes like *HFDF v. Biden* were not as strong and durable as they should have been, and many issues remain unresolved.

Some plaintiffs who stuck to their guns, like Idaho church deacon Gabriel Rench, arrested in the summer of 2020 for gathering with others to worship outside while refusing to wear a mask, were finally vindicated. In February 2023, Senior United States District Judge Morrison C. England, Jr. ruled that Rench "should never have been arrested in the first place."[1141] But even these vindications were often based on technicalities rather than the fundamental rights which were violated. Gabriel Rench and his co-worshippers were arrested in violation of their First Amendment rights to freedom of speech, free exercise of religion, and right to expressive association (and more). Yet their arrest was ruled wrongful simply because the arresting police and prosecuting city officials had failed to honor their own mask mandate carve-outs for the First Amendment.

Kicking this can down the road until the next hysteria will condemn future generations to suffer *guaranteed* abuses on similar grounds in the future. Good COVIDcrisis precedents exist alongside bad ones, often providing enough citations for judges to rule in whichever direction they are so inclined in the future. Nevertheless, many well-articulated and well-fought legal actions challenging compulsory masking now exist in publicly available forms, and can be consulted by any future litigants willing to take the time to read them. Those wins against compulsory masking in definitively completed cases like *Green v. Alachua County* in Florida have become all the more important. Other COVIDcrisis mask cases not mentioned in this book are still in progress, and the arguments they involve are

---

1141   *Rench v. City of Moscow*, No. 3:21-CV-00138, Document 40, p. 6 (District of Idaho, February 1, 2023), available online: https://casetext.com/case/rench-v-city-of-moscow or https://gaberench.com/wp-content/uploads/2023/02/40-Memo-and-Order-2-1-23.pdf

nowhere near exhaustive. More strong arguments against compulsory masking are still waiting to be thoroughly researched, articulated, tried and (ultimately) won by determined plaintiffs.

Part 5 Conclusion and general thoughts for those considering legal action

Personally, the experience of the COVIDcrisis has convinced me that a dedicated Medical Freedom Amendment is needed to more fully secure the fundamental right to not be subjected to a medical intervention in the name of some greater good. Even without such an amendment, though, we have seen that many other long-established components of the Bill of Rights still provide substantial direct and indirect protections for individual liberties when it comes to issues like compulsory masking.

For anyone who feels called to use any formal mechanism — including the legal system — to push back on compulsory masking, the next few pages are intended to be a summary of experiences and pitfalls others have encountered to provide you with a leg up and hopefully save you a few dozen (or a few hundred) hours of work. As stated previously, this is not legal advice or practicing law.

The first step in pushing back against a mask mandate using any formal mechanism involves figuring out exactly where the mandate in question came from and exactly what it says. Fortunately, this is not nearly as difficult as it sounds, though it *is* time-consuming. When someone tells you some mask requirement is "the law," do not assume they are correct. As one commentator aptly put it, "Laws are passed by legislative bodies. Around face masks there are 'orders,' 'policies,' 'statements,' 'guidances,' 'letters,' 'protocols,' and many other official sounding words that do not mean the same as 'laws.'"[1142] Most mask mandates precisely cite the laws they rely on, and since those laws are publicly available,

---

1142   Allan Stevo, *Face Masks in One Lesson*, Crafting 52 (2020), p. 6.

it becomes a simple matter to read them and find exactly what they do and do not authorize.[1143] Do not give the issuing authorities *any* benefit of the doubt when it comes to the accuracy of their statutory interpretations. Look up the statutes they cite and read them for yourself. They are publicly available by law. Cases like *HFDF v. Biden* show how important an analysis of statutory minutiae can be to the outcome of a case, and how prone to overreaching interpretations authorities tend to be when that kind of interpretation stands to increase their power.

Do your research before filing anything. Search for a lawyer before going *pro se*. If you do go *pro se*, expect to be trading a *lot* of your time for any monetary savings. As you might expect, success is never guaranteed, but your chances are directly proportionate to the amount of work you are willing to put in. Whether or not finances dictate that you end up going *pro se*, a good starting point is to read the initial complaints and rulings of successful challenges to compulsory masking, probably including one or more from cases covered in this book. Most of them are publicly available, whether on CourtListener, Justicia, individual plaintiff websites, or are linked from news articles (some are linked directly in the footnotes of this book). Don't stop there. Review the motions to dismiss, the responses to the motions to dismiss, the appellate briefs and anything else that you think may be helpful. Looking at failed challenges can also be useful. During the COVIDcrisis, many challenges to compulsory masking simply did not get a fair shake, and besides, failure tends to teach more than success.

Make sure you know where you are filing and what statutes you are filing claims under. Making a basic error like filing under the wrong statute, or filing a federal claim in state court, or a filing a state claim in federal court would get your efforts off on the wrong foot (and a number of *pro se* challenges to compulsory masking foundered on this — e.g. *Denis v. Ige*). For example, The Eleventh Amendment bars suits against state officials on the basis of state law being brought in *federal* courts. 42 U.S.C. § 1983 is the statute that permits federal courts to hear claims for injunctive relief brought against state officials under federal law: "constitutional claims against state officials are § 1983 actions, whereas claims against federal officials fall under Bivens."[1144] Considerations like this take time and mental effort to sort out correctly.

1143   For example, Texas Governor Abbott's July 2020 Mask Mandate specifically locates its authority in Sections 418.011 and Section 418.173 of the Texas Government Code, and Massachusetts Governor Baker's May 2020 Mask Mandate cites "sections 7, 8, and 8A of Chapter 639 of the Acts of 1950" for authority to issue the mandate, and General Law chapter 11 section 30 (G.L. c. 11 § 30) for the Department of Public Health, and local boards of health enforcement.

El Paso Mayor Dee Margo's mask mandate listed Texas Government Code §418.1015(a) and El Paso City Code Chapter 2.41.010 and 2.48.020.

1144   *Denis v. Ige et al*, No. 1:2021cv00011 - Document 62, p. 9 (D. Haw. 2021), available online: https://law.justia.com/cases/federal/district-courts/hawaii/hidce/1:2021cv00011/152658/62/

It is easy to be an armchair general or backseat driver after the fact. The truth is that lawyers drafting challenges to compulsory masking during the COVIDcrisis performed yeoman's service under incredibly difficult conditions, even when not successful. For many of these lawyers, it was akin to a general practitioner MD suddenly finding themselves dropped into the emergency room during a mass casualty event with no trauma surgeons available or likely to be coming anytime soon. Many of the lawyers who were willing to challenge mandates came from other specialties and had to get spun up on the relevant law while drafting the complaints and dividing their limited time among other challenges to the multiple outrageous overreach measures all clamoring for their attention. Prior cases on masks were mostly obscure and indirectly applicable, and previous mask cases that *were* directly applicable had to be located and read carefully, which took hours. Once a suit was finally filed, judges in general tended to disproportionately default to "experts" and other professionals or governmental institutions, and the poisoned legal atmosphere of the COVIDcrisis meant that most legal challenges to compulsory masking were brought in covertly (or overtly) hostile courts. Even good arguments were met with derision, whether they were about the law or the science involved. Jeff Childers recalled:

> My arguments were met by lazy, scoffing lawyers who didn't bother doing any research of their own — who mostly dismissed my cases as just being about awful Ku Klux Klan behavior. Wrong! Not one single case cited in my brief was a Klan case. I left all the many Klan mask cases out of my brief, on purpose, knowing my opponents would get too distracted by the racial issue to wrestle with the underlying logic, even though the reason Klan members shouldn't mask in public is exactly the same reason nobody else should either. When I pointed out that to my opponents there were TONS of mask cases having nothing to do with the Klan, it was like their brains suddenly malfunctioned, sparked out, and everyone just moved on to their next arguments. It was vexing.[1145]

Despite the frustrations and many failures, even unsuccessful legal challenges to compulsory masking served a useful function. Failed lawsuits are like failed scientific hypotheses or range-finding artillery rounds. They still provide valuable information that helps rule out what doesn't land on the mark so that you or some future challenger can try a different approach that *does* work or find a more fruitful channel for your efforts (legislative reform, for example).

1145   Jeff Childers, "SWITCHEROO," *Coffee and Covid* (Blog), March 4, 2023, online: https://www.coffeeandcovid.com/p/switcheroo-saturday-march-4-2023
Note: All the appellant briefs leading up to the oral arguments can be downloaded directly from the Florida First District Court of Appeal website: https://1dca.flcourts.gov/Oral-Arguments/Briefs-for-Appeals-Scheduled-for-Oral-Argument/20-1661

Moreover, even in cases where failure is practically a foregone conclusion, it is important to make the bar as high as possible for the other side, and not all of the rulings that dismissed legal challenges to compulsory masking were illegitimate or purely pretextual. If you or someone you know is thinking about challenging masks in court, put in the effort to thoroughly prepare. Even in normal times, poorly framed initial complaints can be fatal for otherwise viable claims, and judges in cases challenging masks were rarely inclined to do the plaintiffs any discretionary favors. Citing references to substantiate allegations of fact was essential, and vague generalizations were called out as grounds for dismissal. Lawsuits targeting the lowest level of mandate the plaintiffs encountered seem to have been more successful on average than those which aimed higher. Remember that successes like those in Florida, Ohio, and Illinois were aimed at county, local public health, and school district mandates, respectively. By contrast, multiple lawsuits that aimed at governors' statewide mask mandates were dismissed because the requested relief would be too "speculative" (after all, if the governor's mandate was enjoined, a municipality or private actor could still theoretically impose one).

As with many (but not all) civil rights cases throughout American history, pre-emptive challenges against mask mandates without a showing of direct harm tended to get dismissed. Some courts said the claims were not reviewable due to a lack of subject matter jurisdiction (depending on exactly what challenge was brought). Many plaintiffs challenging compulsory masking were ruled to have no standing to sue, often by failure to state a cause of action or failure to state a claim, such as failing to allege either personal intent to violate a specific mask mandate by entering a public space without face coverings or failing to allege having been personally punished for refusing to wear a mask. The more the plaintiffs had directly suffered from the enforcement of the mask mandates, the further their lawsuits typically got.

The initial complaints and responses to the motion to dismiss also had to be thorough. In at least one case, raising a new argument for the first time at oral argument was considered a form of forfeiting that argument.[1146] Many (though by no means all) mask challenges were low on scientific references. When references *were* cited to contest the efficacy of compulsory masking, they tended to be secondary sources rather than primary scientific articles. This made it easy for judges inclined to uphold compulsory masking to claim that the science was on the side of masks (though, as we've seen,

---

1146   "During oral argument for this appeal, an amicus supporting plaintiffs offered another argument as to why this claim remains live—namely, that Resurrection School's principal admitted to violating the mask mandate and thus potentially could be subject to prosecution in the future. But arguments in support of justiciability can be forfeited. See *California v. Texas*, ___ U.S. ___, 141 S. Ct. 2104, 2116, 210 L.Ed.2d 230 (2021); *Glennborough Homeowners Ass'n v. U.S. Postal Serv.*, 21 F.4th 410, 414 (6th Cir. 2021). And this argument was forfeited because it was raised for the first time at oral argument."
Resurrection School v. Hertel, 35 F.4th 524, *530 (2022), available online: https://scholar.google.com/scholar_case?case=5344911665347l6699&q=resurrection+school+v+hertel&hl=en&as_sdt=4000003

plenty of these judges set the evidential bar for defending mandates scandalously low). Lawsuits that included direct quotes from whoever officially put out the mandate, either written or spoken (e.g., at a press conference) typically got farther than those which didn't — especially when those quotes pointed to an ulterior or mixed motive on the part of whoever was mandating the masks. Evidence that showed the public health officials were not acting in good faith, or being inconsistent, arbitrary and capricious, or disproportionately burdening one group (like churches) made things especially clear.

Some First Amendment speech claims challenging masks failed to allege an actual burden on speech (either speaking or expressive conduct). First Amendment religious claims against masks needed to articulate the core religious beliefs involved and explicitly allege a substantial burden on those beliefs through compulsory masking. A Free Exercise claim against masks, *at a minimum*, needed to identify the plaintiff's religion and describe its tenets that conflicted with wearing a mask. Failure to do so made dismissal easy, as happened in *Denis v. Ige* in Hawaii:

> "The Complaint does not even identify his religion. It certainly does not
> describe the tenets of that religion or explain how the Mask Mandates affect
> any religious practice. Without more, this court cannot determine whether the
> Mask Mandates coerce Denis to act in a way that is contrary to his beliefs. Even
> if Denis had sufficiently alleged a substantial burden on his exercise of religion,
> he has failed to allege that the Mask Mandates were an irrational response to
> COVID-19. 'The right to exercise one's religion freely . . . does not relieve an
> individual of the obligation to comply with a valid and neutral law of general
> applicability… Denis has never asserted that the Mask Mandates are not neutral
> laws of general applicability. Accordingly, the Mask Mandates only violate the
> Free Exercise Clause if they fail to survive rational basis review[.]'"[1147]

As we can see from the example above, it was also important to specifically allege that the mask mandate was not neutral and generally applicable. Absent this specific allegation, most courts *automatically* "found" that mask mandates *were* neutral and generally applicable and then applied rational basis scrutiny.[1148] Even when plaintiffs made this allegation, most courts ultimately insisted that the mandates *were* neutral and generally applicable and applied rational basis scrutiny anyway, but at least alleging that the mask mandates were not neutral and generally applicable pushed the

---

1147   *Denis v. Ige et al*, No. 1:2021cv00011 - Document 62, p. 23-24 (D. Haw. 2021), available online:
https://law.justia.com/cases/federal/district-courts/hawaii/hidce/1:2021cv00011/152658/62/

1148   *Massie v. Pelosi*, 590 F. Supp. 3d 196, *230 (D.C. 2022). Available online:
https://scholar.google.com/scholar_case?case=735226917287193397&hl=en&as_sdt=4000003&scfhb=1

bar (and effort) necessary to do this that much higher, and built more support to help prepare for the inevitable appeal.

As we have seen, it is possible to successfully challenge a mask mandate even under what is essentially a rational basis standard. Alleging that mask mandates *per se* are irrational is a non-starter because "rational" in "rational basis" is defined so broadly that it includes "intuitive", and "no evidence needed," but alleging that a *specific* mask mandate is irrational or "arbitrary and capricious" — at least as implemented — *can* be viable. This is a subtle but critical distinction. Every COVIDcrisis mask mandate was self-contradictory or violated common sense in multiple ways. What took effort was spotlighting and spelling out those internal contradictions. The key pitfall to avoid was not to argue that *all* mask mandates are arbitrary and capricious, but that the *particular mandate being challenged* was arbitrary and capricious. This lesser burden is much more feasible to meet. Judges who think mask mandates *in general* are rational and justified may still be open to the idea that the mask mandate *immediately in front of them* is irrational and unjustified.

Remember that challenges brought against State and local governments under the Federal Constitution's Bill of Rights only have a chance at success insofar as they allege violations of rights which have been specifically incorporated against the states under the Fourteenth Amendment. Remember also that challenges to compulsory masking brought under the Due Process Clause of the United States Constitution were generally less effective than challenges brought under the state constitutions or due process requirements from specific statutes, as in the Ohio and Illinois cases we reviewed earlier. Each State has its own bill of rights, some of which, like Florida's, contain provisions protecting certain individual rights that are stronger than those in the United States Constitution. Jeff Childers' critical 2021 win against masks came because of *Florida's* constitution.

Procedural and statutory lines of attack were often the most effective options available to challenge compulsory masking, as in the successful statutory due process challenge brought by attorneys Brant Hadaway and George Wentz against masking Illinois schoolchildren. Masking was a form of modified quarantine because it involved devices intended to limit disease transmission, and had been imposed and then re-imposed without any of the due process required by quarantine laws. Many state and local health codes contain detailed due process procedures and guarantees. When courts were reticent about Fourteenth Amendment and State Due Process Clause challenges, these procedural due process requirements were sometimes able to act as surrogates, though they needed to be read carefully, because due process protections for custodial detention and non-custodial orders often differed. The Fourteenth Amendment should not be left out, but it benefits greatly from some local assistance, at least where challenges to masks are concerned.

Any legal win against compulsory masking is a major achievement and cause for celebration, but some wins are more decisive and durable than others. In a sense, it was actually a *good* thing that purely science-based challenges to masking were legal non-starters, because contesting compulsory masking based purely on the science implicitly granted the utilitarian premise that individual rights turn on a numbers game — that they are contingent on how many people may benefit from violating them. If you find yourself arguing purely on the science of masks, you've likely already implicitly conceded the most important ground. On the other hand, because the science of masks is so far removed from the deeper arguments that are already entrenched and get people riled up, sometimes arguing purely on the science can be an effective foot in the door. Beliefs about The Science™ can be as dogmatic as the most fanatical religions, but people's beliefs about what science *says* are generally more open to change than their beliefs about what they should do *based* on what the science says.

In theory, an administrative or procedure-based win against masks could simply have the net effect of making mask mandates more robust by forcing the issuing authority to dot all their *i*'s and cross all their *t*'s. In many cases during COVID, the mandating authority was simply given a procedural pass because of the ongoing "emergency," and the mandate was allowed to remain in force while going through the processes required to make it kosher. This lenient deference is what Supreme Court Justice Alito referred to when he wrote that: "At the dawn of an emergency — and the opening days of the COVID-19 outbreak plainly qualify — public officials may not be able to craft precisely tailored rules…. Thus, at the outset of an emergency, it may be appropriate for courts to tolerate very blunt rules."[1149]

Judges were subjected to the same psychological blitzkrieg that everyone else was. On top of this, people with credentials tend to have a pro-credential bias. Courts are no exception, and default to the assumption that any emergencies have been declared with good reason. Judges who imposed mask mandates in their own courts often looked very dubiously — if not with outright hostility — on anything which even indirectly implied they may have been wrong to do so. Still, as we have seen, a portion of judges resisted the mass hysteria just as a portion of people among the general population did, and panic will wear off eventually as long as it doesn't penetrate too deeply. None of the potential drawbacks should prevent using administrative and procedural grounds to challenge masks. Sometimes these are the only tools available, and an administrative or procedural win can still be decisive in practice. Even a temporary reprieve from masking is beneficial. Also, if the only authority that issued the mandate is also its primary (or only) proponent, then stopping that mandate may be all that is needed. Even if it is simply a matter of doing more work to make their mandates stick, if the issuing authority is too lazy or has too many other demands on its time to put in that work, they may just

---

1149   *Calvary Chapel Dayton Valley v. Sisolak*, 140 S. Ct. at 2605 (Alito, J., dissenting), available online: https://casetext.com/case/calvary-chapel-dayton-valley-v-sisolak

decide it's not worth the effort of dotting all those *i*'s and crossing all those the *t*'s. If you are willing to put in more effort than the opposition, then you already have a big advantage.

Administrative and procedural challenges to compulsory masking also have a number of advantages. Procedure-based challenges can sometimes be the most straightforward. If one can show that the issuing authority simply did not have the legal power to do so, then that's that. Authorities that may *have* the power often have a list of procedures they have to follow if the rules they put out are to be valid, and it is usually a pretty black-and-white question as to whether or not they did that in the case of their mask mandate. The CDC's habit of bypassing notice and comment is a prominent example of how authorities often failed to follow their own rulemaking procedures when mandating masks. Furthermore, if an authority has overstepped their legitimate domain to issue a mask mandate, then the odds are good that they have done so by encroaching on some other institutional authority's territory, as when federal agency mask mandates encroached on state prerogatives. Even if the second authority generally holds the bad philosophy that allows compulsory masking, they may not agree that compulsory masking is warranted in this particular case, and they are more than likely to jealously guard their powers and prerogatives, which means their interests and yours may both align against the particular mask mandate you are opposing. Also, from a judge's point of view, it is a much more comfortable decision to say "Authority B cannot mandate masks" while leaving it an open question as to whether or not Authority A can mandate them.

The most successful lawyers and plaintiffs set the bar for the judge to rule in their favor as low as possible. A good starting point was not seeking any monetary redress. In *Denis v. Ige*, the plaintiff sued for over 600 million dollars. That was a mistake.

Smart lawyers and plaintiffs made it clear the judge knew what ruling in their favor would *not* amount to. No judge in 2020 was (and only a rare judge is at any time) willing to make a finding of fact that contradicts an organization like the CDC. In *Green v. Alachua County*, Jeff Childers made sure the judges were aware of exactly how low the bar for ruling in favor of his client actually was. He started by making it clear that his case was not about whether or not masks actually work. Childers also emphasized that granting a temporary injunction in his client's favor did not amount to striking down the mask mandate, because it was still possible the mask mandate would survive a hearing on the merits. He repeatedly pointed out that showing Alachua County lacked authority to require face masks was not the same thing as striking down *all* mask mandates, because the higher authorities of the Governor or the State Legislature could still potentially have the authority to mandate masks. Jeff Childers' reminders about these things apparently did not fall on deaf ears, because the appellate court's ruling was careful to highlight this distinction: "To be clear, we are not saying that the mask mandate in fact was unconstitutional. If, however, Green persists in his challenge to some new mask mandate that the county adopts, the trial court would have to start its analysis with this presumption of unconstitutionality."[1150] In other words, we're not saying the county's mask mandate is unconstitutional. We're saying it's *presumptively* unconstitutional. For that stage of the case, this was the correct call, and *presumptively* unconstitutional is often all that is needed to stop measures based in fear, bad science, manipulative psychology, and bad philosophy.

Another good example comes from the challenge to masks out of Minnesota: "This Court need not reach any conclusion about the effectiveness of face coverings in Minnesota, but it should conclude that absent evidence that masks are actually necessary to slow the spread of COVID-19, there is no compelling state interest in forcing people to wear masks of wildly varying types."[1151] Note that finding that the mandating authority failed to provide enough evidence that masks work and are necessary is not the same thing as finding that masks do not work at all. It is also far less of a stretch for any court to reach than a finding that masks do not work.

At the time of this writing, the CDC and other federal agencies are still very much exploring alternative avenues that they can use to reimpose mask mandates, and there are plenty of judges who would be happy to rule in their favor. The legal duels over compulsory masking have abated, but they are far from over. In a sense, they will never be over, because the mentality underlying compulsory masking mutates, renews, and resurfaces in every generation. Trying to forcibly inoculate the entire population against it by imposing ideological conformity would be immoral even if it could be done effectively. The only real answer is to strengthen enough individuals on a voluntary basis that a sort of societal immunity prevents the next panic from gaining the same foothold compulsory masking did during the COVIDcrisis. In the final section of this book, we will look at ways you can personally make a difference in how the final denouement on masks turns out, as well as to how this general cycle plays out the next time around.

1150   *Green v. Alachua County*, 323 So. 3d 246, *254 n 6 (Fla. Dist. Ct. App. 2021). Available online: https://law.justia.com/cases/florida/first-district-court-of-appeal/2021/20-1661.html

1151   *Snell v. Walz*, No. A21-0626, appellant brief pdf page 59 (Minn. July 14, 2023), available online: https://www.umlc.org/wp-content/uploads/2021/08/download1.pdf

Part 6:

Fighting Back

What next?

Congratulations! You now know far more about all aspects of facemasks than any of the doctors, politicians, public health officials, or other authorities who pushed, prodded, cajoled, coerced, and finally forced them on you, your friends, and your loved ones. In the process of acquiring this knowledge, we have examined some of the deceptive games played using The Science™, how the so-called "experts" actually know nothing like the amount they purport to, and how their rule-making is often based more on visceral guesswork and moral assumptions no better than those of anyone else (in fact, typically worse). While it is impossible for one person to do this level of research on more than a few topics, having seen how manipulative tricks are deployed in one area will give you some cross-immunity when the same techniques are tried in other areas. You are now more readily able to discern and resist the telltale hallmarks of manipulation elsewhere.

So, what now? In this final chapter, I offer some closing thoughts and suggestions. Ultimately, though, what use you make of this knowledge is up to you. It will be different for every person.

The Most Important Battleground is Your Own Mind

Simply having this knowledge is beneficial. You are now far more able to resist and contest compulsory masking at every level if and when it tries to make a comeback — not just at the scientific or policy level, but at the more fundamental levels of the psychology and philosophy involved. Having read this book, you are now more able to help free those still trapped by the fear and false beliefs that led to compulsory masking. Even more importantly, you are personally better able to withstand and effectively oppose the philosophy and mentality behind compulsory masking when it manifests in other domains.

Compulsory masking is *currently* in retreat, but the mentality that led to the self-inflicted disaster that was the COVIDcrisis cannot be eliminated. It is like an endemic virus. The specific symptoms and manifestations vary over time, but it will never completely go away because the attitudes and beliefs that give rise to it are renewed in many people each generation. Moreover, the court penchant for mooting so many of the challenges to compulsory masking just as they were gaining momentum means that the issue is guaranteed to return in either our lifetimes or those of our children unless we take action now to put additional barriers in place.

What I have been referring to as the "masking mentality" is simply another term for a tendency present to one extent or another in all humans (myself included). It has many other names. Masking was simply the way this mentality manifested during the COVIDcrisis. The core trait of the masking mentality is what German Lutheran pastor Dietrich Bonhoeffer referred to as "stupidity," or "folly" (depending on which English translation of Bonhoeffer's writings you read). Pastor Bonhoeffer is best known for his courageous opposition to the Third Reich, for which he suffered arrest and execution. What is popularly known as "Bonhoeffer's Theory of Stupidity," comes from a letter essay he wrote to friends and family in December 1942. After 10 years of living under the rule of the National Socialist German Workers' Party, pastor Bonhoeffer wrote:

> "Stupidity is a more dangerous enemy of the good than malice. One may protest against evil; it can be exposed and, if need be, prevented by use of force… Against stupidity we are defenseless. Neither protests nor the use of force accomplish anything here; reasons fall on deaf ears; facts that contradict one's prejudgment simply need not be believed — in such moment the stupid person even becomes critical — and when facts are irrefutable they are just pushed aside as inconsequential, as incidental. In all this the stupid person, in contrast to the malicious one, is utterly self-satisfied and, being easily irritated, becomes dangerous by going on the attack. For that reason, greater caution is called for when dealing with a stupid person than with a malicious one. Never again will we try to persuade the stupid person with reasons, for it is senseless and dangerous."[1152]

To Bonhoeffer, stupidity:

> "[I]s in essence not an intellectual defect but a human one. There are human beings who are of remarkably agile intellect yet stupid, and others who are intellectually quite dull yet anything but stupid… The impression one gains is not so much that stupidity is a congenital defect but that, under certain circumstances, people are made stupid or that they allow this to happen to them… it would seem that stupidity is perhaps less a psychological than a sociological problem… The process at work here is not that particular human capacities, for instance, the intellect, suddenly atrophy or fail. Instead, it seems that under the overwhelming impact of rising power, humans are deprived of their inner independence and, more or less consciously, give up establishing an autonomous position toward the emerging circumstances. The fact that the stupid person is often stubborn must not blind us to the fact that he is not independent. In conversation with him, one virtually feels that one is dealing not at all with him as a person, but with slogans, catchwords, and the like that have taken possession of him. He is under a spell, blinded, misused, and abused in his

---

1152 Dietrich Bonhoeffer, "After Ten Years" (Dec. 1942), reprinted in *Dietrich Bonhoeffer: Letters and Papers From Prison*, John W. DeGruchy, editor, Barbara and Martin Rumschedit, translators, Minneapolis: Fortress Press (2010), p. 43, available online: https://archive.org/details/letterspapersfro0008bonh/page/42

very being. Having thus become a mindless tool, the stupid person will also be capable of any evil and at the same time incapable of seeing that it is evil."[1153]

As Bonhoeffer observed, this stupidity can arise from virtuous traits and good "pro-social" motives:

"Who can deny that in obedience, duty and calling we Germans have again and again excelled in bravery and self-sacrifice? But the German has preserved his freedom… by seeking deliverance from his own will through service to the community… The trouble was, he did not understand his world. He forgot that submissiveness and self-sacrifice could be exploited for evil ends."[1154]

But Bonhoeffer also found "a grain of consolation in these reflections on human folly. There is no reason for us to think that the majority of men are fools under all circumstances."[1155] The less-optimistic corollary to this is that all of us are stupid fools in at least *some* situations, and to make matters worse, we are least likely to recognize those situations when we are in the middle of them. Different people are stupid in different situations. A single person can be a pristine example of how to avoid this trap in one area of their life but fall headlong into it in another area. I make no claim to be an exception to this rule. The same held true during the COVIDcrisis. The COVIDcrisis was *built* on this type of stupidity. Many people easily perceive and decry this failing in others while completely failing to notice it in themselves (here, too, I make to claim to be an exception). This is one of the reasons why, even on unexamined utilitarianism, we ought to vigorously protect *everyone's* individual liberties, so that those people who get it right can have the chance to persuade everyone else who is being stupid. If someone is really being stupid, their stupidity will, on its own, incur plenty of natural consequences if they persist in being stupid. Piling artificial legal consequences on top of the natural ones is just vindictive and only adds unnecessary suffering.

Bonhoeffer also wrote, "we know that to sow and to nourish mistrust is one of the most reprehensible things, and that, instead, trust is to be strengthened and advanced wherever possible."[1156] Whether the

1153   Dietrich Bonhoeffer, "After Ten Years" (Dec. 1942), reprinted in *Dietrich Bonhoeffer: Letters and Papers From Prison*, John W. DeGruchy, editor, Barbara and Martin Rumschedit, translators, Minneapolis: Fortress Press (2010), p. 43-44, available online: https://archive.org/details/letterspapersfro0008bonh/page/44

1154   Dietrich Bonhoeffer, "After Ten Years" (Dec. 1942), reprinted in *Prisoner for God: Letters and Papers from Prison*, Eberhard Bethge, editor, Reginald H. Fuller, translator, New York: The Macmillan Company (1959), p. 16, available online: https://archive.org/details/DietrichBonhoefferLettersFromPrison/page/n19

1155   Dietrich Bonhoeffer, "After Ten Years" (Dec. 1942), reprinted in *Prisoner for God: Letters and Papers from Prison*, Eberhard Bethge, editor, Reginald H. Fuller, translator, New York: The Macmillan Company (1959), p. 19, available online: https://archive.org/details/DietrichBonhoefferLettersFromPrison/page/n21

1156   Dietrich Bonhoeffer, "After Ten Years" (Dec. 1942), reprinted in *Dietrich Bonhoeffer: Letters and Papers From Prison*, John W. DeGruchy, editor, Barbara and Martin Rumschedit, translators, Minneapolis: Fortress Press

pretext is political ideology or public health, sowing mistrust is a go-to play for totalitarian regimes throughout human history. Anyone could be an informer who denounces you. Compulsory masking successfully sowed and nourished mistrust like few other measures in history — pushing the message that you should be very afraid because anyone could be carrying a virus that could infect and kill you.

The masking mentality sees dissent as an existential threat. It seeks to shut down debate through force. This tendency is part of human nature and has not changed for thousands of years. It shows up in multiple different contexts. It is as old as the Bible. When the Babylonian King Nebuchadnezzar was besieging Jerusalem, the prophet Jeremiah urged surrender:

> "This is what the LORD says: 'Whoever stays in this city will die by the sword, famine, or plague, but whoever goes over to the Babylonians will live. They will escape with their lives; they will live.' And this is what the LORD says: 'This city will certainly be given into the hands of the army of the king of Babylon, who will capture it.'" Jeremiah 38:2-4 (NIV)

The response of those who disagreed with Jeremiah was not debate, setting an example by their courage defending the walls, or even to simply expel him from the city. Instead, "the officials said to the king, 'This man should be put to death. He is discouraging the soldiers who are left in this city, as well as all the people, by the things he is saying to them. This man is not seeking the good of these people but their ruin" (Jeremiah 38:2-5, NIV). How many times were similar accusations flung at those who refused to comply with one or another of the ostensibly anti-COVID measures? Emotional-blackmail was used to push compulsory masking, and vituperation was systematically directed at those who (quite rightly) objected and resisted.

The masking mentality involves a mindset of fear. It confuses what is theoretically *possible* with what is likely, or certain. The masking mentality habitually conflates sins of omission with sins of commission, as though refusing to do something *for* someone is the same as doing something *to* them. The masking mentality has a knee-jerk authoritarian reflex; it follows rules simply because they are rules. The masking mentality is characterized by non sequiturs: "It is good for *me* to wear a mask, therefore *everyone* should do it," or "This works, therefore it should be mandatory." The masking mentality involves outsourcing one's critical reasoning. Václav Havel's description of totalitarianism applies: "an essential aspect of this ideology is the consignment of reason and conscience to a higher authority. The principle involved here is that the center of power is identical with the center of truth."[1157] The

---

(2010), p. 47, available online: https://archive.org/details/letterspapersfro0008bonh/page/46

1157   Václav Havel, *The Power of the Powerless.* International Journal of Politics October, 1978, p. 4. Available Online: https://www.nonviolent-conflict.org/resource/the-power-of-the-powerless/

masking mentality will often disparage an unappealing explanation by arguing that the explanation *itself* needs to be explained before it can be accepted. It is the sort of mentality that says, "Until you can show me *why* masks don't work to stop disease transmission, you should wear one."

Under the masking mentality, few, if any, sources are cited to bolster confident pronouncements that: "The evidence is overwhelming." When sources *are* cited, they're typically secondary sources at best, most often tertiary ones. Nevertheless, personal judgement must capitulate to authority, no matter how well that personal judgment is informed by original sources and primary evidence. People operating under the masking mentality may actually be very well-versed in what their particular source of authority says, but may brush off, entirely ignore, or be unable to support other even more basic assumptions, such as whether their authority should be trusted in the first place. They may, for example, be able to quote CDC recommendations chapter and verse, but be completely unaware of the massive body of literature which the CDC ignored when coming up with its recommendations, and they simply take for granted that the CDC should be everyone's ultimate authority on individual preventative medicine in the first place. Confirmation bias, motivated reasoning, and cognitive dissonance are integral components of the masking mentality. Under the masking mentality, actions that are fundamentally ritual and superstition are couched in the language of science and then mandated. Gut fear gets the final say. The rituals are intended to alleviate primal fears, but in reality they perpetuate and exacerbate those fears.

Being an "amateur" does not make someone automatically wrong, and being an "expert" does not make someone automatically right, but the masking mentality default rebuttal is *ad-hominem* attack based on credentials and motivations. The worst motivations are assigned to those who disagree about matters of fact, such as whether or not unmasked and unvaccinated people are truly a threat to those around them. Those operating under the masking mentality seek to place the entire burden of proof on the opposing side while denying their own burden of proof. One way this is done is to subtly frame a debate as a proposition in which all one side has to do is play the skeptic instead of a question where both sides have to defend their position. It is the difference between being willing to debate the question: "Should people be forced to wear masks?" vs. demanding: "prove to me that you should not have to wear a mask!"

Just because mask mandates are gone does not mean the masking mentality is gone. Widespread compulsory masking has already happened twice, first during the Spanish Flu and now during the COVIDcrisis. During the intervening century, dehumanizing treatment of prisoners and unexamined "best practices" for "evidence based" infection control in the medical field served as institutional reservoirs for masking. This second time around, the masking mentality really got its hooks into a large subset of the population. The second time *happened* because the first time was never repudiated and guarded against.

The only way to keep the masking mentality from achieving the same dominance it did in 2020 is to strengthen and reaffirm the intellectual, moral, and spiritual principles that provide immunity — not by trying to force belief as though it were a compulsory mask or vaccine, but by persuasion, example, and an unyielding defense of our own individual boundaries. This can only take place, *voluntarily*, at an individual level. If even 20% of the population had resolutely opposed compulsory masking, refusing to comply under any circumstances, compulsory masking and everything that followed from it would have collapsed before it even got going, and much of the harm that followed afterwards would have been avoided. This is analogous to dealing with a virus that cannot be eliminated. When you cannot eliminate a virus, the next best option is to strengthen as many people as possible to achieve pre-existing immunity or to recover without assistance. Those who started out immune or recover on their own can provide palliative care for those who need help to recover. The more people that recover, the less vulnerable those who are perpetually susceptible to the social influences involved become. When it comes to the mentality that the COVIDcrisis was built on, you cannot prevent exposure and you cannot guarantee preexisting immunity, so the best defense is to strengthen as many individuals as possible, and to vigorously uphold the individual liberties that those with innate or acquired immunity can exercise to help others recover. The more you bolster your own innate immunities, the more you help those around you.

Asch's Findings and Milgram's Exit Interviews

When no alternatives are visible, group pressure and groupthink can easily produce a self-reinforcing mass hysteria, delusion, or mass formation.[1158] If anyone wonders how cultures like the Aztecs could come to institutionalize practices like human sacrifice, the COVIDcrisis was our practical introduction to the first steps in that process. In one sense, the findings of Solomon Asch, Stanley Milgram, and their successors on obedience and conformity are a complete black pill, but variations in their basic experiments also revealed ways in which the forces of group influence can be mitigated or reversed by those willing to stand on their convictions. In Asch's original experiment, when the test subject had a partner who agreed with their answers, conformity to the group error decreased by 2/3, from 30% of test subjects to just 10%. When you refuse to conform to a group error by wearing a mask, you help others by filling that role. When Solomon Asch inverted his experiment, using a majority of test subjects, but adding a single accomplice who consistently dissented by giving wrong answers, the majority did not take the dissenter seriously. They even found him amusing: "contagious laughter spread through the group at the droll minority." However, when Asch increased the number of accomplices giving the wrong answer to at least three, he found that: "the attitude of derision in the majority turns to seriousness and increased respect."[1159] Vocal, determined minorities can shift the majority. This is why totalitarianism demands 100% conformity and compliance, and the presence of this demand for 100% conformity and compliance is a dead giveaway.

1158   Thank you to Dr. Mattias Desmet and Dr. Robert Malone for popularizing this term.

1159   Solomon E. Asch, "Effects of group pressure upon the modification and distortion of judgments." *Groups, Leadership and Men: Research in Human Relations*, Harold Guetzkow, ed. United States: Carnegie Press (1951), p. 177-190, p. 189, available online: https://www.google.com/books/edition/Groups_Leadership_and_Men/aPRGAAAAMAAJ?hl=en

In the 2012 study examining peer influence on people's beliefs about the morality of torture which we looked at in our review of mask psychology,[1160] those with pre-existing strongly held moral convictions were far less likely to change their stance on torture than were their peers who held no such beliefs. This was especially true when their moral pre-commitments were informed by religious belief. This type of conviction was exhibited by at least two of the subjects in Stanley Milgram's original experiments. These two men were part of the minority that refused to proceed past the 150-volt shock when the "learner" demanded to be released, and they held to their refusal in spite of the lab-coated authority figure's promptings.

In his book, Milgram describes the reasons which both of these men gave for how they were able to resist the inhumane orders from an authority figure. One of the men was a teacher of Old Testament liturgy. When asked: "What in your opinion is the most effective way to strengthen resistance to inhumane authority?" he answered, "If one had as one's ultimate authority God, then it trivializes human authority."[1161] The other man was an electrical engineer by the name of Jan Rensaleer. He stopped giving the shocks for two reasons. First, he knew from personal experience how the victim must feel. He said, "I know what shocks do to you. I'm an electrical engineer, and I have had shocks... and you get real shook up by them — especially if you know the next one is coming." Second, he refused to assign any responsibility for his actions to anyone else in the experiment. "When asked who was responsible for shocking the learner against his will, he said, 'I would put it on myself entirely.'"[1162]

"*I* am responsible." Breaking the diffusion of responsibility is one of the ten steps that Dr. Zimbardo listed to help resist unwanted influences in his book *The Lucifer Effect: Understanding How Good People Turn Evil*. During COVID, there was a lot of equivocation over what "individual responsibility" meant. Advocates for mandates argued that diffusion of responsibility is exactly what people who *refused* to wear masks or get vaccinated were wrongfully doing. Mandate proponents argued that refusing to wear a mask (and later, refusing to get a vaccine) was refusing to take responsibility for one's part in public health by potentially spreading infections the refusers did not even know they had. This was and is simply wrong. Setting aside the general rule that, by definition, transient infections with no

1160   Aramovich, N.P., B.L. Lytle, and L.J. Skitka, Opposing torture: Moral conviction and resistance to majority influence. Social Influence, 2012. 7(1): p. 21-34. https://doi.org/10.1080/15534510.2011.640199, available online: https://www.tandfonline.com/doi/full/10.1080/15534510.2011.640199

1161   Milgram, S., *Obedience to Authority*. 1974, New York: Harper and Row, p. 49, available online: https://archive.org/details/obediencetoautho0000milg/page/n7/

1162   Milgram, S., *Obedience to Authority*. 1974, New York: Harper and Row, p. 51-52, available online: https://archive.org/details/obediencetoautho0000milg/page/n7/

Note: Humans, especially tyrants, intuitively recognize this, and that is why totalitarian regimes the world over single out religions like Christianity for special attention and persecution.

symptoms are not major public health threats, refusing to hold other people responsible for one's own personal health, and vice versa, is the indispensable first step in taking *real* responsibility for one's part in the overall public health. *Every* human action has the potential to spread germs, and we depend on microbes for our very existence. If we could somehow achieve aseptic perfection and eliminate the spread of all germs, all we would get for our trouble would be a sterile world devoid of any organic life.[1163] There are *multiple* ways to reduce the risk of disease. Who is so presumptuous that they can insist that one person who manages to achieve a 100% reduction in their personal risk of a given disease is doing more to benefit public health than another person who achieves a somewhat lower reduction in risk from *multiple* diseases by generally healthy living or some other method? Medical and public health recommendations are just as prone to fads as any other academic discipline. Failing to make use of the latest obsessive public health craze does nothing "to" anyone else. At the most, it is refusing to do something "for" someone, and that is very different.

You are responsible for those things which you can control, not for every possible outcome your actions or non-actions could possibly lead to. "I am responsible" does not mean you are responsible for someone else's overall well-being or for every bad outcome in the chain of events your actions happen to be part of, as the crude consequentialism behind mask mandates assumed. Rather, it means that your actions, your life, and your health are ultimately your own responsibility and not someone else's. The same applies to other people. Regardless of where, why, when, or how SARS-CoV-2 originated, ultimately the only ones answerable for our individual responses to it were *ourselves*. The worst harms from COVID did not come from SARS-CoV-2, but from our *response* to SARS-CoV-2, and were *self-inflicted*, not just on ourselves, but on our children and grandchildren.

1163   I am indebted to Allan Stevo (*Face Masks in One Lesson*, 2020) for his uniquely eloquent expression of this point. Personally, the COVIDcrisis obsession with infection control reminds me very much of the general revulsion for organic life evinced by the villains in C.S. Lewis' science fiction novel *That Hideous Strength*.

What you do matters

When it came to masking and other COVID lockdown measures, seemingly trivial acts made a difference. Mutual encouragement and evidence-based sparring in various forums were not wastes of time. The people pushing masking and other measures were just as human as anyone else, and many (if not most) were even more sensitive to social pressure and averse to confrontation. However, verbal dissent can only do so much. In Milgram's experiments, virtually all of the subjects dissented verbally, but only a minority stopped administering the shocks. In a way, having the option of verbal dissent acted as a pressure release valve that helped enhance behavioral compliance. Of the minority of dissenters in Milgram's experiments who went beyond verbal dissent and refused to continue, none took further steps beyond nonparticipation to challenge the experiment's existence. Philip Zimbardo's assistant professor (and future wife) Christina Maslach laid out the ramifications:

> "Suppose it was not an experiment; suppose Milgram's 'cover story' were true, that researchers were studying the role of punishment in learning and memory and would be testing about one thousand participants in a host of experiments to answer their practical questions about the educational value of judiciously administered punishment. If you disobeyed, refused to continue, got paid, and left silently, your heroic action would not prevent the next 999 participants from experiencing the same distress. It would be an isolated event without social impact unless it included going to the next step of challenging the entire structure and assumptions of the research. Disobedience by the individual must get translated into systemic disobedience that forces change in the situation or agency itself and not just in some operating conditions. It is too easy for evil situations to co-opt the intentions of good dissidents or even heroic rebels by giving them medals for their deeds and a gift certificate for keeping their opinions to themselves."[1164]

1164 Philip Zimbardo. *The Lucifer Effect*, New York: Random House Publishing Group, 2007, Kindle Edition, p. 458-459.

To be fair to Milgram's subjects, some might very well have taken concrete steps against the experiments if the ruse had not been revealed to them immediately at the conclusion of their session, when they were presented with the "victim" who they saw to be healthy, friendly, and in good spirits. The point remains, however, that direct *behavioral* challenges to compulsory masking and other COVIDcrisis overreach by nonparticipation were (and remain) absolutely indispensable. The little things mattered, even if it was just a little bit of passive-aggression like making a habit of not putting on a mask until directly asked to do so. Those who were willing to go the next step and further challenge the existence of compulsory masking performed a service for everyone. These individuals sacrificed to do so, whether by giving up time, money and energy, paying a fine, suffering an arrest, losing a job, being denied access to school, or various forms of social humiliation and ostracism. This is true even in those cases where the individuals challenging compulsory masking were obnoxious or belligerent about it. People who refused to wear masks were (at best) derided as being dangerously irresponsible, and treated accordingly. Some endured quietly, but many, very understandably (and not entirely inaccurately), perceived attempts to pressure them to wear a mask as an aggressive violation of an intensely personal boundary — bordering on a common law form of battery — and responded accordingly.

Many of the authors and thinkers whose work we have examined also offer valuable insights into what we can do on an individual basis to inoculate ourselves against situational, psychological, social, and spiritual forces pushing us onto a majority path to immoral actions or becoming compliant and complicit in a way that we later look back on with guilt, shame, and a desire to rationalize. In his book, *The Lucifer Effect*, Philip Zimbardo suggests 10 statements that, if internalized, help us resist situational forces which may be pushing us to do the wrong thing. He also provided prescient warnings that too many people ignored during COVID:

> "The Mephistophelian tempter will argue that his power to save you depends
> upon all the people making small sacrifices of this little right or that small
> freedom. Reject that deal. Never sacrifice basic personal freedoms for the promise
> of security because the sacrifices are real and immediate and the security is a
> distant illusion."[1165]

In his book, *Influence: The Psychology of Persuasion*, Robert Cialdini provides an example of how to consciously define ourselves in such a way that we are more likely to stay open-minded and less likely to fall for the type of behavioral manipulations employed during the COVIDcrisis. The Czech Statesman Václav Havel and the dissident Russian writer Aleksandr Solzhenitsyn also offer valuable

---

1165   Philip Zimbardo. *The Lucifer Effect*, New York: Random House Publishing Group, 2007, Kindle Edition, p. 455

insights into how totalitarian systems function, and how the smallest actions or *non*-actions can have a paradigm-changing cumulative impact for good.

The only person we can control is ourself. All of us, if we are being honest, will admit even *that* control is, at best, incomplete. The most accessible and the most important actions we can take are internal. To the extent that our internal house is out of order, our external efforts will be undermined, and may even be vulnerable to misdirection or misuse. A good first step is to consciously determine your values, where your own personal boundaries lie, and what you are willing to do, pay, or sacrifice to hold to them. Are you willing to sacrifice time? Money? Are you willing to deal with dirty looks from strangers? Awkward social situations? Worse? When confronted by something like widespread COVID masks and lockdowns, prioritize enough time and mental effort to figure out where your own personal lines in the sand are. This includes both where they *actually* are and where you *want* them to be. Determine what you are able to do and what you are *willing* to do to push back, and then do that thing, however small, to the best of your ability. Gradual successive steps most easily lead downwards, but they can also lead upwards. As Aleksandr Solzhenitsyn pointed out, if "civil disobedience *a la Gandhi*," is too much, then the simplest key to liberation is "a personal nonparticipation in lies… We are not called upon to step out onto the square and shout out the truth, to say out loud what we think — this is scary, we are not ready. But let us at least refuse to say what we do *not* think!"[1166] Follow through and make habits of whatever positive actions (or non-actions) you are willing and able to take. Step back and re-evaluate from time to time. I speak from experience when I say that your own personal breaking point can and will change over time — in a good way! You will get stronger if you figure out what small steps you are willing to take in the direction you want to go and then follow-through.

If you are new to this sort of thing, start small. Even as minor an act as refusing to put on a mask until someone directly tells you to do so can have a large cumulative effect if enough people do it. Courage and personal boundaries are like muscles — they get stronger the more you make use of them. At any rate, such minor acts set good internal precedents. My own assessment is that individuals can make the most predictable impact by starting closest to home, inside their own minds, and working outwards and upwards from there. Florida lawyer Jeff Childers' advocacy mantra is "local, local, local" for a good reason, though your own circumstances may compel you to start bigger. Oftentimes, the most selfless and effective thing you can do is simply to implacably hold your own personal line in the sand. Letting any observers know what you are doing and *why* you are doing it will make your stand more effective.

1166   Aleksandr Solzhenitsyn, *Live Not by Lies*, (1974), available online: https://www.solzhenitsyncenter.org/live-not-by-lies

In this book, we have looked in depth at many of the manipulations, fallacies, tricks, and "nudges" used to push masks. I cannot claim to be immune to these things when I encounter them in other areas of my own life, but that's part of the overall point. No one is immune — not me, not you, not doctors, not public health officials, and *especially* not politicians! (If anything, politicians may be *more* susceptible because they earn their living by their skill in recognizing and appropriating social trends.) These vulnerabilities, by their very nature, live in our personal blind spots, and we're better at spotting them in others than in ourselves (which is also part of human nature). However, with conscious effort, we *can* shrink our personal blind spots and diminish the extent to which we engage in defective thinking. We can stay alert for cues that influence tactics are being used to manipulate us. They're usually pretty easy to spot once you learn to recognize them, especially when they are used *en masse*. Brute coercion, moving goalposts, and gaslighting are all, of course, giant red flags.

When all else fails, weigh the opposing arguments as best you can, stand firm on fundamental principles, and let other people do the same. When in doubt and push comes to shove, do not let yourself be pushed into making exceptions based on "emergency" and necessity. Be open to reevaluating your positions, but remain indomitable in the defense of your personal boundaries. At a minimum, I recommend taking the time to make a conscious addition to your self-definition. If you make a mental note or internal commitment, that is better than nothing. If you write your commitment down, simply having done so and the physical reminder you produce will make you adhere to it more closely and consistently. Handwriting works better than typing because the additional time, effort, and personalization enhance psychological commitment. If you share your commitment with others, you are even more likely to follow it when the going gets tough (though you do run the risk of having your failure to live up to one or more of your commitments pointed out to you in the heat of the moment — such is life).

I share my own personal resolution here as an example and perhaps a jumping off point for anyone else inclined to make one. It is drawn from all the works we have looked at together, including those in the Appendix. It is perpetually in-progress, and I do not claim to always live up to it (in fact, I routinely fall short). It is as much a statement of aspiration as it is a statement of present reality, but in adhering to it as best I can, I become less vulnerable to groupthink, mass hysteria, and compromising sacred values. Commitment to an ideal does not mean perfection in pursuit of that ideal. Strive for excellence, not perfection (some internal rebuke for failure may be warranted, but there is a point where self-flagellation becomes counterproductive, unhealthy, or self-indulgent). Here is my personal resolution:

I hold ideas, but I do not become my ideas. My views are based on the information, evidence, arguments, and experiences I have had to-date. I can and do modify my views as I encounter more or better information, evidence, arguments, and experiences. I am willing to evaluate information, evidence, and arguments even when those things come from sources that I regard as unfriendly or antagonistic. I am willing to admit to error. I do not affirm what I know or believe to be untrue. I defend my personal boundaries. I consistently second-guess my motivations and my reasoning. I do not blindly follow authority, nor do I consign my reason and my conscience to any human authority. I value my integrity more than I value group acceptance. Even if I am ordered to do something, I will still bear ultimate responsibility for having done it. I can and do turn off my autopilot, pause, take a step back, and critically evaluate a situation, especially when something does not seem right. I do not allow the present moment to dominate me — my present actions are informed by my memory of the past and desires for the future. I do not sacrifice fundamental principles and freedoms for safety. I can and do oppose unjust rules.

When you take a stand to oppose things like compulsory masking, there are *at least* four broad objectives that you can achieve regardless of what anyone else does, assuming you do not quit:

- You can make the arguments.
- You can set an example.
- Dealing with you forces the other side to actually get their unstated arguments and assumptions out in the open, where they can be addressed directly.
- You can enhance informed consent, one way or the other. (I am big on informed consent, and if you've read this book, you probably are, too. Where coercion is present, informed consent is not.)

As you begin to enforce your boundaries, do your best not to hand would-be tyrants an easy win. Sometimes, even when physically fighting back is more than morally justified, the most effective way to win is still to take a punch and then turn the other cheek to take another one without backing down. Even when you *know* your rights are being violated, the authorities who are in the process of violating them will give each other every benefit of the doubt and seize on any excuse to justify each other's actions (up to and past the point of reason). If you are going against the herd, you can expect the opposite of that, at least at first. Expect to have your motives, maturity, intellect and integrity routinely referred to as worse than garbage. It's wrong. It's not fair. It's not how the law *itself* says it is supposed to be applied, but that is still how it works in situations like the COVIDcrisis, so be strategic, sober-minded, and self-disciplined. When you exercise your rights in a context where an authority is determined to violate them and thinks they have the moral and legal high ground, then you give yourself the best fighting chance for ultimate vindication if you assert your rights in a manner that that makes it as obvious as possible to every semi-objective outside observer that the authorities violating your rights are in the wrong. This may require immense patience, forbearance, and self-discipline, because it means getting repeatedly kicked around and not reciprocating, despite your legitimate right of self-defense.

An outstanding example of this is the conduct of the civil rights student protestors we discussed earlier whose First Amendment rights were eventually vindicated by the Supreme Court in *Edwards v. South Carolina*. The Supreme Court made clear that it was their conduct in the face of blatantly unfair treatment that made the difference:

> "Not until they were told by police officials that they must disperse on pain of arrest did they do more. Even then, they but sang patriotic and religious songs after one of their leaders had delivered a 'religious harangue.' There was no violence or threat of violence on their part, or on the part of any member of the crowd watching them…

> "If, for example, the petitioners had been convicted upon evidence that they had violated a law regulating traffic, or had disobeyed a law reasonably limiting the periods during which the State House grounds were open to the public, this would be a different case."[1167]

Even then, the winning students still had to endure through a *two-year* process of repeatedly losing at the local court level and the appellate court level, with fines or imprisonment adding injury to injury and no guarantee of final success. A common thread in court cases that resulted in landmark decisions upholding Americans' essential individual liberties was that the vast majority of the time, the person whose liberty was finally vindicated had lost at the level of one or more lower courts.

Don't despair. It's ok to get tired. It's ok to get discouraged. It's a marathon, not a sprint. It's ok to take a mental health break to remind yourself why you are fighting. If you've reached the point past which you are not willing to go to resist, it may even be ok to step back. One thing you must *not* do, however, is despair. Even after a final court adverse ruling, the passage of new laws and ordinances or the repeal of old ones can easily shift the legal landscape, creating new avenues and reopening old

---

1167   *Edwards v. South Carolina*, 372 U.S. 229, 236 (1963), available online:
https://tile.loc.gov/storage-services/service/ll/usrep/usrep372/usrep372229/usrep372229.pdf

ones. Quitting in despair is the only guaranteed way to lose, and sometimes a tactical loss is necessary to pave the way for a strategic victory.

If you are resisting a mask mandate within almost any setting, you are likely to find yourself having to pursue whatever administrative process for accommodation is in place. Don't expect immediate success. In fact, expect to lose at the lowest levels and have to appeal. There is almost always an appeal process, even when it is not obvious at first glance. *Do* expect it to be a massive timesink, but don't use that expectation to lower the quality of your work. Your job is to exhaust any administrative remedies and make the bar for denying your accommodation request as high as possible. Authorities are human too, and have other issues on their plates. It is even possible they secretly loathe the mandates they are enforcing, and are just waiting for a good challenge to materialize to provide them with an excuse to rescind it. (However, don't ever assume this is the case unless you have strong evidence. It's much more likely that the imposing authorities have internalized the enforcement of the mask mandate as an extension of their own personal authority, and will see any challenge to the mandate as a personal attack.)

> *"Always imagine a future time when today's deed will be on trial and no one will accept your pleas of 'only following orders,' or 'everyone else was doing it.'"*
>
> — Philip Zimbardo, *The Lucifer Effect*, 2007[1168]

Whatever you are opposing, do not fall for attempts to guilt you into complying with something that your conscience is uneasy with or revolts against. It is usually pretty easy to tell the difference between the unease that comes from a pricked conscience and simple fear. Standing up for your individual rights is not selfish, and when you uphold one person's rights, including your own, you uphold everyone's. We all know going against the herd is scary, and for good reason. Don't beat yourself up by shooting for perfection or expecting to convert people on the first try. In most cases, the best you can do is simply to raise the bar that someone's cognitive dissonance has to vault over to reach their current conclusion. *Do* be strategic about where you put your efforts. There is a certain percentage of the population your efforts are just going to be wasted on. Another subset of people will be naturally more amenable to giving you a genuine hearing. Sometimes the right crucial piece of information enables a lightning-fast *volte-face*, but for most of us, coming around takes time and happens gradually by degrees; gradual successive steps moving either upwards or downwards. Even if you manage to wear someone down, they may very well resent you to the extent you did so. Note

---

1168   Philip Zimbardo. *The Lucifer Effect*, New York: Random House Publishing Group, 2007, Kindle Edition, p. 453

that this does *not* mean you should not be persistent. The line between what comes across as being persistent vs. trying to grind someone down varies from person to person, but persistence is essential.

Acknowledging error, even privately to ourselves, is difficult and painful. Solomon Asch observed that once a subject had yielded to the group, he continued to yield more than previously independent subjects:

> "[H]aving once committed himself to yielding, the individual finds it difficult and painful to change his direction. To do so is tantamount to a public admission that he has not acted rightly. He therefore follows the precarious course he has already chosen in order to maintain an outward semblance of consistency and conviction."[1169]

The mere act of not wearing a mask under the same circumstances where masks were previously worn can't help but be a symbolic repudiation of previous mask wearing. The behavioral psychology we examined earlier when looking at the work of Robert Cialdini predicted what we all saw happen as masking started to lose ground in hearts and minds while advocates doubled down:

> "Because the only acceptable form of truth had been undercut by physical proof, there was but one way out of the corner for the group. It had to create another type of proof for the truth of its beliefs: social proof… because the physical evidence could not be changed, the social evidence had to be. Convince, and ye shall be convinced."[1170]

1169   Solomon E. Asch, "Effects of group pressure upon the modification and distortion of judgments." *Groups, Leadership and Men: Research in Human Relations*, Harold Guetzkow, ed. United States: Carnegie Press (1951), p. 177-190, available online: https://www.google.com/books/edition/Groups_Leadership_and_Men/aPRGAAAAMAAJ?hl=en

1170   Robert B. Cialdini, Influence, New and Expanded (p. 142-143). HarperCollins. Kindle Edition. 2021.

Part 6 Conclusion

In *Federalist* No. 8, Alexander Hamilton warned about how a perpetual state of war leads to the gradual erosion of a people's essential liberties.

> The inhabitants of territories, often the theatre of war, are unavoidably subjected to frequent infringements on their rights, which serve to weaken their sense of those rights; and by degrees the people are brought to consider the soldiery not only as their protectors, but as their superiors. The transition from this disposition to that of considering them masters, is neither remote nor difficult; but it is very difficult to prevail upon a people under such impressions, to make a bold or effectual resistance to usurpations supported by the military power.[1171]

What Hamilton did not anticipate was that metaphorical wars could achieve the same result, whether the "war" is a war on terrorism, a war on drugs, a war on poverty, or a war on a particular disease. The pretext of increasing safety has always been a useful tool for expanding power, but over the last 150 years, a new public health soldiery and aristocracy has quietly developed. The COVIDcrisis brought this class to the forefront and revealed just how much power and influence it had already accrued.

Masking during the COVIDcrisis was so divisive because it embodied differences in fundamental belief that have always coexisted in America. Tensions stemming from irreconcilable disagreements so fundamental cannot resolve completely, even between people of good will who genuinely love each other. It is the same tension experienced by members of families who hold different religious beliefs (sometimes even different beliefs within the same religious tradition). These disagreements may take a backseat to more pressing concerns, or be relegated to the periphery of awareness such that

---

1171   "The Federalist No. 8, [20 November 1787]," *Founders Online,* National Archives, https://founders.archives.gov/documents/Hamilton/01-04-02-0160.
[Original source: *The Papers of Alexander Hamilton*, vol. 4, *January 1787–May 1788*, ed. Harold C. Syrett. New York: Columbia University Press, 1962, pp. 326–332.]

you can almost forget about them, but they come back to the forefront during a crisis the moment their real-life implications become unavoidable. On a practical level, the options for dealing with differences in belief this divergent are either force, persuasion, or learning to live with the tension. In this multiple-choice test, compulsory masking and other mandatory COVIDcrisis measures were not just a wrong answer, but the *worst* answer. They physically forced the beliefs of one segment of the population onto everyone else's faces.

People who got it right on masks rarely did so based on chance alone, and attempts to frame the issue that way are mistaken at best and a form of gaslighting at worst. People who rightly perceived the underlying ideology and struggled against the COVIDcrisis tyranny, whether they did so from the start or came around slowly, deserve to celebrate that fact. At the same time, we also need to be careful for the simple reason that it is entirely possible for even luminaries like John Adams to be instrumental in the War of Independence and then sign the Alien and Sedition acts in complete violation of the First Amendment less than 20 years afterwards. People in the minority that turned the tide this time need to maintain their vigilance and introspection because they could very easily find themselves on the side they now oppose when it comes to another future issue. This is why we need to vigorously reaffirm and defend the unalienable rights that all individuals possess. The next time something like this happens, we may find ourselves in the place of those people who took a long time to come around during COVID. We need to protect the rights of those people who followed the herd this time, because *they* may be the ones setting the example that helps restore *us* the next time history starts rhyming.

We have learned from painful experience that compliance is only a virtue depending on what we are complying *with*, and that complying with demands and "rules" made by bullies, fear-mongers, or would-be humanitarian tyrants only makes things worse. Firm adherence to principles, especially those Natural Law principles of individual liberty recognized by the Bill of Rights, are essential to immunizing ourselves against falling prey to panic. Even when we personally get something wrong, if we honor the liberty of others to set a contrary example to what we think is the best course of action, we *still* benefit even on a utilitarian outlook, because this creates voluntary control groups composed of subjects who can cross over at any time. If we are right, the other group serves as additional evidence and a warning, who we help by our example. But for those times when *we* are in the wrong, this other group will provide an essential example that can help restore *us*.

However you decide to oppose compulsory masking and the mentality that gave birth to it, whether by lobbying representatives in your state legislature to add involuntary masking to your state's tort claims act,[1172] debating masks on the street, or simple stubborn noncompliance, Godspeed. If we fail to uphold and defend the principles of individual liberty right now, then at some point people will argue for compulsory masking again with some new disease in spite of its glaring failure with COVID. Plenty of people are still willing to argue that masks will work against the next disease *du jour*, or that masking in the context of COVID was eventually overturned not because it was immoral and ineffective, but because COVID-19 was not *enough* of an emergency or we didn't mask hard enough. We, our children, and our grandchildren, and future generations, will be forcibly masked and worse, on a recurring basis, until enough of us on an individual level resolve — at the very least with regard to masks and anything else that we know to be based in lies and manipulation — in the words of Aleksandr Solzhenitsyn, "Let their rule hold not through me!"[1173]

1172   Suggestion inspired by Wendy E. Parmet, "Quarantining the Law of Quarantine: Why Quarantine Law Does Not Reflect Contemporary Constitutional Law" (2018). *Wake Forest Journal of Law & Policy*, Vol. 9, No. 1, p. 32 (2019), Northeastern University School of Law Research Paper No. 342-2019, Available at SSRN: https://ssrn.com/abstract=3341375

1173   Aleksandr Solzhenitsyn, *Live Not by Lies*, (1974), available online: https://www.solzhenitsyncenter.org/live-not-by-lies

Appendix:

A Christian Case
Against Compulsory Masking

Why include this at all?

Compulsory masking was and remains controversial because it gets at the heart of so many people's sacred values — mine included. I was initially hesitant to include this section. It is an obnoxious and frustrating reality that when you put forward a religious argument or motivation for or against something, a certain subset of antagonistic interlocutors will point to the mere *existence* of that religious argument or motivation as somehow undermining or invalidating either the substance or the sincerity of your other arguments on the subject. Alternatively, they may point to the existence of secular arguments as evidence to doubt or suspect the sincerity or intellectual integrity of your religious arguments. Such is rhetoric and human nature. I hope and trust that all readers who have made it to this point will instantly perceive and reject the bad reasoning inherent in such *ad-hominem* non-counterarguments.

Two driving internal reasons (which tend to alternate in predominance) ultimately prompted me to include this material. First, I believe that no comprehensive work on masks which fails to address this aspect of masking would be worthy of the title. Second, without the sincerely held religious beliefs articulated and defended in this section, this book would not exist. Nothing short of sincerely held religious belief could have induced me to subject myself to the grueling process of writing this book or to any of the personal background events related to the subject matter (which are ongoing at the time of this writing). All of the foregoing material in this book *should* have been enough to get me to say "no" to masks when I stood to suffer something more than dirty looks, a fine, marital friction, or getting kicked out of Walmart by a manager and police officer without my groceries for refusing to put one on (yes, that happened). But the prior contents of this book were not, on their own, enough to get me to do this. I was still too afraid — not of COVID, but of the formal and informal punitive measures used to impose and enforce masking. What finally induced me to refuse *all* mask-wearing and other PPE use in excess of what was common in my profession and the general public prior to 2020, even with much more on the line, were the sincerely held religious beliefs that I lay out here.

In this appendix, I am not speaking on behalf of any organization or denomination. These are my own personal religious beliefs. Christians have historically gone to war with each other (multiple times) over basic doctrinal issues, so it should come as no surprise to anyone that Christians will likewise vehemently split over a subsidiary issue like masks. I expect that no reader will 100% concur with the content of this section, but I strongly suspect that most readers, including those of other faiths, will find at least a few things within it that resonate with them. My views are actually nuanced enough to potentially validate even those who perceived an individual moral obligation to wear a mask.

I do not pretend to be anything other than a layman, and I have no formal theological training. But even laymen can make solid arguments when they put in the time and effort. I consider this appendix to be what is known in the Christian tradition as an *apologia* — a reasoned defense of one's beliefs. It is a specifically Christian second half to the more generalized philosophy section earlier in this book. Insofar as these arguments are specifically directed at anyone, it is at other Christians. That being said, if you are an atheist, an agnostic, a non-Christian theist, or coming from another faith tradition entirely, I think you are still likely to find that one or more of these arguments mirror some of your own sincerely held beliefs. Even if not, you may still find this appendix worth your time in order to gain insight into some of the motivations and reasons that your friends or acquaintances may have had for opposing compulsory masking.

As one snarky but insightful internet blogger pithily observed: "many people *have* ideas, but those who become the more committed of the zealots or dogmatists often *become* their ideas."[1174] This section is not intended in the way of lecturing or talking down to anyone, though in places it certainly does come down heavily against particular *ideas* that a lot of people hold. I hope the previous discussion of separating oneself from one's ideas will help to mitigate any offense. My primary goal is to articulate and defend the sincerely held religious beliefs that brought me to this point. In general, I don't consider myself a "good" Christian. Sincere? Yes. Good? Not if I'm being completely honest. But in many ways, my personal flaws, failures, and shortcomings are part of the point. My personal duty exceeded that of many people in the COVIDcrisis because I *always* knew better than to wear a mask, even in March 2020. But while I resisted wearing a mask in many settings, I did it anyway for far too long in one other, and I still wrestle with whether I could have or should have pushed back even harder than I did overall. Peter denied Jesus three times. Based on my experience and the self-knowledge I've gained since March 2020, I am willing to bet good money that I would have done worse than Peter, were I in his position. Thankfully, Christianity is predicated on the notion that only one man was (and is) really and truly good, and that His sacrifice is sufficient to cover the wrongdoing of all the rest of us.

1174   El Gato Malo (Pseudonym), "free thought as existential threat," *bad cattitude* (blog), September 29, 2022, online: https://boriquagato.substack.com/p/free-thought-as-existential-threat?publication_id=323914&post_id=75138778&is-Freemail=false

One quick bookkeeping note on the Biblical translations cited: I quote from multiple different translations of the Bible in this section. I grew up reading primarily the 1984-2011 edition of the New International Version (NIV). In this chapter, I primarily cite using the more word-for-word English Standard Version (ESV), though I also occasionally cite to the New International Version (NIV) based on either personal familiarity or for maximal clarity on the point being made. For anyone inclined to dispute my readings based on a different translation, the very existence of such a disagreement is a win-win, because it means that more time is being spent reading the verses in question and the Bible in general.

— Philip Traugott Buckler, DDS, August 2023

Differing moral duties

As we saw in the philosophy section of this book, moral values and duties exist objectively, apart from individual or societal beliefs, but where do these values and duties originate? What grounds them? From a Christian perspective, our objective moral values and duties originate in God. The classic Euthyphro dilemma asks the question: "Does God create The Good or does God recognize The Good?" Phrased another way: does God will something because it is good, or is something good because God wills it? The Christian answer to the Euthyphro dilemma is that this is simply a false dilemma because there is a third option: namely, God wills something because *He* is good — not because He is recognizing an external standard or arbitrarily declaring something to be good. In Christian theology, God's nature is what defines goodness; the degree to which something aligns with God's nature is the degree to which it is good. Thus, God's commands, which constitute our moral duties, are a necessary expression of his nature, and since God's nature is The Good, it is impossible that He should will things such as murder or cruelty be good. This is the traditional Christian theologian's response to the Euthyphro dilemma, and this is what I hold to as well. On a practical level, this means that while consideration of the consequences can be relevant (even important) to deciding which moral principle is applicable in a situation, utilitarian and consequentialist systems of ethics are morally deficient and often incompatible with Christian practice. Though the practical applications of utilitarianism and Christianity can overlap to some extent, compulsory masking is one of many places where they radically diverge.

Moral values and duties exist objectively, but having objective existence does not necessarily imply that something is also absolute — applying to every person in the exact same way regardless of circumstance. To be sure, there are some moral duties that are both universal *and* absolute, such as every human's moral duty to worship God alone. But most day-to-day moral values and duties are situation-*relative* — they supervene on individuals differently according to their individual circumstances. Put another way, God lays a number of general moral values and duties on all mankind, for example in the Ten Commandments, but practical application of those general moral duties often varies situationally, depending on the individuals and circumstances involved (not always, but often). The fact that many moral duties laid on individuals are circumstantial does not imply those duties are merely subjective.[1175] Objective does not imply absolute. Relative does not imply subjective.

God frequently lays special moral duties on particular individuals, depending on who they are and the circumstances they are in. He can do this through the generalized channels of Natural Law, Conscience, and Scripture, but also through other means. Too many Biblical examples to count show that God calls different believers to forego or engage in different specific practices depending on their individual situation. Thus, it is entirely possible that even within the same household, one individual could have a moral duty to wear a mask, another could have a moral duty before God to refuse *all* mask-wearing regardless of the consequences, and another could fall somewhere in-between, wearing a mask in some situations but adamantly refusing in others. The Apostle Paul's instructions to believers in Romans 14 regarding differing day-to-day practices that: "Each one should be fully convinced in his own mind," are not just for trivialities. Again, moral values and duties have objective existence, but they supervene on individuals differently according to circumstances.

Repeatedly throughout the Bible, in various times, places, and contexts, God calls, assigns, or outright *orders* groups and individuals to either engage in — or *refrain* from engaging in — acts which have great practical and/or symbolic significance. The Supreme Court of the United States explicitly recognized and cited this in the landmark case *West Virginia State Board of Education v. Barnette et al.* (1943):

> "Early Christians were frequently persecuted for their refusal to participate in ceremonies before the statue of the emperor or other symbol of imperial authority... The Quakers, William Penn included, suffered punishment rather than uncover their heads in deference to any civil authority."[1176]

1175   It is worth repeating: we must take care not to mix up two crucial pairs of opposites: absolute vs. relative, and objective vs. subjective. If something is absolute, it applies regardless of circumstances, whereas something that is relative is circumstance-dependent. If something exists objectively, it does so independently of people's beliefs, whereas things that are relative vary depending on people's beliefs and opinions. Again, something being objective does not imply that it is also absolute. Likewise, something being relative does not imply that it is subjective.

In the author's experience, internet lay-atheists tend to make the mistake of demonstrating that moral values and duties often supervene on individuals differently according to circumstances (i.e. are not absolute), and then triumphantly conclude that this means moral values and duties have no objective existence. A complimentary version of this category error that frequently occurs among Christians is to accurately perceive the objective existence of a moral duty which is specific to oneself, and then erroneously conclude that it applies absolutely to everyone else.

1176   *West Virginia Board of Education v. Barnette et al.*, 319 U.S. 624, 633 n 13 (1943), available online: https://www.loc.gov/item/usrep319624/

Incidentally, the undergraduate university that I attended from 2001 to 2005 was named after another Quaker leader, George Fox

In the Biblical record, God routinely confers individualized moral duties, and sometimes these moral duties are very peculiar. Below are just *some* of the symbolic acts God has required of believers in the Bible in various times and contexts. Compared to several of these, refusing to wear a mask should barely raise an eyebrow:

- Refraining from cutting hair, drinking wine, or going near a dead body — even that of an immediate family member (Numbers 6:1-8).

- Refusing to eat foods proscribed in the Israelite dietary restrictions found in Exodus, Leviticus, Numbers and Deuteronomy.

- Eating foods proscribed in the Israelite dietary restrictions found in Exodus, Leviticus, Numbers, and Deuteronomy (Acts 10:9-16).

- Eating or refusing to eat food sacrificed to idols (1 Corinthians 8).

- Circumcising (Genesis 17:9-14) or refusing to circumcise (Galatians 5:1-2).

- Marrying a prostitute, having children with her, and naming the daughter by that marriage "No Mercy" (or "Not Loved," depending on which translation you read) (Hosea 1:2-6).

- Refraining from mourning for a dearly loved wife when she died (Ezekiel 24:15-18).

- Buying property from a countryman at full price when that property is under occupation by enemy armed forces and your country's national destruction is a foregone conclusion and there is no way you can reasonably expect to exercise your claim of ownership (Jeremiah 32: 6-15).

- Traveling to deliver a message without eating bread, drinking water, or returning by the same route you came (1 Kings 13).

The most commonly known example of individualized moral duty was front-and-center during the birth of the United States. From the earliest days of the Church straight through to the present, God has placed upon a subset of Christians the moral duty to affirm the high value He places on human life by engaging in various degrees of pacifism. At the same time, God has called, assigned, or at least permitted other contemporaneous Christian groups and individuals to take human life in certain other narrow contexts, such as defense of oneself, one's family, and one's country. These duties *may* last for the duration of a person's whole life, but they may also change depending on the person's circumstances.

Since God assigns highly individualized and varying moral duties even in matters of basic survival like diet and self-defense, it should have come as no surprise that God would likewise place similarly varying moral duties on individuals in the context of COVID-19 or other diseases with regard to interventions like masks and vaccines — even individuals within the same families. It may be one person's duty before God to do everything possible to avoid a disease, while at the same time it is another person's duty to forego one or more prevention measures that others consider to be basic and essential. Also, these moral duties are not the same for every disease, even in the same individual.

I am a non-denominational Protestant Christian,[1177] and I hold the following religious beliefs regarding mask-wearing for COVID. I do not doubt that there are additional Christian beliefs precluding masking which could be articulated, and this list is not exhaustive, but these are the arguments that were (and remain) the most decisive for me, personally. Any single one of them is enough to preclude wearing masks in the context of COVID-19. A few are so straightforward they do not require additional explanation, but I will expand on the rest in the sections that follow before providing some closing thoughts.

- Because wearing a mask is a form of symbolic speech, doing so causes me to violate the Commandment not to lie - not to bear false witness against my neighbor (Exodus 20:16, c.f. Proverbs 12:22). In my particular case, because I know so much about all aspects of masks, wearing a mask or even remaining silent on the issue is a failure to testify when called to do so in violation of Leviticus 5:1: "If anyone sins because they do not speak up when they hear a public charge to testify regarding something they have seen or learned about, they will be held responsible" (NIV Translation); "If anyone sins in that he hears a public adjuration to testify, and though he is a witness, whether he has seen or come to know the matter, yet does not speak, he shall bear his iniquity" (ESV Translation).

- When I wear a mask, I am violating the second-greatest commandment to "love your neighbor as yourself" (Matthew 22:38-40; Mark 12:29-31), because when I wear a mask, I am not doing unto others as I want them to do unto me (Luke 6:31).

- Mask-wearing is contrary to language found throughout the Bible (more on this later, but too many references to list in a single bullet point).

- In Biblical terms, compulsory masking is forcing everyone with unblemished skin to

---

1177   I say "non-denominational" because I am not familiar enough with the finer doctrinal differences between most of the denominations to accurately designate which one I am at this time (other than to rule out being a Catholic, Calvinist, or Seventh Day Adventist). I strongly suspect that I fall somewhere in the region of Lutheran or Baptist, but I am not certain, and the time that I could have spent refining my understanding of denominational differences was instead channeled into learning everything about masks — a tradeoff which I am personally not entirely happy with.

behave like lepers, symbolically shouting "unclean, unclean!"

- I do not for one moment believe that Jesus would have worn a mask.

- Compulsory masking is inconsistent with Christian practice throughout history - even during the worst plagues.

- Regardless of whether or not it met the legal definition of an establishment of religion, during COVID, public health *did* take on the nature of a performative secular religion, and I refuse to participate in rituals which promote this novel idolatry, of which mask-wearing was and remains the most visible.

- Attempts to push or defend compulsory masking on Biblical grounds all fail miserably.

- Matters of religion are no less important than matters of public health. Coercion in matters of religion is utterly immoral. Therefore, coerced masking in the name of public health is utterly immoral.

- Compliance with mask-wearing makes me complicit in the widespread constitutional violations perpetrated in the name of public health, and thereby causes me to violate an oath I took to defend the United States Constitution — an oath which I took with a Biblically based understanding (c.f. Numbers 30:2).

- Adhering to a Biblically based recognition of the role of empirical evidence in decision-making precludes compulsory masking, as do Biblical judicial principles.

I am not opposed to *all* mask wearing or to *anyone* wearing a mask, nor do I think it is *every* Christian's moral duty to refuse to wear a mask in the context of COVID-19. Rather, I believe that there is a subset of Christians upon whom God has laid a special moral duty to refuse to wear a mask in the context of COVID-19, and that I am part of this group. Whether (and to what extent) anyone else falls into this group is between them and God. People have different breaking points, and God knows our breaking points better than we do ourselves. He takes our circumstances and breaking points into account when assigning our individualized moral duties.[1178] One moral duty which I do not believe that God has ever laid on *anyone*, however, is to force masks or other interventions like vaccines onto

---

1178   In one of the military vaccine cases, Judge Steven Merryday astutely listed some of the many factors that God can bring to a individual's moral awareness, altering their moral duties: "The Assistant Commandant says nothing about the development of the vaccine or the religious concepts of, for example, accepting a personal benefit from evil, assisting someone in profiting from evil, cooperating in evil, appropriation of evil, de-sensitization to evil, moral contamination by intimacy with evil, ratification of evil, complicity with evil, or other considerations undoubtedly familiar to a theologian and likely familiar to a thoughtful religious lay person.

others in such a way that it brings them into conflict with their own duties to God. God may *permit* such conflicts, but that does not mean He causes or approves of them. What He asks of each of us on an individual level is to earnestly seek his will and then follow it to the best of our ability.

Navy Seal 1 v. Austin, 8:21-cv-02429, Document 111, p. 14-15 (M.D. Fla.), Available online: https://storage.courtlistener.com/recap/gov.uscourts.flmd.395057/gov.uscourts.flmd.395057.111.0_4.pdf

additional case documents available: https://www.courtlistener.com/docket/60650721/navy-seal-1-v-biden/

Bearing false witness through wearing a mask

Every time I put on a mask in a manner exceeding that which was common in my profession and the general public prior to 2020, I violate the commandment not to lie — not to bear false witness against my neighbor. As we have established, "speech" is more than mere words. Speech also consists of the symbolic actions, gestures and displays which people make on a daily basis, often without even thinking about them. Symbolic speech is a major component of nonverbal communication. As early as the Mosaic law, there is a recognition that performing the requirements of statutes gives them a form of affirmation and confirmation, as in Deuteronomy 27:26 (ESV): "Cursed be anyone who does not confirm the words of this law by doing them." Because wearing a mask is a form of symbolic speech, in my case, doing so causes me to violate the divine prohibition against lying — i.e., not to bear false witness against my neighbor. (c.f. Proverbs 12:22 (ESV): "Lying lips are an abomination to the Lord, but those who act faithfully are his delight"; and Exodus 20:16 (ESV): "You shall not bear false witness against your neighbor.") "Bearing false witness" in a Biblical sense is not limited to legal settings. Martin Luther, in his Small Catechism, elaborates that not bearing false witness means "that we do not falsely deceive, betray, or slander our neighbor."[1179] As we saw in Part 4, there is a point at which compliance becomes complicity. For me, at least, putting on a mask outside of direct patient treatment passes that inflection point. I am also unwilling to lend any tacit confirmation or legitimacy to the mentality and tyranny inherent in compulsory masking or the other COVIDcrisis measures by my participation.

Wearing a mask constitutes a broad-spectrum symbolic affirmation of multiple propositions. Many of these propositions, I know for a fact or sincerely believe to be false, and wearing a mask effectively communicates them more powerfully than any verbal utterance. The fact that messages communicated via symbolism are just as dependent on the recipient's understanding as they are in any other form

---

1179   Paul Rydecki, *Luther's Small Catechism: An Introduction to the Catholic Faith.* Las Cruces, NM: Paul Rydecki (2019).

of communication does not imply that no specific message has been sent or received. Wearing a mask symbolically conveys multiple statements about my internal beliefs to those around me. These include (but are not limited to):

- Affirmation that masks have efficacy in mitigating the spread of respiratory viral infections.
- Affirmation of the false belief that because it is *possible* any given individual *may* be infectious, that therefore *every* individual should be assumed to be infectious and treated as such until proven healthy. (To me, this is as bad as affirming that someone should be considered guilty of practicing witchcraft or having unwittingly committed manslaughter until they prove otherwise.)
- Affirmation that I fear those around me, that I believe them to be dangerously contagious, and that they have good grounds for holding the same beliefs about me.
- Affirmation that I believe in the moral and legal validity of mask-wearing requirements, orders, directives, and guidelines.
- Affirmation that I believe there is imminent danger present which the mask I am wearing will mitigate in some way.
- Affirmation that my beliefs which run contrary to wearing masks are not strong enough or sincere enough to warrant defying the authorities who are wrongfully mandating masks.
- Affirmation that the truth I know is less important than the preservation of my own temporal comforts which depend on wearing a mask.

Wearing a mask is not loving my neighbor

When I wear a mask outside of direct patient treatment, I am violating the second-greatest commandment to "love your neighbor as yourself," (Matthew 22:38-40; Mark 12:29-31) because when I wear a mask, I am not doing unto others as I want them to do unto me (Luke 6:31). I want my neighbor to critically examine what he is told by authority figures, to: "not believe every spirit" (1 John 4:1, ESV). Jesus does not tell us to *never* judge ourselves or others. Instead, what He tells us is, "Do not judge by appearances, but judge with right judgment" (John 7:24), and He warns us that we are to always keep in mind that: "with the measure you use, it will be measured to you" (Matthew 7:2). I want my neighbor to evaluate risks in perspective, and to implement the risk tradeoffs that he deems most appropriate, living life to the fullest, extending that same accommodation and toleration to myself and others. I want my neighbor to engage in normal life activities without fear, even though individual risks vary for *every* normal life activity. I want my neighbor to not fear being around other people. If my neighbor thinks I am wrong on some point, I want him to articulate his arguments (in a public forum if necessary). If I am wrong on some issue, I want my neighbor to attempt to correct me, not by coercion, but through persuasion, debate, and *example*, living and acting consistently with his beliefs, and so I strive to do the same in this case.

By its very nature, Jesus' command implies individual liberty rather than collectivism, because the way one person wants to be treated is not necessarily the same as the way another person wants to be treated. The Biblical injunction is to love our neighbors *as* ourselves, not *more* than ourselves. The Golden Rule must not be misstated. The Golden Rule *is*: "Do unto others as *you* would have them do unto *you*." The Golden Rule is *NOT* "Do unto others as *they* would have you do unto *them*." The difference is subtle, but profound. In the real formulation, the individual must take others' preferences into account when acting, but others' preferences are not determinative, whereas in the un-Biblical mal-formulation, others' preferences are used to *dictate* one's actions. Christ's command is also very much NOT "do unto others as they would have done unto you," nor is it "do unto others as they did unto you," though this last one is probably the hardest variant to resist on a day-to-day basis.[1180]

For me, loving my neighbor involves the opposite of leaving them physically and emotionally isolated in the name of safety under "lockdown," and it precludes ever hiding my face from them by wearing a mask outside of direct patient treatment (when my face is 18 inches away above their open mouths and I am using a dental handpiece to generate tooth dust and flying globs of plaque in the process of tooth repair and restoration). While sincere Christians can come to opposing conclusions about whether or not to wear a facemask, many, myself included, can't help but see something subtly diabolic in reasoning which twists our best impulses to care for and protect one another into strong pressure to send out signals of distance, displeasure, fear, and alienation to those around us. This, in combination with the other reasons articulated in this book, makes it an intolerable violation of conscience for me to wear a mask according to post-2020 "guidelines" and "recommendations" issued by public health agencies. That goes double for "guidelines" and "recommendations" put out by manifestly corrupt and incompetent public health agencies with conflicts of interest like the CDC.

Even secular psychology, including the work of Dr. Philip Zimbardo, has shown the un-person-ing, "deindividuating" effect that mask-wearing tends to generate both within the wearer and towards those observed wearing them, as well as the dangerously slippery slope inherent in the methods used to enforce compliance and conformity. As Dr. Zimbardo pointed out:

> "… a key ingredient in transforming ordinarily nonaggressive young men into
> warriors who can kill on command is first to change their external appearance…
> For the young men, it becomes easier to do so if they first change their
> appearance, altering their usual external façade by putting on military uniforms
> or masks or painting their faces."[1181]

Based on the conceptualization of masks liberating hostile impulses that was illustrated in William Golding's novel *Lord of the Flies*, Dr. Zimbardo said: "I had conducted research showing that research participants who were 'deindividuated' more readily inflicted pain on others than did those who felt more individuated."[1182] When I assist in our mutual deindividuation, desensitization, and un-person-ing by my compliance with masking directives, I am not loving my neighbor but harming him.

1180   Also, do not do unto others what you would never allow them to do unto you.
1181   Philip Zimbardo. *The Lucifer Effect*, New York: Random House Publishing Group, 2007, Kindle Edition, p. 304.
1182   Philip Zimbardo. *The Lucifer Effect*, New York: Random House Publishing Group, 2007, Kindle Edition, p. 24.

The argument, "Love thy neighbor, mask up" was deployed with clinical intentionality, and a group of authors published in the *Proceedings of the National Academy of Sciences* even quantified the resultant increase in support (and presumably compliance) for masks among the target group of evangelicals (in the study, support for masks rose from 34% to 43%).[1183] During the push for masks, I often read arguments like the following: "… if you are a Christian, you should be wearing a mask — even if you do think they are useless… it all boils down to this: it doesn't matter what you believe about wearing masks — it matters what those around you believe about wearing masks."[1184] This argument in favor of mask-wearing is utterly unbiblical, as though your neighbor's belief and emotional or physical comfort can determine what is right. One Christian ethicist, Steffen Flessa, in the process of arguing that loving our neighbor makes vaccination a Christian duty, made a common misapplication of the principle that: "a purely self-benefit decision is unacceptable for a Christian."[1185] This misapplication was also common with regard to masks. It is a misapplication because "love your neighbor" and warnings against self-interest cut both ways. What is and is-not a purely self-benefit decision for each individual can only be fully known by God. A decision to be vaccinated or wear a mask can just as easily be a purely self-benefit decision as a resolution to *refuse* those things. Yet many Christians and people of other faiths were and remain willing to incur social opprobrium, discrimination, job loss, and even legal penalties by refusing masking or vaccination. These are clearly not "purely self-benefit" decisions. Rather than assuming these individuals are erroneously perceiving their moral duties, or assuming they are making a cynical, cowardly, selfish, or irrational risk tradeoff, these Christian brothers and sisters' decisions and moral stands should be given the same respect and benefit of the doubt regarding loving motivation and accurate moral awareness that was extended to those who enthusiastically participated in mask-wearing and vaccination. I believe that this is at least part of what the Apostle Paul was referring to when he says in Romans 14:4 (ESV): "Who are you to pass judgment on the servant of another? It is before his own master that he stands or falls."

1183   Stephanie L. DeMora, Jennifer L. Merolla, Brian Newman, B., and Elizabeth J. Zechmeister "Reducing mask resistance among White evangelical Christians with value-consistent messages." *Proceedings of the National Academy of Sciences U.S.A.* 2021, 118:e2101723118. doi: 10.1073/pnas.2101723118, available online: https://www.pnas.org/doi/10.1073/pnas.2101723118
Reported on by J.D. Warren, "'Love thy neighbor, mask up' White evangelicals respond to religious pro-mask messages; study authors say lessons apply to vaccines," *UC Riverside News*, May 12, 2021, online: https://news.ucr.edu/articles/2021/05/12/love-thy-neighbor-mask

1184   Murphy, K., NC pastor: "Jesus would wear a mask," 2021, *The Charlotte Observer*. Online https://www.charlotteobserver.com/opinion/article242898806.html

1185   Steffen Flessa, "Vaccination Against COVID-19 as a Christian Duty? A Risk-Analytic Approach," *Christian Journal for Global Health*, 2021. **8**(2): p. 2-15. https://journal.cjgh.org/index.php/cjgh/article/download/611/975/

Steffen Flessa went on to argue that "It would be insufficient to base love on emotions and reduce it to avoid[ing] direct harm to the people in front of me."[1186] This is true as far as it goes, but it is every bit as insufficient to base "love" on the risk profile of one disease and reduce it to one or two infection control measures. Reducing "love" to a medical intervention with the physical, psychological, legal, and spiritual baggage of masks is even worse. Love does not necessarily entail shielding someone from all risks and suffering, and loving others sometimes entails doing things that make them very uncomfortable. Refusing to subtly validate a loved one's fears about a particular virus by putting on a mask is just one of many examples. Sometimes one of the best ways to alleviate a loved one's anxiety is to persistently demonstrate just how far from reality their fears have brought them. Also, never forget that the person refusing to wear a mask is not the only moral actor in this drama. If it is selfish, inconsiderate, and unloving to refuse to make use of a medical intervention for someone, it is even *more* selfish, inconsiderate, and unloving to *force* a medical intervention on someone else against their will, and outsourcing that force to a third party like the government does not make it any *less* selfish and unloving.

It is not loving to cater to an untrue belief by pretending that it is true. That is dysfunctional enabling. A psychiatrist who makes an elaborate show of searching his office for surveillance devices before beginning counseling does nothing to help a patient who is suffering from paranoia — such actions merely validate that paranoia. If you have a friend who is convinced that investing their life savings in a particular stock is a good idea, and you have knowledge or even reasonable grounds to suspect the investment is a *bad* idea, it is not loving to reinforce their belief by purchasing some of that stock yourself, even if they ask you to do so. In cases where you and your neighbor both vehemently disagree about the factual basis of certain beliefs, the most loving thing is to tolerate their beliefs and practices while vigorously challenging them by argument and example. In the case of COVID or any other disease, this may include wearing or not wearing a mask, and vaccinating or refusing vaccination.

Christ did not command his followers to be safe at all costs, nor did he himself live that way. "For whoever would save his life will lose it, but whoever loses his life for my sake will find it" (Matthew 16:25, Mark 8:35, Luke 9:24, ESV). Christianity entails that temporal safety for oneself or others *cannot* be a preeminent superseding value or virtue. In Christianity, the beliefs and comfort of those around you must not be allowed to dictate your own beliefs or behavior in a way that deviates from what you know or believe to be morally right and true, and it is ultimately your own *individual* beliefs and *personal* response to Jesus that determine your eternal outcome. Christianity *assumes* that

---

1186   Flessa also makes the same basic error of conflating inaction with active endangerment when he says: "the endangerment of people can represent a breach of love, even if its materialization is subject to a certain probability." Likewise, "imposing" something is by definition an action. Erroneously conflating the two is either bad reasoning or rhetorical manipulation.

more often than not, our moral beliefs will conflict with those of many others, that we will suffer consequences for this, and that we are not supposed to let those consequences prevent us from doing what we know or believe to be right. Many, if not the vast majority, of those who refused to wear masks for COVID were loving their neighbors as themselves by being willing to endure the social censure (and worse) that came along with their refusal in order to resist what they knew to be unbiblical coercion, panic-promoting immoral governmental overreach, emotional blackmail, non-evidence-based interventions, and a foolish reliance on "experts" with conflicts of interest.[1187]

Refusing to wear a mask is never doing something "to" someone. At worst, it is refusing to do something "for" someone. As psychologists Henry Cloud and John Townsend point out in their book, *Boundaries*:

> "Remember the landowner's words in the parable of the workers in the vineyard: 'Don't I have the right to do what I want with my own money?' (Matthew 20:15). The Bible says that we are to give and not be self-centered. It does not say that we have to give whatever anyone wants from us. We are in control of our giving."[1188]

This directive applies to forms of assistance as well as monetary gifts. Paul writes to Christians: "Each of you should give what you have decided in your heart to give, not reluctantly or under compulsion, for God loves a cheerful giver" (2 Corinthians 9:7, ESV). Imposing a particular form of "love" as a legal obligation twists it into a form of oppression, corrupting its exercise and contaminating it with resentment.[1189] Compulsory masking and its ilk put a new twist on the parable of the Good Samaritan: "Everyone is your neighbor and you *will* stop to render assistance in the prescribed way — or else."

1187   See also Janet E. Smith's published rebuttal to M. Terese Lysaught's National Catholic Reporter article arguing that "Catholics seeking 'religious' exemptions from vaccines must follow true church teaching on conscience."
Janet E. Smith, "Opinion: The Fake Theology Behind Vaccine Mandates," *Crisis Magazine*, September 27, 2021, available online: https://crisismagazine.com/opinion/the-fake-theology-behind-vaccine-mandates
responding to M. Therese Lysaught, "Catholics seeking 'religious' exemptions to vaccines must follow true church teaching on conscience," *National Catholic Reporter, Opinion, Guest Voices*, September 21, 2021, available online: https://www.ncronline.org/news/opinion/catholics-seeking-religious-exemptions-vaccines-must-follow-true-church-teaching

1188   Henry Cloud and John Townsend. *Boundaries: When To Say Yes, How to Say No* (p. 241). Zondervan. Kindle Edition.

1189   For a fuller discussion of this point, I direct readers to essays by Paul Kingsnorth and David Cayley, discussing the work of philosopher Ivan Illich.
Paul Kingsnorth, "What Progress Wants," *The Abbey of Misrule* (Blog), May 24, 2022, online: https://paulkingsnorth.substack.com/p/what-progress-wants
discussing David Cayley, "'The Apocalypse Has Begun': Ivan Illich and Rene Girard on Anti-Christ," DavidCayley. com (Blog), January 2, 2016, online: https://www.davidcayley.com/blog/2016/1/2/the-apocalypse-has-begun-ivan-illich-and-ren-girard-on-anti-christ

Actually, measures like compulsory masking went a step beyond this by requiring that people stop and render aid in the prescribed manner even when it was unclear whether there was anybody lying hurt by the side of the road in the first place.[1190] This sort of boundary violation is bad enough when friends and family do it without getting the law involved. Recall C. S. Lewis' warning that "a tyranny sincerely exercised for the good of its victims may be the most oppressive."[1191] When you actively refuse to wear a mask, you may not be doing something *for* someone that they want you to do, but you are certainly not doing anything *to* them, and you may very well be doing something else for them that they need even more.

1190   And by suggesting that every would-be Samaritan was also a potential bandit.

1191   Clive Staples Lewis, *God in the dock: essays on theology and ethics*, Grand Rapids, Michigan: William B. Eerdmans Publishing Company (1970), p. 292. Available online: https://archive.org/details/godindockessayso0000lewi/page/292/mode/2up

Wearing a mask is contrary to Biblical language

Mask-wearing is contrary to ubiquitous iconic Biblical language and imagery. Throughout the Bible and especially in the Psalms, the symbolic act of covering or hiding one's face is used as a way of indicating distance, displeasure, and alienation from another. For instance:

- "Why do you hide your face and count me as your enemy?" Job 13:24 (ESV)
- "… as one from whom men hide their faces he was despised, and we esteemed him not." Isaiah 53:3 (ESV)
- "O Lord, why do you cast my soul away? Why do you hide your face from me?" Psalm 88:14 (ESV)
- "Why do you hide your face? Why do you forget our affliction and oppression?" Psalm 44:24 (ESV)
- "Hide not your face from me, lest I be like those who go down to the pit." Psalm 143:7 (ESV)
- "How long, O Lord? Will you forget me forever? How long will you hide your face from me?" Psalm 13:1 (ESV)
- "Hide not your face from me. Turn not your servant away in anger, O you who have been my help. Cast me not off; forsake me not, O God of my salvation! Psalm 27:9 (ESV)
- "By your favor, O Lord, you made my mountain stand strong; you hid your face; I was dismayed." Psalm 30:7 (ESV)
- "Do not hide your face from me in the day of my distress! Incline your ear to me; answer me speedily in the day when I call!" Psalm 102:2 (ESV)

Often, the gesture of hiding one's face is provoked by wrongdoing on the part of those from whom one's face is hidden:

- "Then they will cry to the Lord, but he will not answer them; he will hide his face from them at that time, because they have made their deeds evil." Micah 3:4 (ESV)

- "your iniquities have made a separation between you and your God, and your sins have hidden his face from you so that he does not hear." Isaiah 59:2 (ESV)

Conversely, *showing* one's face is used as an indication of favor, mutual intimacy, safety, and blessing:

- "The LORD make his face to shine upon you and be gracious to you." Numbers 6:25 (ESV)
- "Make your face shine on your servant; save me in your steadfast love!" Psalm 31:16 (ESV)
- "Thus the Lord used to speak to Moses face to face, as a man speaks to his friend." Exodus 33:11 (ESV)
- "Now therefore, O our God, listen to the prayer of your servant and to his pleas for mercy, and for your own sake, O Lord, make your face to shine upon your sanctuary, which is desolate." Daniel 9:17 (ESV)

Seeking God's face is used to describe part of the process of repentance: "If my people who are called by my name humble themselves, and pray and seek my face and turn from their wicked ways, then I will hear from heaven and will forgive their sin and heal their land" 2 Chronicles 7:14 (ESV). This repentance is then followed by a restoration of fellowship, again described in terms of hiding and seeing another's face:

- "'In overflowing anger for a moment I hid my face from you, but with everlasting love I will have compassion on you,' says the LORD, your Redeemer." Isaiah 54:8 (ESV)
- "Then man prays to God, and he accepts him; he sees his face with a shout of joy, and he restores to man his righteousness." Job 33:26 (ESV)
- "And I will not hide my face anymore from them, when I pour out my Spirit upon the house of Israel, declares the Lord God." Ezekiel 39:29 (ESV)
- "For then you will delight yourself in the Almighty and lift up your face to God." Job 22:26 (ESV)
- "There are many who say, 'Who will show us some good? Lift up the light of your face upon us, O Lord!'" Psalm 4:6 (ESV)
- "For the Lord is righteous; he loves righteous deeds; the upright shall behold his face." Psalm 11:7 (ESV)
- "As for me, I shall behold your face in righteousness; when I awake, I shall be satisfied with your likeness." Psalm 17:15 (ESV)
- "Make your face shine upon your servant, and teach me your statutes." Psalm 119:135 (ESV)
- "And we all, with unveiled face, beholding the glory of the Lord, are being transformed into the same image from one degree of glory to another." 2 Corinthians 3:18a (ESV)
- "They will see his face, and his name will be on their foreheads." Revelation 22:4 (ESV)

While covering one's face is occasionally used in a more neutral cultural context (for example: Genesis 24:65, where Rebekah veils herself when first meeting her future husband Isaac), every such cultural reference can be matched by one that is less positive in nature, as in Genesis 38:15 (ESV): "When Judah saw her, he thought she was a prostitute, for she had covered her face." In the book of Esther, figurative language becomes literal, when Haman's face is physically covered after the Persian king orders his execution after his plot to exterminate all the Jews in Persia is exposed and foiled: "As the word left the mouth of the king, they covered Haman's face… So they hanged Haman on the gallows that he had prepared for Mordecai" (Esther 7:8-10, ESV). Elsewhere, the covering of one's face is used to evoke moral blindness, as in Job 9:24 (ESV): "The earth is given into the hand of the wicked. He covers the faces of its judges[.]"

In 2 Corinthians 3:13-18, the Apostle Paul uses the illustration of a *face covering* — a veil worn by Moses — as a symbol of mental dullness and continuing adherence to the old covenant, which was replaced and superseded by the new covenant of Christ. Christians are "not like Moses, who would put a veil over his face to keep the Israelites from gazing at it while the radiance was fading away. But their minds were made dull, for to this day the same veil remains when the old covenant is read. It has not been removed, because only in Christ is it taken away" (2 Corinthians 3:13-14, NIV). "But," says Paul, "whenever anyone turns to the Lord, the veil is taken away" (2 Corinthians 3:16, NIV). Since the veil has been taken away by Christ in the new covenant, believers, "with *unveiled faces*" (emphasis added) are able to directly behold and then reflect more and more of God's glory through the ongoing process of transformation produced by Christ's presence in their lives (2 Corinthians 3:18, NIV).[1192]

Given the frequent, consistent, and powerful use of such imagery throughout the Bible, it should come as no surprise that *many* sincere Christians (myself included) are highly resistant to *any* instructions which require them to cover their faces around others except under carefully limited and narrowly defined circumstances. The Bible's pervasive language recognizes and affirms a very real hardwired part of human nature conferred by God. The multiple human authors who contributed to the Bible over more than a millennium through inspiration by the Holy Spirit made remarkably consistent use of this imagery for a reason. Humans were created as relational beings, and a negative reaction to things that undermine or subvert this essential need — things like wearing a mask — is hardwired into us.

Compulsory masking was ultimately driven by primal fears, flawed intuitions, coercive impulses, herd mentality, an unexamined utilitarian ethic, and (I believe) "the spiritual forces of evil" referenced

---

1192   Not incidentally, America's founders were especially fond of a quote from this same passage: "where the Spirit of the Lord is, there is liberty" (2 Corinthians 3:17, KJV).

by the Apostle Paul in Ephesians 6:12[1193] — which Christians are assigned to oppose. Compulsory masking in the name of COVID-19 not only conflicts with my personal religious beliefs, but it is indefensible even on a utilitarian ethic, because masks have been *decisively* shown to lack *any* real efficacy in stemming the spread of COVID and other respiratory viruses, masks have numerous non-trivial side-effects, and compulsory masking violates *multiple* God-given, unalienable individual human rights recognized in the Constitution of the United States.

1193   "For our struggle is not against flesh and blood, but against the rulers, against the authorities, against the powers of this dark world and against the spiritual forces of evil in the heavenly realms" Ephesians 6:12 NIV).

Wearing a mask is contrary to historical Christian practice

Under the COVIDcrisis safety-*über-alles* public health mindset, care for the sick as it has been given from the earliest days of the Church was prohibited and subject to prosecution. Reasoning under the 2020 lockdown and masking mentality would condemn Christian actions during the 3rd century Plague of Cyprian in Alexandria as "unloving" (or at least fatally wrong-headed). As the third- and fourth-century A.D. Church historian, Eusebius, wrote:

> "Most of our brethren showed love and loyalty in not sparing themselves while helping one another, tending to the sick with no thought of danger and gladly departing this life with them after becoming infected with their disease. Many who nursed others to health died themselves, thus transferring their death to themselves. The best of our own brothers lost their lives in this way — some presbyters, deacons, and laymen — a form of death based on strong faith and piety that seems in every way equal to martyrdom. They would also take up the bodies of the saints, close their eyes, shut their mouths, and carry them on their shoulders. They would embrace them, wash and dress them in burial clothes, and soon receive the same services themselves."[1194]

Mask-wearing has never been a part of congregant worship practice throughout the last 2,000 years of Christian history (including many plagues next to which COVID-19 is barely a blip on the radar. Nor was it part of any pre-Christian Jewish practice prior to the birth of Christ. In no account of major plagues in Church history do medical masks feature prominently, including (but not limited to) the Antonine Plague, the Plague of Cyprian, the Plague of Justinian, or the Bubonic Plague known

---

1194   Eusebius and P.L. Maier, *Eusebius--the church history: a new translation with commentary.* 1999, Grand Rapids, MI: Kregal Publications. Book 7, Chapter XXII, section 7, available online: https://archive.org/details/eusebiusthechurc00euse/page/268/mode/2up

as the Black Death.[1195] For example, when a plague in 590 A.D. turned Rome into a "desert," and Pope Pelagius II died from it "screaming in agony,"[1196] his successor, Gregory the Great led seven processions through the streets of Rome for three days, with prayers and hymns.[1197] What the new Pope did *not* do was mandate any kind of masks. Lack of precedent does not necessarily imply that it is *wrong* to wear a mask, but it certainly lends no *support* to doing so.

1195   Plague doctor costumes do not count.

1196   Shelley, Bruce L. *Church History in Plain Language: Fifth Edition* (p. 203). Zondervan Academic. Kindle Edition

1197   Shelley, Bruce L. *Church History in Plain Language: Fifth Edition* (p. 204). Zondervan Academic. Kindle Edition

The secular religion of public health

As mentioned earlier in this book, in 2020 the statue of Christ the Redeemer in Rio de Janeiro, Brazil, was lit at night to make it appear that Jesus was wearing a mask, along with the hashtag "MascaraSalva," #MaskSaves.[1198] A year later, in 2021, the statue of Christ was again illuminated. This time with a message reading "Vaccine Saves," and included photography showing enthusiastic cultists wearing KN95 masks outdoors.[1199] I appreciate satire and irony as much as the next man, but I still couldn't help being concerned by substitutions of Anthony Fauci's name to change the phrase "In God we trust" to "In Fauci we trust." Public Health in the context of COVID took on the nature of a performative religion that emphasizes salvation by works, and I refuse to participate in rituals which promote this novel idolatry, of which mask-wearing is the most visible.

To the extent that they are rituals widely believed to be efficacious and essential to the public good and general welfare, mask-wearing and vaccination in modern society are serving the same function that public worship served in the ancient Roman Empire. Those who refuse to participate are seen as not just lacking in civic virtue, but as direct threats, as though inaction is now a form of aggression, to be treated accordingly.[1200] As historian Bruce L. Shelley explains in his *Church History in Plain Language*:

> "The main cause of the hatred of early Christians in Roman society lies in their distinctive lifestyle… simply by living according to the teachings of Jesus, Christians were a constant unspoken condemnation of the pagan way of life. It was not that Christians went about criticizing and condemning and

---

1198   Allen Kim, "Face mask projected onto Brazil's famous Christ the Redeemer statue," *CNN Travel*, May 4, 2020, available online: https://edition.cnn.com/travel/article/rio-redeemer-statue-trnd/index.html

1199   Lisa Shumaker, "Brazil's Christ the Redeemer statue lights up for vaccine equality," *Reuters*, May 17, 2021, available online: https://www.reuters.com/world/americas/brazils-christ-redeemer-statue-lights-up-vaccine-equality-2021-05-17/

1200   The Romans at least had a ready-to-hand pretextual justification for equating Christian failure to engage in such rituals as a form of incipient sedition or revolution, because such actions presaged the Jewish revolt of AD 66-70: "Finally, in AD 66 the Jews revolted, signaling their intent by refusing to perform the daily sacrifice for the emperor." Shelley, Bruce L.. *Church History in Plain Language: Fifth Edition* (p. 31). Zondervan Academic. Kindle Edition.

disapproving, nor were they consciously self-righteous and superior. It was simply that the Christian ethic in itself was a criticism of pagan life."[1201]

Christian refusal to offer sacrifices to the gods was seen as an overt threat to the general welfare. Moreover, Christian insistence that there was an authority above Caesar was seen as a threat to Imperial rule, rather like how military servicemembers who refused masks or vaccines were seen as an existential threat to "good order and discipline" and had their accommodation requests summarily denied, often followed by various direct and indirect "disciplinary" actions. A person wearing a mask who believes that masks work, that all rules should be presumed lawful until proven otherwise, or that their mask-wearing is a manifestation of them being a good person, will perceive an unmasked face as an attack on their body, their worldview, their character, their intelligence, or all four at once, and react accordingly. This is one of the major unstated underlying reasons why unmasked and unvaccinated individuals were subjected to such duress at the height of the COVID hysteria.

Church history is among my many historical interests, and my recreational research from prior to 2020 included several books and lectures on various aspects of the subject.[1202] My recreational reading also included psychology, fear, *actual* plagues, how people judge risk, and episodes of mass psychosis like the Jonestown collective suicide. This background made me more sensitive than average to the implications of compulsory masking and other COVID policies. Furthermore, the impulses and mentalities that were given free rein during the COVIDcrisis have parallels in the history of the Christian Church, and these parallels constitute some of the most infamous episodes in Church history: the two specific ones I will discuss here are The Inquisition[1203] and the sale of indulgences. As Professor Thomas Madden explains in his lectures on the history of the Inquisition:

1201   Shelley, Bruce L.. *Church History in Plain Language: Fifth Edition* (p. 50). Zondervan Academic. Kindle Edition.

1202   Of particular relevance to this discussion:

Shelley, B.R., *Church History in Plain Language: Fourth Edition.* 2013. Grand Rapids, MI: Zondervan Academic.

Madden, T.F., *Heaven or Heresy: A History of the Inquisition*, in *The Modern Scholar.* 2007

Madden, T.F. *From Jesus to Christianity: A History of the Early Church*, in *The Modern Scholar.* 2005

Madden, T.F. *Christianity at the Crossroads: The Reformation of the Sixteenth and Seventeenth Centuries*, in *The Modern Scholar.* 2008.

1203   It is actually a misnomer to speak of "The Inquisition." The caricatured popular image of "The Inquisition" has been filtered through centuries of acrimonious post-Reformation debates, pamphleteering, literary trope-mining, and Hollywood sensationalism and anti-Catholic slant. It thus bears less resemblance to history than does Disney's Pirates of the Caribbean franchise. However, I still use the term here for rhetorical purposes to tap into the popular understanding, mental images, and nigh-universal moral outrage which are still justly applicable to those very real abuses which did occur. I refer readers seeking a more accurate view of this episode in church history to the work of scholars like Thomas Madden. Perhaps it would have been better for me to refer to examples like the case of Jan Hus, or how Martin Luther at times had expectations of arrest and execution which were very understandable (even reasonable) under his own circumstances. Any defects in my choice of illustration notwithstanding, the point about coercion being

"In one form or another, all societies have heresies. There are beliefs that are considered by the consensus of a society — whether those beliefs be religious, or political or social — are considered to be repugnant... From the Medieval point of view, of course, what mattered most was not the things of this world, but the things of the next world, and therefore heresy for them was not simply dissent or what we would today refer to as "religious diversity" but rather heresy was a disease, it was an infection, it was something that came into the society, a society bound together by common religious beliefs and tore apart at that society. More importantly, it not only doomed the heretic to eternal damnation, but, they believed, it also threatened those around him with eternal damnation."[1204]

"The overriding sense one gets from reading these inquisitors' manuals, is that above all, the inquisitors saw themselves as the doctors of souls — almost as a physician that was being brought in because of a sickness. The person who was a heretic was the immediate person who was sick. He or she was the one who had a spiritual sickness that would ultimately claim not their body — not their physical life — but their eternal life; and so the action of the inquisitor was one of pastoral care and it was considered to be an act of charity, one in which one brother is helping another brother to heal them of this sickness. But it was also otherwise seen as a healing of a societal sickness as well — of heresy — of ensuring that the sickness did not spread, because heresy was a sickness of great contagion, and therefore would be deadly to other souls... from their writings, at least, and from the writings of others who watched them work, we can tell that their primary interest was in getting rid of what they saw as this dangerous infection in the Body of Christ."[1205]

Christians would affirm that everyone everywhere has an absolute, objective moral duty to worship God and no one else. Yet one of America's founding principles is that this is at least one moral duty which should *not* be a legal duty (those of a more pluralistic or atheistic bent who do not believe that such a moral duty even exists will also certainly agree that worshiping the God of the Bible should

wrong in matters of religion remains valid and should still resonate with most sympathetic readers, despite any friendly in-house disagreements about which example would be best to use. If anything, many Christians' experience from 2020 onwards may make them more empathetic to the very real social influences and good intentions that inquisitors labored under. "There but for the grace of God..."

1204   Madden, T.F., Heaven or Heresy: A History of the Inquisition, in The Modern Scholar. 2007, Recorded Books, Lecture 2, 1:27-3:05

1205   Madden, T.F., Heaven or Heresy: A History of the Inquisition, in The Modern Scholar. 2007, Recorded Books, Lecture 7, 2:37-3:55, 7:35-7:55.

not be a legal duty). At the final judgement, every knee will bow and every tongue will proclaim that Jesus is LORD, but in the meantime, God created all humans at enough of an epistemic distance to allow each one of us the free choice of whether or not to acknowledge and worship Him. Spending eternity with God is an incommensurable good. Accepting Jesus as Lord and savior is 100% effective for salvation, and the worst side effects are, in the words of St. Paul, "light momentary affliction" (2 Corinthians 4:17). But even taking all this as gospel truth, it is *still* wrong to even attempt to compel such belief because God has endowed every man, created in His image, with free will, and with the right of free choice.

Christians spent centuries trying to legally enforce religious belief, and a lot of areas around the world *still* do this to one extent or another. Does the Church now regard those episodes in its history with pride? Do we consider the centuries when Christians tried to turn the moral duty to worship God into a legal duty to be eras of justice, moderation, and good public policy? Are those countries that still try to do similar things flourishing utopias? Do we consider those modern countries which have enshrined Sharia law to one extent or another to be amongst the best places to live? Of course not! The opposite, in fact. We very properly look back on episodes like The Inquisition, the Wars of Religion, or coerced conversions in the New World with aversion and revulsion.

Today, it is practically axiomatic (at least in western countries) that coercion in matters of religion is — and always *was* — *wrong*. There are, however, two key distinctions between the spiritual virus of heresy and a physical virus like SARS-CoV-2, which make compulsory masking even *less* defensible than coercion in matters of religion. First, the historical Inquisition generally provided its accused better due process than today's politicians, public health officials, and busybully bureaucrats have done in matters relating to COVID.[1206] Second, heretics tend to spread their beliefs intentionally, whereas those who temporarily carry a physical virus almost never do. The eternal death produced by heresy is, from a Christian perspective, incomparably worse than the worst outcome of any physical illness. Inquisitors were, in effect, spiritual public health officials.[1207] The Inquisitors were right in that heresy *is*, in a sense, a spiritual virus, and "infection" can and does cause eternal death. From a Christian perspective, matters of religion are at least as important as matters of public health. By the core rationale espoused during COVID, in light of the eternal consequences of heresy and the intentional nature of its spread, medieval persecutors of heresy had far *stronger* moral justification for their actions than their modern public health analogues had for persecuting the unmasked and

---

1206   *See generally*, Madden, T.F., Heaven or Heresy: A History of the Inquisition, in The Modern Scholar. 2007, Recorded Books.

1207   In fact, based on my reading of the work of medieval historian Thomas Madden, the Papal Inquisition gave accused heretics *more* benefit of the doubt, due process, and leniency than unvaxxed and unmasked got during the COVIDcrisis.

unvaccinated. Yet all such actions were still wrong. It took Western Civilization hundreds of years — and thousands of lives — to internalize the fact that coercion is wrong in matters of religion. Many areas of the world still repudiate this proposition.

If it is wrong to use State force and legal compulsion either to stop a spiritual virus or to attempt to spread a belief which is necessary for eternal life, how much *more* immoral is it to use force or compulsory measures to stop a physical pathogen even as bad as the Black Plague — much *less* a virus like SARS-CoV-2 with a less than 0.15% infection fatality rate.[1208] Even if COVID really *had* been a second coming of the Black Death, that would have meant there was even *less* justification for *any* church to close its doors, because with so many people being ushered into eternity, the need for the gospel of eternal life to be delivered in every way possible — including in-person *face-to-face* — would be all the *greater*, and social distancing be *damned*! Matters of religion are no less important than matters of public health. Coercion in matters of religion is utterly immoral. Therefore, coercion is immoral in matters of public health.

Additionally, the emotional-blackmail reasoning used to push compulsory masking is eerily reminiscent of Johann Tetzel's indulgence sales tactics that goaded Martin Luther into posting his 95 Theses and kick-starting the Protestant Reformation. Indulgences were sold as a way of exchanging earthly money for spiritual credit that, once purchased, could be used on behalf of oneself or relatives — including dead relatives — to reduce time in the torments of purgatory before proceeding onwards to Heaven. You can see the potential for emotional blackmail, and Johann Tetzel was one of the most effective in the business of selling indulgences:

> "Don't you hear the voices of your wailing dead parents and others who say, 'Have mercy upon me, have mercy upon me, because we are in severe punishment and pain. From this you could redeem us with a small alms and yet you do not want to do so.' Open your ears as the father says to the son and the mother to the daughter, 'We have created you, fed you, cared for you, and left you our temporal goods. Why then are you so cruel and harsh that you do not want to save us, though it only takes a little? You let us lie in flames so that we only slowly come to the promised glory.'"[1209]

1208   John P. A. Ioannidis, "Reconciling estimates of global spread and infection fatality rates of COVID-19: An overview of systematic evaluations." *European Journal of Clinical Investigation*, 2021.
https://dx.doi.org/10.1111/eci.13554

1209   Hans J. Hillerbrand, ed., *The Reformation: A Narrative History Related by Contemporary Observers and Participants*. New York, 1964, pp. 41-43. Retrieved from:
https://archive.org/details/reformationnarra00hill/page/42/mode/2up?view=theater

This is, at its core, the same basic argument that was used to "sell" masks and vaccines to avoid COVID. The parallels between Johann Tetzel's sales pitch for indulgences and the public health browbeating to get people to participate in masking and vaccination during COVID are obvious. In modern terms, Tetzel's slogan was: "We have a moral obligation to buy indulgences to protect our family members," and "We don't have the right to keep people in purgatory." Compare this to: "We don't have the right to infect others," with all its underlying fallacious reasoning relating to causality and ultimate responsibility. This is just one of many such abuses of conscience that Martin Luther was referring to when he spoke at the Diet of Worms: "Is it not evident that the human laws and doctrines of the popes, entangle, torture, martyr the consciences of the faithful[?]"[1210] It's your fault your family members are still suffering in purgatory. It's your fault your family got infected. The difference is that in the modern episode of COVID, public health officials and politicians took things one step further by using force under color of law, rather than mere guilt and social pressure. So here (again) modern public health officials managed to do *worse* than one of the most embarrassing episodes in Church history.

At the Diet of Worms, when he was standing before Charles V, ruler of the Holy Roman Empire, Martin Luther declared: "Unless I am convicted by Scripture and plain reason — I do not accept the authority of popes and councils, for they have contradicted each other — my conscience is captive to the Word of God. I cannot and I will not recant anything, for to go against conscience is neither right nor safe."[1211] Popes and councils still come out of this assessment better off than the public health

---

See also J. E. Kapp, Schauplatz des Tetzelischen Ablass-Krams (Erfurt, 1717), pp. 46ff. A free German translation is found in Dar martin Luther's Sammtliche Schriften, herausgeg. Von John. Gg. Walch, vol. 15 (St Louis, 199), pp. 340-1. An English translation of another Ttezel sermon is found in Translations and Reprints from the Original Sources of European History (Philadelphia, 1894ff.), vol. II, No. 6, pp. 9-10.

1210   Martin Luther, address at the Diet of Worms, 17 April 1521, quoted in Merle d'Aubigné, J. H. (Jean Henri), 1794-1872 *History of the reformation in the sixteenth century*, G.P. Putnam & Sons: New York, 1872 (2009 Facsimile Ed. Powder Springs Press: Georgia), p. 203, available online: https://archive.org/details/historyofreforma0000merl_s7r1/page/202

Also quoted in Roland H. Bainton, *Here I Stand: A Life of Martin Luther*, Abingdon Press: Nashville (1950; 2013 reprint ed.), p. 185, available online: https://archive.org/details/hereistandlifeof0000bain/page/184

1211   Quoted in Roland H. Bainton, *Here I Stand: A Life of Martin Luther*, Abingdon Press: Nashville (1950; 2013 reprint ed.), p. 185, available online: https://archive.org/details/hereistandlifeof0000bain/page/184

Martin Luther gave his address at the Diet of Worms in German and then again in Latin. English translations obviously have variant wordings that do not affect the substance of what he said. An alternate version is included below for the sake of completeness.

> "I cannot subject my faith either to the pope or to councils, because it is clear as day, that they have often fallen into error, and even into great self-contradiction. If, then, I am not disproved by passages of Scripture, or by clear arguments, - if I am not convinced by the very passages which I have quoted, and so bound in conscience to submit to the Word of God, *I neither can nor will retract anything*, for it is not safe for a Christian to speak against his conscience."

establishment during COVID. Popes and councils contradicted *each other*. During the COVIDcrisis, politicians, public health officials, and all manner of authorities contradicted not only each other, but repeatedly and shamelessly contradicted *themselves* when imposing mandates, using The Science™ and moral posturing as a cloak to conceal what was at bottom political expediency and the self-interest that they tried to project onto those who resisted their diktats. Even in their worldly domains of the Papal States, the most corrupt, worldly Renaissance Popes did not even *attempt* to impose the kind of measures common during COVID.

We moralized COVID in the same simplistic way that we disparage moralization in other contexts: "If you hadn't done something wrong, this wouldn't be happening to you." The diabolical twist, though, was the communalization of the penalty. Instead of something bad happening to *you* because *you* did something wrong, with COVID it became, "something bad happened to *me* because you did something wrong." A lot of people took it a step further: "Something bad happened to someone I care about because you did something wrong." This let them frame themselves as being in the position of altruistic protector of the weak against the villainous non-compliant. All based on abstract delusions of risk in a total absence of any direct causation.

Long before COVID, The Science™ was elevated to a religion in certain circles. The eugenics movement was just one of the results. Christian writer G. K. Chesterton saw this a century before COVID, and addressed it in his 1922 book, *Eugenics and Other Evils*:

> "The thing that really is trying to tyrannise through government is Science. The thing that really does use the secular arm is Science. And the creed that really is levying tithes and capturing schools, the creed that really is enforced by fine and imprisonment, the creed that really is proclaimed not in sermons but in statutes, and spread not by pilgrims but by policemen—that creed is the great but disputed system of thought which began with Evolution and has ended in Eugenics. Materialism is really our established Church; for the Government will really help it to persecute its heretics. Vaccination, in its hundred years of experiment, has been disputed almost as much as baptism in its approximate two thousand. But it seems quite natural to our politicians to enforce vaccination; and it would seem to them madness to enforce baptism.

Martin Luther, address at the Diet of Worms, 17 April 1521, quoted in Merle d'Aubigné, J. H. (Jean Henri), 1794-1872 *History of the reformation in the sixteenth century*, G.P. Putnam & Sons: New York, 1872 (2009 Facsimile Ed. Powder Springs Press: Georgia), p. 204, available online: https://archive.org/details/historyofreforma0000merl_s7r1/page/204

"I am not frightened of the word "persecution" when it is attributed to the churches; nor is it in the least as a term of reproach that I attribute it to the men of science. It is as a term of legal fact. If it means the imposition by the police of a widely disputed theory, incapable of final proof—then our priests are not now persecuting, but our doctors are. The imposition of such dogmas constitutes a State Church—in an older and stronger sense than any that can be applied to any supernatural Church to-day. There are still places where the religious minority is forbidden to assemble or to teach in this way or that; and yet more where it is excluded from this or that public post. But I cannot now recall any place where it is compelled by the criminal law to go through the rite of the official religion."[1212]

Regardless of whether courts were willing to recognize it, during COVID, public health became a religion in all but name, with masking and vaccination as its chief rituals, and those who refused to participate were persecuted accordingly.

1212   Gilbert Keith Chesterton, *Eugenics and Other Evils*, Cassell and Company: London (1922), p. 76-77, available online: https://archive.org/details/cu31924013462555/page/n87/

Attempts to push masking using the Bible fail miserably

Multiple misguided arguments were put forward attempting to establish a Biblical basis from which to push masking-wearing on objectors. All such arguments fall apart on close examination.

Often, I see 1 Corinthians Chapter 8 applied to argue that Christians who are comfortable without masks should put them on for the sake of their "weaker" fellow believers. In this passage, the Apostle Paul points out that there are times when, for the sake of their weaker brothers, Christians with strong faith are called to limit actions they would otherwise be free to take. One guest author for CrossWalk. com applied this passage to masks like so: the Apostle Paul:

> "… sees that the only faithful response to this matter is to limit his own freedom for the sake of others. This is the only faithful response to the face mask matter at hand, because this is the response of Jesus… Let us take up the call of the cross, and willfully empty ourselves for the sake of another. Let us emulate Paul as he emulates Christ. And, if a refusal to wear a mask causes a brother or sister to fall into sin, may we consider wearing them as a way to express agape[1213] love, instead."[1214]

Another author writes: "For the sake of unity in our time and place, all worshippers should wear masks and keep social distance in public worship. This will do no harm. It is also the right thing to do in order to build up our community with love."[1215]

1213   *Agape*, pronounced agápē ("ah-ga-pay"), is an ancient Greek term for the highest form of love.

1214   Reverend Norman, K. "What Would the Apostle Paul Say about Wearing Face Masks?" *CrossWalk*, 2020. https://www.crosswalk.com/special-coverage/coronavirus/what-would-the-apostle-paul-say-about-wearing-face-masks.html

1215   Taylor-Troutman, A. "Do no harm: Wear your mask," *The Presbyterian Outlook*, 2021 https://pres-outlook.org/2021/05/do-no-harm-wear-your-mask/

There are two fatal problems with this reasoning. First, it cuts both ways depending on how you frame "weak" vs. "strong". One could just as easily exhort readers: "If *wearing* a mask causes a brother or sister to fall into sin, may we consider going *without* them as a way to express agape love, instead." If you start out by defining the people who are concerned about wearing masks as the "weak" brothers, the onus instead falls on those "strong" Christians who are wearing masks to take their masks *off*. Moreover, if this passage *must* be applied to masks, I would argue that this alternative framing better fits the context in which Paul was writing, because in the context in which Paul is writing, the "weaker" Christians were those trying to *not* do something (they were concerned about eating food which had been sacrificed to idols). In the same way that Christians in the early church did not want to lend even an indirect sense of approval or legitimacy to idol-worship by eating food which had been sacrificed to idols, many Christians today, myself included, vehemently resist wearing masks because of concerns over the symbolic legitimacy which doing so might lend to mask-wearing, compulsory masking, and other activities, such as the regime of social-distancing, widespread fear, and elevation of public health to a *de facto* secular religion.

Second, even if one insists on applying 1 Corinthians 8 in the way that the authors cited above do, Paul's words and actions in other areas of his ministry clearly indicate that he considered his guidance in 1 Corinthians 8 (and Romans 13-14) to be situation-dependent. Contrast 1 Corinthians 8 with Galatians 2 (ESV), where Paul inveighs against: "false brothers secretly brought in—who slipped in to spy out our freedom that we have in Christ Jesus, so that they might bring us into slavery — to them we did not yield in submission even for a moment, so that the truth of the gospel might be preserved for you." Clearly, then, while there are times when Christians may be individually called to rein in the exercise of their freedom for the sake of others, there are plenty of other instances when they are called to not give an inch even if that stand makes those around them uncomfortable. Even more relevant, Galatians 2 was written in the context where Paul called out Peter for limiting his own freedom based on the opinions of other believers who were separating themselves for reasons of ritual purity: "before certain men came from James, he was eating with the Gentiles; but when they came he drew back and separated himself…" (Galatians 2:12, ESV). Ritual purity was the more charitable way of characterizing most of the response measures to COVID like spaced out stickers on the floor, directional lanes in supermarkets, and arbitrary occupancy limitations. The ritual component of mask-wearing should be obvious to all readers by this point, and the bottom line is that Christian love for others does not necessarily require doing something like wearing a mask even if that non-action makes others uncomfortable, offends them, or makes them feel unsafe. In fact, Christian love for others may very well *require* refusing actions like mask-wearing, despite the odium that such a non-action may incur.

Another common pseudo-Biblical argument ran like this: "… if you are a Christian, you should be wearing a mask — even if you do think they are useless… it all boils down to this: it doesn't matter what you believe about wearing masks — it matters what those around you believe about wearing masks."[1216] This argument in favor of mask-wearing is utterly unbiblical, as though your neighbor's belief and emotional or physical comfort can determine what is right. Does God act this way? Does the being whose very nature is love use how people *feel* about a particular action as the primary criteria for choosing His actions? If the Bible is anything to go by, clearly, He does not. This being the case, loving others does not necessarily entail doing *only* things which they like. If God avoided everything that caused us distress, or did everything possible to prevent us from having to endure pain, privation, or fear, we would be (at best) nothing but pampered pets rather than the morally mature spiritual beings we are meant to be. We justly criticize parenting styles that give children everything they want, and our best friends are those who will tell or show us truths we don't necessarily want to hear. Those Christians who refuse to wear a mask cannot all be tarred with the same brush as being unloving, because it may well be that refusing to wear a mask is actually the *best* way for them to show love to those around them.

One particularly grating assertion coming from people who ought to know better is: "What would Jesus do? He'd wear a mask."[1217] Not for one second do I buy this claim. Insofar as anyone can hypothesize about what Jesus would have done, the weight of the evidence suggests that He would have refused to wear a mask, at least to the extent that having worn a mask would have reinforced the legitimacy of pseudo-legal Pharisaic traditions. As mentioned in our earlier discussion of symbolic speech, Jesus did not follow public health directives against touching lepers (Matthew 8:2-3), and on at least one occasion, declined to participate in the standard hygiene practice of handwashing before eating at a dinner party in order to make a broader point that many around him considered highly offensive (Luke 11:37-41). The gospel writer who related this incident was Luke, the physician.

Another time, some Pharisees tried to call out Jesus' disciples for violating the traditions of the elders (i.e., "best practices") by not washing their hands before eating. They asked Jesus, "Why don't your disciples live according to the tradition of the elders instead of eating their food with defiled hands?" (Mark 7:5, NIV). Rather than reprimanding His disciples for not washing their hands before eating, Jesus reprimanded the Pharisees for ranking their traditions too highly (Matthew 15:1-20, Mark 7:1-23). Jesus retorted: "Isaiah was right when he prophesied about you hypocrites; as it is written:

1216   Murphy, K., NC pastor: "Jesus would wear a mask," 2021, *The Charlotte Observer*. Online https://www.charlotteobserver.com/opinion/article242898806.html

1217   Lorenz, M., Karl, "What would Jesus do during COVID-19? He would wear a mask," 2020, *The Dallas Morning News*. Online https://www.dallasnews.com/opinion/commentary/2020/07/26/what-would-jesus-do-during-covid-19-he-would-wear-a-mask/

'These people honor me with their lips, but their hearts are far from me. They worship me in vain; their teachings are merely human rules.' You have let go of the commands of God and are holding on to human traditions" (Mark 7:6-8). Jesus then told the surrounding crowd: "Nothing outside a person can defile them by going into them. Rather, it is what comes out of a person that defiles them" (Mark 7:15, NIV). Jesus' disciples asked what he meant, and the examples that Jesus gave in his answer about what defiles a person, making them unclean, had nothing whatsoever to do with infectious disease. Jesus said, "What comes out of a person is what defiles them. For it is from within, out of a person's heart, that evil thoughts come — sexual immorality, theft, murder, adultery, greed, malice, deceit, lewdness, envy, slander, arrogance and folly. All of these evils come from inside and defile a person" (Mark 7:20-23, NIV).

Compare this to the COVIDcrisis masking mentality, where getting infected by a virus was presumed to be a moral failing unless all the public health rites and traditions had been observed to the letter. Handwashing (especially in communal meals) has a *much* stronger evidential link to disease control than does wearing a mask, which has been a practice and tradition in certain medical settings only since the late 1800s (at the earliest). If "safety first" (for Himself or for those around Him) had been Jesus' top priority or that of His followers, Christianity would have been stillborn. Nor did Jesus "social distance" from people with communicable diseases like leprosy. On the contrary, as again reported by both Matthew and Luke, the physician: "a leper came to him and knelt before him, saying, 'Lord, if you will, you can make me clean.' And Jesus stretched out his hand and touched him, saying, 'I will; be clean.' And immediately his leprosy was cleansed" (Matthew 8:2-3, ESV; see also Luke 5:12-14).

Compulsory masking should be as unthinkable as a Biblical priest telling someone with unblemished skin to behave like a leper, visually shouting, "Unclean! Unclean!" (Leviticus 13:45). Indeed, in Biblical terms, that is exactly what compulsory masking is, and I do not for one moment believe that Jesus would have condoned it by his participation. Based on these incidents, what do you think Jesus would have said to a Pharisee, a priest, a doctor, or a public health official who told someone with unblemished skin that they should behave like a leper? What do you think Jesus would have told such people if they had *forced* people with unblemished skin to behave like lepers? I think Jesus would have had *very* harsh words for anyone making such recommendations — and even harsher words for those issuing such *mandates*. I think it is far *more* plausible that Jesus would have responded to today's public health officials in a similar vein to how He critiqued the scribes and Pharisees: "They tie up heavy burdens, hard to bear, and lay them on people's shoulders, but they themselves are not willing to move them with their finger. They do all their deeds to be seen by others" (Matthew 23: 4-5, ESV).

I, for one, do *not* think that Jesus would have worn a mask. The Bible specifically says that Jesus came to "deliver all those who through fear of death were subject to lifelong slavery" (Hebrews 2:15,

ESV), and wearing masks for COVID *promoted* lifelong slavery through both fear of death *and* public health totalitarianism. By contrast, the Bible says that "It is for freedom that Christ has set us free" (Galatians 5:1, NIV). Some might argue against refusing masks by referring to 1 Peter 2:16 (ESV), that Christians are to: "Live as people who are free, not using your freedom as a cover-up for evil, but living as servants of God." In response, I say, refusing to wear a mask is not evil, never has been evil, and a substantial subset of Christians have a moral duty before God to *refuse* to wear masks.

At least one pastor argued that Jesus' example favors mask-wearing on the reasoning that "Jesus didn't insult or belittle people who were afraid. He went out of his way to come alongside them. Jesus went out of his way to show… that their lives mattered to him… He didn't mock those who were afraid, he relieved their fears."[1218] Therefore, this pastor reasons, Christians should wear masks. There is an *element* of truth to this argument, but it is a partial truth at best. When Jesus and his disciples were in a boat that "was being swamped by the waves" in a "great storm on the sea," "they went and woke him, saying, 'Save us, Lord; we are perishing.' And he said to them, 'Why are you afraid, O you of little faith?'" (Matthew 8:24-26, ESV); see also Mark 4:37-40). Jesus certainly didn't do any name-calling, but he also did not validate his disciples' fears, even though having one's fishing boat in the process of being swamped during a great storm is a far more reasonable and immediate cause for fear than an apparently healthy person within six feet not wearing a mask. Yet Jesus still rebuked his disciples for their lack of faith. So, would Jesus have worn a mask for COVID? I see no good reason to think so, and I see *multiple* good reasons to think he would have *refused* to do so, and possibly would have rebuked others for doing so.

Another supposedly Biblical argument in favor of mask-wearing is that "Asserting rights is all about me, but wearing a mask is all about my care and concern for the feelings and well-being of others"[1219] Neither half of this statement is entirely true. Each half is *possibly* true, but not *necessarily* true, and that is what makes it so insidious. Asserting and defending rights that you share in common with other people benefits you, but it *also* benefits everyone else as well, and this remains true even, or *especially*, when others either choose not to assert their rights or cannot do so for themselves. This follows a strong Biblical injunction: "Open your mouth for the mute… defend the rights of the poor and needy" (Proverbs 31:8-9, ESV). It may very well be that a stand for your own rights and those of others is taken contrary to personal interest and in the face of personal loss and social ostracism. Conversely, the primary impetus behind one's decision to wear a mask may simply be a matter of personal convenience and a desire to avoid getting hassled, with care and concern for the feelings of

---

1218   Kate Murphy, NC pastor: "Jesus would wear a mask," 2021, *The Charlotte Observer*: Online
https://www.charlotteobserver.com/opinion/article242898806.html

1219   Rob Schenck, *Wearing a Mask Is Biblical*. 2020: Online
https://sojo.net/articles/bible-says-wear-mask-Christians-church-coronavirus

others being a distant secondary motivation. Wearing a mask can be simply going along to get along, doing good works to be seen by men - which Jesus specifically warned against when He said: "Beware of practicing your righteousness before other people in order to be seen by them, for then you will have no reward from your Father who is in heaven" (Matthew 6:1, ESV; see also Matthew 23:5). When you assert and defend your God-given, unalienable rights, you are asserting and defending everyone else's God-given, unalienable rights as well — even those of people who hate and revile you for doing so. This is actively loving not just your neighbor, but loving your *enemy* as well!

Two verses in particular showed up frequently in debates on COVID mandates. The first of these is when Jesus said, "render to Caesar the things that are Caesar's, and to God the things that are God's" (Luke 20:25, ESV). The second is when the Apostle Paul said, "Let every person be subject to the governing authorities. For there is no authority except from God and those that exist have been instituted by God" (Romans 13:1, ESV). Typically, those citations were followed by very strong assertions: "Listen to me, believer—God has established that authority. Your response in such matters is to obey. Anything else, is rebellion to God and His established authorities."[1220] Checkmate, anti-maskers! I respond that it is simply false to conflate human authority — the things that are "Caesar's" — with the Law of God to the extent which many Christians have done since March 2020 (especially as pertaining to mask mandates). What is "legal," and what is lawful or just in the Biblical sense are often two different things. The Bible is constantly warning against perverting justice through legal mechanisms, and these injunctions occur right out of the gate: "You shall not fall in with the many to do evil, nor shall you bear witness in a lawsuit, siding with the many, so as to pervert justice, nor shall you be partial to a poor man in his lawsuit" (Exodus 23:2-3, ESV). The apostles' earliest preaching regarding Jesus' trial and execution takes for granted that legal mechanisms do not always confer moral legitimacy on an authority's actions and that "just following orders" may still be wrong: "this Jesus, delivered up according to the definite plan and foreknowledge of God, you crucified and killed by the hands of lawless men" (Acts 2:23, ESV).

As America's Founders rightly understood, the instruction to: "Be subject for the Lord's sake to every human institution" (1 Peter 2:13, ESV) is not one of the absolute moral duties, nor is it limited to mere matters of worship or preaching. Biblically permissible acts of disobedience range from working within the system to obtain a dietary accommodation (Daniel 1), avoidance and deception (the midwives in Exodus 1) or even failure to comply with a legally mandated gesture of respect, as in the case of Queen Esther's cousin Mordecai:

1220   Ron Cantor, *Does The Bible Speak to Mask-Wearing? Yes!* | God TV. 2020, GodTV: Online
https://godtv.com/does-the-bible-speak-to-mask-wearing-yes/

> "And all the king's servants who were at the king's gate bowed down and paid homage to Haman, for the king had so commanded concerning him. But Mordecai did not bow down or pay homage. Then the king's servants who were at the king's gate said to Mordecai, "Why do you transgress the king's command?" And when they spoke to him day after day and he would not listen to them, they told Haman[.]" (Esther 3:2-4, ESV).

Situational moral imperatives may even require direct refusal in more stringent contexts where authorities are given a freer hand, like the military. Issuing authorities might view any disobedience in such contexts as existential threats to "good order and discipline," but this would be erroneous. In these cases, it is the orders being issued that are the real source of any harm to "good order and discipline," because of how they force inescapable moral conflicts on the recipients.[1221]

As we saw in our discussion covering St. Thomas Aquinas and St. Augustine, an unjust law is no law at all. The only way in which an unjust law *might* incur any moral obligation is indirectly — when resisting or disobeying it would bring about greater evils than obeying it.[1222] This was the favored indirect route by which so many people were coerced against their will into making use of measures like lockdowns, compulsory masking and compulsory vaccination — by weaponizing their other moral duties against them to get compliance. Everyone who wore a mask or got the COVID vaccine against their will to keep their job to provide for their family knows *exactly* how this worked. Faced with this pair of bad choices, resisting to the utmost or gambling one's health to provide for one's family were *both* honorable and godly choices, and in the final analysis should have been between each individual and *God*. It is one thing when outside circumstances like a virus or natural disaster put people to this sort of choice, but doing so artificially is quite another. Actively putting people to this sort of choice is just one of many despicable aspects of the coercive public health response to COVID.

1221   c.f. 1 Samuel 14:45, when King Saul ordered the men of his Army not to eat during post-battle pursuit and wanted to execute his own son for disobeying, but was overruled by his own soldiers. *See also* 2 Samuel 11:11, when King David instructed his soldier Uriah to go home and rest while the rest of his unit was in the field but Uriah refused to do so.

1222   "This argument is true of a law that inflicts unjust hurt on its subjects. The power that man holds from God does not extend to this: wherefore neither in such matters is man bound to obey the law, provided he avoid giving scandal or inflicting a more grievous hurt." Aquinas also says that some laws like those that induce to idolatry or anything else directly contrary to God's Divine Law must not be obeyed at all.
Saint Thomas Aquinas. *The Summa Theologica*, First Part of the Second Part, Treatise on Law, Question 96, Article 4, "On The Power of Human Law, Whether Human Law Binds a Man in Conscience," Reply to Objection 3. Catholic Way Publishing. Kindle Edition. Alternative translation available online:
http://www.sophia-project.org/uploads/1/3/9/5/13955288/aquinas_law.pdf

During the French Revolution, when Quakers asked that their pacifist beliefs be respected, the *comte de Mirabeau*, a leader in the French National Assembly, voiced the common counterargument:

> "Don't you think the defence of yourselves and your neighbours to be a religious duty also? Otherwise you would surely be overwhelmed by tyrants! Since we have gained liberty for you as well as for ourselves, why would you refuse to preserve it?

> If your brethren in Pennsylvania had been settled nearer its savage inhabitants, would they have allowed their wives, children, and old people to be slaughtered rather than resist? And aren't stupid tyrants and ferocious conquerors equally savages?…

> Whenever I meet a Quaker I intend to say to him: My brother, if you possess the right to be free, you have also an obligation to prevent anyone from making you a slave. Loving your neighbour, you must not allow a tyrant to destroy him: to do so would be the same to kill him yourself.[1223]

The *comte*'s final objection, in particular, was a false equivalency common during COVID. It is simply false that not preventing a tyrant from destroying one's neighbor is the same as killing him yourself. A public health permutation of the *comte de Mirabeau's* argument was dominant during the COVID crisis, and every bit as wrong: "Loving your neighbor, you must not allow a virus to infect him: to do so would be the same to infect or kill him yourself." This is simply false, even assuming that you have that much influence over which viruses do or do not infect your neighbor. To be sure, when "Caesar" decides to intrude on a domain where he has no rightful authority, God lays a duty on many Christians to resist tyrants in the way the *comte de Mirabeau* described, but God does not place that same duty on *every* believer to the same extent, or call them to resist tyranny in the same *way*. Some believers may be called to physically fight, some may be called to help the sick, while others are called to serve God in some completely different way.

To a certain extent, I can see where the impulse to see inaction as direct harm comes from. There is at least one place in the Bible which could be read that way (at least at first glance), James 4:17.[1224] The

---

1223   Peter Brock, "Conscientious Objection in Revolutionary France," *The Journal of the Friends Historical Society*, 1995. 57(2), page 171, see also footnote 22 on page 180, available online: https://journals.sas.ac.uk/fhs/article/view/3500

1224   See also Ezekiel 33:1-9

NIV translation of this verse reads: "If anyone, then, knows the good they ought to do and doesn't do it, it is sin for them" (NIV). The ESV translation reads: "So whoever knows the right thing to do and fails to do it, for him it is sin" (ESV). That certainly sounds like an endorsement for the moral equivalency of action and inaction, but applying this to justify universal masking is simply wrong. James does not say "the good that *everyone* ought to do" or anything of the sort. The key phrase is "the good *they* ought to do" (emphasis added). The good that someone knows they ought to do is *individualized* to themselves depending on their circumstances and individual characteristics. In matters of things like masks and vaccines, who am I to presume to know what the good that someone else ought to do is? I, Dr. Philip Buckler, may have a duty to oppose compulsory masking with everything I have, but for all I know, my family member has a duty to just comply. Likewise, no family member, doctor, politician, or public health bureaucrat can presume to know that I or anyone else has a moral duty to wear a mask either. Moreover, even in this verse, failing to do a particular good that one ought to is *still* not the same as doing the opposite evil. Even assuming a duty to protect or shield someone, *failing* to do that duty is not morally equivalent to assaulting them yourself.

People often have difficulty perceiving exactly what their moral duty is, leading to the common confusion between our ability to *know* our moral duties (moral epistemology) vs. the *existence* of those moral duties (moral ontology). To further complicate matters, sometimes our uncertainty over our moral duty is used as a semi-conscious dodge when we know our moral duty and simply do not want to do it. Most people have an innate sense that their culpability for not doing their moral duty is diminished if they didn't actually *know* what they ought to do. Jesus himself implies this in Luke 12:47-48 (ESV): "And that servant who knew his master's will but did not get ready or act according to his will, will receive a severe beating. But the one who did not know, and did what deserved a beating, will receive a light beating." One duty which I do *not* believe God has placed on *any* Christian (with the *possible* exception of parents for their children or guardians of those unable to make a decision on their own behalf), is to *force* any medical intervention on another against their will, or to aid, abet, or otherwise facilitate such things being done. My position and understanding of this principle shifted as a direct result of seeing the outrageous, inexcusable medical tyranny perpetrated in the name of public health during COVID.

The authors of the United States Constitution recognized the distinction between human law and the Higher Law which human law must reflect in order to retain moral and legal legitimacy, and they appealed to this Higher Law and the unalienable Rights with which men are endowed by their Creator to justify their own forcible separation from England: "to secure these rights, Governments are instituted among Men… that whenever any Form of Government becomes destructive of these ends, it is the Right of the People to alter or to abolish it[.]" As James Madison affirmed, men are, first and foremost, subjects of the Governor of the Universe, and civil society does not supersede this claim.[1225] One of the original proposals for the great seal of the United States included Benjamin Franklin's suggestion that it include the words "Rebellion to Tyrants is Obedience to God," and Thomas Jefferson added this phrase to his own personal seal at least as early as 1790.[1226] It is my firm conviction that no part of "Caesar's" domain includes compulsory public health measures like masking, lockdowns, and vaccine mandates. When it comes to compulsory masking and all other such measures, "Caesar" was attempting to take what *is* not his, never *was* his, and never *should* be his. Whenever "Caesar" involves himself outside of his *narrow* legitimate domain, regardless of how good his motives are or how effective the intervention is, God lays a moral duty on some (or many) individual believers to push back.

Unfortunately, many people, especially in positions of authority, tend to have an attitude which I think is best exemplified in a statement by Pliny the Younger. Pliny was a Roman aristocrat who lived from approximately 61 to 113 A.D. He is not as well-known these days, but he is a good writer and his letters are entertaining (among other things, he gives an eyewitness account of what it was like to live through the eruption of Mount Vesuvius in A.D. 79). Later in Pliny's life, the Roman emperor Trajan made him governor of the province of Bithynia and Pontus. In one of his letters to the emperor while serving as governor, Pliny made a very revealing statement. He said, "stubbornness and inflexible obstinacy surely deserve to be punished."[1227] A lot of people might agree with that in a knee-jerk kind of way, and this authoritarian mentality was given full rein during the COVIDcrisis. But as details surrounding this letter of Pliny's show, that belief is both mistaken and harmful. Stubbornness and inflexible obstinacy are not inherently good or bad. They become vice or virtue depending on the objects to which they are directed and the context in which they are exercised. The context of that particular statement by Pliny was part of his explanation to the emperor why he had some Christians executed:

1225   James Madison, "Memorial and Remonstrance against Religious Assessments, [ca. 20 June] 1785," *Founders Online,* National Archives, https://founders.archives.gov/documents/Madison/01-08-02-0163. [Original source: *The Papers of James Madison*, vol. 8, *10 March 1784–28 March 1786*, ed. Robert A. Rutland and William M. E. Rachal. Chicago: The University of Chicago Press, 1973, pp. 295–306.]

1226   Jefferson, T., From Thomas Jefferson to Richard Gem, 4 April 1790, in Founders Online. 1790, Princeton University Press: Online https://founders.archives.gov/documents/Jefferson/01-16-02-0169 p. 297-298.

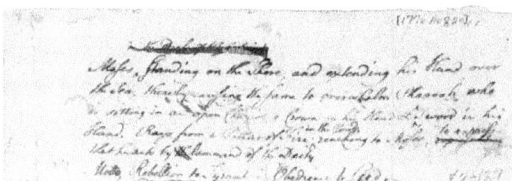

Moses in the Dress of a High Priest[*] standing on the Shore, and extending his Hand over the Sea, thereby causing the same to overwhelm Pharoah who is sitting in an open Chariot, a Crown on his Head & a[ᵇ] Sword in his Hand. Rays from a Pillar of Fire in the Clouds[ᶜ] reaching to Moses, expressing to express that he acts by the Command of the Deity

Motto, *Rebellion to Tyrants is Obedience to God*.

ᵃ The words "in the Dress of a High Priest" are inserted with a caret and deleted.
ᵇ The article "a" is inserted with a caret.
ᶜ The words "in the Clouds" are inserted with a caret.

1227   Pliny *Letters* 10.96-97, available online: https://sourcebooks.fordham.edu/source/pliny1.asp

"Meanwhile, in the case of those who were denounced to me as Christians, I have observed the following procedure: I interrogated these as to whether they were Christians; those who confessed I interrogated a second and a third time, threatening them with punishment; those who persisted I ordered executed. For I had no doubt that, whatever the nature of their creed, stubbornness and inflexible obstinacy surely deserve to be punished… Those who denied that they were or had been Christians, when they invoked the gods in words dictated by me, offered prayer with incense and wine to your image, which I had ordered to be brought for this purpose together with statues of the gods, and moreover cursed Christ--none of which those who are really Christians, it is said, can be forced to do--these I thought should be discharged. Others named by the informer declared that they were Christians, but then denied it, asserting that they had been but had ceased to be, some three years before, others many years, some as much as twenty-five years. They all worshipped your image and the statues of the gods, and cursed Christ."[1228]

Most laws and moral issues that Christians are faced with are not nearly as cut-and-dry as this. Usually, moral choices are derived from broader principles and applied to individual day-to-day situations, sometimes after specifically seeking God's will for especially difficult choices. Medical interventions fall into this category. During the inoculation controversies of the 18th Century, Christian writer John Newton (best known for being the author of the song "Amazing Grace") discussed the moral controversy surrounding medical interventions for a far more deadly disease — Smallpox. In a letter from the summer of 1777, during America's Revolutionary War, Newton's summary of two valid Christian positions on the critical subject of inoculation remains just as relevant and applicable to masks and vaccines for COVID (or any other disease) today as it was when he wrote it. One of Rev. Newton's friends wrote to him asking for advice about whether a parishioner should change her plans or get inoculated for smallpox. When reading Newton's response, note that the positions he summarizes are *individual* choices based on *individual* situations and *individual* understanding of God's will for *individual* lives:

"I am not a professed advocate for inoculation: but if a person who fears the Lord should tell me, I think I can do it in faith, looking upon it as a salutary expedient, which he in his providence has discovered, and which therefore appears my duty to have recourse to, so that my mind does not hesitate with respect to the lawfulness, nor am I anxious about the event; being satisfied,

---

1228   Pliny *Letters* 10.96-97, available online: https://sourcebooks.fordham.edu/source/pliny1.asp

that whether I live or die, I am in that path in which I can cheerfully expect his blessing; I do not know that I could offer a word of dissuasion.

"If another person should say, My times are in the Lord's hands; I am now in health, and am not willing to bring upon myself a disorder, the consequences of which I cannot possibly foresee. If I am to have the small-pox, I believe he is the best judge of the season and manner in which I shall be visited, so as may be most for his glory and my own good; and therefore I choose to wait his appointment, and not to rush upon even the possibility of danger without a call. If the very hairs of my head are numbered, I have no reason to fear that, supposing I receive the small-pox in a natural way, I shall have a single pimple more than he sees expedient; and why should I wish to have one less?'... therefore I am determined, by his grace, to resign myself to his disposal... If a person should talk to me in this strain, most certainly I could not say, Notwithstanding all this, your safest way is to be inoculated...

"I dare not advise: but if she can quietly return at the usual time, and neither run intentionally into the way of the small-pox, nor run out of the way, but leave it simply with the Lord, I shall not blame her. And if you will mind your praying and preaching, and believe that the Lord can take care of her without any of your contrivances, I shall not blame you: nay, I shall praise him for you both."[1229]

Many Christians have argued that *not* wearing a mask would be presenting a bad Christian witness to unbelievers. However, every argument adduced to support this position could be easily flipped to argue that *wearing* a mask would be presenting a bad Christian witness. I think Christians who make these arguments are, at best, erroneously generalizing their own individual moral duties to all other believers. Remember that Jesus said: "Woe to you, when all people speak well of you, for so their fathers did to the false prophets" (Luke 6:26).

1229   John Newton, "Letters to the Rev. Mr. R ****. Letter IX, June 3, 1777," *The Works of John Newton*, 6 vols., Reprint: (Edinburgh: The Banner of Truth Trust, 1985), Volume 2, p. 129-131, Originally published: (London: Hamilton, Adams & CO, 1824), available online: https://archive.org/details/worksofjohnnewto0002newt/page/128/

Appendix Conclusion: My oath

Finally, in my particular case, compliance with mask-wearing made me complicit in the widespread Constitutional violations perpetrated in the name of public health, thereby causing me to violate my oath to defend the United States Constitution — an oath which I took with a Biblically based understanding in 2006 and again in 2018.

- "If a man vows a vow to the Lord, or swears an oath to bind himself by a pledge, he shall not break his word. He shall do according to all that proceeds out of his mouth" (Numbers 30:2, ESV).
- "O Lord, who shall sojourn in your tent? Who shall dwell on your holy hill? He who walks blamelessly and does what is right and speaks truth in his heart.... who swears to his own hurt and does not change;" (Psalm 15:1-4, ESV).

As I stated earlier, the Constitution of the United States lacks an emergency clause not because of accident or omission, but by design - from conscious intent, and it is my firm conviction that my oath to defend the Constitution, taken before God, requires (at least in my case) refraining from actions that overtly lend intellectual credence and moral legitimacy to measures like mandatory masking. The God of Abraham, Isaac, and Jacob, and those unalienable rights to Life, Liberty, and the Pursuit of Happiness with which He has endowed all mankind in supersedence of *any* government are a greater good which we have repeatedly sacrificed on the idolatrous altar of safety and public health since March 2020 and even long prior to that.

The mentality and philosophy underlying compulsory masking has existed as long as mankind, and it will endure until Christ returns. The best we can do is strengthen ourselves and others of the current and future generations to successfully oppose it in whatever guises it next manifests. Dietrich

Bonhoeffer's theory of stupidity is the most well-known part of his essay-letter "After Ten Years," but the rest of Bonhoeffer's essay contains a great deal that was relevant and applicable, not just in his own era, but in many others, including ours:

> "In other times it may have been the task of Christianity to testify to the equality of all human beings; today it is Christianity in particular that should passionately defend the respect for human boundaries and human qualities. The misinterpretation that it is a matter of self-interest, or the cheap allegation that it is an antisocial attitude, must be resolutely faced. They are the perennial reproaches of the rabble against order."[1230]

Necessity and self-preservation are always the justification used for immoral overreach and tyranny, humanitarian or otherwise, including compulsory medical interventions. These rationales are just as abusable in a medical context as in any other. As Bonhoeffer put it: "the suspension of God's commandments on principle in the supposed interest of earthly self-preservation acts precisely against what this self-preservation seeks to accomplish."[1231] "The world *is* simply ordered in such a way that a profound respect for the absolute laws and human rights is also the best means of self-preservation."[1232] Every reader of this book has a God-given *right* to refuse to wear a mask, and many — including myself — have a God-given *duty* to refuse to do so.

The COVIDcrisis was never something that we could comply our way out of, and the more everyone complied, the worse it got. It was only after enough people started to say "no" – that the situation began to improve. To many people, this trajectory was clear from the beginning, because that is the nature of "emergency" power throughout history. In my case, I had a conviction from the moment of the first mask mandates that I ought not to comply in any context. By January 2022, this conviction reached a crisis point. Knowing what I know, and believing what I believe, made wearing a mask in *any* manner exceeding what was common in my profession and the general public prior to 2020 an act of cowardice on my part, and the Bible specifically states that cowards get thrown into the Lake of Fire (Revelation 21:8). I am certain that many people, not just Christians, experienced similar moral

---

1230   Dietrich Bonhoeffer, "After Ten Years" (Dec. 1942), reprinted in *Dietrich Bonhoeffer: Letters and Papers From Prison*, John W. DeGruchy, editor, Barbara and Martin Rumschedit, translators, Minneapolis: Fortress Press (2010), p. 47, available online: https://archive.org/details/letterspapersfro0008bonh/page/46

1231   Dietrich Bonhoeffer, "After Ten Years" (Dec. 1942), reprinted in *Dietrich Bonhoeffer: Letters and Papers From Prison*, John W. DeGruchy, editor, Barbara and Martin Rumschedit, translators, Minneapolis: Fortress Press (2010), p. 45, available online: https://archive.org/details/letterspapersfro0008bonh/page/44

1232   Dietrich Bonhoeffer, "After Ten Years" (Dec. 1942), reprinted in *Prisoner for God: Letters and Papers from Prison*, Eberhard Bethge, editor, Reginald H. Fuller, translator, New York: The Macmillan Company (1959), p. 21, available online: https://archive.org/details/DietrichBonhoefferLettersFromPrison/page/n23

imperatives to resist various measures being imposed in the name of public health, including masks and vaccines. This book, and everything surrounding its writing, has been part of my own efforts to fulfill the requirements of that moral imperative and my own personal duty before God. Thank you for taking the time to accompany me by reading.

Acknowledgements

In writing this book, I have been acutely conscious that I am standing on the shoulders of giants, past and present. I gratefully acknowledge my indebtedness to the work of those I have cited, as well as to the millions of everyday, unsung heroes who pushed back on compulsory masking, in whatever way they could, however small. Doing justice to the heroes of the COVIDcrisis would take many books far longer than this one.

> To every doctor, dentist, nurse, therapist, healthcare provider, lawyer, congressman, podcaster, and youtuber who spoke out against compulsory masking and what followed —

> To every student from grade school through college who resisted or refused to wear a mask —

> To all the state governors and attorneys general who did their own due diligence and worked to put a stop to the abusive masking of their citizens —

> To all those who sacrificed their time, money, and peace of mind to push back against compulsory masking (and what followed after) in some way, however small —

> thank you from the bottom of my heart.

In thanking specific individuals, I limit myself to those people with whom I have personally spoken or communicated. Upon reviewing this list, I realized that the only person on it who I had met prior to 2020 was my cousin, Daniel Emerick. Had it not been for COVID and the surrounding hysteria, I very likely would have gone the rest of my life without knowing these people, and my life would have been far poorer for it. The fact that I can recognize these people by name is, for me, a major silver lining to the events of 2020 and beyond.

Dr. Terry Lakin: his listening ear, wise advice, encouragement, and prayers have been indispensable to me, as was his personal example, completely in keeping with the highest traditions of loyalty, duty, respect, selfless service, honor, integrity, and personal courage. From the evening he got an unexpected phone call from an agitated dentist who he'd never heard of, his experience and gracious support have made a bigger difference than he knows.

Daniel Emerick: his moral compass and support have been unwavering from the start of COVID, not to mention his hours of video production, encouragement, beta-reading, and local leadership in the fight to return to liberty and a true normal. Americans like him and his wife, Jenna, are at the heart of the nascent reformation and restoration flowing just beneath the surface.

Rev. Paul Rydecki: his courage, support, Biblical feedback, and wise counsel (all often on short notice), have been personally vital. The unmasked Christmas service in his church in December 2020 will live as a bright spot in my memory for the rest of my life.

Tracy Hollister: her encouragement, optimism, advice, feedback, introductions, and timely news have been an indispensable asset, especially in the completion of this book during the many moments when my motivation needed a pick-me-up.

Professor David Clements: his eloquence and fearless example were, and remain, an inspiration. His feedback and advice were crucial for me at several forks in the road of background events which led to the completion of this book.

Jeff Childers: at the time of this writing we have only spoken once, and he was not involved in (or aware of) this book until after its completion, but the legal section of this book nevertheless owes him a particular debt. He not only led by example, but his appellate brief in the First District of Florida brought many relevant cases and arguments to my attention for the first time. His consistent and cogent articulation of multiple arguments on his Coffee and COVID Substack have sharpened my own thinking "as iron sharpens iron."

There are many more: **Daniel New**, **Corey Zinman**, **Mark Joseph**, **Lucas Wall**, **Sarah Smith**, and others. For those people who deserve recognition but whom I have failed to list, that is a deficiency is on my part, not theirs. To all these, and many more, I proffer my sincere gratitude.

Finally, my wife **Vanessa**. Apart from myself, she bore the next highest opportunity cost for the overall creation of this book, and that is not even counting all the associated background events. Apart from being a wonderful mother to our two sons, her decision to maintain unrestricted, unmasked, contact between our firstborn (and, at the time, only) son and his grandparents throughout the COVIDcrisis will forever be to her (and their) credit. I do not know what opportunity cost is normal for an author's spouse, but if my own experience of writing this book is anything to go by, to-date, hers has been quite a bit higher than average. It is my sincere and prayerful hope that at the end of our lives, when all is said and done, we will both be able to say (without hesitation) that we would do all of this again.

Going Forward

The fight over compulsory masking is far from over. If you enjoyed this book, please help me improve its visibility by leaving a review on Amazon or with another vendor. It does not need to be long or eloquent. Even 1 or 2 simple sentences would make a difference, and be a big help.

For more information and free extras, including full-resolution versions of the charts, tables, and pictures found in this book, visit www.thebookonmasks.com.

Subscribe for the latest news, extras, and supplemental releases!

Thank you for reading.

www.ingramcontent.com/pod-product-compliance
Lightning Source LLC
Chambersburg PA
CBHW052349210326
41597CB00038B/6301